Mine Hazards Identification, Prevention and Control

Mine Hazards Identification, Prevention and Control

Editors

Xiangguo Kong
Dexing Li
Xiaoran Wang

MDPI

Basel • Beijing • Wuhan • Barcelona • Belgrade • Novi Sad • Cluj • Manchester

Editors
Xiangguo Kong
Xi'an University of Science and Technology
Xi'an, China

Dexing Li
China University of Mining and Technology
Xuzhou, China

Xiaoran Wang
China University of Mining and Technology
Xuzhou, China

Editorial Office
MDPI
St. Alban-Anlage 66
4052 Basel, Switzerland

This is a reprint of articles from the Topical Collection published online in the open access journal *Sustainability* (ISSN 2071-1050) (available at: https://www.mdpi.com/journal/sustainability/topical_collections/ZAP59IWIFS).

For citation purposes, cite each article independently as indicated on the article page online and as indicated below:

Lastname, A.A.; Lastname, B.B. Article Title. *Journal Name* **Year**, *Volume Number*, Page Range.

ISBN 978-3-0365-9430-9 (Hbk)
ISBN 978-3-0365-9431-6 (PDF)
doi.org/10.3390/books978-3-0365-9431-6

Contents

About the Editors

Xiangguo Kong

Xiangguo Kong is an associate professor at Xi'an University of Science and Technology. Dr. Kong earned his Ph.D. from China University of Mining and Technology. During his Ph.D. (2015.09–2018.12), he was funded, as a visiting Ph.D. student, by the China Scholarship Council (CSC) to study in Professor Jim Underschultz's research group at the University of Queensland, Australia. Now, Dr. Kong's research interests mainly include (1) coal and rock gas dynamic disasters, (2) geophysical signals response of coal and rock fractures, and (3) identification and monitoring of urban underground engineering disasters. Dr. Kong is responsible for the National Natural Science Foundation Youth Program, the China Postdoctoral Science Foundation General Program, the Shaanxi Provincial Natural Science Basic Research Program, and the Outstanding Youth Program of Xi'an University of Science and Technology. In recent years, he has published more than 30 papers in high-level domestic and international SCI/EI journals, such as Composite Structures, Engineering Frame Mechanics, and Energy, and four papers have been selected for ESI high citation. Dr. Kong is a guest editor for the "Sustainability" journal and an Editorial board member of the International Journal of Mining Science and Technology. He also reviews over twenty international journals, including Fuel, Journal of Natural Gas Science and Engineering, and Fractals.

Dexing Li

Dexing Li is a postdoctoral fellow at China University of Mining and Technology. He earned his bachelor's degree in Safety Engineering from Hubei Polytechnic University in 2015 and completed his Ph.D. in Safety Science and Engineering at China University of Mining and Technology in 2021. Currently, he is involved in postdoctoral research at the School of Safety, China University of Mining and Technology. His primary research areas include mining engineering safety, tunneling, underground engineering disaster prevention, safety monitoring, and early warning technology. He has led research projects, including the National Natural Science Foundation of China for Young Scientists, the China Postdoctoral Science Foundation General Program, and the Guangdong Provincial Foundation for Basic and Applied Basic Research. He has also participated as a key researcher in the National Natural Science Foundation of China Key Projects, General Projects, and National Key Research and Development Program topics. He has published more than 30 papers in journals, including the "International Journal of Rock Mechanics and Mining Sciences", "Engineering Geology", "Construction and Building Materials", and "Geophysics", among others. He is also a guest editor for the "Sustainability" journal and a reviewer for more than twenty international journals.

Xiaoran Wang

Xiaoran Wang is an assistant research fellow at the State Key Laboratory for Fine Exploration and Intelligent Development of Coal Resources, China University of Mining and Technology (CUMT). Dr. Wang earned his Ph.D. degree in Safety Science and Engineering at CUMT. During his Ph.D. (2017.10–2020.07), he was funded, as a visiting Ph.D. student, by the China Scholarship Council (CSC) to study in Professor Joseph F. Labuz's research group at the University of Minnesota, America. Now, Dr. Wang's research interests mainly include (1) rock/coal dynamic disasters monitoring and control; (2) rock fracture mechanics; (3) source characterization using acoustic emission moment tensor; and (4) discrete element simulation, etc. Based on those research interests, Dr. Wang is responsible for the National Natural Science Foundation Youth project, the sub-project of the National Key ix Research and Development Program (China and the United States intergovernmental Science and technology cooperation project), the State Key Laboratory Director Fund, and the"Sailing Plan" for Youth Research of CUMT. As a key member, he also participated in developing National Major Scientific Research Instruments and the National Key Research and Development Program. In recent years, he has published over 30 SCI/EI papers in international journals, such as Rock Mechanics and Rock Engineering, Engineering Fracture Mechanics, and Theoretical and Applied Fracture Mechanics. He also has authorized/applied for around ten national invention patents and five software copyrights. Xiaoran Wang is the guest editor of the Sustainability journal and an Editorial board member of the International Journal of Mining Science and Technology and the Journal of China University of Mining and Technology. He also serves as a guest reviewer for several international journals.

Preface

With the depletion of shallow coal resources, coal mining has entered the deep stage. The occurrence environment of deep coal and rock masses is significantly different from that of shallow areas. Therefore, new characteristics will also emerge in the occurrence mechanism of coal mine disasters, such as rock bursts, coal and gas outbursts, and coal spontaneous combustion in deep areas. Therefore, further research on the formation mechanism, occurrence characteristics, monitoring methods, and identification methods of deep mine disasters is of great significance for promoting the stable and sustainable development of the mining industry. Then, with the invitation of sustainability journal, the Topical Collection, titled "Mine Hazards Identification, Prevention, and Control," was issued, which aimed to provide an opportunity for researchers around the globe to conduct a broader scientific and technological discussion on such advances to improve the prevention and control level of the disasters encountered during underground coal mining. This Topical Collection contains 27 articles, concentrating on basic experiments, modeling, numerical simulation, and field tests of mine disasters. As guest editors, we are very grateful for the valuable research results of the authors. These research results will provide broader ideas for researchers in the field of mine safety, as well as theoretical guidance and practical solutions for disaster prevention and control in mining enterprises.

Due to limited time and understanding, there may still be some issues with publishing the paper. Readers are kindly requested to provide criticism and correction.

Xiangguo Kong, Dexing Li, and Xiaoran Wang
Editors

sustainability MDPI

Article

Precursor Signal Identification and Acoustic Emission Characteristics of Coal Fracture Process Subjected to Uniaxial Loading

Xiangguo Kong [1,2], Mengzhao Zhan [1,2,*], Yuchu Cai [1,3], Pengfei Ji [1,2], Di He [1,2], Tianshuo Zhao [1,2], Jie Hu [1,2] and Xi Lin [1,2]

1 College of Safety Science and Engineering, Xi'an University of Science and Technology, Xi'an 710054, China
2 Key Laboratory of Western Mine and Hazard Prevention, Ministry of Education of China, Xi'an 710054, China
3 School of Chemistry and Chemical Engineering, Xi'an University of Science and Technology, Xi'an 710054, China
* Correspondence: 22220226139@stu.xust.edu.cn

Abstract: In deep underground mine engineering, the critical warning signals before the sudden failure of coal are crucial to predict coal or rock dynamic catastrophes and to help the coal industry grow sustainably. Therefore, with the objective of accurately identifying the precursor signals of coal fracture, a uniaxial compression test was adopted. Tests were performed on multiple sets of raw coal samples, and acoustic emission (AE) technology was used to capture the deformation and destruction courses of the coal samples. Furthermore, the signal intensity of AE energy was discussed. Based on the critical slowing down theory, the AE energy sequence was processed. The results indicate that there are significant discrepancies in the strength of coal affected by initial pore fissures. During the whole loading process, the AE energy signals showed obvious stage characteristics, and there was a high risk of rapid coal energy storage during the unstable rupture development (URD) stage, which predicted the imminent destruction of the coal. The variance mutation point that was not affected by the lag step selection was easier to identify than that of the autocorrelation coefficient, and the precursor points were all in the URD stage, which is more accurate than using the AE cumulative energy curve slope.

Keywords: uniaxial compression; acoustic emission; critical slowing down; precursor signals

Citation: Kong, X.; Zhan, M.; Cai, Y.; Ji, P.; He, D.; Zhao, T.; Hu, J.; Lin, X. Precursor Signal Identification and Acoustic Emission Characteristics of Coal Fracture Process Subjected to Uniaxial Loading. *Sustainability* **2023**, *15*, 11581. https://doi.org/10.3390/su151511581

Academic Editor: Chaolin Zhang

Received: 17 May 2023
Revised: 25 June 2023
Accepted: 25 July 2023
Published: 26 July 2023

1. Introduction

In recent years, China has promoted a green shift in its energy structure and encouraged the energy sector to follow a sustainable development path. As the basic power of China, coal will still occupy a dominant position in China's energy for a considerable period of time in the future [1,2], and mining coal resources safely and efficiently is the key to ensure the sustainable development of the coal industry [1]. With the increasing of the number and capacity of deep coal mines in the last few years, the deep mining of coal resources in China has become an inevitable trend. Coal or rock dynamic catastrophes such as coal burst, roof pressuring, and coal and gas outburst are becoming increasingly serious, causing an enormous threat to the safe production of coal mines [3,4] and seriously hindering the sustainable development of the coal mining industry. Therefore, research of effective coal or rock dynamic catastrophe monitoring methods has become an important scientific issue that will be helpful to guarantee the sustainable development of coal as the main energy source in the energy structure.

The extension of cracks in coal rocks is an extremely complex phenomenon that is accompanied by the emergence, extension and penetration of microcracks during the loading process, causing the deformation of coal rocks at the same time [5,6]. Numerous studies have shown that strain is an important index of coal rock fracture evolution [5,7].

Munoz proposed a non-contact optical uniaxial compressive strain measurement method based on 3D DIC to measure the strain field formed by compressed sandstone more accurately [8]. Ma et al. analyzed the evolution law of strain field on a loaded coal or rock surface. It is believed that the evolution of microcracks in coal or rock is related to the statistical index of strain field [9]. Hou et al. studied the practical micro-scale deformation and crack evolution law of in situ coal seams through simulation experiments [10]. Based on the relationship between strain and fracture development, many scholars have conducted studies from another aspect.

During the process of loading coal or rock, there is the plastic deformation of coal/rock [11,12], associated with the release of energy, part of which is released as elastic waves, namely acoustic emission (AE) [13,14]. These acoustic emission signals provide an effective index for the coal fracture process [15,16] because they reflect the location, formed mechanism and releasing energy of coal rock micro-cracks [17,18]. As a means of non-destructively monitoring and characterizing the development of internal fractures in quasi-brittle materials [19], one of the areas of acoustic emission studies in the laboratory is the time-varying parameters of damage generation [20,21], determining how microcracks accumulate under external loading [22,23]. In the course of destroyed coal or rock, AE means can capture a variety of parameters, including impact times, ringing counts, energy, etc. [24,25], and the occurrence of coal or rock dynamics catastrophes can be predicted [26,27]. Although the genesis of dynamic catastrophes can be forecast by AE as a concomitant signal in coal or rock fracture, its effectiveness is not stable. In the last few years, scientists have discovered that when complex dynamic systems approach a tipping point before a mutation occurs, there is a phenomenon with predictive significance called critical slowing down [28,29]. The critical slowing phenomenon has enormous potential in disclosing whether a sophisticated dynamic system is susceptible to ruinous mutations. Nowadays, it has been used to predict ecological problems [28], earthquakes [29] and climate disasters [30,31]. Therefore, to further identify the precursors of disaster occurrence accurately and effectively is a key issue in the future research field of coal engineering for coal rock dynamics disasters.

Based on the previous research, acoustic emission technology has achieved many excellent results with regard to early warning potential rupture indicators [32,33] in the monitoring and application [34] of rock [35,36], but there are few research results on the critical slowing down of coal specimens [37,38]. Therefore, the research on critical slowing down of coal is particularly important for forewarning and prevention of dynamic catastrophes in the coal mining industry. Through uniaxial compression experiments, AE signals were captured during the fracture process of coal. The critical slowing down theory was used as the basic principle, and the AE signal generated by coal failure was processed. It was concluded that the variance and autocorrelation coefficient characterized the critical slowing phenomenon appearing in the later stage of coal fracture. The sudden increase phenomenon can be regarded as an effective precursor of coal instability destruction.

2. Experiment

2.1. Experimental System

The experimental system is shown in Figure 1 and is composed of the loading control subsystem and the AE monitoring subsystem. In the loading control subsystem, the experimental platform of impact mechanics at high and low frequency was used. The output maximum force is 3000 kN, the output maximum impact force is 1000 kN, the impact reaction time is <0.3 s, the precision of the test machine is 1%, the speed of piston displacement is 1–50 mm/min, and the loading rate of the sample machine is 0.02–2% FS. There are force and displacement control methods that can be used to carry out uniaxial compression, tensile, cyclic loading and other mechanical experiments. In the acoustic emission monitoring subsystem, a Micro-II Express acoustic emission monitoring and data analysis instrument was adopted that could collect the AE data of 24 probes at the same time.

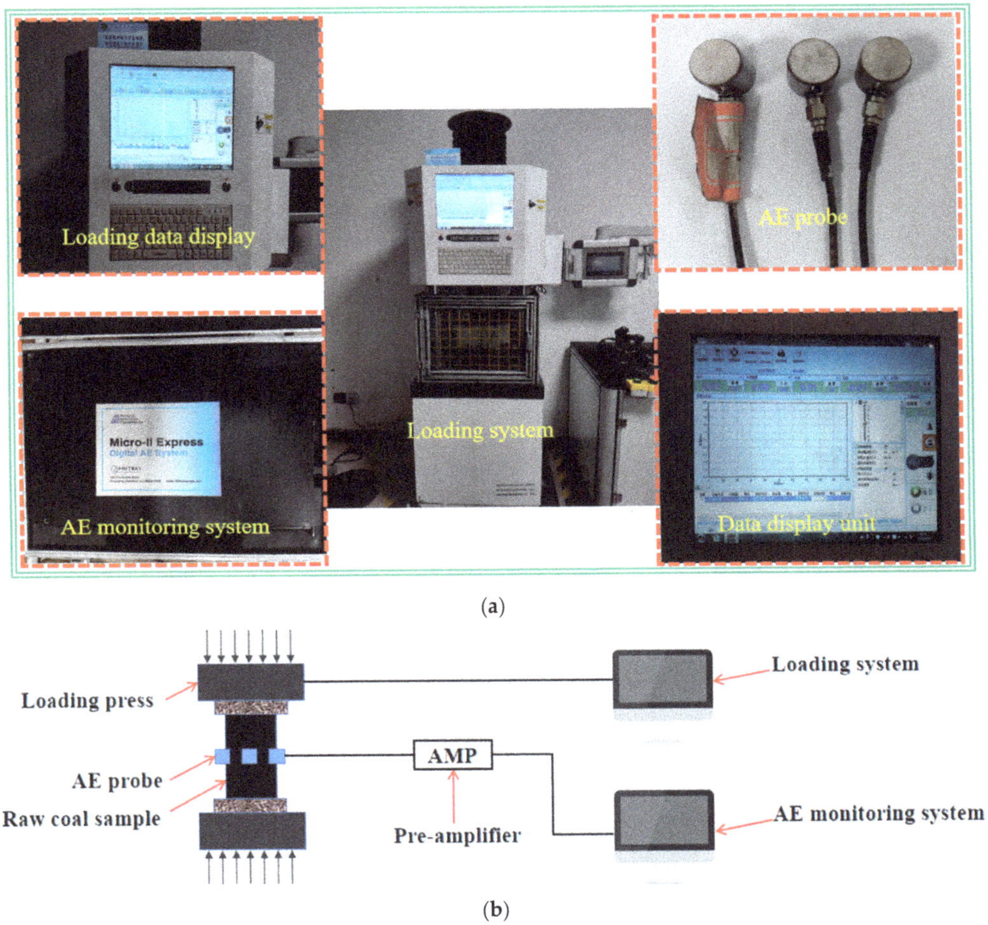

Figure 1. Experimental system. (**a**) Experimental equipment; (**b**) experimental composition.

2.2. Experimental Method

During the uniaxial compression loading process, the acoustic emission monitoring data were collected simultaneously. The experimental loading method was force-controlled loading (the loading rate of 50 N/s), in which the stress–strain was recorded automatically. Six acoustic emission probes (model was NANO-30, peak frequency was 300 kHz) were pasted on the surface of each coal sample to measure AE signal when it was damaged under uniaxial compression. The amplification ratio of the pre-amplifier was set at 40 dB, and the threshold was set at 40 dB. The signal sampling rate was 1 MHz and the sampling length was 1024 points (the first 1/4 of which are pre-triggered) to ensure the coupling between the sample and each AE probe. When the coal sample was placed on the experimental platform, a stress of 0.5 kN could be preloaded first.

2.3. Experimental Material

To avoid the contingency of experimental phenomena, the uniaxial compression experiments shown in Figure 2 had three groups of rectangular standard coal samples with a length × width × height of 50 mm × 50 mm × 100 mm. The parallelism of both ends of the samples was no more than 0.05 mm, and the axial deviation was no more than 0.25 mm. In order to verify the validity of the precursors of coal fracture based on the critical slowing down theory for samples with different damage conditions, coal samples with different internal fractures and damage conditions were selected for the test.

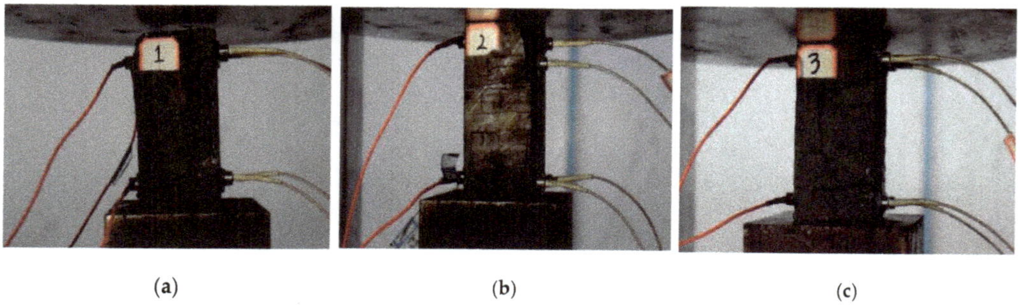

Figure 2. Test samples. (**a**) Sample 1; (**b**) sample 2; (**c**) sample 3.

3. Results

3.1. Mechanical Properties of Coal in Loading Process

The deformation and failure states of coal samples vary during loading due to their own internal damage and primary fracture conditions [39,40]. Figure 3 shows the stress–strain curve, which has typical compaction, elastic, strengthening and post-peak phases.

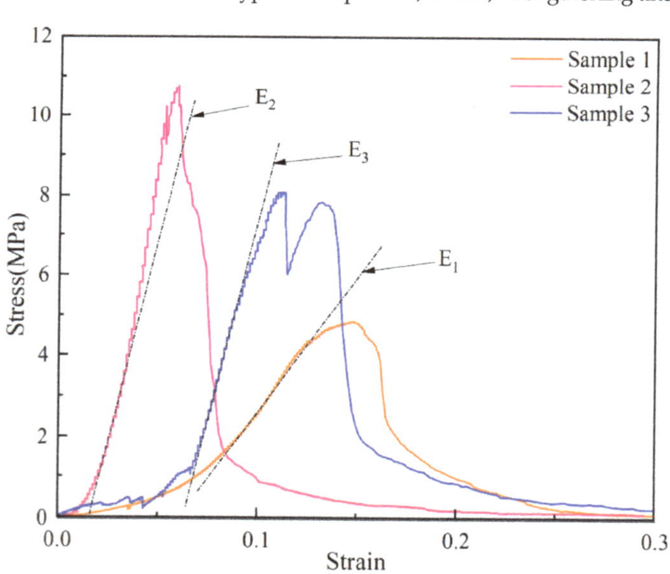

Figure 3. Stress–strain curves.

The strength of coal is significantly different among raw coal samples taken from the same coal source and is influenced by its pores and fissures. However, the better the integrity of the coal samples, i.e., no obvious surface fissures, the higher the strength, the longer the elastic phase, the more obvious the stress drop in the post-peak phase, and the shorter the damage time. On the contrary, the more significant the softening characteristics of the post-peak stress, the longer the damage time. The elastic modulus was calculated according to the stress–strain relationship in the elastic phase, and the uniaxial compressive strength was calculated at the time of coal destruction. The elastic modulus, uniaxial compressive strength and size of breaking strain of the coal samples used in the test affected by fissures were obtained statistically as shown in Table 1, and the statistical results show that the elastic modulus and uniaxial compressive strength increased with the increase in coal body integrity and the breaking strain decreased with the increase in coal body integrity.

Table 1. The mechanical properties of coal samples.

	Uniaxial Compressive Resistance Intensity/MPa	Elastic Modulus/MPa	Failure Strain
Sample 1	4.840	61.094	0.149
Sample 2	10.760	264.151	0.059
Sample 3	8.096	221.951	0.111

3.2. Evolution Laws of AE Energy

Compressive force was applied to the coal samples until they were destroyed. Figure 4 is the curve of AE energy, cumulative energy and stress with time. According to the relation between AE energy and time, the AE signals can be assigned to five phases: pore-fracture compaction (PFC) phase, elastic deformation (ED) phase, stable development of microelastic fractures (SDMF) phase, unstable fracture development (UFD) phase and posterior rupture (PR) phase.

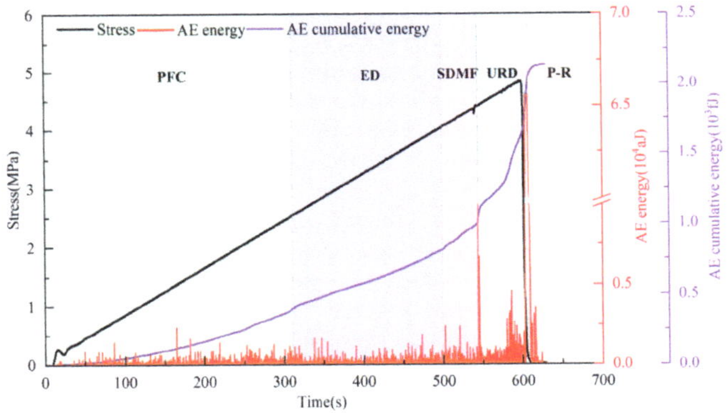

(**a**)

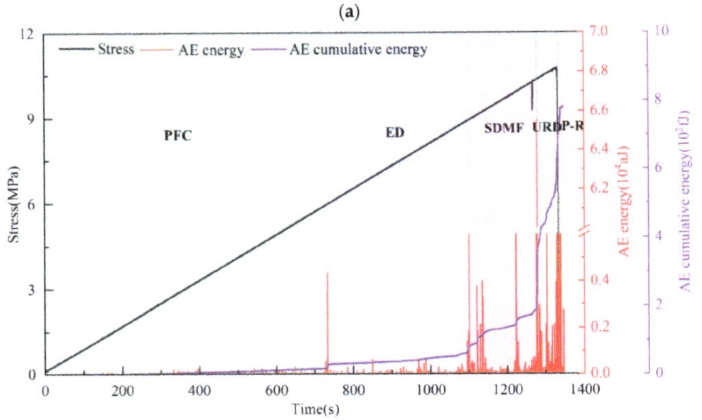

(**b**)

Figure 4. *Cont.*

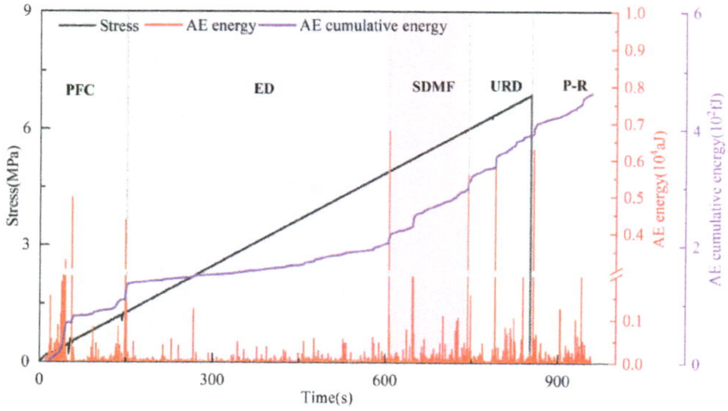

(**c**)

Figure 4. Acoustic emission energy, accumulated energy and stress with time during coal failure. (**a**) Sample 1; (**b**) Sample 2; (**c**) Sample 3.

The AE energy of the three samples show the same variation characteristics in the corresponding stages. The AE energy is weak at the incipience of loading, and the original pores or damage inside the coal become compacted by the exterior load. With the increase in stress, a little AE energy is generated, and the cumulative energy has a slow growth trend. In this phase, the original damage or pores are compacted and begin to produce new micro-cracks inside the samples. The AE energy grows steadily until a sudden change occurs at some point, and similarly, there is a tendency for the accumulated AE energy to grow abruptly. In this phase, new micro-cracks start to emerge, grow and expand inside the samples, then connect with the original pores to form larger cracks. When the samples reach the stress maximum, the peak AE energy appears, and then samples destabilize and break. However, due to the existence of local micro-cracks in the destroyed coal, the AE energy does not suddenly disappear, but gradually decreases until it tends to be stable.

When sample 1 is in the PFC phase, the stress in this phase accounts for 0–51.5% of the peak stress, after which the coal enters the ED phase. In this phase, stress accounts for 51.5–84.0% of the peak stress. The frequency of the AE signal is higher than that in the PFC phase, and when entering the SMFD phase, the stress in this phase only accounts for 84–90% of the peak stress. After a short time, the coal sample starts to enter the later loading stage, which is called the URD stage. The stress in this stage reaches 90–100% of the peak stress, and the peak stress is followed by the PR stage.

There are similar physical characteristics between sample 1 and sample 2, so the generation of the acoustic emission signal also has a similar rule; in the PFC stage, stress accounted for 0–55.3%; in the ED stage, stress accounted for 55.3–82.7%; in the SDMF stage, stress accounted for 82.7–96.3%; and in the URD stage, stress accounted for 96.3–100%.

The acoustic emission law of sample 3 is affected by its internal pore cracks, and it is slightly different from the other two samples. The AE signal is more pronounced at the beginning of loading by the original damage. In the PFC phase, the stress accounts for 0–16.7%; in the ED phase, stress accounts for 16.7–62.8%; in the SDMF phase, stress accounts for 62.8–76.9%; and in the URD phase, stress accounts for 76.9–100%.

In summary, the AE signals caused by coal failure correspond well to the whole loading process with their significant variation in features, and the accumulated energy of AE can more intuitively reflect these stage characteristics.

3.3. Cumulative Acoustic Emission Energy Risk Index

The cumulative AE energy is the cumulative value of AE rupture events, which can be used as a characteristic index of coal rupture AE events during the loading process [41]. The analysis of its variation law aims to analyze the evolutionary features of the energy accumulation in the loading course of the coal sample so as to provide a new index for the fracture precursor signal and offer a reference for the prediction of coal fractures.

According to the change law of the cumulative AE energy curve (Figure 4), it can be seen that the fracture of coal under load is a process of energy accumulation and then release. The PFC and ED stage and SDMF and URD phase in fracture development are all part of the energy accumulation process. When the energy accumulation is up to a certain critical value, the energy releases as coal breaks. Based on the experience of previous studies [41,42], the cumulative AE energy can be used as an index to determine the risk of coal rupture. The slope of the cumulative AE curve of each stage was obtained by the linear fitting method and the definition of the risk index P_1 is the ratio of PFC to ED slope in the cumulative curve (Figure 4) while P_2 is the ratio of SDMF to ED slope in the cumulative curve and P_3 is the ratio of URD to ED slope in the cumulative curve. The calculation formula of P_1, P_2 and P_3 is:

$$P_1 = \frac{K_1}{K_2} \tag{1}$$

$$P_2 = \frac{K_3}{K_2} \tag{2}$$

$$P_3 = \frac{K_4}{K_2} \tag{3}$$

where K_1 is the slope of the PFC stage; K_2 is the slope of the ED stage; K_3 is the slope of the SDMF stage; and K_4 is the slope of the URD stage.

The calculation results of P_1, P_2 and P_3 are shown in Table 2.

Table 2. Loaded rupture risk determination indexes P_1, P_2 and P_3.

	K_1	K_2	K_3	K_4	P_1	P_2	P_3
Sample 1	1.30	2.17	4.93	16.20	0.59	2.27	7.46
Sample 2	0.03	0.14	0.64	9.06	0.21	4.57	64.71
Sample 3	0.71	0.19	0.59	1.21	3.74	3.11	6.37

It can be seen from Table 2 that the range of risk index P_1 is between 0.21 and 3.74, that of P_2 is between 2.27 and 4.57, and that of P_3 is between 6.37 and 64.71. (1) When $P < 2.27$, the PFC phase has similar fluctuation characteristics to the ED phase and has a weak risk of rupture. (2) When $2.27 < P < 6.37$, this indicates that the coal in the SDMF phase has a medium risk of fracture, and the acoustic emission energy appears "convex" in shape. (3) When $P > 6.37$, this is the time period of rapid energy storage of coal fracture. At this time, the possibility of coal fracture is greatly increased and has a high risk of fracture. Therefore, effective prediction should be carried out in the URD phase to reduce the occurrence of disasters caused by coal fracture.

4. Discussion

4.1. Critical Slowing down Theory

In natural dynamic systems, if the phase state changes, i.e., when the system transitions from the old phase to a new phase, a dispersion rise and fall phenomenon favoring the formation of the new phase will occur near the proximity point, and this dispersion rise and fall is characterized by increased amplitude, elongated time, slower recovery of perturbations, and less ability to recover to the old phase, which is called the critical slowing down phenomenon [38,43], In dynamic systems, this phenomenon is usually characterized as a phenomenon of increasing variance and autocorrelation coefficients of the covariates. Therefore, the computational analysis of the variance and autocorrelation coefficients of acoustic emission characteristic parameters with time can provide a precursor prediction method for the loaded rupture of coal samples.

Variance is a characteristic mass that represents the deviation of data from the mean value x in a sample, denoted as s^2; s is called the mean square error, and the calculation formulas are:

$$S^2 = \frac{1}{n}\sum_{i=1}^{n} (x_i - \overline{x})^2 \tag{4}$$

$$S = \sqrt{\frac{1}{n}\sum_{i=1}^{n} (x_i - \overline{x})^2} \tag{5}$$

An autocorrelation coefficient is a statistical quantity that represents the relevance between different times of the same argument. The autocorrelation coefficient with lag length (j) is marked as $r(j)$ so as to judge the possibility of predicting x_{i+j} by x_i. For variable x, the autocorrelation coefficient with lag length (j) is:

$$r(j) = \frac{1}{n-j}\sum_{i=1}^{n-j} \left(\frac{x_i - \overline{x}}{s}\right)\left(\frac{x_{i+j} - \overline{x}}{s}\right) \tag{6}$$

s is the mean square error of n-length time series, and s is obtained by (5).

Firstly, it is assumed that there is a periodic variable Δt with a recovery speed of λ. In a simple model of autoregression, it can be represented as:

$$x_{n+1} = e^{\lambda \Delta t} x_n + s\varepsilon_n \tag{7}$$

where x_n is the deflection distance of system state from disturbance state to equilibrium state. If λ and Δt are not determined by x_n, the course can be simplified to a one-step autoregressive model:

$$x_{n+1} = \alpha x_n + s\varepsilon_n \tag{8}$$

Among them, the autocorrelation coefficient $\alpha = e^{\lambda\Delta t}$, and the pair (8) autoregressive process is examined by square deviation:

$$Var(x_{n+1}) = E(x_n^2) + (E(x_n))^2 = \frac{s^2}{1-\alpha^2} \tag{9}$$

Generally, as the system approaches the critical point, the recovery rate of small perturbations becomes slower and slower. When it approaches the critical point, the reparatory rate λ tends to 0 and the autoregressive item α tends to 1 [44,45]. The variance of (9) tends to infinity, so the variance can also be used as an index to determine whether the system reaches the critical state. In summary, when the complex multi-dynamic system approximates the critical point, the reparatory rate of the perturbation gradually decreases and approaches 0. At this time, the autocorrelation coefficient and variance increase of the disturbance information are two important indexes for testing the critical slowing down phenomenon.

4.2. The Effect on Different Window or Lag Length on Critical Slowing Down

Before calculating the autocorrelation coefficient and variance, we first need to select the appropriate size of window or lag length, because the change and stability situation of the autocorrelation coefficient and variance are closely relevant to the window or lag length size. As shown in Figure 5, a certain unit that needs to be calculated for a sequence is referred to as the window length, and the lag step represents the length of the lag sequence to be slid backward to obtain a new sequence of window length, starting from the selected window length sequence [38,43]. To investigate the effect on different window lengths or lag steps on the critical slowing down characteristics, there are different window length and lag steps selected for comparative analysis.

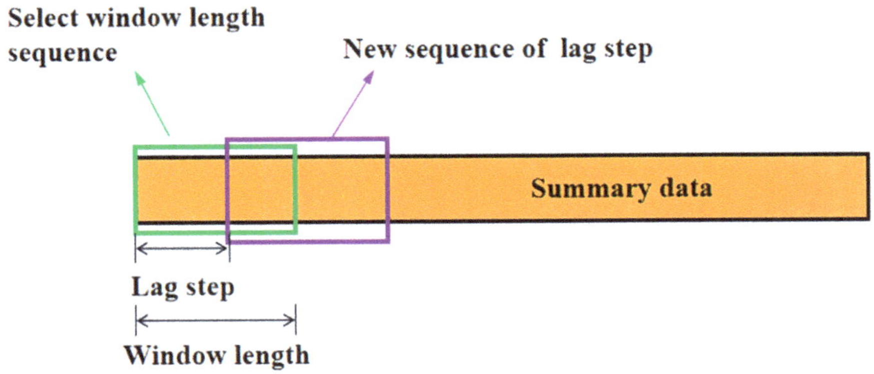

Figure 5. Window length and lag step.

Due to the large amount of raw test data, the window length and the lag step should be chosen accordingly as a larger value. Firstly, the window length is 3000, and the lag steps are 500, 1000 and 1500. Then, the lag step is 1000, and the window lengths are 2000, 3000 and 4000. The effects of different lag steps and window lengths on the autocorrelation coefficient and variance are compared separately in Figure 6.

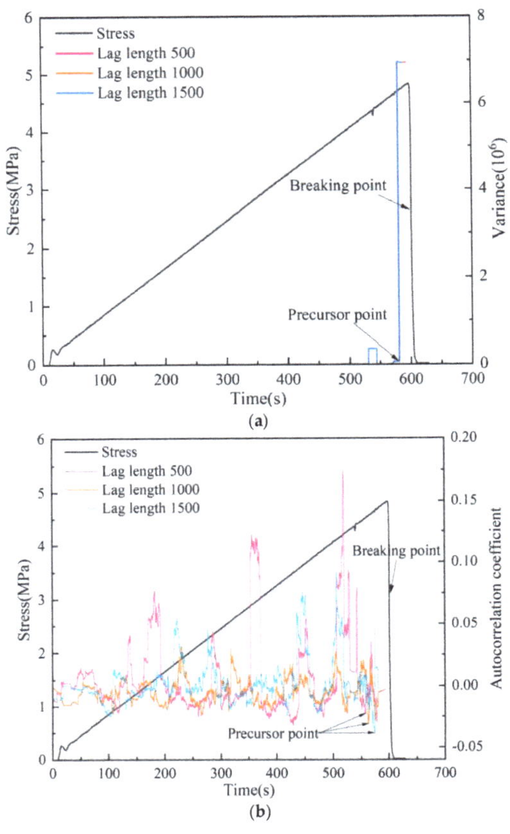

Figure 6. *Cont.*

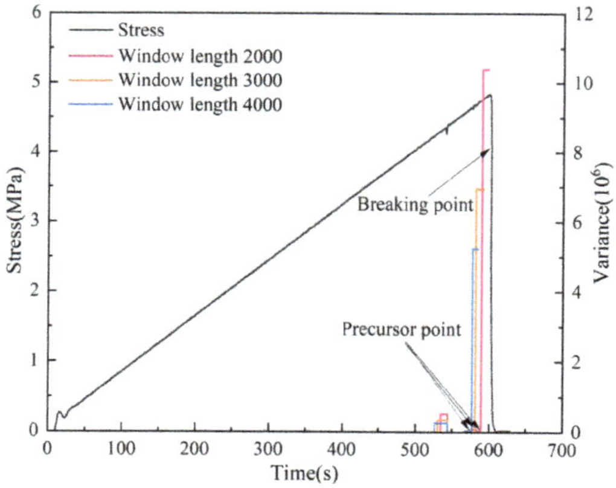

(**c**)

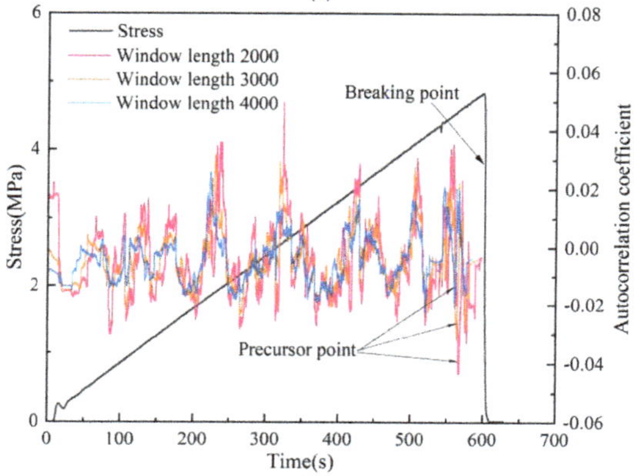

(**d**)

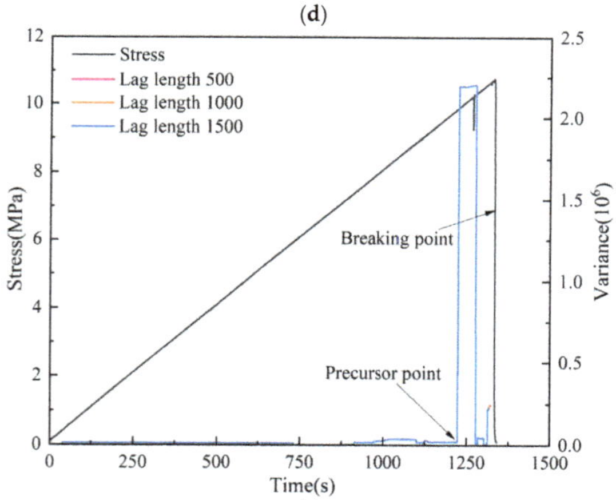

(**e**)

Figure 6. *Cont.*

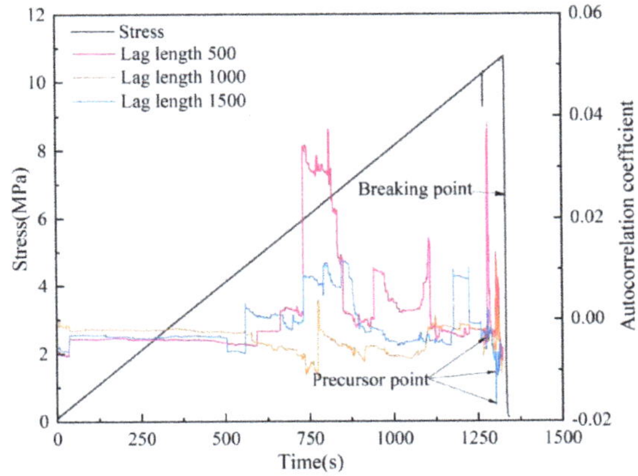

(**f**)

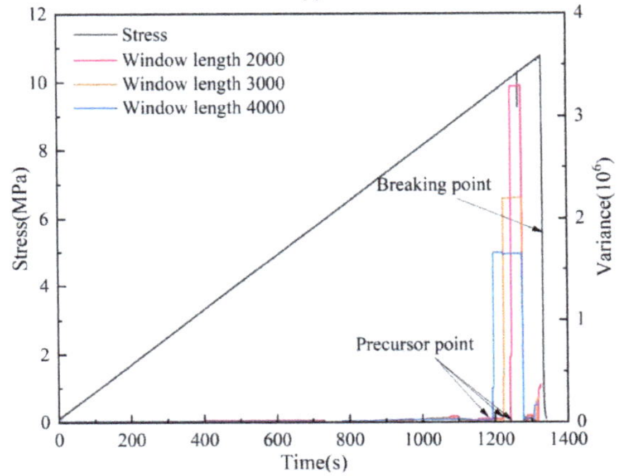

(**g**)

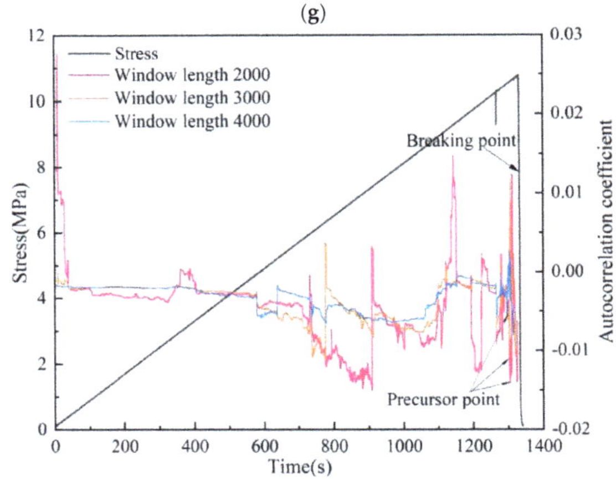

(**h**)

Figure 6. *Cont.*

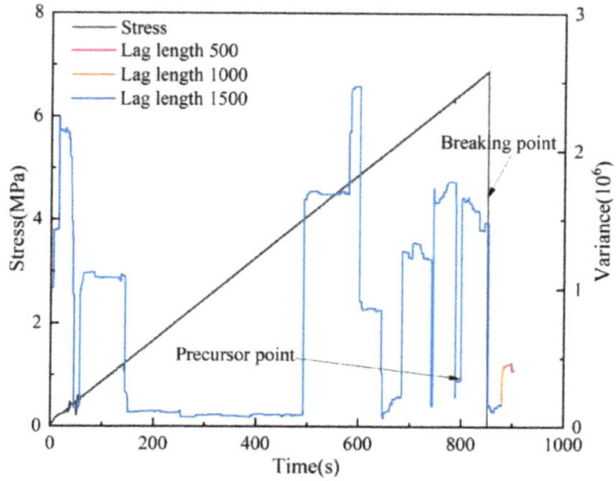

(i)

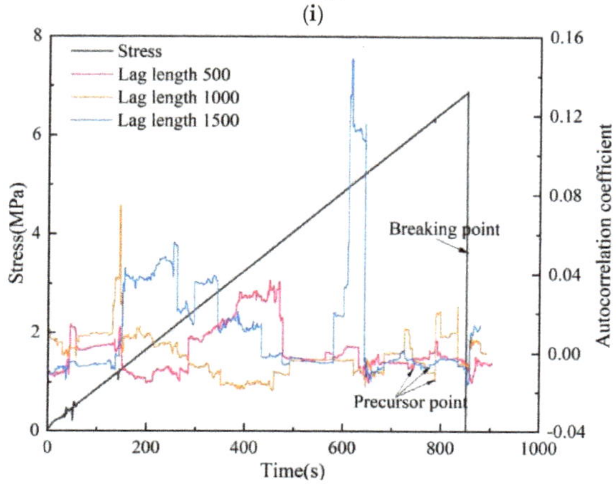

(j)

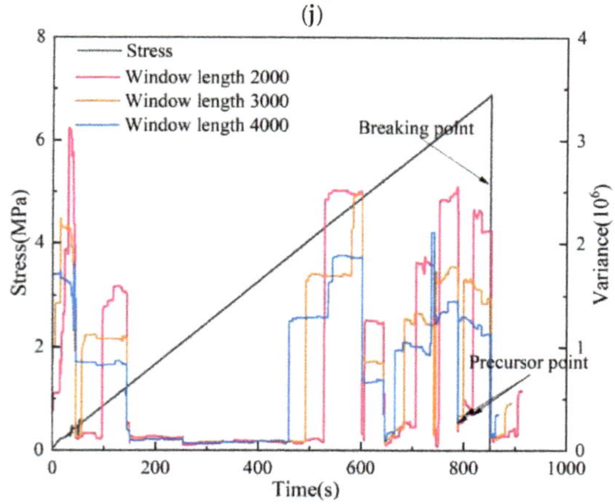

(k)

Figure 6. *Cont.*

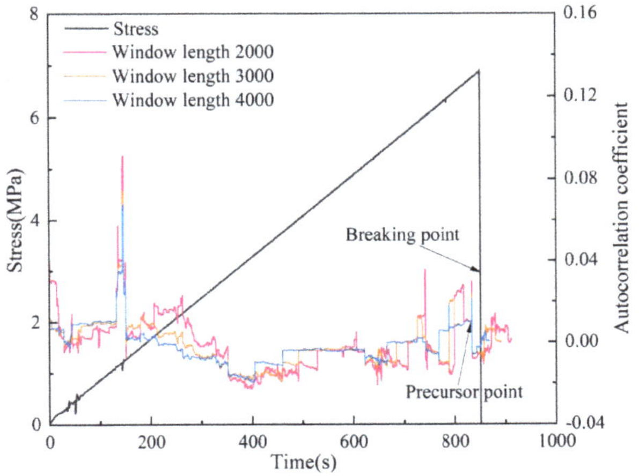

(l)

Figure 6. Comparison of variance and autocorrelation coefficient curve. (**a**) Sample 1 variance curves with different lag steps; (**b**) Sample 1 autocorrelation coefficient curves with different lag steps; (**c**) Sample 1 variance curves with different window lengths; (**d**) Sample 1 autocorrelation coefficient curves with different window lengths; (**e**) Sample 2 variance curves with different lag lengths; (**f**) Sample 2 autocorrelation coefficient curves with different lag steps; (**g**) Sample 2 variance curves with different window lengths; (**h**) Sample 2 autocorrelation coefficient curves with different window lengths; (**i**) Sample 3 variance curves with different lag steps; (**j**) Sample 3 autocorrelation coefficient curves with different lag steps; (**k**) Sample 3 variance curves with different window lengths; (**l**) Sample 3 autocorrelation coefficient curves with different window lengths.

The window length was taken as 3000 to compare the effect of variance for different lag steps. As shown in the Figure 6a,e,i, the variance curves relative to the lag steps of 500, 1000 and 1500 show the same trajectory with time and reunite together, and the time points of sudden increase are basically the same, that is, the variance curve does not vary with the variation in lag length.

The window length was taken as 3000 to compare the effect of the autocorrelation coefficient for different lag steps. The autocorrelation coefficient curves with different lag steps of the same window length are shown in Figure 6b,f,j. It can be seen that different lag steps have a certain influence on the autocorrelation coefficient, but this change does not show the overall regularity. There is no obvious correlation between the fluctuation range and the choice of lag step size, but the trend of local change is consistent, meaning that the autocorrelation coefficient has independent variation characteristics under the new sequence of lag step size.

A lag step of 1000 was taken to compare the effect of different window lengths on the variance. Figure 6c,g,k show the same lag length variance curves with different window length. As shown in the figure, the variance curves with various window lengths of 2000, 3000 and 4000 show the same change rule and affect the change range of variance; in particular, the increase in the inflection point of the variance curve reduces with the growth in window length.

A lag step of 1000 was taken to compare the effect of different window lengths on the autocorrelation coefficient. As can be seen in Figure 6d,h,l, when the lag step lengths are equal, the change trend of the autocorrelation coefficient curve corresponding to different window lengths is relatively consistent, and the volatility range of the curve gradually reduces with the growth in window length or even tends to be stable; that is, to determine the length of the new series, the larger window length series is selected, with a smaller effect on the autocorrelation coefficient.

In summary, although the precursor points of critical slowing can be successfully captured by both variance and autocorrelation coefficient, however, compared with the variance, the autocorrelation coefficient presents a messy change process with more peaks, which is very easy to be judged as a pseudo-signal. Therefore, the variance is more definite than the autocorrelation coefficient precursor signal, so the variance can be used as the main criterion of the precursor signal and the autocorrelation coefficient as an auxiliary criterion.

4.3. Compare the Precursor Characteristics of Each Index

Usually, the sudden increase in acoustic emission signal level and the increase in slope of the accumulated energy curve of acoustic emission are often shown before the destabilization damage of coal [46], which is because the microfracture is sprouting and developing through, so the precursor point of each index characterization must be controlled in the URD stage if the precursor signal of coal rupture can be effectively identified.

Based on the load, cumulative energy of acoustic emission, variance and autocorrelation coefficient versus time, the magnitude of the load values corresponding to the precursor signals characterized by the cumulative energy, positive sequence variance and autocorrelation coefficient are summarized from the loads and compared with the load at the damage point. The degree of damage load at the precursory signal point is represented by each index before the failure of the coal sample is analyzed. According to the effect of different window or lag length on the critical slowing, the lag step of 1000 and window length of 3000 were selected to calculate the critical slowing down precursor index. As shown in Figure 6, among the three samples, the strength corresponding to the appearance of the damage precursor signal of sample 1 is 4.651 MPa and the peak strength is 4.840 MPa, which is about 96.1% of the peak strength; the strength corresponding to the appearance of the damage precursor of sample 2 is 9.895 MPa and the peak strength is 10.736 MPa, which is about 92.2% of the peak strength; and the strength corresponding to the appearance of the damage precursor of sample 3 is 6.472 MPa and the peak strength is 6.904 MPa, which is about 93.7% of the peak strength. The precursor signals all appear in the URD stage and have sufficient validity. From Figure 4, the precursor signal is characterized by a steep change in the slope of the accumulative energy curve of AE. The strength corresponding to the appearance of the damage precursor signal of sample 1 is 4.452 MPa and the peak strength is 4.836 MPa, which is about 90.8% of the peak strength; the strength corresponding to the appearance of the damage critical precursor of test coal sample 2 is 8.626 MPa and the peak strength is 10.736 MPa, which is about 80.4% of the peak strength; and the strength corresponding to the appearance of the sample 3 damage critical precursor is 5.676 MPa and the peak strength is 8.092 MPa, which is about 82.2% of the peak strength. The precursor signals of sample 2 and sample 3 appear in the SDMF stage. Although they can indicate the damage of coal to some extent, the early warning characteristics are not obvious. The calculation results show that the precursor signal characterized by variance and autocorrelation coefficient is more accurate and effective.

From a time perspective, the precursor signal appeared 24 s earlier for sample 1, 25 s earlier for sample 2 and 50 s earlier for sample 3 after the critical slowing treatment, accounting for 4.0%, 1.9% and 5.8% of the total loading time, respectively, all of which were in the URD stage. The precursor signals characterized by the cumulative energy curve were 45 s, 54 s and 103 s earlier, accounting for 7.5%, 4.1% and 12.1% of the total loading time. This shows that the occurrence time of the precursor signal represented by the critical slowing down is neither too early to become a false signal nor too late to cause difficulties in early warning work. Therefore, on the basis of the critical slowing theory, effective measures are taken to deal with the fracture of coal with good timeliness in the precursor of failure, which has a certain use as a reference for enriching the AE monitoring technology of coal failure.

5. Conclusions

(1) The strengths of coal samples are significantly different due to the influence of internal pore fractures. The elastic moduli and uniaxial compressive strengths grow with increase in coal integrity, while the failure strain decreases.

(2) AE signals are generated during the destruction of the coal, and the whole loading process can be divided into five stages: PFC, ED, SDMF, URD, and P-R. The accumulated AE energy can reflect the characteristics of these stages more intuitively.

(3) Based on the acoustic emission signal, a risk index of acoustic emission parameters was established. According to P_1, P_2 and P_3, the risk was divided into three levels: weak risk, medium risk and high risk. At a high risk index, the possibility of coal fracture is greatly increased and there is a high risk of fracture. Therefore, effective prediction should be carried out in the URD stage to reduce the occurrence of disasters caused by coal fracture. Furthermore, this conclusion was obtained for uniaxial compression conditions. The study of other loading conditions as well as samples is necessary to verify this result.

(4) The critical slowing precursor point characterized by variance is clearer and more accurate and provides an early warning response to coal rupture at the URD stage. In underground coal mining engineering operations, effective measures must be taken to prevent and manage coal rupture after capturing the precursor signals at this stage so as to reduce the losses caused by coal–rock power disasters. This can serve as a reference for enriching the AE monitoring technology of coal failure.

Author Contributions: Conceptualization, X.K.; methodology, X.K., Y.C., P.J., D.H., T.Z., J.H. and X.L.; software, M.Z.; validation, M.Z.; formal analysis, M.Z.; investigation, M.Z.; resources, M.Z.; data curation, M.Z.; writing—original draft preparation, M.Z.; writing—review and editing, M.Z.; visualization, M.Z.; supervision, X.K.; project administration, X.K.; funding acquisition, X.K. All authors have read and agreed to the published version of the manuscript.

Funding: This research was funded by Xiangguo Kong, the grants National Natural Science Foundation of China (51904236), Natural Science Basic Research Program of Shaanxi (2020JQ-756), China Postdoctoral Science Foundation (2021M693879) and Excellent Youth Program of Xi'an University of Science and Technology.

Conflicts of Interest: The authors declare no conflict of interest.

References

1. Li, X.L.; Cao, Z.Y.; Xu, Y.L. Characteristics and trends of coal mine safety development. *Energy Sources Part A Recovery Util. Environ. Eff.* **2021**, 1–19. [CrossRef]
2. Lin, H.F.; Ji, P.F.; Kong, X.G.; Li, S.G.; Long, H.; Xiao, T.; Li, B. Experimental study on the influence of gas pressure on CH_4 adsorption-desorption-seepage and deformation characteristics of coal in the whole process under triaxial stress. *Fuel* **2023**, *333*, 126513. [CrossRef]
3. Lin, H.F.; Ji, P.F.; Kong, X.G.; Li, S.G.; Dou, G.D.; Li, K. Precise hole placement pattern and engineering practice of pre-pumping coal seam gas by down-seam drilling. *J. Coal* **2022**, *47*, 1220–1234.
4. Qin, L.; Li, S.G.; Zhai, C.; Lin, H.F.; Zhao, P.X.; Shi, Y.; Bing, Y. Changes in the pore structure of lignite after repeated cycles of liquid nitrogen freezing as determined by nitrogen adsorption and mercury intrusion. *Fuel* **2020**, *267*, 117214. [CrossRef]
5. Ma, W.B.; Chen, Y.L.; Yi, W.; Guo, S.C. Investigation on crack evolution behaviors and mechanism on rock-like specimen with two circular-holes under compression. *Theor. Appl. Fract. Mech.* **2022**, *118*, 103222. [CrossRef]
6. Zhou, D.; Ye, Y.C.; Hu, N.Y.; Wang, W.Q.; Wang, X.H. Crack evolution of soft-hard composite layered rock-like specimens with two fissures under uniaxial compression. *Front. Struct. Civ. Eng.* **2021**, *15*, 1372–1389. [CrossRef]
7. Ma, W.Q.; Wang, J.T.; Li, X.X.; Wang, T.X. Crack evolution and acoustic emission characteristics of rock specimens containing random joints under uniaxial compression. *Acta Geophys.* **2021**, *69*, 2427–2441. [CrossRef]
8. Munoz, H.; Taheri, A.; Chanda, E.K. Pre-peak and post-peak rock strain characteristics during uniaxial compression by 3D digital image correlation. *Rock Mech. Rock Eng.* **2016**, *49*, 2541–2554. [CrossRef]
9. Ma, S.P.; Zhou, H. The evolution of strain field on specimen surface during rock failure process Characteristics study. *J. Rock Mech. Eng.* **2008**, *27*, 1667–1673.
10. Hou, C.L.; Jiang, B.; Li, M.; Song, Y.; Cheng, G.X. Micro-deformation and fracture evolution of in-situ coal affected by temperature, confining pressure, and differential stress. *J. Nat. Gas Sci. Eng.* **2022**, *100*, 104455. [CrossRef]
11. Lockner, D.A.; Byerlee, J.D.; Kuksenko, V.; Ponomarev, A.; Sidorin, A. Quasi-static fault growth and shear fracture energy in granite. *Nature* **1991**, *350*, 39–42. [CrossRef]

12. Cox, S.J.D.; Meredith, P.G. Microcrack formation and material softening in rock measured by monitoring acoustic emissions. *Int. J. Rock Mech. Min. Sci. Geomech. Abstr.* **1993**, *30*, 11–24. [CrossRef]
13. He, M.C.; Miao, J.L.; Feng, J.L. Rock burst process of limestone and its acoustic emission characteristics under true-triaxial unloading conditions. *Int. J. Rock Mech. Min. Sci.* **2010**, *47*, 286–298. [CrossRef]
14. Zhao, A.H.; Ma, Y.K.; Liu, J.; Chen, D.Z.; Yuan, H.Y.; Fu, M. Energy characteristics of AE signal frequency band during coal rupture in mines with gas. *China Saf. Sci. J.* **2020**, *30*, 115–121.
15. Ishida, T.; Labuz, J.F.; Manthei, G.; Meredith, P.G.; Nasseri, M.H.B.; Shin, K.; Yokoyama, T.; Zang, A. ISRM suggested method for laboratory acoustic emission monitoring. *Rock Mech. Rock Eng.* **2017**, *50*, 665–674. [CrossRef]
16. Zang, A.; Wagner, F.C.; Stanchits, S.; Dresen, G.; Andresen, R.; Haidekker, M.A. Source analysis of acoustic emissions in Aue granite cores under symmetric and asymmetric compressive loads. *Geophys. J. Int.* **1998**, *135*, 1113–1130. [CrossRef]
17. Aker, E.; Kühn, D.; Vavryčuk, V.; Soldal, M.; Oye, V. Experimental investigation of acoustic emissions and their moment tensors in rock during failure. *Int. J. Rock Mech. Min. Sci.* **2014**, *70*, 286–295. [CrossRef]
18. Wang, X.R.; Liu, X.F.; Wang, E.Y.; Li, X.L.; Zhang, X.; Zhang, C.; Kong, B. Experimental research of the AE responses and fracture evolution characteristics for sand-paraffin similar material. *Constr. Build. Mater.* **2017**, *132*, 446–456. [CrossRef]
19. Burud, N.B.; Kishen, J.M.C. Investigation of long memory in concrete fracture through acoustic emission time series analysis under monotonic and fatigue loading. *Eng. Fract. Mech.* **2023**, *277*, 108975. [CrossRef]
20. Triantis, D.; Kourkoulis, S.K. An alternative approach for representing the data provided by the acoustic emission technique. *Rock Mech. Rock Eng.* **2018**, *51*, 2433–2438. [CrossRef]
21. Wang, X.R.; Wang, E.Y.; Liu, X.F. Damage characterization of concrete under multi-step loading by integrated ultrasonic and acoustic emission techniques. *Constr. Build. Mater.* **2019**, *221*, 678–690. [CrossRef]
22. Lavrov, A. The Kaiser effect in rocks: Principles and stress estimation techniques. *Int. J. Rock Mech. Min. Sci.* **2003**, *40*, 151–171. [CrossRef]
23. Lockner, D.A. The role of acoustic emission in the study of rock fracture. *Int. J. Rock Mech. Min. Sci. Geomech.* **1993**, *30*, 88–899. [CrossRef]
24. Backers, T.; Stanchits, S.; Dresen, G. Tensile fracture propagation and acoustic emission activity in sand stone: The effect of loading rate. *Int. J. Rock Mech. Mining Sci.* **2005**, *42*, 1094–1101. [CrossRef]
25. Wang, E.Y.; He, X.Q.; Liu, Z.T.; Li, Z.H. Study on frequency spectrum characteristics of acoustic emission in coal or rock de-formation and fracture. *J. China Coal Soc.* **2004**, *29*, 289–292.
26. Xiao, C.X.; Jin, C.; Ding, X.; Pan, Y. Experimental study on rock burst tendency of coal with different moisture content based on acoustic emission time-frequency signals. *J. China Coal Soc.* **2018**, *43*, 931–938.
27. Yang, D.; Hu, J.; Ma, S.; Zeng, P. Analysis of dynamic fracture of granite after uniaxial recompression predamaged by high confining pressure cyclic loading based on acoustic emission. *Eng. Fract. Mech.* **2022**, *266*, 108414. [CrossRef]
28. Carpenter, S.R.; Brock, W.A. Rising variance: A leading indicator of ecological transition. *Ecol Lett.* **2006**, *9*, 311–318. [CrossRef]
29. Yan, R.; Jiang, C.S.; Zhang, L.P. Study on critical slowing down phenomenon of radon concentrations in water before the Wenchuan Ms 8.0 earthquake. *Chin. J. Geophys.* **2011**, *54*, 1817–1826.
30. Hao, W.; Wei, H.; Yan, P.C. Using the principle of critical slowing down to discuss the abrupt climate change. *Acta Phys. Sin.* **2013**, *62*, 556–565.
31. Dakos, V.; Scheffer, M.; van Nes, E.H.; Brovkin, V.; Petoukhov, V.; Held, H. Slowing down as an early warning signal for abrupt climate change. *Proc. Natl. Acad. Sci. USA* **2008**, *105*, 14308–14312. [CrossRef] [PubMed]
32. Hao, T.X.; Li, F.; Tang, Y.J.; Zhao, L.; Wang, Z. Infrared precursor of pre-cracked coal failure based on critical slowing down. *Geomat. Nat. Hazards Risk* **2022**, *13*, 1682–1699. [CrossRef]
33. Loukidis, A.; Stavrakas, I.; Triantis, D. Non-extensive statistical mechanics in acoustic emissions: Detection of upcoming fracture in rock materials. *Appl. Sci.* **2023**, *13*, 3249. [CrossRef]
34. Wei, Y.; Li, Z.H.; Kong, X.G.; Zhang, Z.B.; Wang, J.L.; Cheng, F.Q. Critical slowing characteristics of uniaxial compression damage in sandstone. *J. Coal* **2018**, *43*, 427–432.
35. Zhu, X.; Liu, H.X.; Hu, J.W.; Fan, J. Experimental study on the characteristics of critical slowing precursors of sandstone damage acoustic emission. *Geotechnics* **2022**, *43*, 164–172.
36. Li, D.X.; Wang, E.Y.; Li, Z.H.; Ju, Y.; Wang, D.; Wang, X. Experimental investigations of pressure stimulated currents from stressed sandstone used as precursors to rock fracture. *Int. J. Rock Mech. Min. Sci.* **2021**, *145*, 104841. [CrossRef]
37. Wang, X.R.; Wang, E.Y.; Liu, X.; Zhou, X. Micromechanisms of coal fracture: Insights from quantitative AE technique. *Theor. Appl. Fract. Mech.* **2021**, *114*, 103000. [CrossRef]
38. Kong, X.G.; Wang, E.Y.; Hu, S.B.; Li, Z.; Liu, X.; Fang, B.; Zhan, T. Critical slowing down on acoustic emission characteristics of coal containing methane. *J. Nat. Gas Sci. Eng.* **2015**, *24*, 156–165. [CrossRef]
39. Yuan, S.X.; Jiang, T.; Lei, J.H.; Cui, C.H. Experimental study on fracture characteristics of rock-like material with prefabricated cracks under compression shear. *Sci. Rep.* **2022**, *12*, 2809. [CrossRef]
40. Sun, H.; Chen, S.J.; Jin, A.B.; Zhu, D.F. Uniaxial compressive strength characteristics and crack evolution law of fractured rock samples. *J. Northeast. Univ. Nat. Sci. Ed.* **2022**, *43*, 404–413.
41. Zhang, X.; Tang, J.P.; Pan, Y.S.; Yu, H. Experimental study on intensity and energy evolution of deep coal and gas outburst. *Fuel* **2022**, *324*, 124484. [CrossRef]

42. Zhang, X.; Tang, J.P.; Yu, H.H.; Pan, Y. Gas pressure evolution characteristics of deep true triaxial coal and gas outburst based on acoustic emission monitoring. *Sci. Rep.* **2022**, *12*, 21738. [CrossRef]
43. Kong, X.G.; Wang, E.Y.; He, X.Q.; Li, D.; Liu, Q. Time-varying multifractal of acoustic emission about coal samples subjected to uniaxial compression. *Chaos Solitons Fractals* **2017**, *103*, 571–577. [CrossRef]
44. Scheffer, M.; Bascompte, J.; Brock, W.A.; Brovkin, V.; Carpenter, S.R.; Dakos, V.; Held, H.; Van Nes, E.H.; Rietkerk, M.; Sugihara, G. Early-warning signals for critical transitions. *Nature* **2009**, *461*, 53–59. [CrossRef]
45. Ramos, O. Criticality in earthquakes. Good or bad for prediction? *Tectonophysics* **2009**, *485*, 321–326. [CrossRef]
46. Zhang, Z.B.; Wang, E.Y.; Zhang, H.T.; Bai, Z.M.; Zhang, Y.H.; Chen, X. Research on nonlinear variation of elastic wave velocity dispersion characteristic in limestone dynamic fracture process. *Fractals* **2023**, *31*, 2350008. [CrossRef]

sustainability

MDPI

Article

Characteristics of Pressure Stimulated Current and Damage Evolution of Granite under Progressive Uniaxial Loading

Dexing Li [1,2,3,*], Enyuan Wang [1,*], Jianhua Yue [2], Manman Li [4], Li Li [3], Dongming Wang [1] and Wei Liang [3]

1 School of Safety Engineering, China University of Mining and Technology, Xuzhou 221116, China
2 School of Resources and Geosciences, China University of Mining and Technology, Xuzhou 221116, China
3 Shenzhen Urban Public Safety Technology Research Institute Co., Ltd., Shenzhen 518000, China
4 Xuzhou Jinglan New Material Technology Co., Ltd., Xuzhou 221000, China
* Correspondence: ldx3180@cumt.edu.cn (D.L.); weytop@cumt.edu.cn (E.W.)

Abstract: The application of load on rock materials stimulates a weak current known as Pressure Stimulated Current (PSC). This study focuses on investigating the damage evolution of granite rocks through the analysis of PSC responses. Uniaxial loading experiments were conducted on granite samples, and the accompanying PSC was measured in real-time. The relationship between PSC characteristics and mechanical behaviors of granite was examined to explore precursory information related to granite failure. The damage evolution of granite was assessed using a damage variable defined as the cumulative charge (time integral of PSC). The results clearly demonstrate a close correlation between the variation of PSC and the mechanical behaviors of rock. Specifically, during the compaction and elastic deformation stages, PSC exhibits a slow and linear increase. However, once deformation enters the plastic stage, PSC demonstrates an accelerated upward trend. Additionally, it was observed that a stress drop coincides with an abnormal increase in PSC, which is followed by a rapid decay. The fluctuation observed after the abnormal increase in PSC during the accelerated growth phase can serve as a precursor of rock failure. Furthermore, the cumulative charge quantity effectively correlates with the damage process of granite samples. The stress–strain curve obtained from a theoretical constitutive model, established based on the damage variable represented by normalized cumulative charge, aligns reasonably well with the experimental results, affirming that the defined damage variable accurately reflects the damage evolution process of rocks. It is hypothesized that PSCs are carried by electrons within the rocks, which are stimulated by electron diffusion during deformation and experience a sharp increase upon rock fracturing. The research findings hold theoretical significance for predicting rockburst incidents using the PSC method.

Keywords: rock mechanics; weak current; damage variable; rock burst; precursor information

Citation: Li, D.; Wang, E.; Yue, J.; Li, M.; Li, L.; Wang, D.; Liang, W. Characteristics of Pressure Stimulated Current and Damage Evolution of Granite under Progressive Uniaxial Loading. *Sustainability* **2023**, *15*, 14526. https://doi.org/10.3390/su151914526

Academic Editor: Edoardo Bocci

Received: 4 September 2023
Revised: 2 October 2023
Accepted: 2 October 2023
Published: 6 October 2023

1. Introduction

Rockburst, a geological disaster resulting from the sudden release of accumulated elastic deformation energy along the free face of underground engineering, poses a significant threat to human lives and property [1]. The occurrence process and severe consequences of rockburst make it a global problem in deep underground engineering, hindering efficient development and utilization of underground space and resources [2–4]. The process of rockburst evolution primarily involves the accumulation of damage in rock mass, with sudden rock failure being the direct cause [5–7]. Therefore, studying the damage evolution process of rocks is crucial for rockburst prediction.

During the deformation and fracture of rocks under loading, a weak current known as Pressure Stimulated Current (PSC) can be generated [8–13]. The variation of PSC is closely associated with the mechanical behaviors of rock materials. Stavrakas et al. [10] observed a significant increase in PSC when stress exceeds approximately 0.6 times the peak stress of marble specimens, with the peak current being directly proportional to the loading rate. Triantis et al. [11] suggested that PSC arises from changes in Young's modulus rather than the loading rate. Kyriazopoulos et al. [14] proposed that the magnitude of PSC is directly proportional to the strain rate, and fracture occurrence is accompanied by a sudden increase in PSC. Freund et al. [12,15–17] found that weak current generated from igneous rocks increases rapidly even at low stress levels and is influenced by the silicate particle content of rocks. Li et al. [13,18] observed that PSC exhibits different responses in various deformation stages of rock materials and increases with the strain rate. Additionally, Li et al. [19] discovered different patterns of PSC changes in rocks under low and high loading rates. Moreover, PSC serves as a reliable indicator of the damage evolution in rock materials and exhibits precursor responses to failure. Triantis et al. [20] detected clear indications of the breaking stress of cement mortar based on PSC, which can be used for predicting final failure. Vallianatos and Triantis [21] introduced non-extensive statistical mechanics (NESM) to study the characteristics of PSC from marble and amphibolite, and the behavior of the Tsallis q-parameter indicates that the rock fracturing is a subadditive process with hierarchically constrained dynamics. Subsequently, Stergiopoulos et al. [22] studied PSCs from mortar beams using NESM and found that the change of q-parameter can be used to predict the failure and instability of cement mortar materials. Li et al. [13] identified an accelerated increase in PSC as a precursor of rock failure under progressively increasing loads, while pulsed fluctuations in PSC serve as precursors to creep rupture. Stavrakas et al. [23] investigated the spatiotemporal evolution of damage in marble specimens under uniaxial compression using PSC and acoustic emission (AE) techniques, demonstrating that both PSC and AE can be employed as pre-failure indicators. Triantis et al. [24] conducted a study investigating the relationship between PSC and AE in brittle materials near fracture load levels. Their findings revealed that the precursor to failure based on weak currents precedes that of AE. Similarly, Loukidis et al. [25] conducted research on the correlation between AE and PSC in the vicinity of fractures in cement mortars during uniaxial compressive loading, yielding similar results. Li et al. [26] conducted a study on the mechanical behaviors of coal under concentrated load conditions and examined the response of PSCs in this context. Based on their findings, they proposed principles for using the PSC technique in predicting rock dynamic disasters. Subsequently, Li et al. [18] successfully measured weak currents in an underground coal mine using their specially developed mine-used weak current monitoring equipment. Their investigation revealed that weak currents exhibit an advance response to mine seismicity. Therefore, the use of the PSC technique for rockburst prediction in underground engineering is considered feasible.

The damage evolution of rocks can be quantitatively characterized by a damage variable, which has been extensively studied in the literature [27,28]. This variable can be defined based on mechanical property parameters and emitted physical signals, enabling a comprehensive understanding of the damage process [29,30]. In the context of the strain equivalence hypothesis, Lemaitre [31] proposed an expression for the damage variable defined by elastic modulus. Jin et al. [32] defined the damage variable for rocks based on energy dissipation and provided a theoretical calculation formula, as well as a method for determining the damage threshold. They utilized this approach to study the damage evolution law of rock materials. Liu et al. [33] introduced a damage variable for rocks based on the normalized cumulative ring-down count of AE signals. They further developed a damage model for coal-rock under uniaxial compression, which facilitated the investigation of the damage evolution law. Li et al. [34] defined the damage variable for sandstone based on electrical resistivity and proposed a comprehensive damage variable capable of effectively reflecting and describing the damage evolution process of rock samples subjected to progressive loading. Gong et al. [30] established a damage constitutive model for brittle rock under uniaxial compression by utilizing a damage variable derived from the linear energy dissipation law. However, the quantitative description of damage evolution in rocks using a damage variable defined by PSCs has not been reported in the literature.

This knowledge gap hinders a deep understanding of the generation mechanism of PSCs in rocks and limits the broader application of the PSC technique in rockburst prediction.

In this paper, the response characteristics of PSCs from granite samples under a progressive uniaxial load are investigated. The damage variable based on cumulative charge is defined, and the damage evolution formula is derived, allowing for further analysis of the damage evolution process. The precursor information of rock failure based on PSCs is determined. Furthermore, a potential generation mechanism of PSCs from rocks during deformation and fracture is proposed. Lastly, the application of the PSC method in rockburst prediction is discussed.

2. Material and Method

2.1. Specimen Preparation

The granite blocks used in this study were obtained from a mine located in Shandong, China. To prepare the specimens, twelve cylindrical samples with a diameter of 50 mm and a height of 100 mm were cored and cut from the blocks, following the standards set by the International Society for Rock Mechanics [35]. After the cylindrical samples were obtained, the surfaces of each specimen were meticulously ground to achieve a flat and uniform surface. This step was crucial to ensure proper contact and alignment during the subsequent experimental procedures.

2.2. System and Equipment

The experimental system was comprised of a loading system, a weak current measuring system, and an electromagnetic shielding system (Figure 1). To apply the load, the YAW4306 compression testing machine from MTS Industrial Systems (China) Co., Ltd., Shanghai, China was utilized. The loading process was controlled, and mechanical parameters such as stress and strain were recorded using the Power Test V3.3 software from SANS (Shanghai) Enterprise Development Co., Ltd., Shanghai, China. The weak current measuring system included a Keithley 6517B electrometer from Tektronix (China) Co., Ltd., Shanghai, China, a computer, data acquisition software, a tri-coaxial cable, and electrodes. To minimize the interference of surrounding electrical noise, the experiments were conducted in an electromagnetic shielding room.

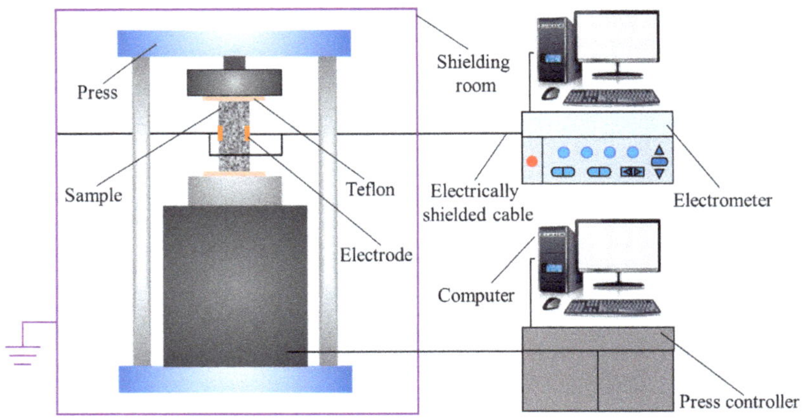

Figure 1. Experimental system and equipment.

2.3. Experimental Procedures and Schemes

Figure 1 illustrates the experimental setup used in this study. Two oblong copper electrodes, measuring 30 mm × 20 mm, were affixed to the side of a granite specimen using conductive silver paint. Insulating tape was used to secure the electrodes in place and prevent detachment. The electrodes were connected to the electrometer via a tri-coaxial cable equipped with alligator clips. To provide electrical insulation, the sample was positioned between two polytetrafluoroethylene plates with an 80 mm diameter and 2 mm thickness. Prior to each test, the press was adjusted to ensure that the upper insulation sheet was approximately one millimeter away from the loading head. A load of 200 N was then applied at a rate of 0.2 mm/min and maintained until the current reached a stable state. The experimental scheme involved subjecting the granite specimens to a progressive uniaxial load at constant rates of 0.1 MPa/s, 0.2 MPa/s, and 0.5 MPa/s until the failure occurred.

3. Results and Analysis

3.1. PSC Response Characteristics

Due to the similarity of the results, the results of Specimens GLS01, GLS02, and GLS03 was selected for analysis. Figure 2 illustrates the variation of stress and PSC with time throughout the entire uniaxial loading process for these three specimens. Notably, the PSC variations from these specimens exhibit similar trends. Based on the PSC-time curve, the variation of PSC can be categorized into two distinct stages:

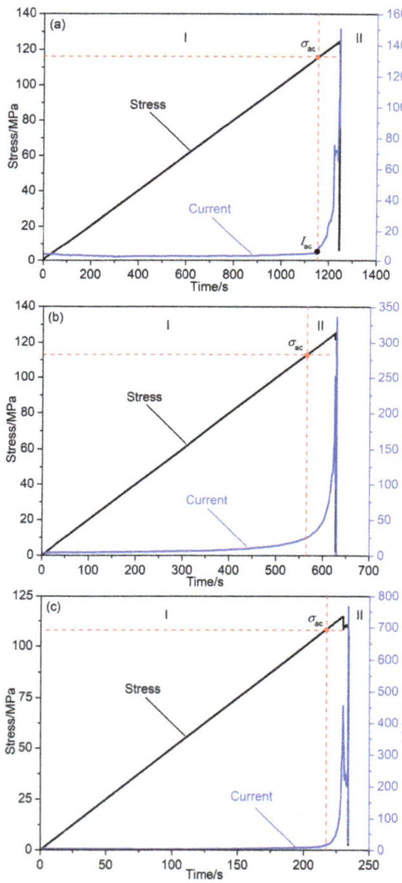

Figure 2. Temporal variation of stress and PSC of Specimen (**a**) GSL01, (**b**) GSL02 and (**c**) GSL03. Phases (I) Slow Increase and (II) Accelerating Increase.

(I) Slow Increase Phase: During this phase, the PSC values are initially very small and exhibit a gradual and slow increase. The increase in PSC is approximately linear with the increase in stress.

(II) Accelerating Increase Phase: In this phase, the PSC demonstrates an accelerated increasing trend, eventually reaching its peak value at the point of failure. Subsequently, the PSC abruptly decreases.

To differentiate between these two phases, a stress value called the Current Accelerating Stress (σ_{ca}) is identified, and the stress represents the stress at which the transition from phase I to phase II occurs. The specific values of σ_{ca} for the three specimens are listed in Table 1. By analyzing the stress and PSC variations over time for the selected specimens (GLS01, GLS02, and GLS03), the distinct phases of PSC evolution can be observed, providing valuable insights into the behavior of PSCs during the uniaxial loading process.

Table 1. Characteristic stress and strain of the three specimens.

Specimen	σ_{ca}/MPa	σ_y/MPa	ε_y/%
GLS01	117.2	116.4	0.720
GLS02	112.9	113.1	0.852
GLS03	107.5	108.2	0.755

According to the stress–strain curve shown in Figure 3, the deformation of rocks before failure can be divided into three stages: compaction stage, elastic deformation stage, and plastic deformation stage [3]. Each stage is characterized by a different shape of the stress–strain curve. In the compaction stage, the stress–strain curve is concave, indicating a gradual increase in stress with relatively small strain. In the elastic deformation stage, the stress–strain curve is linear, indicating that the material exhibits elastic behavior with stress and strain being proportional to each other. Finally, in the plastic deformation stage, the stress–strain curve becomes convex, indicating a rapid increase in strain with a relatively small increase in stress.

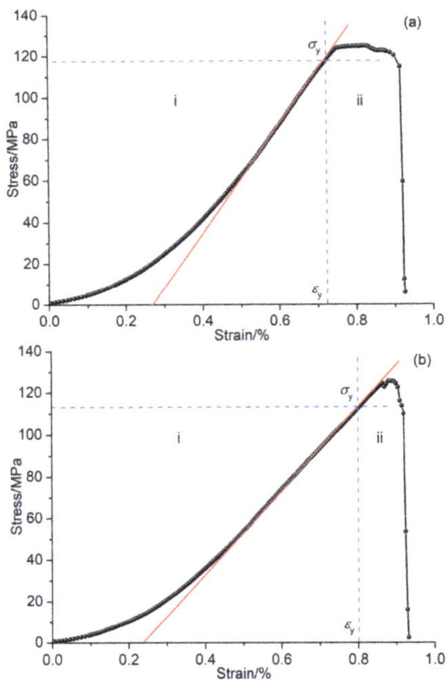

Figure 3. *Cont.*

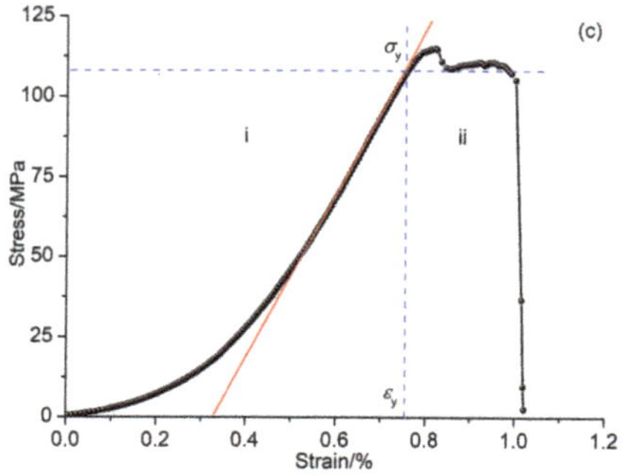

Figure 3. Stress–strain curve of Specimen (**a**) GSL01, (**b**) GSL02 and (**c**) GSL03 during the whole loading. (i) Compaction and elastic deformation stage and (ii) Plastic deformation stage. The red line marks the elastic deformation stage.

The yield stress (σ_y) is used to distinguish between the elastic and plastic deformation stages, with the yield strain (ε_y) corresponding to this transition. Table 1 provides the values of the yield stress (σ_y) and the Current Accelerating Stress (σ_{ac}) for specimens GSL01, GSL02, and GSL03. It is observed that the yield stress values are approximately equal to the corresponding Current Accelerating Stress values for all three specimens. This suggests that the point at which the PSC starts to accelerate (σ_{ac}) is close to the yield stress point (σ_y).

Based on these observations, it can be concluded that the PSC increases slowly and linearly with stress during the compaction and elastic deformation stages. However, it accelerates during the plastic deformation stage. Therefore, the accelerated increase in PSC can be considered as an indication of the onset of the plastic deformation stage.

3.2. *Damage Evolution of Granite Analyzed by PSC*

3.2.1. Analytical Expression of Damage Variable

The damage variable, denoted as D, represents the proportion of failed micro elements to the total micro elements in a material. It quantifies the extent of damage or loss of bearing capacity in the material. Kachanov [27] defined the damage variable as the ratio of the area where the material has lost its bearing capacity to the initial area of the material. The damage variable can be calculated by comparing the total area of failure elements on the bearing surface (A_d) with the undamaged or nondestructive area (A), as shown in Figure 4. The calculation equation for the damage variable is as follows:

$$D = \frac{A_d}{A} \quad (1)$$

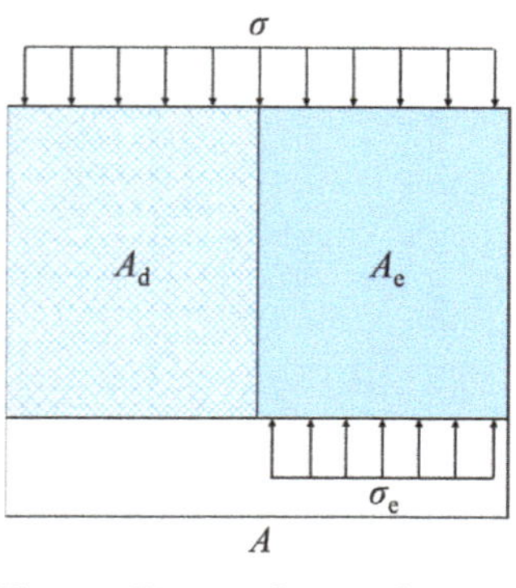

Figure 4. Diagram showing the stress state of an elastoplastic damaged body element.

It is important to note that the damage variable typically ranges between 0 and 1, where D = 0 represents an undamaged material with full bearing capacity, and D = 1 indicates complete failure or loss of bearing capacity.

Due to the loss of bearing capacity in the damaged part (A_d), the applied stress is entirely borne by the undamaged area. This relationship can be expressed by the following equation:

$$\sigma A = \sigma_e A_e \tag{2}$$

where σ is the applied stress, σ_e is the stress in the undamaged area, A_e is the undamaged area, and A is the total area. Then

$$\sigma = \sigma_e(1 - D) \tag{3}$$

Assuming that the stress and strain of the undamaged part of the element follow Hooke's law of linear elasticity, the relationship between stress and strain can be expressed as

$$\sigma_e = E\varepsilon \tag{4}$$

where E is the Young's modulus of the material, and ε is the strain.

By combining Equations (3) and (4), the following relationship can be derived as

$$\sigma = (1 - D)E\varepsilon \tag{5}$$

The application of load to the rock induces the movement of charges, leading to the generation of a weak current. If the cumulative charge of complete destruction of the whole section A of the material is Q_0, the charge per unit area of destruction (Q_w) can be calculated by dividing Q_0 by the area of the destructed region (A_d). This can be expressed as

$$Q_w = \frac{Q_0}{A} \tag{6}$$

When the damaged area of the section reaches A_d, the charge accumulation Q_d is calculated using the following formula:

$$Q_d = Q_w A_d = \frac{Q_0}{A} A_d \tag{7}$$

Then

$$D = \frac{Q_d}{Q_0} \tag{8}$$

In many cases of loading tests, the compression testing machine may stop working before the granite samples are completely damaged. This could be due to the machine's insufficient stiffness or the different failure conditions set for various samples. Therefore, the traditional definition of the damage variable needs to be modified to account for this incomplete damage [33]. The modified damage variable can be expressed as follows:

$$D = D_u \frac{Q_d}{Q_0} \tag{9}$$

where D_u is the critical damage variable, and Q_0 is the cumulative charge quantity when the damage variable reaches D_u. The value of D_u can be determined using the following formula:

$$D_u = 1 - \frac{\sigma_r}{\sigma_f} \tag{10}$$

where σ_f represents the peak stress and σ_r represents the residual stress.

The damage constitutive model for the rock samples under uniaxial compression, derived from cumulative charges, can be expressed as follows:

$$\sigma = (1 - D)E\varepsilon = \left(1 - D_{\mathrm{u}}\frac{Q_{\mathrm{d}}}{Q_0}\right)E\varepsilon \tag{11}$$

3.2.2. Damage Evolution Law of Granite

The quantity of charge flowing through a section of a rock specimen in a certain period of time can be calculated by integrating the current with respect to time, which corresponds to the area under the current-time curve, as shown in Figure 5. The accumulated charge (Q) can be calculated as follows:

$$Q = \int_{t_1}^{t_2} I(t)\mathrm{d}t \tag{12}$$

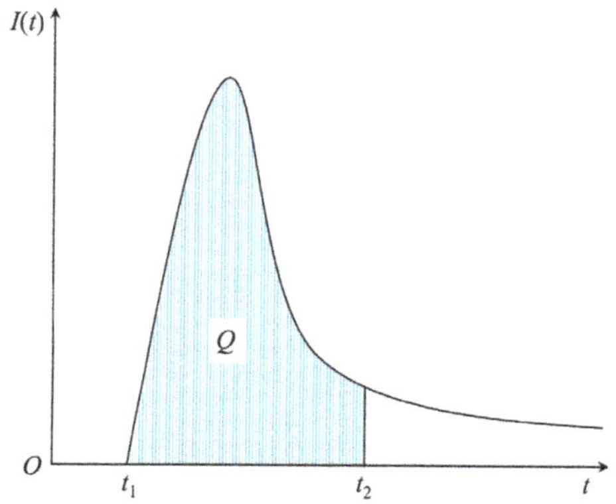

Figure 5. Schematic diagram of charge calculation.

The quantity of cumulative charges (Q) in the entire process of rock damage can be calculated by summing up the charges at each time step. This can be expressed as:

$$Q_0 = \int_{0}^{t_{\mathrm{f}}} I(t)\mathrm{d}t \tag{13}$$

where I(t) is the current at time t, and t_{f} is the time it takes for the granite specimen to fail finally.

The cumulative charge Q_{d} at any time t_{i} during loading can be calculated by

$$Q_{\mathrm{d}} = \int_{0}^{t_{\mathrm{i}}} I(t)\mathrm{d}t \tag{14}$$

According to the calculations using Equation (14), the real-time cumulative charge quantity of Specimens GLS01 and GLS02 during the entire loading process was obtained. The variation of cumulative charges with time is depicted in Figure 6. From the figure, it can be observed that the cumulative charge initially exhibits a linear and steady increase. As the loading progresses, the cumulative charge shows an accelerated growth rate. This behavior is consistent with the change in current, indicating a correlation between the cumulative charge and the current. These findings provide valuable insights into the relationship between cumulative charge and the mechanical behavior of the specimens under loading conditions.

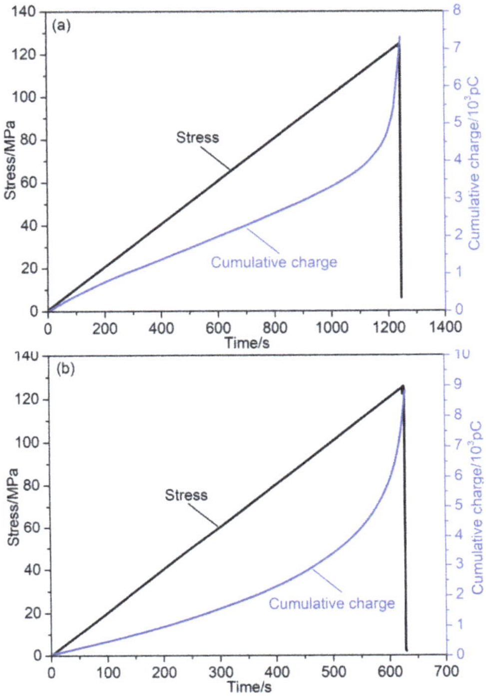

Figure 6. Variation of stress and cumulative charge with time of Specimen (**a**) GSL01 and (**b**) GSL02.

According to the provided data, the residual stresses of Specimens GLS01 and GLS02 are 5.96 MPa and 2.09 MPa, respectively. Using Equations (10) and (13), the parameters D_u and Q_0 are calculated, and their values are listed in Table 2. By applying Equation (9), the damage variable at any moment can be calculated. The scatter plot of the damage variable (D) with respect to strain (ε) is shown in Figure 7. From the figure, it can be observed that the scatter curve exhibits an inflection point. Before the inflection point, the D-ε curve is concave, indicating a slower increase in the damage variable with strain. After the inflection point, the curve becomes convex, suggesting a more rapid increase in the damage variable with strain.

Table 2. The typical stresses and damage variable of Specimen GSL01 and GSL02.

Specimen	σ_f/MPa	σ_r/MPa	D_u/s	Q_0/pC	ε_i
GLS01	124.9	5.96	0.952	7309.5	0.0074
GLS02	125.6	2.09	0.983	8815.6	0.0087

Therefore, the damage evolution process can be analyzed in two stages, which are divided by the inflection point. The corresponding strain values (ε_i) for Specimen GLS01 and GLS02 are calculated to be 0.0074 and 0.0087, respectively. These observations highlight the non-linear behavior of the damage evolution process and provide valuable insights into the deformation characteristics of granite specimens under uniaxial compression.

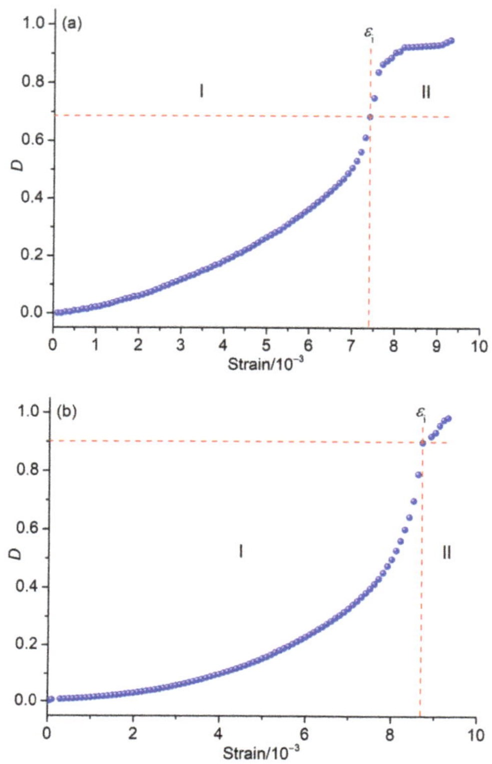

Figure 7. Scatters of damage variable with strain of Specimens (**a**) GSL01 and (**b**) GSL02.

Based on the scatter plot of the damage variable (D) with respect to strain (ε) for Specimen GLS01 (Figure 7a), a piecewise fitting is performed to approximate the damage evolution equation (Figure 8). The deduced equation is as follows:

$$D = \begin{cases} 0.08046e^{-288\varepsilon} - 0.08089 \; (0.0000 \leq \varepsilon \leq 0.0074) \\ -7.94026 \times 10^{11} e^{-3891\varepsilon} + 0.93218 \; (0.0074 \leq \varepsilon \leq 0.0093) \end{cases} \quad (15)$$

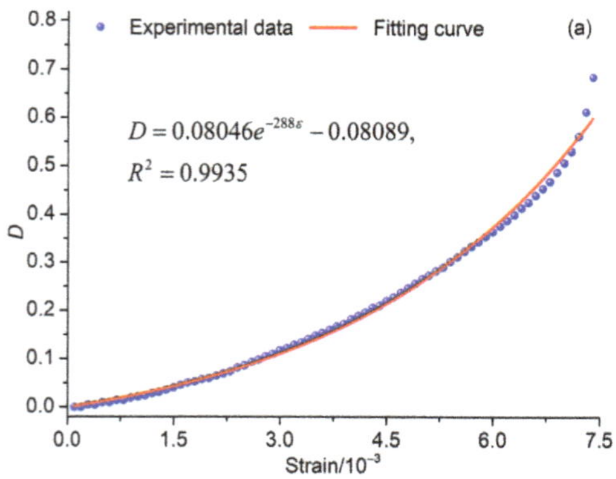

Figure 8. *Cont.*

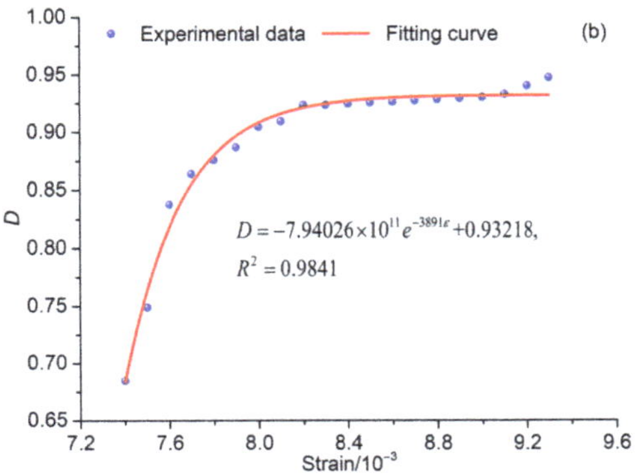

Figure 8. Piecewise fitting of D-ε scatters of Specimen GSL01 during the whole damage process. (**a**) $0 < \varepsilon \leq 0.0074$; (**b**) $0.0074 \leq \varepsilon \leq 0.0093$.

Similarly, based on the scatter plot of the damage variable (D) with respect to strain (ε) for Specimen GLS02 (Figure 7b), a piecewise fitting is performed to deduce the damage evolution equation (Figure 9). The equation is as follows:

$$D = \begin{cases} 0.04173e^{312\varepsilon} + 4.46085 \times 10^{-12}e^{2866\varepsilon} - 0.04526\,, \ (0 \leq \varepsilon \leq 0.0087) \\ 152.35638\varepsilon - 0.4373\,, \ (0.0087 \leq \varepsilon \leq 0.0093) \end{cases} \quad (16)$$

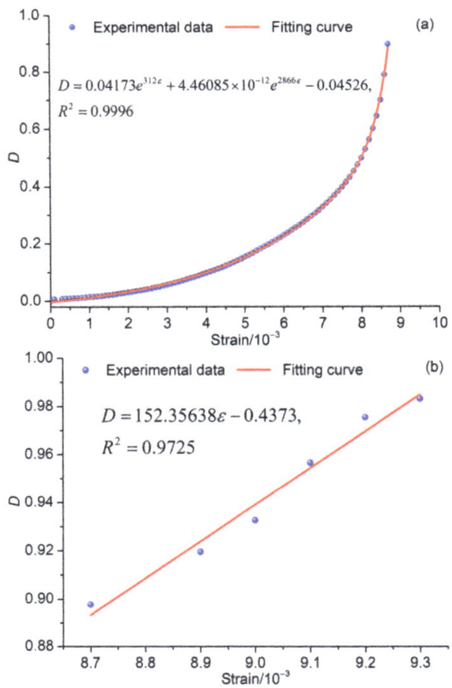

Figure 9. Piecewise fitting of D-ε scatters of Specimen GSL02 during the whole damage process. (**a**) $0 < \varepsilon \leq 0.0087$; (**b**) $0.0087 \leq \varepsilon \leq 0.0093$.

3.2.3. Damage Constitutive Model for Granite

The elastic moduli of Specimens GSL01 and GSL02 were determined by performing linear fitting on the stress–strain curves obtained during the elastic deformation stage. The calculated elastic modulus values for Specimens GSL01 and GSL02 are 30,523 MPa and 22,316 MPa, respectively. By incorporating Equations (11) and (15), the one-dimensional damage constitutive models for Specimens GSL01 and GSL02 based on cumulative charges can be expressed as follows:

$$\sigma = \begin{cases} 30523\varepsilon(1.08089 - 0.08046e^{-288\varepsilon})\ (0.0000 \leq \varepsilon \leq 0.0074) \\ 30523\varepsilon(7.94026 \times 10^{11}e^{-3891\varepsilon} + 0.06782)\ (0.0074 \leq \varepsilon \leq 0.0093) \end{cases} \quad (17)$$

$$\sigma = \begin{cases} 23316\varepsilon(1.04526 - 0.04173e^{312\varepsilon} - 4.46085 \times 10^{-12}e^{2866\varepsilon}),\ (0 \leq \varepsilon \leq 0.0087) \\ 23316\varepsilon(1.4373 - 152.35638\varepsilon),\ (0.0087 \leq \varepsilon \leq 0.0093) \end{cases} \quad (18)$$

As depicted in Figure 10, the trend of the theoretical stress–strain curve obtained from the constitutive model aligns with that of the experimental curve. This indicates that the defined damage variable is reasonable and effectively captures the damage progression in granite samples. However, there exists a disparity between the peak limit strains (ε_p) obtained from the theoretical and experimental curves. To quantify this difference, the stress–strain curve in the elastic deformation stage is extended with a dotted line (Figure 10), intersecting the horizontal axis at the point with coordinates (ε_0, 0). In this paper, ε_0 is defined as the ideal initial strain. The peak strain offset ($\Delta\varepsilon_p$) between the theoretical and experimental failure limit strains can be calculated using the following equation:

$$\Delta\varepsilon_p = \varepsilon_{pe} - \varepsilon_{pt} \quad (19)$$

where ε_{pe} represents the experimental failure limit strain, and ε_{pt} represents the theoretical failure limit strain.

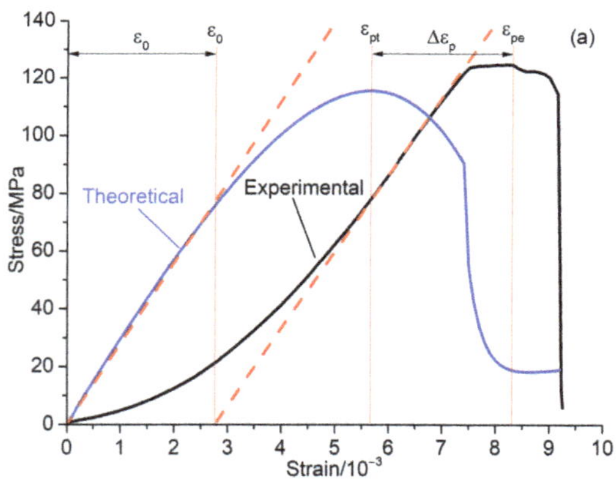

Figure 10. *Cont.*

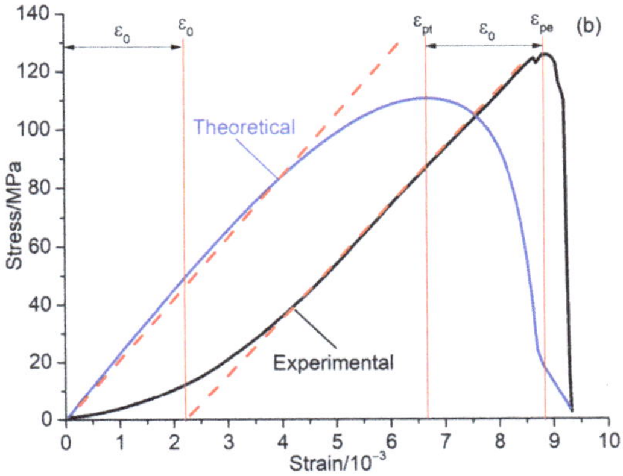

Figure 10. Theoretical and experimental stress–strain curve of Specimen (**a**) GLS02 and (**b**) GLS03.

The strain parameters of Specimen GSL01 and GSL02 are provided in Table 3. It can be observed that the ideal initial strain is very close to the peak strain offset, indicating a strong correlation between the difference in peak limit strains and the ideal initial strain. During the initial stages of loading, the rock undergoes compaction, where primary fractures gradually close, leading to an increase in deformation modulus until reaching the elastic deformation stage. This concave-shaped stress–strain curve in the elastic deformation stage results in the presence of an ideal initial strain. Therefore, the deviation between the theoretical and experimental strain-stress curves is primarily influenced by the closure of primary fractures. Based on the analysis, it can be concluded that the damage variable defined by cumulative charge is reasonable and effectively reflects the damage evolution process of granite specimens.

Table 3. Typical strain parameters of Specimen GSL01 and GSL02.

Specimen	ε_{pt}	ε_{pe}	$\Delta\varepsilon_p$	ε_0
GLS01	0.00660	0.00885	0.00225	0.00220
GLS02	0.00695	0.00960	0.00265	0.00330

4. Discussion

4.1. Mechanism of PSCs from Granite during Damage

The generation of weak current in granite indicates the presence of carriers within the rock. According to the dielectric conduction theory, there are three forms of carriers: ions, electrons, and holes [36].

Ionic conductivity requires the presence of freely moving ions, which typically occurs in aqueous solutions or molten states. However, in our loading experiments on dry granite at room temperature, there is no condition for the free movement of ions, suggesting that ions are not the carriers of the weak current in this case. Freund [37] and Scoville et al. [16] proposed that p-holes, generated by the loss of an electron by oxygen atoms in silicate minerals, are the main carriers of current in igneous rocks. This theory was also applied to explain the weak current observed in stressed coals [26]. The formation of p-holes involves the fracture of chemical bonds, which requires the absorption of energy. The rapid increase in the weak current suggests the quick formation and movement of a large number of p-holes, indicating the absorption of significant energy that cannot be released during the initial compression stage. Therefore, it is important to note that the main carriers of the weak current are not p-holes.

The above analysis and discussion show that the main carriers of the weak current in rocks are electrons. While a small number of electrons naturally exist in rocks, a significant quantity of electrons can be generated through friction, stress-induced polarization, and crack propagation [26]. These processes provide an ample supply of carriers for the weak current in rocks.

The current is defined as the rate of flow of charge and can be calculated using the formula [36]:

$$I = \frac{dq}{dt} \tag{20}$$

where I is the current, dq is the change in charge, and $\mathrm{d}t$ is the change in time.

In Figure 11a, when the rock is not loaded, it is in a thermal equilibrium state with free electrons in chaotic motion, resulting in electrical neutrality. However, it should be noted that the distribution of electrons follows the tip effect, where electrons tend to concentrate at the tip of a crack, and the electron density increases with the stress experienced by the crack tip. This means that the electrons in the primary microcracks are not uniformly distributed but rather concentrated at the tip of the crack, resulting in a higher electron density at the crack tip.

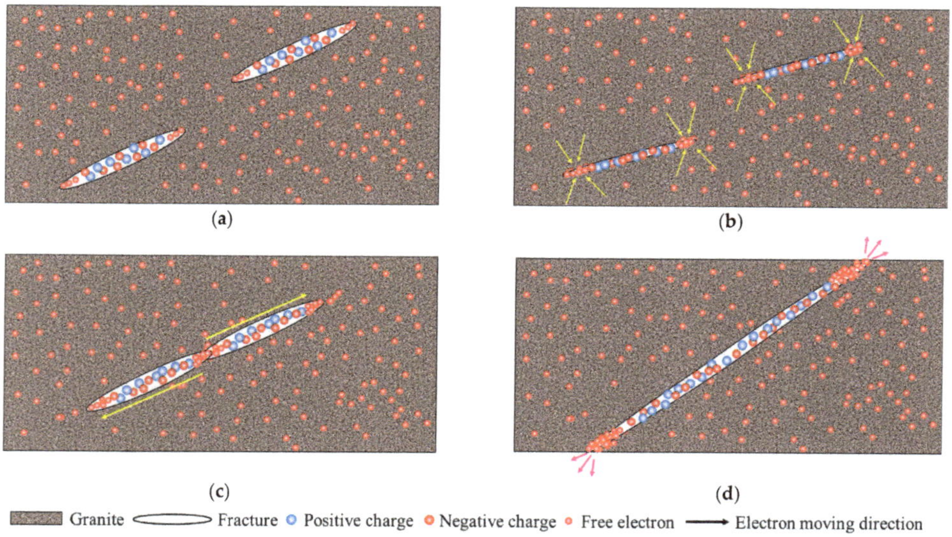

Figure 11. Schematic diagram showing a mechanism of PSCs from rock during deformation and fracture. (**a**) Unstressed state; (**b**) Crack closure; (**c**) Crack propagation; (**d**) Crack coalescence.

In Figure 11b, as the rock is loaded and the cracks start to close, both the original and newly generated free electrons are enriched towards the tip of the primary cracks. This leads to a reduction in charge density in the vicinity of the crack, causing remote charges to flow towards the low-density region to replenish the charge imbalance. This results in the generation of a diffusion current. During the compaction stage, where primary crack closure occurs gradually, the process of charge diffusion is slow. Similarly, during the elastic deformation stage, where the stress at the crack tip does not reach the critical value for crack propagation, the formation and flow of free charges remain slow. Therefore, PSC exhibits a slow and linear increase during the compaction and elastic deformation stages.

As the stress continues to increase, the primary microcracks start to propagate, entering the plastic deformation stage. In this stage, the release of free charges becomes rapid as cracks grow and move towards newly generated crack tips (Figure 11c). With increasing stress, both the number of new cracks and the speed of crack growth gradually increase, accelerating the process of charge transfer and release. As a result, the PSC increases with stress at an accelerated rate during the plastic deformation stage.

In Figure 11d, when the rock fractures due to internal crack coalescence, the accumulated charges at the crack tip are released rapidly, causing a sudden increase in current.

This is often referred to as a transient current. Additionally, as the final failure of the rock approaches, the activity of internal crack coalescence becomes more intense, leading to fluctuations in abnormal currents. These current fluctuations can serve as precursors to rock failure.

Overall, the observations in Figure 11 demonstrate that the distribution and movement of electrons in the rock are influenced by stress and crack propagation, leading to the generation and transfer of weak currents. These currents provide valuable insights into the damage evolution process and can serve as indicators of rock failure.

4.2. Application of PSC in Rockburst Prediction

Rockburst, as the name suggests, is a sudden and violent failure of rock mass that occurs during mining and tunnelling activities. The ultimate trigger for a rockburst event is the sudden failure of the rock, which results in the instantaneous release of stored elastic energy [38]. The process of rockburst involves the progressive aggravation of deformation and accumulation of damage in the rock mass.

Before the occurrence of a rockburst, a rock typically undergoes three stages of deformation: the compression stage, elastic deformation stage, and plastic deformation stage. During the compression and elastic deformation stages, the rock experiences deformation and accumulates elastic energy. In these stages, there is minimal crack generation and release of elastic energy, making it unlikely for a rock failure to occur. It is only when the rock enters the plastic deformation stage that local fractures start to develop, accompanied by the generation and propagation of cracks. This stage is critical, as it leads to the failure of the rock under increasing stress. Therefore, the prediction of rockburst events primarily relies on monitoring the deformation of the rock mass and associated physical signals. By closely monitoring the deformation patterns and physical signals, such as changes in stress, strain, AEs, electromagnetic radiations and other relevant parameters, it becomes possible to identify the onset of the plastic deformation stage and the potential for rockburst [39]. Early detection and prediction of rockburst can help mitigate the risks associated with these events and ensure the safety of mining and tunnelling operations.

Based on our experimental results, the damage variable defined by cumulative charge proves to be a reasonable indicator that effectively reflects the damage evolution process of rocks. Therefore, the variation in PSC can be utilized for rockburst prediction. In particular, the behavior of PSC during different deformation stages provides valuable insights. During the compaction and elastic deformation stages, PSC exhibits a slow and gradual increase. However, in the plastic deformation stage, there is an accelerated increase in PSC, which indicates that the rock deformation has entered a critical phase. This continuous acceleration in PSC can serve as an early signal of the transition to plastic deformation.

To further analyze the relationship between stress and PSC, we zoomed in on the stress and PSC curves during the period leading up to final failure, as shown in Figure 12. It can be observed that when a stress drop occurs, there is a sharp increase in PSC followed by a rapid decrease, indicating that the fracture is accompanied by an abnormal surge in PSC. Subsequently, PSC fluctuates at a high level and demonstrates a responsive behavior to stress variations. When the final failure occurs, PSC experiences another surge, reaching its peak value. It is noteworthy that after the failure, PSC drops to the background value. In contrast, when a local fracture occurs, PSC may drop slightly but continues to vibrate at a high value, significantly higher than the background level. Therefore, the post-abnormal increase fluctuation in PSC during the accelerated increase process can be regarded as a precursor to rock failure or rockburst.

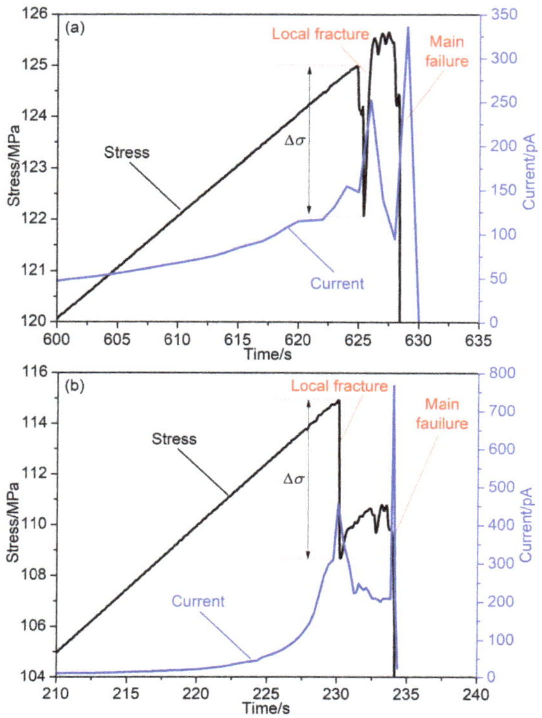

Figure 12. Variation of stress and current with time before rock failure. (**a**) GSL02; (**b**) GSL03.

Considering the strong anti-interference capability and early response to mine seismicity, the PSC technique holds promising potential for broader applications in rockburst prediction [18]. By utilizing the PSC technique, it becomes possible to enhance the ability to detect and forecast rockburst events, contributing to improved safety measures in mining and tunnelling operations.

5. Conclusions

The main conclusions can be summarized as follows:

(1) PSC can be generated from granite during deformation, and its variation effectively reflects the different deformation stages of the rock. During the compaction and elastic deformation stages, PSC shows a gradual increase with stress. However, after the deformation enters the plastic stage, PSC exhibits an accelerated increase in response to stress.

(2) PSC exhibits precursory behavior prior to granite failure. A stress drop is accompanied by a sudden and significant increase in PSC, which then decays rapidly. The growth rate and amplitude of the PSC associated with the main fracture are greater compared to those observed during a local fracture. Following the occurrence of the main failure, PSC experiences a sudden drop to pre-loading levels. In the case of a local fracture, PSC decreases slightly and continues to vibrate at a high value. The fluctuation following an abnormal increase in PSC during the accelerated increase process can serve as a precursor to rock failure.

(3) The cumulative charge calculated based on PSC effectively reflects the damage process of granite samples. The stress–strain curve obtained using a theoretical constitutive model, established according to the defined damage variable based on cumulative charge, is in good agreement with the experimental results. This indicates that

the damage variable defined by cumulative charge accurately reflects the damage evolution process of granite.

(4) The carriers of PSC in rocks are primarily electrons, and their distribution follows the tip effect, with charge enrichment towards the crack tip. The generation mechanism of weak currents during rock damage involves the transfer, accumulation, and release of free charges. During the deformation process, currents are stimulated due to charge diffusion resulting from density differences. When a fracture occurs, currents are stimulated by the rapid release of accumulated charges.

Author Contributions: Conceptualization, D.L., E.W. and J.Y.; Methodology, D.L. and J.Y.; Software, W.L.; Validation, L.L.; Formal analysis, M.L. and D.W.; Investigation, D.L. and D.W.; Data curation, M.L. and W.L.; Writing—original draft, D.L.; Writing—review & editing, E.W. and J.Y.; Supervision, E.W.; Project administration, L.L.; Funding acquisition, D.L. All authors have read and agreed to the published version of the manuscript.

Funding: This research was funded by the National Natural Science Foundation of China (52204257), Jiangsu Funding Program for Excellent Postdoctoral Talent (2022ZB512), the GuangDong Basic and Applied Basic Research Foundation (2022A1515110132), China Postdoctoral Science Foundation (2022M713371) and the Fundamental Research Funds for the Central Universities (2022QN1014).

Institutional Review Board Statement: Not applicable.

Informed Consent Statement: Not applicable.

Data Availability Statement: Data is available on request from the authors.

Conflicts of Interest: The authors declare no conflict of interest.

References

1. Kaiser, P.K.; McCreath, D.R.; Tannant, D.D. *Canadian Rockburst Support Handbook*; Geomechanics Research Centre, Laurentian University: Sudbury, ON, Canada, 1996; p. 314.
2. Li, X.B.; Zhou, J.; Wang, S.F.; Liu, B. Review and practice of deep mining for solid mineral resources speed. *Chin. J. Nonferrous Met.* **2017**, *27*, 1236–1262.
3. Li, D.X.; Wang, E.Y.; Kong, X.G.; Ali, M.; Wang, D.M. Mechanical behaviors and acoustic emission fractal characteristics of coal specimens with a pre-existing flaw of various inclinations under uniaxial compression. *Int. J. Rock Mech. Min. Sci.* **2019**, *116*, 38–51. [CrossRef]
4. Feng, G.L.; Feng, X.T.; Xiao, Y.X.; Yao, Z.B.; Hu, L.; Niu, W.J.; Li, T. Characteristic microseismicity during the development process of intermittent rockburst in a deep railway tunnel. *Int. J. Rock Mech. Min. Sci.* **2019**, *124*, 104135. [CrossRef]
5. Keneti, A.; Sainsbury, B.A. Review of published rockburst events and their contributing factors. *Eng. Geol.* **2018**, *246*, 361–373. [CrossRef]
6. Li, C.C. Principles and methods of rock support for rockburst control. *J. Rock Mech. Geotech.* **2021**, *13*, 46–59. [CrossRef]
7. Luo, S.; Gong, F.Q. Evaluation of energy storage and release potentials of highly stressed rock pillar from rockburst control perspectives. *Int. J. Rock Mech. Min. Sci.* **2023**, *163*, 105324. [CrossRef]
8. Vallianatos, F.; Tzanis, A. Electric current generation associated with the deformation rate of a solid: Preseismic and coseismic signals. *Phys. Chem. Earth* **1998**, *23*, 933–939. [CrossRef]
9. Vallianatos, F.; Triantis, D.; Tzanis, A.; Anastasiadis, C.; Stavrakas, I. Electric earthquake precursors: From laboratory results to field observations. *Phys. Chem. Earth* **2004**, *29*, 339–351. [CrossRef]
10. Stavrakas, I.; Anastasiadis, C.; Triantis, D.; Vallianatos, F. Piezo stimulated currents in marble samples: Precursory and concurrent-with-failure signals. *Nat. Hazards Earth Syst. Sci.* **2003**, *3*, 243–247. [CrossRef]
11. Triantis, D.; Stavrakas, I.; Anastasiadis, C.; Kyriazopoulos, A.; Vallianatos, F. An analysis of pressure stimulated currents (PSC), in marble samples under mechanical stress. *Phys. Chem. Earth* **2006**, *31*, 234–239. [CrossRef]
12. Freund, F.T.; Takeuchi, A.; Lau, B.W. Electric currents streaming out of stressed igneous rocks—A step towards understanding pre-earthquake low frequency EM emissions. *Phys. Chem. Earth* **2006**, *31*, 389–396. [CrossRef]
13. Li, D.X.; Wang, E.Y.; Li, Z.H.; Ju, Y.Q.; Wang, D.M.; Wang, X.Y. Experimental investigations of pressure stimulated currents from stressed sandstone used as precursors to rock fracture. *Int. J. Rock Mech. Min. Sci.* **2021**, *145*, 104841. [CrossRef]
14. Kyriazopoulos, A.; Anastasiadis, C.; Triantis, D.; Brown, C.J. Non-destructive evaluation of cement-based materials from pressure-stimulated electrical emission—Preliminary results. *Constr. Build. Mater.* **2011**, *25*, 1980–1990.
15. Takeuchi, A.; Lau, B.W.S.; Freund, F.T. Current and surface potential induced by stress-activated positive holes in igneous rocks. *Phys. Chem. Earth* **2006**, *31*, 240–247. [CrossRef]

16. Scoville, J.; Sornette, J.; Freund, F.T. Paradox of peroxy defects and positive holes in rocks Part II: Outflow of electric currents from stressed rocks. *J. Asian Earth Sci.* **2015**, *114*, 338–351. [CrossRef]
17. Freund, F.; Ouillon, G.; Scoville, J.; Sornette, D. Earthquake precursors in the light of peroxy defects theory: Critical review of systematic observations. *Eur. Phys. J. Spec. Top.* **2021**, *230*, 7–46.
18. Li, D.X.; Wang, E.Y.; Feng, X.J.; Wang, D.M.; Zhang, X.; Ju, Y.Q. Weak current induced by coal deformation and fracture and its response to mine seismicity in a deep underground coal mine. *Eng. Geol.* **2023**, *315*, 107018. [CrossRef]
19. Li, M.; Lin, Z.J.; Shi, S.L.; Wang, D.M.; Lu, Y.; Li, H.; Ye, Q.; Zhang, X.N. Experimental research on influence mechanism of loading rates on rock pressure stimulated currents. *Int. J. Min. Sci. Technol.* **2023**, *33*, 243–250. [CrossRef]
20. Triantis, D.; Stavrakas, I.; Kyriazopoulos, A.; Hloupis, G.; Agioutantis, Z. Pressure stimulated electrical emissions from cement mortar used as failure predictors. *Int. J. Fract.* **2012**, *175*, 53–61. [CrossRef]
21. Vallianatos, F.; Triantis, D. A non-extensive view of the Pressure Stimulated Current relaxation during repeated abrupt uniaxial load-unload in rock samples. *Europhys. Lett.* **2013**, *104*, 68002. [CrossRef]
22. Stergiopoulos, C.; Stavrakas, I.; Triantis, D.; Vallianatos, F.; Stonham, J. Predicting fracture of mortar beams under three-point bending using non-extensive statistical modeling of electric emissions. *Phys. A* **2015**, *419*, 603–611.
23. Stavrakas, I.; Kourkoulis, S.; Triantis, D. Damage evolution in marble under uniaxial compression monitored by pressure stimulated currents and acoustic emissions. *Fract. Struct. Integr.* **2019**, *13*, 573–583. [CrossRef]
24. Triantis, D.; Pasiou, E.D.; Stavrakas, I.; Kourkoulis, S.K. Hidden affinities between electric and acoustic activities in brittle materials at near-fracture load levels. *Rock Mech. Rock Eng.* **2022**, *55*, 1325–1342.
25. Loukidis, A.; Tzagkarakis, D.; Kyriazopoulos, A.; Stavrakas, I.; Triantis, D. Correlation of acoustic emissions with electrical signals in the vicinity of fracture in cement mortars subjected to uniaxial compressive loading. *Appl. Sci.* **2023**, *13*, 365. [CrossRef]
26. Li, D.X.; Wang, E.Y.; Ju, Y.Q.; Wang, D.M. Laboratory investigations of a new method using pressure stimulated currents to monitor concentrated stress variations in coal. *Nat. Resour. Res.* **2021**, *30*, 707–724. [CrossRef]
27. Kachanov, M. Effective elastic properties of cracked solids: Critical review of some basic concepts. *Appl. Mech. Rev.* **1992**, *45*, 304–335. [CrossRef]
28. Voyiadjis, G.Z.; Kattan, P.I. A comparative study of damage variables in continuum damage mechanics. *Int. J. Damage Mech.* **2009**, *18*, 315–340. [CrossRef]
29. Xiao, J.Q.; Ding, D.X.; Jiang, F.L.; Xu, G. Fatigue damage variable and evolution of rock subjected to cyclic loading. *Int. J. Rock. Mech. Min. Sci.* **2010**, *47*, 461–468. [CrossRef]
30. Gong, F.Q.; Zhang, P.L.; Xu, L. Damage constitutive model of brittle rock under uniaxial compression based on linear energy dissipation law. *Int. J. Rock. Mech. Min. Sci.* **2022**, *160*, 105273. [CrossRef]
31. Lemaitre, J. How to use damage mechanics. *Nucl. Eng. Des.* **1984**, *80*, 233–245. [CrossRef]
32. Jin, F.N.; Jiang, M.R.; Gao, X.L. Defining damage variable based on energy dissipation. *Chin. J. Rock Mech. Eng.* **2004**, *23*, 1976–1980.
33. Liu, B.X.; Huang, J.L.; Wang, Z.Y.; Liu, L. Study on damage evolution and acoustic emission character of coal-rock under uniaxial compression. *Chin. J. Rock Mech. Eng.* **2009**, *28*, 3234–3238.
34. Li, S.C.; Xu, X.J.; Liu, Z.Y.; Yang, W.M.; Liu, B.; Zhang, X.; Wang, Z.C.; Nie, L.C.; Li, J.L.; Xu, L. Electrical resistivity and acoustic emission response characteristics and damage evolution of sandstone during whole process of uniaxial compression. *Chin. J. Rock Mech. Eng.* **2014**, *33*, 14–23. [CrossRef]
35. Barton, N. Suggested methods for the quantitative description of discontinuities in rock masses. International Society for Rock Mechanics. *Int. J. Rock Mech. Min. Sci. Geomech. Abstr.* **1978**, *15*, 319–368.
36. Fischer-Cripps, A.C. *The Electronics Companion*; CRC Press: Boca Raton, FL, USA, 2004.
37. Freund, F.T. Pre-earthquake signals: Underlying physical processes. *J. Asian Earth Sci.* **2011**, *41*, 383–400.
38. Kong, X.G.; He, D.; Liu, X.F.; Wang, E.Y.; Li, S.G.; Liu, T.; Ji, P.F.; Deng, D.Y.; Yang, S.R. Strain characteristics and energy dissipation laws of gas-bearing coal during impact fracture process. *Energy* **2022**, *242*, 123028. [CrossRef]
39. Qiu, L.M.; Zhu, Y.; Song, D.Z.; He, X.Q.; Wang, W.X.; Liu, Y.; Xiao, Y.Z.; Wei, M.H.; Yin, S.; Liu, Q. Study on the nonlinear characteristics of EMR and AE during coal splitting tests. *Minerals* **2022**, *12*, 108. [CrossRef]

sustainability

Article

Joint Characterization and Fractal Laws of Pore Structure in Low-Rank Coal

Yuxuan Zhou [1,2,*], Shugang Li [1,2,*], Yang Bai [1,2], Hang Long [1,2], Yuchu Cai [1,2] and Jingfei Zhang [1,2]

[1] School of Safety Science and Engineering, Xi'an University of Science and Technology, Xi'an 710054, China
[2] Key Laboratory of Coal Resources Exploration and Comprehensive Utilisation, Ministry of Natural Resources, Xi'an 710054, China
* Correspondence: zyx@stu.xust.edu.cn (Y.Z.); lisg@xust.edu.cn (S.L.)

Abstract: The pore structure of low-rank coal reservoirs was highly complex. It was the basis for predicting the gas occurrence and outburst disasters. Different scale pores have different effects on adsorption–desorption, diffusion, and seepage in coalbed methane. To study the pore structure distribution characteristics, which are in different scales of low-rank coal with different metamorphism grade, the pore structure parameters of low-rank coal were obtained by using the mercury injection, N_2 adsorption, and CO_2 adsorption. These three methods were used to test the pore volume and specific surface area of low-rank coal in their test ranges. Then, the fractal dimension method was used to calculate the fractal characteristics of the pore structure of full aperture section to quantify the complexity of the pore structure. The experimental results showed that the pore volume and specific surface area of low-rank coal were mainly controlled by microporous. The pore fractal characteristics were obvious. With the influence of coalification process, as the degree of coal metamorphism increases, fluctuations in the comprehensive fractal dimension, specific surface area, and pore volume of the pore size range occur within the range of R_{max} = 0.50% to 0.65%.

Keywords: low-rank coal; pore structure; comprehensive fractal dimension; fractal law

Citation: Zhou, Y.; Li, S.; Bai, Y.; Long, H.; Cai, Y.; Zhang, J. Joint Characterization and Fractal Laws of Pore Structure in Low-Rank Coal. *Sustainability* **2023**, *15*, 9599. https://doi.org/10.3390/su15129599

Academic Editor: Cun Zhang

Received: 14 May 2023
Revised: 4 June 2023
Accepted: 12 June 2023
Published: 15 June 2023

1. Introduction

Coal bed methane (CBM), as clean energy, will promote the sustainable development of global energy supplying [1]. The reserves of CBM will depend on the safety and reliability of national energy. Low-rank coal is of complex pore structure, especially rich in microporous, which control the adsorption capacity of CBM, This is the basis for evaluating coal mine disasters [2–4]. Therefore, it is of great significance to identify the distribution of the pores about low-rank coal.

The pore structure plays a crucial role in various fields such as coal mining, gas extraction, and hazard prediction [5–7]. The size and connectivity of pores directly impact the permeability and extraction rate of gas [8]. Larger pores and improved connectivity facilitate enhanced gas permeability and extraction efficiency while reducing gas pressure [9]. Simultaneously, larger pore surfaces provide more adsorption sites for gas adsorption. This increases the chances of gas molecules coming into contact with the pore walls, thereby enhancing gas adsorption capacity [10]. The pore structure also influences the flotation kinetics of coal. It affects the diffusion rate and adsorption rate of the flotation reagents, as well as the stability of the foam, thereby impacting the flotation speed and recovery rate of coal [11]. By conducting ventilation borehole monitoring of gas content within coal seams, it becomes possible to assess the gas composition and concentration and predict potential hazards associated with the pore structure [12] because pores can be used as gas storage and migration channels. When a large amount of gas accumulates in the coal seam, larger pores and better connectivity help to improve gas permeability [13]. However, at the same time, it will increase the speed and scale of gas accumulation, thus increasing the risk of gas explosion [14]. Secondly, the mining disturbance will make the coal dust suspended in the air, and the fire source may cause the coal dust explosion. The pore size and distribution characteristics directly affect the suspension capacity of coal dust and the risk of coal dust explosion. Smaller pores may be more likely to suspend coal dust and form explosive dust clouds, increasing the aerodynamic risk during mining [15]. Therefore, the study of pore structure plays an irreplaceable role throughout the entire process of coal mining and utilization.

There are many experimental methods for testing the pore structure, including gas adsorption, visualization, and capillary pressure methods [16]. The pore structure characteristics of coal are mainly tested using scanning electron microscopy [17], gas adsorption [18,19], mercury injection (MIP) [20], nuclear magnetic resonance (NMR) [21,22], X-ray diffraction (XRD) [23], CT imaging (X-CT) [24], and small angle scattering (SAXS) [25]. The pore structure has been studied by many scholars using the above means. Using scanning electron microscopy, XRD, etc., the pore morphology and connectivity of coal rock bodies can be obtained visually [26–29]. However, the pore characteristic data observed by these methods are mainly used for qualitative characterization, and it is difficult to analyze quantitatively.

To obtain a quantitative characterization of the pore structure, mercury injection, gas adsorption, and small-angle scattering methods were introduced. Obtaining pore size parameters through fluid intrusion into the pore of coal bodies is widely recognized because of their wide measurement range and high measurement accuracy [30–32]. The low-temperature N_2 adsorption method and low-pressure CO_2 adsorption method were employed to measure the characteristics of mesopores and micropores, respectively [33]. After obtaining the pore structure parameters, the introduction of fractal dimension calculation provides a quantitative characterization of the complexity of the pore network [34].

Due to the limitations of the type of pore structure testing method, different experiments could only characterize the pore distribution at a certain scale but cannot fully reflect the pore structure characteristics. In recent years, scholars have successively attempted to combine multiple testing methods to jointly characterize the pore structure features and pore distribution of coal [35–38]. However, some scholars have only tested mesopores and macropores [39], while others have not quantified the complexity of the pore structure [40,41].

This study is to complete a more comprehensive joint characterization of a full aperture about the pore structure in low-rank coal and to quantify the pore structure complexity. Eight low-rank coal samples from the northern Shaanxi coalfield were used to investigate the pore joint characterization. Pore structure characteristics are evaluated using mercury injection, low-temperature N_2 adsorption, and CO_2 adsorption, to study the causes of pore structure complexity in low-rank coal. The study had significant implications for CBM reservoir and transport patterns.

2. Materials and Methods

2.1. Sample Preparation

Primary coal from eight different mines in the northern Shaanxi coalfield were selected for the experiments, from Bailiang 5# coal seam (BL), Liangshuijing 4-3 coal seam (LSJ), Daljuta (DLT), Ningtiaota 3-1 coal seam (NTT), Jianxin 4-2 coal seam (JX), Ruineng 401 coal seam (RN), Xiaozhuang 4# coal seam (XZ) and Huangling 2# coal (HL). The distribution of coal mines was shown in Figure 1.

The samples were collected at the newly exposed coal wall, sealed, and brought back to the laboratory. The coal samples were crushed by grinding, and 2.4–4 mm coal samples were screened for mercury injection. In addition, 0.18–0.25 mm coal samples were screened for low-temperature N_2 adsorption and carbon dioxide adsorption experiments. An amount of 10 g of each sample was weighed to set aside. The experimental coal samples were analyzed according to ISO 11722:2013 [42] *(Solid mineral fuels—Hard coal—Determination of moisture in the general analyzed test sample by drying in nitrogen)* and ISO 1171:2010 [43] *(Solid mineral fuels-Determination of ash)*, and the maximum reflectance of the specular group was determined according to the national standard GBT6948-2008 [44] *(Method of determining microscopically the reflectance of vitrinite in coal)*. The test results were shown in Table 1.

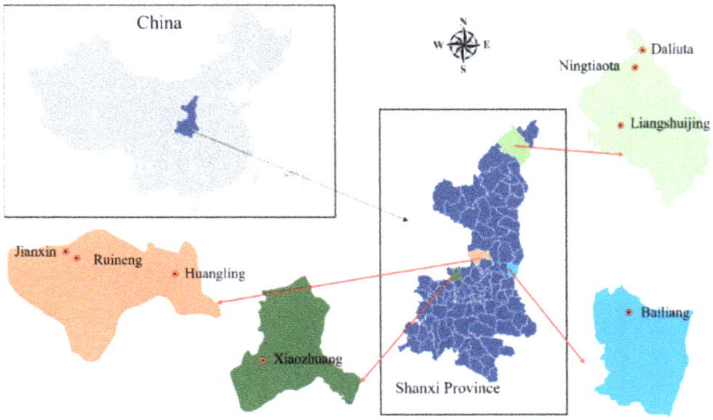

Figure 1. Distribution map of mining areas.

Table 1. Results of coal quality analysis.

Sample ID	Proximate Analysis (%)				R_{max}/%
	M_{ad}	A_{ad}	V_{daf}	FC_{ad}	
BL	0.81	27.36	12.72	59.43	0.33
LSJ	4.55	4.69	30.73	61.64	0.47
DLT	6.89	6.95	31.49	57.32	0.49
NTT	5.26	4.14	32.76	59.78	0.53
JX	2.94	9.05	36.66	52.70	0.55
RN	2.46	13.7	30.86	54.08	0.60
XZ	3.90	5.74	33.08	58.79	0.66
HL	2.1	4.86	31.03	62.76	0.68

M_{ad} is Moisture, V_{daf} is Volatile fraction, A_{ad} is Ash, FC_{ad} is Fixed carbon, and R_{max} is Maximum vitrinite reflectance.

The low-rank coal samples with volatile matter were ranging from 12.72% to 36.66% (10% to 40% volatile matter for low-rank coal) [45]. The specular group emissivity characterized the degree of deterioration of the coal samples (maximum specular group reflectance R_{max} < 0.68% for low-rank coal), the lower degree of coal deterioration with the lower the R_{max}.

2.2. High-Pressure Mercury Injection Experiments

The pore structure of the experimental coal samples was analyzed using an AutoProe IV 9510 fully automatic mercury injection [46,47]. The test system diagram is shown in Figure 2a.

Coal samples were prepared by weighing about 3 g, and it was dried at 70 °C for 8 h and vacuumed in the dilatometer. The experimental temperature was 298 K, the maximum working pressure was 414 MPa, and testable pore size ranges from 3 nm to 1000 µm. During high-pressure mercury injection experiments, the native pore structure of the coal sample was damaged as the mercury solution was pressed in. The experiment had a greater advantage for mesoporous and macropore tests but a greater deviation for microporous and mesoporous tests. The specific test conditions were as follows: in/out mercury contact angle 130°, surface tension 0.48 N/m, and expansion gauge volume 0.5 cc. The relationship between inlet pressure and pore size can be obtained using the Washburn equation [48]. The pore surface area parameter can be determined using the theoretical model proposed by Rootare [49].

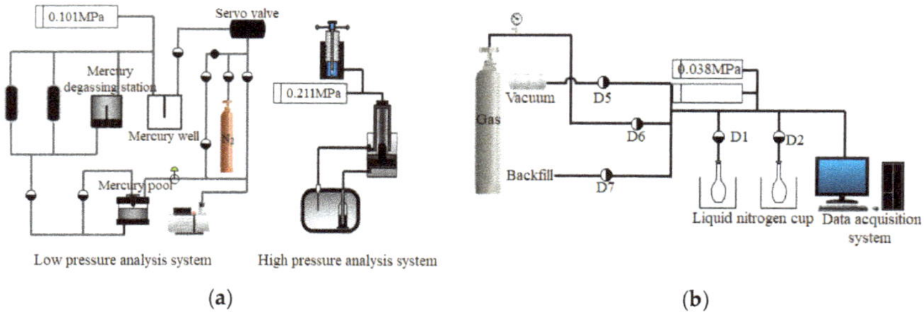

Figure 2. Experimental system. (**a**) High-pressure mercury injection experiments; (**b**) Low-temperature N_2 adsorption and carbon dioxide adsorption experiments.

2.3. Low Temperature N_2 Adsorption Experiments and Low Pressure CO_2 Adsorption Experiments

The pore structure of experimental coal samples was analyzed by ASAP 2460 specific surface area and porosity tester [50]. The test system diagram is shown in Figure 2b. The maximum working pressure is 133 MPa, and pore size test ranges from 0.35 nm to 500 nm. The prepared coal samples were weighed to approximately 200 mg and dried at 70 °C for 8 h before the experiments were carried out.

Low-temperature nitrogen adsorption experiments are carried out at liquid nitrogen temperature (77 K). The amount of nitrogen adsorbed on the solid surface depends on the relative pressure of the nitrogen (P/P_0). P is partial pressure, and P_0 is saturation steam pressure. When P/P_0 is between 0.05 and 0.35, the relationship between adsorption and relative pressure is in accordance with the BET equation, and it is the basis for the determination of the specific surface area of powder materials by low-temperature nitrogen adsorption. When $P/P_0 \geq 0.40$, nitrogen agglomerates in microporous. The parameters such as pore volume, pore size distribution, and specific surface area can be determined according to BJH or DFT theoretical models.

Low-pressure CO_2 adsorption experiments were carried out at saturation temperature (273 K). Testable pore sizes in the range of less than 2 nm. The principle of carbon dioxide adsorption experiments was similar to low-temperature N_2 adsorption experiments. However, carbon dioxide molecules are smaller and diffuse at a faster rate, it had a greater saturation pressure at saturation temperature and can be tested on micropore.

In order to ensure the accuracy of the experimental data, each group of experiments were tested three times. The final data used were the average of the three groups of experiments. The relative error of each group is less than 5%.

3. Results and Discussion

3.1. Adsorption Curves and Pore Distribution Characteristics

The article used the IUPAC pore classification method. Pore was classified as micropore (<2 nm), mesopore (2 nm to 50 nm), and macropore (>50 nm).

The mercury injection experiments focused on the pore characteristics of the mesoporous and macropore. The mercury intrusion and exit curves are drawn based on the experimental parameters of coal samples (Figure 3).

The opening degree of pore in coal samples could be reflected by different curves with mercury entry and exit. All eight low-rank coal samples showed significant mercury injection hysteresis loops, indicating that the open-pore spaces are more developed and the inter-pore connectivity is better. The efficiency of mercury exit was calculated to be 30.85–48.34%. The mercury exit efficiency is medium, indicating that there are both open pore and semi-open pore in coal samples, and the pore connectivity is good. Mercury exit efficiency of low-rank coal is less than 40% with R_{max} = 0.60% and R_{max} < 0.50%. And there were in the range of 40–50% with R_{max} = 0.50–0.60% and R_{max} > 0.60%. The calculations show that the pore connectivity and openness of the more highly metamorphosed low-rank coal are better than less metamorphosed low-rank coal.

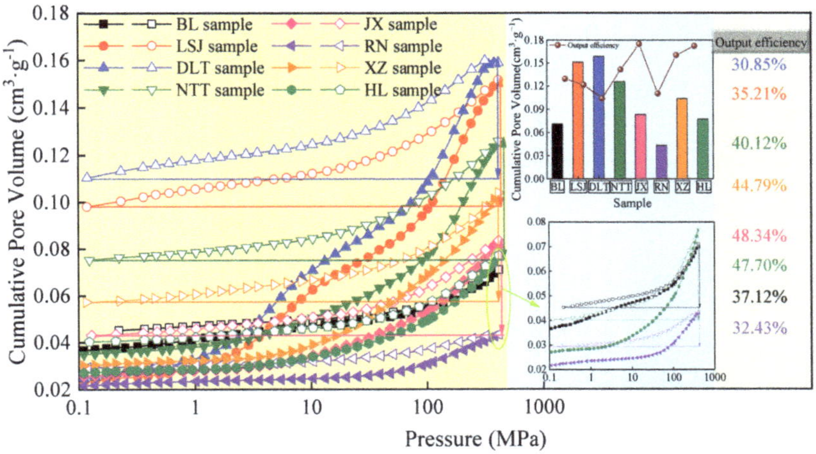

Figure 3. Mercury entry and withdrawal curves of the mercury intrusion experiments.

When the pressure is greater than 10 MPa, the compression effect of the coal matrix is obvious [51,52]. Affected by this effect, the reflection of the real pore structure is biased. With the increase of pressure, the deformation is more obvious, and the deformation of pore under this pressure must be considered [53,54]. It is necessary to modify the experimental data of mercury injection with pressure above 10 MPa, and the corresponding pore size is about 135 nm.

$$\begin{cases} V_{xPi} = V_{cPi} - ZV_m(P_i)P_i \\ V_m(P_i) = V_m - \frac{dV_{cPi}}{dP_i}P_i \\ Z = \frac{1}{V_m}\left(K - \frac{\Delta V_P}{\Delta P}\right) \end{cases} \quad (1)$$

where P_i is the mercury inlet pressure corresponding to the pore volume to be modified (>10 MPa), MPa; V_{Pi} is the hole volume test value corresponding to attack pressure P_i, cm^3/g; V_{XPi} is the corrected pore volume corresponding to the inlet mercury pressure P_i, cm^3/g; $V_m(P_i)$ is the coal matrix volume after the change of mercury injection pressure, cm^3/g; V_m is the volume of coal matrix, cm^3/g; Z is compression coefficient of coal matrix; ΔV_P is the cumulative pore volume value of mercury entry section to be corrected, cm^3/g; ΔP is the difference between the maximum value and the minimum value of the mercury injection pressure to be corrected, MPa; K is the slope of the high-pressure stage (>10 MPa).

The curves were obtained by fitting a linear regression to the mercury feed curve (Figure 4).

The pore volume and specific surface area data were corrected for the high-pressure section of the mercury injection experiment (Table 2).

Table 2. Correction results for mercury injection experimental data.

Simple ID	K	V/cm^3·g^{-1}	V_X/cm^3·g^{-1}	S/cm^3·g^{-1}	S_X/cm^3·g^{-1}
BL	1.300×10^{-4}	0.0024	0.0015	0.1750	0.2828
LSJ	7.694×10^{-4}	0.0153	0.0045	1.0897	3.6836
DLT	8.477×10^{-4}	0.0154	0.0036	1.1764	4.9916
NTT	5.225×10^{-4}	0.0095	0.0029	0.5989	1.9913
JX	3.903×10^{-4}	0.0069	0.0013	0.4157	2.2026
RN	6.259×10^{-4}	0.0010	0.0001	0.0642	0.0647
XZ	5.782×10^{-4}	0.0100	0.0036	0.5840	1.6372
HL	2.952×10^{-4}	0.0053	0.0020	0.3212	0.8500

K is the experimental result for pore volume of pore > 50 nm in diameter, V_X is the corrected pore volume for pore > 50 nm in diameter, S is the experimental result of the specific surface area of pore with a pore size > 50 nm, S_X is the specific surface area after correction for pore size > 50 nm pore.

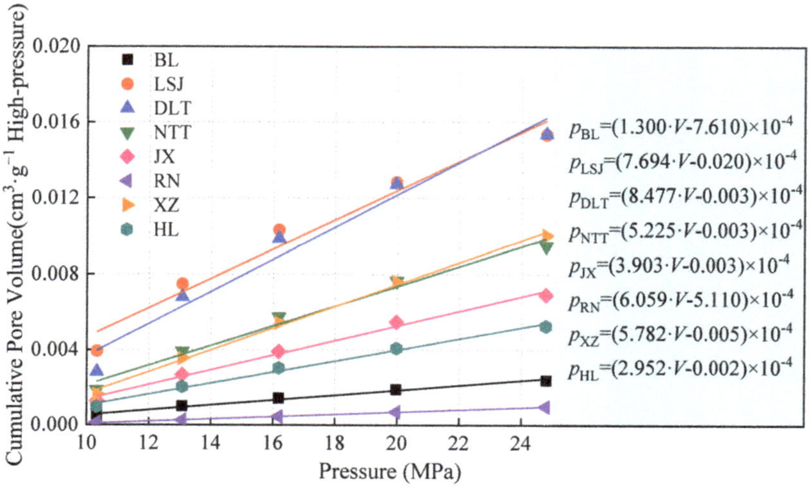

Figure 4. Linear regression of the high-pressure section of the mercury injection curve.

Low-temperature nitrogen adsorption experiments are better for mesopore [55]. The adsorption isotherms were plotted from the low-temperature N_2 adsorption experimental data for each coal sample (Figure 5). LSJ, DLT, and NTT coal samples at the start of the low-pressure section with the curve were biased towards the Y axis. These three coal samples had a strong interaction with liquid nitrogen, with more microporous present in this category and a strong adsorption potential within the microporous. The starting segment was type I according to the IUPAC isotherm classification. The starting adsorption curve of BL, JX, RN, XZ, and HL coal samples was type III, which had a weak interaction force with liquid nitrogen.

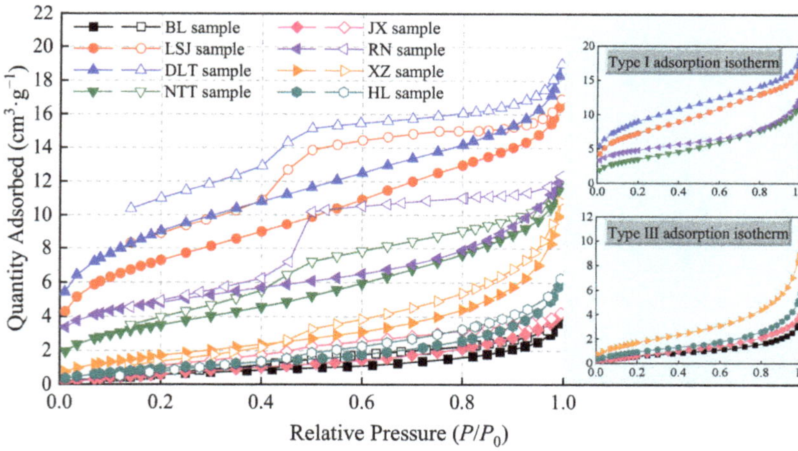

Figure 5. Low-temperature nitrogen adsorption isotherms.

CO_2 adsorption experiments are more capable of testing microporous. In the microporous stage, CO_2 was presented as a monolayer adsorption or microporous filling on the coal surface. Therefore, the adsorption and desorption curves overlap (Figure 6). The shape of the isothermal sorption curve for low-rank coal samples was fitted by the Langmuir equation. The adsorption volume increased rapidly, and the adsorption curve is clearly raised upwards in the low relative pressure region. It tends to be straight in the high-pressure region. The adsorption capacity shows a general trend of decreasing with increasing coal deterioration, except for the BL, XZ, and HL coal samples.

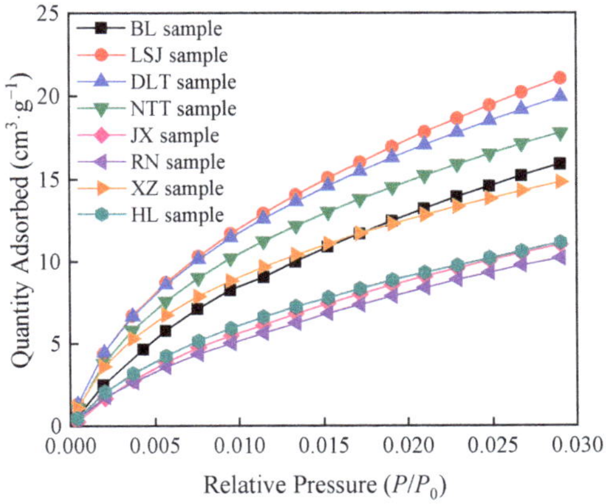

Figure 6. CO_2 adsorption isotherms.

The pore structure parameters tested by mercury pressure, low-temperature N_2 adsorption, and CO_2 adsorption experiments are shown in Table 3.

Table 3. Test parameters for mercury injection, low-temperature N_2 adsorption, and low-pressure CO_2 adsorption experiments.

Sample ID		BL	LSJ	DLT	NTT	JX	RN	XZ	HL
Mercury injection experiment	Porosity/%	8.39	15.81	16.31	13.37	9.33	6.62	11.28	8.47
	Average pore size/nm	19.90	14.10	16.80	12.90	12.90	16.70	13.40	12.80
	Pore volume/cm^3·g^{-1}	0.0710	0.1515	0.1588	0.1257	0.0835	0.0434	0.1038	0.0772
	Pore specific surface area/m^2·g^{-1}	14.2830	42.9030	37.8090	39.0020	25.8020	10.3630	31.069	24.1220
Low pressure nitrogen adsorption experiment	Average pore size/nm	11.08	4.85	4.56	6.43	7.80	6.64	10.57	10.21
	BJH pore volume/cm^3·g^{-1}	0.0059	0.0223	0.0239	0.0172	0.0063	0.0153	0.0160	0.0092
	BET pore specific surface area/m^2·g^{-1}	2.5756	25.8364	31.9735	12.8436	3.1947	16.6833	6.6900	3.9026
Carbon dioxide adsorption experiment	Pore volume/cm^3·g^{-1}	0.0206	0.0271	0.0281	0.0244	0.0133	0.0128	0.0213	0.0133
	Pore specific surface area/m^2·g^{-1}	132.856	168.463	153.955	141.948	94.137	85.012	115.587	102.615

The mercury injection experiments test data showed that the porosity of the eight low-rank coal samples ranged from 6.62% to 16.31%. The DLT coal sample had the largest porosity, and the RN coal sample was the smallest, with a porosity difference of 9.69%. The selected coal samples have a wide range of pore coverage. The average pore size ranged from 12.8 nm to 19.9 nm. The difference between BL coal sample with the largest average pore size, and HL coal sample with the smallest was 1.55 times. The difference between DLT coal sample with the maximum pore volume, and RN coal sample with the minimum value is 3.66 times. The difference between LSJ coal sample with the maximum specific surface area and RN coal sample with the minimum is 4.14 times.

The low-temperature N_2 adsorption experiments test data showed that the average pore size of the eight low-rank coal samples ranged from 4.56 nm to 11.08 nm. The average is small overall compared to the mercury injection experiment. The BL sample had the largest pore size, consistent with the results of the mercury injection test, and differed by a factor of 2.43 from DLT sample, which had the smallest average pore size. The specific surface area measured by BET method is smaller than that obtained by the mercury injection test, and the overall variation is greater. The difference between DLT sample with the largest specific surface area and the smallest BL coal sample was 12.39 times. The pore volumes measured by BJH method are also smaller than those obtained by the mercury injection test. The difference between DLT sample with the largest pore volume and the smallest BL sample was 4.05 times.

The low-pressure CO_2 adsorption experiments test data showed that pore volume and specific surface optimum of low-rank coal microporous were calculated based on DFT model analysis. The distribution is consistent with the results of the mercury injection experimental tests. The DLT sample had the largest pore volume and RN sample had the smallest pore volume, with a difference of 2.20 times. The LSJ coal sample has the maximum specific surface area, and RN coal sample has the least specific surface area, with a difference of 1.98 times. The variation in pore volume and specific surface area of the micropore section varies less.

3.2. Joint Characterization of the Pore Structure of Full Aperture of Low-Rank Coal

Due to the different test principles and ranges of the three pore structure tests, it may have some errors in the test results for the same pore size. Using the experimental data above, the least difference or overlap in test results for the same pore size range was selected to connect and characterize the pore structure of the full aperture of the low-rank coal. The position of the pore joint should be within the test range of the corresponding test method. The three experimental data were bridged at 1.46~1.66 nm and 45.47~46.06 nm to plot the distribution of pore volume and pore specific surface area (Figures 7 and 8).

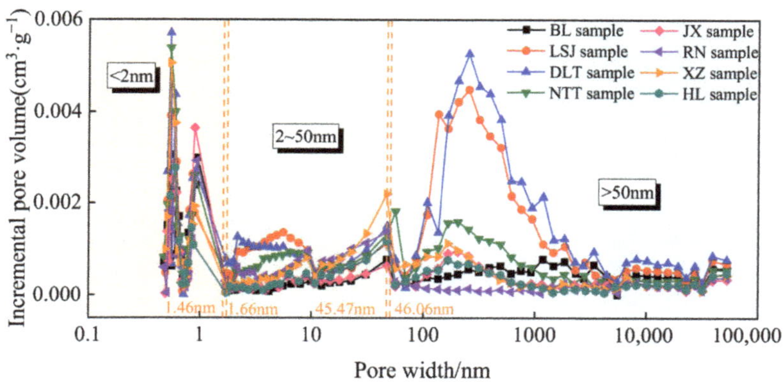

Figure 7. Pore volume distribution feature of full aperture.

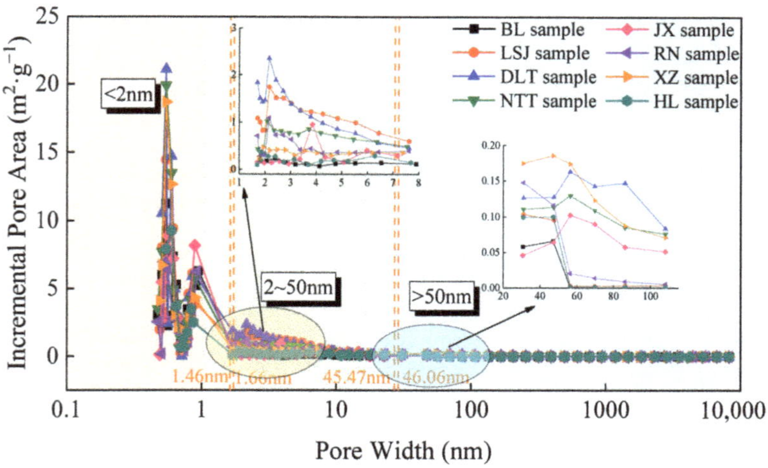

Figure 8. Specific surface distribution feature of full aperture.

From Figure 7, the pore space developed in the low-rank coal is mainly concentrated in the microporous stage. In the microporous section, the increase in pore volume was greater between 0.5 and 0.7 nm, and it reached a maximum around 0.55 nm. After 0.55 nm, the pore volume gradually decreases and shows a multi-peaked distribution. Small fluctuations in pore volume variation in the mesoporous section, with a tendency for the pore volume to increase as the pore size increases. The low-rank coal in the large pore section shows a trend of increasing and then decreasing between pore sizes of 100 and 1000 nm, with pore volumes decreasing gradually and fluctuating less after greater than 1000 nm.

From Figure 8, The specific surface area development of low-rank coal is mainly concentrated in the microporous stage. It had a large difference compared with the mesoporous and macropore stages. The peak of the microporous phase occurs around 0.55 nm, after which the specific surface area gradually decreases and shows a multi-peak distribution. The specific surface area of the mesoporous and macropore stages tends to decrease more steadily with the increase in pore size.

The pore structure characteristics are shown in Table 4. The specific surface area of the eight coal samples is mainly controlled by micropores. This can also be seen from Figure 8. The proportion of micropores reached 79.73–96.56%. The change rule of pore volume ratio is more complicated. The largest proportion of micropores is JX coal sample, accounting for 53.10%. The largest proportion of mesopores is RN coal sample, accounting for 46.81%. The largest proportion of macropores is DLT coal sample, accounting for 51.81%. As the degree of metamorphism increases, the coal is subjected to different stages of coalification. The evolution of pore space in different pore sizes appears to be significantly different due to the influence. Pore volume is more significantly influenced by coalification. The pore-specific surface area also has the same evolutionary characteristics as the pore volume. As the trends in specific surface area in the full aperture are mainly controlled by the microporous, the magnitude of change in the mesopore and macropore sections of the data is its slightest. The change rule is obvious. The change in pore volume with the increase in pore size is more complicated. Thus, this discussion focuses on the trends in the pore volume of low-rank coal.

Table 4. Pore structure characteristics of full aperture section.

Sample ID	Pore Volume/cm^3·g^{-1}	Proportion of Pore Volume/%			Specific Surface Area/m^2·g^{-1}	Proportion of Specific Surface Area/%		
		V_{mic}	V_{mes}	V_{mac}		S_{mic}	S_{mes}	S_{mac}
BL	0.0391	52.36	10.26	37.39	63.6561	96.56	2.99	0.46
LSJ	0.0839	32.00	22.08	46.80	100.6521	81.02	15.29	3.69
DLT	0.0963	28.70	19.48	51.81	106.0858	79.73	15.22	5.04
NTT	0.0522	44.93	26.96	28.11	85.8990	86.26	11.26	2.49
JX	0.0257	53.10	19.31	27.59	42.9067	94.68	0.01	5.31
RN	0.0274	47.56	46.81	5.64	45.8984	82.71	17.13	0.17
XZ	0.0478	41.94	23.99	34.07	70.9783	90.25	7.33	2.42
HL	0.0252	41.85	27.13	31.02	35.7180	89.05	8.56	2.39

V_{mic} is volume of micropore, V_{mes} is volume of mesopore, V_{mac} is volume of macropore. $S_{mic}$is the specific surface area of micropores, S_{mes} is the specific surface area of mesopores, and S_{mac} is the specific surface area of macropore.

The microporous in low-rank coal were smaller pore formed by the stacking of macromolecular structures in the coal. In addition, there would exist some interlayer pore. The process of microporous evolution has similar characteristics to mesopore. The fracture of functional groups, branched chains, etc., in the macromolecular structure of the coal produces gas and forms some of the microporous. During this process, the increase in aromatic ring sheet layers of the microporous pore caused the aromatic sheet layers to stack up against each other as interlayer pore increases and thus the volume of the microporous pore increases.

The evolution of the mesopore is mainly controlled by the stacking of macromolecules such as coal molecular chains and aromatic ring lamellae. As the coalification process proceeds, the aromatization of the coal gradually increases and the condensation of the aromatic thick ring system further increases. At higher levels of metamorphism, some branched chains in coal are synthesized into aromatic rings and reduced. The macromolecular structure is more compact, and the molecular spacing is reduced, resulting in a reduction in mesopore pore volume. The pore volume of low-rank coal mesopore evolves in a similar trend to that of large pore with coalification but changes more smoothly during the coalification jump stage.

The change in macropore has the most significant effect on the change with pore volume in the full aperture. Excluding the effect of coalification jump stage on pore volume, the pore volume tends to increase and then decrease as the degree of metamorphism increases. The turning point of the change occurs at $R_{\max} = 0.50\%$, when the first coalification jump begins. With the end of the coalification jump and the onset of the cracking reaction in the coal body, the volume of macropore pore space begins to decline.

3.3. Fractal Dimension Calculation of Coal Pore

Low-rank coal has strong heterogeneity and complex pore structure. Fractal geometric characteristics could be used to study the irregularity of pore structure and surface and to characterize the adsorption capacity of low-rank coal. The pore fractal dimension obtained by mercury intrusion experimental data was calculated by thermodynamic model [56]. The Frenkel–Halsey–Hill (FHH) model was used to calculate the pore fractal dimension obtained from the experimental data of low-temperature N_2 adsorption [57]. Combined with the data obtained by fitting the two models, the comprehensive fractal dimension is calculated to characterize the complexity of the pore structure of low-rank coal.

3.3.1. Fractal Dimension Based on a Thermodynamic Model

During the mercury injection process, the amount of incoming mercury gradually increases as the pressure increases, resulting in constant increase in the surface energy of the pore [58]. The increase in the surface energy of mercury entering the pore is equal to the work exerted on the mercury by the external environment. The incremental pore surface energy in the whole system is consistent with the work done by the surroundings, thus giving the following Equation (2)

$$\mathrm{d}W_n = \sigma \cos \alpha \mathrm{d}S \quad (2)$$

where W_n is the surface energy (J), σ is the mercury surface tension (0.48 N/m), α is the contact angle between the mercury and the solid surface (130°), and S is the specific surface area of the pore (m^2).

After the correction of Equation (2), the following equation was obtained:

$$\ln \frac{W_n}{r_n^2} = D_f \ln \frac{V_n^{1/3}}{r_n} + \ln C \quad (3)$$

where r_n is the pore diameter (nm), V_n is the pore volume (m^3), and D_f is the fractal dimension of the pore surface area.

According to Equation (3), the fractal dimension was calculated for low-rank coal samples with pore size greater than 50 nm, and a fitting model of thermodynamic fractal was obtained (Figure 9).

The fitting results are shown in Table 5.

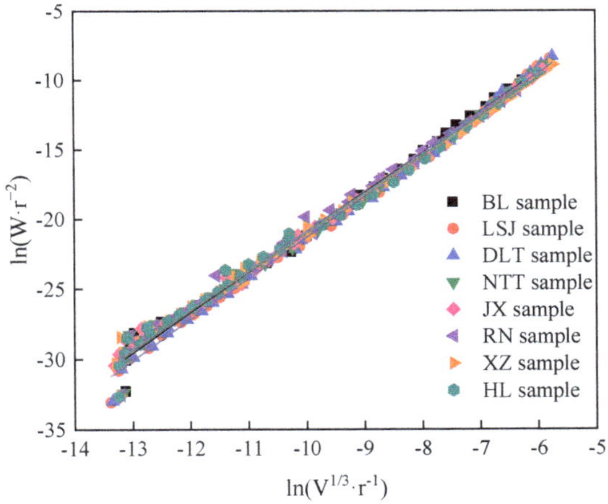

Figure 9. Thermodynamic model fitting curve.

Table 5. Fractal dimension that is based on the thermodynamic model.

Sample	Fitting Equations	$R_1{}^2$	D_1
BL	$y = 2.8737x + 7.7737$	0.9925	2.8737
LSJ	$y = 2.9219x + 7.9515$	0.9953	2.9219
DLT	$y = 2.9397x + 8.1395$	0.9953	2.9397
NTT	$y = 2.7982x + 6.9738$	0.9926	2.7982
JX	$y = 2.7646x + 6.6578$	0.9919	2.7646
RN	$y = 2.8261x + 7.4171$	0.9894	2.8261
XZ	$y = 2.7474x + 6.4992$	0.9918	2.7474
HL	$y = 2.7631x + 6.6941$	0.9907	2.7631

3.3.2. Fractal Dimension Based on FHH Model

The FHH model, first proposed by Frenkel, Halsey and Hill, describes the theory of multilayer adsorption of gas molecules in porous media and is relatively simple to calculate [59]. On a fractal surface in a capillary condensation region and non-homogeneous porous solids, Avnir established the FHH equation of gas adsorption theory by studying the adsorption of gas molecules [60].

$$\ln V = C + (D_h - 3)\ln[\ln(\frac{P_0}{P})] \quad (4)$$

where V is the amount of gas adsorbed at relative air pressure (cm^3/g). P_0/P is the relative pressure, C is a constant, and D_h is the value of the fractal dimension of the porous material.

According to Equation (4), the fractal dimension of low-rank coal samples with mesoporous was fitted by the FHH model, shown in Figure 10. Fitting equation and fractal dimension as shown in Table 6. At $\ln[\ln(P_0/P)]$,the value of lnV changed significantly, so piecewise fitting fractal dimension values are presented here. The fitting degrees were high, which above 0.93. The fractal dimension ranged from 2.4794 to 2.8123, with an average of 2.6656. The fractal characteristics were obvious, and the pore structure of this section was highly complex.

The fractal dimension of low-rank coal samples with microporous was also fitted by the FHH model, as shown in Figure 11.

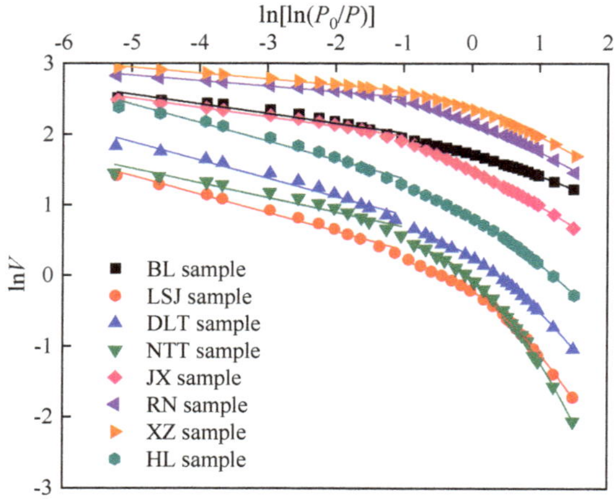

Figure 10. FHH model fitting curve of the mesoporous.

Table 6. Fractal dimension of the mesoporous based on FHH model.

Sample	Relative Pressure < 0.7			Relative Pressure > 0.7			D_2
	Fitting Equations	R_{21}^2	D_{21}	Fitting Equations	R_{22}^2	D_{22}	
BL	$y = -0.1367x + 1.8853$	0.9367	2.8633	$y = -0.2832x + 1.6999$	0.9889	2.7168	2.8066
LSJ	$y = -0.2642x + 0.0942$	0.9862	2.7358	$y = -0.7645x - 0.3073$	0.9526	2.2355	2.5326
DLT	$y = -0.2589x + 0.6009$	0.9615	2.7411	$y = -0.6817x + 0.1944$	0.9688	2.3183	2.5694
NTT	$y = -0.2082x + 0.4790$	0.9326	2.7918	$y = -0.9773x - 0.1844$	0.9474	2.0227	2.4794
JX	$y = -0.1339x + 1.8445$	0.9673	2.8661	$y = -0.4811x + 1.4708$	0.9941	2.5189	2.7251
RN	$y = -0.0779x + 2.4487$	0.9686	2.9221	$y = -0.3852x + 2.1473$	0.9854	2.6148	2.7973
XZ	$y = -0.0833x + 2.5326$	0.9872	2.9167	$y = -0.3402x + 2.3229$	0.9713	2.6598	2.8123
HL	$y = -0.2632x + 1.1156$	0.9772	2.7368	$y = -0.5951x + 0.7743$	0.9836	2.4049	2.6020

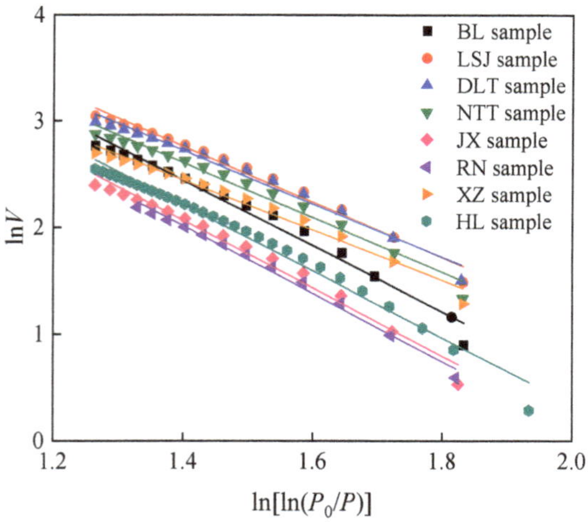

Figure 11. FHH model fitting curve of the microporous.

Fitting equation and fractal dimension are shown in Table 7. The fitting degrees of micropore section were high, which was above 0.98. The fractal dimension is 1.0084 to 1.8414, and the average value is 1.4107.

Table 7. Fractal dimension of the microporous based on FHH model.

Sample	Fitting Equations	R_3^2	D_3
BL	$y = -2.7544x + 6.3004$	0.9894	1.2456
LSJ	$y = -2.4086x + 6.1353$	0.9911	1.5914
DLT	$y = -2.3040x + 5.9513$	0.9889	1.6960
NTT	$y = -2.3447x + 5.8928$	0.9872	1.6554
JX	$y = -2.8999x + 6.1255$	0.9893	1.1001
RN	$y = -2.9962x + 6.2010$	0.9951	1.0084
XZ	$y = -2.1586x + 5.4703$	0.9889	1.8414
HL	$y = -2.8524x + 6.2028$	0.9942	1.1476

3.4. Relationship between the Fractal Dimension of Low-Rank Coal and the Degree of Coal Metamorphism

The calculated fractal dimensions were concentrated between 2.4 and 3.0, and the correlation degree is 86.09–98.30%, with high correlation and good fractal characteristics. The fractal dimension of RN sample joint pore greater than 50 nm pore size segment was greater than 3. It had been suggested that there may be compressive damage to the fractures and pore in the coal as a result of high-pressure mercury injection. It is also possible that the coal seam is highly metamorphosed or highly fractured and deformed. The high-pressure stage (>10 MPa) and pore sizes greater than 50 nm have been corrected for compression. It was unlikely that the first cause of the fractal dimension was greater than 3. In conjunction with the above experimental analysis, possible due to a high degree of fracture deformation in RN coal sample.

As a porous, non-homogeneous solid, coal has different fractal characteristics at different pore-size sections. In order to better characterize quantitatively the complexity of the experimental coal samples and their rough surface, the comprehensive fractal dimension was calculated. The comprehensive fractal dimension was obtained by weighing the pore volume ratios of different pore size sections as weights and summing the fractal dimensions of different pore size sections.

$$D_t = \sum D_i \, T_i \quad (5)$$

where D_t is the integrated fractal dimension of the coal, D_i is the fractal dimension corresponding to the i pore size section, T_i s the pore volume ratio corresponding to the i pore size section (%), and i is the i Pore Size Section and is a positive integer.

The pore volume share of microporous, mesopore and macropore was calculated. Based on the range of applications of the fractal model discussed above, the integrated fractal dimension is calculated for the full aperture section. The integrated fractal dimension of the full aperture section of the coal sample is calculated from Equation (5). The results are shown in Table 8.

Table 8. Comprehensive fractal dimensions of full aperture.

Sample	Microporous Stage			Mesopore Stage			Macropore Stage			Synthesis of Fractal Dimensions
	R_1^2	D_1	V_1/%	R_2^2	D_2	V_2/%	R_3^2	D_3	V_3/%	
BL	0.9894	1.2456	52	0.9569	2.8066	10	0.9925	2.8737	38	1.9916
LSJ	0.9911	1.5914	32	0.9726	2.5326	21	0.9953	2.9219	47	2.4397
DLT	0.9889	1.6960	29	0.9645	2.5694	19	0.9953	2.9397	52	2.5087
NTT	0.9872	1.6554	45	0.9386	2.4794	27	0.9926	2.7982	28	2.1979
JX	0.9893	1.1001	53	0.9782	2.7251	19	0.9919	2.7646	28	1.8749
RN	0.9951	1.0084	48	0.9754	2.7973	46	0.9894	2.8261	6	1.9404
XZ	0.9889	1.8414	42	0.9807	2.8123	24	0.9918	2.7474	34	2.3825
HL	0.9942	1.1476	42	0.9798	2.6020	27	0.9907	2.7631	31	2.0411

During coal formation, coalification does not evolve linearly but undergoes several jumps. Low-rank coal pore is subject to more complex changes by coalification [61,62]. The aliphatic, alicyclic functional groups, and side chains are shed from the aromatic layer to form methane-based volatiles when the coalification was at a R_{max} of 0.50% to 0.60% [63]. The asphaltene caused by asphalting shows that the first jump begins.

During the first coalification jump, the pore structure is controlled by the dissociation and polymerisation of functional groups and aromatic structure, and the trend of change fluctuates considerably. The phenomenon is most evident in the microporous stage. Here, it is shown that, from R_{max} = 0.50%, a substantial reduction in pore volume and specific surface area occurs, and the comprehensive fractal characteristics of the full aperture diminish sharply (Figure 12).

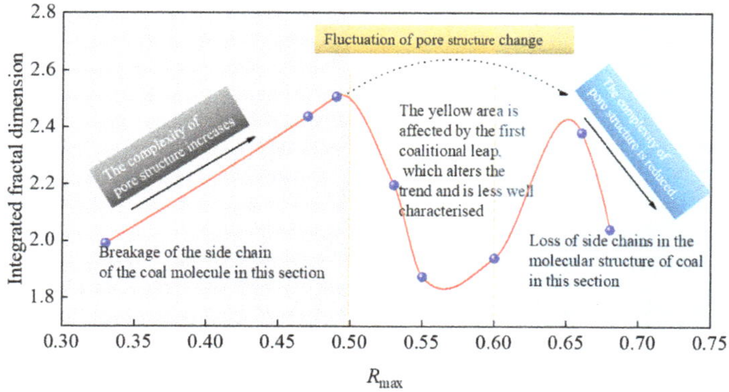

Figure 12. Trends in integrated fractal dimensionality with coal evolution.

As the coalification of low-rank coal increases, the evolution of the pore structure of low-rank coal can be broadly divided into two stages, excluding the coalification jump stage. The first stage is before the start of the first coalification jump (R_{max} < 0.30%), The chemical reactions in coal are dominated by the formation of hydrocarbons, the destruction of coal molecular chains and aromatic ring lamellae, the gradual increase in pore volume, and the increase in specific surface area of coal. After the end of the first coalification jump, the coalification reaction gradually changed to a predominantly cracking reaction. At this stage, moisture, volatile matter, hydrogen, and oxygen content gradually decrease. The organic molecules are gradually arranged in a regular manner. Furthermore, as the degree of polymerisation increases, the carbon content gradually increases. The side chains of the coal molecules decrease, and the layer space decreases, resulting in a gradual decrease in the pore volume and specific surface area of the coal.

In order to study the development and complexity of the pore volume and specific surface area of low-rank coal, the pore volume and specific surface area variation graphs were obtained, as shown in Figures 13 and 14.

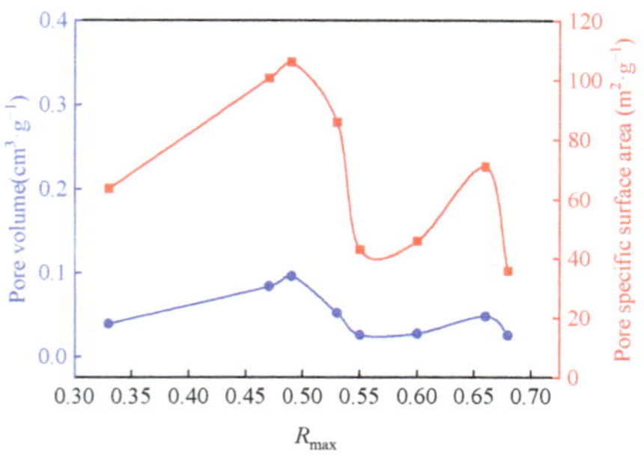

Figure 13. Trend in pore volume and specific surface area with coal evolution.

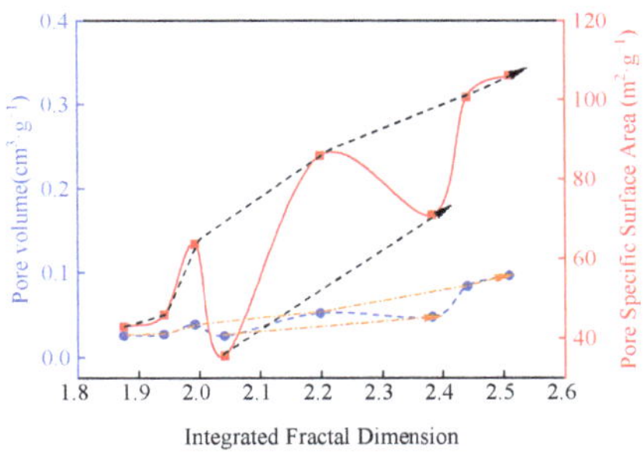

Figure 14. Trend in pore volume and specific surface area with integrated fractal dimensionality.

From Figure 13, the trends in pore volume and specific surface area with coal evolution are similar to those in integrated fractal dimensionality with coal evolution. The trend in specific surface area coincides well with the trend in integrated fractal dimension with an increasing degree of metamorphism. As the specific surface area is mainly controlled by the microporous, the complexity of the pore structure depends more on the microporous.

From Figure 14, there is an overall trend of increasing pore volume and specific surface area as the number of integrated fractal dimensions increases. The larger the pore volume or specific surface area, the more complex the pore structure is. The pore volume and specific surface area before the completion of the coalification jump are larger than those after the completion of the coalification jump. In these two stages, the pore volume and specific surface area increase with the increase of pore structure complexity.

The study shows that the change in specific surface area with R_{max} is most consistent with the change in total fractal dimension with R_{max}. To verify the accuracy of the experimental law, some experimental data in the literature are cited, and the results are basically consistent with the law obtained in this paper (Figure 15). The law of pore structure variation with coalification mentioned in this paper can be confirmed by subsequent research.

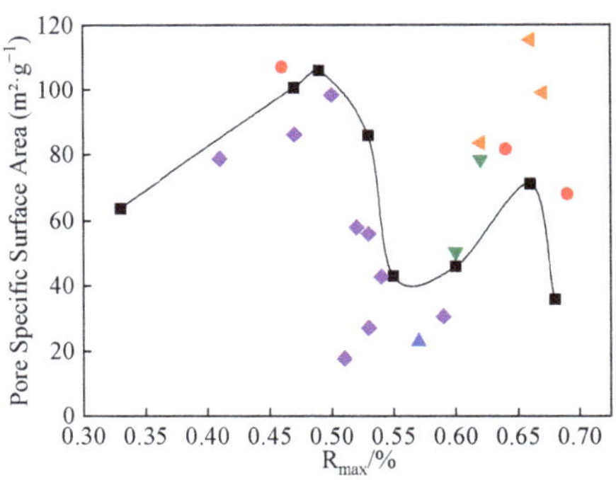

Figure 15. Trend in specific surface area with coal evolution [63–67].

4. Conclusions

In mercury injection experiments, the pore volume and specific surface area of low-rank coal depend mainly on the mesopore. Coal samples with R_{max} between 0.40% and 0.55% also have a large number of developed pores in the large pore section. In low-temperature N_2 adsorption experiments, it shows that pore volume and specific surface area were more developed in 2 nm to 50 nm mesopore section. In CO_2 adsorption experiments, pore volume and specific surface area of low-rank coal depend on the microporous and show an "increasing–decreasing–increasing" trend at 0.4 nm–0.55 nm–0.7 nm–0.9 nm.

The pore structure characteristics of full aperture were characterized using low-pressure CO_2 adsorption to characterize the microporous, low-temperature N_2 adsorption, and mercury compression. It is bridging at 1.46–1.66 nm and 45.47–46.06 nm, respectively. The pore-specific surface area of the full aperture was mainly controlled by micropores. The specific surface area of micropores accounts for 79.73–96.56% of the full aperture. The change in pore volume is more complicated due to the influence of coalification jump. The change rule needs specific analysis.

The pore volume of the full aperture of low-rank coal is mainly controlled by the macropore via the segmental union pore. The specific surface area is mainly controlled by micropores, which are more effectively controlled. The pore fractal characteristics of full aperture in low-rank coal vary between the different degrees of metamorphism, with the fractal dimension ranging from 1.8749 to 2.5087. The macropore fractal features are most pronounced, with fractal dimensions ranging from 2.7474 to 2.9397. Mesopores are next in line, with fractal dimensions ranging from 2.4794 to 2.8123. The micropores are most affected by coalification and have the weakest fractal characteristics, with fractal dimensions of 1.0084 to 1.8414.

The fractal characteristics of low-order coal fluctuate in R_{max} = 0.50–0.60% stage as the degree of metamorphism increases by the coalification jump. In R_{max} =0.30–0.50% stage, the pore structure complexity of low-rank coal increases with the vitrinite reflectivity. After the first coalification jump, the complexity of the pore structure in low-rank coal decreases with the vitrinite reflectivity.

Author Contributions: Investigation, Data curation, Writing-original draft preparation, Formal analysis, Y.Z.; Conceptualization, Methodology, Validation, funding acquisition, S.L.; Methodology, Resources, Y.B.; Methodology, Validation, Supervision, H.L.; Data curation, Formal analysis, Resources, Y.C.; supervision, Writing—review & editing, J.Z. All authors have read and agreed to the published version of the manuscript.

Funding: This research was funded by National Natural Science Foundation of China, (grant No. 52074217, funder: Li S). China Postdoctoral Science Foundation (grant No. 2022MD713795, funder: Bai Y). Natural Science Foundation of Shaanxi Province (grant No. 2022JQ-347, funder: Bai Y). Shaanxi Provincial Department of Education Youth Innovation team (grant No. 22JP048, funder: Bai Y). The authors would like to thank these founda-tions for the financial support.

Institutional Review Board Statement: Not applicable.

Informed Consent Statement: Not applicable.

Data Availability Statement: The experimental data can be provided upon request to the corresponding author.

Conflicts of Interest: The authors declare no conflict of interest.

References

1. Song, Y.; Jiang, B.; Li, M.; Hou, C.; Xu, S. A review on pore-fractures in tectonically deformed coals. *Fuel* **2020**, *278*, 118248.
2. Kaji, R.; Hishinuma, Y.; Nakamura, Y. Low temperature oxidation of coals: Effects of pore structure and coal composition. *Fuel* **1985**, *66*, 297–302. [CrossRef]
3. Lin, H.; Ji, P.; Kong, X.; Li, S.; Long, H.; Xiao, T.; Li, B. Experimental study on the influence of gas pressure on CH_4 adsorption-desorption-seepage and deformation characteristics of coal in the whole process under triaxial stressmain. *Fuel* **2023**, *333*, 126513. [CrossRef]
4. Zhang, J.; Lin, H.; Li, S.; Yang, E.; Ding, Y.; Bai, Y.; Zhou, Y. Accurate gas extraction (AGE) under the dual-carbon background: Green low-carbon development pathway and prospect. *J. Clean. Prod.* **2022**, *377*, 134372. [CrossRef]
5. Zhang, L.; Huang, M.; Xue, J.; Li, M.; Li, J. Repetitive mining stress and pore pressure effects on permeability and pore pressure sensitivity of bituminous coal. *Nat. Resour. Res.* **2021**, *30*, 4457–4476. [CrossRef]
6. Si, L.; Li, Z.; Kizil, M.; Chen, Z.; Yang, Y.; Ji, S. The influence of closed pores on the gas transport and its application in coal mine gas extraction. *Fuel* **2019**, *254*, 115605. [CrossRef]
7. Saqib, A.S.; Jurgen, F.B.; Muhammad, U.K. Optimization of gob ventilation boreholes design in longwall mining. *Int. J. Min. Sci. Technol.* **2020**, *30*, 811–817.
8. Yang, L.; Ai, L.; Xue, K.; Ling, Z.; Li, Y. Analyzing the effects of inhomogeneity on the permeability of porous media containing methane hydrates through pore network models combined with CT observation. *Energy* **2018**, *163*, 27–37. [CrossRef]

9. Davudov, D.; Moghanloo, R.G.; Zhang, Y. Interplay between pore connectivity and permeability in shale sample. *Int. J. Coal Geol.* **2020**, *220*, 103427. [CrossRef]
10. Yao, Y.; Liu, D.; Tang, D.; Tang, S.; Huang, W. Fractal characterization of adsorption-pores of coals from North China: An investigation on CH4 adsorption capacity of coals. *Int. J. Coal Geol.* **2008**, *73*, 27–42. [CrossRef]
11. Wang, S.; Xia, Q.; Xu, F. Investigation of collector mixtures on the flotation dynamics of low-rank coal. *Fuel* **2022**, *327*, 125171. [CrossRef]
12. Zhang, H.; Pera, L.S.; Zhao, Y.; Sanchez, C.V. Researches and applications on geostatistical simulation and laboratory modeling of mine ventilation network and gas drainage zone. *Process Saf. Environ.* **2015**, *94*, 55–64.
13. Kong, X.; Wang, E.; Liu, Q.; Li, Z.; Li, D.; Cao, Z.; Niu, Y. Dynamic permeability and porosity evolution of coal seam rich in CBM based on the flow-solid coupling theory. *J. Nat. Gas Sci. Eng.* **2017**, *40*, 61–71. [CrossRef]
14. Bosikov, I.I.; Martyushev, N.V.; Klyuev, R.V.; Savchenko, I.A.; Kukartsev, V.V.; Kukartsev, V.A.; Tynchenko, Y.A. Modeling and complex analysis of the topology parameters of ventilation networks when ensuring fire safety while developing coal and gas deposits. *Fire* **2023**, *6*, 95. [CrossRef]
15. Balovtsev, S.V.; Skopintseva, O.V.; Kolikov, K.S. Aerological risk management in preparation for mining of coal mines. *Sustain. Dev. Mt. Territ* **2022**, *14*, 107–116. [CrossRef]
16. Li, H.; Shi, S.; Lin, B.; Lu, J.; Ye, Q.; Lu, Y.; Wang, Z.; Hong, Y.; Zhu, X. Effects of microwave-assisted pyrolysis on the microstructure of bituminous coals. *Energy* **2019**, *187*, 115986.1–115986.14. [CrossRef]
17. Deirieh, A.; Casey, B.; John, T.G.; Xu, G.P. The integration of magnifications: A novel approach to obtain representative information about the pore space of mudrocks from SEM images. *Appl. Clay Sci.* **2018**, *154*, 73–82. [CrossRef]
18. Du, B.; Liang, Y.; Tian, F.; Guo, B. Analytical Prediction of Coal Spontaneous Combustion Tendency: Pore Structure and Air Permeability. *Sustainability* **2023**, *15*, 4332. [CrossRef]
19. Jagiello, J.; Thommes, M. Comparison of DFT characterization methods based on N_2, Ar, CO_2, and H_2 adsorption applied to carbons with various pore size distributions. *Carbon* **2004**, *42*, 1227–1232. [CrossRef]
20. Liu, K.; Mirzaei-Paiaman, A.; Liu, B.; Ostadhassan, M. A new model to estimate permeability using mercury injection capillary pressure data: Application to carbonate and shale samples. *J. Nat. Gas Sci. Eng.* **2020**, *84*, 103691. [CrossRef]
21. Anderson, R.J.; McNicholas, T.P.; Kleinhammes, A.; Wang, A.; Liu, J.; Wu, Y. NMR Methods for Characterizing the Pore Structures and Hydrogen Storage Properties of Microporous Carbons. *J. Am. Chem. Soc.* **2010**, *132*, 8618–8626. [CrossRef]
22. Yan, J.; He, X.; Zhang, S.; Feng, C.; Wang, J.; Hu, Q.; Cai, J.; Wang, M. Sensitive parameters of NMR T-2 spectrum and their application to pore structure characterization and evaluation in logging profile: A case study from Chang 7 in the Yanchang Formation, Heshui area, Ordos Basin, NW China. *Mar. Pet. Geol.* **2020**, *111*, 230–239. [CrossRef]
23. Li, Z.; Xu, Y.; Fan, L.; Kang, W.; Cheng, B. Fabrication of polyvinylidene fluoride tree-like nanofiber via one-step electrospinning. *Mater. Des.* **2016**, *92*, 95–101. [CrossRef]
24. Li, Y.; He, S.; Deng, X.; Xu, Y. Characterization of macropore structure of Malan loess in NW China based on 3D pipe models constructed by using computed tomography technology. *J. Asian Earth Sci.* **2018**, *154*, 271–279. [CrossRef]
25. Liu, Y.R.; Paskevicius, M.; Sofianos, M.V.; Parkinson, G.; Li, C.Z. In situ SAXS studies of the pore development in biochar during gasification. *Carbon* **2020**, *172*, 454–462. [CrossRef]
26. Ni, H.Y.; Liu, J.F.; Pu, H.; Chen, X. Microscopic Pore Structure and Improved Permeability Characterization of CO_x Argillite Based on SEM Images. *Int. J. Geomech.* **2021**, *21*, 04021144. [CrossRef]
27. Ni, H.; Liu, J.; Huang, B.; Pu, H.; Meng, Q.; Wang, Y.; Sha, Z. Quantitative analysis of pore structure and permeability characteristics of sandstone using SEM and CT images. *J. Nat. Gas Sci. Eng.* **2021**, *88*, 103816. [CrossRef]
28. Li, H.; Dong, S.; Qin, N.; Liu, J.; Yu, Y.; Zhang, Z.; Wu, M.; Chen, Z. On automatically extracting the statistical information of pores in heterogeneous materials from the SEM morphology. *J. Porous Media* **2021**, *24*, 83–100. [CrossRef]
29. Ilzig, T.; Schumacher, D.; Wilhelm, M.; Gunther, S.; Odenbach, S. Image data analysis of high resolution μCT data for the characterization of pore orientation and pore space interconnectivity in freeze cast ceramics. *Mater. Charact.* **2021**, *174*, 110966. [CrossRef]
30. Clarkson, C.R.; Solano, N.R.; Bustin, R.M.; Bustin, A.M.M.; Chalmers, G.R.L.; He, L.; Melnichenko, Y.B.; Radlinski, A.P.; Blach, T.P. Pore structure characterization of North American shale gas reservoirs using USANS/SANS, gas adsorption, and mercury intrusion. *Fuel* **2013**, *103*, 606–616. [CrossRef]
31. Zhang, N.; He, M.; Zhang, B.; Qiao, F.C.; Sheng, H.; Hu, Q.H. Pore structure characteristics and permeability of deep sedimentary rocks determined by mercury intrusion porosimetry. *J. Earth Sci.* **2016**, *4*, 670–676. [CrossRef]
32. Guo, F.; Pang, X.; Chen, D.; Liu, G.; Pang, H.; Pang, B.; Li, M. Characterization of micropore structure of a middle to deep clastic reservoir: The oligocene reservoir in the Nanpu Sag, Bohai Bay Basin, China. *J. Pet. Sci. Eng.* **2021**, *202*, 108567. [CrossRef]
33. Zhang, J.; Wei, C.; Luo, J.; Lu, G.; Quan, F.; Zheng, K.; Peng, Y. Volume and Surface Distribution Heterogeneity of Nano-pore in Coal Samples by CO_2 and N_2 Adsorption Experiments. *Acta Geol. Sin.-Engl. Ed.* **2020**, *94*, 1662–1675. [CrossRef]
34. Lqbal, M.A.; Rezaee, R.; Smith, G.; Ekundayo, J.M. Shale lithofacies controls on porosity and pore structure: An example from Ordovician Goldwyer Formation, Canning Basin, Western Australia. *J. Nat. Gas Sci. Eng.* **2021**, *89*, 103888.

35. Han, H.; Dai, J.; Guo, C.; Zhong, N.; Pang, P.; Ding, Z.; Chen, J.; Huang, Z.; Gao, Y.; Luo, J. Pore Characteristics and Factors Controlling Lacustrine Shales from the Upper Cretaceous Qingshankou Formation of the Songliao Basin, Northeast China: A Study Combining SEM, Low-temperature Gas Adsorption and MICP Experiments. *Acta Geol. Sin.-Engl. Ed.* **2021**, *95*, 585–601. [CrossRef]
36. Zong, Y.T.; Yu, X.L.; Zhu, M.X.; Lu, S.G. Characterizing soil pore structure using nitrogen adsorption, mercury intrusion porosimetry, and synchrotron-radiation-based X-ray computed microtomography techniques. *J. Soils Sediments* **2015**, *15*, 302–312. [CrossRef]
37. Wang, F.Y.; Yang, K.; Cai, J.C. Fractal Characterization of Tight Oil Reservoir Pore Structure Using Nuclear Magnetic Resonance and Mercury Intrusion Porosimetry. *Fractals* **2018**, *26*, 1840017. [CrossRef]
38. Lin, H.; Bai, Y.; Bu, J.; Li, S.; Yan, M.; Zhao, P.; Qin, L. Comprehensive Fractal Model and Pore Structural Features of Medium- and Low-Rank Coal from the Zhunnan Coalfield of Xinjiang. China. *Energies* **2020**, *13*, 7. [CrossRef]
39. Okolo, G.N.; Everson, R.C.; Neomagus, H.W.; Roberts, M.; Sakurovs, R. Comparing the porosity and surface areas of coal as measured by gas adsorption, mercury intrusion and SAXS techniques. *Fuel* **2015**, *141*, 293–304. [CrossRef]
40. Zhu, Q.; Yang, Y.; Lu, X.; Liu, D.; Li, X.; Zhang, Q.; Cai, Y. Pore Structure of Coals by Mercury Intrusion, N_2 Adsorption and NMR: A Comparative Study. *Appl. Sci.* **2019**, *9*, 1680. [CrossRef]
41. Xu, J.; Cheng, X.; Peng, S.; Jin, Q.; Cheng, F.; Lou, D.; Zhang, F.; Li, F. Depositional environment and hydrocarbon potential of Early Permian coals deposit in the Huanghua Depression, Bohai Bay Basin. *Ore Geol. Rev.* **2023**, *154*, 105315. [CrossRef]
42. *ISO 11722:2013*; Solid Mineral Fuels—Hard coal—Determination of Moisture in the General Analysis Test Sample by Drying in Nitrogen. ISO Copyright Office: Geneva, Switzerland, 2013.
43. *ISO 1171:2010*; Solid Mineral Fuels-Determination of Ash. ISO Copyright Office: Geneva, Switzerland, 2010.
44. *GBT6948-2008*; Method of Determining Microscopically the Reflectance of Vitrinite in Coal. The Central People's Government of the People's Republic of China: Beijing, China, 2008.
45. Schmitt, M.; Fernandes, C.P.; da Cunha Neto, J.A.; Wolf, F.G.; dos Santos, V.S. Characterization of pore systems in seal rocks using nitrogen gas adsorption combined with mercury injection capillary pressure techniques. *Mar. Pet. Geol.* **2013**, *39*, 138–149. [CrossRef]
46. Katz, A.J.; Thompson, A.H. Prediction of rock electrical conductivity from mercury injection measurements. *J. Geophys. Res.-Solid Earth* **1987**, *92*, 599–607. [CrossRef]
47. Washburn, E.W. Note on a method of determining the distribution of pore sizes in a porous material. *Proc. Natl. Acad. Sci. USA* **1921**, *7*, 115–116. [CrossRef]
48. Rootare, H.M.; Prenzlow, C.F. Surface areas from mercury porosimeter measurements. *J. Phys. Chem.* **2002**, *71*, 2733–2736. [CrossRef]
49. Li, K.; Zhou, M.; Liang, L.; Jiang, L.; Wang, W. Ultrahigh-surface-area activated carbon aerogels derived from glucose for high-performance organic pollutants adsorption. *J. Colloid Interface Sci.* **2019**, *546*, 333–343. [CrossRef]
50. Friesen, W.I.; Mikula, R.J. Mercury porosimetry of coals Pore volume distribution and compressibility. *Fuel* **1988**, *67*, 1516–1520. [CrossRef]
51. Liu, S.M.; Zhang, R.; Karpyn, Z.; Yoon, H.; Dewers, T. Investigation of Accessible Pore Structure Evolution under Pressurization and Adsorption for Coal and Shale Using Small-Angle Neutron Scattering. *Energy Fuels* **2019**, *33*, 837–847. [CrossRef]
52. Kong, X.; Li, S.; Wang, E.; Wang, X.; Zhou, Y.; Ji, P.; Shuang, H.; Li, S.; Wei, Z. Experimental and numerical investigations on dynamic mechanical responses and failure process of gas-bearing coal under impact load. *Soil Dyn. Earthq. Eng.* **2021**, *142*, 106579. [CrossRef]
53. Kong, X.; He, D.; Liu, X.; Wang, E.; Li, S.; Liu, T.; Ji, P.; Deng, D.; Yang, S. Strain characteristics and energy dissipation laws of gas-bearing coal during impact fracture process. *Energy* **2022**, *242*, 123028. [CrossRef]
54. Zhang, Z.; Du, J.; Zhu, H. Surface Area and Pore Distribution of Sponge Palladium by Low Temperature Nitrogen Adsorption Method. *Chin. J. Rare Metal* **2011**, *35*, 411–416.
55. Cai, Y.; Liu, D.; Yao, Y.; Li, J.; Liu, J. Fractal Characteristics of Coal Pores Based on Classic Geometry and Thermodynamics Models. *Acta Geol. Sin.* **2011**, *85*, 1150–1162. [CrossRef]
56. Lu, G.; Wang, J.; Wei, C.; Song, Y.; Yan, G.; Zhang, J.; Chen, G. Pore fractal model applicability and fractal characteristics of seepage and adsorption pores in middle rank tectonic deformed coals from the Huaibei coal field. *J. Pet. Sci. Eng.* **2018**, *171*, 808–817. [CrossRef]
57. Zhang, B.; Li, S. Determination of the surface fractal dimension for porous media by mercury porosimetry. *Ind. Eng. Chem. Res.* **1995**, *34*, 1381–1386. [CrossRef]
58. Pandey, R.; Harpalani, S. An imaging and fractal approach towards understanding reservoir scale changes in coal due to bioconversion. *Fuel* **2018**, *230*, 282–297. [CrossRef]
59. Avnir, D.; Jaroniec, M. An isotherm equation for adsorption on fractal surfaces of heterogeneous porous materials. *Langmuir* **1989**, *5*, 1431–1433. [CrossRef]
60. Daniels, E.J.; Aronson, J.L.; Altaner, S.P.; Clauer, N. Late Permian age of NH_4-bearing illite in anthracite from eastern Pennsylvania: Temporal limits on coalification in the central Appalachians. *Geol. Soc. Am. Bull.* **1994**, *106*, 760–766. [CrossRef]
61. Liu, Y.; Zhang, C.; Zhang, Y. Evolution mechanism of methane adsorption capacity in vitrinite-rich coal during coalification. *J. Nat. Gas Sci. Eng.* **2021**, *96*, 104307. [CrossRef]

62. Ibarra, J.; Muñoz, E.; Moliner, R. FTIR study of the evolution of coal structure during the coalification process. *Org. Geochem.* **1996**, *24*, 725–735. [CrossRef]
63. Ding, L.; Zhao, M.; Wei, Y.; Wang, A. Impact from Medium to Low Ranked Coal Vitrinite Micropore Structure on Methane Adsorptivity. *Coal Geol. China* **2021**, *33*, 17–21.
64. Yang, F.; He, D.; Ma, D.M.; Duan, Z.H.; Tian, T.; Fu, D.L. Multi-scale joint characterization of micro-pore structure of low-rank coal reservoir. *Lithol. Reserv.* **2020**, *32*, 14–23.
65. Yan, H.; Yang, F.; Duan, Z.; Tian, T.; Fu, D. Microscopic Pore Structure Sharacteristics Coal Microscopic Pore Geometry Features and Its Impact Factors in Zhuanjiao Exploration Area, Huanglong Coalfield. *Coal Geol. China* **2022**, *34*, 19–25.
66. Yang, G.; Qin, Y.; Wu, C.; Li, S. Pore structure features of Xishanyao Formation low rank coal in Hoxtolgay Basin of Xinjiang. *Coal Sci. Technol.* **2017**, *45*, 123–130.
67. Lu, J.; Fu, X.; Kang, J. Quantitative and comprehensive characterization of full pore size in medium and low rank coal in southern margin of Junggar basin. *China Sci.* **2022**, *17*, 62–71.

sustainability

MDPI

Article

Study on the Effect of External Air Supply and Temperature Control on Coal Spontaneous Combustion Characteristics

Changkui Lei [1,*], Xueqiang Shi [2,*], Lijuan Jiang [3], Cunbao Deng [1], Jun Nian [1] and Yabin Gao [1]

[1] College of Safety and Emergency Management Engineering, Taiyuan University of Technology, Taiyuan 030024, China
[2] School of Environment and Safety Engineering, North University of China, Taiyuan 030051, China
[3] Faculty of Business and Economics, University of Pécs, 7626 Pécs, Hungary
* Correspondence: leichangkui@tyut.edu.cn (C.L.); shixueqiang@nuc.edu.cn (X.S.)

Abstract: Coal spontaneous combustion in underground mine goaf has a great impact on coal mining. The temperature-programmed experiment is a commonly used and effective method for studying the characteristics of coal spontaneous combustion. Aiming at the problem that the numerical simulation of coal spontaneous combustion characteristics under the condition of external air supply and temperature control in a temperature-programmed experiment is insufficient, a multi-physical field coupling numerical model of coal spontaneous combustion in the temperature-programmed experiment is established. The variation characteristics of coal temperature, oxygen, and oxidation products under external air supply and temperature control were studied. The results show that the numerical simulation results are consistent with the experimental results. With the increase in temperature, the volume fractions of oxygen and carbon dioxide decrease and increase, respectively. As the air supply volume increases, the oxygen volume fraction at the outlet increases, and the peak value of the oxygen volume fraction change rate exhibits a "hysteresis" feature, and the time corresponding to the peak value increases. Moreover, the temperature change rate increases. With the increase in the heating rate, the peak value of the oxygen volume fraction change rate increases and shows an "early appearance" characteristic, at the same time, the maximum coal temperature displays a linear increase trend.

Keywords: coal spontaneous combustion; air supply volume; heating rate; oxygen volume fraction; numerical simulation

Citation: Lei, C.; Shi, X.; Jiang, L.; Deng, C.; Nian, J.; Gao, Y. Study on the Effect of External Air Supply and Temperature Control on Coal Spontaneous Combustion Characteristics. *Sustainability* **2023**, *15*, 8286. https://doi.org/10.3390/su15108286

Academic Editors: Baoqing Li, Xiangguo Kong, Dexing Li and Xiaoran Wang

Received: 6 March 2023
Revised: 14 April 2023
Accepted: 12 May 2023
Published: 19 May 2023

1. Introduction

Coal spontaneous combustion is one of the main disasters that affect coal mine safety [1]. It not only causes resource waste and equipment damage but also leads to gas and coal dust explosions, triggering secondary disasters and causing serious casualties and property losses [2]. Due to the concealment and difficulty in accessing the interior of the goaf, it is one of the most prone locations for coal spontaneous combustion [3,4]. Once the conditions of oxidation reaction are met, namely, oxygen and a good thermal storage environment, coal will spontaneously undergo an exothermic reaction, and the temperature rise of coal caused by the oxidation reaction will feedback to accelerate its reaction process, eventually achieving extremely hazardous and violent combustion [5,6]. Therefore, the problem of coal spontaneous combustion in goaf has always been an urgent problem that troubles the safety production of mines [7]. However, in the actual production process of mines, avoiding oxidation reactions of residual coal in goaf is the key to preventing coal spontaneous combustion, which mainly involves analyzing the environment of coal spontaneous combustion in goaf. Therefore, coal spontaneous combustion and its influencing factors are one of the important topics of fire science research [8,9]. By investigating the characteristics of coal spontaneous combustion under different environmental conditions, it is possible to have a more comprehensive understanding of the oxidation laws of coal spontaneous combustion, which has important guiding significance for the prevention and control of coal spontaneous combustion in goaf.

In view of the coal oxidation reaction and its characteristics, a lot of studies have been conducted in recent years, including the use of advanced experimental instruments and self-developed equipment to study the mechanism and characteristics of its spontaneous combustion [10–12]. Deng et al. [13] tested the heat release intensity and oxygen consumption rate of the coal samples with a temperature-programmed device and calculated the change law of gas products of coal spontaneous combustion. Qi et al. [14] built a similar experimental platform for coal smoldering simulation and studied the smoldering characteristics of coal fire under different air supply rates. Wen et al. [15] compared and analyzed the characterization parameters of coal spontaneous combustion through two different experimental systems, the large-scale coal spontaneous combustion experiment, and the isothermal difference leading experiment. Beamish et al. [16] studied the influence of moisture content, initial temperature, coalbed methane content, and reactive pyrite content on the self-heating rate of low-temperature coal through an adiabatic furnace test. Further, Zhang et al. [17] established a calculation method for the coal spontaneous combustion period based on pure oxygen adiabatic oxidation experiment. Jia et al. [18] employed the temperature-programmed experimental system to conduct oxidation experiments on coal samples with six oxygen concentrations and studied the gas products of coal samples at different temperature stages.

In addition, with the development of computing technology, numerical simulation has gradually become an effective means to predict the risk of coal spontaneous combustion and optimize industrial conditions [19,20]. Meanwhile, it also provides a reliable method for studying the coal spontaneous combustion process and external coal spontaneous combustion conditions in goaf [21,22]. Shi et al. [23] used numerical simulation and experimental methods to explore the characteristics of temperature evolution in large coal spontaneous combustion silo and obtained the movement law of high temperature points. Li et al. [24] adopted numerical simulation to study the compound hazard area of coal spontaneous combustion and gas and obtained the change characteristics of the hazard area under the change of relevant parameters. Ma et al. [25] studied the distribution of oxygen concentration in adjacent goafs by numerical simulation method and determined the risk zone of spontaneous combustion in adjacent goafs. Yuan et al. [26] adopted the obtained experimental data to simulate the oxidation and temperature rise process of different coal samples in the longwall goaf with three-dimensional CFD, and studied the influence of coal surface area and reaction heat on the natural heat release process. Liu et al. [27] developed a new transient model under the framework of multi-physical coupling to simulate the process of coal spontaneous combustion during stopping mining in longwall goaf.

However, there are few studies on the influence of external conditions, especially external air supply, temperature control, etc. on coal spontaneous combustion characteristics. Ren et al. [28] used a C80 experimental system to study the low-temperature oxidation heat flux of pulverized coal under different oxygen concentrations to investigate the risk of spontaneous combustion of pulverized coal. Yan et al. [29] conducted a programmed temperature rise experiment to test the oxidation process of coal samples at different oxygen concentrations. Therefore, current research on the impact of the external environment on coal spontaneous combustion mainly focuses on controlling oxygen concentration, rather than directly regulating the gas supply. In addition, the analysis of the impact of temperature control on coal spontaneous combustion is mainly reflected in the analysis of the heating rate on the thermal kinetics of coal oxidation. For example, Zhang et al. [30] studied the thermodynamic parameters of coal spontaneous combustion at different heating rates before and after the ignition temperature. Sabat et al. [31] studied the thermal degradation characteristics, kinetics, and thermodynamics of different grades of low-rank coal at different heating rates through thermogravimetric analysis. Meanwhile, Wang et al. [32] employed a thermogravimetric simulation test bench to study the pyrolysis characteristics and kinetics of bituminous coal at different heating rates.

In this paper, the characteristics of coal spontaneous combustion under different external air supplies and temperature controls are studied by using a numerical simulation method combined with a self-developed temperature-programmed coal spontaneous combustion experiment. The characteristics of reactants, products, and temperature changes during the coal oxidation reaction were obtained. This study is of great significance for understanding the characteristics of coal spontaneous combustion and preventing its occurrence in goaf.

2. Experiment and Methodology

2.1. Coal Sample

To study the oxidation process of coal spontaneous combustion, lignite from the Wantugou coal mine, in Shaanxi Province, China, was used as the experimental coal sample. Fresh coal samples are collected from the in situ and transported to the laboratory through oxygen isolation. After being crushed by a jaw crusher, coal samples with a particle size of 3–5 mm are screened out as experimental samples. For the experiment, 1 kg coal samples with the same particle size range were selected. Table 1 shows the results of proximate and element analyses of the coal sample.

Table 1. The proximate and element analyses of coal sample.

Proximate Analysis				Ultimate Analysis				
$M_{ad}/\%$	$V_{ad}/\%$	$A_{ad}/\%$	$FC_{ad}/\%$	$C_{daf}/\%$	$H_{daf}/\%$	$O_{daf}/\%$	$N_{daf}/\%$	$S_{daf}/\%$
11.75	35.78	13.03	39.44	65.36	1.25	12.66	1.91	0.52

Note: M_{ad}, V_{ad}, A_{ad}, and FC_{ad} correspond to air-dried moisture, volatile matter, ash, and fixed carbon, respectively; C_{daf}, H_{daf}, O_{daf}, N_{daf}, and S_{daf} correspond to dry ash-free carbon, hydrogen, oxygen, nitrogen, and sulfur, respectively.

2.2. Experiment

The experimental equipment is a self-developed temperature-programmed device that leads the oxidation reaction of coal under additional heating conditions. Figure 1 presents the schematic diagram of the temperature-programmed system used in the experiment, which includes three parts, namely the gas supply system, the temperature control system, and the gas analysis system.

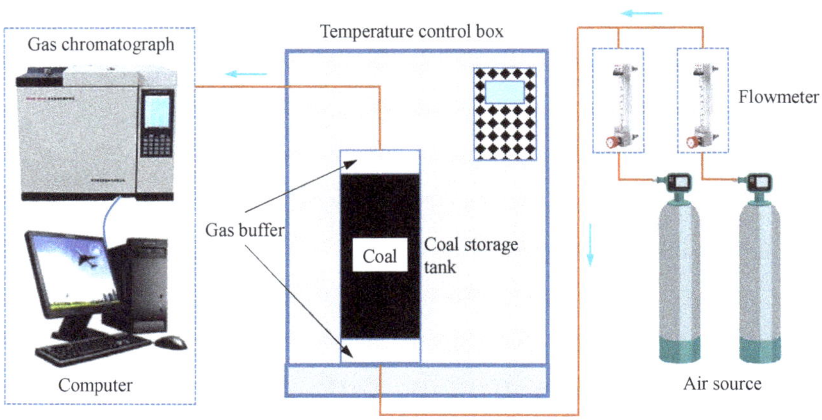

Figure 1. Schematic diagram of the temperature-programmed experiment system.

To prevent heat loss and maintain constant heating conditions, the coal storage tank is placed in the temperature-programmed heating box. The temperature-programmed heating box is automatically controlled by a silicon-controlled regulator with a furnace space of 50 × 40 × 30 cm. The sample tank for coal storage is made of stainless steel, with a height and radius of 220 mm and 50 mm, respectively. To obtain a more uniform oxygen supply environment, 20 mm gas buffer space is set at the upper and lower parts of the coal storage tank. In the experiment, nitrogen and oxygen were premixed at 79:21 to simulate the air environment. In order to keep the intake temperature basically the same as the temperature of the coal sample, a 2 m long copper tube is coiled in the temperature-programmed heating box, and the air flow is preheated through the coil before entering the coal sample.

Before the experiment, the coal storage tank was continuously ventilated for half an hour to ensure that the gas in the coal storage tank is constant, and the gas entering the coal storage tank was controlled by the flow control valve and float flowmeter. In addition, in order to ensure that the difference in oxygen concentration between the inlet and outlet of the coal sample tank can be within the resolution range of the gas chromatograph used in mining, the experimental gas supply flow rate should be 41.8–190.0 mL/min. Therefore, the experimental gas supply flow rate is set to 120 mL/min, and the test is stopped after the coal is heated to 180 °C at a heating rate of 0.3 °C/min. The gas produced in the experiment was analyzed by the SP-2120 gas chromatograph. K-type thermocouples are set at the center of the coal body and the wall of the coal storage tank to measure the temperature.

2.3. Numerical Model

Coal oxidation reaction involves multiple physical processes, such as fluid flow, heat transfer, mass transfer, and chemical reaction. To solve the parameter coupling law of multiple physical fields, the multi-physical field coupling software COMSOL Multiphysics (COMSOL, Stockholm, Sweden) was applied to solve the reaction process. According to the actual experimental conditions, the numerical model of programmed temperature rise of coal spontaneous combustion is established considering the momentum, mass, energy transfer, and chemical reaction process. For the heat transfer during coal oxidation and heating, the energy equation is as follows [33,34]:

$$(\rho C_p)_{\text{eff}} \frac{\partial T}{\partial t} + \rho C_p \vec{u} \cdot \nabla T + \nabla \cdot \vec{q} = Q \tag{1}$$

where ρ is the density of gas, kg/m^3; C_p is the specific heat of gas at constant pressure, $C_p = 2.8\,T + 50$ according to the experiment, J/kg·K; $(\rho C_p)_{\text{eff}}$ is the average effective volume heat capacity at atmospheric pressure considering the characteristics of solids and gases, J/m^3·K; T is the temperature, K; $\vec{q}$ is heat conduction flux, W/m^2; $\vec{u}$ is the velocity vector, m/s; and t is time, s.

The Brinkman equation is employed to solve the fluid flow in the coal void [35,36]:

$$(1-\varepsilon)\frac{\partial \rho_c}{\partial t} = -V_c \tag{2}$$

$$\frac{\partial \varepsilon \rho}{\partial t} + \nabla \cdot (\rho \vec{u}) = Q_{br} \tag{3}$$

$$\rho \frac{\partial \vec{u}}{\partial t} = \nabla \cdot [-p\vec{\mathrm{I}} + \mu \frac{1}{\varepsilon}(\nabla \vec{u} + (\nabla \vec{u})^{\mathrm{T}}) - \frac{2}{3}\mu \frac{1}{\varepsilon}(\nabla \cdot \vec{u})\vec{\mathrm{I}}] - \frac{Q_{br}}{\varepsilon^2}\vec{u} \tag{4}$$

where ε is the porosity of the coal sample; ρ_c is the density of the coal sample, kg/m^3; V_c is the coal oxidation reaction rate, kg/m^3·s; Q_{br} is the mass source, kg/m^3·s; and μ and p are dynamic viscosity and pressure, respectively, kg/m·s, Pa.

Considering the generation and consumption of various substances in the coal oxidation process, Equation (5) is used to solve for the material transfer:

$$\frac{\partial(\varepsilon c_j)}{\partial t} + \nabla \cdot \vec{\mathrm{J}}_j + \vec{u} \cdot \nabla c_j = V_j \tag{5}$$

where $\vec{J}_j$ is mass flux, kg/m^2·s; c_j represents gas concentration, mol/m^3; and the subscript j indicates the type of gas.

Table 2 provides the parameter values that need to be assigned to the numerical model mentioned above. Based on the actual situation on site and corresponding thermophysical experiments, the physical parameters of coal were obtained, and it is assumed that the parameters in the table are fixed values and do not change with temperature and other conditions.

Table 2. Physical parameters used in numerical simulation.

Parameters	Description	Value
p_0	Pressure (atm)	1
T_0	Initial temperature (°C)	30
R_s	Specific gas constant (J/kg·K)	287
γ	Specific heat	1.4
k_c	Thermal conductivity (W/m·K)	0.21
ρ_c	Coal density (kg/m^3)	1200
$C_{p,c}$	Specific heat capacity (J/kg·K)	1000

3. Results and Discussion

3.1. Numerical Simulation Verification

To verify the reliability of the numerical simulation method, the simulation results and experimental data are compared and analyzed. Figure 2 shows the changes in oxygen and carbon dioxide volume fraction with temperature under the condition that the initial and boundary conditions of the numerical simulation are consistent with those of the experiment. By comparing the simulation results of oxygen and carbon dioxide volume fractions with the experimental results, it is found that the average error is 2.3% and 4.6%, respectively. Thus, the results of the experiment and simulation are in good agreement, which demonstrates that the acquisition of numerical simulation parameters and the numerical simulation method can reasonably characterize the coal oxidation and heating process. In addition, with the increase in coal temperature, the oxygen volume fraction at the outlet gradually decreases and the carbon dioxide volume fraction gradually increases, which indicates that the oxygen consumption gradually increases with the increase in temperature [37].

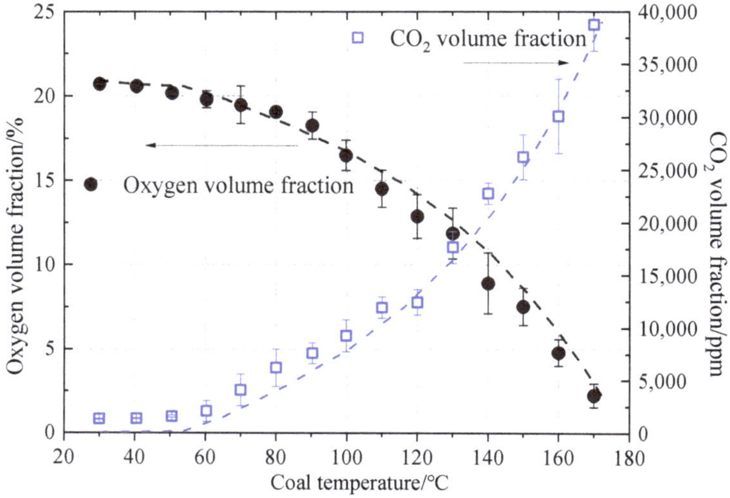

Figure 2. Changes of oxygen and carbon dioxide at the outlet with temperature during coal oxidation.

3.2. Influence of Inlet Air Volume on Coal Spontaneous Combustion Characteristics

As is well known, air leakage in the working face has a significant impact on the spontaneous combustion of coal in the goaf and is the main cause of coal spontaneous combustion in the goaf [38]. To further study the influence of inlet air volume on the coal spontaneous combustion process, the inlet air volume was changed in the numerical simulation, and the programmed heating rate was set to 0.375 °C/min. Figure 3 shows the oxygen volume fraction and its change rate at the outlet under different air volume conditions. As can be seen from Figure 3a, with the increase in air supply, the oxygen volume fraction at the outlet also shows an increasing trend, which is due to the slow oxidation reaction of coal at low temperature and the small consumption of oxygen. Take the fourth hour as an example, the oxygen gas integral value at 144 mL/min is 1.77 times higher than that without ventilation. Figure 3b shows that the peak value of the oxygen volume fraction change rate at the outlet is "hysteresis" with the increase in air volume. The "hysteresis" feature is that the increase in air supply promotes the heat loss inside the coal body, and the peak value of the coal oxidation reaction is delayed. The above results prove that heat is the main factor that dominates the reaction of coal at the initial stage of low-temperature oxidation [39].

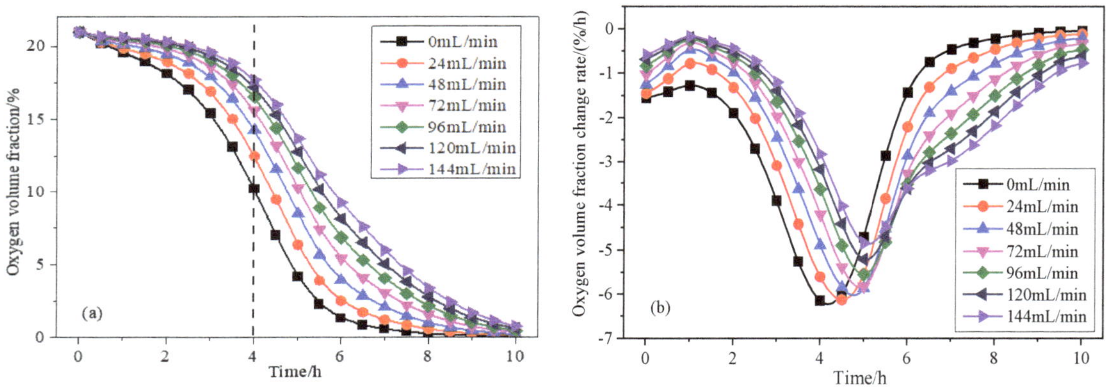

Figure 3. Oxygen volume fraction and its change rate at the outlet under different air volumes: (**a**) oxygen volume fraction; (**b**) oxygen volume fraction change rate.

Figure 4 gives the relationship between the peak time of the oxygen volume fraction change rate at the outlet and the flow rate. With the increase in flow rate, the peak time of the change rate increases. Further, based on the method of numerical fitting, Equation (6) shows the relationship between them.

$$y = -1.36e^{-x/64} + 5.38 \quad (6)$$

With the further increase in the flow rate, it can be predicted that the peak time of the change rate will not change, which highlights the leading role of oxygen, that is, under a certain low-temperature oxidation rate of coal, temperature dominance will change to oxygen dominance.

Figure 5 shows the maximum coal temperature and its change rate under different air supply conditions. It can be seen from Figure 5a that the maximum coal temperature before 7.5 h is always lower than the leading temperature of programmed temperature rise. Except for air volume of 0 and 24 mL/min, the maximum coal temperature under other air supply conditions exceeds the leading temperature of programmed temperature rise after 7.5 h. This shows that the oxidation reaction of coal after 7.5 h is the main reason for the temperature increase, and sufficient oxygen supply promotes the rapid coal oxidation reaction, while the heat release of coal oxidation at air volume of 0 and 24 mL/min is less than the heat dissipation, so their heat cannot be accumulated.

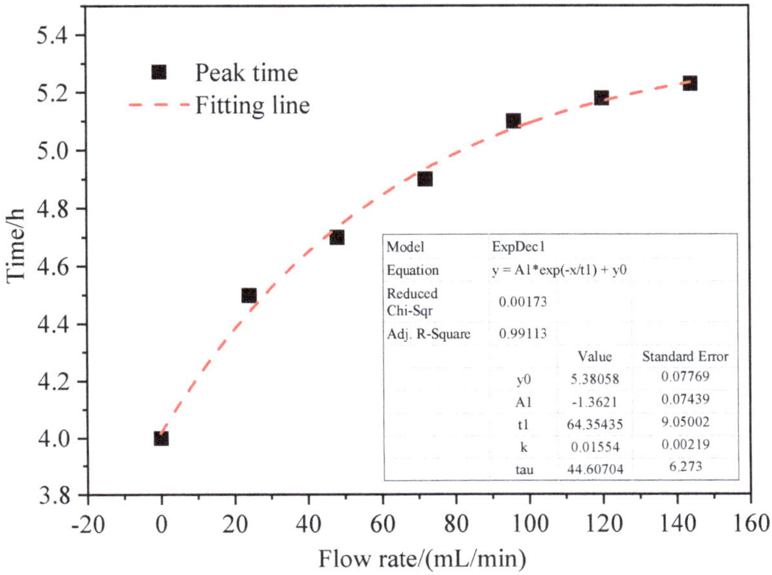

Figure 4. The relationship between the peak time of oxygen volume fraction change rate at the outlet and flow rate.

From Figure 5b, it can be seen that except for the air volume of 0 and 24 mL/min, the maximum temperature change rate of coal under other air supply conditions exceeds the temperature change rate adjusted by the programmed temperature rise system within 4 h, and as time goes on, the larger the air supply volume is, the more obvious the increase is. This reflects that after 4 h, the oxidation heat release of coal is greater than the heat dissipation, resulting in heat accumulation. Furthermore, it also shows that the greater the air supply, the more intense the chemical reaction of the spontaneous oxidation process of coal, and the greater the heat production. However, the oxidation heat release of the coal sample at air volume of 0 and 24 mL/min is lower, and its temperature change rate changes slightly with the increase in time, which is mainly due to the insufficient air supply and the inability to provide sufficient oxygen for the coal oxidation reaction. Therefore, oxygen conditions are a key factor in the occurrence of coal spontaneous combustion [40,41]. Only by minimizing air leakage and oxygen supply in the goaf can we prevent the occurrence of coal spontaneous combustion in the goaf.

In order to more intuitively reflect the relationship between coal spontaneous combustion temperature and air supply flow, Figure 6 shows the maximum coal temperature at 10 h under different air supply conditions. The maximum coal temperature under the condition of air volume of 0 and 24 mL/min is lower than the leading temperature of programmed temperature rise. With the increase in air supply volume, the maximum coal temperature shows an upward trend, and the two are positively correlated as a whole. Equation (7) is obtained by linear fitting of the maximum coal temperature at 10 h with different air supply conditions.

$$y = 0.675x + 288 \tag{7}$$

When the air volume is less than 24 mL/min, the result does not conform to a linear change. The air volume exceeds 24 mL/min, and the result conforms to the characteristics of linear change. The internal reason for this different trend is that the oxygen content determines the completeness of the coal oxidation reaction. Therefore, in the process of prevention and control of coal spontaneous combustion in the mine goaf, the control of air leakage in the goaf is crucial, which is the key measure to prevent the occurrence of coal spontaneous combustion [42].

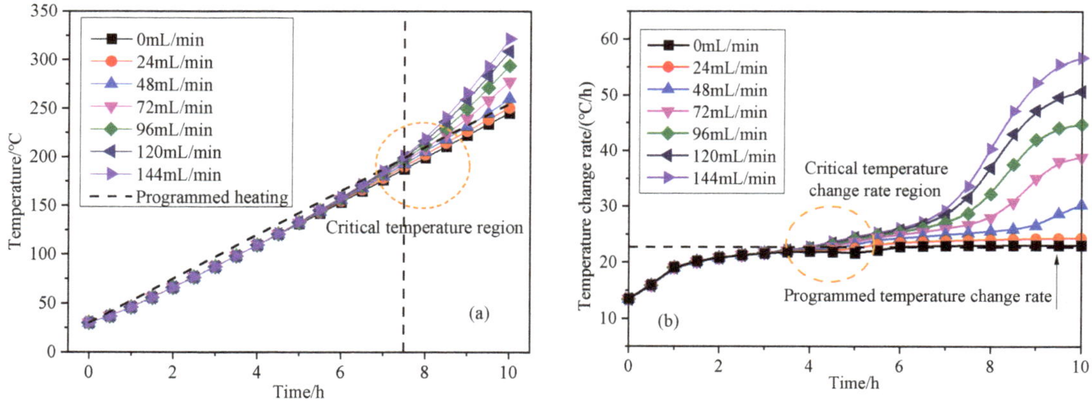

Figure 5. Maximum coal temperature and its change rate under different air supply conditions: (**a**) coal temperature; (**b**) coal temperature change rate.

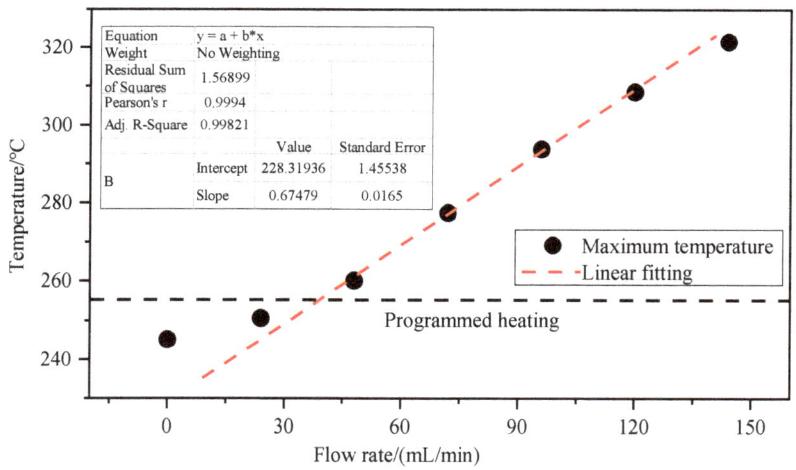

Figure 6. Maximum coal temperature at 10 h under different air supply conditions.

Figure 7 displays the cloud diagram of oxygen volume fraction and temperature distribution under different air flow conditions at the 10th hour. With the increase in air flow, the area with high oxygen volume fraction moves upward, for example, the line representing 1% oxygen volume fraction moves upward, as shown in Figure 7a. The distribution of oxygen volume fraction reflects that with the migration of oxygen in coal, oxygen is consumed due to the low-temperature oxidation reaction of coal. Obviously, the higher the air flow, the higher the oxygen volume fraction in the coal body. The distribution of oxygen volume fraction in the coal body presents stratified characteristics.

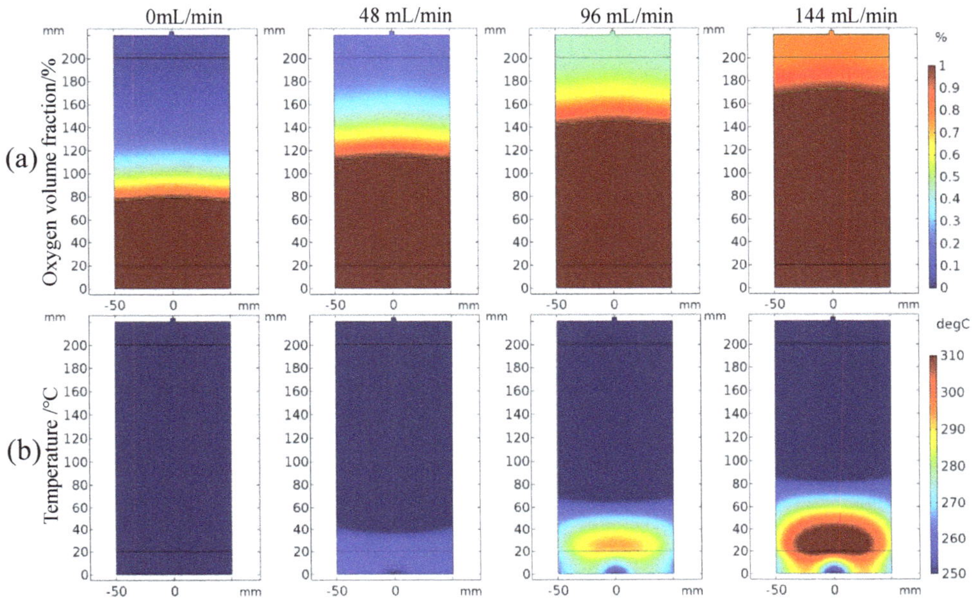

Figure 7. Cloud diagram of oxygen volume fraction and temperature distribution under different air flow conditions: (**a**) oxygen volume fraction; (**b**) coal temperature.

From Figure 7b, with the increase in air flow, the internal temperature of coal is getting higher and higher. Interestingly, the area with a higher temperature in the coal body is located at the inlet of the air. This is because the low-temperature oxidation reaction of coal and oxygen is related to the concentration of oxygen (see Equation (2)). The higher the oxygen, the more intense the coal oxidation reaction. Due to the influence of thermal buoyancy and air flow, the temperature in the coal body shows an ellipse, the center of the high-temperature area is upward, as shown in the example of 144 mL/min.

3.3. Effect of Heating Rate on Coal Spontaneous Combustion Characteristics

The thermal environment outside the coal body also has a great influence on the coal spontaneous combustion process. Therefore, the effect of heating rate on coal spontaneous combustion characteristics is investigated by numerical simulation. The heating rate is taken as a variable and the air supply for programmed heating is controlled to 120 mL/min. Figure 8 presents the outlet oxygen volume fraction and its change rate under different heating rates. It can be seen from Figure 8a that with the increase in heating rate, the decrease in oxygen volume fraction at the outlet is intensifying. The oxygen volume fraction at a heating rate of 0 °C/min has little change with time, which is 20.63% at 10 h, but it is only 0.16% when the heating rate is 0.45 °C/min. Moreover, from Figure 8b that with the increase in heating rate, the peak value of the change rate of oxygen volume fraction increases, it is characterized by "early appearance", that is, the higher the heating rate, the earlier the peak of oxygen volume fraction change rate appears, which indicates that increasing the heating rate will promote the low-temperature oxidation intensity of coal. This is mainly because coal spontaneous combustion is originally an oxidative exothermic process, and the increase in heating rate further reduces the reaction threshold at the low-temperature stage of coal spontaneous combustion, thereby accelerating the oxidation reaction.

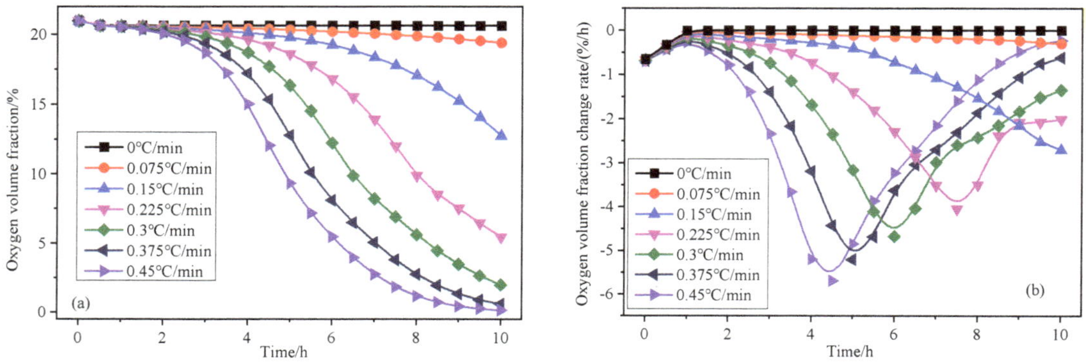

Figure 8. Oxygen volume fraction and its change rate at the outlet under different heating rates: (**a**) oxygen volume fraction; (**b**) oxygen volume fraction change rate.

Furthermore, Figure 9 presents the relationship between the peak time of the oxygen volume fraction change rate at the outlet and the heating rate. Since the peak value of outlet oxygen volume fraction change rate corresponding to 0, 0.075, and 0.15 °C/min does not appear within 10 h, only the temperature rise rate greater than 0.225 °C/min is listed. From Figure 9, with the increase in heating rate, the peak time decreases, reflecting the characteristic of "early appearance", and Equation (8) is obtained by exponential fitting.

$$y = 17.6e^{-x/0.15} + 3.63 \quad (8)$$

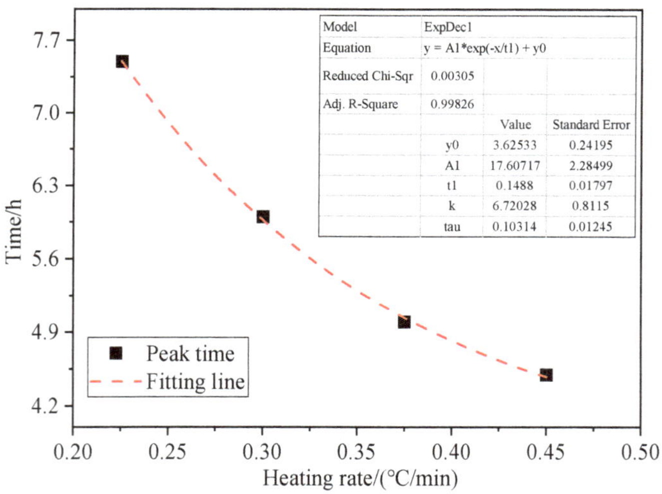

Figure 9. The relationship between the peak time of the outlet oxygen volume fraction change rate and the heating rate.

Figure 10 shows the maximum temperature value and its change rate of coal under different heating rates. On the whole, the higher the heating rate, the faster the temperature rises, indicating that the external thermal environment promotes the coal oxidation process. Meanwhile, except for heating rate 0 °C/min, under other heating conditions, the maximum coal temperature and its change rate always change from lower than the programmed temperature to higher than the programmed temperature, and in the later stage of coal oxidation, the change rate of temperature rise increases and the coal temperature rises faster. This indicates that under the effect of programmed temperature rise, the heat accumulation of coal low-temperature oxidation exceeds the heat dissipation, and the net heat of coal is positive. Therefore, in the process of prevention and control of coal spontaneous combustion disaster in the mine, the early low-temperature stage is the key to treatment [43]. Once the coal temperature exceeds a certain critical value, usually the dry crack temperature, its oxidation reaction increases rapidly, and spontaneous combustion may occur in a very short time.

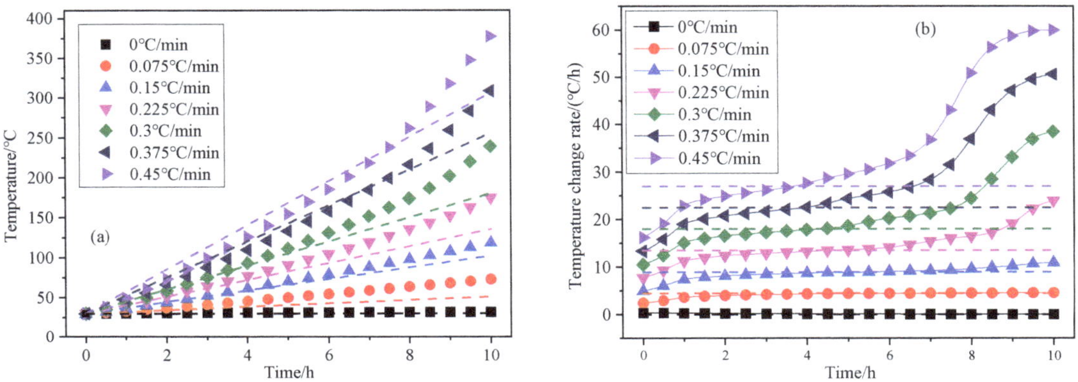

Figure 10. Maximum coal temperature and its change rate under different heating rates: (**a**) coal temperature; (**b**) coal temperature change rate.

Figure 11 shows the maximum coal temperature at different heating rates at 10 h. With the increase in heating rate, the maximum coal temperature value presents a linear increase trend, and the linear fitting result is shown in Equation (9). Therefore, in the process of mine production, attention should be paid to controlling the ventilation temperature to prevent it from intensifying the coal oxidation reaction.

$$y = 774.5x + 14 \quad (9)$$

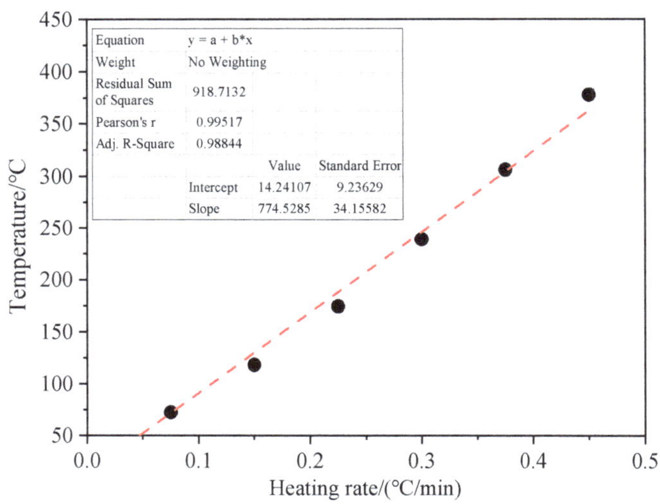

Figure 11. Maximum coal temperature at 10 h under different heating rates.

Figure 12 gives the cloud diagram of oxygen volume fraction and temperature distribution under different heating rate conditions at the 10th hour. According to Figure 12a, the oxygen volume fraction in the coal storage tank decreases with the increase in the heating rate. As the air enters the coal body from the inlet and then is discharged from the outlet, this process consumes a large amount of oxygen due to the influence of coal oxidation, resulting in the lowest oxygen volume fraction at the outlet. Especially when the heating rate is 0.45 °C/min, the oxygen volume fraction in the coal body is less than 4%, and the higher oxygen volume fraction area is located in the inlet gas buffer zone.

With the increase in heating rate, the coal temperature in the coal storage tank is getting higher and higher, as shown in Figure 12b. The coal temperature distribution corresponding to the heating rate of 0 and 0.15 °C/min is relatively uniform, which indicates that the oxidation reaction inside the coal is weak under the above conditions, and the change of the coal temperature depends on the external heat transfer. The temperature distribution of coal corresponding to the heating rate of 0.3 and 0.45 °C/min is uneven, and the high-temperature zone is located at the air inlet. This is because the oxygen volume fraction here is high, which can fully meet the demand of coal oxidation reaction, and the oxidation reaction here is more full and intense. In addition, due to the rapid consumption of oxygen in the coal body at the entrance, the temperature distribution in the upper part of the coal body is characterized by a high surrounding area and low center at the heating rate of 0.3 and 0.45 °C/min. This reflects that under the conditions of heating rates 0.3 and 0.45 °C/min, the temperature of the upper part of the coal body is mainly affected by heat transfer rather than coal oxidation.

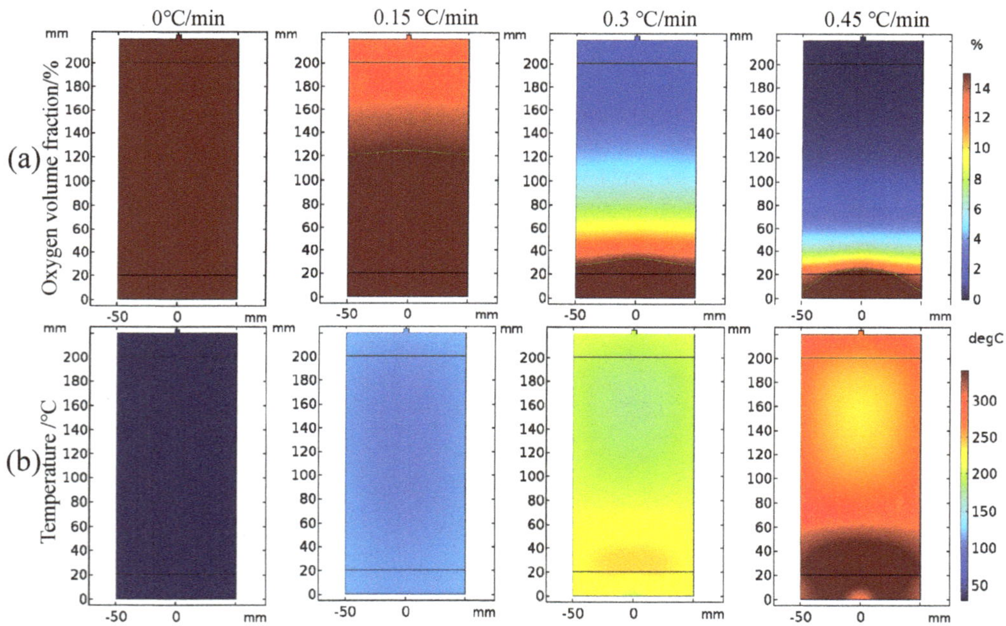

Figure 12. Cloud diagram of oxygen volume fraction and temperature distribution under different heating rates conditions: (**a**) oxygen volume fraction; (**b**) coal temperature.

4. Conclusions

(1) A programmed temperature rise model of coal spontaneous combustion is established using the multi-physical field coupling method to study the characteristics of coal spontaneous combustion under different external air supplies and temperature controls. The average error between simulation results of oxygen and carbon dioxide volume fractions and experimental results is only 2.3% and 4.6%, respectively. The acquisition of numerical simulation parameters and the numerical simulation method

can reasonably characterize the coal oxidation process. The research results have important practical significance for the prevention and control of coal spontaneous combustion in goaf.

(2) With the increase in air supply volume, the oxygen volume fraction at the outlet increases, and the peak value of the change rate of oxygen volume fraction at the outlet presents the "hysteresis" characteristic. As the air flow increases, the area with high oxygen volume fraction moves upward and the internal temperature of coal is getting higher, and linearly related to the air supply flow. After the air supply flow is greater than 24 mL/min, the maximum coal temperature exceeds the leading temperature of the programmed temperature rise after 7.5 h, its change rate exceeds the programmed temperature rise rate after 4 h, and the oxidation reaction is further intensified.

(3) As the heating rate increases, the decrease in oxygen volume fraction at the outlet intensifies, and the peak value of the oxygen volume fraction change rate presents the "early appearance" characteristic. The maximum coal temperature and its change rate always change from below the programmed leading temperature (heating rate) to above the programmed leading value, and in the later stage of coal oxidation, their values increase rapidly, and the risk of coal spontaneous combustion also increases further.

Author Contributions: Conceptualization, C.L. and X.S.; methodology, C.L.; software, C.L. and X.S.; validation, L.J., C.D. and J.N.; writing—original draft preparation, C.L.; writing—review and editing, C.L. and X.S.; visualization, C.L.; supervision, C.D. and Y.G.; funding acquisition, C.L. All authors have read and agreed to the published version of the manuscript.

Funding: This research was funded by the National Natural Science Foundation of China, grant numbers 52204229 and 52274220; the Basic Research Program in Shanxi province, grant number 20210302124349; and the Scientific and Technological Innovation Programs of Higher Education Institutions in Shanxi, grant number 2021L054.

Institutional Review Board Statement: Not applicable.

Informed Consent Statement: Not applicable.

Data Availability Statement: Not applicable.

Conflicts of Interest: The authors declare no conflict of interest.

References

1. Ma, T.; Zhai, X.; Xiao, Y.; Bai, Y.; Shen, K.; Song, B.; Hao, L.; Ren, L.; Chen, X. Study on the influence of key active groups on gas products in spontaneous combustion of coal. *Fuel* **2023**, *344*, 128020. [CrossRef]
2. Li, D.; Wang, E.; Kong, X.; Ali, M.; Wang, D. Mechanical behaviors and acoustic emission fractal characteristics of coal specimens with a pre-existing flaw of various inclinations under uniaxial compression. *Int. J. Rock Mech. Min.* **2019**, *116*, 38–51. [CrossRef]
3. Xu, K.; Li, S.; Liu, J.; Lu, C.; Xue, G.; Xu, Z.; He, C. Evaluation cloud model of spontaneous combustion fire risk in coal mines by fusing interval gray number and DEMATEL. *Sustainability* **2022**, *14*, 15585. [CrossRef]
4. Shi, X.; Zhang, Y.; Chen, X.; Zhang, Y.; Rui, L.; Guo, R.; Zhao, T.; Deng, Y. Numerical simulation on response characteristics of coal ignition under the disturbance of fluctuating heat. *Combust. Flame* **2022**, *237*, 111870. [CrossRef]
5. Ren, S.; Ma, T.; Zhang, Y.; Deng, J.; Xiao, Y.; Zhai, X.; Zhang, Y.; Song, Z.; Wang, C. Sound absorption characteristics of loose bituminous coal porous media with different metamorphic degrees. *Fuel* **2023**, *332*, 126091. [CrossRef]
6. Kong, X.; He, D.; Liu, X.; Wang, E.; Li, S.; Liu, T.; Ji, P.; Deng, D.; Yang, S. Strain characteristics and energy dissipation laws of gas-bearing coal during impact fracture process. *Energy* **2022**, *242*, 123028. [CrossRef]
7. Liang, Y.; Zhang, J.; Wang, L.; Luo, H.; Ren, T. Forecasting spontaneous combustion of coal in underground coal mines by index gases: A review. *J. Loss Prev. Proc.* **2019**, *57*, 208–222. [CrossRef]
8. Sahu, A.K.; Joshi, K.A.; Raghavan, V.; Rangwala, A.S. Comprehensive numerical modeling of ignition of coal dust layers in different configurations. *Proc. Combust. Inst.* **2015**, *35*, 2355–2362. [CrossRef]
9. Zeneli, M.; Nikolopoulos, A.; Nikolopoulos, N.; Grammelis, P.; Karellas, S.; Kakaras, E. Simulation of the reacting flow within a pilot scale calciner by means of a three phase TFM model. *Fuel Process. Technol.* **2017**, *162*, 105–125. [CrossRef]
10. Joshi, K.A.; Raghavan, V.; Rangwala, A.S. An experimental study of coal dust ignition in wedge shaped hot plate configurations. *Combust. Flame* **2012**, *159*, 376–384. [CrossRef]

11. Zhang, Y.; Yang, C.; Li, Y.; Huang, Y.; Zhang, J.; Zhang, Y.; Li, Q. Ultrasonic extraction and oxidation characteristics of functional groups during coal spontaneous combustion. *Fuel* **2019**, *242*, 287–294. [CrossRef]
12. Zhao, J.; Deng, J.; Wang, T.; Song, J.; Zhang, Y.; Shu, C.; Zeng, Q. Assessing the effectiveness of a high-temperature-programmed experimental system for simulating the spontaneous combustion properties of bituminous coal through thermokinetic analysis of four oxidation stages. *Energy* **2019**, *169*, 587–596. [CrossRef]
13. Deng, J.; Xiao, Y.; Li, Q.; Lu, J.; Wen, H. Experimental studies of spontaneous combustion and anaerobic cooling of coal. *Fuel* **2015**, *157*, 261–269. [CrossRef]
14. Qi, G.; Lu, W.; Qi, X.; Zhong, X.; Cheng, W.; Liu, F. Differences in smoldering characteristics of coal piles with different smoldering propagation directions. *Fire Safety J.* **2018**, *102*, 77–82.
15. Wen, H.; Wang, H.; Liu, W.; Cheng, X. Comparative study of experimental testing methods for characterization parameters of coal spontaneous combustion. *Fuel* **2020**, *275*, 117880. [CrossRef]
16. Beamish, B.B.; Theiler, J. Coal spontaneous combustion: Examples of the self-heating incubation process. *Int. J. Coal Geol.* **2019**, *215*, 103297. [CrossRef]
17. Zhang, D.; Yang, X.; Deng, J.; Wen, H.; Xiao, Y.; Jia, H. Research on coal spontaneous combustion period based on pure oxygen adiabatic oxidation experiment. *Fuel* **2021**, *288*, 119651. [CrossRef]
18. Jia, X.; Wu, J.; Lian, C.; Rao, J. Assessment of coal spontaneous combustion index gas under different oxygen concentration environment: An experimental study. *Environ. Sci. Pollut. Res.* **2022**, *29*, 87257–87267. [CrossRef]
19. Shi, X.; Chen, X.; Zhang, Y.; Zhang, Y.; Guo, R.; Zhao, T.; Liu, R. Numerical simulation of coal dust self–ignition and combustion under inclination conditions. *Energy* **2022**, *239*, 122227. [CrossRef]
20. Tang, Y.; Zhong, X.; Li, G.; Yang, Z.; Shi, G. Simulation of dynamic temperature evolution in an underground coal fire area based on an optimised Thermal-Hydraulic-Chemical model. *Combust. Theory Model.* **2019**, *23*, 127–146. [CrossRef]
21. Chen, X.; Shi, X.; Zhang, Y.; Zhang, Y.; Ma, Q. Numerical simulation study on coal spontaneous combustion: Effect of porosity distribution. *Combust. Sci. Technol.* **2023**, *195*, 472–493. [CrossRef]
22. Wen, H.; Yu, Z.; Deng, J.; Zhai, X. Spontaneous ignition characteristics of coal in a large-scale furnace: An experimental and numerical investigation. *Appl. Therm. Eng.* **2017**, *114*, 583–592. [CrossRef]
23. Shi, X.; Zhang, Y.; Chen, X.; Zhang, Y. Effects of thermal boundary conditions on spontaneous combustion of coal under temperature-programmed conditions. *Fuel* **2021**, *295*, 120591. [CrossRef]
24. Li, Z.; Xu, Y.; Liu, H.; Zhai, X.; Zhao, S.; Yu, Z. Numerical analysis on the potential danger zone of compound hazard in gob under mining condition. *Process. Saf. Environ.* **2021**, *147*, 1125–1134. [CrossRef]
25. Ma, L.; Guo, R.; Wu, M.; Wang, W.; Ren, L.; Wei, G. Determination on the hazard zone of spontaneous coal combustion in the adjacent gob of different mining stages. *Process. Saf. Environ.* **2020**, *142*, 370–379. [CrossRef]
26. Yuan, L.; Smith, A.C. Numerical study on effects of coal properties on spontaneous heating in longwall gob areas. *Fuel* **2008**, *87*, 3409–3419. [CrossRef]
27. Liu, W.; Qin, Y.; Shi, C.; Guo, D. Dynamic evolution of spontaneous combustion of coal in longwall gobs during mining-stopped period. *Process. Saf. Environ.* **2019**, *132*, 11–21. [CrossRef]
28. Ren, L.; Deng, J.; Li, Q.; Ma, L.; Zou, L.; Laiwang, B.; Shu, C. Low-temperature exothermic oxidation characteristics and spontaneous combustion risk of pulverised coal. *Fuel* **2019**, *252*, 238–245. [CrossRef]
29. Yan, H.; Nie, B.; Liu, P.; Chen, Z.; Yin, F.; Gong, J.; Lin, S.; Wang, X.; Kong, F.; Hou, Y. Experimental investigation and evaluation of influence of oxygen concentration on characteristic parameters of coal spontaneous combustion. *Thermochim. Acta* **2022**, *717*, 179345. [CrossRef]
30. Zhang, Y.; Zhang, Y.; Li, Y.; Shi, X.; Che, B. Determination of ignition temperature and kinetics and thermodynamics analysis of high-volatile coal based on differential derivative thermogravimetry. *Energy* **2022**, *240*, 122493. [CrossRef]
31. Sabat, G.; Gouda, N.; Panda, A.K. Effect of coal grade and heating rate on the thermal degradation behavior, kinetics, and thermodynamics of pyrolysis of low-rank coal. *Int. J. Coal Prep. Util.* **2022**, 1–19. [CrossRef]
32. Wang, Z.; Xu, M.; Fu, X.; Cao, J.; Li, C.; Feng, M.; Li, K. Study on pyrolysis characteristics and kinetics of bituminous coal by thermogravimetric method. *Combust. Sci. Technol.* **2022**, *194*, 558–573. [CrossRef]
33. Ejlali, A.; Mee, D.J.; Hooman, K.; Beamish, B.B. Numerical modelling of the self-heating process of a wet porous medium. *Int. J. Heat Mass Transf.* **2011**, *54*, 5200–5206. [CrossRef]
34. Xia, T.; Zhou, F.; Wang, X.; Zhang, Y.; Li, Y.; Kang, J.; Liu, J. Controlling factors of symbiotic disaster between coal gas and spontaneous combustion in longwall mining gobs. *Fuel* **2016**, *182*, 886–896. [CrossRef]
35. Zheng, Y.; Li, Q.; Zhu, P.; Li, X.; Zhang, G.; Ma, X.; Zhao, Y. Study on multi-field evolution and influencing factors of coal spontaneous combustion in goaf. *Combust. Sci. Technol.* **2023**, *195*, 247–264. [CrossRef]
36. Wu, D.; Vanierschot, M.; Verplaetsen, F.; Berghmans, J.; Van den Bulck, E. Numerical study on the ignition behavior of coal dust layers in air and O_2/CO_2 atmospheres. *Appl. Therm. Eng.* **2016**, *109*, 709–717. [CrossRef]
37. Lei, C.; Deng, J.; Cao, K.; Xiao, Y.; Ma, L.; Wang, W.; Ma, T.; Shu, C. A comparison of random forest and support vector machine approaches to predict coal spontaneous combustion in gob. *Fuel* **2019**, *239*, 297–311. [CrossRef]
38. Deng, J.; Lei, C.; Xiao, Y.; Cao, K.; Ma, L.; Wang, W.; Laiwang, B. Determination and prediction on "three zones" of coal spontaneous combustion in a gob of fully mechanized caving face. *Fuel* **2018**, *211*, 458–470. [CrossRef]

39. Zhang, Y.; Shu, P.; Deng, J.; Duan, Z.; Li, L.; Zhang, L. Analysis of oxidation pathways for characteristic groups in coal spontaneous combustion. *Energy* **2022**, *254*, 124211. [CrossRef]
40. Wang, C.; Deng, Y.; Xiao, Y.; Deng, J.; Shu, C.; Jiang, Z. Gas-heat characteristics and oxidation kinetics of coal spontaneous combustion in heating and decaying processes. *Energy* **2022**, *250*, 123810. [CrossRef]
41. Lei, C.; Jiang, L.; Bao, R.; Deng, C.; Wang, C. Study on multifield migration and evolution law of the oxidation heating process of coal spontaneous combustion in dynamic goaf. *ACS Omega* **2023**, *8*, 14197–14207. [CrossRef] [PubMed]
42. Kong, B.; Li, Z.; Yang, Y.; Liu, Z.; Yan, D. A review on the mechanism, risk evaluation, and prevention of coal spontaneous combustion in China. *Environ. Sci. Pollut. Res.* **2017**, *24*, 23453–23470. [CrossRef] [PubMed]
43. Lei, C.; Deng, J.; Cao, K.; Ma, L.; Xiao, Y.; Ren, L. A random forest approach for predicting coal spontaneous combustion. *Fuel* **2018**, *223*, 63–73. [CrossRef]

sustainability

MDPI

Article

Study on the Pseudo-Slope Length Effect of Buried Pipe Extraction in Fully Mechanized Caving Area on Gas Migration Law in Goaf

Pengxiang Zhao [1,2,3,4], Xingbao An [1,2,*], Shugang Li [1,2,3], Xinpeng Kang [1,2], Yitong Huang [5], Junsheng Yang [5] and Shikui Jin [4,6]

1 College of Safety Science and Engineering, Xi'an University of Science and Technology, Xi'an 710054, China
2 Key Laboratory of Western Mine Exploitation and Hazard Prevention Ministry of Education, Xi'an University of Science and Technology, Xi'an 710054, China
3 Western Engineering Research Center of Mine Gas Intelligent Drainage for Coal Industry, Xi'an 710054, China
4 Xinjiang Uygur Autonomous Region Coal Science Research Institute, Urumqi 830000, China
5 Liuhuanggou Coal Mine, Yankuang Xinjiang Mining Co., Ltd., Changji 831100, China
6 Xinjiang Coal and CBM Engineering Technology Research Center, Urumqi 830000, China
* Correspondence: 20220226069@stu.xust.edu.cn

Abstract: To study the law of gas transportation in mining areas, Fluent numerical simulation software was applied to examine the influence of different pseudo-slope lengths (PSL) on gas concentration in a U-ventilated working area under no-extraction conditions. Based on this, numerical simulation experiments were conducted on the buried pipe extraction arrangement parameters. The simulation found that when there was no extraction, the PSL had an impact on the airflow in the extraction area, which caused the airflow in the extraction area to be disordered, causing gas to accumulate locally at the working area. When the buried pipe depths (BPDs) and PSLs of the working area worked together, the gas concentration of the working area was lower when the inlet air influence zone and the extraction influence zone were through; otherwise, gas concentration accumulation occurred at the working area. The research results showed that when the PSL was at 25 m and BPD was at 20 m, the gas concentration at the working area was not abnormal, and the gas concentration in the upper corner was lower. By adjusting the PSL and BPD of the test working area, the maximum gas concentration in the upper corner was reduced to 0.46% and the maximum gas concentration in the return air outlet was reduced to 0.41%. The experimental and practical results provide important reference values for coal and gas co-mining.

Keywords: buried pipe extraction; inclined thick coal seam; pseudo-slope length; u-shaped ventilation; upper-corner gas concentration

Citation: Zhao, P.; An, X.; Li, S.; Kang, X.; Huang, Y.; Yang, J.; Jin, S. Study on the Pseudo-Slope Length Effect of Buried Pipe Extraction in Fully Mechanized Caving Area on Gas Migration Law in Goaf. *Sustainability* **2023**, *15*, 6628. https://doi.org/10.3390/su15086628

Academic Editor: Adam Smoliński

Received: 12 February 2023
Revised: 8 March 2023
Accepted: 3 April 2023
Published: 13 April 2023

1. Introduction

China is one of the few countries in the world where coal is the main energy source. Currently, the coal mined underground accounts for 81.8% of the total coal [1,2]. Coal seams generally have high gas content and low permeability. The rapid increase in frequent mining disturbances and unloading gas emission intensity is caused by the continuous increase in the mining intensity. This led to frequent gas overrun problems in the return airway and upper corner. It hindered the safety production of the mine [3–5]. The concentration of gas extracted during coal mine production is less than 30%. This value is lower than the national standard for gas use. Many coal mines still adopt direct emptying treatment, resulting in more than 20 billion cubic meters of gas being discharged into the atmosphere every year during coal mining operations. The amount of gas pollution discharged into the atmosphere is tens of times more than that of CO_2. At present, the rapid development of clean coalbed methane resources to replace the demand for coal energy has become increasingly urgent. After more than ten years of development, the theory of coal and gas co-mining has achieved a more comprehensive technical standard. However, the geological environment of the coal seams in China is complex. The existing theoretical technology still cannot support the safe and efficient co-mining of coal and coalbed methane.

To ensure safe and efficient production in mines, scholars had established different experimental models for various factors inside the mining area based on different research purposes, and have researched the evolution of laws related to wind volume, temperature field, and pressure field [6–9], etc. Using computational fluid dynamics simulation, Chang et al. [10] studied the amount of air supply to the working area and the negative ventilation pressure used for gas discharge in the roof channel to obtain the most suitable ventilation pressure. Shao et al. [11] simulated the gas control effect of "U + L" and "Y + L" ventilation modes and optimized the main ventilation factors of the "U + L" ventilation mode by orthogonal test and fuzzy evaluation method. The results showed that the gas concentration of the "U + L" ventilation mode was lower than that of the "Y + L" mode. The optimized "U + L" ventilation mode effectively reduced the gas concentration in the upper corner and returned it to the air tunnel. Yang et al. [12] investigated the effect of the gas concentration distribution pattern of the "Y + HLDR" ventilation method under different air supply conditions with the help of COMSOL. The results showed that with an increase in the air supply, the gas concentration near the working area decreased, and the high-concentration gas was directed to the deeper part of the mining area. Li et al. [13] simulated the gas concentration and flow field under different wind speeds by constructing a cavity collapse rock model in the mining area. Liang et al. [14] built a two-dimensional distribution model of permeability in the mining area based on different collapse patterns. Wang et al. [15] combined FLAC3D and Fluent simulation methods to obtain the flow field of air leakage in the mining area. Wang et al. [16] studied the flow field distribution and air leakage in the extraction zone under different conditions for the long-walled working area. As for field tests, Zhu et al. [17] qualitatively analyzed the non-synchronous correlation characteristics of gas transport in the direction of airflow at the working area and proposed an algorithm for identifying gas monitoring data anomalies based on spatiotemporal correlation analysis. Yu et al. [18] studied the asymmetry of the flow field in the mining area. The location of the gas aggregation zone varies with the ventilation parameters, which can be used as an evaluation index to study the degree of air leakage in the mining area. Zhang et al. [19] analyzed the development height of fracture zone in an inclined coal seam by using FLAC and used Fluent simulation to optimize the arrangement parameters of a high-level directional long borehole, instead of a high extraction lane, for efficient extraction. Brigida et al. [20] processed the dynamic data of gas concentration by using an optimization algorithm to obtain the cyclic nonlinear distribution of mine gas subject to perturbation. Dzhioeva et al. [21] studied the relationship between mine longwall distance and gas as a function of distance and used polynomial regression to analyze the results in order to improve the prediction reliability of the dynamic distribution of gas.

Concerning the problem that gas concentration in the upper corner of high gas mines easily exceeds its limit, many scholars have analyzed the possible causes from different angles and have carried out field verifications [22,23]. Li et al. [24] established the physical model of the mining area and have proposed a combined drainage technology of buried and sprinkled pipes in the upper mine corner, proving the effectiveness of the technology through Fluent simulation experiments. Chen et al. [25] studied the effects of air intake and advancement speed on the gas concentration in the upper corner and obtained the best airspeed of 2 m/s for the air intake lane and the best speed of 3 m/d for the working area mining. Liu et al. [26] compared the gas pressure distribution and methane concentration distribution under the conditions of no extraction measures, buried pipe extraction in the upper corner, and gas extraction holes at different locations with the help of numerical simulations. The results showed that the gas extraction hole can be used to discharge a large amount of gas and can effectively control gas transport. Chai [27] established different arrangement parameters of high extraction lanes for the problem of gas over-limit in the upper corner during the rapid advance of the extra-thick coal seam. In addition, based on a field verification, we realized the safety requirements of gas control at the working area. Xiong et al. [28] proposed secondary extraction under double stress interference and divided the coal body of the working area into three zones based on extraction concentration: high efficiency zone, effective zone, and original extraction zone. Among them, the average extraction concentration in the high-efficiency zone exceeded 20%.

With a shortage of oil and natural gas, China relies heavily on coal for energy. The coal demand would be more than 50% of the total energy consumption in the long term. Moreover, the energy demand is increasingly aggravated as China is in the stage of rapid development. The demand for coal is bound to increase further if the vigorous development of clean energy is not accelerated. At the same time, the safety problems in the process of coal production have not been effectively curbed. These problems have become an important obstacle to the healthy development of ecological civilization in China [29]. With the westward shift of China's coal mining center of gravity, the proportion of mining of thick coal seams is gradually increasing. The Xinjiang region is rich in coal resources, featuring large coal seam thickness, large coal seam inclination, and high gas content. Due to the limitation of mining technology, a large amount of coal remains in the mining area, easily leading to the abnormal accumulation of gas in the mining area. As an essential means to adjust the movement of the hydraulic bracket in the large inclination coal seam, the PSLs performed a decisive role in the wind flow diameter and flow field distribution at the working area. Meanwhile, as an essential means to solve the gas anomaly in the upper corner, buried pipe extraction has the advantages of a small engineering volume, short working period, simple technology, and sound effects. When the amount of gas gushing from the mining area was small, the buried pipe extraction method could be used alone. When the amount of gas pouring out from the mining area is large, it can be used in combination with other extraction methods. Liuhuanggou Coal mine in Changji area of Xinjiang Province is selected as the research background. In this study, Fluent numerical simulation software was used to study the influence of the length of the slope and the depth of the buried pipe on the gas concentration in mining area and working area. The optimal pseudo-slope length and buried pipe depth were determined and verified in the field. The research results provide important theoretical guidance for gas disaster prevention and control in the upper corner of inclined thick coal seam comprehensive release mining.

2. Experimental Design

2.1. Overview of the Experimental Working Area

The experimental area used in this study was located in Liuhuanggou Coal Mine, Changji City, Xinjiang Uygur Autonomous Region. Coal was mainly extracted from 4-5 seams. The test working area was mined by the method of long-walled backward caving. The maximum and minimum depth of the orbiting groove on the working area was 557 m and 435 m, respectively. The top and minimum depth of the belt chute was 662.5 m and 486.1 m, respectively. The strike length of the test area is 3250 m, the width of the working area is 180 m, the average coal seam inclination is 24°, and the average mining height is 6.15 m. The mining of inclined coal seams was often accompanied by the problem of upward movement and downward movement of the conveyor. When the slippage was severe, it caused the bracket to dump and could not effectively control the roof, which deteriorated the working area conditions and even caused roofing accidents. At the same time, it also caused the safety exit width of the lower chute to be insufficient, which brought safety risks. The upward movement of the conveyor will cause overload with the conveyor and, in severe cases, will cause the upper chute to be pedestrian. Generally, the PSL is increased to ensure the normal movement of the hydraulic support and to provide more functional space for the headstock. However, the increase in the PSL will affect the gas change between the frames and the upper corner, resulting in abnormal gas concentration at the working area. According to the actual conditions of the test working area, the PSL was adjusted between 18 m and 37 m, respectively. Four groups of 20 m, 25 m, 30 m, and

35 m PSLs were selected for the study. The PSL layout of different working areas are shown in Figure 1.

2.2. Model Building

2.2.1. Geometric Modeling

The dimensions of the working area at the site served as the basis for the model for the numerical simulation. The working area inclination length is 180 m. The coal seam inclination angle is 24°. The relic coal seam height is 2 m. The thickness of the direct top was used to compute the theoretical height of the fall zone. Calculate the height of the fall band according to Equation (1).

$$m_z = \frac{h-\Delta}{N_\mathrm{p}-1} \quad (1)$$

where m_z is the height of the bubble fall zone, m; h is the thickness of the coal seam, m; Δ is the filling thickness due to coal relics, $\Delta = h \times (1 - v) \times N_\mathrm{m}$, m; c is the total extraction rate, taken as 85%; N_m is the coefficient of coal breakage and swelling of the fallen top coal, taken as 1.1; and N_p is the coefficient of fragmentation and expansion of coal rock seam, taken as 1.2.

Combining with Equation (1), according to the lithological characteristics of the leading mining area, the theoretical bubble fall zone height of this working area was calculated at 22.8 m. According to Table 1, the range of the fissure zone was 92–130 m, and the height of the fissure zone was taken as 110 m.

Table 1. Table of equations for the maximum height of the rift zone.

Lithology	Applicable to "3 m<$\sum M \leq$12 m" Coal Seam
Hard (40~80 MPa, quartz sandstone, limestone, conglomerate)	$H_\mathrm{L} = \frac{100\sum M}{0.15\sum M+3.12} \pm 11.18$
Medium-hard (20~40 MPa, sandstone, muddy tuff, shale)	$H_\mathrm{L} = \frac{100\sum M}{0.23\sum M+6.10} \pm 10.42$
Soft (10~20 MPa, mudstone, muddy shale)	$H_\mathrm{L} = \frac{100\sum M}{0.31\sum M+8.81} \pm 8.21$

Note: M is the mining height, m; H_L is the height of the rift zone, m.

According to the lithological characteristics of the main mining area, the height of the theoretical bubble fall zone in the mining area was calculated at 22.8 m and the height of the fissure zone was 110 m, using the above equation.

The workbench-Design Modeler was used to build the geometric model. The mesh module was used to unstructured the model, and the tetrahedral mesh was used to delineate the fluid domain. The geometric model is shown in Figure 2. The model was divided into three different densities of meshes for mesh validation. The total number of grid 1 was 81,371, and the number of working area grids was 7162. The total number of grid 2 was 1,947,318, and the number of working area grids was 400,709. The total number of grid 3 grids was 14,148,963, and the number of working area grids was 3,111,898.

Figure 3 shows the wind speed distribution of nine measurement points arranged along the working area tendency under different grid densities. As the grid density increased, the difference between the results calculated for the density of grid 2 and that of grid 3 was small. Grid extraneous verification showed that the medium-density grid satisfies the irrelevance requirement of the calculation results. The PSLs of 20 m, 30 m, and 35 m were computed using the same mesh-partitioning method.

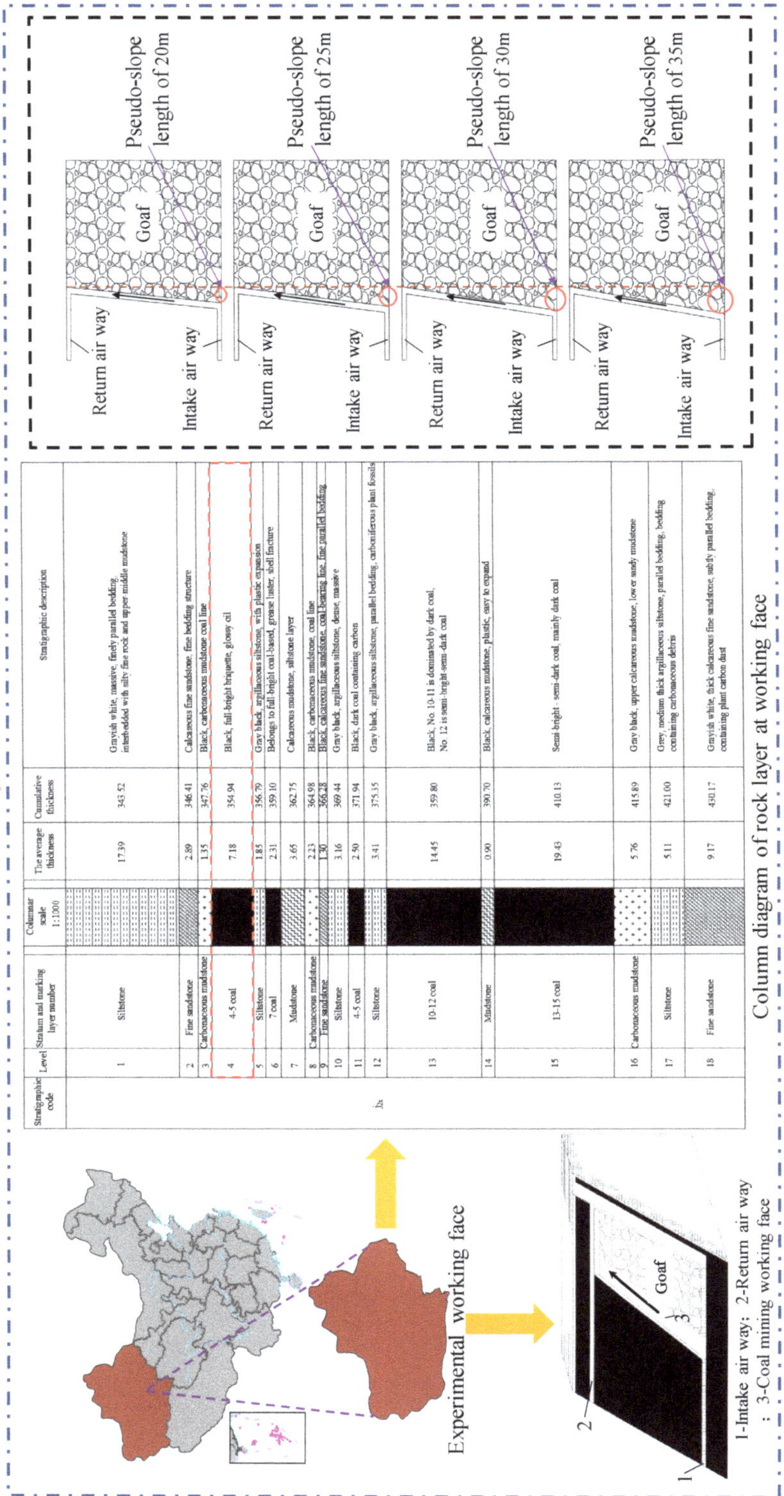

Figure 1. Working area layout diagram.

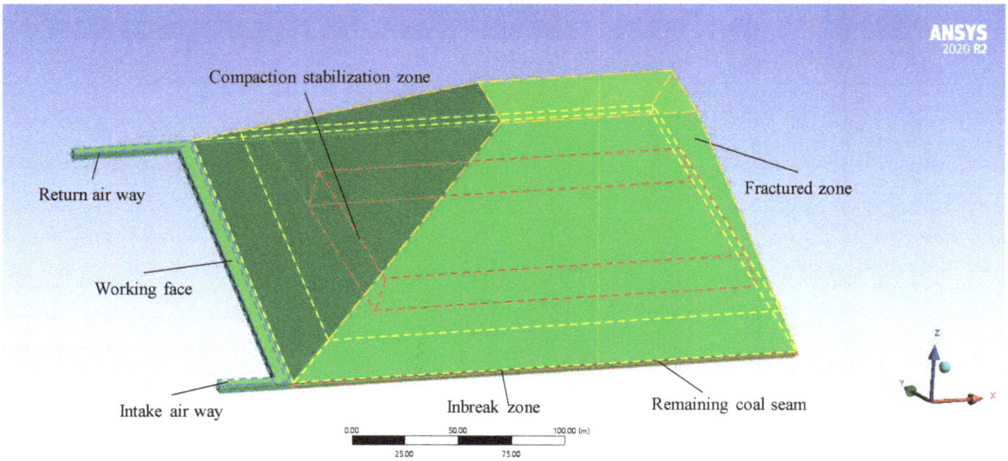

Figure 2. Geometric model diagram.

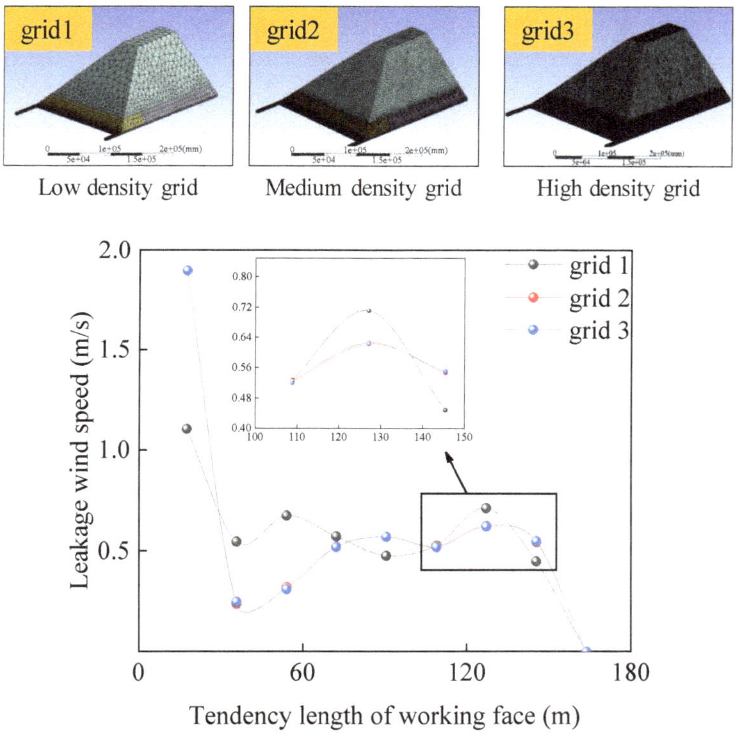

Figure 3. Wind speed along the inclination direction of the working area at different grid densities.

2.2.2. Boundary Conditions and Parameter Setting

Several constraints and assumptions must be made to simplify the numerical simulation analysis while ensuring the reliability of the simulation results in the real world. Only methane, nitrogen, and oxygen were considered in the species transport model. In the computational domain, gas was a continuous and incompressible medium. The flow process did not consider the energy exchanged, such as heat transfer.

(1) Flow model selection

The Reynolds number was obtained by substituting the field monitoring parameters into Equation (2). The model *RNG-k-ε* was chosen because it added a term to the ε equation to improve the accuracy of the flow. Additionally, it provided an analytical formulation of the turbulent Prandtl number and allowed for higher accuracy and confidence compared to the standard *k-ε* model. The fluid was set as a mixed phase of air and methane.

$$Re = \frac{\rho v d}{\mu} \tag{2}$$

where Re is the Reynolds number; v is the flow velocity of the fluid, m/s; ρ is the density of the fluid, kg/m^3; μ is the coefficient of viscosity, kg/(m·s); and d is the characteristic length, m.

(2) Source item settings

After the advancement of the header workings, the relic coal is the primary source of gas gushing out from the mining area. Under the influence of mining, the gas gushed out from the adjacent coal seam under the main mining seam after unloading pressure. Finally, it diffused and floated through the legacy coal seam to the mining area. Integrate the gas outflow from the coal deposit, the adjacent seam, and the surrounding rock to the source of porous media in the bottom 2 m of the model. It was calculated that the total gas outflow from the mining area was 6.15 m^3/min, and the gas quality source term in the mining area was calculated according to Equation (3).

$$Q_v = \frac{Q_g \rho_g}{V} \tag{3}$$

where Q_v is the gas mass source term gushing volume, kg/(m^3·s); Q_g is the absolute gas gushing volume, m^3/s; ρ_g is the gas density, 0.716 kg/m^3; V is the model volume of the mining area, m^3; and $Q_v = 6.15 \times 0.716/(2.7 \times 10 - 6) = 1.6e^{-6}\text{kg}/(\text{m}^3{\cdot}\text{s})$.

(3) Basic flow equations [30,31]

① Conservation of mass Equation:

Equation (4) depicts the mass conservation equation, also known as the continuity equation.

$$\frac{\partial \rho}{\partial t} + \nabla(\rho v) = S_m \tag{4}$$

where t is the time, s; and S_m is the mass added to the continuous phase by the dispersive secondary term and the defined source, kg/(m^3·s).

② Conservation of momentum equation:

$$\frac{\partial}{\partial t}(\rho v) + \nabla(\rho v v) = -\nabla p + \nabla(\tau) + \rho g + F \tag{5}$$

where p is the static pressure, Pa; τ is the stress tensor, Pa; ρg is the gravitational force; and F is the external force, N.

③ Component mass conservation equation:

$$\frac{\partial}{\partial t}(\rho c_s) + \nabla(\rho v c_s) = \nabla(D_s \text{grad}(\rho c_s)) + S_s \tag{6}$$

where c_s is the volume fraction of component s; D_s is the diffusion coefficient, m^2/s; and S_s is the mass of the component produced by the chemical reaction per unit time, kg/(m^3·s).

④ Considered the extraction area as a porous medium and add momentum sources, which included viscous loss term and inertial loss term.

$$S_i = \sum_{j=1}^{3} E_{ij}\mu v_j + \sum_{j=1}^{3} F_{ji}\frac{1}{2}\rho v_{mag} v_j \quad (7)$$

where S_i is the source of the "i" (x, y, z) momentum equation, N/m^3; μ is the molecular viscosity; v_{mag} is the mode of the velocity vector; v_j is the velocity component in the "x, y, z" direction, m/s; and E and F are predefined matrices.

⑤ The type of flow in the model is turbulent, and the solution equations are given in Equations (8) and (9) using the "*RNG-k-ε*" model in Fluent.

$$\frac{\partial}{\partial t}(\rho k) + \frac{\partial}{\partial x_i}(\rho k u_i) = \frac{\partial}{\partial x_i}\left(\alpha u_{iff}\frac{\partial k}{\partial x_j}\right) + G_k + G_b - \rho\varepsilon - Y_M + S_k \quad (8)$$

$$\frac{\partial}{\partial t}(\rho\varepsilon) + \frac{\partial}{\partial x_i}(\rho\varepsilon u_i) = \frac{\partial}{\partial x_j}\left(\alpha_\varepsilon u_{eff}\frac{\partial\varepsilon}{\partial x_j}\right) + G_{1\varepsilon}\frac{\varepsilon}{k}(G_k + C_{3\varepsilon}G_b) - C_{2\varepsilon}\rho\frac{\varepsilon^2}{k} - R_\varepsilon + S_\varepsilon \quad (9)$$

where G_k is the turbulent kinetic energy generated by the velocity gradient, J; G_b is the turbulent kinetic energy generated by buoyancy, J; and k is the turbulent kinetic energy, J. Y_M is the contribution of pulsating expansion in compressible turbulence to the total dissipation rate; α_k, α_ε are the inverse effective Prandtl numbers of "k", "ε", respectively; S_k, S_ε are the defined source terms; k source term in $kg/m{\cdot}s^3$; ε source term in $kg/m{\cdot}s^4$; and $C_{1\varepsilon}$, $C_{2\varepsilon}$, and $C_{3\varepsilon}$ are constants.

(4) Porosity setting of mining area module

The mining area model was divided into two modules, fracture zone, and fallout zone. The porosity of the fracture zone was divided according to the "O" circle theory, and the porosity was calculated according to Equation (10) [32]. The average porosity P_0 distribution of the sub-domains was shown in Table 2.

$$\varphi(x,y) = 1 + \frac{\left(1 + e^{-0.15*(\frac{l_y}{2} - |y|)}\right)\left[1 - \frac{h_d}{h_d + H - (H - h_d(N_{P_b} - 1))\left(1 - e^{-\frac{x}{2l}}\right)}\right] - 1}{1 + \sigma_0^{-1}\beta_1\gamma\left(\frac{l_y}{2} - y\right)sin\theta} \quad (10)$$

where l_y is the tendency width of the mining area, m; h_d is the thickness of the direct top, m, take 3 m; N_{P_b} is the coefficient of rock fragmentation and expansion after the direct top damage, take 1.1; l is the length of the basic fixed broken rock, m, take 15 m; γ is the bulk weight of the fallen rock, N/m^3, the bulk weight of the fallen gangue is generally 2×10^4~3×10^4; β_1 is the regression coefficient, the fallen rock is Mudstone is taken −0.028; and is the coal seam inclination, 24°.

Table 2. Porosity assignment table for each part of goaf.

	Remaining Coal Seam	Inbreak Zone				Fractured Zone
		Natural Accumulation Area	Influence Area of Coal Wall on the Return Wind Side	Compaction Stabilization Zone	Inlet Side Coal Wall Influence Area	
$P_0(\%)$	0.2	0.4	0.25	0.13	0.22	0.2

(5) Parameters of the inlet and return air lanes

The airflow was inflowed from the inlet lane. The inlet air speed was 3.5 m/s. The inlet air mass fraction was 23%, which was converted to a molar fraction for calculation purposes. The return air lane was set as a free outlet. The hydraulic diameter was taken as the equivalent diameter of the lane. The calculation was made according to Equation (11).

$$de = \frac{2ab}{(a+b)} \tag{11}$$

where *de* is the equivalent diameter of the roadway, m; and *a* and *b* are the geometric length and width, m, respectively.

2.3. Monitoring Point Layout Program

Gas concentration monitoring at the working area was calculated via grid nodes. Gas concentration was monitored by arranging to measure surfaces, measuring lines, and measuring points in the calculation domain. The working area domain was divided into three measurement surfaces that were parallel to the base plate, and the measurement points on each surface were organized using the point cloud measurement method. The number of point clouds was set to 1000. There was a total of 12 measuring lines that were spaced 0.3 m, 2.8 m, 5.3 m, and 7.8 m from the coal wall, respectively. Each group of measurement lines was separated into upper, medium, and lower spatial heights. At the end of the return air tunnel, we set up a 2 m by 2 m by 2 m grid and nine grid measurement stations to keep an eye on the gas concentration in the upper corner. The measurement arrangement was shown in Figure 4.

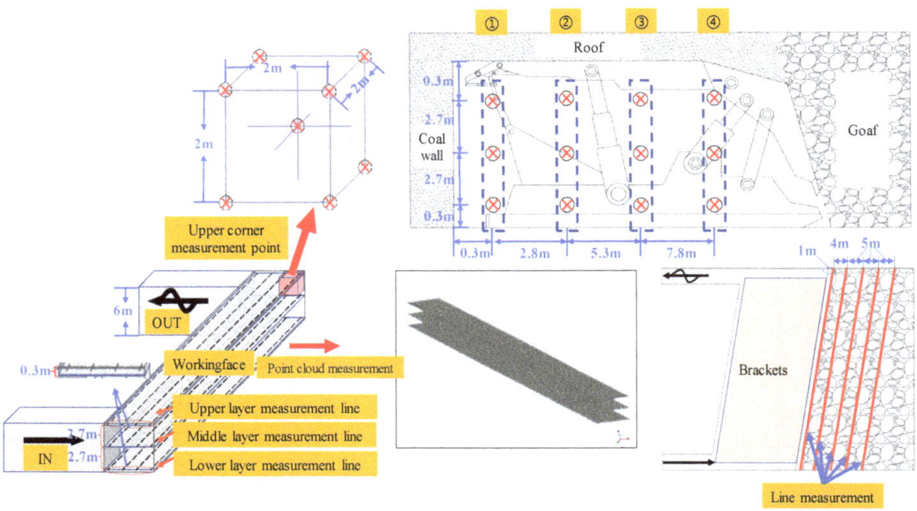

Figure 4. Working area measuring point layout.

3. Gas Concentration Distribution Pattern under Different PSLS

3.1. Variation in the Working Area's Gas Spewing Location while Using Various PSLs

The flow field vector lines at each location inside the mining region are shown in Figure 4 under the influence of various PSLs, respectively. Under various PSLs, the flow field lines in the mining region exhibited significant variances. The wind in the mining region flowed into the working area from the upper corner when the PSL was 20 m and 25 m. When the PSL was 30 m, two air leakage spots on the working area were visible, one in the A region and the other in the upper corner. The B area and the top corner of the working area were the two primary air leakage locations when the PSL was 35 m, with the B area being closer to the inlet tunnel than the A area, as shown in Figure 5a–d. In summary, as the PSLs increased to more than 30 m, the air leakage point drifted toward the inlet tunnel. The wind transported high gas concentrations from the mining region to the working area, abnormally concentrated at the wind leakage area.

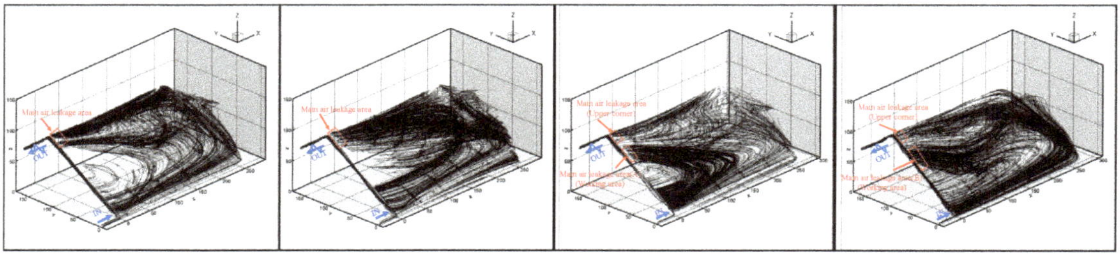

(**a**) (**b**) (**c**) (**d**)

Figure 5. Airflow vector map of goaf under different PSLs. (**a**) PSL-20 m; (**b**) PSL-25 m; (**c**) PSL-30 m; (**d**) PSL-35 m.

The disorder and uniformity in the arrangement of measurement points characterize the three-dimensional grid point cloud measurement method. The experimental results show that the measured gas concentration can reflect the spatial distribution of gas concentration under different PSLs.

Figure 6 shows the analysis of gas concentration distribution in working areas at different levels. The experimental results showed that PSLs have a great influence on the location of gas accumulation zone in the working area (concentration of gas accumulation zone ranges from 0.8% to 1.0%, and over 1.0% of the abnormal gas zone). The overall gas concentration in the working area was low, and gas concentrated in the upper corner under PSL-20 m, PSL-25 m, and PSL-30 m. At a PSL of 35 m, gas accumulated in the upper corner and at the working area. Gas builds up in the PSL's 20 m and 25 m upper corners at the middle and upper levels. Gas builds up at the upper corner and central portion of the working area at PSL of 30 m and 35 m. The working area's gas concentration gathering region was moved nearer the input tunnel as the PSLs increased. The gas accumulation zones (expressed as the distance from the inlet lane, m) and peak gas concentration (%) at different PSLs' working areas are shown in Table 3.

Table 3. Distribution of gas anomalous areas and gas concentration peaks in working area under different PSLs.

	PSL 20 m		**PSL 25 m**		**PSL 30 m**		**PSL 35 m**	
	Gas Concentration Anomaly Zone (m)	**Maximum Gas Concentration (%)**	**Gas Concentration Accumulation Area (m)**	**Maximum Gas Concentration (%)**	**Gas Concentration Anomaly Zone (m)**	**Maximum gas Concentration (%)**	**Gas Concentration Anomaly Zone (m)**	**Maximum Gas Concentration (%)**
Upper Level	84.9~159.7	1.76%	156.4~160.6	0.98%	64.1~161.1	3.05%	23.6~163.1	4.90%
Middle Level	-	-	160.1~161.1	0.81%	65.3~73.6 149.4~160.1	1.46% 1.58%	29.9~126.6 155.4~163.1	3.28% 2.57%
Lower level	-	-	-	-	-	-	35.2~102.3 160.2~161.7	2.19% 1.88%

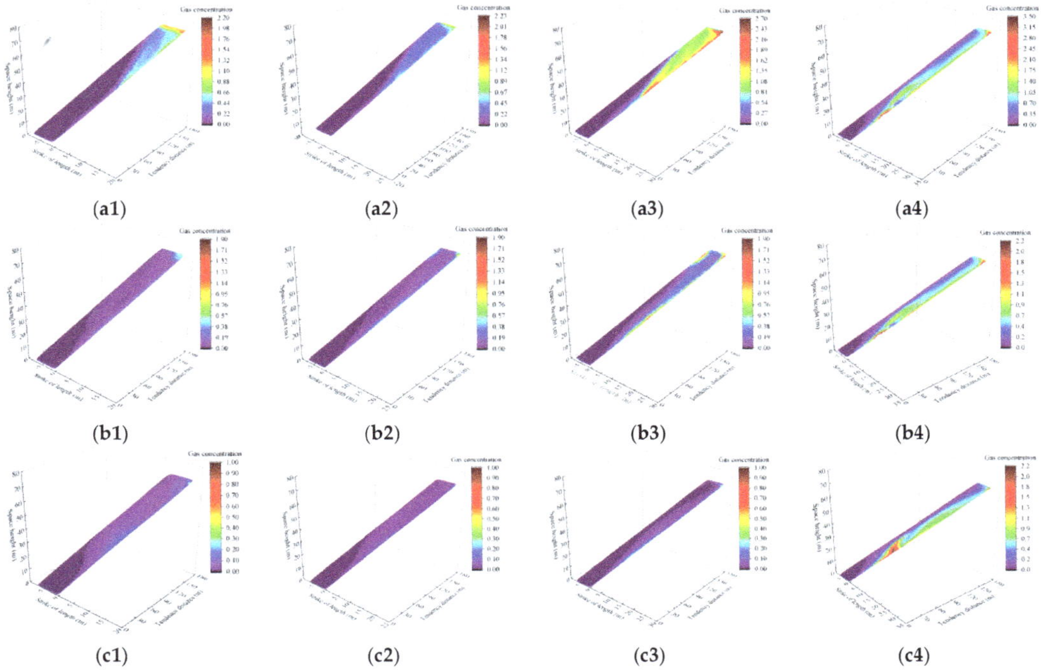

Figure 6. Gas cloud map of the working area under different PSLs. (**a1–a4**) show the gas concentration clouds at the high level of the working area at pseudo-slope lengths of 20 m, 25 m, 30 m, and 35 m, respectively; (**b1–b4**) show the gas concentration clouds at the middle level of the working area at pseudo-slope lengths of 20 m, 25 m, 30 m, and 35 m, respectively; (**c1–c4**) show the gas concentration clouds at the low level of the working area at pseudo-slope lengths of 20 m, 25 m, 30 m, and 35 m, respectively.

In order to reflect the spatial distribution of gas concentration in the working area, four groups of measuring lines are arranged according to the working area. According to Formula (12), the standard deviation of gas concentration at different positions in the tilt direction of the working area was calculated. The gas concentration error zone diagram in the working area was drawn, as shown in Figure 7a–d. The gas concentration error band chart reflected the variation characteristics of the spatial trend average gas concentration data supported by the comprehensive working area. According to the change rate of average gas concentration and gas anomaly in the working area, the gas concentration curve in the working area was divided into three zones. G_0 was the gas concentration area of the working area, G_1 was the gas growth area of the working area, G_2 was the gas anomaly area of the working area. Zone G_0 was located near one side of the intake tunnel. As the dip Angle of coal seam was large and the space height of air inlet roadway was low, the gas in the mining area will be transported to the side of air return roadway under the action of floating. A small amount of gas overflowing from the mining area to the working area will be quickly carried away by fresh air with high velocity, so the gas area at the side of the intake lane was 0. In G_1 zone, the spatial height of the working area increased, resulting in higher gas concentration in the mining area. Due to the influence of ventilation resistance, the wind speed in the working area decreased and the dilution effect on gas was weakened, resulting in the increase in gas concentration in the area. In the space 7.8 m away from the coal wall, there exists the G_2 zone, which was an abnormal gas concentration zone in the working area, and most of the gas gush from the mining area will be diluted by fresh air and taken out of the working area. However, when a large amount of gas gushing out, the flow of wind cannot dilute the gas in time, leading to the occurrence of gas anomalies in different positions of the working area far from the coal wall.

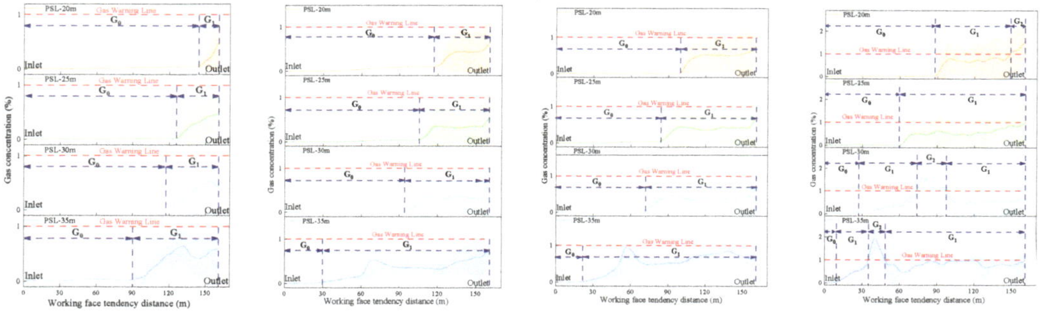

(a) (b) (c) (d)

Figure 7. Distribution of Gas Concentration in Space with Different PSLs. (**a**) Distance to coal wall 0.3 m. (**b**) Distance to coal wall 2.8 m. (**c**) Distance to coal wall 5.3 m. (**d**) Distance to coal wall 7.8 m.

Under different PSLs conditions, the working area's gas distribution pattern differed. Overall, the G_0 zone extended further away from the coal wall. The range of the G_0 zone gradually decreased as the PSLs increased, while the scope of the G_1 zone gradually expanded toward the side of the inlet tunnel. Different PSLs factored occasionally had gas concentration abnormalities at 0.3 m and 2.8 m from the coal wall, but the average gas concentration value did not exceed 1%. A gas concentration of 0.94% at 5.3 m from the coal wall, PSL of 35 m, and 56 m, respectively, from the inlet airway. The G_2 zone appeared in the upper corner at 7.8 m from the coal wall and a PSL 20 m. When the PSL was 30 m or 35 m, the G_2 zone appeared on the working area, which was 78.1 m and 45.1 m away from the incoming wind tunnel, respectively. The gas concentration at different heights in inclined coal seams varied greatly due to the wind flow's dual action and the gas's rise and float. When the PSL was 20 m, 25 m, or 30 m, the vertical distribution of gas was more stable at 0.3 m, 2.8 m, and 5.3 m from the coal wall, respectively. When the PSL was 30 m, it was 7.8 m from the coal wall, and the vertical distribution of gas in the G_1 area was less stable in the range of 61.6~91.6 m from the inlet airway. When the PSL was 35 m, the vertical distribution of gas concentration in the G_1 area was disordered. An abnormal area of gas concentration appeared due to severe air leakage from the working area. The tendency lengths of gas anomalies that appeared in the working area under different PSLs were fitted to obtain the spatial distribution of gas in the working area under different PSLs, as shown in Figure 8. The average gas area of the working area was below the curve, and the mining area's gas-gushing influence area was above the curve. When the PSL was 20 m and 25 m, the gas gushing out of the mining area impacted the area dividing the line in line with the linear distribution, as shown in Equations (12) and (13). When the PSL was 30 m and 35 m, the high-concentration gas gushed from the mining area to the working area in advance. At this point, the gas gushing out of the mining area was influenced by the area dividing line in line with $y = y_0 + Ae^{-x/t}$, as shown in Equations (14) and (15).

$$Y = 144.66478 - 7.2749X \quad (12)$$

$$Y = 130.77093 - 8.5809X \quad (13)$$

$$Y = 119.73454exp(-X/5.29068) + 2.82769 \quad (14)$$

$$Y = 112.10118exp(-X/1.35962) + 15.60497 \quad (15)$$

where Y is the distance from the gas emergence point to the inlet tunnel, m; and X indicates the distance from the coal wall, *m*.

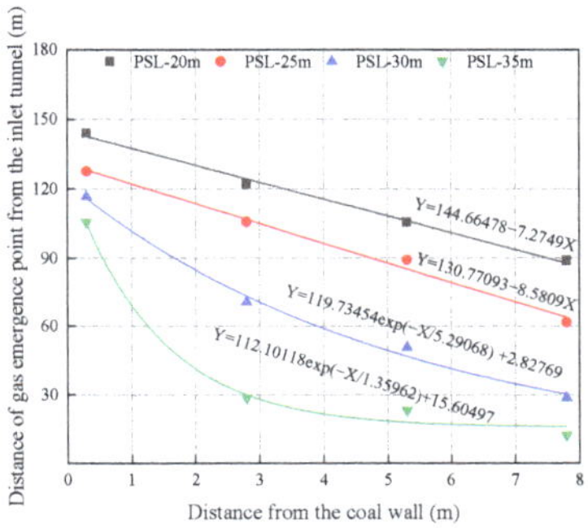

Figure 8. Spatial distribution map of the gas-containing area in working area under different PSLs.

When the PSLs were short, the mining area's air leakage point was on the return wind side. The wind flow carries away a small amount of gas gushing out from the working area, and the gas concentration at the working area was 0. The area of the gas emergence domain at the working area was calculated by integration as $W_{20m} = 575.99\ m^2$, $W_{25m} = 719.86\ m^2$, $W_{30m} = 965.27\ m^2$, and $W_{35m} = 1201.22\ m^2$. The results showed that as the PSLs increased, as did the area affect by gas gushing from the working area.

3.2. Gas Concentration Distribution in the Upper Corner at Various PSLs

Most of the fresh air used in coal mining operations entered the working area from the inlet roadway and carried the coal wall and support overflow gas out from the return air side. The high concentration of gas was brought into the work by the small air flow from the return air side, and the small air flow entered the mining area from the intake side. In the working area, the fluid domain was turbulent, so it was easy to generate eddy currents at the angle between the working area and the road surface. In the upper corner, the vortex created an airflow blind spot, encouraging the gas to build up, causing gas anomalies. Under different PSLs, the upper corner gas concentration fluctuated more obviously, as shown in Figure 9. Figure 9 shows the error bands for gas concentrations determined by nine measuring points located on different PSLs in the upper corners of the layout. The solid line at the top corner shows the average gas concentration. The dashed line was the fitting curve of upper corner gas concentration increasing with PSLs. Error tape was the shaded area. With the increase in PSLs, the upper corner gas concentration decreased gradually. The average gradient of gas concentration in the upper corner of PSLs was 0.0856%/m, 0.010%/m, and 0.0066%/m, respectively, in the range of 20~35 m. The inflection point of airflow of different PSLs was different in mining area. The results showed that the leakage area and the upper corner vortex intensity vary with the pressure difference. In other words, with the increase in PSLs, the size of the shadow region gradually increased, and the concentration of each measuring point fluctuated obviously. As mentioned above, the upper corner gas concentration gradually decreased as the PSLs increased. However, at 35 m PSL, the average gas concentration in the upper corner still reached 0.89%, and the gradient of the decrease in the upper corner gas concentration decreases gradually. Therefore, there is an effective range for PSLs to reduce the upper corner concentration. The increase in PSLs would increase the length of the working area, reduce the area of the extraction area, increase the gas concentration in the extraction area, and leads to gas gushing out of the working area. The effect of reducing the upper corner gas concentration became progressively worse. Therefore, it was of great significance to find the optimal PSLs for gas disaster management in the working area.

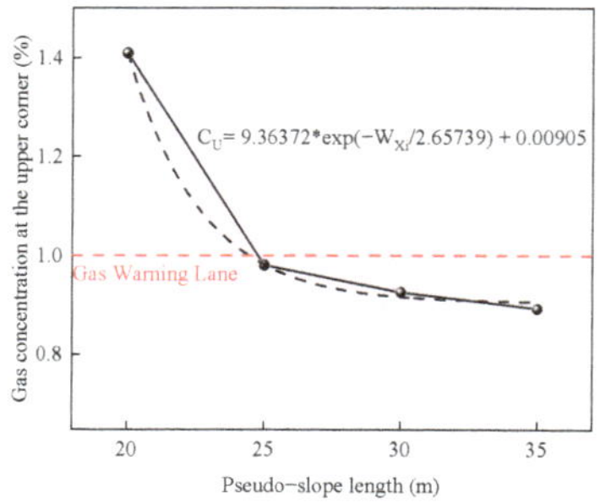

Figure 9. Gas concentration in the upper corner under different PSLs.

4. The Effect of the Coupling Mechanism of the BPDs of the Extraction Port and PSLs on the Gas Concentration Distribution in the Working Area

4.1. Influence of BPDs and PSLs on the Gas Concentration Distribution in the Extraction Zone

The gas concentration on the return side of the "U" ventilation of the inclined coal seam was high. The working area might have had abnormal gas concentration. Although increasing the air volume at the working area can reduce the gas concentration to a certain extent. Due to site conditions, equipment and operation costs, and wind speed limitations, the air volume cannot be increased indefinitely. Therefore, if wind discharged alone cannot solve the problem of gas overload, effective extraction measures must be taken. Buried pipe extraction, as an auxiliary gas extraction method, was used to solve the problem of gas overload in the upper corner. The specific practice was to lay a large diameter steel pipe in the outer helper of the return airway near the bottom plate. A tee containing a combination of valves was installed on the pipe at regular intervals. The gas extractor was connected when the working area was advanced to a certain distance. The top of the extractor was inserted vertically into the roof plate. When it was in the best extraction position in the extraction area, the valve of the extraction port was opened to extract gas from the extraction area. When the extraction port continued to penetrate deeper into the extraction zone, the next three-way valve was opened so that the extraction port was always in the best extraction position. The gas in the upper corner was often abnormal for the high gas concentration on the return side of the "U" ventilation. The effect of BPDs and PSLs on the gas concentration distribution in the extraction area was simulated by Fluent. Figure 10 shows the geometric model of the extraction zone for buried pipe extraction. According to the site situation, the buried pipe extraction pipe was proposed to be a 0.4 m cylindrical pipe. The height of the BPD was 2.6 m from the roof, and the negative pressure of the buried pipe extraction port was 8 kPa.

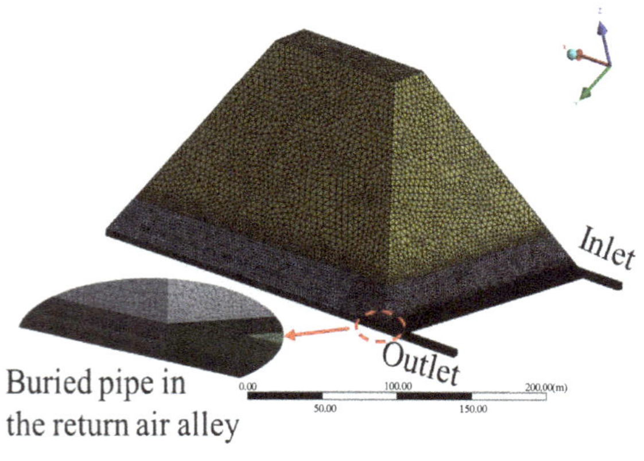

Figure 10. Geometric model of the extraction zone under buried pipe extraction.

To study the influence of BPDs and PSLs on gas concentration distribution, Fluent was used to simulate BPDs of 10 m, 20 m, 30 m, and 40 m at PSLs of 20 m, 25 m, 30 m, and 35 m, respectively. The complete results are shown in Figure 11a–d.

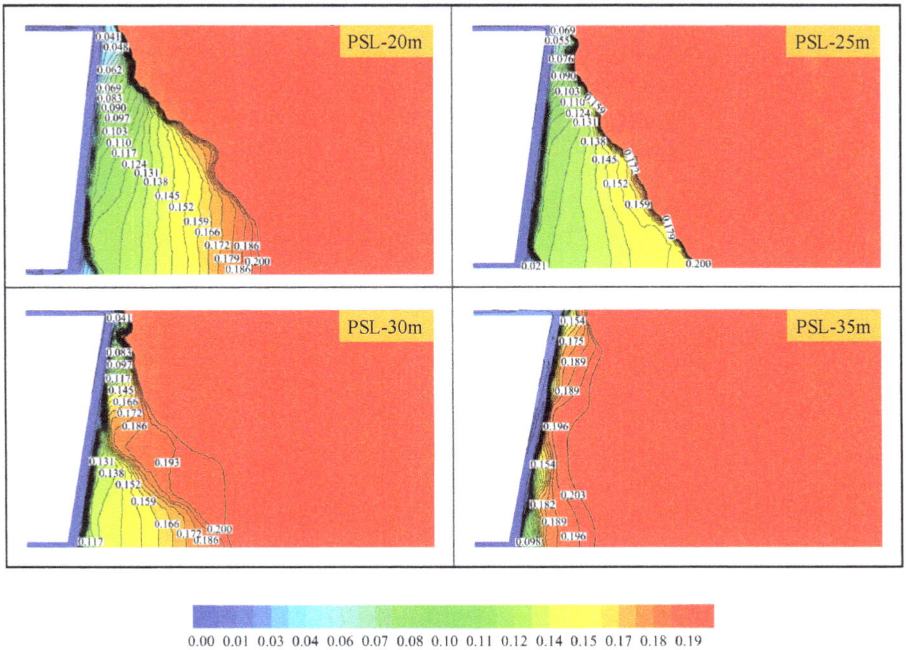

(**a**)

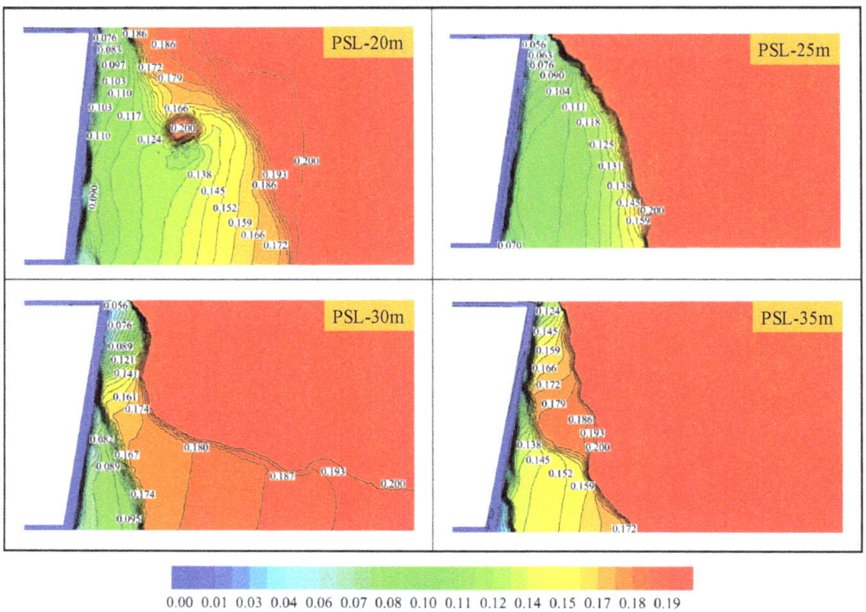

(**b**)

Figure 11. *Cont.*

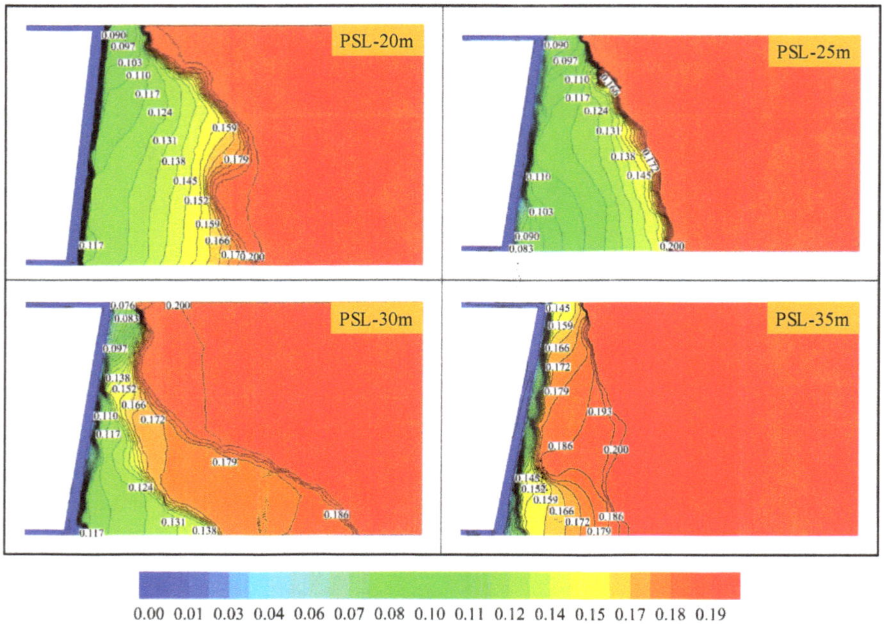

(**c**)

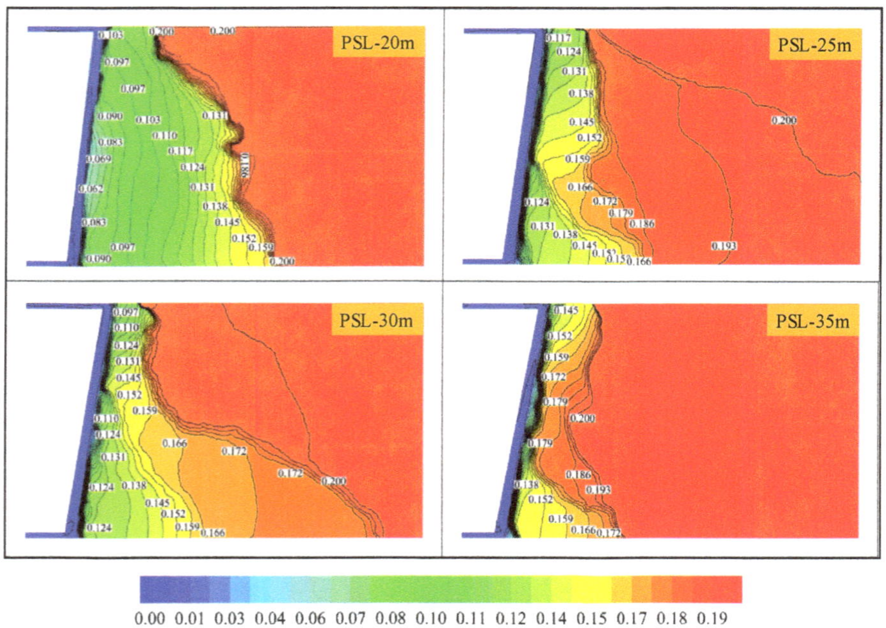

(**d**)

Figure 11. Cloud map of gas concentration distribution in the extraction zone under the coupling effect of different PSLs and BPDs: (**a**) BPD—10m; (**b**) BPD—20m; (**c**) BPD—30m; (**d**) BPD—40m.

Figure 12 shows the simulation results of gas concentration distribution at different PSLs and BPDs in the roof height of the return air tunnel. When the PSL was 20 m, the influential zone of the inlet airway and the suitable location of the buried pipe extraction under different BPD conditions were all through in the mining area, forming a low gas area channel. When the PSL was 25 m, the influential area of the incoming wind lane and the adequate size of the buried pipe extraction failed to form a penetration when the BPD was 40 m, and the high concentration gas area in the mining area spread to the working area. When the PSL was 30 m and 35 m, the influential zone of the inlet airway and the suitable location of the buried pipe extraction failed to form a penetration under different BPDs, and the high gas concentration area in the extraction area tended to spread to the working area. With the increase in BPDs, the low-concentration gas area within the extraction radius in the A (buried pipe influence area) area gradually increased. However, influenced by the PSLs, the location of low gas concentration in the B (wind-influenced place) area gradually decreased. When the PSL was 25 m and the BPD was 40 m, the penetration of the air inlet influence zone and buried pipe extraction influence zone in the mining area was interrupted, and high gas appeared in the middle of the two zones and gathered near the working area. A measuring line was arranged at the height of the top plate in the return tunnel to monitor the gas concentration distribution along the return side direction.

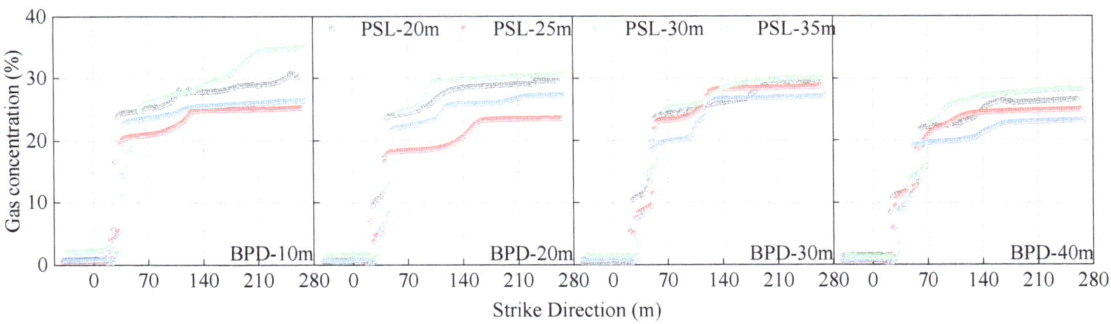

Figure 12. Gas concentration distribution pattern on the return side.

When comparing the gas concentration extraction on the inlet and return side of the mining area under the joint action of PSLs and BPDs, the results showed that the extraction port was closer to the working area, affected by air leakage, and could not maximize the negative pressure, and it was easy to form a vortex on the return side. When the buried pipe extraction port was 10 m from the working area, the gas concentration on the return wind side decreased rapidly in the range of 22~58 m and slowly in the field of 58 m~270 m. In the case of the same burial depth of the extraction port, the gas concentration on the return wind side decreased, then increased when the PSL increased. The negative pressure of the buried pipe has limited influence on the deep part of the compacted mining area. It can only extract gas locally, so its concentration remained unchanged within a certain distance. With the increased PSLs and the BPDs, the joint action of the leakage wind flow and damaging pressure extraction wind flow tended to be smooth, and the low-concern gas forms a penetration near the working area. The unloaded gas was transported to the return wind side under the influence of wind flow and discharged by the negative pressure of the BPD. As the extraction port's depth increased, the buried pipe extraction's guiding effect decreased. The low gas penetration area formed by the negative extraction pressure and the leakage wind flow gradually disappears, and the high concentration gas accumulates near the working area. Although the gas concentration in the direction of the return wind side was low, the gas management of the upper corner has yet to play a noticeable effect. Therefore, when BPDs and PSLs work together, the PSL was challenging to exceed 25 m, and the BPD-20 m was more reasonable.

4.2. Influence of the Combined Effect of BPDs and PSLs on the Distribution Pattern of Gas Concentration in the Upper Corner

To study the pattern of high-concentration gas gushing out from the mining area to the working area under different PSLs and BPDs. The fourth set of measurement lines monitored the gas concentration distribution at the working area near the extraction zone. The test results were shown in Figure 13a–d, where different colors were blocked to represent different gas concentration intervals.

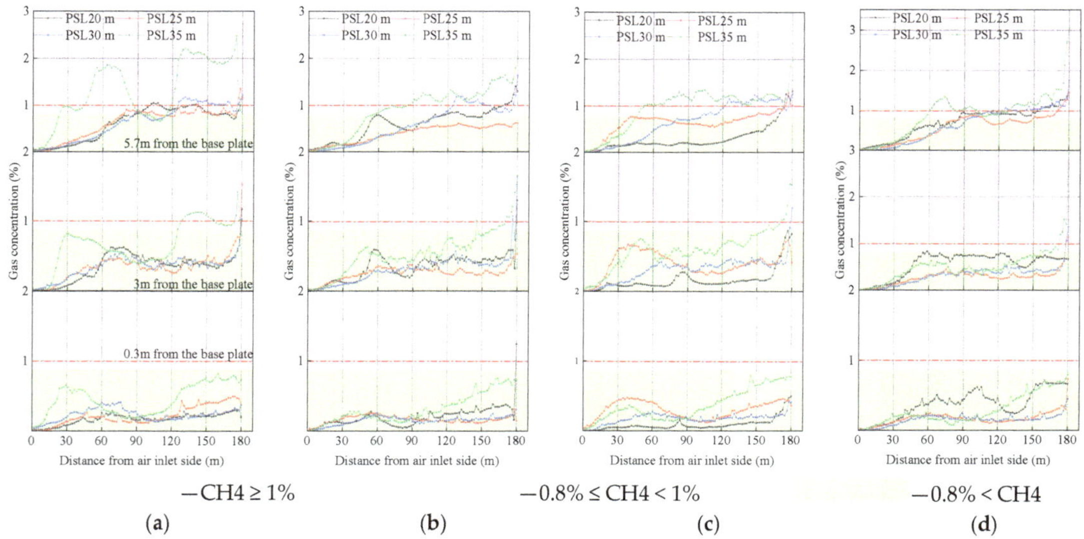

Figure 13. Distribution of gas concentration in the working area of the near mining area. (**a**) BPD—10 m; (**b**) BPD—20 m; (**c**) BPD—30 m; (**d**) BPD—40 m.

Gas exhibits significant variability in the spatial height of the working area under the action of rising floats. When the extraction port was 10 m from the working area, the high concentration of gas in the deep part of the extraction area gathered near the working area under the influence of the negative pressure of buried pipe extraction. Under the intervention of air leakage, the gas in the extraction area gushed out to the working area. With the increase in PSLs, the gas concentration at the working area also increased gradually, and there were more gas concentration abnormalities at the working area when the PSL was 35 m. When the extraction port was buried 20 m deep, gas did not accumulate at the working area when the PSL was 20 m. With the increase in the extraction opening depth, the extraction effect of the upper corner gradually decreased. Due to the guiding influence of wind flow, gas accumulation occurs in the upper corner under different working area PSLs and gas accumulation occurs in the working area.

Figure 14 shows the effect of different PSLs and BPDs on the concentration of extracted gas. Due to the low concentration of gas gathering on the return side at the PSLs of 25 m and the influence of air leakage, the air leakage diluted and carried away. Therefore, the extraction concentration was the lowest at the PSLs of 25 m for different extraction opening depths. When the extraction port was buried 10 m deep, the extraction effect was poor, and the concentration of extracted gas was low, about 13%. The extraction port in this area was shallow and influenced by the leakage wind flow, so the gas was diluted and carried away, which led to low extraction concentration and could not effectively prevent the gas in the mining area from gushing out to the working area. When the extraction port was buried 20 m deep, the extraction concentration reached 17%. The effect of managing the abnormal gas accumulation in the corner of the working area was remarkable. The gas concentration in the upper corner was the lowest, about 0.43%, as shown in Figure 15. When the extraction port was 30 m inside the extraction area, the extraction gas concentration was more significant, about 20%. Although the gas extraction concentration was higher in the extraction area, the effect of extracting gas on the corner angle on the working area was gradually reduced. When the extraction port was just 40 m inside the extraction area, the extraction concentration was as high as 28%. However, there was an abnormal gas concentration in the upper corner because the extraction port was far from the working area. Although the extraction concentration was high, it had less effect on the gas control of the working area. The results were consistent with Liu. Z's [33] on the effect of BPD on gas concentration in the upper corner.

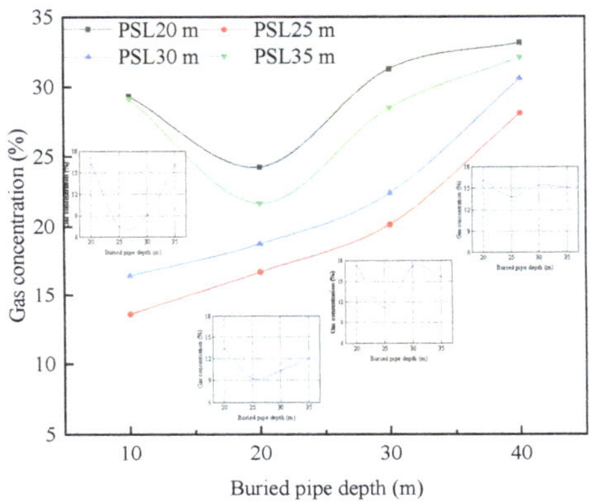

Figure 14. Variation pattern of extracted gas concentration.

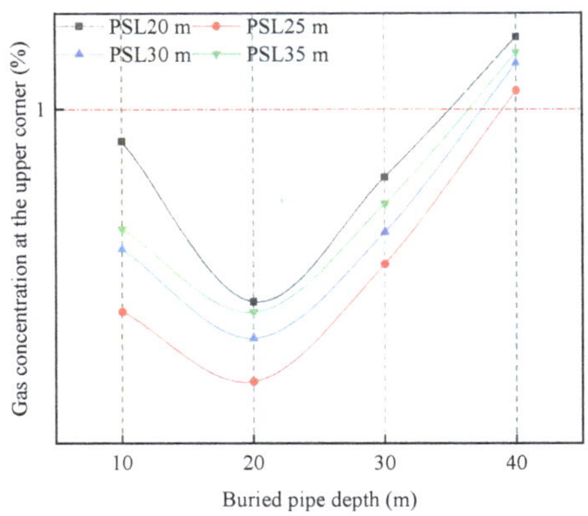

Figure 15. Change pattern of gas concentration in the upper corner.

In sum, at PSL 25 m and BPD 20 m, the influencing area on the inlet side was connected with the influencing area of buried pipe extraction. There was no gas accumulation in the working area at this time, and the gas concentration in the upper corner was low. By adjusting the PSLs and BPDs, the gas accumulation at the working area and upper corner can be effectively prevented during mining and meet the safety production standards. Research on the PSL effect of cooperative extraction, including parameters of high drilling arrangement, buried pipe extraction arrangement, and high extraction lane arrangement, will be carried out in the follow-up study to improve the safety and efficiency of coal and gas co-mining.

5. Engineering Practices

In light of the recent abnormal increase in gas concentration in the main mining area, the effect law of PSLs distance on gas gushing out was investigated. Before the 25th day, the PSLs were at most 30 m. The peak gas concentration was in the upper right-hand corner, with an average gas concentration of 0.89%. As the PSLs increased, the angle between the working area and the inlet and outlet lane changed, affected the flow of wind at the working area. It took time for the gas to rise and float and for the airflow to change. As a result, changing the PSLs did not immediately change the gas concentration at the working area and upper corner. The change in gas concentration had hysteresis. The 13th day saw a change as the PSLs gradually increased. On the 17th day, the corner gas concentration gradually decreased, and the PSL was 25 m. As the PSL continued to grow, the rate of gas concentration reduction in the upper corner decreased. The high gas concentration in the extraction area entered the working area earlier. From day 33 onwards, the gas concentration at the working area increased. On the 40th day, the length of the PSL increased to 35 m, and the high concentration of gas in the mining area continued to enter the working area earlier, reaching its peak at this time. After the 45th day, the PSLs decreased, as did the gas concentration at the working area. Figure 16 shows the pattern of change.

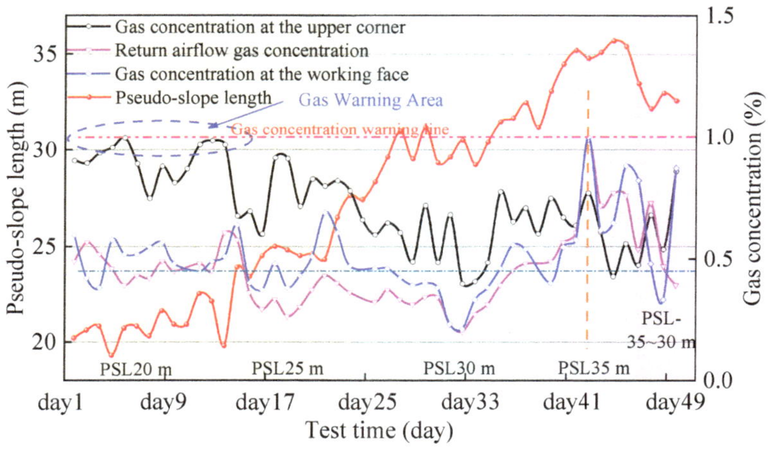

Figure 16. Curve of PSLs Distance and Gas Concentration.

The variation pattern shown in Figure 17 was obtained by conducting continuous field observations of gas concentration along the working area. As the PSLs increased, the gas anomaly area gradually shifted from the 122 #–124 # bracket to the incoming wind side. When the PSLs reached 35 m, the gas anomaly zone appeared slowly at the 103 #–108 # frame. When the PSLs gradually decreased after 44 days, the working area's gas anomaly area shifted back to the backwind lane side.

The buried pipe was buried in the extraction area to extract gas from the upper corner, combined with the numerical simulation experiment results to reduce the problem of local gas accumulation. The PSL was controlled on-site to be about 25 m, and two trips of φ219 mm steel pipes were laid along 700 mm of the bottom plate of the return tunnel. Each journey of steel pipe was connected with the main line in the return wind tunnel, and valves were installed on each BPD to control them, respectively. We installed an upward tee in the pipeline every 12 m and started pumping when the first trip of buried pipe entered 12 m into the mining air. When the first buried pipe entered 24 m into the mining area, we disconnected this buried pipe and started pumping the second buried pipe. We continuously extended the connection forward to ensure that the burial depth of the pumping port was around 20 m, and constantly monitored for 180 days. The buried pipe plan of the working area is shown in Figure 18. As shown in Figure 19, the monitoring results shown that after optimizing the buried depth parameters of the extraction port and the PSLs conditions, the average gas concentration in the upper corner was 0.26%, and the average gas concentration in the return wind tunnel was 0.22%. The maximum gas concentration in the upper corner was 0.46%, and the highest in the return airway was 0.41%. As stipulated in the Coal Mine Safety Regulations, it was far below the warning line of 1% gas concentration in the upper corner, return air tunnel, and tail shaft. By applying the experimental results of numerical simulation to optimize the PSLs of the working area and the BPD in the mining area, the maximum gas concentration in the upper corner and return wind lane can be effectively controlled below 1%. It was practical and feasible for gas prevention and control of the test working area and realizing the efficient production of the tilted coal seam extended release working area.

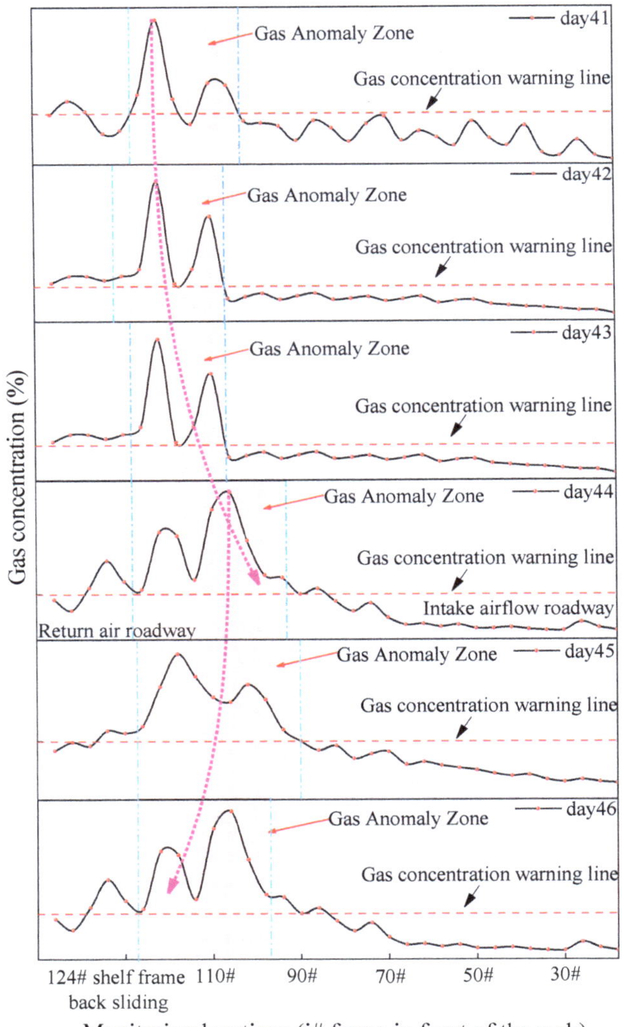

Figure 17. Distribution law of gas concentration along working area from day 41 to day 46.

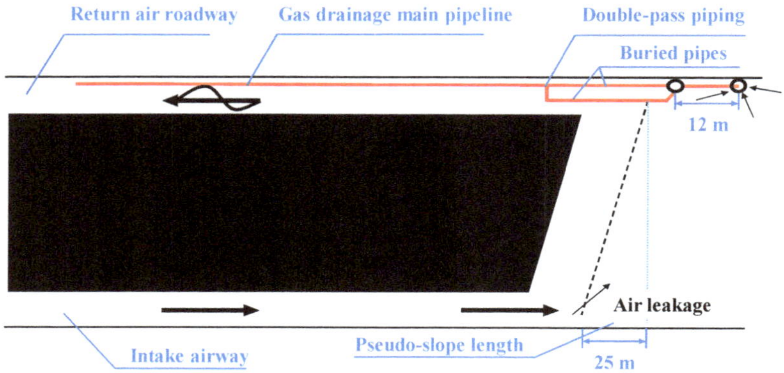

Figure 18. Plan view of buried pipe extraction in the extraction area.

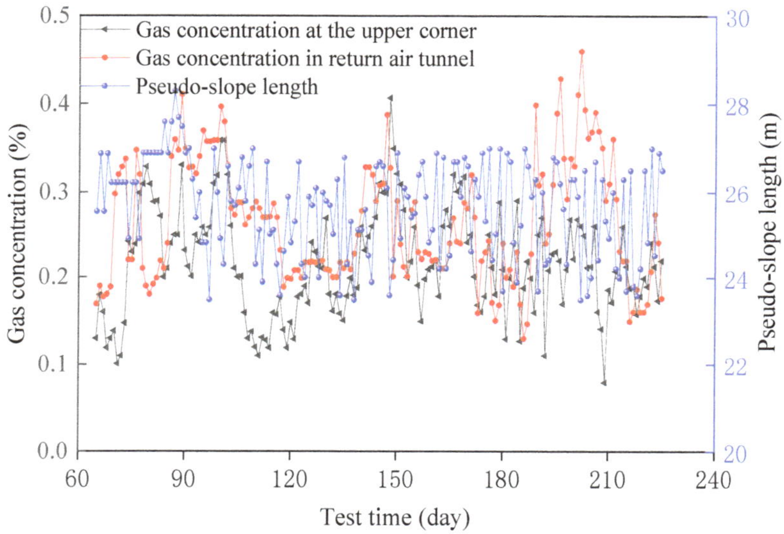

Figure 19. Change the pattern of gas concentration in the upper corner and return the air tunnel after optimization.

6. Conclusions

Through this study, three conclusions were reached:

(1) The influence law of PSLs on the wind flow in the mine area was obtained by Fluent simulation. With the increased PSL, the air leakage area gradually changed from the upper corner to the middle of the working area and the upper corner. The abnormal gas concentration area shifted to the inlet side. In contrast the gas concentration in the upper corner gradually decreased. The numerical simulation results showed that the optimal PSL was 25 m when there was no extraction measure.

(2) When PSLs and BPD work together, the gas concentration in the corner of the working area can be effectively reduced. When the extraction opening was buried shallowly, the negative pressure formed by extraction would form a low gas concentration zone with air leakage. With the increase in BPD, the effect of BPD on the control of gas concentration in the upper corner gradually decreased. The results show that when PSL was 25 m and BPD was 20 m, it had the best effect on the gas control in the upper corner.

(3) The change of PSLs and the depth of the extraction port significantly affected the gas concentration distribution on the working area. When the PSLs exceeded 25 m, the measured gas concentration in the field caused a gas anomaly at the 122# frame, and when the PSLs increased to 35 m, the measured gas concentration in the area caused a gas anomaly at the 108# frame. Therefore, the PSLs should be, at most, 25 m. Combined with the numerical simulation results, the gas concentration in the upper corner gradually decreased with the increase in the PSL. Therefore, the best PSL should be 25 m. When the gas concentration in the upper corner was low, no abnormal gas zone appeared in the working area. Based on the numerical simulation results, the maximum gas concentration in the upper corner was 0.46%, with the extraction port of BPD-20 m arranged in the return air tunnel. The numerical simulation was consistent with the engineering practice results. The gas concentration at the working area was effectively controlled by adjusting the PSLs and BPDs to realize the efficient production of the inclined coal seam extended release working area.

Author Contributions: Conceptualization, P.Z.; methodology, S.L.; data curation, X.K. and X.A.; writing—original draft preparation, X.A.; investigation, Y.H.; writing—review and editing, J.Y.; funding acquisition, S.J. All authors have read and agreed to the published version of the manuscript.

Funding: This research was funded by the National Natural Science Foundation Grants of China (5217-4205 and 5197-4237), the Basic Research on National Natural Science Foundation critical project of China (5173-4007), and the Xinjiang Natural Science Foundation (2019D01B42). The authors are obliged to the Xinjiang Liuhuanggou coal mine for providing field tests supported in this study.

Institutional Review Board Statement: Not applicable.

Informed Consent Statement: Not applicable.

Data Availability Statement: Data available on request due to privacy restrictions.

Conflicts of Interest: The authors declare no conflict of interest.

References

1. Yuan, L. Scientific problem and countermeasure for precision mining of coal and associated resources. *J. China Coal Soc.* **2019**, *44*, 1–9.
2. Yuan, L. Concept of gas control and simultaneous extraction of coal and gas. *J. China Coal* **2010**, *36*, 5–12.
3. Qian, M.; Xu, J.; Wang, J. Further on the sustainable mining of coal. *J. China Coal Soc.* **2018**, *43*, 1–13.
4. Qin, B.; Shi, Z.; Hao, J.; Ye, D.; Liang, B.; Sun, W. Analysis of the Space-Time Synergy of Coal and Gas Co-mining. *ACS Omega* **2022**, *7*, 13737–13749. [CrossRef] [PubMed]
5. Wang, K.; From, F. Coal-gas compound dynamic disasters in China: A review. *Process Saf. Environ. Prot.* **2020**, *133*, 1–17. [CrossRef]
6. Jia, L.; Peng, S.; Xu, J.; Yan, F.; Chen, J.; Wu, B.; Chen, Y. On the evolution mechanism of permeability during gas drainage: Insights from deformation field, gas pressure field and temperature field. *Process Saf. Environ. Prot.* **2022**, *162*, 825–836. [CrossRef]
7. Tang, M.; Jiang, B.; Zhang, R.; Yin, Z.; Dai, G. Numerical analysis on the influence of gas extraction on air leakage in the gob. *J. Nat. Gas Sci. Eng.* **2016**, *33*, 278–286. [CrossRef]
8. Wang, Z.; Ren, T.; Cheng, Y. Numerical investigations of methane flow characteristics on a longwall face Part II: Parametric studies. *J. Nat. Gas Sci. Eng.* **2017**, *43*, 254–267. [CrossRef]
9. Wang, C.; Yang, S.; Li, X. Simulation of the hazard arising from the coupling of gas explosions and spontaneously combustible coal due to the gas drainage of a gob. *Process Saf. Environ. Prot.* **2018**, *118*, 296–306. [CrossRef]
10. Cheng, J.; Li, S.; Zhang, F.; Zhao, C.; Yang, S.; Ghosh, A. CFD modelling of ventilation optimization for improving mine safety in longwall working faces. *J. Loss Prevent. Proc.* **2016**, *40*, 285–297. [CrossRef]
11. Shao, H.; Chen, K.; Zhao, H.; Li, M. Comparison Study of "U + L" and "Y + L" Working Face Ventilation Patterns on Gas Control in a Highly Gassy Mine. *Teh. Vjesn.* **2015**, *22*, 443–452. [CrossRef]
12. Yang, S.; Zhou, B.; Wang, C. Investigation on coal spontaneous combustion in the gob of Y type ventialtion caving face: A case study. *Process Saf. Environ. Prot.* **2021**, *148*, 590–603. [CrossRef]
13. Li, H.; Liu, Z.; Yang, Y.; Zhu, D.; Yang, H.; Wang, W. Study on the evolution law of the gas flow field based on the distribution characteristics of voids in the overlying strata in a goaf. *Arab. J. Geosci.* **2021**, *14*, 1–8. [CrossRef]
14. Liang, Y.; Wang, S. Prediction of coal mine goaf self-heating with fluid dynamics in porous media. *Fire Saf. J.* **2017**, *87*, 49–56. [CrossRef]
15. Wang, G.; Xu, H.; Wu, M.; Wang, Y.; Wang, R.; Zhang, X. Porosity model and air leakage flow field simulation of goaf based on DEM-CFD. *Arab. J. Geosci.* **2018**, *11*, 148. [CrossRef]
16. Wang, Z.; Ren, T.; Zhang, J. Numerical investigations of airflow patterns on a longwall face. *Int. J. Oil Gas Coal Technol.* **2020**, *24*, 321–344. [CrossRef]

17. Zhu, S.; Wang, Y.; Wei, L. Gas monitoring data anomaly identification based on spatio-temporal correlativity analysis. *J. Coal Sci. Eng.* **2013**, *19*, 8–13. [CrossRef]
18. Yu, Z.; Yang, S.; Qin, Y.; Hu, X.; Cheng, J. Experimental study on the goaf flow field of the "U+I" type ventilation system for a comprehensive mechanized mining face. *Int. J. Min. Sci. Technol.* **2015**, *25*, 1003–1010. [CrossRef]
19. Zhang, F.; Wang, G.; Wang, B. Study and Application of High-Level Directional Extraction Borehole Based on Mining Fracture Evolution Law of Overburden Strata. *Sustainability* **2023**, *15*, 2806. [CrossRef]
20. Brigida, V.S.; Golik, V.I.; Dzeranov, B.V. Modeling of Coalmine Methane Flows to Estimate the Spacing of Primary Roof Breaks. *Mining* **2022**, *2*, 45. [CrossRef]
21. Dzhioeva, A.K.; Brigida, V.S. Spatial non-linearity of methane release dynamics in underground boreholes for sustainable mining. *J. Min. Inst.* **2020**, *245*, 522–530. [CrossRef]
22. Zhang, L.; Wu, W.; Bian, Y.; Yang, M.; Luo, H. Influences of Gas Drainage Pipe Positions on Spontaneous Coal Combustion in the Gob: A Case Study of Baode Coal Mine in China. Combust. *Sci. Technol.* **2021**, *194*, 3357–3373. [CrossRef]
23. Zhang, L.; Yang, M.; Luo, H. Parameter Optimization and Application of Gas Extraction Technology by Buried Pipes in Goaf of Multi-roadway U-type Working Face. *Combust. Sci. Technol.* **2022**, 1–18. [CrossRef]
24. Li, Z.; Wang, F.; Ren, S.; Liu, G.; Xiong, H. Gas Distribution Mechanism in Goaf during Combined Drainage of Upper Corner Buried Pipeline and Intubation for Thick Coal Seams. *Lithosphere* **2021**, *2021*, 8308256. [CrossRef]
25. Chen, X.; Dong, X.; Wang, L.; Huang, Z.; Cui, P. Gas Emergence Characteristics of the Upper Corner on the 215101 Mining Working Face of the Yue Nan Coal Mine. *ACS Omega* **2022**, *7*, 25663–25674. [CrossRef]
26. Liu, Y.; Shao, S.; Wang, X.; Chang, L.; Cui, G.; Zhou, F. Gas flow analysis for the impact of gob gas ventholes on coalbed methane drainage from a longwall gob. *J. Nat. Gas Sci. Eng.* **2017**, *36*, 1312–1325. [CrossRef]
27. Chai, J. Study on Gas Control in Goaf of High Gas Coal Seam: A Case Study of Tingnan Coal Mine, China. *Geofluids* **2022**, *2022*, 5262173. [CrossRef]
28. Ding, X.; Zhai, C.; Xu, J.; Yu, X.; Sun, Y. Study on Coal Seepage Characteristics and Secondary Enhanced Gas Extraction Technology under Dual Stress Disturbance. *Sustainability* **2022**, *14*, 15118. [CrossRef]
29. Zhang, Q.; Li, Q.; Fan, X. Current situation and development trend of theories and technologies for coal and CBM co-mining in China. *Nat. Gas Ind.* **2022**, *42*, 130–145.
30. Tutak, M. The Influence of the Permeability of the Fractures Zone Around the Heading on the Concentration and Distribution of Methane. *Sustainability* **2020**, *12*, 16. [CrossRef]
31. Liu, J.; Gao, J.; Yang, M.; Wang, D.; Wang, L. Numerical Simulation of Parameters Optimization for Goaf Gas Boreholes. *Adv. Civ. Eng.* **2019**, *2019*, 3891080. [CrossRef]
32. Wang, S.; Wang, D.; Cao, K.; Wang, S. Distribution law of 3D fracture field of goaf and overlying strata. *J. Cent. South Univ. Sci. Technol.* **2014**, *45*, 833–839.
33. Liu, Z. Analysis of the effect of buried pipe extraction at the integrated working face of sharply inclined extra-thick coal seam. *Coal Sci. Technol.* **2017**, *45*, 74–76.

sustainability

MDPI

Article

Creep Characteristic Test and Creep Model of Frozen Soil

Yawu Shao [1,*], Yonglu Suo [1,2], Jiang Xiao [1,2], Yuan Bai [1] and Tao Yang [1,2]

[1] School of Energy, Xi'an University of Science and Technology, Xi'an 710054, China
[2] Key Laboratory of Department of Education of Western Mine Mining and Disaster Prevention, Xi'an University of Science and Technology, Xi'an 710054, China
* Correspondence: 18103077007@stu.xust.edu.cn; Tel.: +86-186-0943-3660

Abstract: In order to research the creep deformation characteristics of frozen soil and the effect of various influencing factors on creep, indoor uniaxial creep tests were carried out on frozen soil specimens at temperatures of −5, −4, −3 and −2 °C under loads of 0.25, 0.5, and 0.75 σ_t, respectively. The creep deformation characteristics of frozen soil under different temperatures and load conditions are analyzed under unconfined conditions. The results show that under the uniaxial creep test conditions, when the load is low, there is no accelerated creep stage in the creep curve, which belongs to the decaying type creep; when the load is 0.75 σ_t, the creep curve enters the stage of accelerated creep, and the creep turns non-attenuated; temperature is the most important external factor affecting permafrost soil creep, and the proportion of ice is the key internal factor for affecting permafrost soil creep, the temperature is negatively correlated with the proportion of ice and the sensitivity of creep rate to temperature and load increases with the decrease of the proportion of ice; the damage variable D is introduced to modify the creep constitutive equation of the frozen soil, the creep process of frozen soil is well described by the modified creep constitutive equation for frozen soil.

Keywords: frozen soil; creep properties; decay creep; creep constitutive equation

Citation: Shao, Y.; Suo, Y.; Xiao, J.; Bai, Y.; Yang, T. Creep Characteristic Test and Creep Model of Frozen Soil. *Sustainability* **2023**, *15*, 3984. https://doi.org/10.3390/su15053984

Academic Editors: Marco Lezzerini and Chaolin Zhang

Received: 17 July 2022
Accepted: 13 October 2022
Published: 22 February 2023

1. Introduction

The Qinghai-Tibet Plateau is the largest high-altitude permafrost region on earth [1]. In the negative temperature environment, the water in the soil pores condenses and freezes into ice, resulting in frost heave deformation of the soil. When it comes to engineering construction in permafrost regions, changes in the mechanical properties of frozen soil and their consequences must be considered [2,3]. The most essential difference between frozen soil and conventional thawed or unfrozen soil is the rheological properties of ice and unfrozen water [4,5]. The existence of ice makes the soil change from a three-phase system to that of four phases consisting of solid mineral particles, poly crystalline ice, unfrozen water and gas. Creep is one of the most important mechanical behaviors of frozen soil and is defined as the change in deformation with time when the stress is constant.

Research on the creep of frozen soil has always been a frontier topic in the mechanics of frozen soil mechanics. Many scholars have observed microstructure changes during uniaxial creep of frozen soil. The results show that the creep process is the result of the combined action of the development of micro-cracks in the frozen soil, the destruction of the skeleton particle structure and the evolution of other defects under the action of external loads [6–8]. Through systematic experimental research, Zhu [9,10] found that when the temperature of frozen soil is higher than −1 °C, the axial strain lowers sharply with rising soil temperature, and a small temperature change can also cause a large change in strain, and the time for deformation stabilization of frozen soil is usually long, if ever, such as tens of days or even months. Possibly this is due to a large amount of unfrozen water in the frozen soil, and the unfrozen water under pressure in the soil undergoes slow phase changes, migration, infiltration and recrystallization. Mcrobert et al. [11] analyzed the stable creep rate of frozen soil with high ice content in the range of −0.8~−4.0 °C through experiments, and found that under the same conditions, the higher the ice content, the greater the creep rate of frozen soil. Ma et al. [12] studied the creep characteristics of frozen soil with high ice content. The test results show that no matter how much stress and how long the action time is, the creep process of uniaxial compression of frozen clay with high ice content has attenuation characteristics. In order to make the test conditions more in line with the deformation conditions of permafrost soil under climate warming, Zheng et al. [13] carried out a constant load heating test of permafrost soil indoors. It is found that the compressibility of frozen soil at higher temperature is much larger than that at lower temperature, and temperature is the main factor for affecting the compressibility of frozen soil. Eckardt et al. [14] carried out uniaxial creep test research by means of hierarchical loading and obtained the functional relationship between the mechanical parameters of the frozen specimen and soil temperature. Yao et al. [15] carried out uniaxial compression and graded loading creep tests on clay in deep mines under negative temperature conditions. Liu et al. [16] conducted in-depth research on the permafrost layer of Niangmute coal mine and pointed out that the permafrost layer has special creep characteristics. Shang et al. [17] revealed the law of structural deformation and fracture of the overlying permafrost through the stope "thin plate" theory. The above studies have well revealed the creep properties of frozen soil but the research on the creep properties of frozen soil under various influencing factors is far from adequate.

In this study, the overburden frozen soil of Jiangcang coal mine in the permafrost region on the Qinghai-Tibet Plateau was taken as the specimen, and the creep deformation characteristics of frozen soil in the high temperature range and their relationships with various influencing factors were investigated by conducting indoor uniaxial creep tests. A new permafrost creep model was established based on the revised Nishihara model. The distribution characteristics of frozen soil and underlying coal seams in Jiangcang Coal Mine are shown in Figure 1. The research results can provide a theoretical basis for the mining of underlying coal in the Jiangcang coal mine and green mining and reclamation in Muli coalfield mining area.

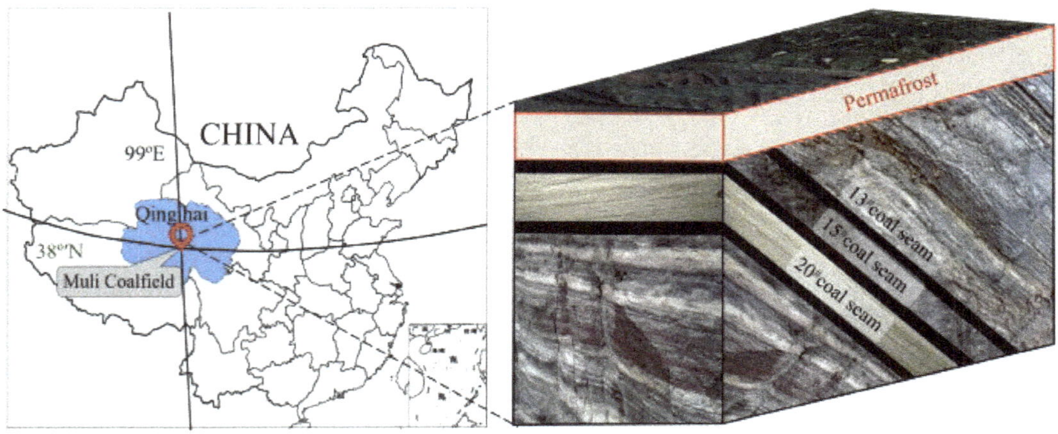

Figure 1. Distribution characteristics of permafrost and underlying coal seams in Jiangcang Coal Mine in the upper Datong river valley, Qinghai Province on the southern flank of the eastern Qilian Mountains on the Qinghai-Tibet Plateau.

2. Experimental Design

2.1. Sample Preparation

The experimental frozen soil was collected from the overlying permafrost layer of Jiangcang Mine in the Muli Coalfield in the headwater area of the Daton River on the east Qinghai-Tibet Plateau. The main components are gravels, loamy sands and ice crystals.

The retrieved soil samples were air-dried, rolled, and passed through the 2.0 mm sieve. A cylindrical standard specimen of Φ50 × 100 mm was prepared, with the moisture content at 23%, and dry density at 2.1 g/cm^3. The production process of frozen soil specimens strictly follows the frozen soil uniaxial compression test method standard for frozen soils. A total of four groups of cylindrical frozen soil specimens were processed in this experiment, and the experimental temperatures were −5, −4, −3 and −2 °C, respectively, to ensure that there are not less than three frozen soil samples under the same conditions. The frozen soil sample used for creep test is named FSC_i-j, where i and j are the absolute value of test temperature and specimen number respectively (For example, the number of the second test piece at −3 °C is FSC_3-b).

2.2. Experimental System

We used the TAW-1000 microcomputer-controlled single-triaxial creep testing machine (Zhongqiao Bolian Instrument Equipment Co., Ltd., Beijing, China). As shown in Figure 2, for the low temperature control, we adopted (Zhejiang Danfu Environmental Technology Co., Ltd. Hangzhou, China) TMS8019-E5000-R40 low temperature circulation device, with the temperature range at −60 to −40 °C, the accuracy of temperature fluctuation is ±0.1 °C. The maximum axial load of the loading system is 1000 kN, and the accuracy is ±1% [18]. The suitable sample size is Φ50 × 100 mm, and the accuracy of deformation measurement is ±0.005 mm.

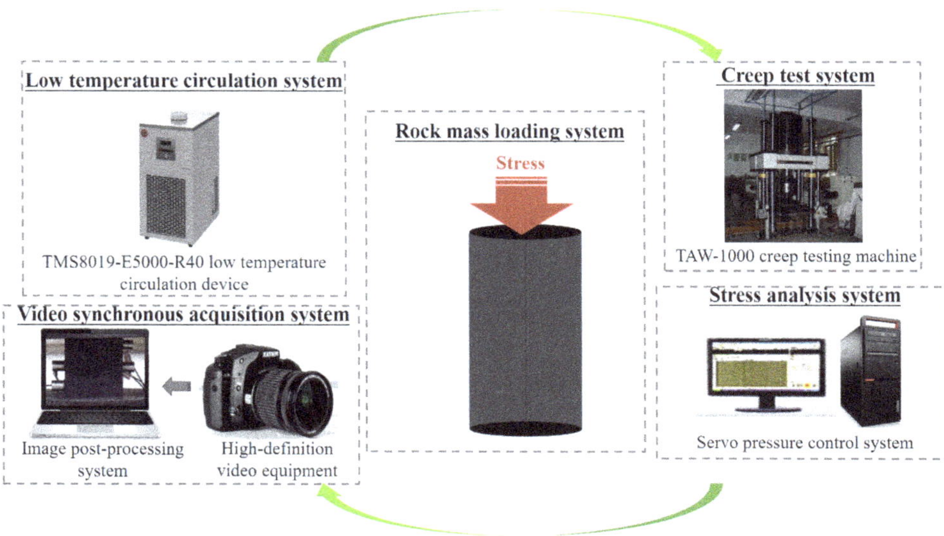

Figure 2. Apparatus and devices for the creep testing of frozen soils.

2.3. Experimental Program

First, a conventional uniaxial compression test was carried out on the frozen soil specimens at −5~−2 °C, and the mechanical parameters of the frozen soil specimens were obtained to provide a basis for determining the creep loading level. The average uniaxial compressive strengths of the four frozen soil specimens at different temperatures are 4.58, 5.18, 5.47 and 6.53 MPa, respectively (Table 1).

Table 1. Results of permafrost uniaxial compression test.

Sample No.	Soil Temperature (°C)	Average Compressive Strength (MPa)
FS_1-1,2,3	−2	4.58
FS_2-1,2,3	−3	5.18
FS_3-1,2,3	−4	5.47
FS_4-1,2,3	−5	6.53

The creep test was carried out on the frozen soil under four different temperature conditions by the method of graded loading. The test load was divided into three grades. The creep loading grades were 0.25, 0.5 and 0.75 σ_t and keep it at 0.75 σ_t until the frozen soil sample is destroyed. The graded loading stress values under different stress states are shown in Table 2.

Table 2. Frozen soil creep test design.

Sample No.	Sample Temperature (°C)	Axial Loading Stress (MPa)		
		0.25 σ_t	0.5 σ_t	0.75 σ_t
FSC_2-a,b,c	−2	1.15	2.29	3.43
FSC_3-a,b,c	−3	1.30	2.59	3.89
FSC_4-a,b,c	−4	1.37	2.74	4.10
FSC_5-a,b,c	−5	1.63	3.27	4.90

3. Creep Test Results and Analysis

Due to the limited space, Table 3 only lists a set of typical creep curves under each test condition for description. The creep behavior of frozen soil can be regarded as the external macroscopic manifestation of the microscopic damage evolution of the internal structure of the sample. In general, the crack development of frozen soil is relatively slow when it is in the initial creep and stable creep stages, and the increase rate of cracks in frozen soil is very rapid when it is in the accelerated creep stage. The soil temperature has a great influence on the failure mode of the frozen soil sample. When the soil temperature is −2 °C, a main oblique crack appears in the frozen soil sample, and some small cracks, around it. At −5 °C~−3 °C, the damage cracks of the frozen soil samples also gradually increased, and tiny cracks appeared. The crack spacing of the samples gradually formed, and the tensile characteristics appeared [19].

In the case of the same soil moisture content and axial stress, the soil temperature of the frozen soil specimens at −5, −4, −3 and −2 °C, respectively, the creep deformation increases with the decrease of soil temperature and the lowest, the creep strain (0.54) at −5 °C increased by 59.26% compared to the highest soil temperature −2 °C (0.86). The higher the temperature, the shorter the specimen in the stable creep stage during the creep process, and the larger the slope of the soil curve in the accelerated creep stage.

As shown in Table 3, when the axial stress is 0.25 σ_t, the axial stress at this stage is mainly borne by the pore pressure of the specimen, and the pores and cracks in the frozen soil are compacted and closed. The lower the temperature, the larger the proportion of ice and the smaller the unfilled porosity. The smaller the compressible space of the sample, the smaller the deformation. When the axial stress is 0.5 σ_t, the initial cracks of the specimen expand and new cracks sprout, and the external load is transferred from the pore pressure to the soil particle skeleton. As soil temperature decreases, the effective stress of the specimen increases. The coupling effect of effective stress is complex, resulting in an insignificant regularity of creep deformation. When the axial stress is 0.75 σ_t, the crack breaks rapidly at this stage, and the axial stress is mainly borne by the soil particle skeleton. When entering the accelerated creep stage, the creep rate of the specimen increases; the lower the temperature is, the later it enters the accelerated creep stage, and the greater the deformation after entering the accelerated creep stage.

Table 3. Creep deformation characteristics under different conditions.

Sample Number	Destruction Form	Binarization	Creep Curve	Brief Description
FSC_2-b		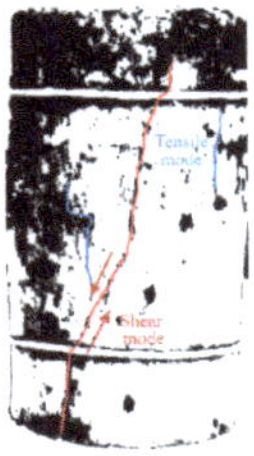	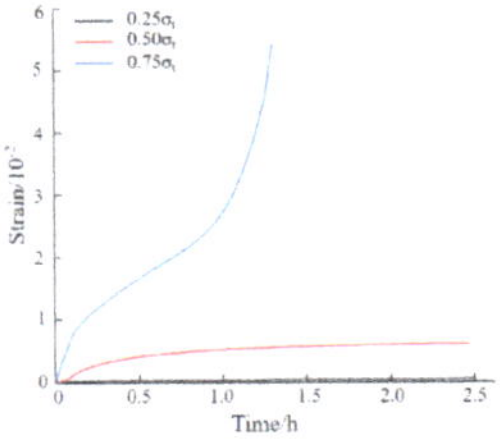	(1) The number of tensile failures increases; (2) The shear failure of the main oblique fissure dominates the failure of the specimen, and a few small fissures appear around it; (3) The maximum axial deformation is 0.54.
FSC_3-c	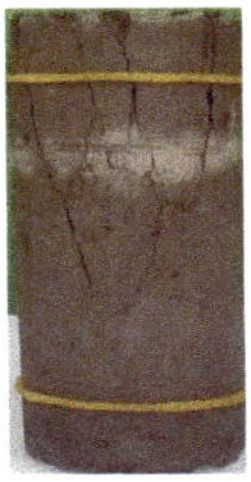	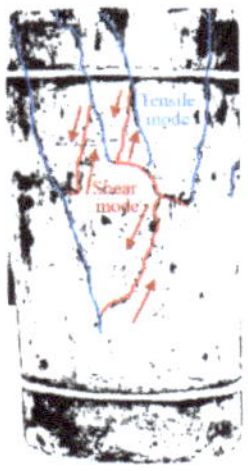	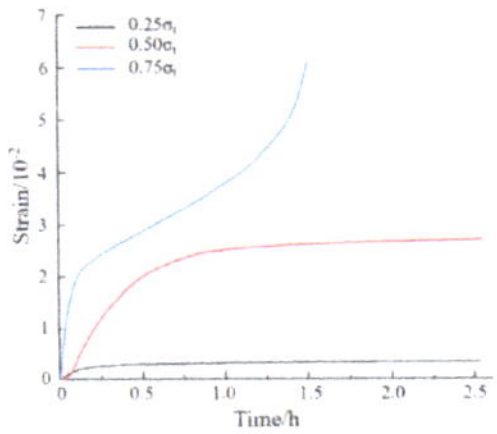	(1) Increased number of tensile and shear failures; (2) The mixed failure of tension and shear leads to the failure of the specimen, and the distribution of cracks is complex; (3) The maximum axial deformation is 0.63.
FSC_4-a		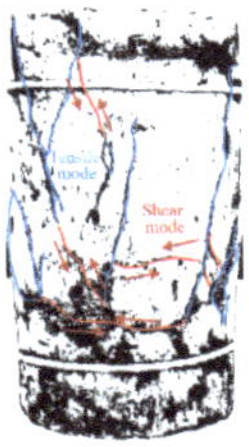	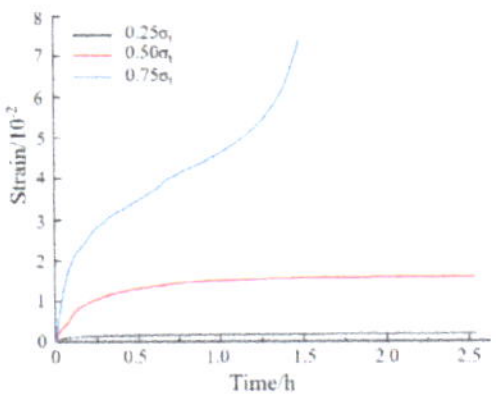	(1) The distribution area of cracks increased, concentrated in the middle of the specimen, and the number of shear failures increased significantly; (2) The mixed failure of tension and shear leads to the failure of the specimen; (3) The maximum axial deformation is 0.78.
FSC_5-b			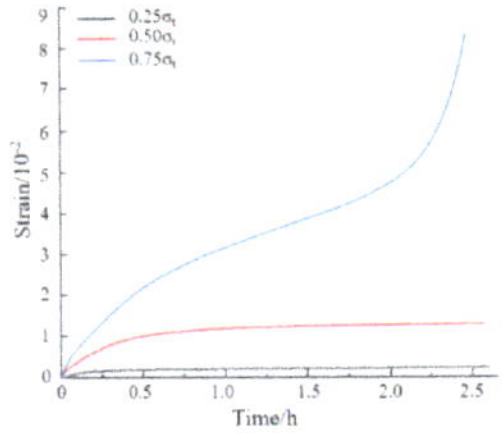	(1) The average track length of cracks increased, concentrated in the middle and upper part of the specimen, and the number of shear failures increased significantly; (2) The mixed failure of tension and shear leads to the failure of the specimen, and the failure form of the specimen is complex; (3) The maximum axial deformation is 0.86.

4. Creep Constitutive Equation of Frozen Soil

The classic Nishihara model is mainly composed of two parts in series: the generalized Kelvin body and the viscoplastic body (Figure 3). This model can describe the visco-elastic-plastic change characteristics during the creep process of frozen soil and can be used for featuring the creep characteristics of geotechnical materials.

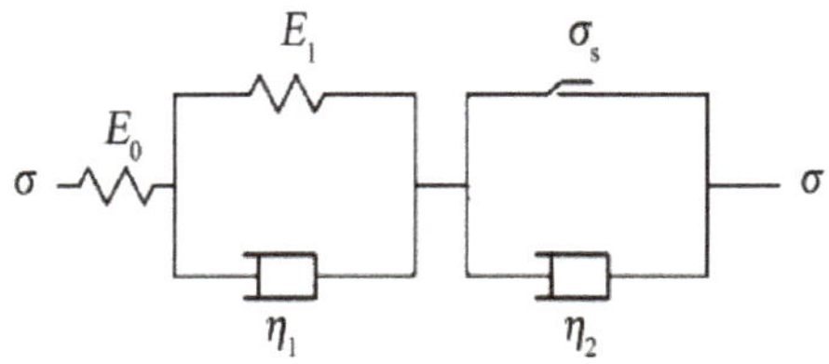

Figure 3. Classical Nishihara model.

The following Formula (1) is the creep equation of this model.

$$\varepsilon = \begin{cases} \frac{\sigma}{E_0} + \frac{\sigma}{E_1}(1 - e^{-\frac{E_1}{\eta_1}t}) & \sigma < \sigma_s \\ \frac{\sigma}{E_0} + \frac{\sigma}{E_1}(1 - e^{-\frac{E_1}{\eta_1}t}) + \frac{\sigma - \sigma_s}{\eta_2}t & \sigma \geq \sigma_s \end{cases} \quad (1)$$

In the formula: E_0 represents the elastic modulus of the Hooke elastomer; η_1 and E_1 represent the viscous coefficient and elastic modulus in the Kelvin model; η_2 represents the viscous coefficient in the viscoplastic body; σ represents the Creep stress; σ_s represents the yield stress.

4.1. Modification of Frozen Soil Creep Constitutive Model

The creep of frozen soil is nonlinear, and the current classical creep models are basically composed of linear elements. The obvious disadvantage is that it can only describe the primary and stable stages of the frozen soil creep process; the accelerated creep phase is not well described. Therefore, some scholars suggest introducing nonlinear elements or damage variables into the creep model for model improvement. By borrowing this method, as shown in Figure 4, this is corrected based on the classical Nishihara model, and the viscous coefficient in the viscoelastic element in the model is corrected as a function of creep time and stress level. The damage variable is considered in the plastic element, and a modified frozen soil creep model is thereupon established [20].

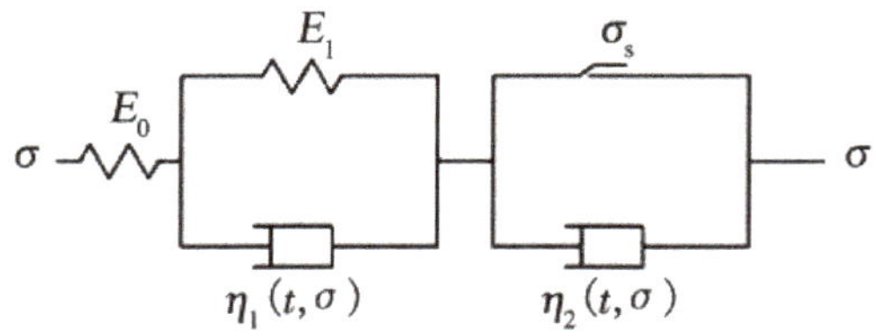

Figure 4. Modified permafrost creep model.

(1) Viscoelastic part

The viscoelastic part of the frozen soil model is mainly composed of nonlinear viscous elements and a spring in parallel. In the creep process of frozen soil, its nonlinear characteristics are significant, and the nonlinear characteristics are affected by the magnitude of the load and its duration at the same time. In this paper, by referring to the relevant literature on the frozen soil creep model [21], the viscosity coefficient is modified as a function of creep time and stress level, which can be expressed as the following expression:

$$\eta_1(t, \sigma) = \eta_{10}\left(\frac{2\sigma_0 - \sigma}{\sigma}\right)\frac{(t/t_0)^{1/3}}{\left[(t/t_0)^{1/3} + 1\right]} \quad (2)$$

In the formula: σ_0 represents the compressive strength of the frozen soil at the same temperature; σ represents the creep stress; η_{10} represents the initial viscosity coefficient of the viscoelastic body; t_0 represents the frozen soil creep time reference value, the unit is *h*, and t_0 = 1 for this frozen soil test.

Derivative of the time in Equation (3), we can get:

$$\dot{\eta}_1(t,\sigma) = \eta_{10}(\frac{2\sigma_0 - \sigma}{\sigma})\frac{\frac{1}{3}(t/t_0)^{-2/3}}{\left[(t/t_0)^{1/3}+1\right]^2} \geq 0 \tag{3}$$

Thus, the constitutive equation of the modified Kelvin style can be obtained as:

$$\sigma = E_1\varepsilon(t) + \eta_1(t,\sigma)\frac{d\varepsilon(t)}{dt} \tag{4}$$

The above Formula (4) can be obtained by definite integration of the separated variables under the condition of constant stress (t = 0, $\varepsilon(t)$ = 0), and finally the creep equation of the unsteady Kelvin body can be obtained as:

$$\varepsilon(t) = \frac{\sigma_0}{E_1}\left[1 - e^{(-\alpha t - \frac{3}{2}\alpha(t/t_0)^{2/3})}\right] \tag{5}$$

In the above expression:

$$\alpha = \frac{E_1}{\eta_{10}}\left(\frac{\sigma}{2\sigma_0 - \sigma}\right) \tag{6}$$

Therefore, the Kelvin style after the above correction can be well-used to describe the decay creep process of frozen soil.

(2) Viscoplastic part

The relevant permafrost creep test research [22] shows that there is a damage effect in the frozen soil. When the stress is greater than a certain critical value, the damage effect of the frozen soil will occur. When it is dominant, the frozen soil will be damaged. Non-attenuated creep occurs. Introducing the damage variable D into the frozen soil creep model can better describe the stable and accelerated creep stages in the permafrost creep process. On the basis of the damage mechanisms, the damage variable D in the frozen soil is defined [23]:

$$E(\sigma,t) = \begin{cases} E_0, \sigma - \sigma_\infty < 0 \\ E_0 e^{-ct^R}, \sigma - \sigma_\infty \geq 0 \end{cases} \tag{7}$$

$$D(\sigma,t) = 1 - \frac{E(\sigma,t)}{E_0} \tag{8}$$

$$D(\sigma,t) = \begin{cases} 0, \sigma - \sigma_\infty < 0 \\ 1 - e^{-ct^R}, \sigma - \sigma_\infty \geq 0 \end{cases} \tag{9}$$

where $D(\sigma,t)$ is the damage variable of the frozen soil; $E(\sigma,t)$ is the function of the elastic modulus changing with time; E_0 is the initial elastic modulus; c, R are the material parameters related to the damage degree of the frozen soil; σ_∞ is the long-term strength of the frozen soil.

According to the definition of effective stress [24]:

$$\sigma' = \frac{\sigma}{1-D} \tag{10}$$

$$\sigma' = \begin{cases} \sigma, \sigma - \sigma_\infty < 0 \\ \sigma e^{ct^R}, \sigma - \sigma_\infty \geq 0 \end{cases} \tag{11}$$

where σ' is the effective stress and σ_∞ is the creep stress.

The accelerated creep stage of frozen soil is described by adopting a viscoplastic element in which the unsteady cohesive body and the plastic body are connected in parallel.

When $\sigma < \sigma_\infty$, the plastic body will not deform; when $\sigma \geq \sigma_\infty$, the plastic body will yield. Therefore, the viscosity coefficient of an unsteady viscous body is expressed as:

$$\eta_2(t,\sigma) = \eta_{20}\left[1 + \frac{2\sigma_\infty}{\sigma}\left(1 - e^{-t/t_0}\right)\right] \tag{12}$$

where η_{20} represents the initial viscosity coefficient of the viscoplastic element.

According to the establishment method of the creep model method, the modified frozen soil creep constitutive equation is obtained as follows:

$$\varepsilon = \frac{\sigma}{E_0} + \frac{\sigma}{E_1}\left[1 - e^{\left(-\alpha t - \frac{3}{2}\alpha(t/t_0)2/3\right)}\right] \qquad \sigma < \sigma_\infty \tag{13}$$

$$\varepsilon = \frac{\sigma}{E_0} + \frac{\sigma}{E_1}\left[1 - e^{\left(-\alpha t - \frac{3}{2}\alpha(t/t_0)2/3\right)}\right] + \frac{\sigma I - \sigma_\infty}{\eta_2(t,\sigma)} \qquad \sigma < \sigma_\infty \tag{14}$$

4.2. Creep Model Parameters

According to the uniaxial creep test results, the modified frozen soil creep model is used for fitting analysis, as shown in Figure 5. The correlation coefficient $R^2 \geq 0.89$ obtained from the fitting is 0.94 on average. The fitted curve is in good agreement with the measured curve as a whole. The initial creep stage is in good agreement with the stable creep stage, and in poor agreement with the accelerated creep stage. At the beginning of the 0.75 σ_t loading stage, with the increase of time, the coefficient of fitting was significantly improved. It shows that the modified creep constitutive model can better describe the decaying and non-attenuating creep processes of frozen soil well.

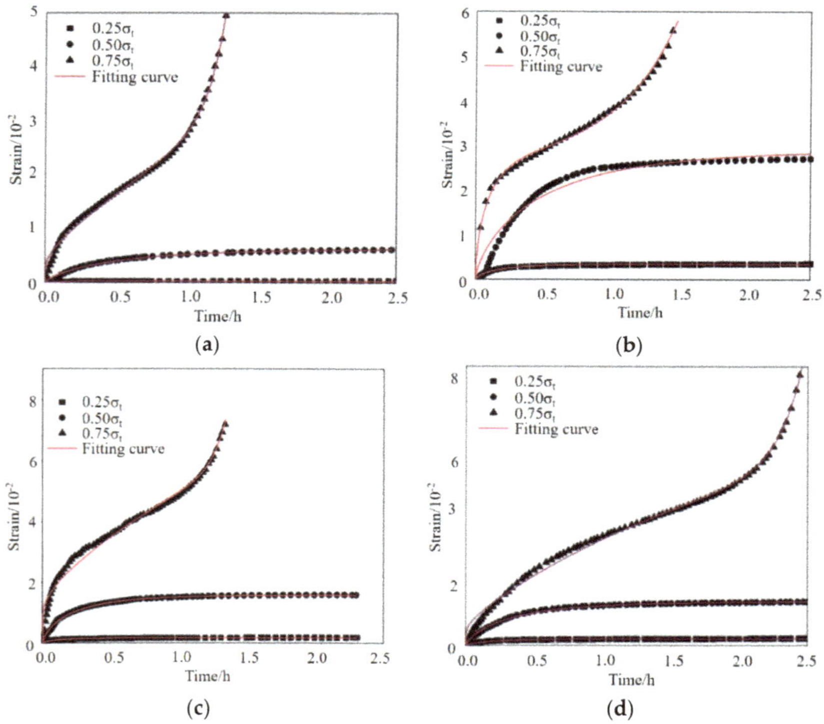

Figure 5. Frozen soil creep test curve fitting, (**a**) FSC_2-b fitted curve; (**b**) FSC_3-c fitted curve; (**c**) FSC_4-a fitted curve; (**d**) FSC_5-b fitted curve.

The initial elastic modulus E_0 represents the deformation resistance at the moment of loading. Table 4 shows that E_0 increases with the decrease of rising temperature but increases with the elevating of the loading level. The progressive elastic modulus E_1 represents the deformation resistance of the soil in the initial creep stage, which is very close to the change trend of E_0. The values of η_{10} and η_{20} represent the initial viscosity coefficient of viscoelastic bodies and viscoplastic body, respectively, which can reflect the deformation rates of soil.

Table 4. Frozen soil creep model parameters.

Sample Number	Temperature/°C	Stress/MPa	c	R	E_0	E_1	η_{10}	η_{20}	R^2
		1.15			2.2188	1.3763	2.0852		0.8945
FSC_2-b	−2	2.29	1.7658	0.3496	2.6575	1.9158	1.8960	8.6866	0.9909
		3.43			2.4864	1.5326	1.5266		0.9950
		1.30			3.0760	1.1810	3.9870		0.9482
FSC_3-c	−3	2.59	1.4622	0.1039	3.5324	1.3213	3.5333	14.7524	0.9971
		3.89			3.1660	1.4639	3.2542		0.9845
		1.37			3.3755	2.2417	6.2449		0.9846
FSC_4-a	−4	2.74	1.3493	0.0808	3.8462	2.7647	5.9199	20.5709	0.9594
		4.10			3.6132	2.4076	5.6791		0.9941
		1.63			4.2981	2.8563	8.7446		0.9805
FSC_5-b	−5	3.27	0.7620	0.0571	5.3059	3.1279	7.4421	26.4141	0.9964
		4.90			5.1398	3.0202	6.2154		0.9948

5. Discussion

In the early stage of this study, many frozen soil basic tests were done to determine the physical and mechanical parameters of the frozen soil specimens. After they were consistent with the frozen soil field specimens of Jiangcang Coal Mine, the permafrost creep test was carried out. Aiming at the deformation and failure characteristics of the overlying frozen soil in the coal mine, the failure characteristics of frozen soil specimens with different stress states, different creep types and different creep stages are mainly studied.

The deformation properties of frozen soil are comprehensively affected by factors such as soil temperature, proportion of soil ice content and load. No matter what the conditions are, the creep type of the final creep curve can be divided into attenuated creep and non-attenuated creep. Attenuated creep only goes through the initial stage and the stable stage, and there is no acceleration stage. The creep rate of the initial stage gradually decreases with elapsing time, and the stable creep curve gradually develops into an approximately straight line, showing a slowly decaying state; the non-attenuated creep can be divided into three stages: initial, stable and accelerating creep stages [25]. In the initial stage, the creep rate gradually decreases with elapsing time until its nadir. In the stable stage, the creep rate remains unchanged at this minimum value. At the accelerating creep stage, the creep rate increases until the frozen soil column fails. When the axial stress is 0.25 σ_t and 0.5 σ_t, the creep curve shows obvious initial and stable creep stages, respectively. The creep deformation rate decreases with passing time, and the creep curve smooths out. The creep type is decaying when the axial stress is 0.75 σ_t, the creep curve goes through the initial creep stage, the stable and the accelerating creep stage, and the creep deformation rate shows a trend of first decreasing and then increasing, showing a curve of a typical non-attenuated creeping.

Temperature, soil ice content and graded load have great influences on the creep properties and failure modes of frozen soil. The higher the soil temperature, the smaller the proportion of soil ice content, the greater the load, the more unstable the mechanical properties of the frozen soil, the greater the creep deformation, the greater the stable creep rate, and the smaller the peak stress of the frozen soil. The lower the soil temperature, the greater the proportion of soil ice content and the smaller the load. As the soil temperature decreases, the proportion of soil ice content and the load increase, the time in the stable creep stage becomes longer, and the specimen enters the longer accelerated creep stage, and the accelerated creep stage is longer. The greater the slope of the curve, the greater the peak value of the frozen soil stress. The above basic research is consistent with the results of most permafrost studies.

However, most scholars believe that the non-attenuating creep of frozen soil has an important influence on the settlement of the foundation and the attenuating creep is ignored. However, both attenuated creep and non-attenuated creeps have a significant impact on the safe production of coal mines. Therefore, this paper analyzes the movement of the coal mine overlying rock and the occurrence of rock pressure according to the mechanical behavior and fracture law of the overlying frozen soil in the coal mine. It goes as follows: The lower the soil temperature, the greater the proportion of soil ice content, the greater the peak stress, the more complex the failure patterns of frozen soil and the more complex the crack evolution path.

With the increase of stress peak value, the failure mode of frozen soil changed from simple shear failure to complex mixed failure, the larger the expansion capacity. At the same time, it shows that temperature is an important factor affecting the mechanical properties and failure form of frozen soil.

6. Conclusions

In order to study the creep deformation characteristics of high-temperature frozen soil and the influence of factors such as temperature and load, this study carried out indoor uniaxial creep tests and the main conclusions are as follows:

(1) Soil temperature is the most important external factor for affecting frozen soil creep, and the proportion of soil ice content is the key internal factor affecting permafrost to the creep. The lower the soil temperature, the higher the proportion of soil ice content in the specimen, the lower the creep rate of frozen soil and the smoother the curve. The greater the number of cracks and the trace length of the specimen, the more complex the failure type. The sensitivity of creep rate to soil temperature and load increases as the proportion of soil ice content decreases.

(2) For the frozen soil specimen under load, the external load in the early stage is mainly borne by the pore pressure and the pores and cracks of the specimen are compacted. With the increase of external load, cracks expand and germinate and the structure bearing external load transfers from pore pressure to soil particle skeleton. The coupling effect of pore pressure and effective stress is complex, leading to the non-obvious regularity of creep deformation. At 0.75σt, the structure bearing the external load transforms into the soil particle skeleton, enters the accelerated creep stage and the failure speed of the specimen increases.

(3) Based on the classical Nishiplasma model, the modified permafrost creep model is obtained, which can better describe the decaying creep and non-attenuating creep processes of frozen soil. The model is relatively simple and has relatively fewer parameters. It can provide a theoretical basis for the mitigation and control of surface subsidence in the mining of underlying coal seams in Jiangcang Coal Mine.

Author Contributions: Conceptualization, Y.S. (Yawu Shao) and Y.S. (Yonglu Suo); methodology, Y.S. (Yawu Shao); software, Y.B.; validation, Y.S. (Yonglu Suo), J.X.; formal analysis, Y.B. (Yuan Bai); investigation, Y.S. (Yawu Shao); resources, Y.S. (Yonglu Suo); data curation, Y.B.; writing—original draft preparation, Y.S. (Yawu Shao); writing—review and editing, Y.S. (Yonglu Suo); visualization, Y.S. (Yawu Shao); supervision, Y.S. (Yonglu Suo); project administration, Y.S. (Yonglu Suo); funding acquisition, T.Y. All authors have read and agreed to the published version of the manuscript.

Funding: This research was financially funded by [Tao Yang] grant number [National Natural Science Foundation of China: (Grant No. 52004200)].

Data Availability Statement: The data analyzed or generated during the research can be provided by the corresponding author upon request.

Conflicts of Interest: The authors declare that they have no conflict of interest.

References

1. Jin, H.J.; Zhao, L.; Wang, S.L.; Jin, R. Thermal Regimes and Degradation Modes of Permafrost along the Qinghai-Tibet Highway. *Sci. China (Ser. D) Earth Sci.* **2006**, *49*, 1170–1183. [CrossRef]
2. Tsytovich, H.A. *The Mechanics of Frozen Ground Soil*; Zhang, C.Q.; Zhu, Y.L., Translators; Science Press: Beijing, China, 1985; pp. 153–158.
3. Qi, J.L.; Ma, W. State-of-art of Research on Mechanical Properties of Frozen Soils. *Rock Soil Mech.* **2010**, *31*, 134–143.
4. Zhang, J.M.; Zhang, J.Y. *Characteristics of Strength and Creep of Frozen Swelling Clay*; Science Press: Beijing, China, 1989; pp. 188–192.
5. Ma, W.; Zhu, Y.L.; Ma, W.T.; Chang, X.X. Analyses of Deformation inFrozen Clayed Soils. *J. Glaciol. Geocryol.* **2000**, *22*, 43–47.
6. Vyalov, C.C. *Rheology of Frozen Soil*; Liu, J.K., Translator; China Railway Press: Beijing, China, 2005.
7. Ma, W.; Sheng, Y.; Wu, Z.W. Stress Change in Ice During the Creep Deformation of Frozen Soil. *Process Nat. Sci.* **1994**, *4*, 921–926.
8. Wu, Z.W.; Ma, W. *Strength and Creep of Frozen Soils*; Lanzhou University Press: Lanzhou, China, 1994; pp. 141–148.
9. Zhu, Y.L.; Zhang, J.Y. Elastic and Compressive Deformation of Frozen Soils. *J. Glaciol. Geocryol.* **1982**, *4*, 29–40.
10. Zhu, Y.L.; Liu, Y.Z.; Xie, X.D. *In-situ Creep Test Research of Ground Ice on Qinghai-Tibet Plateau*; Science Press: Beijing, China, 1983; pp. 124–130.
11. Mcroberts, E.C.; Law, T.C.; Murray, T.K. Secondary Creep Interpretation of Ice-rich Permafrost. In Proceedings of the 5th International Conference on Permafrost, Trondheim, Norway, 2–5 August 1988; pp. 1137–1142.
12. Ma, X.J.; Zhang, J.M.; Chang, X.X.; Zhen, B.; Zhang, Y.M. Experimental Study on Creep of Warm and Ice-rich Frozen soil. *Chin. J. Geotech. Eng.* **2007**, *29*, 848–852.
13. Zheng, B.; Zhang, J.M.; Ma, X.J.; Zhang, J.W. Study on Compression Deformation of Warm and Ice-rich Frozen Soil. *Chin. J. Rock Mech. Eng.* **2009**, *28* (Suppl. S1), 3064–3069.
14. Eckardt, H. Creep Behaviour of Frozen Soils in Uniaxial Compression Tests. *Ground Freez.* **2012**, *26*, 185–195.
15. Yao, Z.M.; Zhang, W.; Guo, M.Y. Analysis of Variable Creep Model in Frozen Clay Considering Temperature Effect. *J. Chang. Acad. Sci.* **2020**, *37*, 81–85+91.
16. Liu, C. Research on the Law of Surface Subsidence in Coal Seam Mining under Permafrost. Ph.D. Thesis, Xi′an University of Science and Technology, Xi′an, China, 2017.
17. Shang, T.L. Research on the Overburden Structure and Its Movement Law in the Mining of Large Dip Angle Coal Seams under Permafrost. Ph.D. Thesis, Xi′an University of Science and Technology, Xi′an, China, 2017.
18. Lai, X.P.; Ren, J.; Shan, P.F.; Zhang, Y.; Zhang, N.; Cui, F. Damage monitoring and energy evolution analysis of self-equilibrium coal under vertical load. *Measurement* **2022**, *203*, 111999. [CrossRef]
19. Shao, Y.W.; Suo, Y.L.; Xiao, J.; Bai, Y.; Yang, T.; Fan, S. Dynamic Evolution Characteristics of Oil–Gas Coupling Fractures and Dynamic Disaster of Coal Mass in Coal and Oil Resources Co-Storage Areas. *Appl. Sci.* **2022**, *12*, 4499. [CrossRef]
20. Wu, M.Y. Theoretical Study on Uniaxial Compressive Strength and Creep of Artificial Frozen Soil. Ph.D. Thesis, Anhui University of Science and Technology, Huainan, China, 2019.
21. Sun, K.; Chen, Z.L.; Chen, J. A Creep Constitutive Relation of Frozen Soil Based on the Modified Nishihara Model. *Rock Soil Mech.* **2015**, *36* (Suppl. S1), 142–146.
22. Yang, Y.G.; Lai, Y.M.; Chang, X.X. Experimental and Theoretical Studies on the Creep Behavior of Warm ice-rich Frozen Sand. *Cold Reg. Sci. Technol.* **2010**, *63*, 61–67. [CrossRef]
23. Hou, F.; Lai, Y.M.; Liu, E.L.; Luo, H.W.; Liu, X.Y. A Creep Constitutive Model for Frozen Soils with Different Contents of Coarse Grains. *Cold Reg. Sci. Technol.* **2018**, *145*, 119–126. [CrossRef]
24. Kachanov, L.M. *Introduction to Continuum Damage Mechanics*; Martinus Nijhoff Publishers: Dordrecht, The Netherlands, 1986.
25. Arenson, L.U.; Springman, S.M.; Sego, D.C. The Rheology of Frozen Soils. *Appl. Rheol.* **2007**, *17*, 1–14. [CrossRef]

sustainability

MDPI

Article

An Engineering Case History of the Prevention and Remediation of Sinkholes Induced by Limestone Quarrying

Zhen Tang [1,2,*], Lei Song [1], Dianqi Jin [2], Ligen Chen [3], Gan Qin [2], Yongjun Wang [2] and Lei Guo [2]

1 State Key Laboratory for Geomechanics and Deep Underground Engineering, China University of Mining and Technology, Xuzhou 221116, China
2 Shenzhen Urban Public Safety and Technology Institute, Shenzhen 518046, China
3 Guangdong Nonferrous Metals Geological Exploration Institute, Guangzhou 510080, China
* Correspondence: tangzhen@szsti.org

Abstract: This paper introduces an engineering case history of the prevention and remediation of sinkholes induced by limestone quarrying in Longmen county, Huizhou city, China, through karst groundwater-air pressure monitoring, the design and construction of a grouting curtain, and grouting effect detection. Based on hydrogeological surveys, the location of the main karst development zones and faults can be accurately delineated by combining geophysical exploration with drilling, providing a basis for curtain setting. According to the interpretation results of geophysical exploration, the monitoring boreholes of groundwater-air pressure were set up, which provided support for mine construction, optimization of prevention and remediation of the sinkhole scheme, and reduction of sinkhole risk. In order to prevent the further expansion of sinkhole hazards, grouting curtain technology was used for engineering treatment of the water inflow points of the quarry. After construction of the grouting curtain was completed, comprehensive detection methods were used to evaluate the grouting effect of the curtain. The results showed that the inflow rate reduced from 3500 to approximately 500 m^3/day, the water plugging effect was significant, and the occurrence of sinkhole hazards was effectively reduced. The monitoring boreholes can capture the changes of groundwater-air pressure within karst conduit systems, and the purpose of monitoring and warning of sinkholes can be achieved by setting an appropriate warning threshold.

Keywords: sinkholes; groundwater-air pressure monitoring; grouting effect detection; monitoring and warning

Citation: Tang, Z.; Song, L.; Jin, D.; Chen, L.; Qin, G.; Wang, Y.; Guo, L. An Engineering Case History of the Prevention and Remediation of Sinkholes Induced by Limestone Quarrying. *Sustainability* **2023**, *15*, 2808. https://doi.org/10.3390/su15032808

Academic Editors: Xiangguo Kong, Dexing Li and Xiaoran Wang

Received: 12 January 2023
Revised: 30 January 2023
Accepted: 31 January 2023
Published: 3 February 2023

1. Introduction

With the acceleration of urbanization and infrastructure construction, the demand for cement has increased sharply. As the main raw material of cement production, the demand for limestone has also increased synchronously. However, frequent unreasonable open-pit mining activities have destroyed the ecological environment around mines, and sinkholes have become the main geological environment problem linked with mine construction.

In karst areas, the high anisotropy of carbonate rocks means that drilling data are unable to truly reflect the karst development characteristics of the mining area and may lead to the wrong conclusion of hydrogeological conditions [1,2]. When the quarrying operation exposes the karst conduits or fracture zones, it is accompanied by many water inflow phenomena. When the groundwater table is lowered around the mining area, local instability of the soil produces soil caves, which eventually leads to the occurrence of sinkholes and other hazards [3]. Since the 1940s, there have been many sinkhole events induced by limestone quarries, and the affected area changes with the water inflow of the quarry. In May 1949, a limestone quarry in Hershey, Pennsylvania, with an average drainage of 6500 gpm, altered groundwater levels over an area of 10 square miles, and about 100 sinkholes formed in the area [4]. In 1986, a limestone quarry in Valley and Ridge

Province was mined to a depth of 60 m below the original groundwater level, resulting in a wide range of groundwater depression cones. The groundwater level at 0.8 km from the quarry was lowered 18 to 24 m, and multiple sinkholes were generated within 1.6 km of the quarry, threatening the operation of local railways [5]. From 1992 to 2008, 353 sinkholes occurred in Chini Village, Huadu District, Guangzhou City, mainly caused by limestone quarrying and large groundwater pumping [6]. Eighty-six sinkholes occurred in the Hejing mining area of Pingnan County, Guangxi Province, China. The zone of dewatering influence was about 9.36 km^2, and the maximum influence radius was about 2.1 km [7]. During 2017 and 2018, more than 30 sinkholes formed in an agricultural region that is located adjacent to a limestone quarry in northeast Thailand, which affected the local agricultural activities and power transmission towers [8]. With the increasing demand for cement materials and the expansion of limestone quarries, the occurrence of sinkholes induced by limestone quarrying is becoming increasingly serious.

In order to effectively protect the ecological environment of the mining area and ensure the safe mining of the quarry, a grouting curtain is often used in mine water inflow treatment engineering. The hydraulic connection on both sides of the curtain is cut off by filling the cracks and karst conduits with cement and other materials. Since the 1960s, the curtain grouting technique has been gradually adopted in mine water inflow treatment engineering. This has promoted the development of the curtain technique and grouting theory, with excellent results [9,10]. Cement–sodium silicate, hot bitumen, coal slurry, modified clay, and polyurethane foam mortar are often used as grouting materials for curtains [11]. The curtain technique mainly includes drilling, grouting, and grouting in special strata [12–16]. Determining the grouting effect of a curtain mainly consists of geophysical analysis, verification boreholes, Lugeon tests, pumping tests, and groundwater-air pressure monitoring on both sides of the curtain [17].

At present, the research on sinkhole hazards induced by limestone quarries mainly focuses on the evolution of the groundwater seepage field and engineering treatment methods, while research on the monitoring and warning of sinkhole hazards in limestone quarry is limited. In addition, quarries usually adopt a negative treatment for sinkholes, without considering the change of groundwater dynamic fluctuations during quarrying, and grouting programs to prevent water inflow often use a single method to detect the grouting effect. This paper introduces the prevention and remediation of sinkholes induced by limestone quarrying through a case study, which provides a reference for similar projects in the future.

2. Research Background

Longmen County is located in the central part of Guangdong Province, with rich mineral resources, mainly including limestone, lead-zinc, and geothermal resources (Figure 1). Longmen County has Guangda, Huarun, Tapai, and other cement enterprises, and is one of the largest cement production bases in Guangdong Province. In 2020, the annual cement output was 15.5453 million tons, accounting for 83.96% of the annual cement output of Huizhou City. There are many expressways in the territory, which are closely connected with the cities in the Pearl River Delta.

The quarry assessed in this paper is located in Longmen County, Huizhou City, China, where limestone is mined for cement. The mining elevation is +231.4 to −30 m, and the mining area consists of 2.6011 km^2. The quarry is divided into north and south pits, which are mined to an elevation of 0 m. The carbonate rocks in the mining area consist of Carboniferous Liujiatang (C_1ylj), Shidengzi (C_1ds), and Huanglong (C_2h) formations. The upper unit (C_2h) is thick-layered crystalline dolomite, which is partially intercalated with dolomitic limestone, and the lower unit consists of thick-layered crystal limestone. The lithology of the Shidengzi Formation is mainly in the form of thick-layered argillaceous limestone, thick-layered carbonaceous limestone, thick-layered crystalline limestone, and thin–medium-layered micritic limestone. The exposed thickness is about 276 m, whichias the main stratum with karst development. Additionally, most of the Liujiatang Formation is covered by Quaternary deposits with a thickness of about 120 m. The upper unit (C_1ylj) contain medium–thick-layered carbonaceous limestone, bioclastic limestone, dolomitic limestone, and calcareous dolomite. The lower unit is mainly composed of argillaceous limestone, calcareous mudstone, argillaceous siltstone, and siltstone, and is locally intercalated with bioclastic limestone. The mining area is located in the uplift end of the axial NE-trending anticline, and the stratigraphic occurrence changes significantly, mainly due to the NNE-trending fault structures (Figure 2).

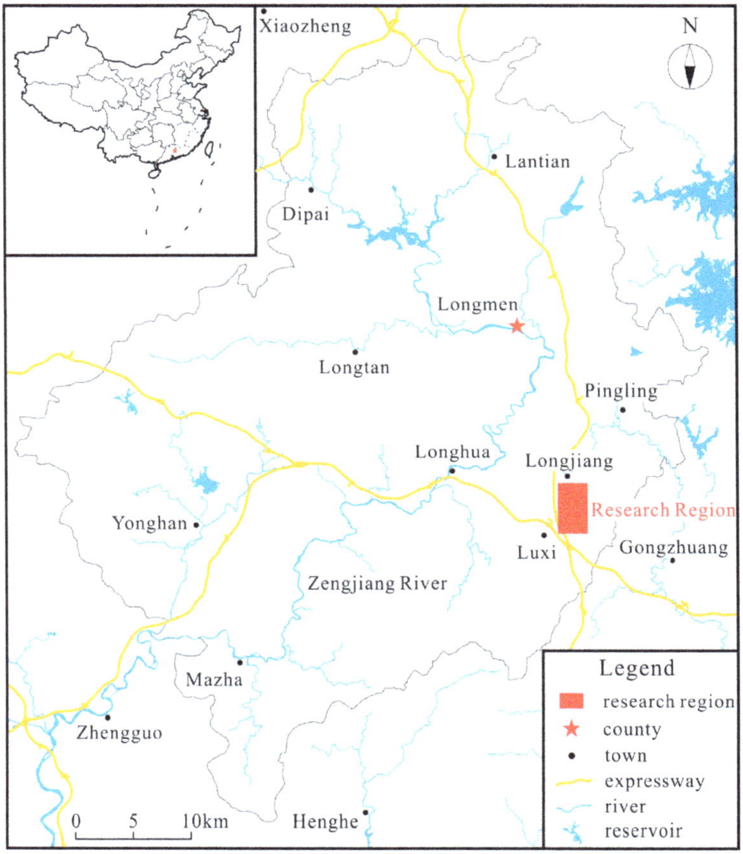

Figure 1. Geographic location of Longmen County.

In mid-October 2018, in the northeastern corner of the south pit of the quarry, the karst fissure zones accidentally broke through and water inflow occurred at a water inflow rate of about 3500 m^3/day. Since 12 November 2018, 17 sinkholes have formed around the mining area, the largest of which is 8 m in diameter and 6 m in depth. Furthermore, many water wells in the nearby village have dried up, and 107 houses have fractured due to mine water inflow (Figure 3). In March 2019, we conducted an emergency investigation, and the results showed that the northeastern corner of the south pit was in the intersection area of faults F1 and F6, and this area was characterized by relatively broken rocks and strong karst development.

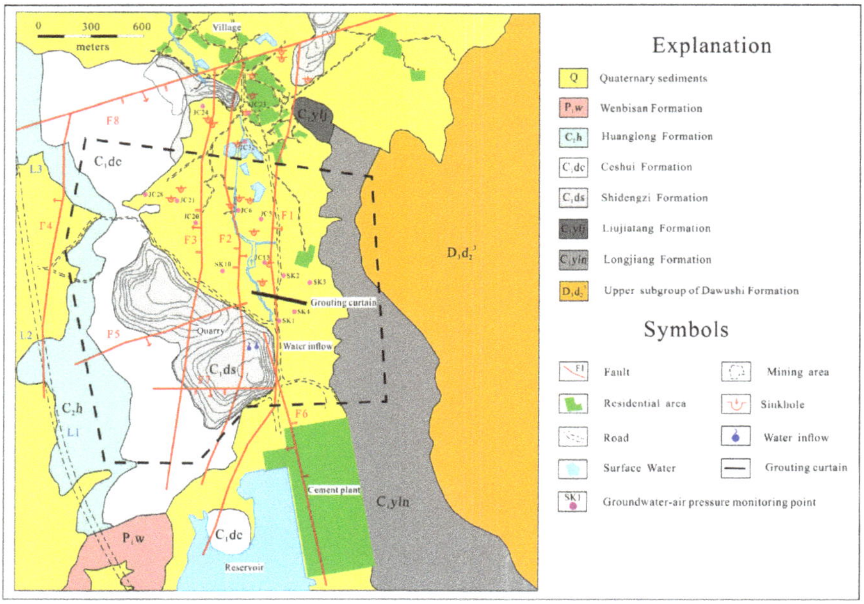

Figure 2. Geologic map of the mining area.

Figure 3. Sinkhole caused by mine water inflow.

3. Karst Groundwater-Air Pressure Monitoring

When the sinkholes formed around the quarry, the quarry companies often ignored the impact of the changes in the groundwater dynamics on the sinkholes. Jiang and Lei proposed a new technique for groundwater-air pressure monitoring in karst fissures and pipeline systems, which could fully reflect the changes in the groundwater-air pressure in karst pipeline fissure systems, as well as explain the sudden changes in karst groundwater-air pressure, and the reasons for these changes [18].

Through the collection of regional geological data and early drilling information from the site, the stratigraphic lithology, geological structures, and hydrogeological conditions were determined. Furthermore, the depth of the monitoring borehole was determined based on the type of exposed karst aquifer, along with its depth and thickness. The boreholes were designed to pass through the karst fissure zones or karst conduits to monitor the dynamic changes in groundwater-air pressure. After the monitoring borehole was drilled, a steel casing was used to isolate the soil aquifer from the karst aquifer. After the sensor was installed, sealing materials were used to seal the borehole orifice to restore the original geological conditions (Figure 4).

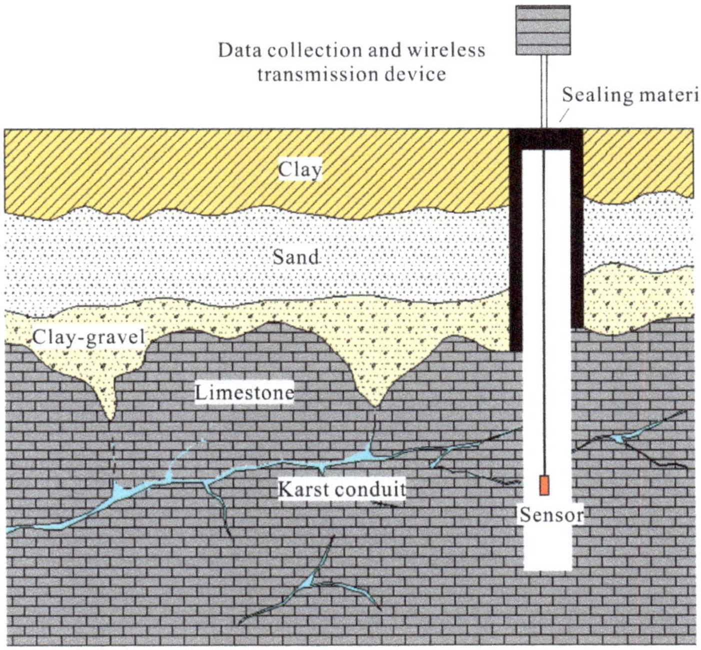

Figure 4. Technique for monitoring the karst groundwater-air pressure.

Groundwater-air pressure monitoring boreholes continuously monitored the abnormal fluctuations in groundwater-air pressure at 10-min intervals during the mining process. Figure 5 shows the monitoring results of the groundwater-air pressure of borehole JC28. It can be seen from Figure 5b that the fluctuation velocity of groundwater pressure caused by rainfall was generally less than 0.5 cm/min, which was not significantly different from that without rainfall.

There were three abnormal fluctuations of groundwater-air pressure, and the sinkhole event could be divided into three stages. Initial stage: the quarry uncovered water-filled karst pipes, which led to the rapid decline of the groundwater level, a large difference in groundwater head was generated, and an initial soil cave was formed. Development stage: under the effect of groundwater head difference, the groundwater flow rate increased, and the soil cave gradually expanded, forming a short-term plugging in the karst pipes, causing abnormal fluctuations in the groundwater-air pressure. Forming stage: under the influence of rainfall during the flood season, the groundwater level rose significantly, the soil was saturated and softened, the seepage pressure increased, and the soil was destabilized, resulting in a sinkhole.

Abnormal groundwater-air pressure fluctuations before and during the sinkhole formation could be captured, and the abnormal values were significantly different from the normal fluctuations; in particular, the abnormal groundwater-air pressure fluctuations during the sinkhole formation could reach tens of times the normal fluctuation values. However, the abnormal groundwater-air pressure fluctuations were not clearly reflected in the monitoring curves of sinkhole hazards induced by loading.

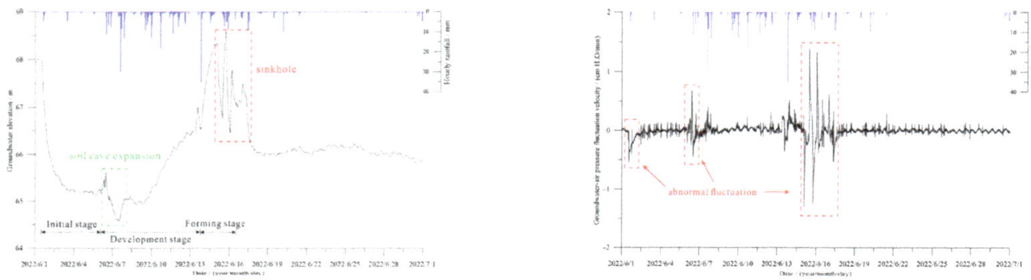

(a) (b)

Figure 5. Groundwater-air pressure monitoring curve of borehole JC28. (**a**) Groundwater elevation curve of borehole JC28. (**b**) Groundwater-air pressure fluctuation velocity curve of borehole JC28.

4. Design of Grouting Curtain

4.1. Location of the Curtain

A grouting curtain was used to seal the groundwater runoff zones at the bottom of the quarry, ensuring the safe mining of the quarry, as well as the safety of the local residents and their property. According to the site survey, the locations of the sinkholes, disappearance of surface water, and house cracks occured in the northern part of the quarry. Thus, the supply sources for the water inflow points could be preliminarily determined. Based on the geophysical test results, and considering the irrigation water of the local residents, the approximate position of the grouting curtain was determined. However, the final location of the curtain was somewhat limited by the terrain, farmland, and houses.

The curtain was constructed perpendicular to the NNE-trending fault, with a length of about 200 m, and the eastern boundary of the curtain was embedded in the sandstone of the Longjiang Formation.

4.2. Grouting Borehole Parameters

A total of 513 sets of borehole data were collected in the study area, and 258 of the boreholes encountered exposed the karst caves (Figure 6). The statistical results of the karst cave development elevation showed that 75% of the karst cave development elevations were between 40 and 70 m, and mainly 20 m below the bedrock surface. The deep karst formations were also mainly developed in the contact zone between the Liujiatang Formation (C_1ylj) and the Shidengzi Formation (C_1ds).

Considering the development elevation of the karst caves and the fact that the quarry will be mined to an elevation of −20 m at a later stage, the bottom elevation of the grouting boreholes was set to −60 m. The grouting boreholes were designed in a single row, with a spacing of 10 m, with 22 grouting boreholes, and a single borehole depth of about 120–150 m. The opening borehole diameter was 130 mm, and the final borehole diameter was 91–75 mm. The designed slurry radius was also at least 7.07 m, and the solid permeability coefficient was 0.06 m/d, forming a 10 m thick curtain wall.

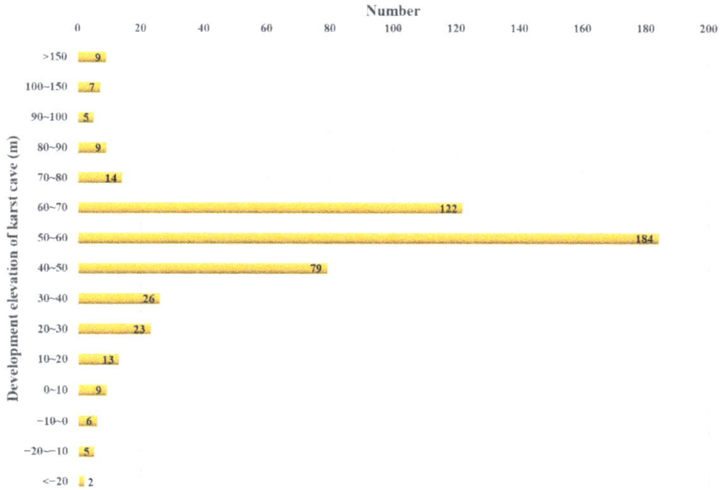

Figure 6. Development elevation of the karst caves.

The karst caves were exposed in grouting boreholes K22, K23, and K24 at roughly the same elevation, and they consisted of filled or semi-filled caves. Water leakage occurred during drilling, and the adjacent boreholes channeled each other during grouting, indicating that the karst caves exposed along the curtain axis were interconnected and were the main runoff zones for groundwater. Additionally, during drilling, two major void zones were discovered. The formations of the west void zones were closely related to the fault position, and the east void zones were developed at the stratigraphic boundary (Figure 7).

Conventional cement grouts were used in the curtain, and the concentration of the slurry was based on the rock permeability obtained from the Lugeon test. The water–cement weight ratios of the grouts were 2:1, 1.5:1, 1:1, and 0.8:1. The additive in the grout was sodium silicate. In order to make the grouts quickly and reduce the influence on the later strength of concrete, the proportion of sodium silicate in the grouts was 1%. Two percent sodium silicate was added in areas requiring large amounts of grouting. The fineness of the cement used for the grouting was such that the sieve allowed for no more than 10% to pass through 80 mm diameter holes. The modulus of the sodium silicate was 2.8–3.4, and the relative density was 1.357–1.453. The grouts were concentrated in the pulping station, pumped to the secondary mixing barrel, and then finally poured into the borehole using a grouting pump.

Once the borehole orifice was closed, the grouts were circulated in the borehole, and the upward and downward directions were connected with segmented grouting. This grouting technique achieved a constant grouting pressure and effectively controlled the diffusion radius of the grouts. The segmented grouting process was conducted as follows. Grouting was carried out after drilling one section, and the next section was drilled after grouting. The above process was conducted in cycles until the designed borehole depth was reached. When large void zones or karst conduits were encountered, the seif−flow grouting method was adopted.

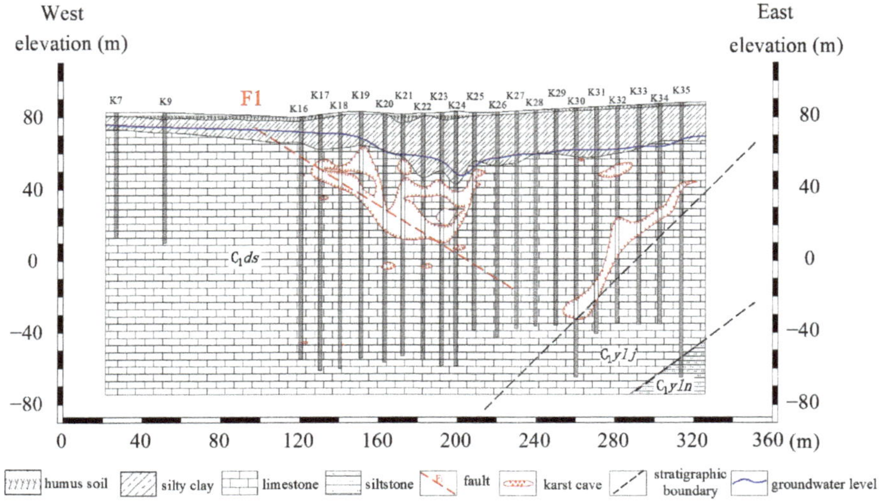

Figure 7. Profile of grouting curtain.

The designed final pressure of the bedrock after grouting was 1.2 MPa, and the curtain grouting was changed by increasing the grout concentrations, step by step. When the injection volume for a certain proportion of grout reached 25 m^3, and the grouting pressure did not increase significantly, nor did the injection rate decrease, the grout concentration was increased. If the grouting pressure did not increase significantly when the grout with the largest specific gravity was used, the amount of sodium silicate was increased from 1% to 2%. Thus, intermittent grouting was needed when the grouting pressure still did not increase. The grouting volume of the intermittent grouting was approximately 150 m^3, and the intermittent time was generally 8 h.

Therefore, the stopping criteria for segmented grouting met the following two criteria:

(1) When the grouting pressure increased uniformly and continuously reaching the designed final pressure, and the slurry suction of the borehole was less than 10 L/min, grouting was conducted for 20–30 min.
(2) After grouting, the Lugeon test was carried out, and the unit permeability was less than 3 lu.

4.3. Implementation of the Data Statistics of the Grouting Curtain

A total of 22 grouting boreholes were used for curtain construction, with a total footage of 2793.28 m. The total thickness of the limestone was 2254.05 m. Nineteen of the boreholes contained karst caves. Thirty-six karst caves were exposed, the total height of the karst caves was 253.08 m, and the average karst rate was 11.23%. The maximum height of the karst cave was 22.5 m, which was exposed in borehole K21, with a maximum karst rate of 37.3%.

This grouting engineering was conducted as an emergency measure to stop the water inflow, and drilling was divided into two sequences for grouting. The total grouting volume of the grouting boreholes was 15,477.46 m^3, and the average unit grouting volume was 7.207 m^3/m. The statistical results of the grouting in each sequence are presented in Table 1, showing that the unit grouting volume of sequence II was significantly lower than that of sequence I. This indicated that the borehole spacing was reasonable, the slurry was effectively diffused, and the sequence grouting effect was significant.

Table 1. Statistical results of the grouting volume.

Sequence	Number of Grouted Sections	Grouting Length (m)	Statistical Data	Elevation Distribution (m)							AUGV (m^3/m)
				< 60	60–40	40–20	20–0	0–−20	−20–−40	>−40	
I	61	1206.81	AUGV (m^3/m)	4.91	3.59	9.24	11.85	10.38	9.46	5.11	8.521
			Grouting length (m)	44.93	175.03	220	220	220	193.25	133.6	
II	56	940.75	AUGV (m^3/m)	1.86	3.08	4.64	4.34	8.01	9.07	0.15	5.521
			Grouting length (m)	28.7	165.9	180	180	180	172.65	33.5	
Total	117	2147.56	AUGV (m^3/m)	3.72	3.34	7.17	8.47	9.31	9.28	4.12	7.207
			Grouting length (m)	73.626	340.93	400	400	400	365.9	167.101	

AUGV refers to the average unit grouting volume.

5. Evaluation of the Grouting Effect of the Curtain

5.1. High-Density Resistivity Method

The high-density resistivity method is based on the differences in electrical conductivity between the rock and the soil, and a stable current field is artificially applied to study the distribution of the conduction current [8,19–21]. By comparing the apparent resistivity section maps before and after grouting, it was found that the apparent resistivity of karst development areas after grouting was lower than before grouting. Thus, the apparent resistivity curve of the original low resistivity anomaly in the horizontal direction became flattened (Figures 8–10).

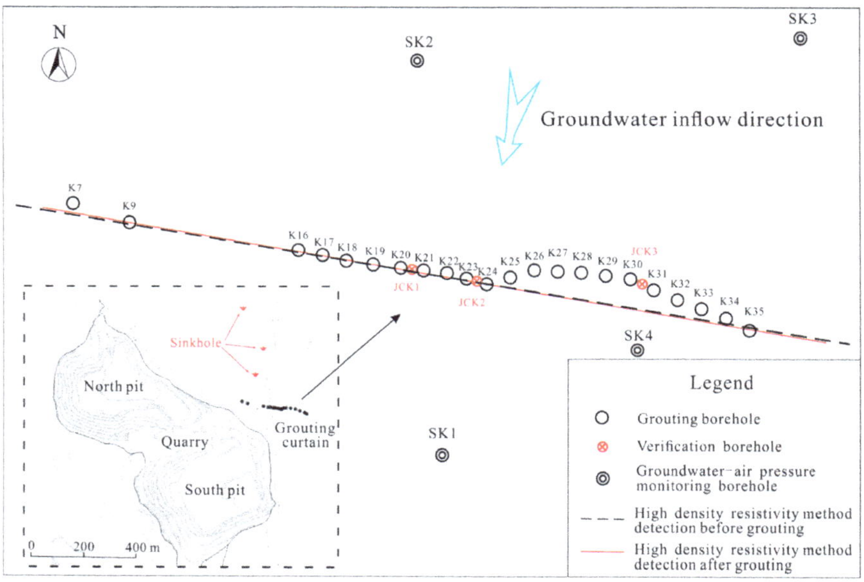

Figure 8. Location of boreholes and survey line for high-density resistivity method.

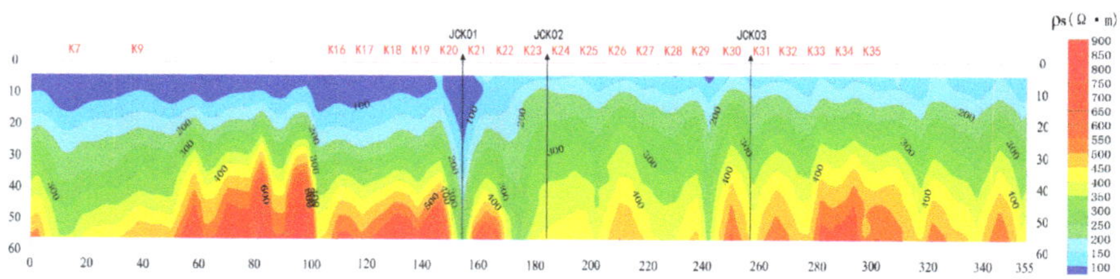

Figure 9. Apparent resistivity section map before grouting.

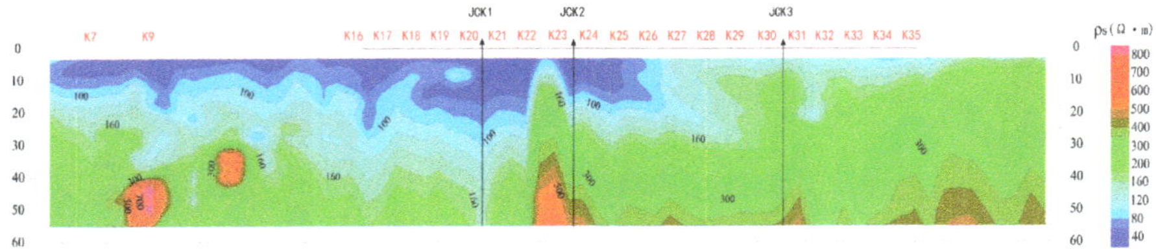

Figure 10. Apparent resistivity section map after grouting.

5.2. Verification Boreholes

Verification boreholes are the most direct method of checking the cross-linking state between the grouting boreholes, specifically to check for the presence of permeable cracks and caves, as well as to assess their filling conditions and distribution in the grouting areas. Based on analysis of the sinkhole locations, karst development distribution, grouting amount, and apparent resistivity section maps, verification boreholes JCK1, JCK2, and JCK3 were drilled between grouting boreholes K20 and K21, K23 and K24, and K30 and K31, respectively. According to the photos of the cores from the verification boreholes, cement slurry stones were found in the fracture zones and karst caves at different depths in the boreholes, and a thin layer of cement was found on many fracture surfaces, indicating that the grout filled the void zones effectively (Figure 11).

Figure 11. Photos of the cores from the verification boreholes.

5.3. Lugeon Test

The Lugeon test uses high pressure to inject water into the borehole to obtain the fracture development and permeable rate of the rock masses based on the water absorption calculation results. The test results provide basic data to evaluate the quality and effect of the grouting curtain. Lugeon tests were carried out in the verification boreholes and the adjacent grouting boreholes.

According to the longitudinal distribution of the permeable rate obtained from the Lugeon tests, the upper rock masses had relatively high permeable rates due to the present of faults, and this was the layer where corrosion cracks and karst caves developed. The permeable rates of the lower rock masses were relatively low, and fractures did not develop, thus allowing it to serve as an aquifuge. The weighted average permeable rates of the verification boreholes were significantly lower than the adjacent grouting boreholes, indicating that the grouts in the grouting areas had a good karst fissure filling effect (Table 2).

Table 2. Lugeon test results of boreholes.

Verification Borehole	Burial Depth of Test Section (m)	q (Lu)	$\bar{q}$ (Lu)	Grouting Borehole	Burial Depth of Test Section (m)	q (Lu)	$\bar{q}$ (Lu)
JCK1	29.00–60.00 60.00–80.00 80.00–100.00	2.445 2.534 2.303	2.170	K19	20.50–40.00 40.00–52.30 52.30–60.50 59.00–82.00 82.00–110.00 112.00–137.03	61.939 30.700 43.089 11.008 2.770 3.632	20.114
	100.00–120.00 120.00–142.95	1.940 1.567		K20	29.60–45.00 45.00–80.00 80.00–100.00 100.00–138.97	128.563 17.841 8.436 1.383	25.847
JCK2	32.00–60.00 60.00–80.00 80.00–100.00	1.717 1.905 1.933	1.481	K23	40.00–47.20 47.20–63.00 63.00–71.00 71.00–100.00 100.00–120.00 120.00–140.74	17.845 21.408 6.262 2.936 2.739 2.238	6.980
	100.00–120.00 120.00–140.74	1.722 0.783		K24	40.00–59.00 59.00–76.00 76.00–100.00 100.00–141.67	19.390 2.757 4.190 0.247	5.175
JCK3	29.60–47.00 47.00–67.00 67.00–87.00	4.814 3.962 2.178	2.643	K30	25.00–45.00 45.00–60.00 60.00–91.00 90.00–120.20 120.00–149.77	29.450 2.817 0.878 14.309 0.729	8.915
	87.00–107.00 107.00–125.83	2.710 0.722		K31	32.70–60.00 60.00–91.00 91.00–106.00 102.00–126.13	3.457 1.239 12.024 1.104	3.637

q refers to the permeable rate, and $\bar{q}$ refers to the weighted average permeable rate.

5.4. Groundwater-Air Pressure Monitoring on Both Sides of the Curtain

There were four groundwater-air pressure monitoring points on both sides of the curtain; SK2 and SK3 were located in the northern part of the curtain, while SK1 and SK4 were located in the southern part of the curtain. The monitoring data showed that (Figure 12):

(1) The groundwater level on both sides of the curtain were obviously different before and after grouting, and the groundwater level on the north side of the curtain was significantly higher than that on the south side. After the curtain grouting was completed, the groundwater level of borehole SK1 rose by 4.2 m, while boreholes SK2 and SK3 rose by 11.2 m and 11.9 m, respectively. The groundwater level difference on both sides of the curtain increased from 8.0 m to 14.9 m.

(2) The response of the groundwater level on both sides of the curtain to the drilling construction was obviously different. In October 2019, the borehole SK4 drilling construction on the south side of the curtain had a great impact on groundwater; the groundwater level of the adjacent monitoring borehole SK1rose and fell sharply, with a maximum variation of 5.3 m, while the groundwater level at the north side of the curtain was nor disturbed.

(3) The response of groundwater level on both sides of the curtain to rainfall were obviously different. After being recharged by rainfall infiltration, the recharge range on the north side of the curtain was large, and the groundwater level had risen, while the groundwater level on the south side of the curtain has slowly decreased due to the drainage of the quarry.

In summary, the curtain cut off the groundwater runoff channel and changed the groundwater recharge conditions on both sides of the curtain, resulting in different dynamic changes in the groundwater level on both sides of the curtain.

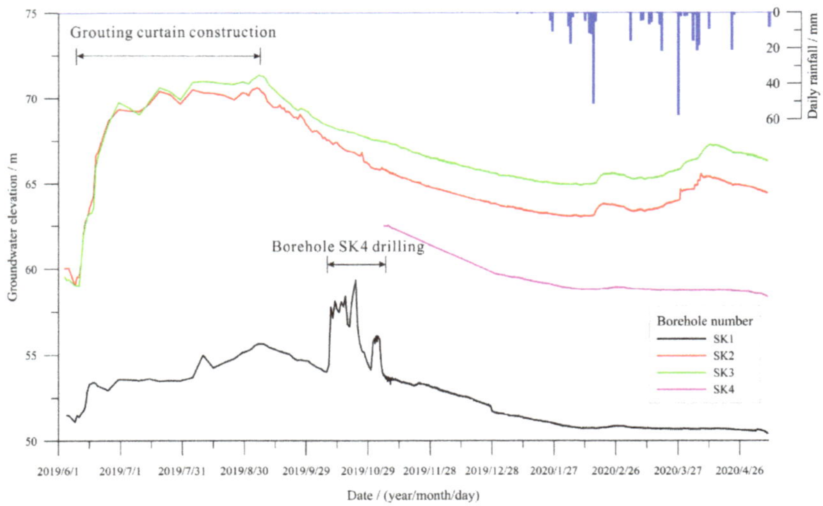

Figure 12. Monitoring curves of groundwater level on both sides of the curtain.

5.5. Water Plugging Effect at Inflow Points

After completion of the curtain, the water channel connected to the main water inflow point was filled with grout, and the soil contained cement slurry (Figure 13). At the remaining two water inflow points, water still flowed out of the fracture flow channels, but the water inflow was reduced from 3500 to 500 m^3/day.

(**a**)

(**b**)

Figure 13. Photos of water inflow points in different periods. (**a**) Photo of water inflow points exposed after pumping in mining pit. (**b**) Photo of main water inflow point after grouting.

6. Discussions

We conducted some seepage deformation tests on the soil of the monitoring borehole and found that the fluctuation velocity of groundwater-air pressure during sinkhole formation has a certain correlation with the critical water flow velocity of soil seepage deformation. We will carry out further research on the direct correlation between these two values, and expect to obtain the setting law of sinkhole monitoring and early warning threshold.

7. Conclusions

The following conclusions were drawn in this paper.

(1) Groundwater-air pressure monitoring is an effective method for monitoring sinkhole hazards, and can capture the abnormal groundwater-air pressure fluctuations before

and during the process of sinkhole formation. The abnormal values of groundwater-air pressure fluctuation velocity can reach tens of times the normal fluctuation values, which can be used as a monitoring index for the monitoring and early warning of sinkhole formation.

(2) Grouting curtains are an effective means of controlling water inflow in quarries, as they can effectively control the hazards posed by water flowing through fractures during the quarrying process and greatly reduce damage to the surrounding geological environment. After the curtain construction was completed, the grouting effect was checked by the variation of the curve shape of the high-density resistivity method physical property parameters before and after grouting, and the permeability of the curtain could be verified by combining the results of verification boreholes and Lugeon test.

(3) The evaluation results show that the curtain cut off the groundwater runoff channel and changed the groundwater recharge conditions on both sides of the curtain, resulting in different dynamic changes in the groundwater level on both sides of the curtain; a significant water level difference formed on both sides of the curtain, and the inflow rate reduced from 3500 m^3/day to approximately 500 m^3/day. However, groundwater will penetrate the karst environment and widen the existing flow paths and create new ones. It is recommended to strengthen groundwater monitoring and make remediation measures based on monitoring results.

Author Contributions: Formal analysis, Z.T.; Investigation, Z.T., L.C., Y.W. and L.G.; Data curation, G.Q.; Writing—original draft, Z.T.; Writing—review & editing, Z.T., L.S. and D.J. All authors have read and agreed to the published version of the manuscript.

Funding: The authors received no financial support for the research, authorship, and/or publication of this article.

Institutional Review Board Statement: Not applicable.

Informed Consent Statement: Not applicable.

Data Availability Statement: The data that support the findings of this study are available from the corresponding author upon reasonable request.

Conflicts of Interest: The authors declare no conflict of interest.

References

1. Kastning, E.H. Quarrying in Karst: Geotechnical estimation of environmental risk. In *Sinkholes and the Engineering and Environmental Impacts of Karst*; American Society of Civil Engineers: Reston, VA, USA, 2008; pp. 704–713.
2. Pan, Z.Y.; Jiang, X.Z.; Lei, M.T.; Guan, Z.D.; Gao, Y.L. Mechanism of sinkhole formation during groundwater-level recovery in karst mining area, Dachengqiao, Hunan province, China. *Environ. Earth Sci.* **2018**, *77*, 799.1–79910. [CrossRef]
3. Wang, H.; Li, L.; Li, J.P.; Sun, D.A. Drained expansion responses of a cylindrical cavity under biaxial in-situ stresses: Numerical investigation with implementation of anisotropic S-CLAY1 model. *Can. Geotech. J.* **2022**, *60*, 198–212. [CrossRef]
4. Foose, R.M. Ground-water behavior in the Hershey Valley, Pennsylvania. *Geol. Soc. Am. Bull.* **1953**, *64*, 623–645. [CrossRef]
5. Kath, R.L.; Mcclean, A.T.; Sullivan, W.R.; Humphries, R.W. Engineering impacts of karst: Three engineering case studies in Cambrian and Ordovician carbonates of the Valley and Ridge Province. *Int. J. Rock Mech. Min. Sci. Geomech. Abstr.* **1996**, *33*, 85A.
6. Luo, R.; Zheng, X.Z.; Yi, S.M. Current situation and prevention of karst surface-collapse geological disaster in Guangzhou Huadu district. *Chin. J. Geol. Hazard Control.* **2012**, *23*, 72–75.
7. Li, T.Y. The Research on the Karstic Water Control of Depression Pits. Ph.D. Thesis, Guangxi University, Nanning, China, 2016.
8. Arjwech, R.; Ruansorn, T.; Schulmeister, M.; Everett, M.E.; Thitimakorn, T.; Pondthai, P.; Somchat, K. Protection of electricity transmission infrastructure from sinkhole hazard based on electrical resistivity tomography. *Eng. Geol.* **2021**, *293*, 106318. [CrossRef]
9. Burston, M.R.; Memon, B.A.; Green, D.S. Reducing Conduit Water Flow into a Quarry in North-Central Alabama: A Case Study. In *Sinkholes and the Engineering and Environmental Impacts of Karst*; American Society of Civil Engineers: Reston, VA, USA, 2008; pp. 640–647.
10. Han, W.W.; Li, S.C.; Zhang, Q.S.; Zhang, X.; Liu, R.T.; Zhang, W.J.e. A comprehensive analysis method for searching weak zones of grouting curtain in mines. *Chin. J. Rock Mech. Eng.* **2013**, *32*, 512–519.

11. Gao, X.T. High Impermeability Composite Curtain Grouting Scheme and Permeability Detection Method. *Mod. Min.* **2020**, *36*, 223–225.
12. Wang, J. A New Technology of Groundwater Flow Cut off by Grout Curtain for Karst Mineral Deposit. *Min. Res. Dev. S* **2006**, *1*, 151–153.
13. Yang, Q.; Gao, G.F.; Han, G.L.; Yao, H.M. Study on Construction Technology of Mine Curtain Grouting for Cave Formation. *Explor. Eng. (Rock Soil Drill. Tunn.)* **2010**, *37*, 67–69.
14. Sun, K.G.; Li, S.C.; Xu, W.P.; Gong, L. On Grouting Techniques for Preventing Water Inflows in Karst Tunnels. *Mod. Tunn. Technol.* **2015**, *52*, 178–183.
15. Yang, Z.; Zhao, Q.; Huang, B.R. Grouting Control Technology and Engineering Application of Strong Dynamic Channel of Mine Curtain. *Min. Res. Dev.* **2020**, *40*, 117–121.
16. Shao, W.; Sun, Q.X.; Xu, X.; Yue, W.H.; Shi, D.D. Durability life prediction and horizontal bearing characteristics of CFRP composite piles in marine environments. *Constr. Build. Mater.* **2023**, *367*, 130116. [CrossRef]
17. Wei, H.M.; Li, X.; Sun, B.T.; Zhou, S.; Niu, J. Exploration and analysis of geophysical methods in curtain grouting water control. *Geophys. Geochem. Explor.* **2021**, *45*, 245–251.
18. Jiang, X.Z.; Lei, M.T. Monitoring technique and its application of karst groundwater-air pressure in karst collapse. *Carsologica Sin.* **2018**, *37*, 786–791.
19. Dong, H.B.; Wang, C.L. Development and application of 2D resistivity imaging surveys. *Earth Sci. Front.* **2003**, *10*, 171–176.
20. Ezersky, M.G.; Eppelbaum, L.V.; Al-Zoubi, A.; Keydar, S.; Abueladas, A.; Akkawi, E.; Medvedev, B. Geophysical prediction and following development sinkholes in two Dead Sea areas, Israel and Jordan. *Environ. Earth Sci.* **2013**, *70*, 1463–1478. [CrossRef]
21. Satitpittakul, A.; Vachiratienchai, C.; Siripunvaraporn, W. Factors influencing cavity detection in Karst terrain on two-dimensional (2-D) direct current (DC) resistivity survey: A case study from the western part of Thailand. *Eng. Geol.* **2013**, *152*, 162–171. [CrossRef]

sustainability

MDPI

Article

Research on Deformation and Failure Law of the Gob-Side Roadway in Close Extra-Thick Coal Seams

Shengquan He [1,2], Le Gao [1], Bin Zhao [3,4,*], Xueqiu He [1], Zhenlei Li [1], Dazhao Song [1], Tuo Chen [5], Yanran Ma [1] and Feng Shen [1]

1 Key Laboratory of Ministry of Education for Efficient Mining and Safety of Metal Mine, University of Science & Technology Beijing, Beijing 100083, China
2 State Key Laboratory of Coal Mining and Clean Utilization, Beijing 100013, China
3 School of Safety Engineering, China University of Mining and Technology, Xuzhou 221116, China
4 China Coal Datong Energy Co., Ltd., Datong 037038, China
5 Department of Mining and Materials Engineering, McGill University, Montreal, QC H3A 0E8, Canada
* Correspondence: zhaobin_tashan@163.com

Abstract: To reveal the deformation and failure law of the gob-side roadway (GSR) and the main influencing factors in close extra-thick coal seams, the research methods of field monitoring, theoretical analysis, and numerical simulation are adopted in this paper. Field monitoring data shows that microseismic events occur and accumulate frequently in the surrounding rock and some overlying key layers of the GSR. Large deformation is experienced in the middle part of roadway near the solid coal side, the middle and upper parts of the roadway near the coal pillar side, and the roadway floor. The overlying strata of the GSR are fractured to form a composite structure as "low-level cantilever beam and high-level masonry beam". The coal pillar is squeezed and effected by the composite beam structure and the rotation moment M, causing serious bulge in middle and upper part of the coal pillar side. The stability of the solid coal side of the roadway is affected by the stress transferred from gangue contact point. Numerical simulation shows that the immediate roof and key layer breakage are induced by the mining of the 30,501 working face. Shear and tension failures happen in the GSR due to overburden subsidence and rotary extrusion. The stress and displacement at the middle and upper of the roadway on the coal pillar side are larger than the other area. Compared with the solid coal side, the coal on the coal pillar side is obviously more fractured, with a lower bearing capacity. The peak stress in the coal pillar shows up 2 m away from the roadway, which is close to the length of bolt support. The mining-induced stress and the stress transferred from gangue contact point are the direct reasons for solid coal bulge beside the roadway. The peak stress on the solid coal side is located 7 m away from the roadway, at the gangue contact point where overburden fractures. The overburden strata loads and the transferred stress near the gangue contact point are transferred from the sides to the roadway floor. Their coupling effect with the in situ horizontal stress acts as the force source for the plastic floor heave.

Keywords: near coal seams; gob-side roadway; microseismicity; overburden fracture; roadway deformation

Citation: He, S.; Gao, L.; Zhao, B.; He, X.; Li, Z.; Song, D.; Chen, T.; Ma, Y.; Shen, F. Research on Deformation and Failure Law of the Gob-Side Roadway in Close Extra-Thick Coal Seams. *Sustainability* **2023**, *15*, 2710. https://doi.org/10.3390/su15032710

Academic Editor: Jianjun Ma

Received: 20 December 2022
Revised: 31 January 2023
Accepted: 31 January 2023
Published: 2 February 2023

1. Introduction

With the increasing demands of coal resources and the construction of resource-saving coal mines, the implementation of narrow coal pillars to arrange working faces has become a more popular choice of more mines [1–3]. However, large deformation and damage of the GSR has also become a large problem, plaguing safe production in mines, especially for close extra-thick coal seams.

Researchers have carried out a lot of research on the deformation and damage laws of GSRs and achieved fruitful results. Yang et al. [4] studied the bearing capacity of narrow coal pillars in the GSR during the process of mining by combining SMP criterion and numerical simulation. Wang et al. [5] studied the fracturing law of overburden during the process of multi-coal mining to determine the relationship between overburden structure failure and GSR deformation using UDEC simulation. Xu et al. [6] analyzed the damage characteristics of the GSR during fully-mechanized caving mining. They concluded that the non-uniform stress distribution during mining is the direct reason for the asymmetric roadway deformation and damage. Liu et al. [7] analyzed the failure mechanism and control technology of the GSR retaining in a short-distance coal seam. The damage and separation of the roof are the main causes of roadway instability. He et al. [8] researched the surrounding rock deformation mechanism during reuse of the gob side entry, and concluded that the surrounding rock failure is mainly affected by the advance abutment pressure. Shi et al. [9] researched the stability of surrounding rock in GSR driving in a deep and thick seam and concluded that the displacement and plastic zone of the surrounding rock of the straight wall semi-circular arch roadway are smaller than the corresponding deformation value of the surrounding rock of the rectangular roadway. Xiong et al. [10] put forward a concept of cyclic damage of the GSR in inclined coal seams, which was verified using numerical simulation. Wang et al. [11], Ma et al. [12], and Gao et al. [13] simulated the stress field, displacement, and plastic zone distribution of rock surrounding roadway when mining coal pillars with different widths. Wang et al. [14] analyzed the relationship between GSR damage and the basic roof fracturing location. They proposed that the roadway stability has a direct relationship with the breaking location of the key block. Additionally, the breaking laws of key blocks were analyzed with the roadway arranged at different positions during the face mining. Mo et al. [15] considered that the strength difference of the surrounding rock would impact the roadway and proposed that the displacement and failure mode of the surrounding rock were sensitive to deformation modulus. Li et al. [16] studied the relationship of in situ stress with roadway deformation and failure and proposed that the roadway layout should be parallel to the maximum horizontal stress. They believed that the concentrated stress at the top corner of the working face should be reduced in advance and the coal seam should be reinforced immediately after excavation. Zhao et al. [17] established a mechanical model of overburden strata on the GSR at the full-mechanized caving face in extra-thick coal seams. The mechanical source of the asymmetric failure of the GSR roof was obtained. Hua et al. [18] reported the mechanical connection between the roadway roof and the gob side roof and obtained the roadway stability mechanism and main roof stability criterion under the dynamic static coupling effect. Zhang et al. [19] studied the deformation and failure law of the GSR in deep mining condition and determined that the face mining and the coal damage are the direct causes of roadway deformation and damage. Guo et al. [20] studied the deformation characteristic of surrounding rock in GSR retaining formed automatically, and determined that the length of roof suspension is the main factor for stress distribution and deformation of the surrounding rock. Zha et al. [21] analyzed the deformation and failure characteristics of the rock surrounding the GSR with narrow coal pillars. The reasonable coal pillar size and roadway excavation time after mining were obtained. They also put forward surrounding rock control technology and effective roadway-side sealing technology. Zhou et al. [22] aimed at the stability of the floor, established a mechanical model to analyze the stability of the roadway floor heave by analogy with the basement heave of the deep foundation pit. It provides a model reference for analyzing the problem of roadway floor heave. Li et al. [23] researched the distribution of in situ stress in a deep roadway by adopting the combination of the stress measurement and comprehensive experimental research method and concluded that the horizontal stress is the key factor for determining the failure depth of the floor. Li et al. [24], Liu et al. [25], Zhang et al. [26], and Xue et al. [27] pointed out that the high stress, low bearing capacity of surrounding rocks and the existing support system are not effective to restrain the rock weathering, which are direct reasons for strength weakening in the deep roadway during the mining process. They proposed that using grouting and high-strength anchor support is an effective measure to ensure roadway stability.

To sum up, our predecessors carried out a lot of research on the deformation and damage laws of GSRs. However, the deformation and failure laws of GSRs and the main influencing factors in close extra-thick coal seams are not clearly revealed.

In this study, a typical GSR of the 30,503 working face in Tashan Coal Mine was investigated to determine deformation and failure characteristics of the GSR via field measurement, theoretical analysis, as well as numerical simulation. The structural characteristics of overburden and its influence on the stress and displacement fields of surrounding rock are obtained. The deformation and failure laws of the GSR and the main influencing factors are revealed. The research results could lay a theoretical foundation for the stability control of the GSR in similar mining conditions.

2. Engineering Background and Roadway Deformation Characteristics

2.1. Temporal and Spatial Evolution Laws of Microseismic Events

#3–5 coal seam of Tashan Coal Mine is extra-thick coal seam, with an average burial depth of 435 m. The Coal Mine is located in Yungang District, Datong City, Shanxi Province, China. The mine field area is 8.146 km^2. The layout of the working face and roadway is shown in Figure 1 (for more details about engineering background please refer to Zhao et al. 2022 [28]). The 30,503 working face passed through the filled roadway on 19 January 2021. To determine the damage characteristics of the GSR before and when the working face passed through the filled roadway, the microseismic monitoring data during these periods were analyzed. The temporal-spatial distribution law of microseismic events near the GSR are shown in Figures 2 and 3. Orlecka-Sikor reported that the concentration area of microseismic events is mainly the stress concentration area [29]. It can be seen from the plan distribution of microseismic events in Figure 2 that the frequency and energy of microseismic events increase significantly as the working face is near the filled roadway, and they are more concentrated near the GSR. This indicates that mining towards the filled roadway area led to the increasing of stress concentration and damage on the gob side. The sectional distribution of microseismic events is shown in Figure 3; the events increase significantly and their range expands along the strike obviously as the work faces get closer to filled roadway, indicating higher mining-induced stress when the working face passed through the filled roadway.

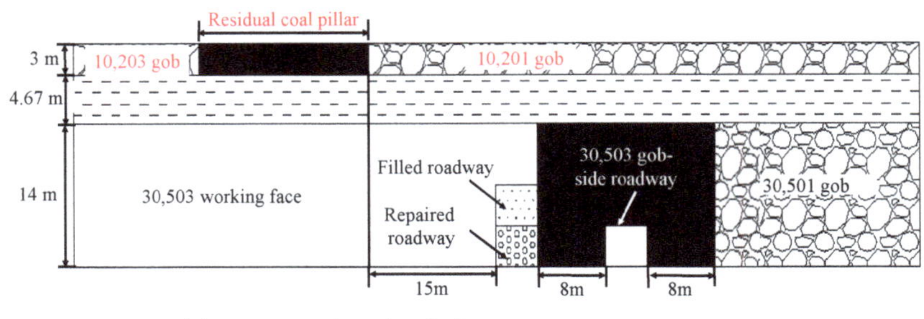

Figure 1. Layout of the 30,503 working face [28].

The overlying microseismic events of the GSR are mainly concentrated in the range 100 m above the coal seam floor. The key layer with thickness of 51 m has frequent fracturing events. The accumulation and frequency of microseismic events in the surrounding GSR, roof, and floor of the adjacent roadway indicate that the area is in a high stress concentration area, which could be the direct cause of large deformation of the roadway.

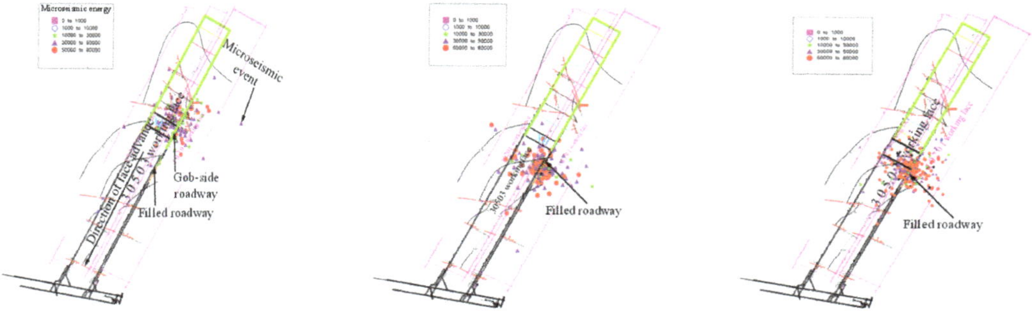

(**a**) November 2020 (**b**) December 2020 (**c**) January 2021

Figure 2. Microseismic events distribution before and during the working face passes through the filled roadway.

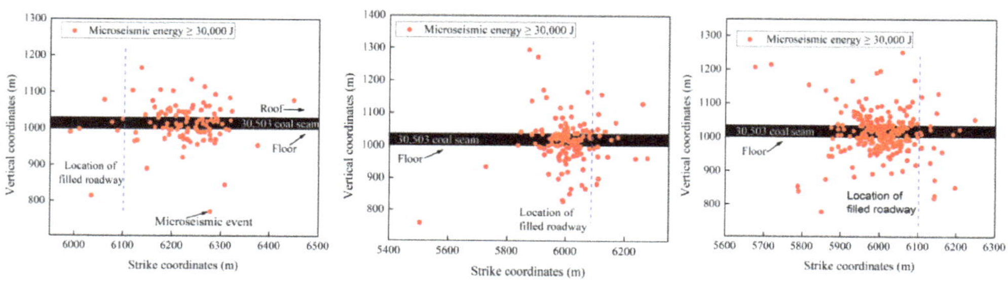

(**a**) November 2020 (**b**) December 2020 (**c**) January 2021

Figure 3. Microseismic events distribution before and during the working face passes through the filled roadway. Blue dotted line is the location of filled roadway.

2.2. Deformation and Damage Characteristics of the Roadway

As shown in Figure 4, the on-site observation shows that the 30,503 working face of the GSR is severely affected by mining. The deformation of the middle part of the solid coal side, the middle and upper part of the coal pillar side, and the bottom of the roadway are considerably large. During mining the 30,503 working face, the GSR received relatively serious deformation and damage. Floor heave begins to appear on the floor within 200 m in front of the working face, reaching 0.5–0.7 m relevant height, causing the concrete laid on the floor to crack. After the roadway floor is repaired, the amount of heave on the floor continuously increases as the working face gets closer. Both sides of the roadway have serious bulge, and the deformation for the middle and upper part of the coal pillar side is more dramatic compared with the solid coal roadside. Although the roadway is supported by bolts, the large pressure still causes the roadway sides to fracture, and embeds bolts and anchor cable trays in the coal seam.

The 30,503 working face of Tashan Coal Mine was mined to the filled road area on 19 January 2021. The deformation data of the two roadsides before and after the working face passes through the filled roadway were monitored to reveal the influence of the filled roadway on the deformation and damage of the GSR, and the results are shown in Figure 5. The maximum deformation of the two sides of GSR is 200 mm before the working face passes through the filled roadway, and the maximum deformation of the GSR is 660 mm when mining to the corresponding area of the filled roadway. It means that the existence of the filled roadway affects the stability of the GSR in the process of coal seam mining. The support strength should be increased in the area where the filled roadway exists.

(**a**) Coal pillar side bulge

(**b**) Solid coal side bulge

Figure 4. Deformation and damage characteristics of GSR.

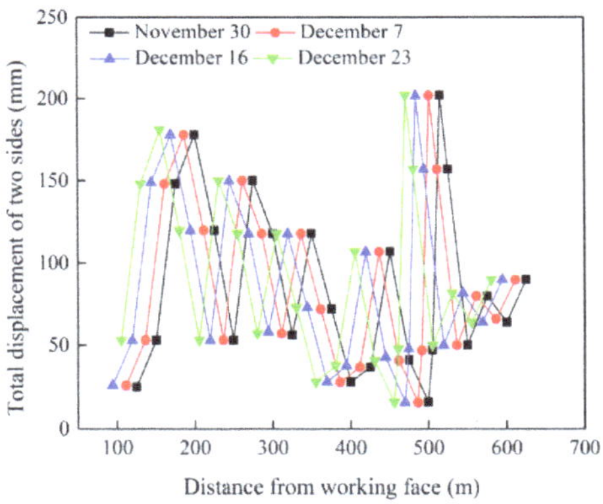

(**a**) Before mining to filled roadway

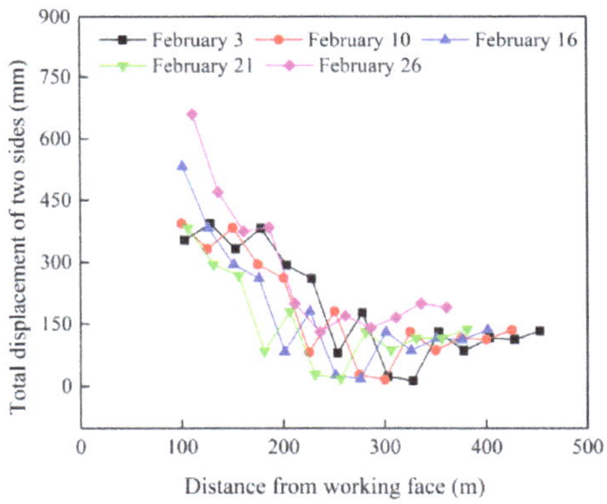

(**b**) Mining to filled roadway

Figure 5. Displacement of two sides of the GSR at different monitoring days.

3. Mechanical Analysis of the Influence of Overburden Structure on the Roadway Stability

3.1. Fracturing Law of Overburden Structure of GSR under Repeated Mining

#3–5 coal seam of the Tashan Coal Mine is typical close extra-thick coal seams. Its distance to the gob of overlying #2 coal seam is only 4.6 m. Since the mining thickness of #3–5 coal seams in one step is 14 m, the gob is greatly increased compared with the #2 coal seam. Therefore, the structural stability of strata overlying the GSR under repeated mining needs to be explored.

There is a 7.9 m thick low-level old roof and a 51.2 m thick high-level old roof overlying the Tashan Coal Mine from borehole histograms. After the #2 coal seam was mined, the low-level old roof was fractured to form a masonry beam structure. After the 30,501 working face is mined, whether the low-level old roof can form a stable structure can be assessed by Formula (1). When the condition is met, the structure can be formed as a cantilever beam [30]:

$$\Delta > \Delta_{max} \tag{1}$$

where Δ is the rotation amount after the rock strata is broken, Δ_{max} is the limit rotation required for the breaking block of the key layer to be hinged to form a stable "masonry beam" structure.

$$\Delta = M_1(1-\mu) - \sum h_i(K_p - 1) > \Delta_{max} = h - \frac{ql^2}{kh\sigma_c} \tag{2}$$

where $\sum h_i$ is immediate roof thickness, K_p is rock fragmentation expansion coefficient, h is the thickness of broken rock stratum, k is dimensionless coefficient, l is periodic step distance of rocks, σ_c is compressive strength of rock, μ is coal caving loss rate, M_1 is coal seam thickness, q is overburden load, $q = \lambda H$, λ is unit weight of rock, H is burial depth of rock.

According to the measured data of Tashan Coal Mine, M_1 is 14 m, $\sum h_i$ is 18.19 m, K_p is 1.2, μ is 15%, the periodic step distance of the low-lever old roof is measured to be 19.8 m, q is 10,625 KN/m^3 and σ_c is 76 MPa. The above parameters are substituted to Formula (1), Δ = 6.44 m, Δ_{max} = 5.33 m. It shows $\Delta > \Delta_{max}$ to satisfy Formula (1). Therefore, the low-level old roof is broken to form a cantilever beam structure when the #3–5 coal seams are mined. The same theoretical calculation is used to get the rotation amount of the high-level old roof is 2.3 m, which is less than the maximum rotation amount of 3.6 m, forming a hinged masonry beam structure. From this, when the #2 coal seam and the 30,501 working face are mined, the overlying strata of the 30,503 GSR will form a "low-level cantilever beam and high-level masonry beam".

The breaking law of overlying rocks when the #2 and #3–5 coal seams are mined separately is analyzed to further determine the fracture law of the overlying high and low old roofs in the GSR under the influence of repeated mining. Firstly, the limit equilibrium theory is used to calculate the distance X_0 between the broken position of the low-level old roof and the coal wall of the working face, when the #2 coal seam is mined [31]:

$$X_0 = \frac{M_2 A}{2\tan\varphi_0}\ln\left[\frac{K\gamma H' + C_0/\tan\varphi_0}{C_0/\tan\varphi_0 + P_0/A}\right] \tag{3}$$

where, M_2 is coal seam mining height, γ is unit weight, A is coefficient of lateral pressure, φ_0 is internal friction angle, K is stress concentration coefficient, C_0 is cohesion, H' is mining depth. P_0 is the support resistance of the support on the upper section level roadway to the lower side.

Combined with the field measured data and Formula (3), it is calculated that the breaking position of the low-level old roof after mining the #2 coal seam is 4.6 m away from the coal wall of the 10,201 gob. The corresponding position of gangue contact point after the breaking of the key block B on the low-level old roof is 15.2 m away from the remaining coal pillars, 26.8 m horizontally away from the 30,501 gob, and 13.8–18.8 m away from the 30,503 GSR. The remaining coal pillar is 28 m away from the narrow coal pillar of the GSR. The broken length of the low old roof is 19.8 m. When the key block of the low old roof fails, the failure position between block C and the adjacent block is 7 m away from the 30,501 gob, located 1 m above the narrow coal pillar with width 8 m and close to the roadway side. The remaining rock blocks with a distance of 12.8 m or shorter will form a cantilever beam structure. Because the narrow coal pillar is affected by the fracture position of the low cantilever beam, stress concentration can be formed on the coal pillar, which will lead to greater damage to the coal pillar. The schematic diagram of the broken structure of the low-level old roof is shown in Figure 6.

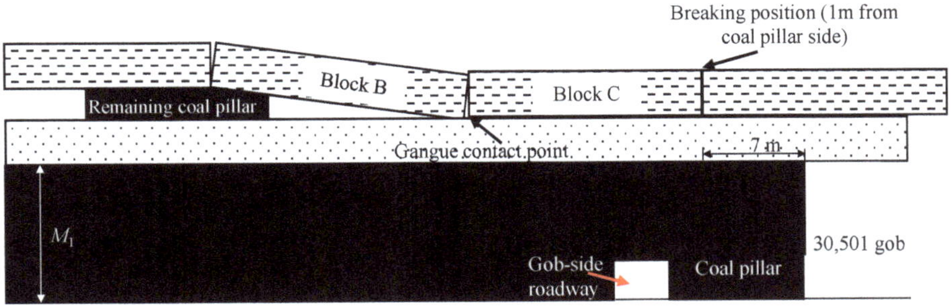

Figure 6. Sectional diagram of the rock layer overlying the GSR after the low-level old roof is broken.

The same calculation method is used to calculate that after mining the 30,501 working face, the periodic breaking length of the high-level old roof is 45.2 m, and the breaking position is 22 m away from the coal wall of the 30,501 gob. The corresponding position of the gangue contact point is 20 m away from the remaining coal pillar and 7–12 m away from the 30,503 GSR after the key block of the high-level old roof is broken. The distance between the breaking point of the high-level old roof and the gangue contact point on the low-level old roof is only 4.8m, according to theoretical calculation results. After the high-level rock beam breaks and sinks, part of the load acts on the gangue contact point, and the other part acts on the coal pillar, roof, and solid coal of the GSR. The load acting on the coal pillar will cause the roadway to deform. The stress around the gangue contact point will be concentrated and transferred to the #3–5 coal seams. The schematic diagram for the overlying rock layer of the GSR after the high-level old roof is broken is shown in Figure 7.

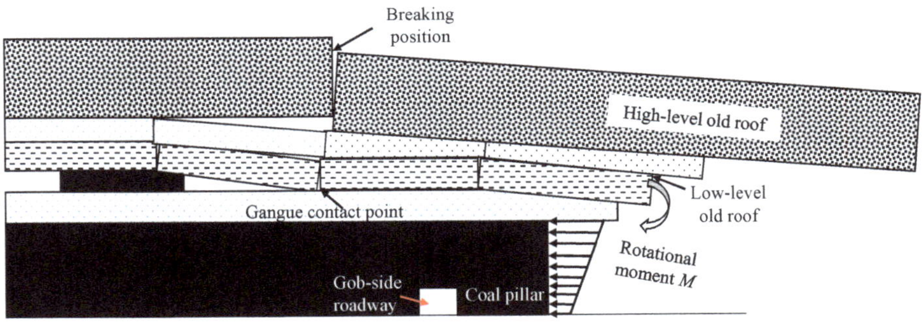

Figure 7. Schematic diagram of rock structure overlying GSR under repeated mining.

When the high-level old roof breaks and sinks on the lower cantilever beam, the cantilever beam will rotate under compressive stress and transfer a part of stress to the lower coal pillar. Combined with the research results of Qu et al. [32], when the high-level old roof breaks and sinks, the coal pillar is not only squeezed by overlying strata, but also is acted upon by the rotational moment M, as shown in Figure 7. The middle and upper part of the coal pillar are more obviously affected by the moment M, which is an important reason for the bulge of the middle and upper part of the coal pillar beside the GSR.

3.2. Stress Transfer Characteristics of Contact Gangue Point under the Action of Overburden Structure

According to the breaking law of overlying stratum of the GSR, the surrounding rock of the adjacent roadway is subject to the compressive stress of the high and low level old roof breaking on the coal body at the solid coal side and the high stress at the gangue contact point after the high and low level old roof breaking. The maximum transferred stress at the contact gangue point where the old roof is broken can be calculated by the following formula [32]:

$$\sigma_{\max} = \frac{\{\gamma(4H - 2L)[2L \tan\alpha + (H - L/2)\tan\alpha] + 2\gamma L^2 \tan\alpha}{4(a + h\cos\varphi\tan(\theta + \varphi) - h\sin\varphi)\tan^2\alpha} \quad (4)$$

where, a is the width of the coal pillar, h is the thickness of the middle layer between the #2 coal seam and the #3–5 coal seams, φ is the inclination angle of the coal seams, α is the gangue angle, L is the inclination length of the gob, and θ is the stress transferring angle.

The length of the stress transferring area at the gangue contact point is r. Taking the coal wall of the 30,501 gob on the right as the starting point, the stress transfer angle θ is shown in Figure 8. The transferred stress continues to increase from 0 on both sides of the contact gangue point and reach to the maximum value σ_{max} at the contact gangue point. The length r can be calculated through the model.

$$r = a + h\cos\varphi\tan(\theta + \varphi) - h\sin\varphi \quad (5)$$

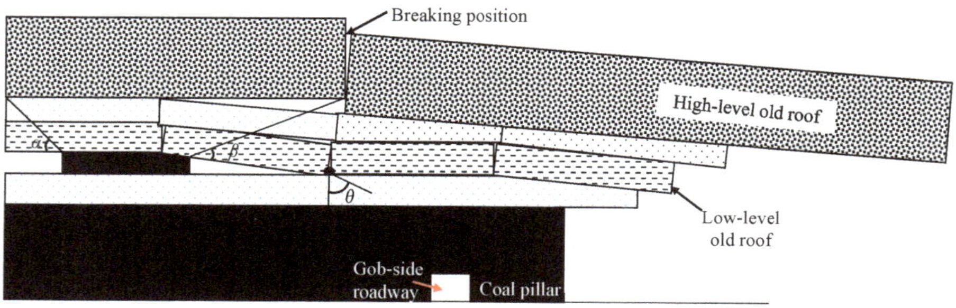

Figure 8. Sectional diagram of the stress transfer model at the contact gangue point.

According to the actual situation of the Tashan Coal Mine, the vertical distance between the #2 coal seam and the #3–5 coal seam is 4.6 m, the width of the narrow coal pillar is 8.0 m, θ is 30°, the dip angle of the coal seam φ is 2°. According to existing research results [32], the gangue angle α is 75° and the rock fracture angle β is 84°. The above data are substituted into Equations (4) and (5) to obtain σ_{max} at the contact gangue point is 23.74 MPa and r is 10.5 m. Combined with theoretical analysis, the breaking position of the old roof is 7–12 m away from the GSR, and the influence range of the contact gangue point stress is 10.5 m. Therefore, the transfer stress of the contact gangue point can affect the stability of the GSR after the high-level old roof breaks and touches the gangue.

Section 2.1 concluded that microseismic events occur frequently in the high and low-level old roof area above the GSR. The failure area of the high and low-level old roof obtained from the theoretical analysis is basically consistent with the microseismic event gathering area monitored on site, which verifies the accuracy of the theoretical analysis results. Simultaneously, the field monitoring shows that the upper and middle part of the coal pillar side of the roadway are seriously bulged compared to other areas, which further verifies the theoretical analysis that the composite structure of low-level cantilever beam and high-level masonry beam squeezed these areas. The stress transferring at the gangue contact point is the main reason for side wall bulge on the solid coal side in the field.

4. Numerical Simulation Research on Deformation and Damage Law of GSR in Close Extra-Thick Coal Seams

4.1. Model Construction and Research Plan

The numerical model is established in FLAC3D modelling software, taking the geological conditions of the Tashan Coal Mine as the engineering background, as shown in Figure 9. The model size is 640 m × 400 m × 130 m (XYZ), with 530,504 grids. The model uses hexahedral mesh. For more details about numerical model and boundary conditions, please refer to Zhao et al. 2022 [28].

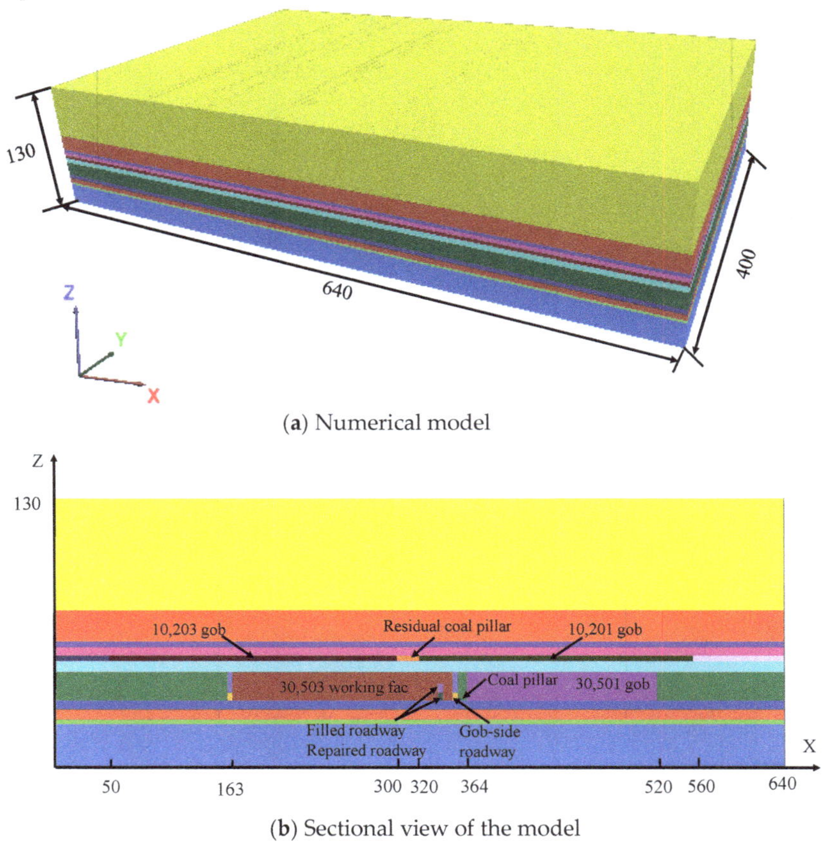

Figure 9. Numerical model.

The Mohr-Coulomb constitutive model [33] is used to reveal the deformation and damage law of the GSR in the close extra-thick coal seams. The physical and mechanical parameters used in the numerical model are from the experimental results of the coal and rock sample in the Tashan Mine [28].

To reflect actual mining conditions, the following mining sequence is simulated: (1) the 30,501 working face is mined with the historical mining sequence of the working face. The stress field and plastic zone of surrounding rock of the GSR of the 30,503 working face are analyzed before mining. (2) The impact of mining on the 30,503 working face stability is studied.

4.2. Distribution Laws of Stress Field and Plastic Zone Surrounding Roadway after GSR Excavation

Modelled excavation is carried out according to the actual mining sequence. The modelled roadway supports are consistent with the actual mine support method. The distribution laws of the stress field and plastic zone for rocks surrounding the GSR after GSR excavation are shown in Figures 10 and 11. As can be seen in Figure 10, the stress in the remaining coal pillar area of the overlying #2 coal seam is highly concentrated, and transferred the stress to the low-level coal mass and upper overlying stratums, with a maximum value of 47 MPa. The main reason for the stress concentration in the coal pillar area is that after the coal masses on the left and right sides of the coal pillar are mined, the overlying strata are in a suspended state, and the overlying stratum is mainly supported by the remaining coal pillars. It can be seen from Figure 11, under the compressive stress of overlying strata, the coal pillar undergoes obvious shear failure, which extends to the lower coal seams. The coal and rock mass near the GSR is affected by the overlying #2 coal seam. The mining of the right side of the 30,501 working face led to obvious pressure relief. Simultaneously, due to the mining of the 30,501 working face on the right side of the GSR, the 4.6 m-thick immediate roof collapsed and drove waste rocks from the mining settlement of the #2 coal seam to further collapse. Moreover, overlying sandstones with 51 m thick also fractured integrally to act on gob and narrow coal pillar. At this time, narrow coal pillars are the main support point of the overlying strata. Under the overlying rock subsidence and rotary extrusion, the stress in the narrow coal pillar area of the #3–5 coal seams increases with a maximum of 17 MPa in the junction area between the cantilever beam and the coal mass. The stress extends to the middle and upper area of the GSR. Shearing and tensioning damage occurred in the GSR due to the subsidence and rotary extrusion of overlying strata. The subsidence and extrusion of overlying key blocks on the coal pillar side is the main reason for the plastic failure of the side, as well as an important cause for the heave occurring in the middle and upper part of the coal pillar side, as obtained in Section 3.1. The vertical stress transmitted by the key blocks acts on the solid coal and coal pillars, which becomes a crucial cause of the floor heave. Based on the above analysis, it can be seen that after the adjacent working faces are mined, the squeezing effect after the key blocks above the GSR sank and broke is an important reason for the deformation and failure of the surrounding rock of the roadway.

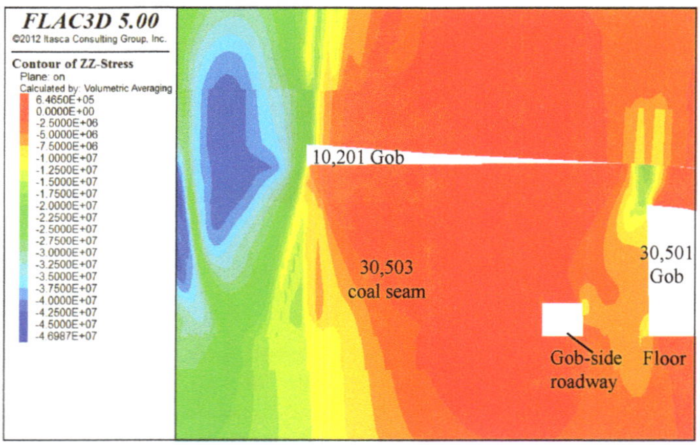

Figure 10. Stress contour of surrounding rocks after GSR excavation.

4.3. Influence Law of Mining Action on Stability of GSR in 30,503 Working Face

4.3.1. Variation Law of Advanced Support Pressure under the Influence of Mining

The field measurement shows that the advanced support pressure is one of the main reasons for the deformation and damage of the rock surrounding the GSR. Figure 12a shows the advanced support pressure distribution in the middle of the coal seams under different mining distances of the 30,503 working face. Figure 12a indicates that there are three stress peaks. The left and right peaks correspond to the stop mining line of overlying coal seam #2 and the open-off cut corresponds to the remaining coal body area, and the middle is the advanced support pressure peak area. When the 30,503 working face is mined 60 m, 80 m, 90 m, 100 m, and 110 m, respectively, the peaks of the advanced support pressure were 20 MP, 28 MPa, 31.2 MPa, 31 MPa, and 29 MPa, respectively. Stress peaks basically occur at the position 20 m ahead of the working face. When the working face is mined to 90–100 m, the advanced support pressure reaches the peak value, which means the working face has been fully mined. When mining to 100 m, the peak value in vertical stress distribution (Figure 12b) and the plastic zone distribution (Figure 12c) are also basically located at the position of 20 m ahead of the working face.

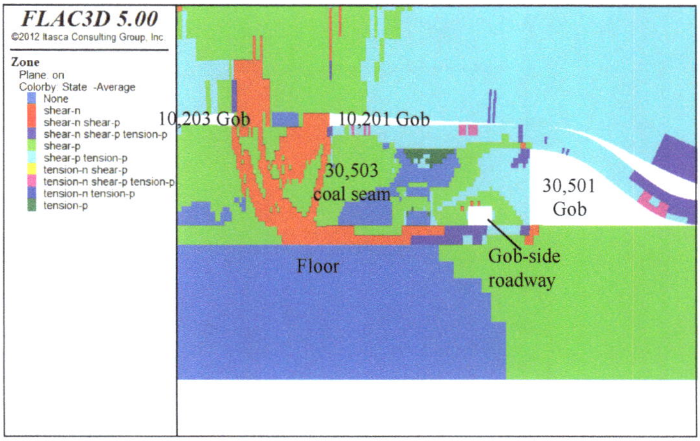

Figure 11. Distribution of plastic zone surrounding the GSR.

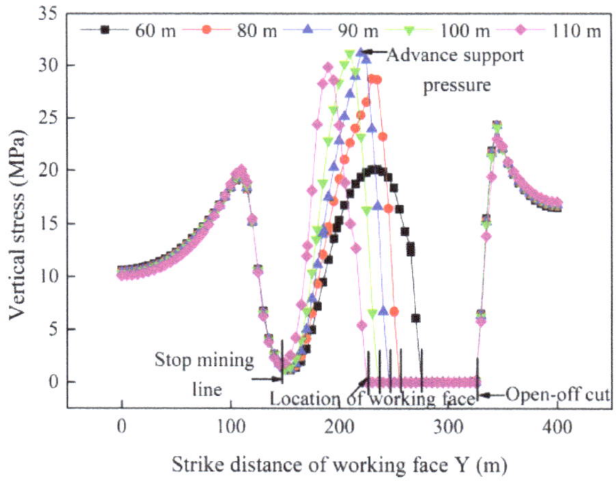

(**a**) Advance support pressure at different mining lengths

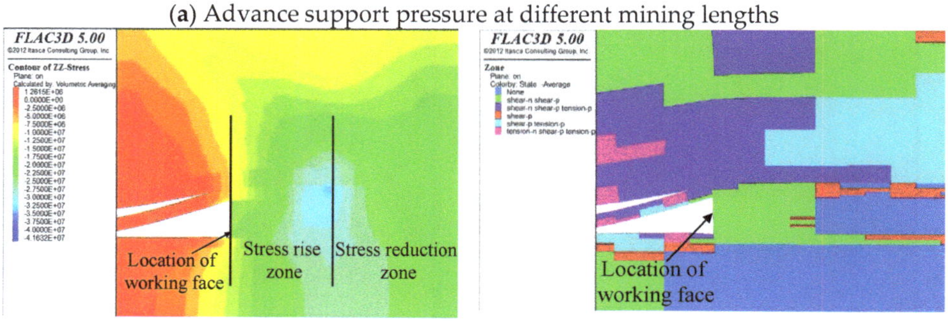

(**b**) Distribution of stress field (**c**) Distribution of plastic zone

Figure 12. Distribution of stress and plastic zone.

A total of six monitoring lines were arranged along the working face on the solid coal and coal pillar sides of the GSR, to further reveal the influence of mining the 30,503 working face on the surrounding rock stability of the GSR. The monitoring lines on the solid coal side are arranged 2 m away from the roadway with an interval of 2 m as lines 1, 2, and 3 sequentially from inside to outside. The coal pillar side is arranged similarly, as shown in Figure 13.

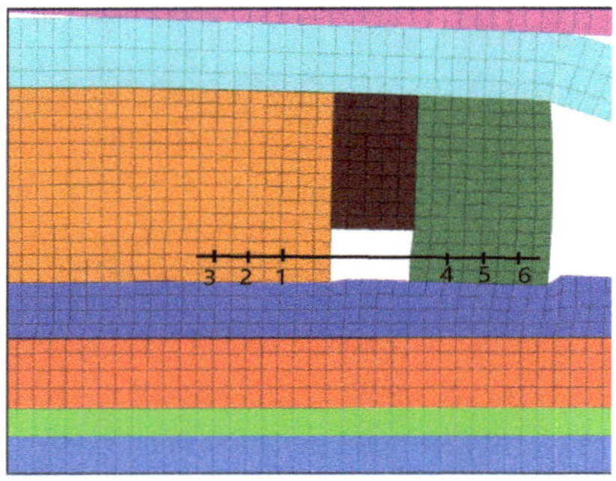

Figure 13. Layout of monitoring lines in two sides of the GSR. Different colors represent different coal and rock beds.

The vertical stress distribution of two roadway sides at different mining distances is simulated as shown in Figures 14–16:

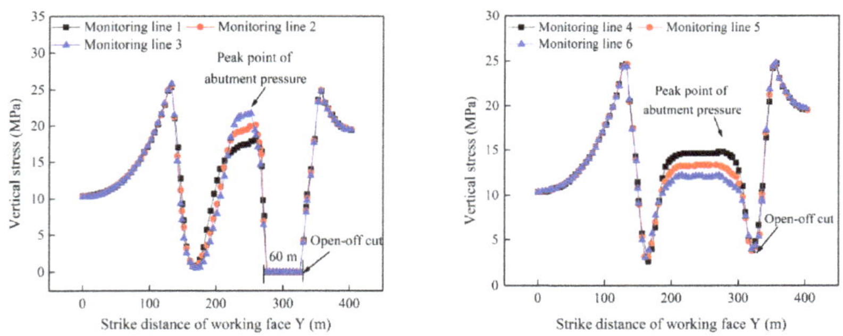

(**a**) Stress distribution on the solid coal side (**b**) Stress distribution on the coal pillar side

Figure 14. Vertical stress distribution when mining at 60 m.

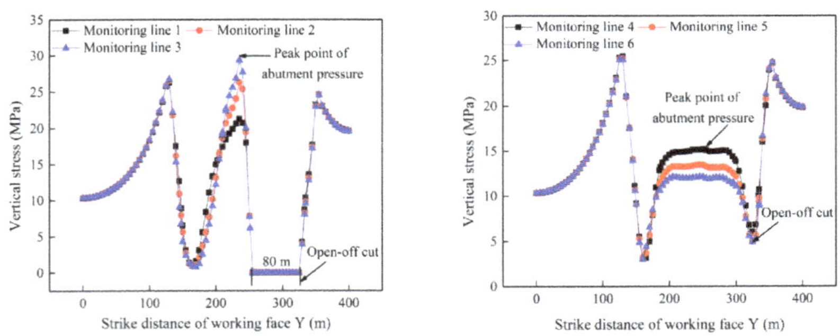

(**a**) Stress distribution on the solid coal side (**b**) Stress distribution on the coal pillar side

Figure 15. Vertical stress distribution when mining at 80 m.

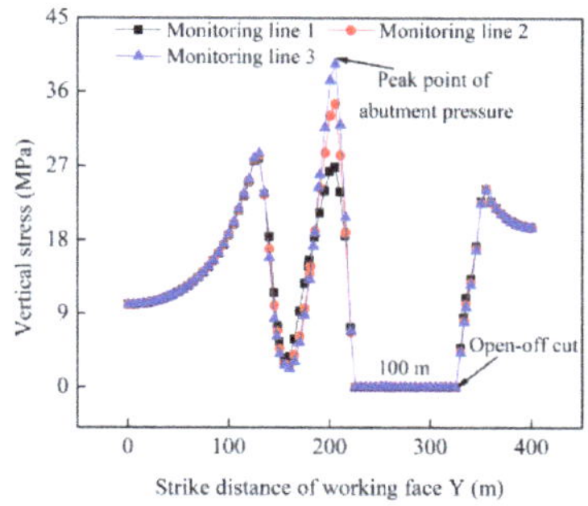

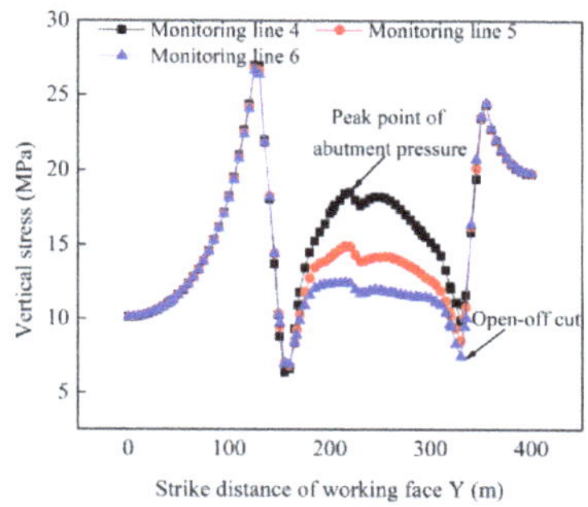

(**a**) Stress distribution on the solid coal side (**b**) Stress distribution on the coal pillar side

Figure 16. Vertical stress distribution when mining at 100 m.

The advance support pressure on the solid coal side of the roadway changes with the continuous advance of the mining face. When the working face is at 60 m, 80 m, and 100 m, respectively, the peak values of the advance support pressure are 21.6 MP, 29.3 MPa, and 39.1 MPa, respectively. With the advance of the working face, the stress peaks basically appear about 20 m in front of the working face, and the stress peaks also increase with the mining length. This is mainly due to the mining of the working face in the proximity of the 30,501 gob. Under the coupling action of the advanced bearing pressure and the lateral bearing pressure, the stress peak value increases continuously.

Comparing supporting pressures on the coal pillar side of the roadway under different mining distances, it can be seen that when the working face is at 60 and 80 m, the distribution trend of the supporting pressure is similar. The peak value of the bearing pressure on the coal pillar is stable at 12.1–15.2 MPa. However, when the working face is mined to 100 m, the peak support pressure is 18.5 MPa. The support pressure on the monitoring line 6 near the gob basically does not change, indicating that the coal in this area is very broken. The supporting pressures along the survey line 4 and line 5 have the increase of 2.77 MPa and 3.7 MPa, respectively, compared with that when the mining is at 80 m. This indicates that the compression effect from the overlying strata on the coal pillar near the gob is more dominant due to mining, which is in line with the theoretical analysis results.

The advanced support pressure variation in two sides of the roadway during mining shows that the peak value basically appears at about 20 m ahead of the working face. Also, the coal on the coal pillar side is significantly more broken than the solid coal side with lower carrying capacity. Within 8 m of the solid coal side, the coal body is more broken as it is closer to the roadway side and the coal pillar is more fragmented as it is closer to the gob. This is an important cause for roadway side walls bulging during mining.

4.3.2. Stress Distribution around the Roadway during Mining

Monitoring lines are arranged surrounding the roadway to further quantify the stress changes of the two sides and the floor of the roadway during mining. The length of the two-side monitoring lines is 8 m. Line A starts parallel to the roof. A total of eight lines are arranged with 1 m interval. Four 6 m long monitoring lines are arranged in the floor with an interval of 1 m. The layout of the monitoring lines is shown in Figure 17. The simulated working face is mined 100 m forward and the stress distribution law on the monitoring section at 0, 20, and 30 m ahead of the working face is extracted. The results are shown in Figure 18.

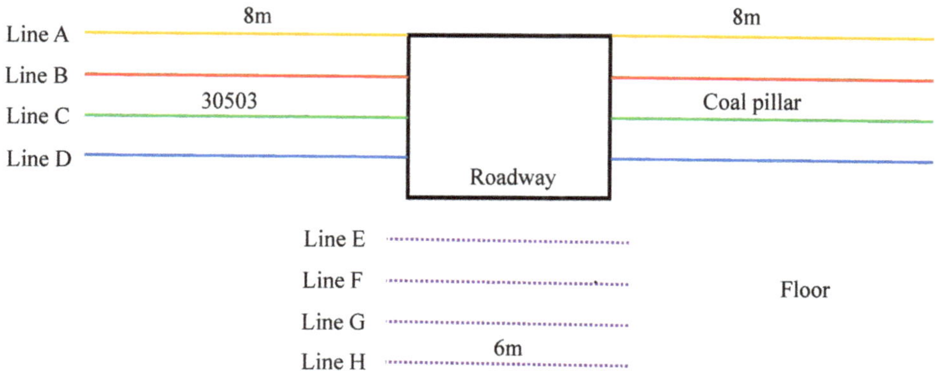

Figure 17. Sectional layout of the roadway monitoring lines.

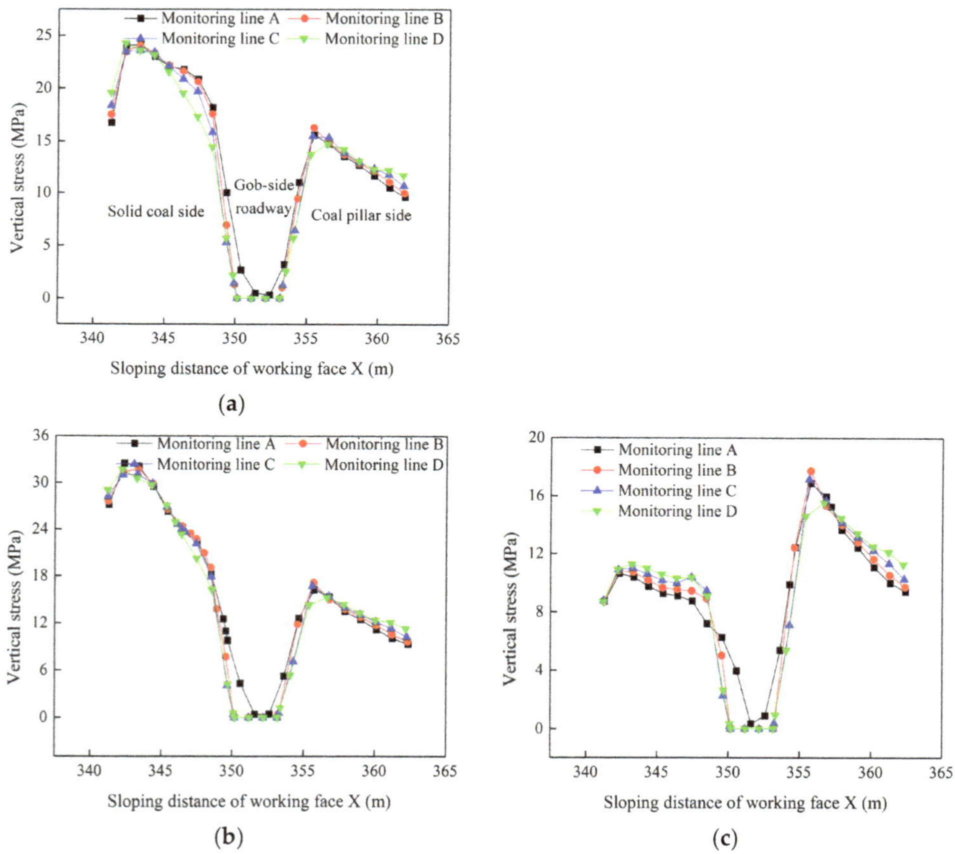

Figure 18. Monitored vertical stresses around the GSR. (**a**) 0 m ahead of the working face, (**b**) 20 m ahead of the working face, (**c**) 30 m ahead of the working face.

Figure 18 shows when the distance of mining to the working face is 100 m, the stress changing trend within 20 m ahead of the working face in solid coal side is basically the same, and its bearing capacity increases as it moves away from the roadway and its peak shows up at 7 m away from the roadway. The peak position is close to the theoretical calculation result in Section 3.1 located at the gangue contact point. The rise of the stress in solid coal side at an advance of 30 m is significantly lower than that within 20 m. It can be seen that the range of 20 m ahead of working face is the main mining affected area. Figure 18 shows that the peak stress on the coal pillar side within 30 m of the working face appears at 2 m near the roadway. 2.2 m bolts have been used for the two sides of roadway. The protection range is just in the peak stress range of the coal pillar side, which is an important reason for the damage and the bulge of bolts on the coal pillar side. The breaking position of the low-level old roof is 1 m away from the coal pillar, as calculated in Section 3.1. The numerical simulation is similar with the theoretical results. It can be seen from the stress distribution within 30 m, the peak stress in the roadway sidewall on the coal pillar side always appears at the position of monitoring lines A, B, and C. This stress distribution characteristic can explain the phenomenon for the heave and serious damage of the middle and upper coal pillar side onsite.

The stress of the solid coal side is significantly larger than that of the coal pillar side within 20 m ahead of the working face. The stress of the coal pillar side is larger at 30 m. This is because the coal pillar within the mining-influenced range is disturbed by repeated mining, yielding a reduced bearing capacity. The mining causes the high-level rock beam to break and in contact with the waste rock. In this case, the solid coal becomes the main bearing structure after the overlying key block is broken with a high stress concentration area appearing on the side. In the area not affected by mining, the coal pillar is relatively complete with the coal pillar as the main bearing structure after the overlying key block is broken. Thus, the stress increases under the compression of broken key block.

Figure 19 shows the monitoring results of the horizontal stress at each monitoring line in the two sides of the roadway. The measured horizontal stress of the solid coal at 20 m and 30 m ahead of the working face is greater than that on the coal pillar side of the roadway. The horizontal stress peaks appear at 2 m from the roadway on the coal pillar side and at 7 m on the solid coal side, which is consistent with the vertical stress peak position. The high horizontal stress on the solid coal side of the roadway is the direct cause for the heave on the solid coal side.

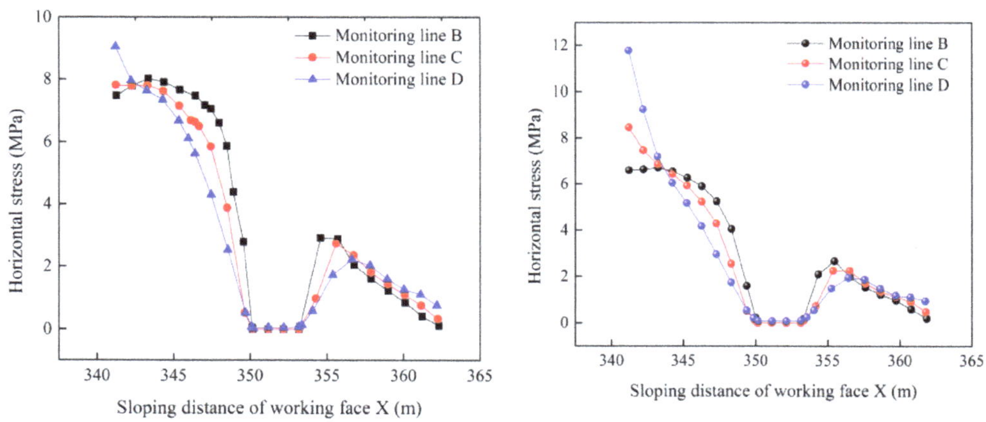

(**a**) 20 m ahead of the working face (**b**) 30 m ahead of the working face

Figure 19. Horizontal stress distribution of two sides of GSR.

When simulated 30,503 working face is at 100 m, the vertical stress distribution of each survey line within 4 m of the roadway floor at 0, 20, and 30 m ahead of the working face is shown in Figure 20. When mining the working face, the stress under the floor presents a "U-shaped" distribution. The stress value on the solid coal side is higher within the mining influence range, while the stress value on the coal pillar side is higher away from the mining influence. The overlying load is transmitted to the roadway floor through the two sides of the roadway, and the high vertical load on the floor is the direct cause for the continuous occurrence of the floor heave.

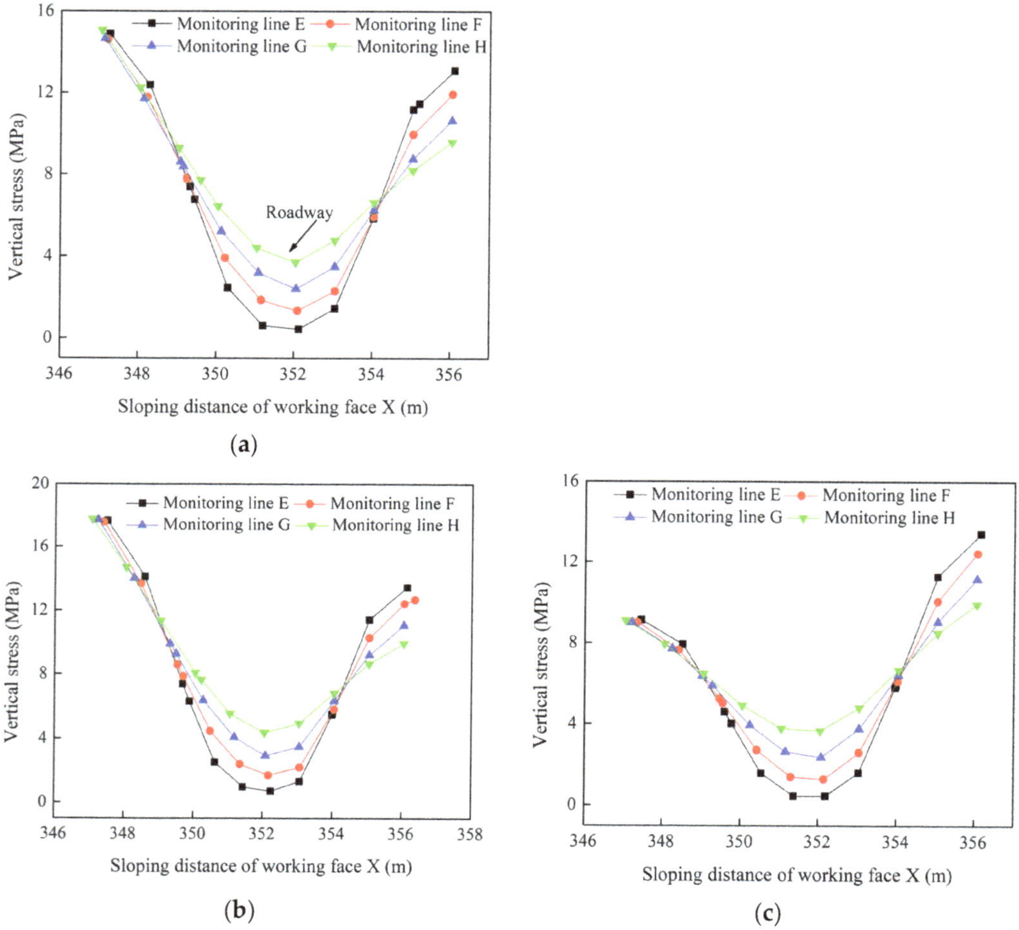

Figure 20. Vertical stress distribution at GSR floor. (**a**) 0 m ahead of the working face, (**b**) 20 m ahead of the working face [28], (**c**) 30 m ahead of the working face.

4.3.3. Displacement Distribution around the Roadway Affected by Mining

When 30,503 working face is at 100 m, the horizontal displacement changes along each survey line in the roadway sides within 30 m ahead of the working face is shown in Figure 21. The changing trend within 30 m is basically the same on two sides of the roadway. The maximum deformation on the solid coal side is 0.69 m, located at 1 m away from the sidewall. The deformation on the gob side increases continuously in coal pillar with a maximum displacement of 1.4 m. The deformations along the monitoring line B and C are larger than that along line D on the coal pillar side, which is basically consistent with the roadway deformation characteristics on site. Based on the above analysis, it can be concluded that the large deformation of the deep coal is an important cause for the occurrence of bulging on coal pillar side of the roadway. Also, there are only 2.2 m bolts for the coal pillar, which cannot control the heave effectively.

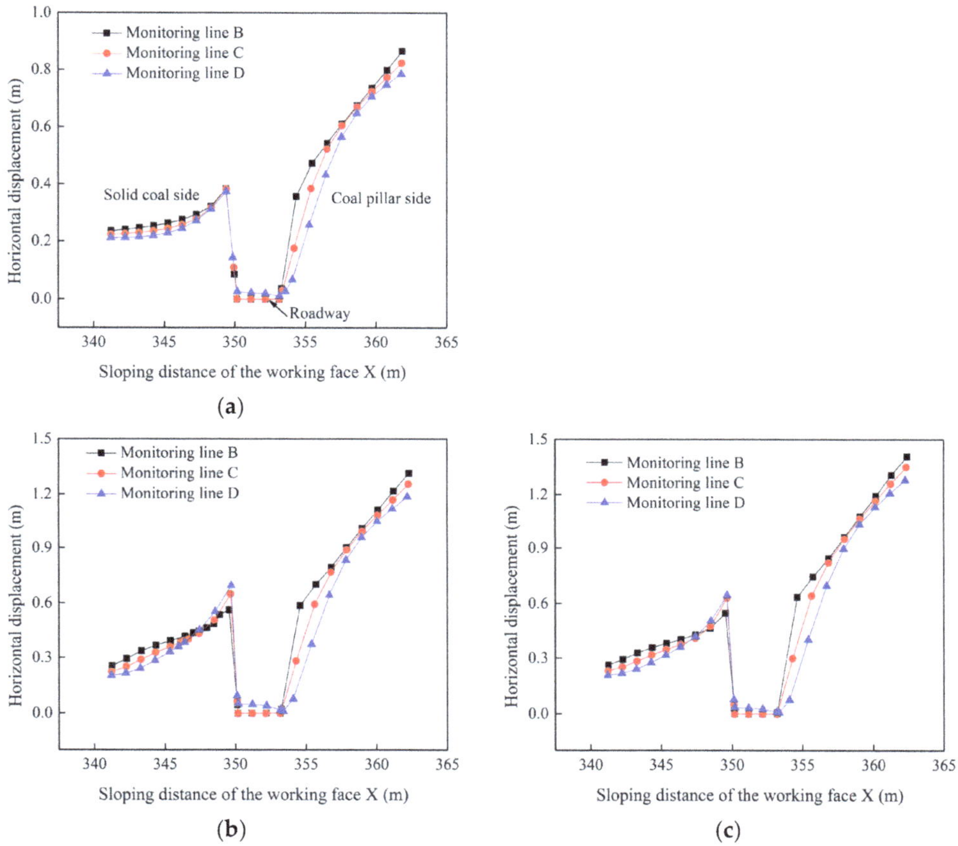

Figure 21. Horizontal displacement distribution of the roadway two sides. (**a**) 0 m ahead of the working face, (**b**) 20 m ahead of the working face [28], (**c**) 30 m ahead of the working face.

According to the research results obtained from the study. The middle part of roadway near the solid coal side, and the middle and upper of the roadway on coal pillar side within 20 m in front of the working face shall be strengthened on the basis of the existing support. Improving the rock bearing capacity on the coal pillar side and using longer bolts could be effective measures to ensure the stability of the GSR.

The vertical displacement changes in the roadway floor within 30 m ahead of the working face when 30,503 working face is at 100 m is shown in Figure 22. The maximum floor heave of 0.41 m appears in the middle roadway 20 m ahead of the working face. Due to high vertical stress in side walls, the roadway continues to heave, and the displacement gradually decreases with increasing distance from the floor. When it is 4 m away from the roadway floor, the displacement is basically negligible. Under the high unbalanced load, the floor is prone to heave plastically under the coupling action of the horizontal compression and the vertical stress. Meanwhile, the vertical stress generated by the breaking of the high and low old roofs acts on rocks surrounding the 30,503 GSR, and the load of overlying rocks is transmitted to the floor through the two sides of the roadway. Under the high stress, two corners of the floor displace downward, and hereby the roadway is squeezed and heave. When the high stress exceeds the ultimate strength of the floor, the floor will be squeezed out and fractured in the roadway.

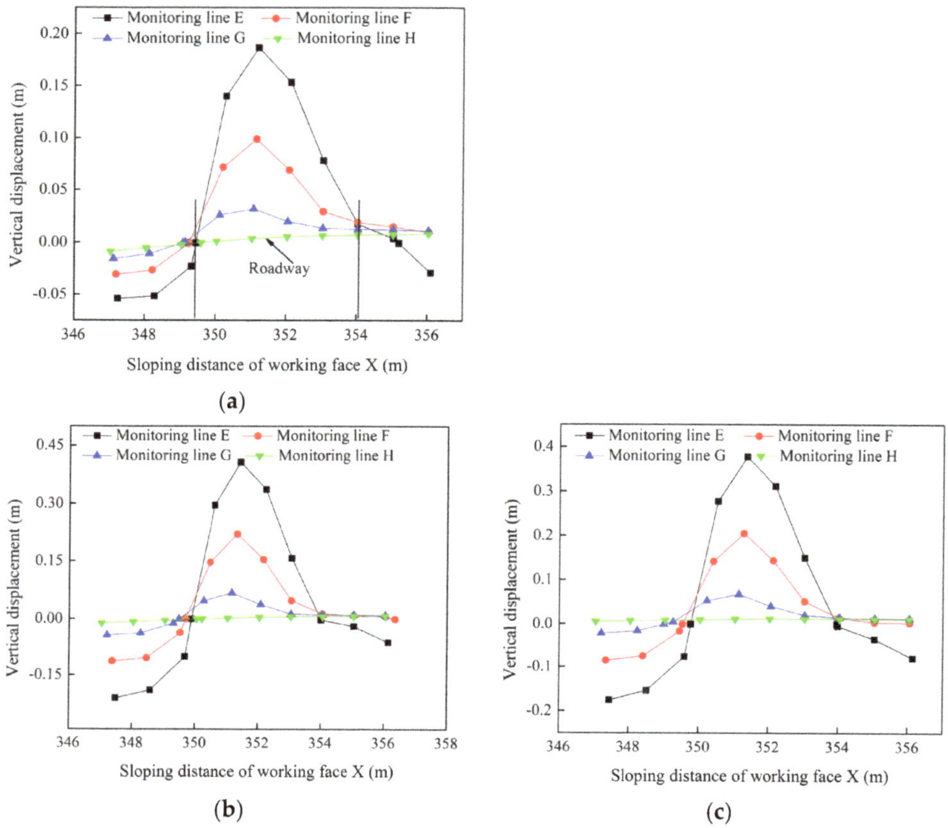

Figure 22. Vertical displacement distribution of the roadway floor. Two black line in (**a**) used to explain the location of roadway. (**a**) 0 m ahead of the working face, (**b**) 20 m ahead of the working face [28], (**c**) 30 m ahead of the working face.

5. Conclusions

(1) The fracturing and subsidence of the key layers accompanied by dense microseismic events have an important impact on the stress field surrounding the GSR. The deformation of the middle roadway on the solid coal side, the middle and upper roadway on the coal pillar side and the bottom of the roadway are relatively large.

(2) Under the influence of mining, the overlying strata of the GSR are broken to form a composite structure as "low-level cantilever beam and high-level masonry beam". The breaking and subsidence of the structures are the main cause for the stress and instability increase in the roadway. The breaking position of the low-level old roof is 1 m away from the coal pillar. The maximum transferred stress at the contact gangue point is 23.7 MPa.

(3) Shear and tension failure happen in the GSR due to overburden subsidence and rotary extrusion. The stress and displacement at the middle and upper coal pillar side of the roadway are relatively large. The peak stress in solid coal side is about 1.8 times that of the coal pillar side. The peak stress on the coal pillar side appears 2 m away from the roadway, which is close to the length of the bolt support. The maximum horizontal displacement on the coal pillar side is about 2 times that of the solid coal side.

(4) The mining-induced stress and the stress transferring at the gangue contact point are the direct reasons for side wall bulge on the the solid coal side. The peak stress in solid coal is 7 m away from the roadway, which is at the gangue contact point where overburden

breaks. The coupling effect of the overburden stress, the transferred stress near the gangue contact point, and the horizontal stress is the force source for the floor heave. The research results are of great significance to guide the targeted roadway support of similar mines.

Author Contributions: Writing original draft, S.H.; Conceptualization, S.H., L.G. and B.Z.; Methodology, S.H., L.G. and F.S.; Funding acquisition, S.H.; Validation, L.G. and T.C.; Investigation, B.Z., Z.L., X.H. and Y.M.; Formal analysis, X.H., Z.L., D.S. and F.S.; Supervision, X.H.; Writing—review &editing, Y.M. and D.S.; All authors have read and agreed to the published version of the manuscript.

Funding: This research was funded by the National Natural Science Foundation of China (No. 52204197), the Postdoctoral Research Foundation of China (Project No. 2021M700371), the Fundamental Research Funds for the Central Universities (Project No. FRF-TP-22-111A1), the Open fund of State Key Laboratory of Coal Mining and Clean Utilization (Project No. 2021-CMCU-KF013).

Institutional Review Board Statement: Not applicable.

Informed Consent Statement: Not applicable.

Data Availability Statement: Data can be obtained by contacting the corresponding author.

Conflicts of Interest: The authors confirm that there are no conflict of interest associated with this publication.

References

1. Han, C.L.; Yang, H.Q.; Zhang, N.; Deng, R.J.; Guo, Y.X. Zoning Control Technology of the gob-side roadway Driving with Small Coal Pillar Facing Mining in a Special Isolated Island Working Face: A Case Study. *Appl. Sci.* **2021**, *11*, 10744. [CrossRef]
2. Mu, H.W.; Wang, A.H.; Song, D.Z.; Su, D.F. Failure Mechanism of the gob-side roadway under Overlying Coal Pillar Multiseam Mining. *Shock. Vib.* **2021**, *2021*, 4403456. [CrossRef]
3. Chen, A. Width Design of Small Coal Pillar of Gob-Side Entry Driving in Soft Rock Working Face and Its Application of Zaoquan Coal Mine. *Adv. Civ. Eng.* **2021**, *2021*, 9999957. [CrossRef]
4. Yang, K.; Gou, P.F. Research on Reasonable Width of Coal Pillars in High Strength Mining Roadway in Wantugou Mine. *Geotech. Geol. Eng.* **2021**, *39*, 2065–2073. [CrossRef]
5. Wang, J.; Ning, J.G.; Tan, Y.L.; Hu, S.C.; Guo, W.Y. Deformation and failure laws of roadway surrounding rock and support optimization during shallow-buried multi-seam mining. *Geomat. Nat. Haz. Risk* **2020**, *11*, 191–211. [CrossRef]
6. Xu, X.H.; He, F.L.; Li, X.B.; He, W.R. Research on mechanism and control of asymmetric deformation of gob side coal roadway with fully mechanized caving mining. *Eng. Fail. Anal.* **2021**, *120*, 105097–105109. [CrossRef]
7. Liu, H.Y.; Zhang, B.Y.; Li, X.L.; Liu, C.W.; Wang, C.; Chen, D.Y. Research on roof damage mechanism and control technology of gob-side entry retaining under close distance gob. *Eng. Fail. Anal.* **2022**, *138*, 106331. [CrossRef]
8. He, F.L.; Xu, X.H.; Qin, B.B.; Li, L.; Lv, K.; Li, X.B. Study on deformation mechanism and control technology of surrounding rock during reuse of gob side entry retaining by roof pre-splitting. *Eng. Fail. Anal.* **2022**, *137*, 106271. [CrossRef]
9. Shi, J.J.; Feng, J.C.; Rui, P.; Zhu, Q.J. The research on stability of surrounding rock in gob-side entry driving in deep and thick seam. *Geotech. Geol. Eng.* **2022**, *40*, 3357–3364.
10. Xiong, X.Y.; Dai, J.; Ouyang, Y.B.; Shen, P. Experimental analysis of control technology and deformation failure mechanism of inclined coal seam roadway using non-contact DIC technique. *Sci. Rep.UK* **2021**, *11*, 20930–20954. [CrossRef] [PubMed]
11. Wang, Y.; Li, Y.M.; Yu, Z.L.; Zhang, H.; Ding, D. Research on the Surrounding Rock Control Technology of Gob-Side Entry with a Narrow Coal Pillar Reserved in a Fully Mechanized Caving Face with Large Mining Height. *Geotech. Geol. Eng.* **2021**, *40*, 285–300. [CrossRef]
12. Ma, Z.Q.; Chen, C.; Liang, X.C.; Chen, A.M.; Song, W.X. Field and numerical investigation on the stability of coal pillars of gob-side entry driving with top coal. *Arab. J. Geosci.* **2020**, *13*, 1193–1206. [CrossRef]
13. Gao, X.; Zhang, S.; Zi, Y.; Khan, S. Study on optimum layout of roadway in close coal seam. *Arab. J. Geosci.* **2020**, *13*, 746–759. [CrossRef]
14. Wang, X.F.; Lu, M.Y.; Gao, Y.H.; Luo, W.B.; Liu, W.G. Structural Mechanical Characteristics and Instability Law of Roof Key Block Breaking in the gob-side roadway. *Adv. Civ. Eng.* **2020**, *2020*, 6682303. [CrossRef]
15. Mo, S.; Sheffield, P.; Corbett, P.; Ramandi, H.L.; Oh, J.; Canbulat, L.; Saydam, S. A numerical investigation into floor buckling mechanisms in underground coal mine roadways. *Tunn. Undergr. Space Technol.* **2020**, *103*, 103497. [CrossRef]
16. Li, H.; Lin, B.Q.; Hong, Y.D.; Gao, Y.B.; Yang, W.; Liu, T.; Wang, R.; Huang, Z.B. Effects of in-situ stress on the stability of a roadway excavated through a coal seam. *Int. J. Min. Sci. Technol.* **2017**, *27*, 917–927. [CrossRef]
17. Zhao, Y.Q.; Li, X.B.; Hou, J.Q. Research on Asymmetric Failure Mechanism and Control Technology of Roadway Along Gob in Extra Thick Coal Seam. *Geotech. Geol. Eng.* **2021**, *40*, 1009–1021. [CrossRef]
18. Hua, X.Z.; Liu, X.; Huang, Z.G.; Yang, P.; Ma, Y. Stability mechanism of non-pillar gob-side entry retaining by roof cutting under the coupled static dynamic loading. *J. China Coal. Soc.* **2020**, *45*, 3696–3708.

19. Zhang, Y.; Xia, X.; Sun, X.M.; Song, P.; Wang, J.; Sun, S.J.; Chen, C. Research on Coupling Control Technology of Constant Resistance in Gob-side Entry Under Mined Gob of Recent Coal Seam Group. *Geotech. Geol. Eng.* **2020**, *38*, 4685–4699. [CrossRef]
20. Guo, P.F.; Zhang, X.H.; Peng, Y.Y.; He, M.C.; Ma, C.R.; Sun, D.J. Research on deformation characteristic and stability control of surrounding rock during gob-side entry retaining. *Geotech. Geol. Eng.* **2020**, *38*, 2887–2902. [CrossRef]
21. Zha, W.H.; Shi, H.; Liu, S.; Kang, C.H. Surrounding rock control of gob-side entry driving with narrow coal pillar and roadway side sealing technology in Yangliu Coal Mine. *Int. J. Min. Sci. Technol.* **2017**, *27*, 819–823. [CrossRef]
22. Zhou, X.M.; Wang, S.; Li, X.L. Research on theory and technology of floor heave control in semicoal rock roadway: Taking longhu coal mine in Qitaihe mining area as an Example. *Lithosphere* **2022**, *2022*, 3810988. [CrossRef]
23. Li, X.L.; Chen, S.J.; Wang, S. Study on in situ stress distribution law of the deep mine taking Linyi Mining area as an example. *Adv. Mater. Sci. Eng.* **2021**, *9*, 5594181. [CrossRef]
24. Li, G.; Ma, F.S.; Guo, J.; Zhao, H.J.; Liu, G. Study on deformation failure mechanism and support technology of deep soft rock roadway. *Eng. Geol.* **2020**, *264*, 105262–105276. [CrossRef]
25. Liu, D.J.; Zuo, J.P.; Wang, J.; Zhang, T.L.; Li, H.Y. Large deformation mechanism and concrete-filled steel tubular support control technology of soft rock roadway-A case study. *Eng. Fail. Anal.* **2020**, *116*, 104721. [CrossRef]
26. Zhang, W.; He, Z.M.; Zhang, D.S.; Qi, D.H.; Zhang, W.S. Surrounding rock deformation control of asymmetrical roadway in deep three-soft coal seam: A case study. *J. Geophys. Eng.* **2018**, *15*, 1917–1928. [CrossRef]
27. Xue, G.Z.; Gu, C.; Fang, X.Q.; Wei, T. A Case Study on Large Deformation Failure Mechanism and Control Techniques for Soft Rock Roadways in Tectonic Stress Areas. *Sustainability* **2019**, *11*, 3510. [CrossRef]
28. Zhao, B.; He, S.; He, X.; Gao, L.; Li, Z.; Song, D.; Shen, F. Research on Deformation and Failure Control Technology of a Gob-Side Roadway in Close Extra-Thick Coal Seams. *Sustainability* **2022**, *14*, 11246. [CrossRef]
29. Orlecka-Sikora, B. The role of static stress transfer in mining induced seismic events occurrence, a case study of the Rudna mine in the Legnica-Glogow Copper District in Poland. *Geophys. J. Int.* **2010**, *182*, 04672. [CrossRef]
30. Zhao, Y.Q. *Study on Narrow Time-Space Interval and Surrounding Rock Control of the Gob-Side Roadway for Fully-Mechanized Top Coal Caving Face in Extra-Thick Coal Seam*; China University of Mining and Technology: Beijing, China, 2019.
31. He, F.L.; Zhang, G.C. Deformation and failure mechanism and control technology of large section coal roadway subjected to severe mining dynamic load. *J. Min. Saf. Eng.* **2016**, *33*, 423–430.
32. Qu, X.K.; Jiang, F.X.; Wang, H.T.; Zhu, S.T.; Zhang, M.; Xu, K.; Zhang, W.T. Research on mechanism of rock burst induced by coal pillar failure in mine gob. *J. Min. Saf. Eng.* **2017**, *34*, 1134–1140.
33. Castelli, F.; Grasso, S.; Lentini, V.; Sammito, M.S.V. Effects of Soil-Foundation-Interaction on the Seismic Response of a Cooling Tower by 3D-FEM Analysis. *Geosciences* **2021**, *11*, 200. [CrossRef]

sustainability

MDPI

Article

Response Characteristics of Weak Current Stimulated from Coal under an Impact Load and Its Generation Mechanism

Dexing Li [1,2,3,*], Enyuan Wang [1,2], Dianqi Jin [3], Dongming Wang [1,2] and Wei Liang [3]

1 Key Laboratory of Gas and Fire Control for Coal Mines (China University of Mining and Technology), Ministry of Education, Xuzhou 221116, China
2 School of Safety Engineering, China University of Mining and Technology, Xuzhou 221116, China
3 Shenzhen Urban Public Safety Technology Research Institute Co., Ltd., Shenzhen 518000, China
* Correspondence: ldx3180@cumt.edu.cn

Abstract: Understanding the response law and mechanism of weak currents stimulated from coal under an impact load is significant for the prediction of coal bumps in deep coal mines. In this paper, the system for the weak current measurement of coal under an impact load is established and the response characteristics of weak currents induced by the deformation of coal under an impact load are investigated. Physical models are established to describe the process of charge transfer and explain the generation mechanism of those currents. The results show that a transient current is stimulated from the coal sample when an impact load is applied, and then, the current decays slowly, tending to be a stable value that is slightly greater than the background current. The weak current flows from the loaded volume to the unloaded volume of the coal and increases with the impact velocity in a negative exponential form. Analysis of weak currents using non-extensive entropy shows that the attenuation of the weak current obeys non-extensive statistical mechanics and the non-extensive parameter q is greater than 2. The carriers are mainly electrons, of which the distribution obeys the tip effect that electrons tend to enrich towards the tip of a crack. The generation mechanism of those weak currents induced by coal deformation is the instantaneous movement of electrons under a density difference caused by the tip effect. Research results can provide a new perspective to understand the electric phenomena of coal under an impact load as well as a new method for coal bump prediction.

Keywords: coal bump; transient current; impact load; mechanism; carriers

Citation: Li, D.; Wang, E.; Jin, D.; Wang, D.; Liang, W. Response Characteristics of Weak Current Stimulated from Coal under an Impact Load and Its Generation Mechanism. *Sustainability* **2023**, *15*, 2605. https://doi.org/10.3390/su15032605

Academic Editor: Gujie Qian

Received: 26 December 2022
Revised: 13 January 2023
Accepted: 29 January 2023
Published: 1 February 2023

1. Introduction

Coal bumps, a kind of underground dynamic disaster induced by the sudden release of elastic deformation energy accumulated in coal mass during mining [1,2], usually cause serious casualties and property loss [3]. A coal bump is closely related to the deformation aggravation of coal mass [4–6], of which the process is always accompanied by the emission of various physical signals, such as acoustic emission (AE), electromagnetic radiation (EMR), surface potential, and charge induction, which can reflect the damage evolution of coal [4,7,8]. Therefore, the measurement of the physical signals emitted from coal during deformation is an important basis for coal bump prediction and early warning [9].

Applying a mechanical load on coal and rock materials can stimulate a weak current [10–16], which is also called the pressure stimulated current (PSC). Stavrakas et al. [11] studied the characteristics of PSC from marbles through laboratory experiments, which showed that an obvious weak current can be generated after the stress reaches $0.6\sigma_f$ (σ_f is the peak stress), and the peak current is considered to be proportional to the loading rate. Afterward, Triantis et al. [13] obtained a more general conclusion on the relationship between PSC and the mechanical behavior of loaded marble specimens: obvious weak currents can be observed only when the applied load exceeds the yield stress, and the generation of weak current from rocks is mainly induced by the change in Young's modulus. Kyriazopoulos et al. [17] believe that the PSC is directly proportional to the strain rate, and the occurrence of fracture is accompanied by the sudden increase in PSC. The experimental results from a study by Freund et al. [14,15,18,19] on weak currents from igneous rocks (granite, anorthosite, gabbro, etc.) show that a weak current is generated the moment a load is applied, and it can increase rapidly to a peak even at a very low stress level. Li et al. [16] believe that PSC response characteristics are closely related to the deformation stages of rock; they also found that the precursors of rock failure based on the PSC response are different for various stress states, such as under progressive loading and during a creep process. In recent years, the weak current technique has been introduced to study the deformation and damage process of coals. He et al. [20,21] measured weak currents of nanoamps on the surface of coal samples under uniaxial compression, confirming that weak currents can also be stimulated from loaded coal. Li et al. [22] investigated the response laws of weak currents stimulated from large-scale coal samples under a concentrated load and proposed the principles of predicting coal bumps using the weak current method. To verify the feasibility of the weak current method in coal bump prediction, Li et al. [23] also conducted field tests in a deep underground coal mine using their self-developed weak current measuring device, filling the gap in the application of the weak current technique in underground engineering. The field test results show that the weak current responds well to mine seismicities and the weak current technique has a broad prospect in coal and rock dynamic disaster prediction due to its advantages of strong anti-interference ability and sensitive response.

Previous research on the weak current stimulated from stressed coals provides a new idea and theoretical basis for coal bump prediction and early warning, but almost all of these studies are focused on the static load condition. In underground mining, the catastrophic failure of coal under the disturbance induced by excavation blasting, mechanical drilling, and roof fracture is becoming one of the major safety risks [24,25]; thus, investigations on the measurement of the physical signals of coal under an impact load are vital for the further understanding of the coal bump evolution process. Liu et al. [26] investigated the burst characteristics of coal under impact loading using an AE system and analyzed the development of fractures in coal samples through an AE count. Feng et al. [27] studied the fracture characteristics of coal and its AE response under dynamic loading, revealing the "double peak" pattern of the dynamic strength of coals. Yang et al. [28] studied the damage evolution characteristics of coal samples under impact loading under different surrounding pressures by measuring the change in ultrasonic wave velocity, through which the internal damage of coal samples was quantitatively characterized. Xu et al. [29] investigated EMR characteristics during the dynamic fracturing progress of gas-bearing coal under impact loadings through laboratory experiments.

So far, the research on the weak current response characteristics of coal under an impact load is still lacking, which limits the improvement of the weak current technique for coal bump prediction and early warning. In the present work, an experimental system for the weak current measurement of coal under an impact load is established, laboratory experiments on coal samples under various impact loads are conducted, and the weak currents are measured synchronously. The response characteristics of the weak current are analyzed and its generation mechanism is investigated, by which the physical models are established to describe the processes of charge transfer and weak current generation.

2. Experimental Details

2.1. Material

Coal blocks were collected from the Zhangminggou coal mine in Shaanxi, China. Cuboid samples measuring $50 \times 80 \times 200$ mm^3 were cored and cut from these blocks in strict accordance with the standards specified by the International Society for Rock Mechanics [30]. The surfaces of each sample were ground flat so that the surface roughness did not exceed ±0.02 mm, and the end face was perpendicular to the axis; the maximum deviation was not greater than 0.25°.

2.2. Experimental System

The experimental system consists of a loading system, a weak current measurement system, and an electromagnetic shielding system (Figure 1). The impact load is supplied by a DK-5621 falling ball impact tester, with the drop height ranging from 0 to 2000 mm, with a minimum scale of 10 mm. The drop of the ball is controlled by a DC solenoid valve with an infrared positioning function. The weak current measurement system consists of a Keithley 6517B electrometer (Tektronix (China) Co., Ltd., Shanghai, China), a computer, a tri-coaxial cable, and two electrodes. The range of this electrometer is from 1 fA to 20 mA, with a minimum resolution of 1 fA. The independently developed data acquisition software, based on the LabVIEW system design platform (National Instruments, Austin, TX, USA), is installed on a personal computer running Windows 10 to acquire, display, and store the current data. To diminish the interference of surrounding electrical noise to the weak currents generated from the coal samples, the experiments were conducted in an electromagnetic shielding room.

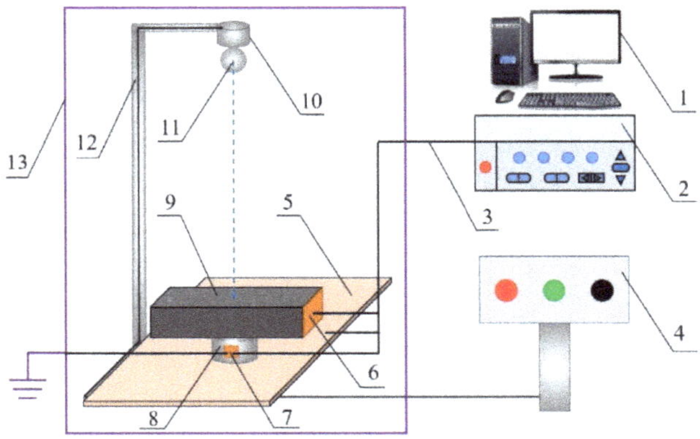

Figure 1. Diagrammatic sketch of impact loading and electrode arrangement. (1) Computer; (2) electrometer; (3) triaxial shielded cable; (4) solenoid valve controller; (5) Teflon sheet; (6,7) copper electrode; (8) stainless steel piston; (9) coal specimen; (10) DC solenoid valve; (11) steel ball; (12) fixed support; (13) electromagnetic shielded room.

2.3. Experimental Scheme

As shown in Figure 1, the sample measuring 50 × 80 × 200 mm^3 is put on a steel piston of 80 mm in diameter that is electrically insulated from the tester by a 3-millimeter-thick Teflon sheet. An oblong copper electrode (80 mm long, 50 mm wide) is attached to the end surface of the unstressed volume, and another copper electrode (30 mm long, 20 mm wide) is attached to the steel piston. The HI (high) side of the electrometer is connected to the electrode attached to the end side of the unstressed volume, while the LO (low) side is connected to the electrode attached to the piston, which is also connected to an electromagnetically shielded room that is grounded. The sampling frequency of 3 Hz for the current measurement has been chosen in the experiments.

Five specimens numbered CIL01, CIL02, CIL03, CIL04, and CIL05 were prepared for impact loading tests. A steel ball weighing 225 g was used for the tests, and eight falling heights of 0.1, 0.2, 0.4, 0.5, 0.6, 0.8, 1.0, and 1.5 m were set to investigate the influence of impact velocity on the magnitude of transient currents. Three groups of parallel tests on Specimens CIL01, CIL02, and CIL03 were carried out; Specimens CIL04 and CIL05 were held in reserve.

3. Results and Discussion

3.1. Weak Current Response Characteristics

Figure 2 shows the variation of weak currents stimulated from Specimen CIL02 with time. It can be seen that before an impact load is applied to the coal sample, the weak current is stable at a certain value, which is called the background current. The moment an impact load is applied, the weak current increases instantly to the peak value; the phenomenon is known as a transient current. After that, the current decreases gradually, tending to be a certain value (marked by the dashed line), which is named the stable current. As shown in Figure 2, after an attenuation of hundreds of seconds, the current seems to be still greater than the stable current, which is verified by the data listed in Table 1. Previous research shows that the weak currents from rock and coal under a static load also decay when the applied load no longer increases or remains constant [12,13,16,19,22], and the decay trend is similar to that shown in Figure 2. The difference is that the stable current is much higher than the background one for the static load condition, while the two are very close for the impact load condition (see Table 1). Due to the fact that the applied load is the driving force of the generation of weak current, the current can also be stimulated, although the load applied on the coal/rock is maintained, making the current stable at a value much higher than the background one. In our experiments, the application of the impact load on the coal samples can be considered instantaneous and is removed subsequently, making the transient current decay without any driving force. Essentially, the weak current stimulated by an impact load tends to be the background value, but this process will take a long time, so the stable current is only a little higher than the background current.

Table 1. Typical currents stimulated from Specimen CIL02 under various impact loadings.

Drop Height/m	Background Current/nA	Peak Current/nA	Stable Current/nA
0.1	0.5	5.4	0.7
0.2	0.6	8.1	1.0
0.4	0.6	10.7	1.0
0.5	1.0	11.9	1.1
0.6	0.4	12.4	0.6
0.8	0.6	15.3	0.6
1.0	1.0	16.4	1.2
1.5	0.5	17.6	0.5

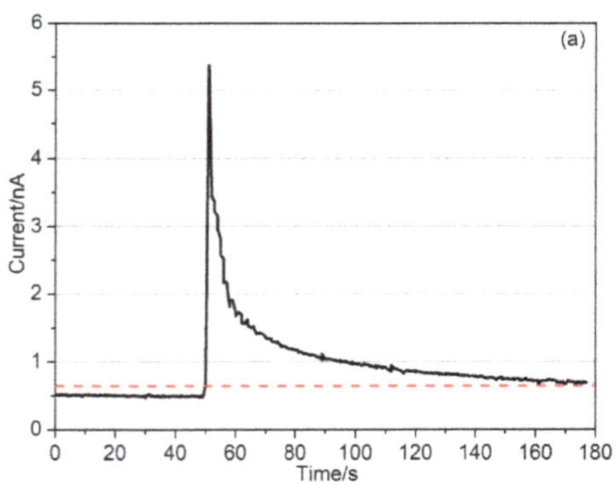

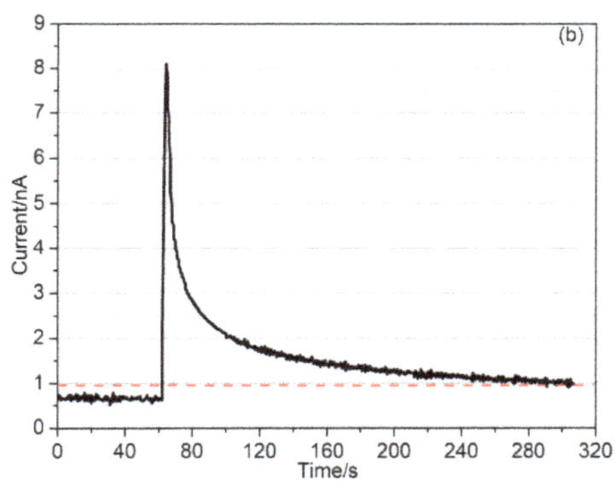

Figure 2. *Cont.*

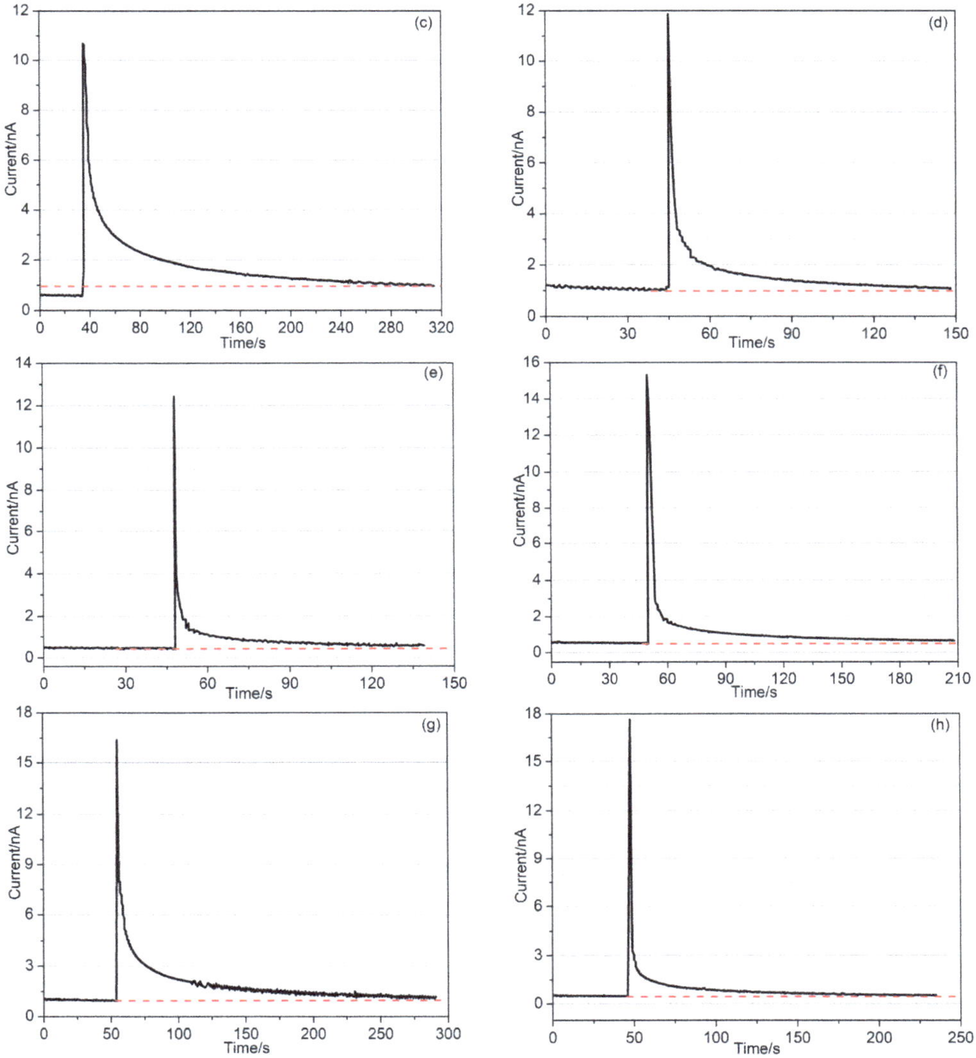

Figure 2. Variation of weak currents with time for Specimen CIL02 for different ball falling heights. (**a**) 0.1 m; (**b**) 0.2 m; (**c**) 0.4 m; (**d**) 0.5 m; (**e**) 0.6 m; (**f**) 0.8 m; (**g**) 1.0 m; (**h**) 1.5 m. The dashed line marks the stable current.

It can be seen from Figure 2 that the variation law of weak currents stimulated from coals under an impact load is consistent, but the peak currents are different. Therefore, it is necessary to investigate the relationship between the falling height of the ball and the intensity of a transient current, which can be calculated by the following formula:

$$I_{tr} = I_p - I_b \quad (1)$$

where I_{tr} is the intensity of a transient current, I_p is the peak current, and I_b is the background current. The intensity of transient currents for the three groups of experiments is listed in Table 2, and the error bar curve is shown in Figure 3. The fitting results of the scatter of the average transient current show that the intensity of transient currents increases exponentially with the falling height of the ball.

Table 2. Intensity of transient current from coal samples under an impact load.

Drop Height/m	Transient Current/nA			
	CIL01	CIL02	CIL03	Average
0.1	3.8	4.9	4.3	4.3
0.2	9.4	7.5	7.4	8.1
0.4	13.1	10.1	10.6	11.5
0.5	14.4	10.9	12.1	12.5
0.6	13.5	12.0	14	13.2
0.8	15.6	14.7	14.8	15.0
1.0	14.7	15.4	17.2	15.8
1.5	19.5	17.1	16.3	17.6

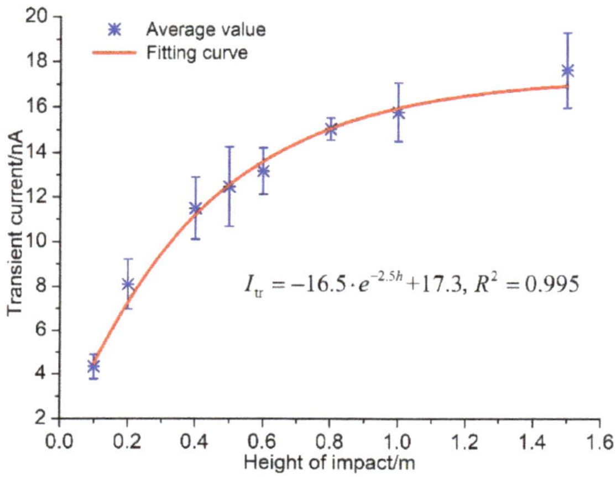

Figure 3. Error bar curve of transient currents with the falling height of the ball (h) for the three groups of tests.

Previous studies show that the weak current stimulated from coals under a static load increases with strain rate [22,30], so it is reasonable to investigate the relationship between the transient current and the strain rate of coal under an impact load.

According to the law of conservation of energy [31], the gravitational potential energy before the ball begins to fall is equal to the kinetic energy the moment it hits the coal sample without considering the effect of air resistance; hence, the energy relationship is

$$\frac{1}{2}mv^2 = mgh \tag{2}$$

where m is the mass of the ball, g is gravitational acceleration, h is the height the ball falls from, and v is the speed of the ball when hitting the coal sample. Then

$$v = \sqrt{gh} \tag{3}$$

The strain rate caused by the impact load is positively correlated with the speed of the ball when striking the coal sample [24], that is

$$\frac{\mathrm{d}\varepsilon}{\mathrm{d}t} \propto v \tag{4}$$

From Equations (3) and (4), we can derive Equation (5):

$$\frac{\mathrm{d}\varepsilon}{\mathrm{d}t} \propto h^{1/2} \tag{5}$$

In order to investigate the relationship between the transient current and the strain rate, the relationship between the transient current and $h^{1/2}$ needs to be determined first; hence, the error bar curve of the transient current, corresponding to the $h^{1/2}$ of Specimens CIL01, CIL02, and CIL03, is drawn in Figure 4. The fitting results to the scatters show that the transient current increases negatively and exponentially with the height that the ball falls from. The relation can be expressed as:

$$I_{\text{tr}} \propto -e^{-h^{1/2}} \tag{6}$$

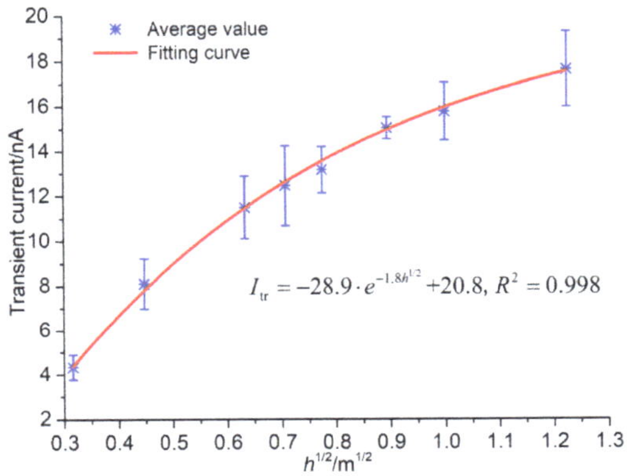

Figure 4. Error bar curve of the transient currents with $h^{1/2}$ for the three groups of tests.

From Equations (5) and (6), we can derive Equation (7):

$$I_{\text{tr}} \propto -e^{-\mathrm{d}\varepsilon/\mathrm{d}t} \tag{7}$$

Equation (7) indicates that the weak current stimulated from coals under an impact load increases negatively and exponentially with the strain rate, which is consistent with that for coals under a static load [22,32].

The application of an impact load on the coal is completed in a very short time, which can be regarded as instantaneous, so the strain rate reaches its maximum instantly. The weak current increases with the strain rate, so the weak current reaches its peak value instantly, generating a transient current. After the application of the impact load, the coal deformation gradually recovers, so the weak current attenuates gradually and, finally, tends to be stable.

3.2. *Attenuation Laws of Transient Currents*

As depicted in Figure 2, the currents exhibit almost an identical attenuation trend. In order to facilitate the analysis of the unity of the attenuation law of the weak current, the time when the current begins to decay is zeroed. As shown in Figure 5, the weak current curves flatten out gradually, indicating that the decay rate of the current decreases. In addition, after hundreds of attenuations, these currents are still greater than the background one, indicating that the attenuation of currents is a slow dissipation process.

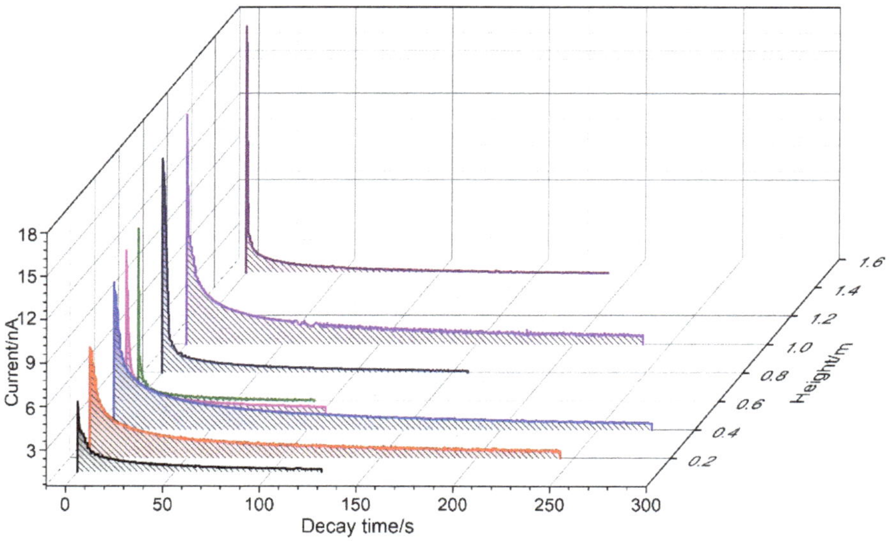

Figure 5. Curves of current attenuation for different ball falling heights for Specimen CIL02.

In 1988, Tsallis C proposed a non-extensive entropy (Tsallis entropy), which is defined as [33]:

$$S_{\mathrm{q}} = k_{\mathrm{B}} \frac{1 - \sum_{i=1}^{W} P_i{}^q}{q - 1} \tag{8}$$

where k_{B} is Boltzmann's constant, P_{i} is a set of probabilities, q is the non-extensive parameter, and W is the total number of possible microscopic configurations. There is a normalization condition where $\sum_{i=1}^{W} P_i = 1$. This kind of statistical mechanics, based on Tsallis entropy, is called non-extensive statistical mechanics, and it will be used to analyze the attenuation laws of the transient current.

Because both the transient currents and the stable currents are different for different ball falling heights, in order to unify the attenuation law of these currents, the decay currents are first normalized using the following formula:

$$\xi(t) = \frac{I_{\mathrm{t}} - I_{\mathrm{s}}}{I_{\mathrm{p}} - I_{\mathrm{s}}} \tag{9}$$

where $\xi(t)$ is the normalized current, I_{t} is the real-time current in the relaxation phase, and I_{s} is the stable current.

If the current relaxation was of the exponential type, then the normalized current $\xi(t)$ would satisfy the following equation [34–36]:

$$\frac{\mathrm{d}\xi}{\mathrm{d}t} = -\beta \cdot \xi \tag{10}$$

Considering the involvement of multi-fractality, a more general equation holds:

$$\frac{\mathrm{d}\xi}{\mathrm{d}t} = -\beta_{\mathrm{q}} \cdot \xi^{\mathrm{q}} \tag{11}$$

where β_{q} is the decay factor that is related to the internal energy of a system.

Equation (11) leads to a generalized q-exponential function:

$$\xi(t) = [1 - (1 - q) \cdot \beta_q \cdot t]^{1/1-q} = \exp(-\beta_q \cdot t) \quad (12)$$

The normalized currents after the withdrawal of the impact load on Specimen CIL02 are calculated by Equation (9). The scatters of ξ, corresponding to different heights, are depicted in Figure 6 and are fitted by using Equation (12). Figure 6 shows that all the fitting curves correspond well to the scatters of current measured in our experiments, and the coefficients of determination, R^2, are all greater than 0.95, showing that the fitting results are very good [37]. In addition, by non-extensive statistical mechanics, $q < 1$ enhances the rare events, while $q > 1$ enhances the frequent events. As shown in Figure 6, the q values are all greater than 2, indicating that the relaxation of these currents obeys the non-extensive statistical mechanics.

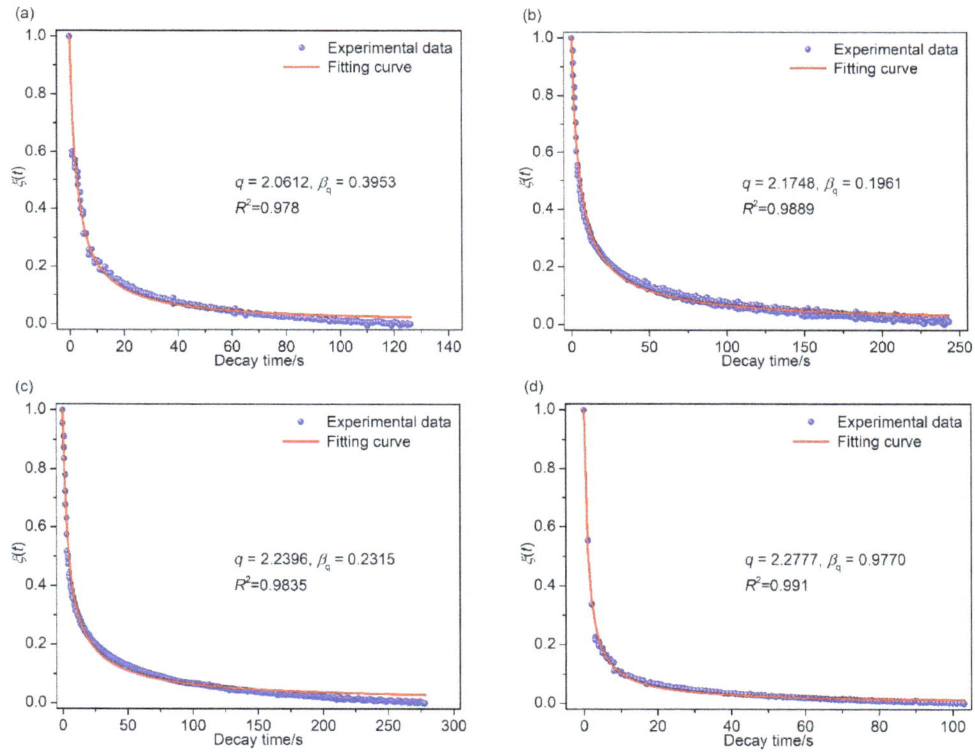

Figure 6. *Cont.*

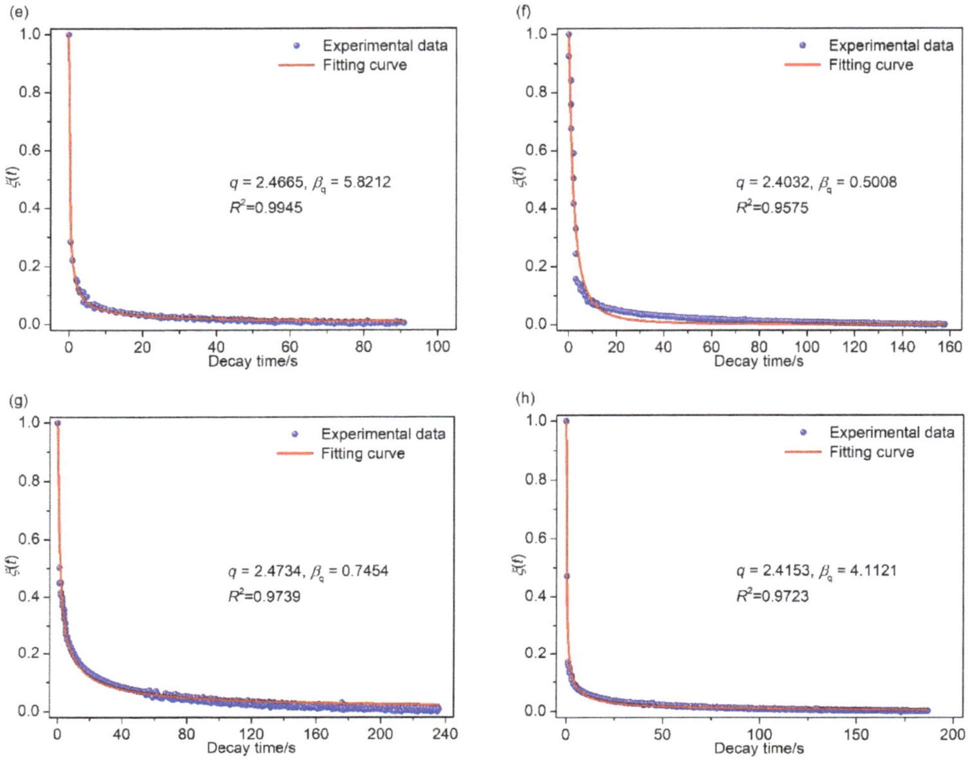

Figure 6. Normalized decaying currents from Specimen CIL02 for different falling heights of the ball. (**a**) 0.1 m; (**b**) 0.2 m; (**c**) 0.4 m; (**d**) 0.5 m; (**e**) 0.6 m; (**f**) 0.8 m; (**g**) 1.0 m; (**h**) 1.5 m.

3.3. Mechanism of Weak Current from Loaded Coals

3.3.1. Carriers in Coal

The fact that the weak current can be stimulated from coals under an impact load shows that there must be carriers in the coal. According to the dielectric conduction theory, there are only three forms of carriers: ions, electrons, and holes [38].

Ionic conductivity is the conductive process of the directional movement of positive and negative ions, so the necessary condition for ionic conductivity is the presence of ions that can move freely [38]. The basis to judge whether there are ions that can move freely in a substance are: (1) whether there are ions and (2) whether there are conditions for ions to move freely. In our experiments, there is neither solution nor molten minerals in the dry coal sample at room temperature, so there is no condition for the free movement of ions. Therefore, it is inferred that the carriers of the weak current generated in the coal samples are not ions.

Freund et al. [18,19] proposed that the p-holes formed by the loss of an electron by the oxygen atom of silicate minerals are the main carriers in igneous rocks. Li et al. [22] introduced the p-hole theory to explain the generation mechanism of weak currents stimulated from coals under a static load and believed that the p-holes are induced by the breaking of covalent bonds within coal molecules. P-hole theory can well illustrate the directionality of weak currents, but there are still some limitations in using this theory. It holds that the p-holes in coals are not primary holes but are generated due to the fracture of chemical bonds. The existing research shows that a weak current can be generated the moment a load is applied to the coal/rock material, and it can still increase rapidly to a peak value even at a low stress level [14–16,18,22]. Considering that current is defined as the quantity of charge passing through the cross-section of a conductor per unit of time [38], the sudden increase in current requires a large quantity of charge in a very short time. If the carriers of those currents are p-holes, then a large number of chemical bonds, such as –O–O–, –C–OH, –C–SH, –C–C_mH_n, in coals need to be broken quickly. However, the breaking of chemical bonds absorbs high energy, but the energy of the ball falling from a height of 0.1 m is too small to provide enough energy to break a large number of chemical bonds, so few p-holes can be activated and the transient current is impossible to stimulate. Therefore, p-holes are not the main carriers of the weak currents from stressed coals.

Based on the above analysis, the carriers of the weak current from loaded coal are electrons. Except for the free electrons that naturally exist in the coal, a large number of electrons can also be generated by friction, stress-induced polarization, and crack propagation, which will provide sufficient carriers for the weak currents from coals [32].

3.3.2. Mechanism and Physical Models

For the electrometer used in our experiments, the sign of the measured current is positive if it flows to the HI side. Our experimental results show that the weak currents measured are all positive, indicating that the current flows to the HI side of the electrometer. Accordingly, the diagram of the equivalent circuit, composed of the coal sample, copper electrodes, wires, and the electrometer, can be obtained, as shown in Figure 7. Considering that the flowing direction of electrons is opposite to that of the current [38], the electrons flow out from the HI side. Due to the movement of electrons being continuous in this circuit, the path that the electron moves along can be depicted in Figure 7, indicating that the electrons flow from the unstressed volume to the stressed one.

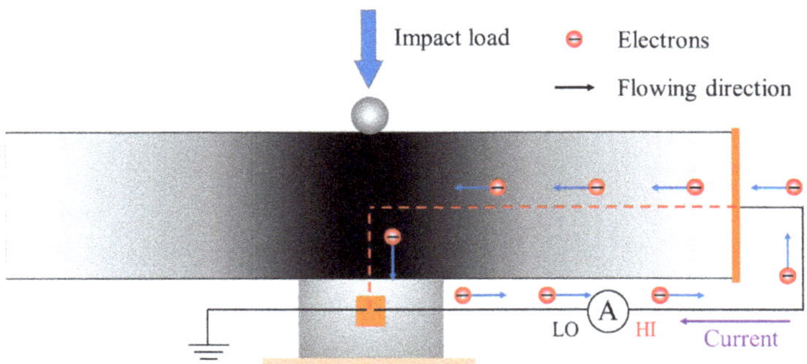

Figure 7. Equivalent circuit diagram showing the flow of the current and the movement of electrons in the coal.

Our previous research shows that the charge distribution in the coal and rock materials obeys the tip effect, which holds that electric charges tend to concentrate at the tip of a crack [30]. Therefore, the charge distribution on the coal surface is illustrated in Figure 8. The distribution of charges is closely related to the surface topography of the coal, and the surface of coal is rough, especially at the friction interface or surface of new cracks, so the charges cannot be evenly distributed. In fact, the negative charges are distributed on the surface of a "sunken" crack, while the positive charges are on that of a "bulge". In addition, under the action of the tip effect, the charges tend to concentrate at the tip location of the coal surface, so the electron density increases along the path to the tip of a "sunken" crack or a "bulge".

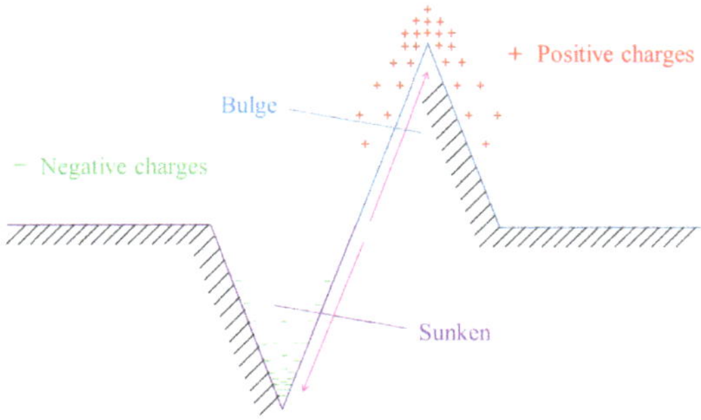

Figure 8. A physical model showing charge distribution on coal surfaces.

The coal is in the thermal equilibrium state when it is not loaded, so the free electrons are in a chaotic motion, representing electrical neutrality. As shown in Figure 9a, before the coal is loaded, the free electrons are evenly distributed in the coal, but when an impact load is applied to the coal (Figure 9b), the primary crack will be closed quickly, which is accompanied by the generation of a large quantity of charges due to friction electrification. Under the action of the tip effect of charges, the original and newly generated free electrons flow to the tip of the fracture space, leading to the reduction of free charge density in the loaded volume of the coal. Because of the charge density difference between the stressed volume and the unstressed one, the electrons in the unstressed volume will diffuse to the stressed volume through the conductive channel to reach a new charge balance in the coal. The diffusion path is depicted by the arrow in Figure 9b.

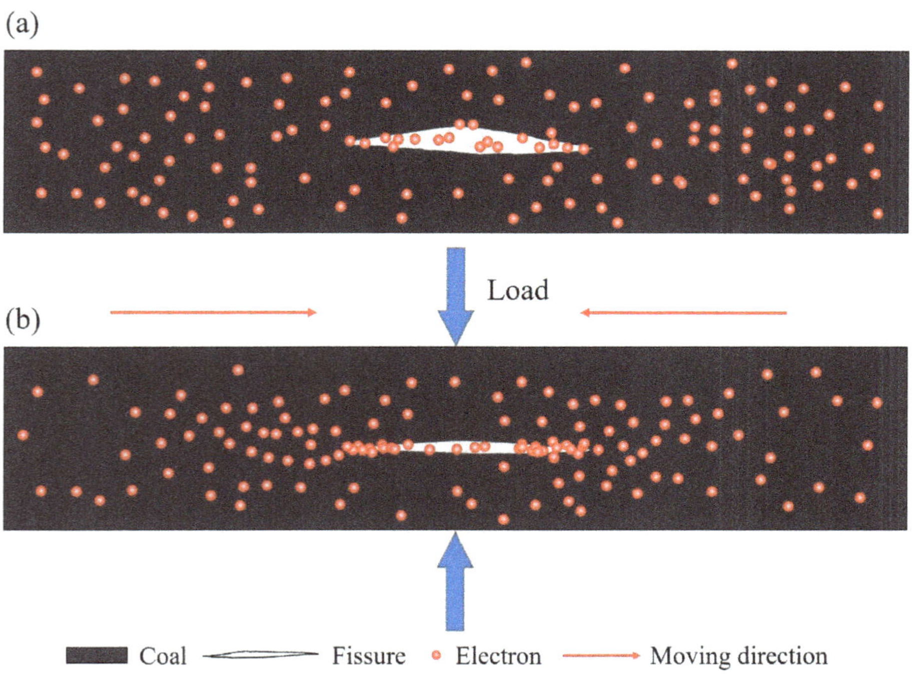

Figure 9. Schematic diagram showing the generation mechanism of currents induced by coal deformation. The coal is (**a**) unloaded and (**b**) loaded.

By definition, the current is calculated by [38]:

$$I = \frac{dq}{dt} \tag{13}$$

where I is the current, q is the charge quantity that flows from a cross-section of a conductor, and t is the time of charge flow.

The application of an impact load on coal can be regarded as instantaneous, making the primary cracks close in a very short time. Under these circumstances, the charge density difference is formed instantly, and the electrons flow instantaneously, so a transient current can be stimulated when an impact load is applied.

4. Conclusions

The contributions of this study can be summarized as follows:

(1) The moment an impact load is applied to coal, a weak current, increasing instantly to the peak value (transient current), is stimulated, and the current flows from the loaded volume to the unloaded one through the coal. The intensity of the transient current increases with impact velocity, which is positively related to the falling height of the ball.

(2) The transient current decays slowly after the withdrawal of the impact load, tending to be a stable value that is slightly greater than the background current before the load application. The attenuation of the transient current lasts hundreds of seconds and obeys non-extensive statistical mechanics, with the non-extensive parameter q greater than 2.

(3) The main carriers of the stimulated weak currents are free electrons. The generation mechanism of the weak currents induced by coal deformation is the instantaneous movement of electrons under a density difference. The closure of primary cracks makes electrons flow toward the tip of the shuttle fracture space under the action of the tip effect, leading to a charge density difference, which makes the electrons in the unloaded volume move to the loaded one to reach a new charge balance, generating the transient currents.

Author Contributions: D.L.: Conceptualization, Methodology, Investigation, Writing-Original Draft, Funding acquisition. E.W.: Writing-Review & Editing, Supervision, Funding acquisition. D.J.: Validation, Project administration. D.W.: Investigation, Formal analysis. W.L.: Software, Data curation. All authors have read and agreed to the published version of the manuscript.

Funding: This work was supported by the National Natural Science Foundation of China (52204257, 51934007), the China Postdoctoral Science Foundation (2022M713371), the Jiangsu Funding Program for Excellent Postdoctoral Talent (2022ZB512), and the GuangDong Basic and Applied Basic Research Foundation (2022A1515110132).

Institutional Review Board Statement: Not applicable.

Informed Consent Statement: Not applicable.

Data Availability Statement: Data is available on request from the authors.

Conflicts of Interest: The authors declare no conflict of interest.

References

1. Keneti, A.; Sainsbury, B.A. Review of published rockburst events and their contributing factors. *Eng. Geol.* **2018**, *246*, 361–373. [CrossRef]
2. Cai, M. Principles of rock support in burst-prone ground. *Tunn. Undergr. Space Technol.* **2013**, *36*, 46–56. [CrossRef]
3. Feng, G.L.; Feng, X.T.; Xiao, Y.X.; Yao, Z.B.; Hu, L.; Niu, W.J.; Li, T. Characteristic microseismicity during the development process of intermittent rockburst in a deep railway tunnel. *Int. J. Rock Mech. Min. Sci.* **2019**, *124*, 104135. [CrossRef]
4. Jiang, Y.D.; Pan, Y.S.; Jiang, F.X.; Dou, L.M.; Ju, Y. State of the art review on mechanism and prevention of coal bumps in China. *J. China Coal Soc.* **2014**, *39*, 205–213.

5. Lawson, H.E.; Tesarik, D.; Larson, M.K.; Abraham, H. Effects of overburden characteristics on dynamic failure in underground coal mining. *Int. J. Rock Mech. Min. Sci.* **2017**, *27*, 121–129. [CrossRef]
6. Sepehri, M.; Apel, D.B.; Adeeb, S.; Leveille, P.; Hall, R.A. Evaluation of mining-induced energy and rockburst prediction at a diamond mine in Canada using a full 3D elastoplastic finite element model. *Eng. Geol.* **2020**, *266*, 105209. [CrossRef]
7. Yuan, L.; Jiang, Y.D.; He, X.Q.; Dou, L.M.; Zhao, Y.X.; Zhao, X.S.; Wang, K.; Yu, Q.; Lu, X.M.; Li, H.C. Research progress of precise risk accurate identification and monitoring early warning on typical dynamic disasters in coal mine. *J. China Coal Soc.* **2018**, *43*, 306–318.
8. Qiu, L.M.; Zhu, Y.; Song, D.Z.; He, X.Q.; Wang, W.X.; Liu, Y.; Xiao, Y.Z.; Wei, M.H.; Yin, S.; Liu, Q. Study on the nonlinear characteristics of EMR and AE during coal splitting tests. *Minerals* **2022**, *12*, 108. [CrossRef]
9. Qiu, L.M.; Liu, Z.T.; Wang, E.Y.; He, X.Q.; Feng, J.J.; Li, B.L. Early-warning of rock burst in coal mine by low-frequency electromagnetic radiation. *Eng. Geol.* **2020**, *279*, 105755. [CrossRef]
10. Vallianatos, F.; Tzanis, A. Electric current generation associated with the deformation rate of a solid: Preseismic and coseismic signals. *Phys. Chem. Earth* **1998**, *23*, 933–939. [CrossRef]
11. Stavrakas, I.; Anastasiadis, C.; Triantis, D.; Vallianatos, F. Piezo stimulated currents in marble samples: Precursory and concurrent-with-failure signals. *Nat. Hazards Earth Syst. Sci.* **2003**, *3*, 243–247. [CrossRef]
12. Stavrakas, I.; Kourkoulis, S.; Triantis, D. Damage evolution in marble under uniaxial compression monitored by pressure stimulated currents and acoustic emissions. *Fract. Struct. Integr.* **2019**, *13*, 573–583. [CrossRef]
13. Triantis, D.; Stavrakas, I.; Anastasiadis, C.; Kyriazopoulos, A.; Vallianatos, F. An analysis of pressure stimulated currents (PSC), in marble samples under mechanical stress. *Phys. Chem. Earth* **2006**, *31*, 234–239. [CrossRef]
14. Freund, F.T.; Takeuchi, A.; Lau, B.W. Electric currents streaming out of stressed igneous rocks—A step towards understanding pre-earthquake low frequency EM emissions. *Phys. Chem. Earth* **2006**, *31*, 389–396. [CrossRef]
15. Freund, F.; Ouillon, G.; Scoville, J.; Sornette, D. Earthquake precursors in the light of peroxy defects theory: Critical review of systematic observations. *Eur. Phys. J.-Spec. Top.* **2021**, *230*, 7–46. [CrossRef]
16. Li, D.X.; Wang, E.Y.; Li, Z.H.; Ju, Y.Q.; Wang, D.M.; Wang, X.Y. Experimental investigations of pressure stimulated currents from stressed sandstone used as precursors to rock fracture. *Int. J. Rock Mech. Min. Sci.* **2021**, *145*, 104841. [CrossRef]
17. Kyriazopoulos, A.; Anastasiadis, C.; Triantis, D.; Brown, C.J. Non-destructive evaluation of cement-based materials from pressure-stimulated electrical emission—Preliminary results. *Constr. Build. Mater.* **2011**, *25*, 1980–1990. [CrossRef]
18. Freund, F.T. Pre-earthquake signals: Underlying physical processes. *J. Asian Earth Sci.* **2011**, *41*, 383–400. [CrossRef]
19. Scoville, J.; Sornette, J.; Freund, F.T. Paradox of peroxy defects and positive holes in rocks Part II: Outflow of electric currents from stressed rocks. *J. Asian Earth Sci.* **2015**, *114*, 338–351. [CrossRef]
20. He, M.; Li, Z.H.; Liu, J.; Liu, Y.J. Experimental study on surface current of coal under uniaxial compression. *J. China Coal Soc.* **2013**, *38*, 966–969.
21. Li, Z.H.; Wang, E.Y.; He, M. Laboratory studies of electric current generated during fracture of coal and rock in rock burst coal mine. *J. Min.* **2015**, *2015*, 235636. [CrossRef]
22. Li, D.X.; Wang, E.Y.; Ju, Y.Q.; Wang, D.M. Laboratory investigations of a new method using pressure stimulated currents to monitor concentrated stress variations in coal. *Nat. Resour. Res.* **2021**, *30*, 707–724. [CrossRef]
23. Li, D.X.; Wang, E.Y.; Yue, J.H.; Zhang, X.; Wang, D.M.; Ju, Y.Q. A weak current technique for coal and rock dynamic disaster prediction and its application. *Chin. J. Rock Mech. Eng.* **2022**, *4*, 764–774.
24. Kong, X.G.; He, D.; Liu, X.F.; Wang, E.Y.; Li, S.G.; Liu, T.; Ji, P.F.; Deng, D.Y.; Yang, S.R. Strain characteristics and energy dissipation laws of gas-bearing coal during impact fracture process. *Energy* **2022**, *242*, 123028. [CrossRef]
25. Kong, X.G.; Wang, E.Y.; Li, S.G.; Lin, H.F.; Xiao, P.; Zhang, K.Z. Fractals and chaos characteristics of acoustic emission energy about gas-bearing coal during loaded failure. *Fractals* **2019**, *27*, 1950072. [CrossRef]
26. Liu, G.J.; Mu, Z.L.; Du, J.L. Investigation of coal burst characteristics under impact loading. *Int. J. Oil Gas Coal Technol.* **2018**, *18*, 118–145. [CrossRef]
27. Feng, X.J.; Ding, Z.; Ju, Y.Q.; Zhang, Q.M.; Ali, M. "Double peak" of dynamic strengths and acoustic emission responses of coal masses under dynamic loading. *Nat. Resour. Res.* **2022**, *31*, 1705–1720. [CrossRef]
28. Yang, S.L.; Yue, H.; Chen, X.L.; Zhai, R.H.; Zhang, S. Experimental study on damage evolution characteristics of coal samples under impact load under different surrounding pressures. *Lithosphere* **2022**, *S11*, 106154. [CrossRef]
29. Xu, X.M.; Wang, Q.; Liu, H.; Zhao, W.W.; Zhang, Y.H.; Wang, C. Experimental investigation on the characteristics of transient electromagnetic radiation during the dynamic fracturing progress of gas-bearing coal. *J. Geophys. Eng.* **2020**, *17*, 799–812. [CrossRef]
30. Barton, N. Suggested methods for the quantitative description of discontinuities in rock masses. International Society for Rock Mechanics. *Int. J. Rock Mech. Min. Sci. Geomech. Abstr.* **1978**, *15*, 319–368.
31. Feynman, R. *The Feynman Lectures on Physics*; Addison Wesley: Boston, MA, USA, 1970; Volume I.
32. Li, D.X. *Study on the Weak Current Effect and Its Mechanism of Stressed Coal during Damage*; China University of Mining and Technology: Xuzhou, China, 2021.
33. Tsallis, C. Possible generalization of Boltzmann-Gibbs statistics. *J. Stat. Phys.* **1988**, *52*, 479. [CrossRef]
34. Vallianatos, F.; Triantis, D. Is pressure stimulated current relaxation in amphibolite a case of non-extensivity? *Europhys. Lett.* **2012**, *99*, 18006. [CrossRef]

35. Vallianatos, F.; Triantis, D. A non-extensive view of the Pressure Stimulated Current relaxation during repeated abrupt uniaxial load-unload in rock samples. *Europhys. Lett.* **2013**, *104*, 68002. [CrossRef]
36. Stergiopoulos, C.; Stavrakas, I.; Triantis, D.; Vallianatos, F.; Stonham, J. Predicting fracture of mortar beams under three-point bending using non-extensive statistical modeling of electric emissions. *Physica A* **2015**, *419*, 603–611. [CrossRef]
37. Li, D.X.; Wang, E.Y.; Kong, X.G.; Ali, M.; Wang, D.M. Mechanical behaviors and acoustic emission fractal characteristics of coal specimens with a pre-existing flaw of various inclinations under uniaxial compression. *Int. J. Rock Mech. Min. Sci.* **2019**, *116*, 38–51. [CrossRef]
38. Fischer-Cripps, A.C. *The Electronics Companion*; CRC Press: Boca Raton, FL, USA, 2004.

sustainability

MDPI

Article

Fracture Evolution of Overburden Strata and Determination of Gas Drainage Area Induced by Mining Disturbance

Yuchu Cai [1,2], Shugang Li [2,3,*], Xiangguo Kong [2,3,*], Xu Wang [4], Pengfei Ji [2,3], Songrui Yang [2,3], Xi Lin [2,3], Di He [2,3] and Yuxuan Zhou [2,3]

1 School of Chemistry and Chemical Engineering, Xi'an University of Science and Technology, Xi'an 710054, China
2 College of Safety Science and Engineering, Xi'an University of Science and Technology, Xi'an 710054, China
3 Key Laboratory of Western Mine and Hazard Prevention, Ministry of Education of China, Xi'an 710054, China
4 Shandong Anke Xingye Intelligent Equipment Co., Ltd., Jinan 250002, China
* Correspondence: lisg@xust.edu.cn (S.L.); kxgtudou7218@xust.edu.cn (X.K.)

Abstract: Overburden strata fracture evolution is critical to dynamic disaster prevention and gas-relief drainage, so it is important to accurately determine the evolution relationships with mining disturbance. In this paper, experiments and numerical simulation were adopted jointly to characterize the time-varying fracture area of overlying strata. The experimental results showed that the roof strata gradually broke and collapsed with coal mining, which indicated the fractures of overburden strata developed in an upward direction. The fracture development causes were explained by numerical simulation, which showed that stress increase exceeded the strength of coal and rock strata, and fractures were formed and expanded. Both experiments and numerical simulation results showed the two sides and the top of fracture areas provided channels and spaces for gas migration and reservoir, respectively. In addition, the breaking angle of overburden strata and the height of fracture areas were analyzed quantitatively. Through microseismic monitoring at the mining site, the fracture scales and ranges of overburden strata were verified by the energy and frequency of microseismic events, which were consistent with the support of maximum resistance. The position of drainage boreholes was considered based on the results of overburden strata fracture evolution. Our study is aimed at promoting coal mining in safety and improving gas drainage with a sustainable approach.

Keywords: fracture evolution; mining disturbance; experimental analysis; numerical simulation; microseismic response

Citation: Cai, Y.; Li, S.; Kong, X.; Wang, X.; Ji, P.; Yang, S.; Lin, X.; He, D.; Zhou, Y. Fracture Evolution of Overburden Strata and Determination of Gas Drainage Area Induced by Mining Disturbance. *Sustainability* **2023**, *15*, 2152. https://doi.org/10.3390/su15032152

Academic Editor: Fuqiang Yang

Received: 5 December 2022
Revised: 18 January 2023
Accepted: 20 January 2023
Published: 23 January 2023

1. Introduction

During coal mining, the overburden strata gradually bend, sink and break [1]. Along the direction of formation height, there are collapse zones, fracture zones and bending subsidence zones formed in overburden strata. Especially in fracture zones, this provides space for gas storage [2–6]. To prevent a gas disaster, it is necessary to conduct gas drainage in this area. The extracted gas can be used as clean energy, which can improve resource utilization efficiency and avoid disasters [7–9]. However, it is a difficult problem to determine the exact location of the overburden fractures caused by mining disturbance.

The processes of fracture initiation, propagation and penetration of coal and rock specimens under different loading conditions have been studied [10–14]. Yang et al. (2020) studied deformation characteristics and fracture evolution of a jointed rock mass, and the relationship between these and joint angles [15]. Li et al. (2022) investigated fracture evolution and failure behavior of granite samples with cross-joints in uniaxial loading experiments, and found that cross-joints affected strength and deformation properties [16]. In cyclic loading experiments, Wang et al. (2022) explored fatigue damage and fracture evolution of red sandstone, and concluded that crack propagation was dominated by tensile cracks [17]. Pirzada et al. (2021) discussed contact area and aperture evolution of natural fractures during the shearing process of rock [18]. Due to the discrete properties of specimens, a large number of tests are required in studies to determine common processes in the evolution of fractures [19–21].

With respect to discrete properties of samples, numerical simulations have been used in the study of coal and rock fracture evolution [22,23]. Han et al. (2020) divided the shearing process of rock-like materials into four stages, including the linear elastic stage, crack strengthening stage, plastic softening stage, and residual strength stage, based on the simulation results [24]. Ju et al. (2019) analyzed stress field evolution during the fracture development process using a numerical solution with discrete elements, which indicated mining effects could be predicted [25]. A large deformation in the mining space was discovered. Vazaios et al. (2019) found that pre-existing joints in the rock mass govern the fracturing mechanism [26]. Further, Park et al. (2022) explained that stress redistribution was the cause of damage evolution and fracture formation [27]. Based on rock slope engineering, Bouissou et al. (2012) investigated fracture evolution and deformation characteristics [28].

To reveal overburden fracture evolution, Yin et al. (2016) built a multi-field coupling coal mine dynamic disaster simulation test system, from which the morphology of fractures in different overburden layers was obtained [29]. Due to different mining conditions, the processes of fracture evolution in overlying rock strata are different. Qiao (2017) researched crack evolution of overlying strata during fully mechanized top-coal caving at a site with a high mining height, and built an empirical formula for the development height of fracture zones [30]. Zhao et al. (2021) studied fracture evolution of overlying strata in a shallow-buried underground mining site with an ultra-high working face (8.8 m), and investigated crack formation mechanisms based on energy dissipation theory [31]. During multi-seam mining conditions, a coal seam with low risk is preferred as the first protective layer for mining [32]. Jiao et al. (2017) analyzed overburden strata movement and fissure evolution in a lower protective mining layer, and found that mining disturbance would affect fracturing of the overburden coal seam [33]. Therefore, fractures of the overlying strata, resulting from repeated mining, would become more developed. With respect to repeated mining of coal seams that are close together, Yang et al. (2022) used fractal dimensions, fracture entropy and fracture rate to quantitatively reflect spatial and temporal characteristics of overburden fractures [34]. With the improvement of support equipment, the mining of extra thick coal seams is taking place, especially in Shaanxi Province, China. Li et al. (2022) summarized the relationship between overburden fractures, mining heights, advancing speeds and dip angles, and established a mathematical model of a compacted area in an elliptical belt affected by the above factors [35]. However, how to determine overburden fracture evolution under these conditions it is not completely clear.

In terms of overburden fracture evolution in an extra-thick coal seam, experiments and numerical simulations were used jointly to characterize the fracture development area. By experiments, the spatial distribution processes of overburden fractures were analyzed across different mining distances, then stress changes were investigated to explain the causes of fracture evolution by numerical simulation. Concerning fracture evolution analysis, Xiao et al. (2022) studied the microseismic responses during overburden fracture evolution, and dense cracks areas were monitored by the energy and frequency of microseismic events [36]. Microseismic monitoring methods were also used to analyze fracture evolution around the mining space in this study. At the mine site, spatial and temporal measurements of microseismic events confirmed the results of experimental and numerical simulations. Our research results will improve determination accuracy of fracture areas, which will provide the basis for gas extraction borehole layout.

2. Engineering Background

In the Shaanxi Binchang Xiaozhuang Mining Co., Ltd., Binzhou city, Shaanxi, China, the main coal seam is 0.8~35.22 m thick, and the average thickness reaches 18.01 m. During coal mining large amount of gas may be emitted. Gas flow into the mining space leads to possible serious gas explosions and other accidents. Therefore, it is necessary to drain gases during coal mining. To improve gas drainage efficiency, it is important to determine the position of extraction boreholes that are closely related to cracks area of overburden stratum.

In terms of this engineering problem, the evolution of cracks in the overburden of an extra-thick coal seam was studied, and the test location was selected at the working face (No. 40205) in Shaanxi Binchang Xiaozhuang Mining Co., Ltd. The strike and dip length of the test working face are 2007 m and 196 m, respectively. This is adjacent to the mined area (No. 40204 working face) at south side, while the north side is the working face of No. 40207, is undisturbed. During coal mining at this site, a fully mechanized top coal caving method was adopted, and the roof was managed by the full caving method. The floor elevation of the coal seam is +370 m~ +386 m. In geological engineering, the strength of the coal and rock strata, and the gas within them, influence mechanical properties, related to the lithology and thickness of overburden strata [37,38]. The lithology and thickness of the upper and lower strata of the coal seam are shown in Table 1, and are decisive factors of overburden fracture evolution.

Table 1. Rock stratum distribution and their ratios of geology and experimental model.

No.	Lithology	Actual Thickness/m	Model Thickness/cm	Ratio	Sand /1 cm	Gypsum /1 cm	Large White Powder/1 cm	Coal Ash/1 cm
33	Conglomerate	52.13	26.0	955	8.66	0.48	0.48	
32	Sandy mudstone	14.80	7.5	955	8.65	0.48	0.48	
31	Sandy mudstone	20.20	10.0	955	8.65	0.48	0.48	
30	Sandy mudstone	33.59	16.5	955	8.65	0.48	0.48	
29	Sandy mudstone	3.6	2.0	955	8.65	0.48	0.48	
28	Coarse gravelly sandstone	3.80	2.0	828	7.69	0.19	0.77	
27	Siltstone	3.95	2.0	837	7.69	0.29	0.67	
26	Coarse gravelly sandstone	0.8	0.5	828	7.69	0.19	0.77	
25	Siltstone	3.95	1.5	837	7.69	0.29	0.67	
24	Siltstone	5.80	3.0	837	7.69	0.29	0.67	
23	Mudstone	3.00	1.5	828	7.69	0.19	0.77	
22	Fine grained sandstone	2.10	1.0	828	7.69	0.19	0.77	
21	Coarse grained sandstone	5.72	2.5	828	7.69	0.19	0.77	
20	Sandy mudstone	9.40	4.5	955	8.65	0.48	0.48	
19	Mudstone	1.84	1.0	828	7.69	0.19	0.77	
18	1# Coal	3.24	1.5	946	4.33	0.38	0.58	4.33
17	Mudstone	1.84	1.0	828	7.69	0.19	0.77	
16	Sandy mudstone	9.10	4.5	955	8.65	0.48	0.48	
15	Sandy mudstone	10.55	5.0	955	8.65	0.48	0.48	
14	3# Coal	1.85	1.0	946	4.33	0.38	0.58	4.33
13	Sandy mudstone	10.55	5.0	955	8.65	0.48	0.48	
12	Sandy mudstone	6.15	3.0	828	7.69	0.19	0.77	
11	Siltstone	3.20	1.5	837	7.69	0.29	0.67	
10	Fine grained sandstone	7.00	3.5	828	7.69	0.19	0.77	
9	Mudstone	1.00	0.5	828	7.69	0.19	0.77	
8	4–1# Coal	1.15	0.5	946	4.33	0.38	0.58	4.33
7	Mudstone	1.00	0.5	828	7.69	0.19	0.77	
6	Fine grained sandstone	3.97	2.0	828	7.69	0.19	0.77	
5	Sandy mudstone	4.00	2.0	955	8.65	0.48	0.48	
4	Coarse gravelly sandstone	2.80	1.5	82	7.69	0.19	0.77	
3	Fine grained sandstone	2.48	1.5	828	7.69	0.19	0.77	
2	Sandy mudstone	2.87	1.5	955	8.65	0.48	0.48	
1	4# Coal	15.00	7.5	946	4.33	0.38	0.58	4.33

3. Experiments

3.1. Experimental Parameters

A test platform (3000 × 200 × 1250 mm) was selected for the coal mining experiments. Based on similarity theory, similar conditions need to be determined by comparison with actual geological conditions in the mine (Table 1). The constants in our experiments are shown in Table 2, and the materials (sand, gypsum, large white powder and coal ash) proportions of the rock strata and coal seams in the experimental model as shown in Table 1.

Table 2. Similarity constants of the model.

Along Coal Seam Direction	Model Size mm × mm × mm	Similarity Constant					
		GeomeTry/α_L	Time /α_t	Bulk Density /A_y	Poisson's Ratio /A_μ	Stress /α_σ	Strength/α_E
Strike	3000 × 200 × 1250	200	14.14	1.5	1.0	300	300

3.2. Experimental Model

The similar experimental materials used in the experiments mainly included sand, gypsum, white powder, and coal ash. When producing the model, mica was evenly spread on each layer as a weak surface layer. The steps for constructing the model were as follows: (1) According to the calculated amount of each layered material in the experimental model, the corresponding ingredients were weighed and loaded in a mixing device. (2) The dry materials were mixed evenly, and then an appropriate amount of water was poured into the prepared materials and mixed immediately to prevent agglomeration. (3) Next, the evenly mixed material was poured into the model support, and tamped with iron blocks to maintain the required bulk density. (4) A wall knife was used to mark natural cracks on the surface of the simulated rock stratum at an interval of about 10–20 mm, and a layer of mica powder was evenly sprinkled on the surface to simulate the layer level, and then smoothed by the wall knife again. (5) The other rock layers were paved in sequence in a similar way until all rock layers were prepared on the test platform. (6) The model was erected and left standing for 2–3 days, then U-shaped steel on the surface of model was removed and the model was placed in a ventilated and dry place. After about 20–30 days, the preliminary treatment of model surface was carried out before coal seam mining experiments. (7) When the thickness of rock stratum on the top boundary could not be simulated, a counterweight was added.

3.3. Experimental Procedure and Results

To eliminate the boundary influence on coal seam excavation, coal pillars (20 cm) were reserved at the left and right sides of the model. Near the left coal pillar, the open-off cut (4.5 cm) was dug and mined at intervals of 3 cm. When the mining was advanced to 65 m, the self-weight of rock stratum and the uncollapsed part of the roof formed a cantilever beam that transmitted pressure on the basic roof, which would otherwise break and collapse under the initial weigh [35,36]. A collapsed morphology is shown in Figure 1a, from which it can be seen that the caving height is 12.9 m, and the left and right breaking angle reach 70.9°and 64.4°, respectively. After coal mining advanced to 95 m, the first periodic weighting occurred on the overlying strata with a weighting step distance of 30 m. From Figure 1b, it can be seen that the caving height was 30 m from the coal seam floor, and the cavity height was 13.1 m, which would act as gas storage area. In addition, the left and right breaking angle decreased to 52.4°and 51.9°, respectively. When the working face was advanced to 131 m, the second cycle of weighting occurred with a weighting step distance of 36 m. The caving height was 53.7 m away from coal seam floor, and the parting fractures tended to develop in an upward direction. In Figure 1c, the breaking angles of the working face side and the cut side were 54.1°and 54.3°, respectively. There were still large parting fractures above the middle of goaf, and the temporary gas storage area reached 10.1 m high. At a mining distance of 161 m, the overall movement of overlying strata produced the third periodic weighting, as shown in Figure 1d. The height of the crack zone reached 64.3 m, when, and the height of caving zone was 37.1 m. Overburden movement caused the central goaf to collapse and become compacted, while the fractures at the cut-off side and the working face side were highly developed, to become the dominant channel of gas migration. With coal mining, the collapse of the overburden rock and the formation of the fracture zone changed regularly and periodically. For example, the fourth periodic weighting occurred at a mining distance of 191 m (Figure 1e) and the fifth cycle of weighting occurred at a mining distance of 239 m (Figure 1f). Fracture evolution, gas migration channels, and storage space, gradually changed with time [2,3].

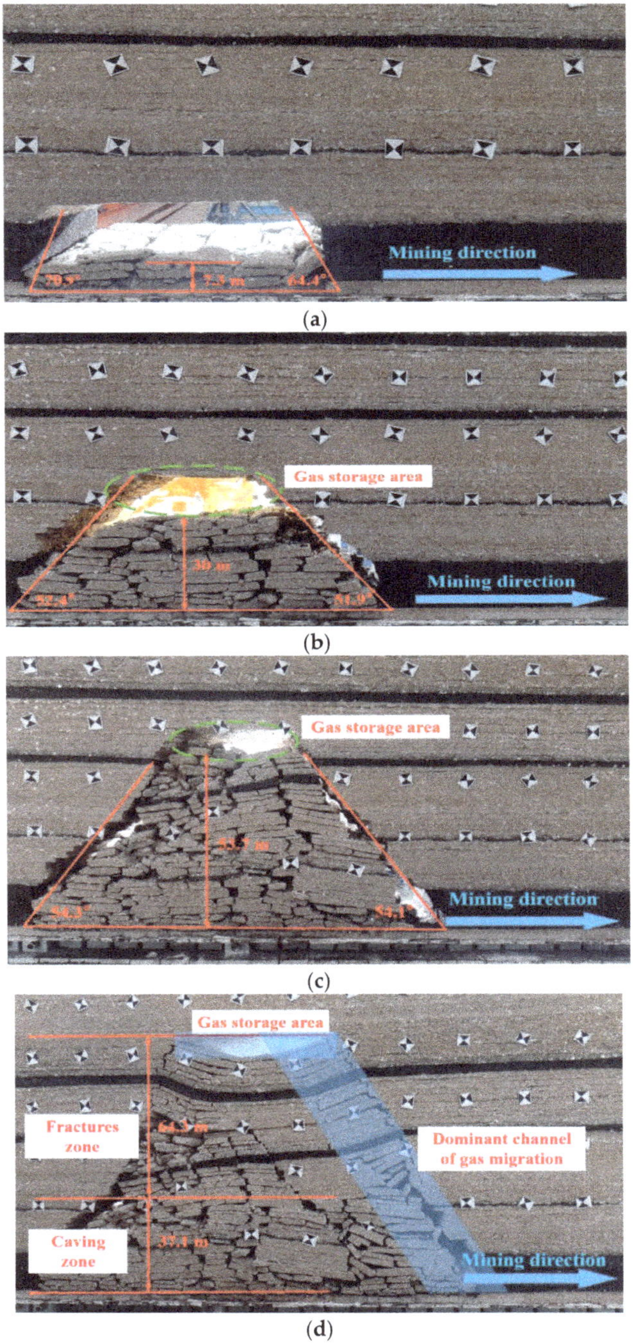

Figure 1. *Cont.*

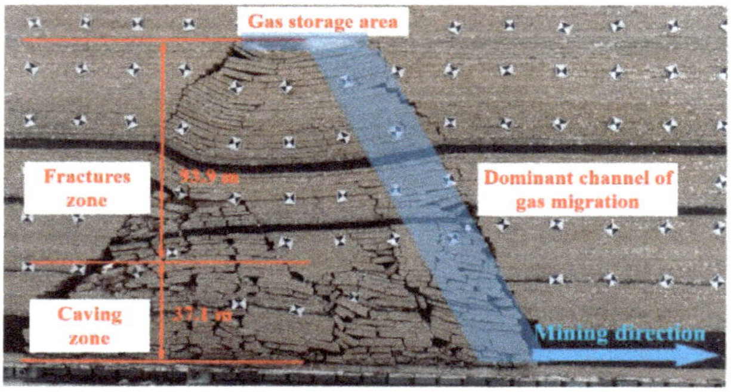

(e)

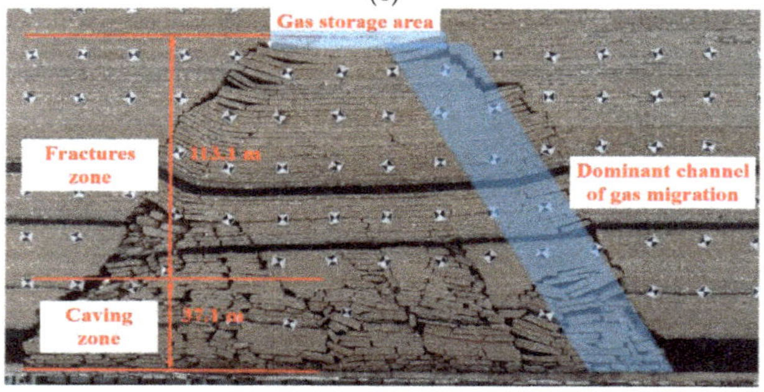

(f)

Figure 1. Fracture evolution, gas migration and storage area formation of overburden strata in experiments. (**a**) Mining distance of 65 m, (**b**) mining distance of 95 m, (**c**) mining distance of 131, (**d**) mining distance of 161 m, (**e**) mining distance of 191 m, (**f**) mining distance of 239 m.

4. Numerical Simulation

Experimental results can provide basic data for studying fracture evolution in the mining overburden, but a large-scale prototype simulation cannot be achieved. To further investigate the evolution of coal mining fractures from large-scale tests, discrete element numerical simulation software was used to study fracture processes [39–42]. In our study, 3DEC was used to establish a numerical model to study spatial fracture distribution and stress characteristics during coal mining.

4.1. Geometric Model, Boundary Conditions and Parameter Setting

Based on the actual geology condition of the coal seam, a geometric model (300 m × 1 m × 300 m) was built. To simulate the weight of rock stratum above the coal seam, 10.94 MPa was exerted on the top boundary. The left and right boundaries of model were set at an equivalent confining pressure of 5.47 MPa. The mechanical parameters of rock and coal stratum, including bulk modulus and shear modulus, are shown in Table 3.

Table 3. Mechanical parameters of rock and coal strata.

Lithology	Rock Stratum						Cleats				
	Density kN/m^3	Bulk Modulus/GPa	Shear Modulus/GPa	Internal Friction Angle/°	Cohesion/MPa	Tensile Strength /MPa	Bulk Modulus /GPa	Shear Modulus/GPa	Internal Friction Angle/°	Cohesion/kPa	Tensile Strength /kPa
Coarse grained sandstone	2410	14.44	12.22	32.00	11.80	6.03	4.55	3.58	32.00	0.01	0.02
Fine grained sandstone	2640	18.60	18.27	28.00	21.13	10.15	4.84	3.11	28.00	0.23	0.20
Mudstone	2420	8.33	5.74	22.00	8.87	4.32	2.06	2.06	22.00	0.10	0.47
Aluminous mudstone	2420	8.33	5.74	22.00	8.87	4.32	2.06	2.06	22.00	0.10	0.47
Sandy mudstone	2220	13.68	12.03	26.00	14.76	7.52	17.78	13.68	26.00	0.45	0.41
4# Coal	1350	0.48	0.23	23.00	4.65	3.01	0.24	0.36	23.00	0.07	0.05
Siltstone	2530	8.06	7.63	31.00	16.77	7.12	16.52	16.12	31.00	0.06	0.08
Conglomerate	2630	16.52	15.25	30.00	16.47	8.09	4.70	3.34	30.00	0.12	0.11

4.2. Simulation Results

(1) Fracture evolution

The results of overburden fracture evolution during coal mining are shown in Figure 2. From Figure 2a, it can be seen that the direct roof collapsed at a mining distance of 29 m, which might be different from the result of experiments. However, the fracture development process was similar to that in experiments. Therefore, the numerical simulation could also reflect overburden movement. When mining continued to advance to 41 m, roof fractures developed and expanded, causing the basic roof of the rock stratum to break and collapse with the initial weighting. As shown in Figure 2b, fracture development layer continued to increase, and the overburden separation space enlarged and further collapsed in a greater range. At a mining distance of 89 m, the overlying rock stratum collapsed again at a larger scale (Figure 2c). With the working face continuing to move forward gradually, it was rare to see widespread collapse, such as at mining distances of 101 m and 165 m (Figure 2d,e). Before the mining distance of 201 m (Figure 2f), the collapse step was maintained at 12~16 m, which was smaller than the result of experiments.

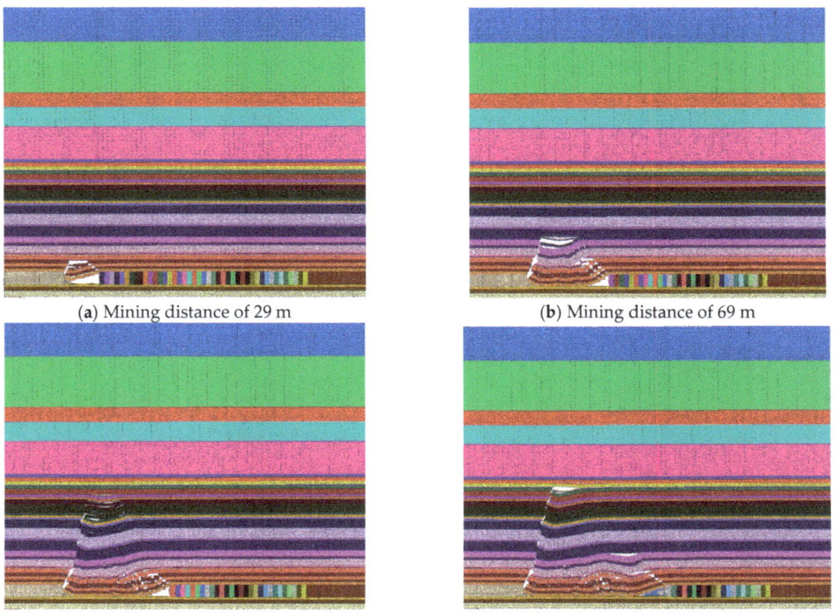

(**a**) Mining distance of 29 m (**b**) Mining distance of 69 m

(**c**) Mining distance of 89 m (**d**) Mining distance of 101 m

Figure 2. *Cont.*

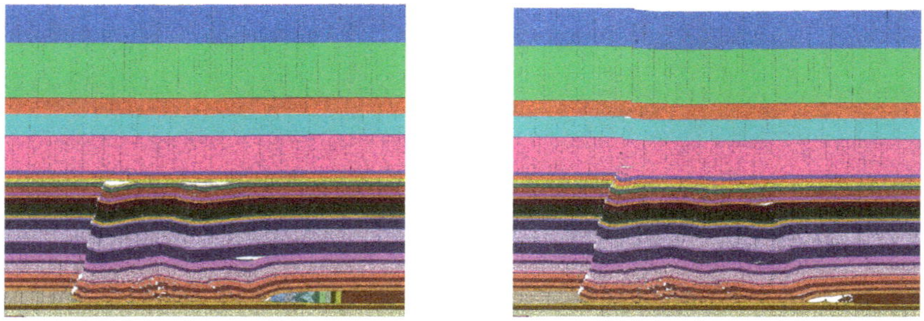

(**e**) Mining distance of 165 m (**f**) Mining distance of 201 m

Figure 2. Fracture evolution of overburden strata in a numerical simulation.

(2) Stress distribution

The fundamental reason for the evolution of overlying rock fractures was stress accumulation, so stress distribution was analyzed during coal mining, as shown in Figure 3. In the mining process of the 40,205 working face, the stress balance of the overburden influenced by the working face was destroyed and redistributed in real time. After the cut-off of the working face was formed, stress occurred on the front and rear coal wall, as shown in Figure 3a. With the advance of the working face, the stress in the goaf gradually decreased. Before the initial weighting, the roof was not broken, so the overlying strata did exert pressure on the floor. The first weighting occurred when the working face was advanced to 41 m, and the overlying strata began to collapse. Then, the goaf was gradually compacted, and the floor stress in the middle of goaf began to increase, as shown in Figure 3b. Comparing the stress results with fracture evolution, stress increase led to fracture development, which decreased stress in its turn. Therefore, stress in the overburden presented alternating cycles of pressurization, relief, and recompression with mining distance (Figure 3c–f). Stress in the goaf behind the working face tended to be stable.

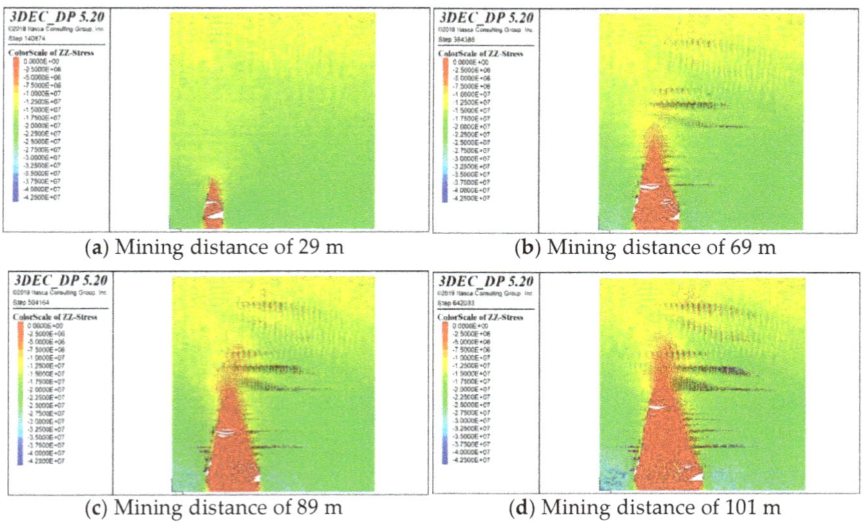

(**a**) Mining distance of 29 m (**b**) Mining distance of 69 m

(**c**) Mining distance of 89 m (**d**) Mining distance of 101 m

Figure 3. *Cont.*

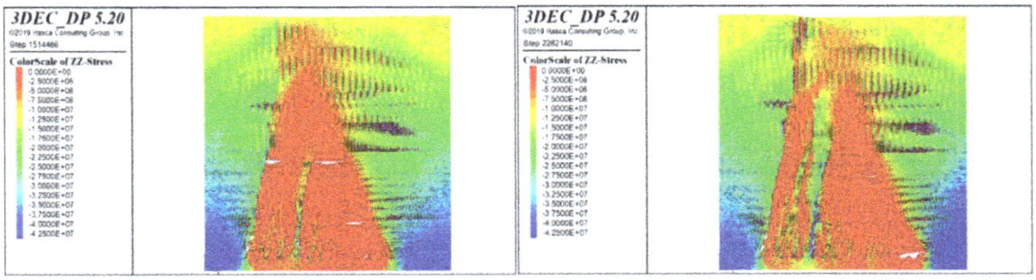

(**e**) Mining distance of 165 m (**f**) Mining distance of 201 m

Figure 3. Stress distribution of overburden strata in numerical simulation.

5. Microseismic Response Characteristics

To improve the accurate location of rock fracture development in overlying layers during coal seam mining, microseismic monitoring was adopted. Multiple methods were used to characterize jointly the development of coal seam mining fracture channels.

5.1. Layout of on-Site Microseismic Sensors

The mine had a KJ551 microseismic monitoring system, and four seismic geophones were laid in the 40,205 working face. The arrangement of the seismic geophones was as uniform as possible in the plane, and their positions were not arranged in one straight line. The layout position moved forward during face mining.

During the installation of seismic geophones, a 2.8 m anchor bolt (one end of the anchor bolt was processed into M20 screw) was used to drill holes in the vertical roadway roof and the sensors fixed with a resin anchoring agent.

5.2. Microseismic Events along the Strike of the Working Face

Abnormal stress leads to fracture evolution, and the process of fracture formation causes microseismic events. What is more, microseismic events within different energies indicated cracks of different scale. At the mine site, microseismic measurements in the mining space were made to determine the fracture range. As shown in Figure 4a, the microseismic response was mainly concentrated between 400 m and 550 m in front of the working face. In the vertical direction, microseismic events occurred about 10–20 m above the roof. Along the dip direction, they were distributed about 20–90 m away from the transport roadway. On 3rd July 2021, the microseismic response changed slightly, but the basic response was consistent. As shown in Figure 4b, the microseismic responses were mainly concentrated between 400–500 m in front of the working face, 7–21 m above the roof, and 0–100 m away from the transport roadway. On 5 July 2021, the microseismic responses were mainly concentrated between 250–550 m in front of the working face, 8–20 m above the roof and 0–100 m away from the transport roadway, as shown in Figure 4c. These results show that the position of microseismic events of high energy indicate instability of the overburden structure, which, due to the influence of advanced stress concentration, was vulnerable to damage. If the working face advanced to this point, the overburden would collapse with mining, and was more likely to form a dominant channel for pressure-relief gas migration. This phenomenon was demonstrated by the microseismic responses on 7 July 2021. On this day, the microseismic events with high energy intensity were mainly concentrated 400–550 m in front of the work face, 7–22 m above the roof, and 0–100 m away from the transport roadway.

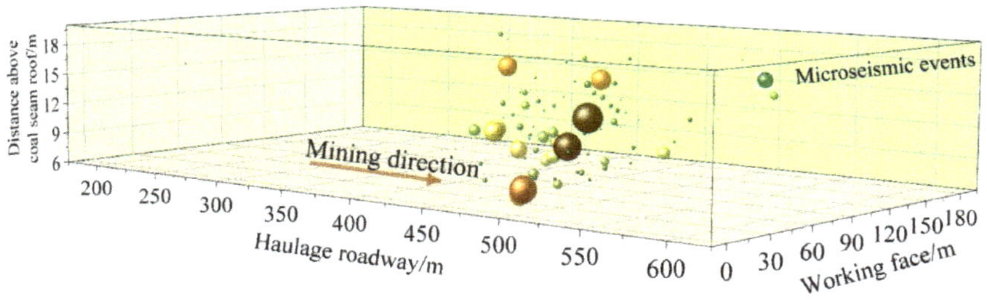

(**a**) 1 July 2021

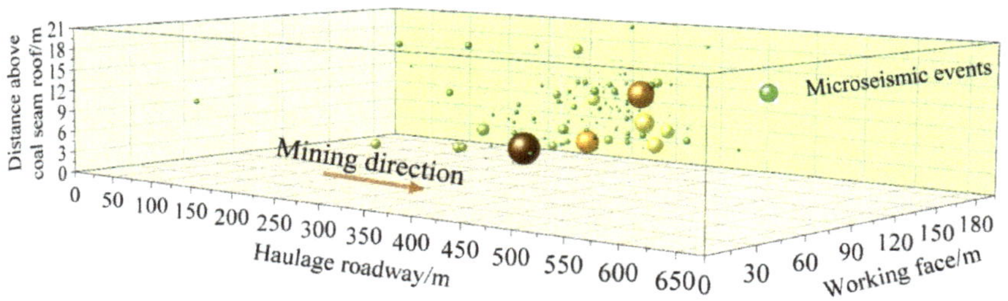

(**b**) 3 July 2021

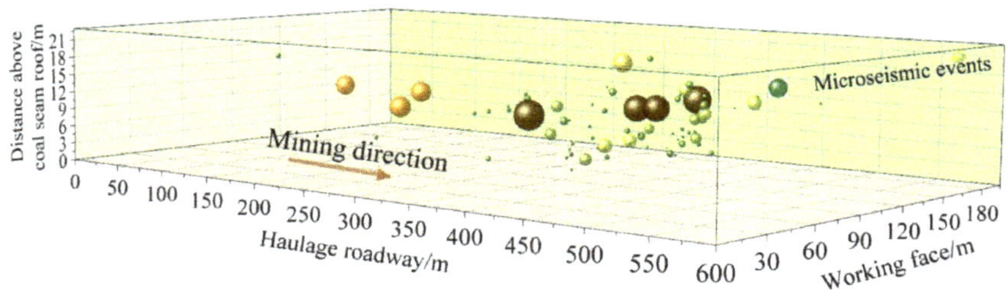

(**c**) 5 July 2021

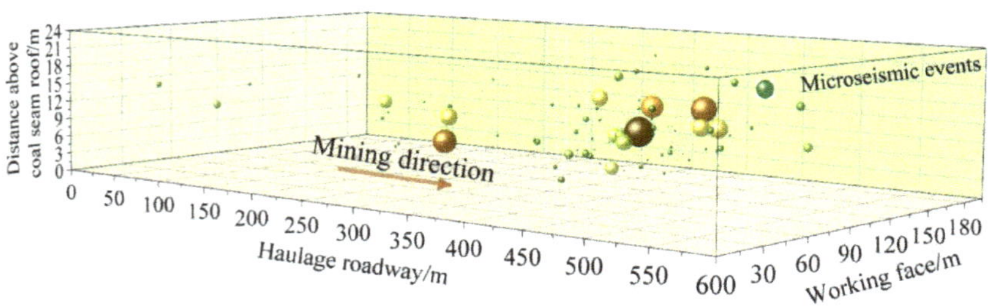

(**d**) 7 July 2021

Figure 4. Microseismic events along the strike of the working face. (**a**) Microseismic events distribution on 1 July 2021; (**b**) Microseismic events distribution on 3 July 2021; (**c**) Microseismic events distribution on 5 July 2021; (**d**) Microseismic events distribution on 7 July 2021. The size of the ball indicates the magnitude of energy values related to microseismic events.

5.3. Microseismic Events along the Dip of Working Face

The distribution of microseismic events in the stope space of the working face can reflect the extent of fracture development. By plotting microseismic events along the dip of working face, the position of fractures can be more accurately ascertained. For example, the microseismic responses were concentrated between 20 and 90 m along the dip of working face, which was consistent with the curve of maximum support working resistance on 1st July 2021, as shown in Figure 5a. It was speculated that due to the mining effects, the support resistance near transportation roadway was than that of the air return lane. The larger stress resulted into more cracks of larger scale, which led to intensive microseismic responses. Figure 5b shows that the maximum support resistance reached a peak 80–90 m away from the transportation roadway, coinciding with the greater energy of the microseismic response at this point. Therefore, microseismic signals may indicate crack ranges and the position of dominant channels. The microseismic events distribution along the dip of working face and support maximum resistance curve were identical on 5 July 2021, as shown in Figure 5c. What is more, the observations were verified by the phenomena on 7 July 2021, as shown in Figure 5d. Intensive and strong microseismic responses are generated in areas severely affected by mining activities. Based on the microseismic monitoring data, distribution of resistance in the support in the working face can be predicted, and the range of dense roof fractures estimated. if the working face advanced to this location, a dynamic disaster would need to be prevented. For gas extraction, it is profitable to determine drilling level, where gas migration channels ensure gas drainage at high concentration.

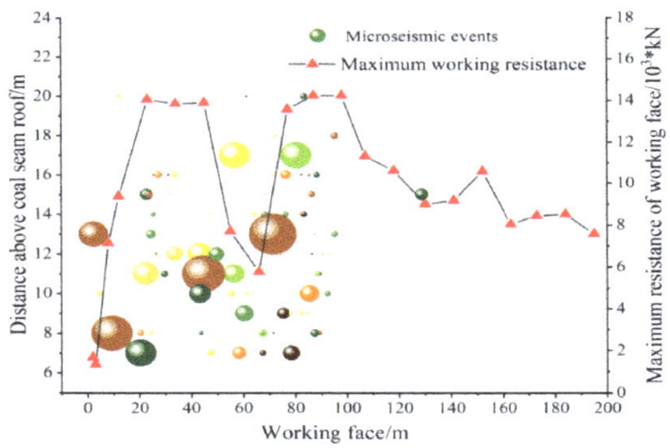

(**a**) 1 July 2021

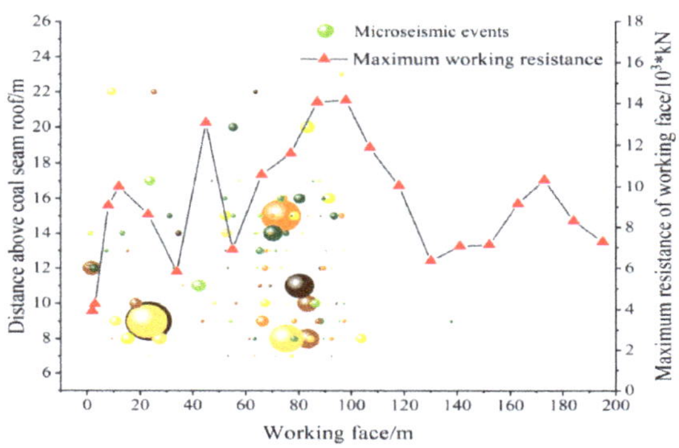

(**b**) 3 July 2021

Figure 5. *Cont.*

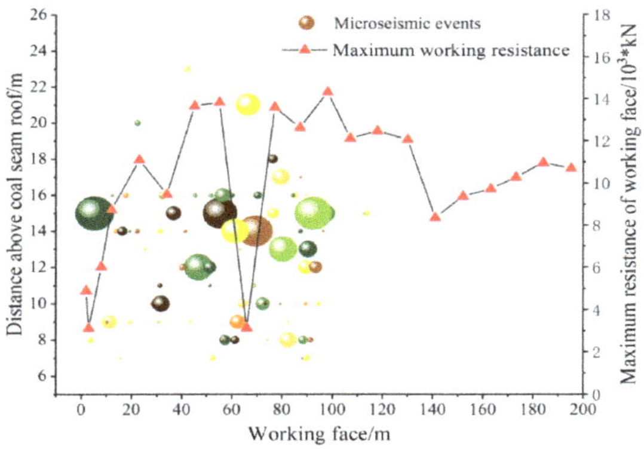

(**c**) 5 July 2021

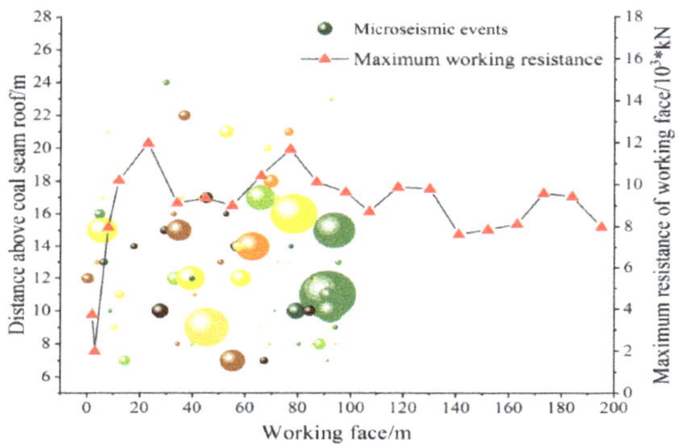

(**d**) 7 July 2021

Figure 5. Microseismic events along the dip of the working face. (**a**) Microseismic events distribution on 1 July 2021; (**b**) Microseismic events distribution on 3 July 2021; (**c**) Microseismic events distribution on 5 July 2021; (**d**) Microseismic events distribution on 7 July 2021. The size of the ball indicates the energy magnitude of microseismic events.

6. Discussion

Based on prediction of overburden fracture evolution, gas drainage boreholes can be designed and arranged in advance. With mining distance, the gas migration channels and storage space change with time, so different drainage methods, such as a buried pipeline, a low level borehole, a high level borehole, a high level suction roadway and directional long drilling maybe recommended at different mining stages (Figure 6), as suggested by Lin et al. (2022) [43]. At the initial stage of mining, there are few overburden cracks, and the gas is close to the bottom of the goaf (Figure 1). Therefore, a pipeline method could be used in this stage. With coal mining, fractures form in overburden strata (Figures 1 and 2), and the drainage boreholes need to be arranged in the roof. As overburden strata develop upward, low-level and high-level boreholes may be adopted successively. For thick coal seam mining, the height of the fractures area are higher than in general coal seam mining. What is more, there is a large amount of gas desorbed by mining fracture in coal seams. This requires a drainage method with pumping capacity. A high-level suction roadway can be applied to decrease gas concentration in a short time. Due to longer service cycles and more secure performance, directional long drilling may be used in gas drainage.

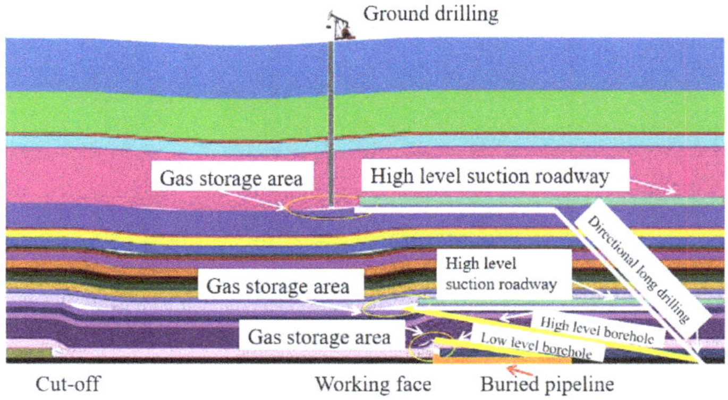

Figure 6. Design and layout of extraction boreholes.

7. Conclusions

To investigate overburden fracture evolution caused by mining disturbance, experiments and numerical simulations were adopted to characterize the development process, and verified by microseismic monitoring. The main conclusions are as follow:

(1) During coal mining, the roof of the coal seam collapses regularly, and overburden fractures developed in an upward direction. With mining distance increasing, the range and height of the overburden fracture areas enlarge, as observed in numerical simulation. From the numerical simulation, stress distribution indicates the stress increases to exceed the bearing capacity of the rock mass, and the overlying strata breaks and collapses, explaining fracture evolution.
(2) Fracture areas near the working face wall and goaf side provide gas migration channels, while the top fractures area of the overlying strata creates a reservoir space for gas. Therefore, overburden fractures can be used to guide drainage borehole design and construction.
(3) Microseismic responses in the mining space display the spatial distribution of overburden fractures, and verified experimental and numerical simulation observations. Energy and frequency characteristics of the seismic sensors reflect fracture scale and range.

Author Contributions: Methodology, Y.C.; formal analysis, Y.C. and X.K.; investigation, Y.C., X.K. and S.L.; resources, S.L. and X.W.; writing—original draft preparation, Y.C. and X.K.; writing—review and editing, P.J. and S.Y.; supervision, X.L., Y.Z. and D.H. All authors have read and agreed to the published version of the manuscript.

Funding: This research was funded by National Natural Science Foundation of China (Grant No. 52074217), Natural Science Basic Research Program of Shaanxi (2020JQ-756), China Postdoctoral Science Foundation (2021M693879), and Outstanding Youth Science Fund of Xi'an University of Science and Technology.

Institutional Review Board Statement: Not applicable.

Informed Consent Statement: Not applicable.

Data Availability Statement: The data used to support the findings of this study are available from the corresponding author.

Conflicts of Interest: The authors declare no conflict of interest.

References

1. Shao, X.; Wu, J.; Zhang, J.; Zhang, J. Study on crack evolution law of overburden strata and stability of interlayer rock in upward coal mining. *Coal Sci. Technol.* **2016**, *44*, 61–66.
2. Zhao, P.; Zhuo, R.; Li, S.; Lin, H.; Shu, C.; Lai, W.; Jia, Y.; Suo, L. Fractal characteristics of gas migration channels at different mining heights. *Fuel* **2021**, *271*, 117479. [CrossRef]
3. Zhao, P.; Zhuo, R.; Li, S.; Shu, C.; Lai, W.; Jia, Y.; Shi, Y.; Suo, L. Analysis of advancing speed effect in gas safety extraction channels and pressure-relief gas extraction. *Fuel* **2020**, *265*, 116825. [CrossRef]
4. Liu, X.F.; Kong, X.G.; Nie, B.S.; Song, D.; He, X.; Wang, L. Pore Fractal Dimensions of Bituminous Coal Reservoirs in North China and Their Impact on Gas Adsorption Capacity. *Nat. Resour. Res.* **2021**, *20*, 4585–4596. [CrossRef]
5. Jia, B.; Chen, Z.; Xian, C. Permeability measurement of the fracture-matrix system with 3D embedded discrete fracture model. *Pet. Sci.* **2022**, *19*, 1757–1765. [CrossRef]
6. Jia, B.; Chen, Z.; Xian, C. Investigations of CO_2 storage capacity and flow behavior in shale formation. *J. Pet. Sci. Eng.* **2022**, *208*, 109659. [CrossRef]
7. Kong, X.; Wang, E.Y.; Liu, X.F.; Chen, L.; Feng, J.; Kong, B.; Li, D.; Liu, Q. Coupled analysis about multi-factors to the effective influence radius of hydraulic flushing: Application of response surface methodology. *J. Nat. Gas Sci. Eng.* **2016**, *32*, 538–548. [CrossRef]
8. Liu, X.; Wang, L.; Kong, X.; Ma, Z.; Nie, B.; Song, D.; Yang, T. Role of pore irregularity in methane desorption capacity of coking coal. *Fuel* **2022**, *314*, 123037. [CrossRef]
9. Kong, X.; He, D.; Liu, X.; Wang, E.; Li, S.; Liu, T.; Ji, P.F.; Deng, D.Y.; Yang, S.R. Strain characteristics and energy dissipation laws of gas-bearing coal during impact fracture process. *Energy* **2022**, *242*, 123028. [CrossRef]
10. Fauchille, A.; Hedan, S.; Valle, V.; Pret, D.; Cabrera, J.; Cosenza, P. Multi-scale study on the deformation and fracture evolution of clay rock sample subjected to desiccation. *Appl. Clay Sci.* **2016**, *132–133*, 251–260. [CrossRef]
11. Yang, N.; Hu, G.; Zhu, J.; Duan, H.; Wang, T.; Li, Y. Evolution of pore-fracture structure and permeability of coal by microwave irradiation under uniaxial compression. *J. Nat. Gas Sci. Eng.* **2022**, *107*, 104759. [CrossRef]
12. Hou, C.; Jiang, B.; Li, M.; Song, Y.; Chen, G. Micro-deformation and fracture evolution of in-situ coal affected by temperature, confining pressure, and differential stress. *J. Nat. Gas Sci. Eng.* **2022**, *100*, 104455. [CrossRef]
13. Xie, H.P.; Lu, J.; Li, C.B.; Li, M.H.; Gao, M.Z. Experimental study on the mechanical and failure behaviors of deep rock subjected to true triaxial stress: A review. *Int. J. Min. Sci. Technol.* **2022**, *32*, 915–950. [CrossRef]
14. Li, D.X.; Wang, E.Y.; Li, Z.H.; Ju, Y.Q.; Wang, D.M.; Wang, X.Y. Experimental investigations of pressure stimulated currents from stressed sandstone used as precursors to rock fracture. *Int. J. Rock Mech. Min. Sci.* **2021**, *145*, 104841. [CrossRef]
15. Yang, S.; Chen, M.; Huang, Y.; Jing, H.; Ranjith, P. An experimental study on fracture evolution mechanism of a non-persistent jointed rock mass with various anchorage effects by DSCM, AE and X-ray CT observations. *Int. J. Rock Mech. Min. Sci.* **2020**, *134*, 104469. [CrossRef]
16. Li, P.; Cai, M.; Gao, Y.; Wang, P.; Miao, S.; Wang, Y. Fracture evolution and failure behavior around an opening in brittle jointed rocks subjected to uniaxial compression. *Theor. Appl. Fract. Mech.* **2022**, *122*, 103651. [CrossRef]
17. Wang, J.; Li, J.; Shi, Z.; Chen, J. Fatigue damage and fracture evolution characteristics of sandstone under multistage intermittent cyclic loading. *Theor. Appl. Fract. Mech.* **2022**, *119*, 103375. [CrossRef]
18. Pirzada, M.; Bahaaddini, M.; Moradian, O.; Roshan, H. Evolution of contact area and aperture during the shearing process of natural rock fractures. *Eng. Geol.* **2021**, *291*, 106236. [CrossRef]
19. Wang, Y.; Deng, H.; Deng, Y.; Chen, K.; He, J. Study on crack dynamic evolution and damage-fracture mechanism of rock with pre-existing cracks based on acoustic emission location. *J. Pet. Sci. Eng.* **2021**, *201*, 108420. [CrossRef]
20. Li, D.X.; Wang, E.Y.; Kong, X.G.; Ali, M.; Wang, D.M. Mechanical behaviors and acoustic emission fractal characteristics of coal specimens with a pre-existing flaw of various inclinations under uniaxial compression. *Int. J. Rock Mech. Min. Sci.* **2019**, *116*, 38–51. [CrossRef]
21. Wang, H.J.; Li, H.Z.; Tang, L.; Li, J.C.; Ren, X.H. Fracturing behavior of brittle solids containing 3D internal crack of different depths under ultrasonic fracturing. *Int. J. Min. Sci. Technol.* **2022**, *32*, 1245–1257. [CrossRef]
22. Sampath, K.; Perera, M.; Elsworth, D.; Matthai, S.; Ranjith, P.; Li, D. Discrete fracture matrix modelling of fully-coupled CO2 flow-Deformation processes in fractured coal. *Int. J. Rock Mech. Min. Sci.* **2021**, *138*, 104644. [CrossRef]
23. Ogata, S.; Yasuhara, H.; Kinoshita, N.; Inui, T.; Nishira, E.; Kishida, K. Numerical analyses of coupled thermal–hydraulic–mechanical–chemical processes for estimating permeability change in fractured rock induced by alkaline solution. *Geomech. Energy Environ.* **2022**, *31*, 100372. [CrossRef]
24. Han, W.; Jiang, Y.; Luan, H.; Liu, J.; Xu, X.; Du, Y. Fracture evolution and failure mechanism of rock-like materials containing cross-flaws under the shearing effect. *Theor. Appl. Fract. Mech.* **2020**, *110*, 102815. [CrossRef]
25. Ju, Y.; Wang, Y.; Su, C.; Zhang, D.; Ren, Z. Numerical analysis of the dynamic evolution of mining-induced stresses and fractures in multilayered rock strata using continuum-based discrete element methods. *Int. J. Rock Mech. Min. Sci.* **2019**, *113*, 191–210. [CrossRef]
26. Vazaios, I.; Vlachopoulos, N.; Diederichs, M. Assessing fracturing mechanisms and evolution of excavation damaged zone of tunnels in interlocked rock masses at high stresses using a finitediscrete element approach. *J. Rock Mech. Geotech. Eng.* **2019**, *11*, 701–722. [CrossRef]

27. Park, K.; Choi, C.; Park, B.; Ji, S. Numerical and experimental investigation of the hydrogeological evolution around a borehole due to stress redistribution in fractured rock. *Tunn. Undergr. Space Technol.* **2022**, *124*, 104461. [CrossRef]
28. Bouissou, S.; Darnault, R.; Chemenda, A.; Rolland, Y. Evolution of gravity-driven rock slope failure and associated fracturing: Geological analysis and numerical modelling. *Tectonophysics* **2012**, *526–529*, 157–166. [CrossRef]
29. Yin, G.; Li, X.; Han, P.; Li, M.; Li, W.; Deng, B. Experimental study on overburden strata fracture evolution law in three dimensional mine-induced stress conditions. *J. China Coal Society.* **2016**, *41*, 406–413.
30. Qiao, X. Failure characteristics and fracture evolution law of overburden of thick coal in fully mechanized sub-level caving mining. *J. Eng. Geol.* **2017**, *25*, 858–866.
31. Zhao, Y.; Ling, C.; Liu, B.; He, X. Fracture evolution and energy dissipation of overlying strata in shallow-buried underground mining with ultra-high working face. *J. Min. Saf. Eng.* **2021**, *38*, 9–18, 30.
32. Kong, X.; Wang, E.; Liu, Q.; Li, Z.; Li, D.; Cao, Z.; Niu, Y. Dynamic permeability and porosity evolution of coal seam rich in CBM based on the flow-solid coupling theory. *J. Nat. Gas Sci. Eng.* **2017**, *40*, 61–71. [CrossRef]
33. Jiao, Z.; Tao, G.; Wang, H.; Lu, Z. Overburden strata movement and fissure evolution in lower protective layer in Jincheng mining district. *J. Min. Saf. Eng.* **2017**, *34*, 85–90.
34. Yang, B.; Yuan, S.; Zheng, D.; Liu, J.; Kang, J.; Ma, L.; Zhou, Y. Spatial and temporal characteristics of overburden fractures due to repeated mining in close distance coal seams. *J. Min. Saf. Eng.* **2022**, *39*, 255–263.
35. Li, S.; Liu, L.; Zhao, P.; Lin, H.; Xu, P.; Zhuo, R. Analysis and application of fracture evolution law of overburden compacted area on fully-mechanized mining face under multiple factors. *Coal Sci. Technol.* **2022**, *50*, 95–104.
36. Xiao, P.; Han, K.; Shaung, H.; Wu, M.; Gao, Z. Similar material simulation test study on evolution law of overburden fracture based on microseismic monitoring. *Coal Sci. Technol.* **2022**, *50*, 48–56.
37. Xue, Y.; Liu, J.; Ranjith, P.G.; Gao, F.; Xie, H.; Wang, J. Changes in microstructure and mechanical properties of low-permeability coal induced by pulsating nitrogen fatigue fracturing tests. *Rock Mech. Rock Eng.* **2022**, *55*, 7469–7488. [CrossRef]
38. Xue, Y.; Ranjith, P.G.; Chen, Y.; Cai, C.; Gao, F.; Liu, X. Nonlinear mechanical characteristics and damage constitutive model of coal under CO2 adsorption during geological sequestration. *Fuel* **2023**, *331*, 125690. [CrossRef]
39. Izadi, G.; Wang, S.; Elsworth, D.; Liu, J.; Wu, Y.; Pone, D. Permeability evolution of fluid-infiltrated coal containing discrete fractures. *Int. J. Coal Geol.* **2011**, *85*, 202–211. [CrossRef]
40. Akhondzadeh, H.; Keshavarz, A.; Al-Yaseri, A.; Ali, M.; Awan, F.; Wang, X.; Yang, Y.; Iglauer, S.; Lebedev, M. Pore-scale analysis of coal cleat network evolution through liquid nitrogen treatment: A Micro-Computed Tomography investigation. *Int. J. Coal Geol.* **2020**, *219*, 103370. [CrossRef]
41. Bertrand, F.; Buzzi, O.; Collin, F. Cleat-scale modelling of the coal permeability evolution due to sorptioninduced strain. *Int. J. Coal Geol.* **2019**, *216*, 103320. [CrossRef]
42. Koyama, T.; Fardin, N.; Jing, L.; Stephansson, O. Numerical simulation of shear-induced flow anisotropy and scale-dependent aperture and transmissivity evolution of rock fracture replicas. *Int. J. Rock Mech. Min. Sci.* **2006**, *43*, 89–106. [CrossRef]
43. Lin, H.F.; Wang, X.; Xu, P.Y.; Kong, X.G.; Shuang, H.Q.; Zhao, P.X. Evolution characteristics and engineering application of pressure relief gas reservoir area in extra-thick coal seam. *Coal Sci. Technol.* **2022**. [CrossRef]

sustainability

MDPI

Article

Simulation Research on Energy Evolution and Supply Law of Rock–Coal System under the Influence of Stiffness

Yanchun Yin [1,2], Xingxue Tang [2,*], Qinwei Ma [1], Zhenan Li [3] and Wei Zhang [4]

[1] School of Aerospace Engineering, Beijing Institute of Technology, Beijing 100081, China
[2] College of Energy and Mining Engineering, Shandong University of Science and Technology, Qingdao 266590, China
[3] Jining Energy Comprehensive Law Enforcement Detachment, Jining 272000, China
[4] Shandong Tangkou Coal Co., Ltd., Jining 272055, China
* Correspondence: bqt2200601014@student.cumtb.edu.cn

Abstract: The energy supply effect caused by the stiffness difference between roofs and sidewalls is an important factor that induces strain coal bursts. In order to quantitatively reveal the energy supply mechanism of strain coal bursts, this paper first establishes a coal burst energy model of the rock–coal system and proposes the calculation formula of coal burst kinetic energy considering supply energy and the stiffness ratio of rock to coal. Then the whole energy evolution law of the rock–coal system with different stiffness ratios is researched by using the numerical simulation method, and the whole process is divided into three stages. With the decrease in the stiffness ratio, the elastic strain energy of the coal changes little, while its kinetic energy is negatively correlated with the stiffness ratio in a power function. Meanwhile, the elastic strain energy and kinetic energy of the rock have power function relations with the stiffness ratio, too. When the rock–coal system is fractured, the kinetic energy of the coal comes from the release of elastic strain energy from the coal and the energy supplied from the rock. The energy supply rate is between 22% and 35% when the stiffness ratio changes from 3.0 to 0.5, and they show a linear relationship, while the supplied energy has a negative power function relationship with the stiffness ratio.

Keywords: coal burst; rock–coal system; stiffness; energy evolution; energy supply

Citation: Yin, Y.; Tang, X.; Ma, Q.; Li, Z.; Zhang, W. Simulation Research on Energy Evolution and Supply Law of Rock–Coal System under the Influence of Stiffness. *Sustainability* **2023**, *15*, 1335. https://doi.org/10.3390/su15021335

Academic Editors: Xiangguo Kong, Dexing Li and Xiaoran Wang

Received: 18 September 2022
Revised: 1 January 2023
Accepted: 6 January 2023
Published: 10 January 2023

1. Introduction

A coal burst is a typical dynamic disaster in underground coal mining. Its frequency and intensity increase with the mining depth, which seriously threatens the safety and production of underground workers [1–3]. In the next five years or even ten years, the prevention and control of coal bursts will be the focus of much research and poses an urgent problem to be solved in mining engineering.

According to the mechanism of coal bursts, scholars have put forward more than ten theories and described several types of coal bursts from different angles [4–6]. Among them, a typical type is a strain coal burst caused by high static stress. A strain coal burst is a sudden event induced by energy accumulation and release, as well as the loss in the dynamic balance of the combined coal and rock mass composed of a roof, sidewall, and floor [7,8]. For a strain coal burst, when the elastic strain energy stored in the coal is greater than the failure dissipation energy, the residual energy will be released in the form of a coal burst [9,10]. However, dynamic damage is unlikely to occur only through the elastic strain energy stored by the coal itself; the surrounding rock mass is required to supply its energy to the coal. According to the stiffness theory [11,12], some experts point out that when the stiffness of the surrounding rock is less than that of the coal, the surrounding rock will supply energy to the damaged coal, thus causing a coal burst [13,14]. Li et al. [15] proposed two concept models to illustrate the energy conversions in a rockburst event, and studied the role of the elastic strain energy released from the surrounding rock mass in rock ejection. Cai et al. [16–18] developed one brittle hard rock testing system with a super high and variable loading stiffness, and studied the influence of loading stiffness on the post-peak stress–strain curves and failure modes of rocks. Zhao et al. [19] developed a rock testing system with changeable stiffness, and researched the mechanical behaviors of rock specimens and testing machines under different loading stiffnesses. The above research reveals the influence mechanism of stiffness on the dynamic failure of the coal (or rock) from the stress–strain curve and the failure characteristics.

A coal burst is a dynamic failure phenomenon driven by energy. Therefore, it is better to reveal the essential characteristics of coal bursts by studying the energy evolution law [20–23]. In the research on the stiffness theory of coal bursts, limited by the test methods, the research on the energy parameters in the current stage is less and generally only involves the analysis of the storage and release of elastic strain energy in the two states of the peak strength point and post-failure [19]. The energy evolution law of the whole process of coal bursts and the energy supply of surrounding rock to the coal are difficult to determine quantitatively, which creates certain limitations for coal burst mechanism research.

In order to quantitatively reveal the stiffness effect on coal burst from the perspective of energy, a rock–coal system model is established by means of numerical simulation, and the energy evolution law and energy supply characteristics of rock and coal under different stiffness combinations are studied. The quantitative relationship between energy parameters and stiffness is obtained.

2. Theoretical Analysis

2.1. Stiffness Theory

The load–deformation curve during the loading process of the testing machine specimen system is shown in Figure 1. When the specimen is damaged after the peak point, the load of the testing machine and the load of the specimen decrease synchronously, the specimen continues to compress, and the deformation of the testing machine recovers. When the stiffness k_{m1} of the testing machine is greater than the slope k_s of the post-peak curve of the specimen, the released elastic strain energy of the testing machine is less than the dissipation energy that causes the fracture of the specimen. In order to fracture the specimen, the testing machine needs to continue to input energy to the specimen. At this moment, the loading process is controllable, and the specimen is statically damaged [24]. When the stiffness k_{m2} of the testing machine is less than the slope k_s of the post-peak curve of the specimen, the released elastic strain energy of the testing machine is greater than the dissipation energy that causes the fracture of the specimen, and the residual energy will be converted into kinetic energy of the specimen, thus causing the dynamic failure of the specimen.

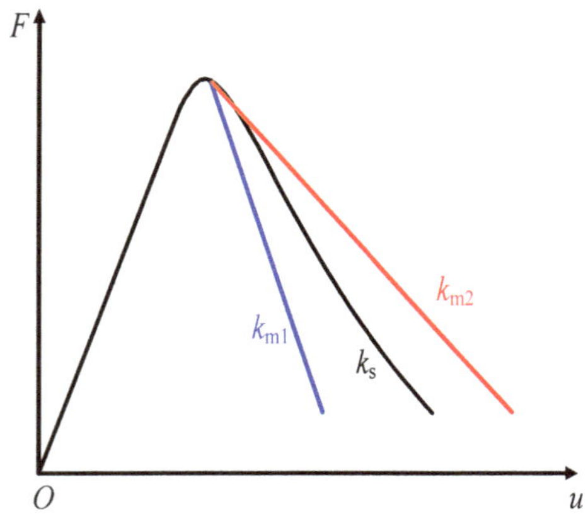

Figure 1. Load–deformation curve of the testing machine specimen system.

2.2. Coal Burst Model of Rock–Coal System

According to the stiffness theory, the sidewall coal of the roadway is the main burst body in the process of coal burst, and the low-stiffness roof supplies energy to the coal. According to the load and deformation relationship between the roof rock and sidewall coal [25,26], it can be simplified as a series-connected rock–coal system model, as shown in Figure 2. The roof stiffness is k_r and the sidewall stiffness is k_c, then the relationship between the sidewall deformation x_c and the roof deformation x_r is:

$$x_r = \frac{k_c}{k_r} x_c \tag{1}$$

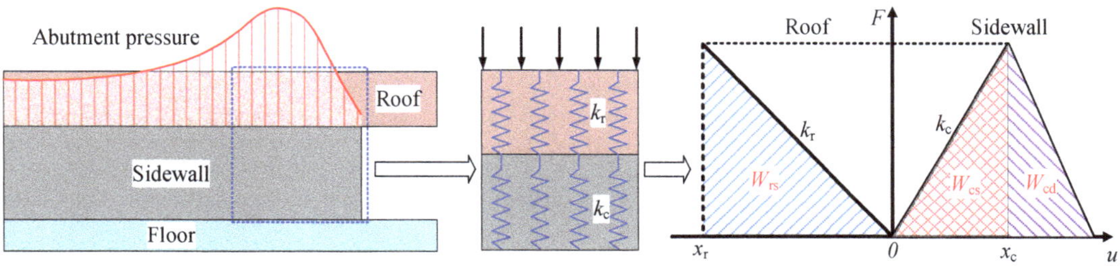

Figure 2. Rock–coal system and their loading–deformation relation.

When the overburdened strata load onto the surrounding rock of the roadway, part of the external input energy is stored in the roof rock, and the other part is stored in the sidewall coal. The elastic strain energy W_{cs} and W_{rs} of the sidewall and roof are, respectively:

$$W_{cs} = \frac{1}{2} k_c x_c^2 \tag{2}$$

$$W_{rs} = \frac{1}{2} k_r x_r^2 = \frac{1}{2} \frac{k_c^2}{k_r} x_c^2 \tag{3}$$

The total elastic strain energy W_{ts} of the rock–coal system is:

$$W_{ts} = \frac{1}{2}\left(k_c + \frac{k_c^2}{k_r}\right) x_c^2 \tag{4}$$

The extreme value of elastic strain energy is certain for the same sidewall coal. The smaller the stiffness k_r of the roof, the greater the extreme value of the elastic strain energy W_{ts} in the rock–coal system. When the rock–coal system transforms from the equilibrium state to the unstable equilibrium state, more elastic strain energy W_{ts} will be released and the coal burst risk will increase.

When the coal reaches the ultimate strength, it will be fractured under the external load, and its stored elastic strain energy W_{cs} will be released. When the sidewall coal is considered separately, part of the elastic strain energy W_{cs} is released in the form of fracture dissipation energy W_{cd}, and the residual energy is converted into kinetic energy W_{ck} [9], which is calculated as:

$$W_{ck} = W_{cs} - W_{cd} \tag{5}$$

For the coal, the greater the elastic strain energy W_{cs}, the smaller the post-peak fracture dissipation energy W_{cd}, and more energy will be released in the form of kinetic energy W_{ck}, showing a strong coal burst tendency.

For the rock–coal system, when the coal is fractured, the elastic strain energy W_{rs} of the roof rock will also be released and a certain amount of energy will be supplied to the coal. Since the strength of rock is generally greater than that of coal, it is assumed that plastic deformation does not occur on the rock. Part of the elastic strain energy W_{rs} released by the rock is released in the form of kinetic energy W_{rk}, and the residual energy is the supply energy $W_{r\text{-}c}$ to the coal. The ratio of the supply energy $W_{r\text{-}c}$ to the elastic strain energy W_{rs} of rock is defined as the energy supply rate α, so the kinetic energy W_{ck} of coal in the rock–coal system is:

$$W_{ck} = W_{cs} - W_{cd} + \alpha W_{rs} = \left(1 + \alpha \frac{k_c}{k_r}\right) W_{cs} - W_{cd} \tag{6}$$

$$\alpha = W_{r\text{-}c} / W_{rs} \tag{7}$$

For the rock–coal system, when the stiffness k_r of the roof decreases, the elastic strain energy W_{rs} stored in the rock increases, and more energy in the rock can be supplied to the coal, which causes greater risk of coal burst.

3. Numerical Simulation

The above theoretical analysis qualitatively reveals the energy supply behavior of the low-stiffness roof to the failure of the sidewall coal, and the energy and stiffness influence mechanism of coal burst. In the following, the numerical simulation method is used to quantitatively study the energy evolution law of rock–coal system under different stiffness combinations.

3.1. Simulation Scheme

The instability and failure of rock–coal system were simulated by Abaqus software. The numerical model of rock–coal system is shown in Figure 3. The lower part of the model is the sidewall coal, which is a rectangular standard specimen with a size of 50 mm × 50 mm × 100 mm. Drucker–Prager failure criterion and shear damage model were selected to simulate the damage and failure process of the coal [27]. The parameters were determined by comparing the stress–strain curve of the coal model with the laboratory test by using trial-and-error method [28]. The specific parameters are listed in Table 1. The stiffness of the coal specimen is 42.5 MN/m.

Table 1. Parameters of the coal.

Density/kg·m^{-3}	Elastic Modulus/GPa	Poisson's Ratio	Angle of Friction/°	Flow Stress Ratio	Dilation Angle/°
1263	1.7	0.21	60	0.778	60
Yield Stress/MPa	**Abs Plastic Strain**	**Fracture Strain**	**Shear Stress Ratio**	**Strain Ratio**	**Displacement at Failure/mm**
20	0.02	0.015	0.33	0.0001	0.001

The roof rock is above the coal, which is a rectangular specimen with a size of 70 mm × 70 mm × 100 mm. In a roadway coal burst, coal sidewall is the main burst body, and the roof rock is the main loading body [29,30]. The strength of rock is greater than that of the coal, and the rock remains in the pre-peak stage when the coal is fractured. Therefore, in the simulation process, the failure of rock was not considered, and the elastic property was applied to the rock with a density of 2000 kg/m^3 and a Poisson's ratio of 0.25 [31]. This paper focuses on the energy evolution law of rock–coal system under different stiffness ratios k_r/k_c of rock to coal, and performs loading tests for different stiffness ratios by changing the elastic modulus of rock. The stiffness ratios k_r/k_c of rock to coal designed in the simulation are 0.5, 1.0, 2.0, and 3.0.

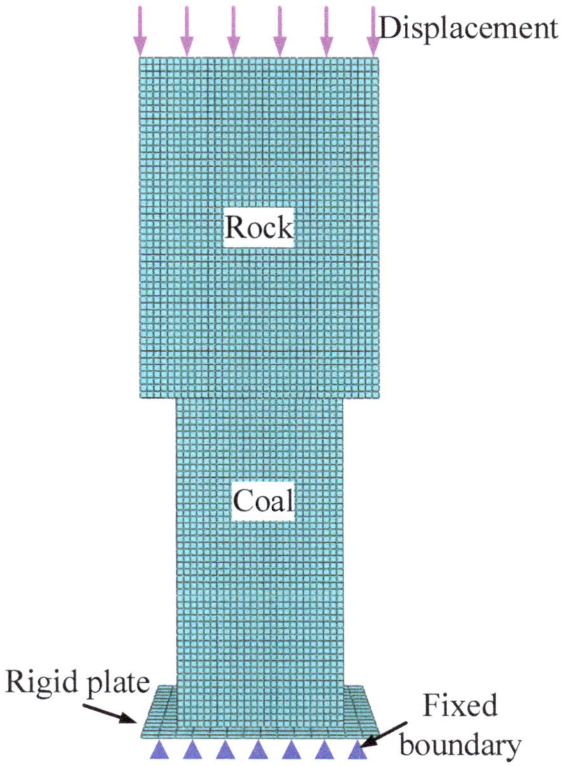

Figure 3. Numerical model of the rock–coal system.

Between the rock and coal, the tangential behavior and normal behavior of the contact surface are defined in the simulation. In order to eliminate the influence of end-face friction on the test results [32], the friction coefficient in the tangential direction is set at zero. In the normal direction, it is set as the "Hard" contact to ensure that there is no model penetration when the rock model and the coal model are in contact with each other.

C3D8R eight-node linear hexahedral element was selected as the finite element mesh type of the model, and Abaqus dynamic display solver was used for calculation. Vertical downward axial displacement was applied to the top of the model for loading, and the rigid plate under the model is fixed. The total axial deformation u of the rock–coal system was recorded.

3.2. *Energy Monitoring*

The energy involved in the whole process of rock–coal system failure includes: elastic strain energy W_{cs}; dissipation energy W_{cd} and kinetic energy W_{ck} of coal; and elastic strain energy W_{rs} and kinetic energy W_{rk} of rock. According to theoretical analysis, the increase in strain energy W_{cs} of coal and elastic strain energy W_{rs} of rock aggravates the dynamic damage degree of the coal [33], while the increase in dissipation energy W_{cd} of coal weakens its dynamic damage degree. The magnitude of kinetic energy W_{ck} of coal is directly related to its dynamic damage degree. Elastic strain energy W_{cs} of coal and elastic strain energy W_{rs} of rock are superposed and transformed into dissipation energy W_{cd} and kinetic energy W_{ck} of coal and kinetic energy W_{rk} of rock. Therefore, the simulation mainly monitors the energy parameters, such as elastic strain energy W_{cs}, dissipation energy W_{cd}

and kinetic energy W_{ck} of coal, and elastic strain energy W_{rs} and kinetic energy W_{rk} of rock. The supply energy $W_{r\text{-}c}$ of rock to coal can be calculated by the following formula:

$$W_{r\text{-}c} = W_{cd} + W_{ck} - W_{cs} \tag{8}$$

In Abaqus, various energy output quantities are provided, such as recoverable strain energy (ALLSE), as well as energy dissipated by rate-independent and rate-dependent plastic deformation (ALLPD) and kinetic energy (ALLKE). In these energy output quantities, ALLSE is the elastic strain of coal or rock, ALLPD is the dissipation energy of coal, and ALLKE is the kinetic energy of coal or rock.

4. Results and Analysis

4.1. Failure Mode

The failure modes of the coal specimens under different stiffness ratios k_r/k_c are shown in Figure 4. When the stiffness ratio k_r/k_c is 0.5, the fragments falling from the specimen are ejected quickly and show obvious burst characteristics. When the stiffness ratio k_r/k_c increases to 1.0, the ejecting velocity of the fragments obviously decreases. When the stiffness ratio k_r/k_c is 3.0 or 2.0, cracks on the post-peak coal specimen are generated slowly, the progressive stripping of the fragments occurs locally in the model, and the failure process is stable. The failure mode and variation law of coal under different loading stiffnesses are similar to those of laboratory testing results [30].

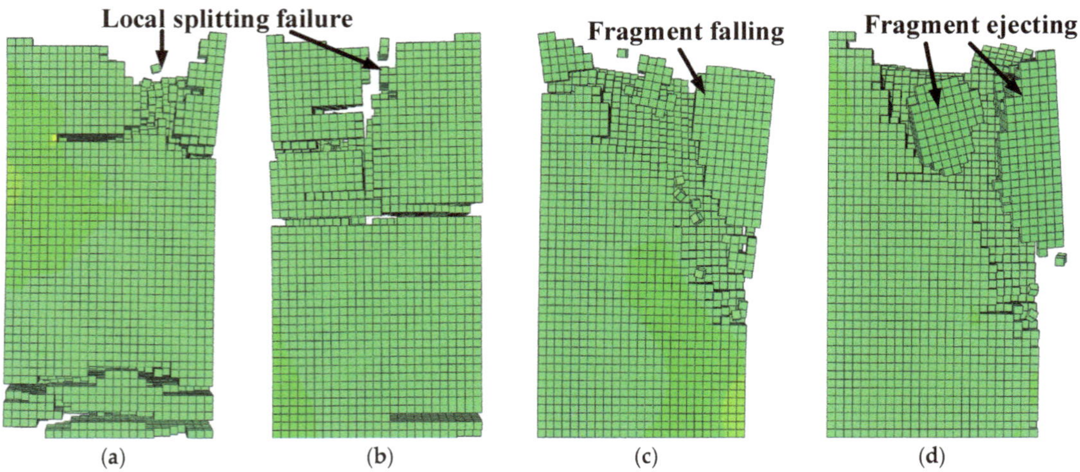

Figure 4. Failure mode of coal specimen under the stiffness ratios of (**a**) 3.0, (**b**) 2.0, (**c**) 1.0, and (**d**) 0.5.

4.2. Strain Energy of Coal and Rock

The evolution curves of the elastic strain energy of coal under different stiffness ratios k_r/k_c are shown in Figure 5a. Under different stiffness ratios k_r/k_c, the extreme value of the elastic strain energy W_{cs} of coal is between 34.2 J and 34.6 J, with little change. When the rock stiffness k_r changes, the extreme value of the elastic strain energy W_{cs} of coal will not change, and the rock stiffness k_r has little effect on the energy storage property of the coal itself. When the elastic strain energy W_{cs} of the coal reaches the extreme value of energy storage, the elastic strain energy curve tends to flatten. At this time, the plastic deformation of the coal occurs and the elastic strain energy W_{cs} does not increase. Under the constant displacement loading speed, when the stiffness ratio k_r/k_c decreases, the duration of the flattened section of the elastic strain energy W_{cs} is shortened. For example, when the stiffness ratio k_r/k_c is 3, the flattened section of the elastic strain energy W_{cs} needs to be continuously loaded by 0.5 mm, and when the stiffness ratio k_r/k_c is 0.5, it is reduced to 0.1 mm.

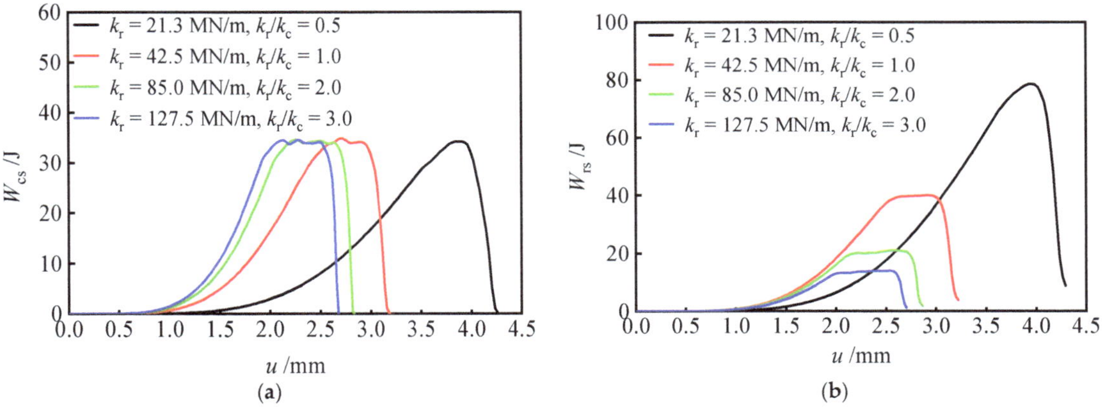

Figure 5. Elastic strain energy evolution curves of (**a**) coal and (**b**) rock.

The elastic strain energy evolution curves of rock are shown in Figure 5b. Combined with the analysis of the elastic strain energy evolution law of coal, when the coal is fractured and its internal elastic strain energy W_{cs} is released, the elastic strain energy W_{rs} of the rock is also released; that is, the deformation rebound phenomenon of the rock occurs [19]. There is an obvious difference between the elastic strain energy W_{rs} of rocks under different stiffness ratios k_r/k_c. When the stiffness ratio k_r/k_c is 0.5, the extreme value of the rock elastic strain energy W_{rs} is about 80 J. When the stiffness ratio k_r/k_c increases to 3.0, the extreme value of the rock elastic strain energy W_{rs} decreases to 13 J. There are also obvious changes in the release rate of rock elastic strain energy. When the stiffness ratio k_r/k_c is 3.0, the release rate of rock elastic strain energy is 11.7 J/s, and when the stiffness ratio k_r/k_c is 0.5, it increases to 21.5 J/s. Compared with rock with high stiffness, rock with low stiffness can accumulate more elastic strain energy, and its energy release rate is faster. According to the stiffness theory, in a rock–coal system, low-stiffness rock will generate a stronger energy supply. The higher elastic strain energy and faster energy release rate of low-stiffness rock are the manifestations of this energy supply.

4.3. Dissipation Energy of Coal

The evolution curves of dissipation energy of coal are shown in Figure 6. Compared with the evolution of coal elastic strain energy in Figure 5a, the dissipation energy W_{cd} starts to increase before the coal elastic strain energy W_{cs} reaches the extreme value, indicating that there are micro-cracks generating in the coal before its failure, and some energy is dissipated. When the elastic strain energy W_{cs} of the coal is fully released and is reduced to 0, the dissipation energy W_{cd} reaches the maximum value, and the coal is completely fractured. With the decrease in the stiffness ratio k_r/k_c, the increasing rate of the dissipation energy W_{cd} of the coal changes little, but the total dissipation energy W_{cd} decreases gradually. When the stiffness ratio k_r/k_c is 3.0, the dissipation energy W_{cd} of coal is 73.8 J, and when the stiffness ratio k_r/k_c is 0.5, the dissipation energy W_{cd} is reduced to 43.3 J. When the roof stiffness k_r decreases, the dissipation energy W_{cd} of coal decreases, which indicates that the rapid rebound of the roof rock leads to a decrease in the coal's plasticity and an increase in brittleness.

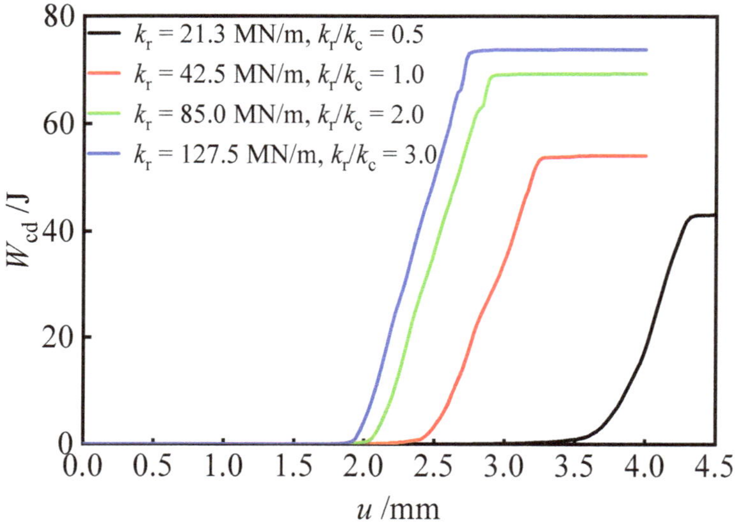

Figure 6. Dissipation energy evolution curves of coal.

4.4. Kinetic Energy of Coal and Rock

The kinetic energy evolution curves of coal and rock are shown in Figure 7. The laboratory tests show that [19,30] the deformation rebound phenomenon obviously occurs when the specimen is fractured, and the rebound speed is negatively correlated with the stiffness of the testing machine; that is, the smaller the stiffness of the testing machine, the faster the deformation rebound speed of the testing machine. For a rock–coal system, the rock is equivalent to a testing machine, and the coal is equivalent to a specimen. When the coal is fractured, the elastic strain energy W_{rs} of the rock is released, and its deformation rebounds. Part of the elastic strain energy W_{rs} is converted into its kinetic energy W_{rk}, and the other part of the elastic strain energy W_{rs} is supplied to the coal specimen and converted into the kinetic energy W_{ck} of the coal. With the decrease in the rock stiffness k_r, the elastic strain energy W_{rs} released by the rock increases when it rebounds, and its kinetic energy W_{rk} increases. The kinetic energy W_{ck} of the coal increases too. When the stiffness ratio k_r/k_c is 3.0, the kinetic energy W_{rk} of the rock is about 11.11 J, and the kinetic energy W_{ck} of the coal is 24.52 J. When the stiffness ratio k_r/k_c is 0.5, the kinetic energy W_{rk} of the rock is 47.69 J and the kinetic energy W_{ck} of the coal is 30.90 J. There is an obvious difference in the energy supply of the rock to coal under the loading of the two kinds of rock stiffnesses. The increase in the kinetic energy W_{rk} of the rock indicates that the rebound speed of the rock is accelerated when the coal is fractured, the loading speed of the deformation rebound of the rock to the coal is faster, and the energy supply is stronger. After the rock stiffness k_r decreases, the dissipation energy W_{cd} of the coal decreases and the kinetic energy W_{ck} increases, indicating that the energy dissipated by the coal decreases under the condition of low-stiffness loading, and more energy is converted into its own kinetic energy W_{ck}. The kinetic energy W_{ck} is a manifestation of the ejecting velocity of coal fragments. The greater kinetic energy W_{ck} of the coal, the faster the ejecting speed of the fragments. The coal shows stronger burst failure characteristics.

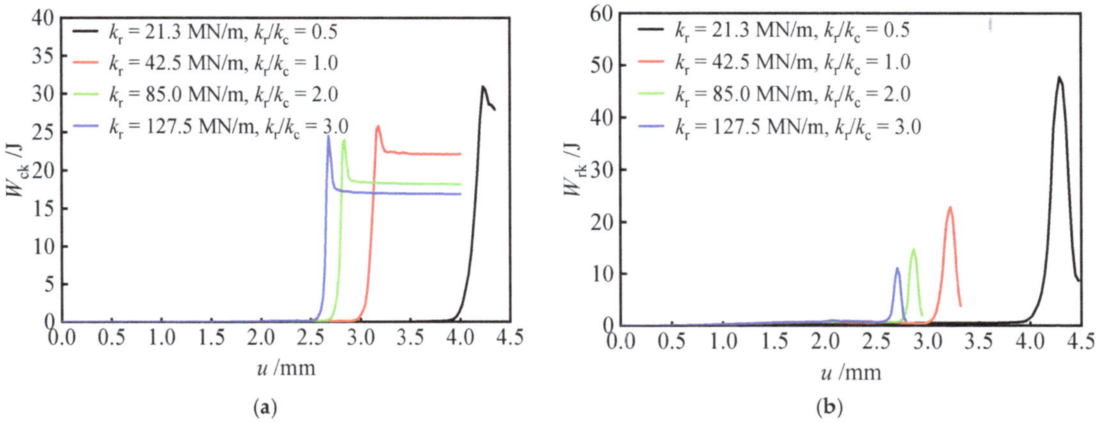

Figure 7. Kinetic energy evolution curves of (**a**) coal and (**b**) rock.

4.5. Energy Evolution Law of Rock–Coal System

When the stiffness ratio k_r/k_c is 0.5, the energy evolution curves of the coal and rock are shown in Figure 8. The energy evolution process can be divided into three stages. In the stage I, the coal and rock bear the load together, and their strain energy gradually increases, converting the external input energy into internal elastic strain energy. The dissipation energy W_{cd} of the coal is zero. In the stage II, before the internal elastic strain energy W_{cs} of the coal reaches its storage limit, the coal will undergo plastic deformation, and part of the external input energy will be dissipated in the form of plastic deformation energy. At this time, the internal elastic strain energy of the coal and rock will continue to increase. In the stage III, the internal elastic strain energy W_{cs} of the coal reaches its energy storage limit and begins to decrease (i.e., energy release). The internal elastic strain energy of the coal and the rock is released synchronously. Part of the elastic strain energy W_{rs} of the rock is converted into its own kinetic energy W_{rk}, and the other part is supplied to the coal. The elastic strain energy W_{cs} of the coal is released in the form of kinetic energy W_{ck} and dissipation energy W_{cd} until the elastic strain energy W_{cs} drops to zero and the coal is completely fractured.

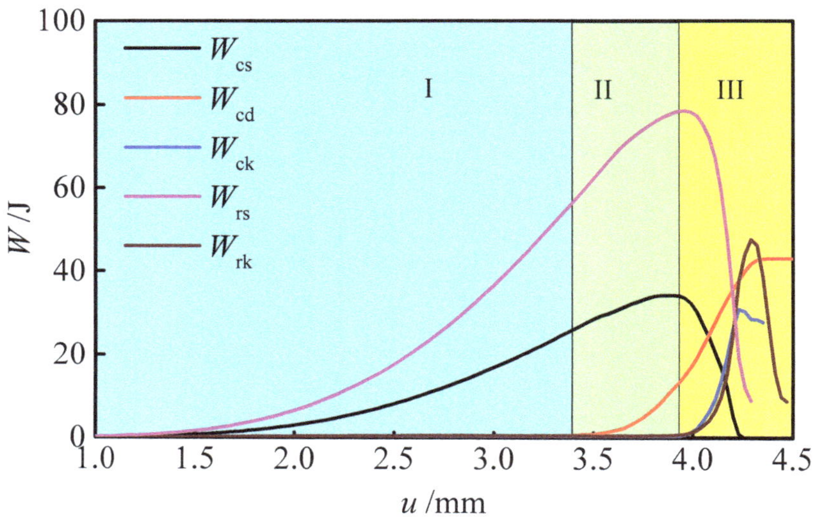

Figure 8. Energy evolution curves of coal and rock.

The relationship between the energy and stiffness ratio is shown in Figure 9. It should be noted that, in order to visually show the changing law of energy parameters of the rock–coal system when the stiffness ratio k_r/k_c is decreased, the stiffness ratio k_r/k_c decreases along the x-axis.

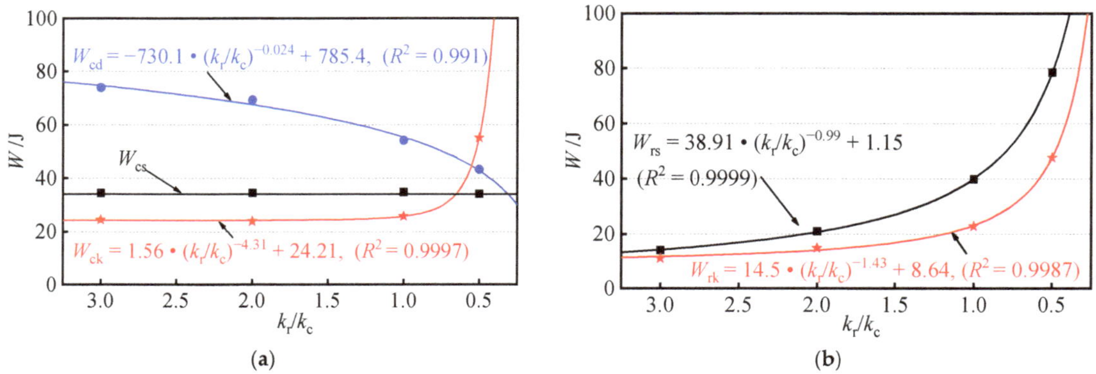

Figure 9. The energy of (**a**) coal and (**b**) rock changing curves with stiffness ratio.

The change in the rock stiffness k_r mainly affects the dissipation energy W_{cd} and kinetic energy W_{ck} of the coal as well as the elastic strain energy W_{rs} and kinetic energy W_{rk} of the rock, and has little effect on the elastic strain energy W_{cs} of the coal. With the decrease in the stiffness ratio k_r/k_c, the dissipation energy W_{cd} of the coal decreases gradually, while the kinetic energy W_{ck} of the coal, the elastic strain energy W_{rs}, and the kinetic energy W_{rk} of the rock increase gradually, all of which have power function relationships with the stiffness ratio k_r/k_c. Compared with other energy, the kinetic energy W_{ck} of coal is more sensitive to the change in the stiffness ratio k_r/k_c. When the stiffness ratio k_r/k_c is greater than one, the kinetic energy W_{ck} of the coal changes slowly, and the influence of the stiffness change on the failure form of the coal is weak. When the stiffness ratio k_r/k_c is less than one, the kinetic energy W_{ck} of the coal increases sharply with the decrease in the stiffness ratio k_r/k_c. When the rock stiffness k_r is less than the coal stiffness k_c, the rock deformation rebound speed is faster and more energy is released, resulting in the accelerated failure speed of the coal after the peak point. The micro-cracks in the coal rapidly expand into the fracture zone in a very short time. The energy dissipated by the coal is reduced, and its kinetic energy W_{ck} is increased.

With the decrease in the stiffness ratio k_r/k_c, the difference between the elastic strain energy W_{rs} and kinetic energy W_{rk} of the rock increases, and more energy is supplied to the coal. The changing curves of the supplied energy $W_{r\text{-}c}$ and energy supply rate α are shown in Figure 10. The supplied energy $W_{r\text{-}c}$ and energy supply rate α increase with the decrease in the stiffness ratio k_r/k_c. The supplied energy $W_{r\text{-}c}$ and the stiffness ratio k_r/k_c have a power function relationship, and the energy supply rate α and the stiffness ratio k_r/k_c have a linear relationship. The lower the rock stiffness k_r, the more energy supply to the coal [19,30].

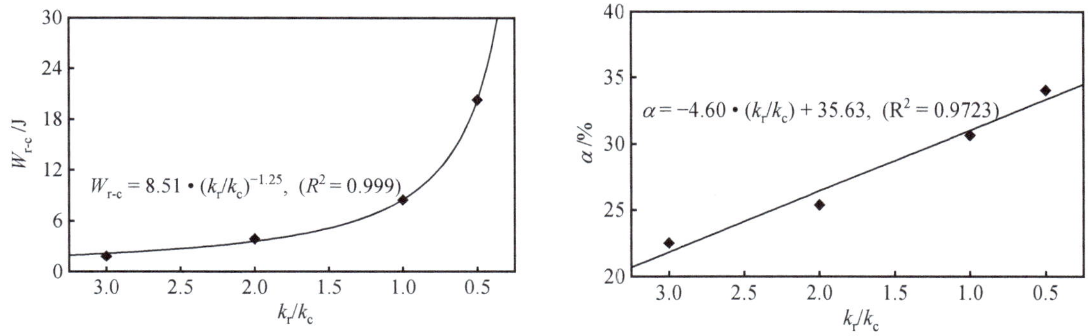

(a) (b)

Figure 10. Changing curves of (**a**) supplied energy and (**b**) energy supply rate.

5. Discussion

In the previous research on the energy mechanism of coal bursts, the strain energy release of coal is the main energy resource. The effect of the surrounding rocks on the coal is mainly reflected in the kinetic energy of the fractured overburdened rock. The commonly used energy criterion is [4]:

$$W_{cs} + W_k > W_{cd} \tag{9}$$

where W_k is the kinetic energy of the fractured overburdened rock.

Based on the stiffness theory and the simulation research above, the roof's energy supply plays an important role in coal bursts. When the stiffness ratio k_r/k_c is 3.0, 2.0, 1.0, and 0.5, the roof's energy supply $W_{r\text{-}c}$ is 1.822 J, 3.871 J, 8.459 J, and 20.293 J. The ratios of the supplied energy $W_{r\text{-}c}$ to the coal elastic strain W_{cs} are 5.30%, 11.25%, 24.59%, and 58.9%, respectively. When the stiffness ratio k_r/k_c is less than the critical value, the roof's energy supply is the key energy source affecting the coal failure mode. Therefore, in the coal burst mechanism, the roof stiffness and energy supply should be considered. Based on Equation (6) and Figure 10b, the coal burst energy criterion can be modified as:

$$k_r/k_c \leq \lambda, (1 + a \cdot k_c/k_r)W_{cs} + W_k > W_{cd} \tag{10}$$

$$k_c/k_r > \lambda, W_{cs} + W_k > W_{cd} \tag{11}$$

The above criterion comprehensively takes into account the stiffness theory and energy theory, and is suitable for coal burst risk assessments of roadways.

6. Conclusions

According to the stiffness theory, one coal burst energy model of a rock–coal system is established, and the numerical simulation method is used to quantitatively reveal the energy supply mechanism of the roof to the sidewall under different stiffness ratios. The main conclusions are as follows:

(1) The stiffness ratio of rock to coal has little effect on the extreme value of the elastic strain energy of the coal, and mainly affects its dissipation energy and kinetic energy. With the decrease in the stiffness ratio, the dissipation energy of the coal after the peak point decreases in a power function, the plasticity decreases, and the brittleness increases. The kinetic energy of coal is negatively correlated with the stiffness ratio in a power function. When the stiffness ratio is less than one, the sensitivity of the kinetic energy of coal to the change in the stiffness ratio increases sharply.

(2) When the coal is fractured, the low-stiffness roof instantaneously generates deformation rebound and kinetic energy, which applies the dynamic load and energy supply behavior to the coal. With the decrease in the stiffness ratio, the release amount

and release rate of the rock elastic strain energy gradually increase. The rock elastic strain energy and kinetic energy have power function relations with the stiffness ratio, respectively.

(3) When a rock–coal system burst failure occurs, the kinetic energy of the coal comes from the release of the elastic strain energy of the coal and the energy supply from the rock. The supplied energy and energy supply rate of the rock to the coal increase with the decrease in the stiffness ratio. The supplied energy and the stiffness ratio have a power function relationship. The energy supply rate and the stiffness ratio have a linear relationship.

Author Contributions: Conceptualization, Y.Y., Z.L. and W.Z.; Data curation, X.T. and Z.L.; Software, X.T. and Q.M.; Writing—original draft, Y.Y. and W.Z.; Writing—review and editing, Q.M. All authors have read and agreed to the published version of the manuscript.

Funding: The study was financially supported by Major Program of Shandong Provincial Natural Science Foundation (NO. ZR2019ZD13), National Natural Science Foundation of China (NO. 52074167), and China Postdoctoral Science Foundation (NO. 2019M660024).

Institutional Review Board Statement: Not applicable.

Informed Consent Statement: Not applicable.

Data Availability Statement: Not applicable.

Conflicts of Interest: The authors declare no conflict of interest.

References

1. Wu, Y.Z.; Gao, F.Q.; Chen, J.Y.; He, J. Experimental study on the performance of rock bolts in coal burst-prone mines. *Rock Mech. Rock Eng.* **2019**, *52*, 3959–3970. [CrossRef]
2. Mottahedi, A.; Ataei, M. Fuzzy fault tree analysis for coal burst occurrence probability in underground coal mining. *Tunn. Undergr. Space Technol.* **2019**, *83*, 165–174. [CrossRef]
3. Yang, X.H.; Ren, T.; Tan, L.H.; Remennikov, A.; He, X.Q. Developing coal burst propensity index method for Austrialian coal mines. *Int. J. Min. Sci. Technol.* **2018**, *28*, 783–790. [CrossRef]
4. Pan, J.F.; Liu, S.H.; Wang, S.W.; Xia, Y.X. A new theoretical view of rockburst and its engineering application. *Adv. Civ. Eng.* **2018**, *2018*, 4683457. [CrossRef]
5. Li, Z.L.; Dou, L.M.; Wang, G.F.; Cai, W.; He, J.; Ding, Y.L. Risk evaluation of rock burst through theory of static and dynamic stresses superposition. *J. Cent. South Univ.* **2015**, *22*, 676–683. [CrossRef]
6. Zhao, T.B.; Guo, W.Y.; Tan, Y.L.; Yin, Y.C.; Cai, L.S.; Pan, J.F. Case studies of rock bursts under complicated geological conditions during multi-seam mining at a depth of 800 m. *Rock Mech. Rock Eng.* **2018**, *51*, 153–1564. [CrossRef]
7. He, M.C.; Ren, F.Q.; Cheng, C. Experimental and numerical analyses on the effect of stiffness on bedded sandstone strain burst with varying dip angle. *Bull. Eng. Geol. Environ.* **2019**, *78*, 3593–3610. [CrossRef]
8. Akdag, S.; Karakus, M.; Taheri, A.; Nguyen, G.; He, M.C. Effects of thermal damage on strain burst mechanism for brittle rocks under true-triaxial loading conditions. *Rock Mech. Rock Eng.* **2018**, *51*, 1675–1682. [CrossRef]
9. Gong, F.Q.; Wang, Y.L.; Wang, Z.G.; Pan, J.F.; Luo, S. A new criterion of coal burst proneness based on the residual elastic energy index. *Int. J. Min. Sci. Technol.* **2021**, *31*, 553–563. [CrossRef]
10. Gong, F.Q.; Si, X.F.; Li, X.B.; Wang, S.Y. Experimental investigation of strain rockburst in circular caverns under deep three-dimensional high-stress conditions. *Rock Mech. Rock Eng.* **2019**, *52*, 1459–1474. [CrossRef]
11. Cook, N.G.W. The failure of rock. *Int. J. Rock Mech. Min. Sci. Geomech. Abstr.* **1965**, *2*, 389–403. [CrossRef]
12. Khosravi, A.; Simon, R. Verification of the CSDS model in estimating the postpeak behavior of hard rocks. *Int. J. Geomech.* **2018**, *18*, 04017166.1–04017166.15. [CrossRef]
13. Qin, S.Q.; Jiao, J.J.; Tang, C.A.; Li, Z.G. Instability leading to coal bumps and nonlinear evolutionary mechanisms for a coal-pillar-and-roof system. *Int. J. Solids Struct.* **2006**, *43*, 7407–7423. [CrossRef]
14. Zhang, D.X.; Guo, W.Y.; Zhang, C.W.; Gong, X.F.; Li, Z.H.; Qiu, Y.; Chen, W.G. A new burst evaluation index of coal-rock combination specimen considering rebound and damage effects of rock. *Geomat. Nat. Hazards Risk* **2020**, *11*, 984–999. [CrossRef]
15. Li, C.C.; Zhao, T.B.; Zhang, Y.B.; Wan, W.K. A study on the energy sources and the role of the surrounding rock mass in strain burst. *Int. J. Rock Mech. Min. Sci.* **2021**, *147*, 104921. [CrossRef]
16. Cai, M.; Hou, P.Y.; Zhang, X.W.; Feng, X.T. Post-peak stress-strain curves of brittle hard rocks under axial-strain-controlled loading. *Int. J. Rock Mech. Min. Sci.* **2019**, *52*, 3959–3970. [CrossRef]
17. Xu, Y.H.; Cai, M. Influence of loading system stiffness on post-peak stress-strain curve of stable rock failures. *Rock Mech. Rock Eng.* **2017**, *50*, 2255–2275. [CrossRef]

18. Hou, P.Y.; Cai, M. Post-peak stress-strain curves of brittle hard rocks under different loading environment system stiffness. *Rock Mech. Rock Eng.* **2022**, *55*, 3837–3857. [CrossRef]
19. Zhao, T.B.; Yin, Y.C.; Tan, Y.L.; Xing, M.L.; Tang, X.X.; Li, C.C. Development of a rock testing system with changeable stiffness and its application in the study on the rock failure mechanical behavior. *Chin. J. Rock Mech. Eng.* **2022**, *41*, 1846–1857.
20. Xu, Y.H.; Cai, M. Influence of strain energy released from a testing machine on rock failure process. *Can. Geotech. J.* **2018**, *55*, 777–791. [CrossRef]
21. Wang, X.Y.; Yin, Y.C.; Xing, M.L.; Zhang, D.D.; Chen, Y.; Wang, E.C. Microsimulation study on energy release and rock block ejection force of granite under different unloading conditions. *Front. Earth Sci.* **2022**, *10*, 909371. [CrossRef]
22. Feng, X.J.; Zhang, Q.M.; Ali, M. 3D modelling of the strength effect of backfill-rocks on controlling rockburst risk: A case study. *Arab. J. Geosci.* **2020**, *13*, 128. [CrossRef]
23. Feng, X.J.; Ding, Z.; Ju, Y.Q.; Zhang, Q.M.; Ali, M. "Double peak" of dynamic strengths and acoustic emission responses of coal masses under dynamic loading. *Nat. Resour. Res.* **2022**, *31*, 1705–1720. [CrossRef]
24. Hudson, J.A.; Crouch, S.L.; Fairhurst, C. Soft, stiff and servo-controlled testing machines: A review with reference to rock failure. *Eng. Geol.* **1972**, *6*, 155–189. [CrossRef]
25. Lu, J.; Yin, G.Z.; Deng, B.Z.; Zhang, W.Z.; Li, M.H.; Chai, X.W.; Liu, C.; Liu, Y.B. Permeability characteristics of layered composited coal-rock under true triaxial stress conditions. *J. Nat. Gas Sci. Eng.* **2019**, *66*, 60–76. [CrossRef]
26. Ma, Q.; Tan, Y.L.; Liu, X.S.; Zhao, Z.H.; Fan, D.Y.; Purev, L. Experimental and numerical simulation of loading rate effects on failure and strain energy characteristics of coal-rock composite samples. *J. Cent. South Univ.* **2021**, *28*, 3207–3222. [CrossRef]
27. Zhang, X.; Wang, J.; Meng, Q.G.; Yu, M.L.; Zhang, Z.J.; Guo, Z.G. Coal rock breaking simulation and cutting performance analysis of disc cutters. *Teh. Vjesn.-Tech. Gaz.* **2021**, *28*, 1755–1761.
28. Han, W.; Jiang, Y.J.; Luan, H.J.; Du, Y.T.; Zhu, Y.G.; Liu, J.K. Numerical investigation on the shear behavior of rock-like materials containing fissure-holes with FEM-CZM method. *Comput. Geotech.* **2020**, *125*, 103670. [CrossRef]
29. Tan, Y.L.; Liu, X.S.; Shen, B.T.; Ning, J.G.; Gu, Q.H. New approaches to testing and evaluating the impact capability of coal seam with hard roof and/or floor in coal mines. *Geomech. Eng.* **2018**, *14*, 367–376.
30. Yin, Y.C.; Zheng, W.W.; Tang, X.X.; Xing, M.L.; Zhang, Y.B.; Zhu, Y.H. Test study on failure and energy supply characteristics of rock under different loading stiffness. *Eng. Fail. Anal.* **2022**, *142*, 106796. [CrossRef]
31. Gercek, H. Poisson's ratio values for rocks. *Int. J. Rock Mech. Min. Sci.* **2007**, *44*, 1–13. [CrossRef]
32. Chen, J.C.; Chemenda, A.I. Numerical simulation of true 3D rock tests with classical and new three-invariant constitutive models focusing on the end effects. *Arab. J. Sci. Eng.* **2020**, *45*, 9367–9378. [CrossRef]
33. Carpinteri, A.; Lacidogna, G.; Corrado, M.; Di Battista, E. Cracking and crackling in concrete-like materials: A dynamic energy balance. *Eng. Fract. Mech.* **2016**, *155*, 130–144. [CrossRef]

sustainability

MDPI

Article

Study on the Characteristics of Coal Ultrasonic Response during Loading and Its Influence Mechanism

Gang Xu [1,*], Tingting Ma [1], Chaofeng Wang [2], Hongwei Jin [1] and Yunlong Wang [1]

[1] College of Safety Science and Engineering, Xi'an University of Science and Technology, Xi'an 710054, China
[2] Xin'an Coal Mine, Henan Dayou Energy Co., Ltd., Luoyang 471842, China
* Correspondence: xugang25193@xust.edu.cn

Abstract: The prediction and prevention· of mine disasters are crucial to identifying the stress and strain state of coal using ultrasonic response characteristics. In this study, ultrasonic testing experiments of primary structure coal samples under uniaxial loading were conducted using a low-frequency rock physics measuring device. Based on the experimental results, the study focused on analyzing the relationship of the stress–strain state of coal samples with the ultrasonic velocity and quality factor of coal samples during stress loading, and exploring the influence mechanism of ultrasonic propagation in coal during stress loading. The results demonstrated that the stress-loading process of coal samples falls into the elastic deformation stage and the plastic deformation stage. In the elastic deformation stage, the ultrasonic velocity and the quality factor of the coal sample increased with the increase in the coal axial strain. In the plastic deformation stage, the ultrasonic velocity and quality factor of coal samples decreased as the axial strain of coal samples increased. Coal porosity was the fundamental factor affecting the coal wave velocity variation and the coal quality factor variation. In the elastic deformation stage, increased coal axial stress was accompanied by decreased coal porosity, contributing to the increase in coal wave velocity and coal quality factor. In the plastic deformation stage, the increase in the coal axial strain increased coal porosity and thus curtailed the wave velocity and quality factor of coal. Significant differences were observed in ultrasonic response characteristics of coal under various stress and strain states. The research results can lay a theoretical foundation for the safe and efficient development of coal resources and the prevention and control of mine disasters.

Keywords: coal porosity; stress loading; ultrasonic velocity; axial strain; quality factor; stress-strain state

Citation: Xu, G.; Ma, T.; Wang, C.; Jin, H.; Wang, Y. Study on the Characteristics of Coal Ultrasonic Response during Loading and Its Influence Mechanism. *Sustainability* **2023**, *15*, 1093. https://doi.org/10.3390/su15021093

Academic Editors: Xiangguo Kong, Dexing Li and Xiaoran Wang

Received: 22 November 2022
Revised: 30 December 2022
Accepted: 4 January 2023
Published: 6 January 2023

1. Introduction

Coal stress and its change are one of the main causes of accidents such as rib spalling, roof fall, rock burst, and coal and gas outbursts. Obtaining the stress–strain state and its variation law of coal can greatly help to reduce various mine disasters in the process of coal mining and realize safe and efficient coal mining.

The ultrasonic detection method mainly focuses on the exploration of geophysical properties. This method mainly emits ultrasonic waves into a coal medium and uses the parameters such as wave velocity, amplitude, and waveform collected after the ultrasonic wave passes through the medium to reversely infer the physical properties of the coal medium [1]. Much research has been done on the relationship between ultrasonic waves and physical properties of coal. Mahmood Karimaei et al. [2] established an exponential relationship between compressive strength and ultrasonic pulse velocity. The research results of Wang Bo et al. [3] show that the ultrasonic velocity is negatively and linearly related to the gas pressure. Kim Wonchang [4] obtained the correlation coefficient between ultrasonic pulse velocity and residual elastic modulus through an ultrasonic pulse experiment. Marta Krzesin'ska [5] applied the ultrasonic velocity measurement method to their study and found that the log (density/velocity) value of bituminous coals is closely related to elastic properties. The above research results provide theoretical support for obtaining the physical properties of coal and play an active role in ensuring coal mining. However, with the increase of coal mining depth, the influence of stress on the physical characteristics of coal is more and more significant, and the influence on coal mining is more and more serious.

Under stress loading, the stress and its changes have a significant effect on the physical properties exhibited by coal and rock mass, affecting the ultrasonic propagation law of the coal. Therefore, investigating the ultrasonic response characteristics of coal can help to identify the coal stress state [6,7], laying a foundation for the prevention and control of mine disasters.

Researchers have conducted many studies to explore the correlation between ultrasonic response characteristics and coal stress [8]. In 1988, Shea and Hanson [9] studied the structural failure law of loaded coal samples in different stages through the test of elastic wave velocity of coal samples. Through a uniaxial loading test, Tiedemann et al. [10], Zhai Xiaojie [11], and Zhao Mingjie et al. [12] revealed that the coal elastic modulus, the density of coal samples, and the compressive strength were obviously and linearly related to their longitudinal wave velocity (LWV) and transverse wave velocity (TWV). They believed that the closing of micro-cracks was the major factor that impacted the axial acoustic characteristics of rocks. Guo Deyong [13], Zhou Feng [14], and Tong Jiqiang et al. [15] suggested that increasing the confining pressure boosted coals' LWV and TWV through ultrasonic detection experiments. Yan Lihong [16] and Wu Jiwen [17] analyzed the relevance of wave velocity to mechanical parameters under tensile conditions through experiments, revealing a good power function correlation between the wave velocity and the tensile strength. Li Qiong et al. [18] paid attention to the impact of pressure on LWV and TWV through experiments, demonstrating that coal samples' LWV and TWV increased with the increase in pressure, and there was a quadratic correlation between them. During loading, rock or coal samples' internal strain exhibited relaxation properties, and the ultrasonic velocity variation could indicate the rock stress relaxation [19]. Engelder and Plumb [20] investigated the changing law of ultrasonic wave velocity under uniaxial loading and performed wave tests for describing the wave velocity–stress relationship. Nur [21] researched the response characteristics presented by ultrasonic wave velocity to the loading stress level. Zheng Guiping et al. [22] confirmed a correspondence relationship of the wave velocity with rock damage variable under uniaxial compression. Liu Xiaofei [23] investigated the laws by which ultrasonic signals propagated through a coal material under different loading conditions, as well as analyzed the changes in ultrasonic parameters such as amplitude, dominant frequency, and velocity. Chen Zhuo [24] explored the response characteristics of coal stress–strain and ultrasonic waves under different confining pressure conditions through triaxial loading of the coal in the laboratory, discovering that the variations of LWV can more accurately reflect the change characteristics of coal pore fissure under different confining pressure conditions compared to TWV.

In conclusion, many studies have focused on the relevance of ultrasonic velocity to the stress of coal, while few paid attention to the relationship between the coals' stress–strain state and their ultrasonic response characteristics during the stress-loading process. Additionally, most of the coal ultrasonic experimental studies only consider the characteristics of acoustic kinematics and only use the acoustic velocity to reflect the coal properties while ignoring the characteristics of its dynamics. Moreover, there are few studies on acoustic attenuation under different loading stress conditions. The stress of underground coal is in different stress and strain states, and the mine disaster is closely associated with the coal stress and strain state. It is particularly critical to obtain the stress and strain state of coal as well as its change trend for predicting and preventing mine disasters. Therefore, the ultrasonic response characteristic experiment under uniaxial loading was performed with the primary structure coal as the research object to obtain the ultrasonic response characteristic law of coal under different stress states. Besides, the ultrasonic velocity variation law and energy attenuation law of coal under uniaxial loading were analyzed considering the experimental results. The study demonstrates the association between ultrasonic kinematics (LWV and TWV), dynamic characteristics (attenuation), and stress and strain under the consideration of the coal anisotropy and lays a theoretical foundation for back calculating the stress–strain state and its changes in coal.

2. Materials and Methods

2.1. Ultrasonic Propagation Theory of Coal during Stress Loading

The physical characteristics exhibited by coal determined the propagation law of ultrasonic waves in coals. The stress and loading process of coal can change their physical properties, thereby affecting the above mentioned propagation. Therefore, the change in the ultrasonic propagation law is the result of the response to the change in the physical properties of coal.

Close et al. believed that coal is a dual structural medium constituted of matrix pores and fractures [25]. Pores in coal include micropores, small pores, mesopores, and macropores, according to their pore size [26]. Following its genesis, the fractures in coal include endogenous, exogenous, and inherited fractures. Generally, the wave velocity and its attenuation characteristics of coal exhibit a close association with the development degree of coal pore fissures. That is to say, the more developed the pores and fissures of coal, the lower the coal wave velocity, and the greater the attenuation coefficient.

The stress–strain relationship of coal during stress loading involves four stages: pore compaction(OA stage), linear deformation(AB stage), micro-fracture propagation(BC stage), and coal failure(CD stage) (Figure 1) [27]. Among them, the coal in the pore compaction and linear deformation stage belongs to elastic deformation, and the coal in the micro-fracture propagation and coal failure stage belongs to plastic deformation. The law and mechanism of the change in pore and fissure in coal caused by elastic deformation and plastic deformation are different, and so is the mechanism of wave velocity change and attenuation characteristics of coal. Therefore, the study analyzed the coal wave velocity change and its attenuation characteristics during stress loading from the perspective of elastic deformation and plastic deformation.

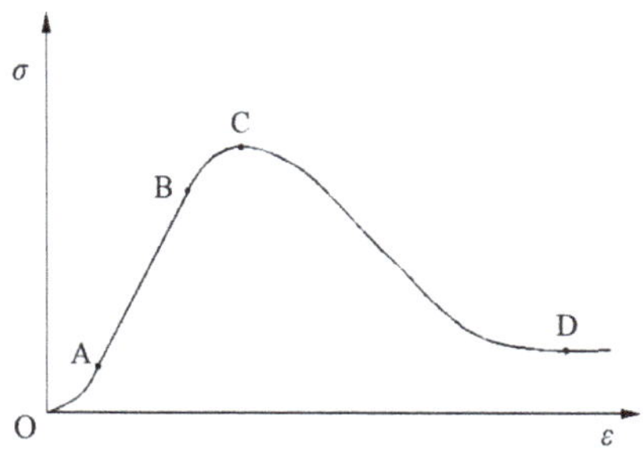

Figure 1. Stress–strain curve of coal.

2.1.1. Theory of Ultrasonic Velocity Change

Coal is a dual structure system composed of matrix pores and fractures which are usually regarded as pores to facilitate the study. Experiments revealed that the relationship between LWV and porosity of coal conforms to the following laws [28,29]:

$$\varphi = be^{aV_p} \tag{1}$$

where φ denotes porosity; a and b are constants; V_p refers to the LWV, i.e., the wave velocity in Equation (1). Since the longitudinal wave and transverse wave had similar propagation laws during stress loading, Equation (1) was also applicable to the transverse wave.

In the elastic deformation stage, the porosity equation expressed in volume strain can be obtained from the definition of porosity [30]:

$$\varphi = 1 - \frac{(1-\varphi_0)}{1+\varepsilon_V} \tag{2}$$

where φ_0 represents the initial porosity, and ε_V indicates the volume strain (negative for volume compression and positive for volume expansion).

Additionally, the volumetric strain–stress relationship of coal is based on the volumetric compressibility coefficient and defined according to the definition of volumetric compressibility coefficient [31]:

$$1+\varepsilon_V = \exp(-C_f\Delta\sigma') \tag{3}$$

where $\Delta\sigma'$ indicates the change of effective stress, C_f denotes the volumetric compressibility coefficient of coal, $C_f = 3(1-2v)/E$, v represents the Poisson's ratio, and E represents the elastic modulus.

By substituting Equation (3) into Equation (2), the relation between porosity and stress can be obtained:

$$\varphi = 1 - \frac{1-\varphi_0}{\exp(-C_f\Delta\sigma')} \tag{4}$$

It can be obtained by substituting Equation (4) into Equation (1) to calculate V_p that

$$V_P = \frac{1}{a}\ln\left\{\frac{1}{b}[1-\frac{1-\varphi_0}{\exp(-C_f\Delta\sigma')}]\right\} \tag{5}$$

Then,

$$V_P = \frac{1}{a}\ln\left\{\frac{1}{b}[1-\frac{1-\varphi_0}{\exp[-\frac{3(1-2v)}{E}\Delta\sigma']}]\right\} \tag{6}$$

Considering that the experiment in this paper was a uniaxial loading test, the change in effective stress during stress loading is $\Delta\sigma' = \frac{1}{3}\Delta\sigma_x$ and $\Delta\sigma_x = \sigma_1 - \sigma_0$; then:

$$V_P = \frac{1}{a}\ln\left\{\frac{1-(1-\varphi_0)\exp[\frac{1-2v}{E}(\sigma_1-\sigma_0)]}{b}\right\} \tag{7}$$

where σ_0 and σ_1 respectively denote the initial stress and the current stress.

In the plastic deformation stage, the change in coal porosity is related to coal damage. The equation below describes the relationship between coal porosity and the coal damage variable [32]:

$$\varphi = \varphi_1 e^{\beta D} \tag{8}$$

where φ_1 represents the porosity at the beginning of coal damage, φ indicates the porosity, β denotes the damage parameter of coal which can be obtained by parameter fitting, and D signifies the damage variable.

The plastic variable is defined as the damage variable [33]:

$$D = \frac{\varepsilon^p - \varepsilon_0^p}{\varepsilon_f^p - \varepsilon_0^p} \tag{9}$$

where ε_0^p denotes the threshold value of damage strain, that is, the plastic strain when the coal starts to damage; ε^p represents the current plastic strain; ε_f^p stands for the plastic strain at the residual strength of coal. ε_0^p, ε^p, and ε_f^p are positive. Considering that the experiment in this paper was a uniaxial stress loading test, various plastic strains were replaced by various axial strains. Specifically, ε_0^p, ε^p, and ε_f^p are the axial strains at the beginning of plastic deformation, during plastic deformation, and after the failure of coal, respectively.

By substituting Equation (8) into Equation (1), LWV V_p at the plastic deformation stage is calculated as:

$$V_P = \frac{1}{a}\ln[\frac{\varphi_1 e^{\beta D}}{b}] \quad (10)$$

2.1.2. Ultrasonic Wave Propagation Energy Attenuation Theory

The main forms of energy attenuation included diffusion attenuation, absorption attenuation, and scattering attenuation when the ultrasonic wave propagated in the medium. Among them, diffusion attenuation is correlated with the distance of ultrasonic propagation [34], and absorption attenuation mainly comprises heat conduction attenuation and viscous absorption attenuation [35]. In the experiment of this study, the coal samples have the same length and same structure type. Meanwhile, the influence of diffusion attenuation and absorption attenuation on ultrasonic is small and thus can be ignored. Hence, the energy attenuation of ultrasonic wave propagation in coal is primarily reflected in scattering attenuation. According to the research results of Urick R J, the scattering attenuation coefficient of ultrasonic wave propagation in coal is [36]:

$$\alpha_s = \varphi\frac{1}{6}(\frac{2\pi f}{v})^4 \bar{r}^3 \quad (11)$$

where α_s denotes the scattering attenuation coefficient of ultrasonic, φ is the coal porosity, $\bar{r}$ indicates the effective radius of pores and fractures in coal, f represents ultrasonic frequency, and v refers to ultrasonic velocity.

According to Equation (11), the scattering attenuation coefficient α_s is majorly associated with the porosity of coal, the effective radius of pores and fractures, and the ultrasonic velocity. During the experiment, the ultrasonic velocity of coal has been measured, the porosity of coal has been obtained in the elastic deformation stage and plastic deformation stage, and the effective radius of pores and fractures in coal is difficult to be directly obtained. The existing research confirms that the porosity and effective radius of coal have the following relationship [37]:

$$\varphi = \frac{12}{\pi}\xi\bar{r}N \quad (12)$$

where ζ denotes the linear density exhibited by cracks in the coal sample, and N represents a coefficient related to the shape of the crack.

Substituting Equation (12) into Equation (11) yields:

$$\alpha_s = \varphi^4\frac{1}{6}(\frac{2\pi f}{v})^4(\frac{\pi}{12\xi N})^3 \quad (13)$$

Considering that the energy attenuation regarding ultrasonic wave propagation in coal is mainly scattering attenuation, the attenuation coefficient can be obtained as:

$$\alpha = \alpha_s = \varphi^4\frac{1}{6}(\frac{2\pi f}{v})^4(\frac{\pi}{12\xi N})^3 \quad (14)$$

Thus, the quality factor of coal during ultrasonic propagation is:

$$Q = 1/\alpha = \frac{1}{\varphi^4\frac{1}{6}(\frac{2\pi f}{v})^4(\frac{\pi}{12\xi N})^3} \quad (15)$$

The parameters ξ and N in Equation (15) are not easy to measure and cannot be effectively verified directly through the laboratory data. Hence, the comparison method is used for verification. During comparison and verification, the ratio of a quality factor at different stages to that at the initial stage is expressed as M with the ultrasonic attenuation coefficient at the beginning of this stage as the standard. The calculation method of M is:

$$M = \frac{Q_b}{Q_a} = \frac{v_b^4 \varphi_a^4}{\varphi_b^4 v_a^4} \tag{16}$$

where M is the ratio of quality factor, Q_b indicates the quality factor of the current state, Q_a denotes the quality factor in the initial state, and v_b and v_a represent the wave velocity in the current state and in the initial state, respectively. φ_b and φ_a denote the porosity in the current state and in the initial state, respectively.

In the elastic deformation stage, substituting Equation (4) into Equation (16) yields:

$$M_1 = \frac{v^4 \varphi_0^4}{\left\{1 - (1 - \varphi_0)\exp\left[\frac{1-2v}{E}(\sigma_1 - \sigma_0)\right]\right\}^4 v_0^4} \tag{17}$$

where M_1 indicates the ratio of quality factor in the elastic deformation stage.

In the plastic deformation stage, substituting Equation (8) into Equation (16) yields:

$$M_2 = \frac{v^4 \varphi_1^4}{\varphi^4 v_1^4} = \frac{v^4 \varphi_1^4}{\left(\varphi_1 e^{\beta D}\right)^4 v_1^4} \tag{18}$$

where M_2 denotes the ratio of quality factor in the plastic deformation stage.

2.2. Experiment

2.2.1. Experimental Equipment

The ultrasonic experiment adopted a rock physics measurement device under low frequency during the stress loading process. Its working principle is illustrated in Figure 2. The device is equipped with a core sample testing system (System 1) and a data acquisition system (System 2).

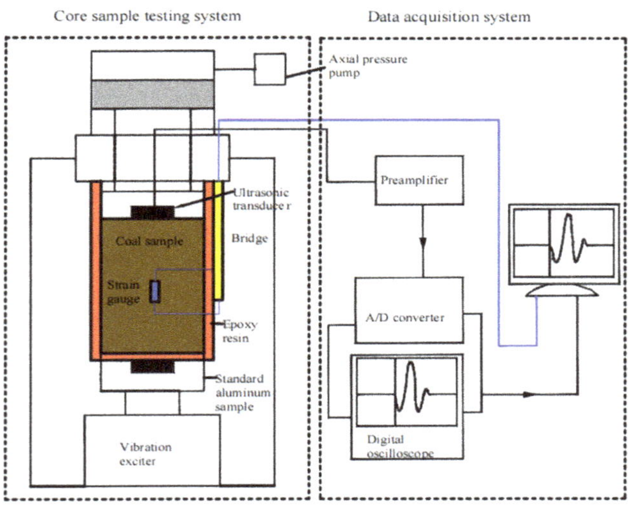

Figure 2. Working principle of low-frequency rock physics measurement device.

System 1, as an essential part of the device, has a semiconductor strain gauge, ultrasonic transducer, standard aluminum sample, servo vibrator, and epoxy resin sheet. System 1 can transmit and receive ultrasonic waves through ultrasonic transducers to realize the ultrasonic measurement of coal media. The frequency range of the ultrasonic measurement is 5 Hz–1 MHz. The axial pressure measurement is realized by applying stress to the test coal sample through a servo vibrator.

System 2, taking charge of recording the change in coal strain and ultrasonic velocity during the experiment, is adopted mainly to achieve high-precision data acquisition, which consists of a control software, signal generator, 12-channel high-precision differential amplifier, servo vibration amplifier, pulse generator receiver, digital oscilloscope, digital acquisition board, embedded computer, and analog-to-digital input board.

Additionally, auxiliary tools such as balances and vernier calipers are also needed during the experiment.

2.2.2. Measurement Principle

The measurement of ultrasonic wave velocity in low-frequency rock physics measurement systems primarily adopted the pulse transmission method [38,39]. Specifically, the measurement was conducted using the polarization filtering method to pick up the initial transverse wave, and then its wave velocity was calculated. Due to the constraints of experimental conditions, the propagation of the ultrasonic transverse wave in the anisotropic thin layer is accompanied by the propagation of fast and slow transverse waves in the medium in the way of orthogonal polarization, so as to weaken the time difference between them reaching the receiver. The detector can only identify the time point of receipt of the transverse wave fast wave, and it is difficult to identify the time point of arrival of the transverse wave slow wave. Therefore, this experiment only focuses on the shear wave velocity given the influence of realistic conditions.

(1) Measurement of wave velocity.

The pulse transmission ultrasonic system was adopted in the experiment. The computer can directly read out the sound wave velocity. The principle of test is expressed as:

$$\begin{cases} V_p = L/(t_p - t_0) \\ V_s = L/(t_s - t_0) \end{cases} \tag{19}$$

where V_p and V_s denote the LWV and TWV, respectively. L represents the distance between the transmitting and receiving transducers, t_p stands for the longitudinal wave propagation time, t_s is the transverse wave propagation time, and t_0 signifies the zero delay of the instrument system.

(2) Measurement of quality factor.

Due to the existence of pores and fissures in coal, ultrasonic waves will inevitably undergo attenuation in coal. The distribution law and structural characteristics of coals' internal microstructure and the ground stress state in the coal can be effectively obtained by studying the attenuation law of ultrasonic waves after passing through the coal sample. Their attenuation characteristics are generally measured by quality factors. A larger quality factor indicates smaller pore and crack size in coal, and slower attenuation. The quality factor is calculated by 2π times the ratio of the energy consumed by ultrasonic propagation attenuation to its own energy within a wavelength [40]:

$$Q = \frac{2\pi E}{\Delta E} \tag{20}$$

where Q denotes the quality factor, E represents the energy it has, and ΔE indicates the decayed energy.

The main methods for determining the quality factor are the time domain method and the frequency domain method. The study employed the amplitude decay method in the time domain to determine the quality factor. The key to this method was the determination of the main frequency and the initial arrival time. The specific steps were detailed as follows. Firstly, the ultrasonic wavelet waveform of the failed coal sample and the ultrasonic waveform passing through the coal sample obtained by the experiment were derived. Then, the more complete transmission wave waveform was intercepted by the first arrival pickup and was subjected to broadband filtering to remove other clutter. Afterward, the frequency point corresponding to the maximum amplitude in the first arrival pickup intercepted waveform was picked up as the main frequency. Next, the root mean square amplitude of the ultrasonic wavelet of failed coal samples and the ultrasonic wavelet passing through coal samples were calculated, respectively. Finally, the obtained main frequency, first arrival time, amplitude, and other parameters were substituted into the formula to obtain the corresponding quality factor value.

The process of determining quality factors by amplitude attenuation method is:

$$A(x) = A_0 \cdot e^{-\alpha x} \tag{21}$$

where A_0 denotes the wavelet amplitude without passing through the coal sample, $A(x)$ indicates the ultrasonic amplitude passing through the coal sample, α represents the attenuation coefficient, and x refers to the length of the medium.

Among them, the attenuation coefficient α can be expressed as:

$$\alpha = \frac{\pi f}{QV} \tag{22}$$

where f and V denote the main frequency and the wave velocity, respectively.

Substituting Equation (22) into Equation (21) yields:

$$A(x) = A_0 e^{\frac{-\pi f x}{QV}} \tag{23}$$

From the first arrival time of ultrasonic wave $t = x/V$, it can be obtained that

$$A(x) = A_0 e^{\frac{-\pi f t}{Q}} \tag{24}$$

After taking logarithms on both sides, Q can be obtained as:

$$Q = -\pi f t / \ln[\frac{A(x)}{A_0}] \tag{25}$$

2.2.3. Coal Sample Collection and Preparation

With the underground raw coal as the research object in the experiment, coal samples were collected per the provisions of Methods for Taking Coal Rock Samples [41] and Classification of Coal Structure [42]. The collected raw coal was primary structural coal. The strike orientations of the fractures in the coal seam are different, suggesting the influence of the anisotropy of the coal seam on the transmission of ultrasonic waves. Hence, attention should be paid to indicating coal samples' strike, tendency, and vertical bedding direction in the coal seam upon the collection of the coal samples on site.

The sampling site was located at 31004 working face of Shanxi Xinyuan Coal Co., Ltd. of Yangmei Group (Jinzhong, China). The working face was located in the Hanzhuang sub-district and mainly mined 3# coal seam. Its coal quality was high-quality lean coal with medium ash and low sulfur. The coal seam was majorly bright coal with developed endogenous fractures. The coal seam contained 1–2 layers of argillaceous gangue, and the thickness was in the range of 0.01–0.04 m (0.02 m on average). The 3# coal seam mined in this working face had steady occurrence and simple structure, and the coal firmness coefficient was 0.51–0.82, meeting the requirements of coal sample collection in this experiment.

The coal with a large volume and relatively complete structure preservation should be picked up as the master sample when coal samples are collected. After picking up the master sample, it should be put into the pre-prepared woven bag. Concurrently, the structure type and direction of the collected coal samples should be indicated on the bag surface. After the master sample was collected, it was transferred to the laboratory and processed into a 100 mm-high cylindrical standard coal sample (50 mm in diameter) by drilling equipment. Additionally, the coal core obtained by drilling was polished to make its flatness less than 0.02% to enable the transducer of the ultrasonic tester to form better contact with the coal core.

The total processing completed three experimental coal samples, including one in the strike, inclination, and vertical bedding direction. The coal sample number is specified as follows for contributing to the statistics and analysis. The direction of the parallel plane cleat, vertical plane cleat, and vertical bedding is defined as the X, Y, and Z direction, respectively, as shown in Figure 3. According to the classification standard of coal structure type, the primary structure coal is class I coal. Hence, experimenters name the three coal samples IX1, IY1, and IZ1 in turn.

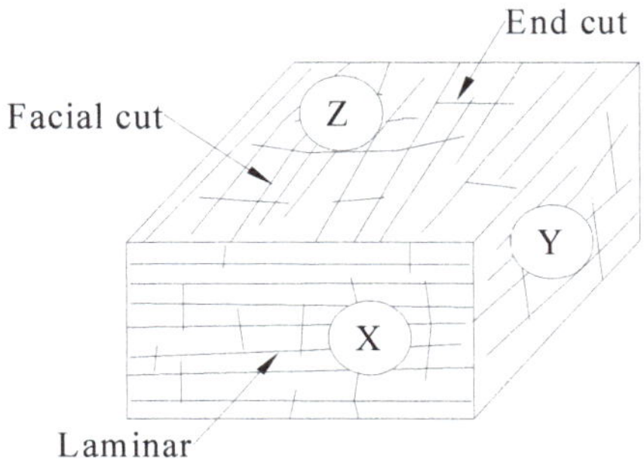

Figure 3. Schematic diagram of coal cleat and bedding.

The coal sample obtained after processing is exhibited in Figure 4. The specific parameters of the coal sample are presented in Table 1.

Figure 4. Physical drawing of coal sample.

Table 1. Statistics of parameters of processed coal samples.

Coal Sample Number	Coal Structure Type	Length (mm)	Diameter (mm)	Fracture Development	$R_{o,max}$	V_{daf} (%)	Bulk Density (t/m^3)	Porosity (%)	Modulus of Elasticity (MPa)	Poisson's Ratio
IX1 IY1 IZ1	Primary structural coal	100.0 99.9 100.1	50.1 49.9 50.1	The bedding is clear without obvious cracks.	1.86	15.7	1.43	3.5	1500	0.3

2.2.4. Experimental Scheme and Steps

In this experiment, there were three coal samples; stress loading was in the form of uniaxial compression; pressure was maintained 5 min after every certain value until the end of the coal failure experiment. Each coal sample experiment's specific steps are described as follows.

1. Clean the sticking position of the coal sample's strain gauge, firmly stick the strain gauge on the coal sample surface with glue, and press it hard to ensure full contact. After the glue is dry and firmly connected, the resistance value is checked with a strain gauge to ensure no open circuit or short circuit.
2. In order to improve the measurement accuracy, the coupling agent is applied to the coal sample and the ultrasonic transducer probe, and put the coal sample into the loading test system. Thus, the coal sample is consistent with the central axis of the transducer probe to ensure stable placement.
3. Connect the outgoing line of the strain gauge with the strain measurement instrument, and obtain the strain of the coal through the output of the strain measurement instrument.
4. Connect the ultrasonic testing instrument following the experimental setting, turn on the computer control terminal and digital oscilloscope power, and debug to assure the normal use of the instrument.
5. Adjust the loading system for controlling axial pressure change, and load the coal sample by stages until it breaks. Stabilize the voltage for 5 min after each new stress value is loaded, read out the corresponding ultrasonic wave velocity, record the first wave amplitude, and export the waveform data on the oscilloscope.
6. Record the loading stress value, axial strain, and ultrasonic response characteristic parameters, and repeat step 5 until the coal sample is broken and the experiment is completed.

3. Results

3.1. Experimental Results

Table 2 lists the experimental results of coal samples in terms of the ultrasonic velocity, quality factor, and axial strain during uniaxial stress loading.

Table 2. Test results of ultrasonic velocity, axial strain, and quality factor of coal samples under uniaxial stress loading.

Loading Step	Axial Stress (MPa)		P-Wave Velocity ($m \cdot s^{-1}$)			S-Wave Velocity ($m \cdot s^{-1}$)			Axial Strain (10^{-4})			Quality Factor Q_p			Quality Factor Q_s		
	X, Y	Z	X	Y	Z	X	Y	Z	X	Y	Z	X	Y	Z	X	Y	Z
1	0	0	1925	1886	1623	1056	959	807	0	0	0	0.2954	0.2846	0.2541	0.2541	0.2489	0.2182
2	0.4	0.5	2114	2013	1785	1129	1131	1056	8.5621	8.1562	10.3216	0.4823	0.4124	0.4015	0.4765	0.4795	0.3695
3	0.9	1	2236	2205	1973	1150	1262	1174	10.5689	9.1564	13.1684	0.6569	0.6386	0.5912	0.6071	0.6025	0.6128
4	1.5	1.5	2280	2342	2003	1296	1364	1203	13.6415	12.6156	14.1235	0.7614	0.7783	0.6962	0.7215	0.7054	0.7218
5	2.2	2.1	2312	2341	2016	1304	1352	1222	15.9501	14.8204	16.2605	0.7962	0.7893	0.7316	0.7428	0.7199	0.7356
6	2.9	2.8	2356	2364	2053	1356	1379	1218	16.7851	16.6925	17.3694	0.8124	0.8201	0.7356	0.7594	0.7259	0.7398
7	3.6	3.7	2392	2368	2123	1375	1378	1230	18.9654	17.1651	18.2068	0.8238	0.8158	0.7465	0.7816	0.7467	0.7695
8	4.2	4.1	2385	2395	2073	1362	1396	1201	19.6258	16.2609	19.2654	0.8417	0.8196	0.7256	0.8119	0.7692	0.7512
9	4.9	4.2	2399	2384	2048	1396	1403	1182	20.6041	17.1656	24.3654	0.8514	0.8214	0.7026	0.8364	0.7697	0.7435
10	5.8	4.3	2416	2399	2046	1381	1399	1174	22.2354	18.1561	27.3691	0.8632	0.8309	0.5865	0.8516	0.7927	0.5216
11	6.6	1.7	2434	2412	1508	1408	1416	765	23.6351	20.1656	35.3614	0.8863	0.8218	0.1369	0.8632	0.7983	0.1689
12	6.8	—	2415	2356	—	1327	1375	—	27.6248	20.2518	—	0.7312	0.7269	—	0.7415	0.6662	—
13	6.9	—	2387	2310	—	1305	1343	—	28.0695	23.6325	—	0.6124	0.6214	—	0.6589	0.5918	—
14	7.2	—	2347	2296	—	1321	1300	—	28.9562	25.1659	—	0.5164	0.514	—	0.5321	0.5238	—
15	1.5	—	1743	1695	—	893	864	—	35.8601	34.8661	—	0.1984	0.1541	—	0.1631	0.1578	—

3.2. Relationship between the Ultrasonic Velocity and Axial Strain of Coal Sample during Stress Loading

Table 2 explains the ultrasonic velocity–axial strain relationship of the coal sample during stress loading (Figure 5).

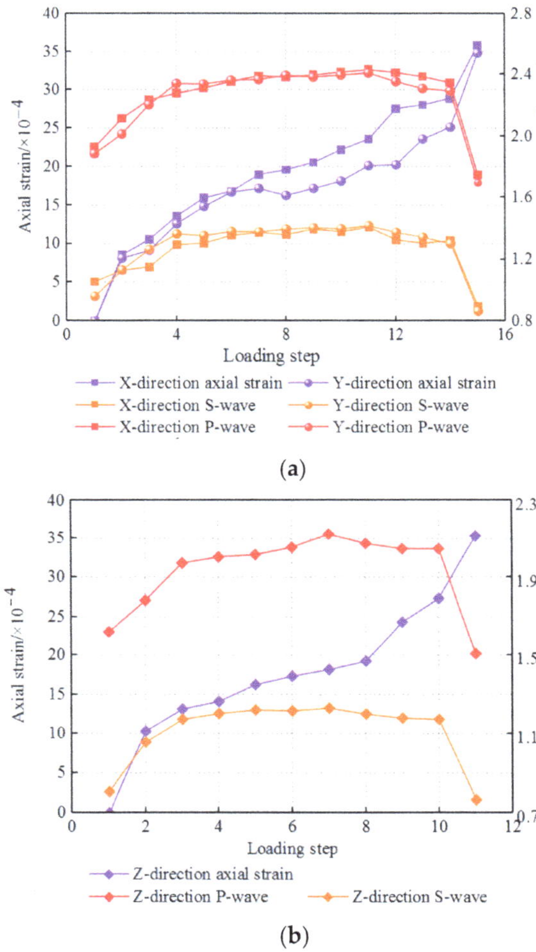

Figure 5. Relationship between coal sample wave velocity and axial strain during stress loading: (**a**) X and Y directions; (**b**) Z direction.

With Figure 5, the analysis on the above relationship assists in drawing the following conclusions:

1. As the loading step increases, the axial strain of the coal sample will be on a rise, while the LWV and TWV will present an increase-to-decrease trend.
2. The wave velocity exhibits strong anisotropy. The coal sample possesses higher LWV than TWV, and the LWV in the Z direction is lower than that in the X and Y directions. Through numerical analysis, the Pearson correlation coefficient of the TWV in the X and Y directions is 0.96501, indicating that the TWV in the X and Y directions is similar. During the whole process of stress loading, the TWV in the X and Y directions of the coal sample showed a change trend of "increase-slow increase-slow decrease-sharply decrease". The TWV in the X direction increases from 1056 m/s to 1296 m/s in the stress range of 0–1.5 MPa. Then, with the increase of axial stress to 6.6 MPa, the TWV in X direction increases slowly from 1296 m/s to 1408 m/s. Then, it slowly decreases to 1321 m/s, and the axial stress is 7.2 MPa. After the coal sample is damaged, the stress value drops sharply to 1.5 Mpa, and the TWV in the X direction decreases sharply to 893 m/s. The TWV in the Y direction increases from 959 m/s to 1364 m/s

in the stress range of 0–1.5 Mpa. Then, with the increase of axial stress to 6.6 MPa, the TWV in Y direction increases slowly from 1364 m/s to 1416 m/s. Subsequently, it slowly begins to decrease to 1300 m/s, and the axial stress is 7.2 MPa. After the failure of the coal sample, the stress value drops sharply to 1.5 MPa, and the TWV in the Y direction decreases sharply to 864 m/s. The Pearson correlation coefficient of the LWV in the X and Y directions is 0.98241, indicating that the LWV in the X and Y directions is similar. In the whole process of stress loading, the LWV of coal samples in X and Y directions showed the change trend of "increase-slow increase-slow decrease-sharply decrease". The LWV in the X direction increases from 1925 m/s to 2280 m/s in the stress range of 0–1.5 MPa; then, as the axial stress increases to 6.6 MPa, the LWV in the X direction slowly increases from 2280 m/s to 2434 m/s, and then slowly decreases to 2347 m/s, and the axial stress is 7.2 Mpa. After the coal sample is destroyed, the stress value drops sharply to 1.5 Mpa, and the LWV in the X direction decreases sharply to 1743 m/s. The LWV in the Y direction increases from 1886 m/s to 2342 m/s in the stress range of 0–1.5 Mpa. Then, with the increase of axial stress to 6.6 Mpa, the LWV in the Y direction increases slowly from 2342 m/s to 2412 m/s, and then decreases slowly to 2296 m/s, and the axial stress is 7.2 Mpa. After the coal sample is destroyed, the stress value drops sharply to 1.5 Mpa, and the LWV in the Y direction decreases sharply to 1695 m/s.

3. According to the stress loading process of coal samples, the changes in axial strain, LWV, and TWV fall into the initial stage, the middle stage, the late stage, and end stage of stress loading.
4. In the early stage of stress loading, the LWV, TWV, and axial strain of coal samples increase. In the middle stage of stress loading, the LWV, TWV, and axial strain of coal samples increase slowly. In the late stage of stress loading, the axial strain of coal samples showed a significant increase trend, and the LWV and TWV of coal samples began to decrease slowly. At the end of the stress loading, the coal sample is broken, the axial strain of the coal sample increases sharply, and the LWV and TWV of the coal sample decreases sharply. The reasons for these trends are as follows: in the early stage of stress loading, the coal structure is compressed, the size of pores and cracks in the coal decreases, the axial strain increases, and the acoustic wave velocity of the coal increases. In the middle of stress loading, with the increase of axial pressure, the compression amplitude of coal structure decreases, so the LWV, TWV, and axial strain of coal sample increase slowly. In the late stage of stress loading, the coal enters the plastic stage, the internal cracks of the coal sample begin to increase, the size of the pores and cracks increases, the axial strain of the coal sample increases, and the LWV and TWV of the coal sample begin to decrease slowly. At the end of the stress loading, because the load of the coal sample exceeds its maximum bearing capacity, a large number of cracks appear in the coal, the axial strain increases sharply, and the wave velocity decreases abruptly.
5. In the initial and middle stages, the ultrasonic velocity and axial strain of coal samples present an uptrend and have a good positive correlation. In the late stage and at the end of stress loading, the axial strain continues to increase, while the wave velocity significantly decreases. The ultrasonic velocity of coal samples shows a negative correlation with the axial strain.

3.3. Relationship between the Axial Strain and Quality Factor of Coal Sample during Stress Loading

Similarly, the relationship between the coal sample quality factor and axial strain during stress loading can be obtained according to Table 2 (Figure 6).

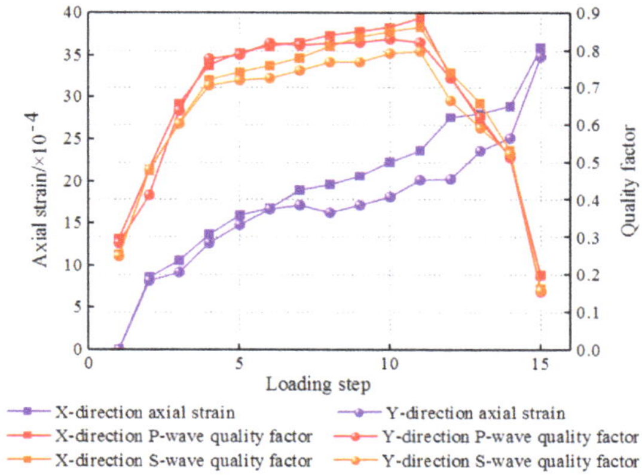

(**a**)

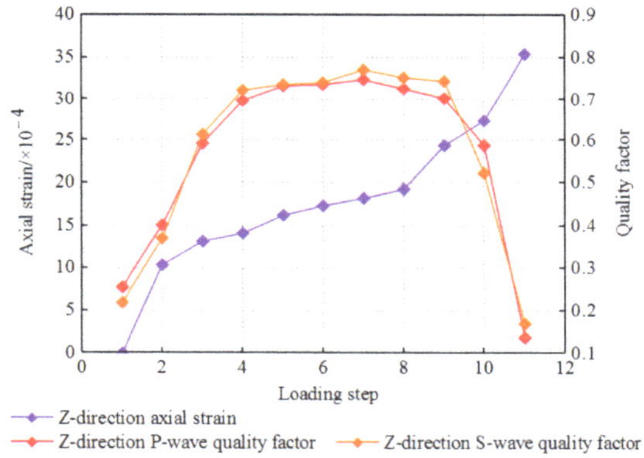

(**b**)

Figure 6. The axial strain-quality factor relationship of coal sample during stress loading: (**a**) X and Y directions; (**b**) Z direction.

Analysis on the above relationship assists in drawing the following conclusions, as shown in Figure 6:

1. The quality factors of coal samples also demonstrate strong anisotropy. The quality factors of coal samples in the Z direction are less than those in the X and Y directions. The quality factors of coal samples in the two directions are similar, and the variation law is the same. In other words, coal samples present a higher wave velocity attenuation amplitude in the Z direction relative to the X and Y directions.
2. The variation trend of coal sample quality factors during stress loading also falls into the initial stage, the middle stage, the late stage, and the end stage of stress loading. In the initial stage, the coal sample quality factor significantly increases. In the middle stage, the quality factor slowly increases. In the late stage, the quality factor significantly decreases. At the end of stress loading, the quality factor sharply decreases.
3. In the initial and middle stage of stress loading, the coal sample quality factor increases as the axial strain increases, indicating that the attenuation amplitude of energy during ultrasonic propagation decreases as the axial strain increases. In the late stage and at the end of stress loading, the axial strain of coal samples keeps increasing, while the quality factor of coal samples sharply decreases, implying that the attenuation range of energy during ultrasonic propagation increases as the axial strain increases.

3.4. Relationship between Coal Sample Wave Velocity and Quality Factor during Stress Loading

Based on Table 2, the coal sample quality factor-wave velocity relationship of coal sample during stress loading can be obtained (Figure 7).

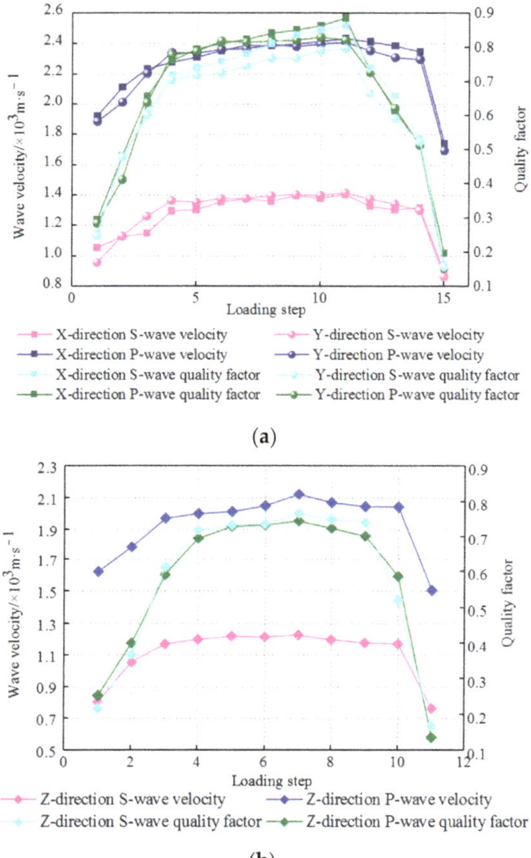

Figure 7. Relationship between coal sample wave velocity and coal sample quality factor during stress loading: (**a**) X and Y directions; (**b**) Z direction.

As revealed in Figure 7, the quality factor-wave velocity relationship of the coal sample during the stress loading process can be concluded as follows. The two parameters have significant stage characteristics, and their variation laws are the same.

4. Discussion

4.1. Analysis of Influencing Factors of Coal Ultrasonic Velocity Change during Stress Loading

Equations (7) and (10) are used to calculate ultrasonic wave velocity in the elastic deformation stage and plastic deformation stage, respectively. Whether Equations (7) and (10) are reliable is further verified with the longitudinal wave data in the X direction as an example.

According to the longitudinal wave data in the X direction in Table 2, Equation (26) can be obtained by fitting according to Equation (1):

$$\varphi = 0.044741915e^{-0.000126133V_P} \tag{26}$$

where $a = -0.000126133$, and $b = 0.044741915$; V_p refers to the LWV, m/s.

The coefficient of Equation (26) is fitted based on the measurement results of this experiment. However, due to the limitation of test conditions, the number of coal samples in this experiment is small, which may lead to a certain error of the coefficient. Therefore, we will focus on improving this problem in subsequent research. The variation data of LWV with stress can be obtained by substituting coal mechanical parameters and fitting parameters a and b into Equation (7). Figure 8 displays the comparison diagram between the calculated value of LWV in the X direction and the experimental value in the elastic deformation stage in combination with the experimental data in Table 2. In Figure 9, the LWV in the X direction demonstrates a linear increase with stress during the elastic deformation phase of the stress loading process, revealing the increase in stress in the elastic deformation stage, the decrease in coal porosity, and the increase in wave velocity. This conclusion is consistent with Yale's research results. Yale [43] also proposed in the research results that the porosity of rock decreases with the increase of stress, which leads to the increase of wave velocity transmitted in rock mass. Therefore, the internal factor affecting the change of coal wave velocity in the elastic deformation stage is porosity, which theoretically explains the root cause of the change of coal wave velocity.

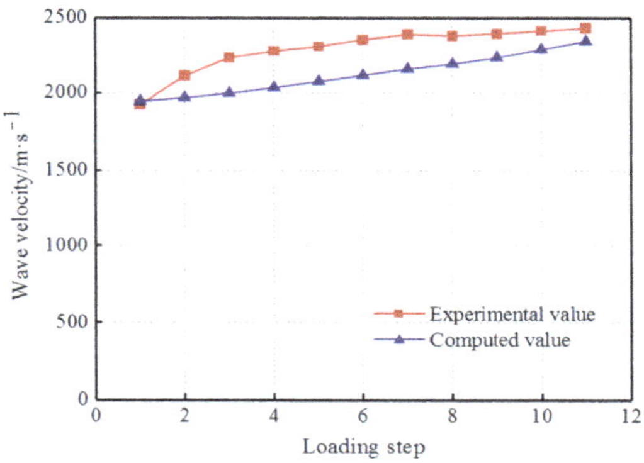

Figure 8. Experimental and calculated values of LWV of coal in the X-direction during the elastic deformation stage.

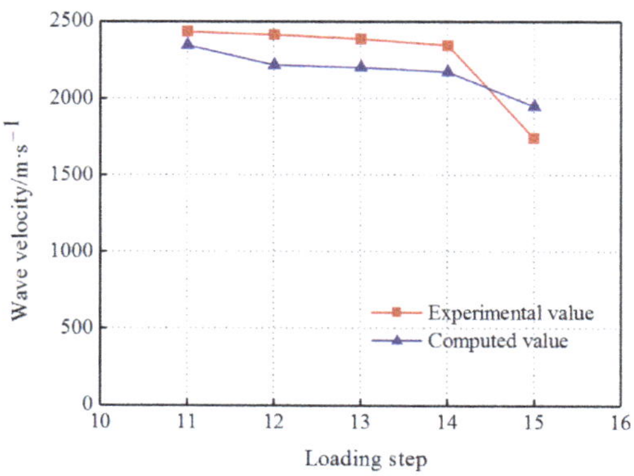

Figure 9. Experimental and calculated values of LWV of coal in the X-direction during the plastic deformation stage.

After calculation, the error rate between the experimental value and the calculated value of the LWV in the X direction of the elastic deformation stage is 1.14% (loading step 1), and the maximum is 10.75% (loading step 4), which shows that it is feasible to calculate the LWV in the elastic deformation stage by this formula.

With the longitudinal wave data in the X direction in Table 2, $\beta = 0.05$ can be obtained by fitting according to Equation (8). Then, the coal mechanical parameters are substituted, and parameters a and b are fitted into Equation (10), so as to acquire the variation data of LWV with strain. With respect to the experimental data in Table 2, the comparison diagram between the calculated value of LWV in the X direction in the plastic deformation stage and the experimental value can be depicted (Figure 9). As observed in Figure 9, the LWV in the X direction decreases exponentially with the strain in the plastic deformation stage during stress loading. Thus, in the plastic deformation stage, the coal axial strain, the damage variable, and the porosity increase, leading to weakened wave velocity. After calculation, the error rate between the experimental value and the calculated value of the LWV in the X direction of the plastic deformation stage is 7.36% (loading step 13), and the maximum is 11.9% (loading step 15), which shows that it is feasible to calculate the LWV in the plastic deformation stage by this formula.

Equation (7) illuminates that the main factor affecting the change of coal wave velocity in the elastic deformation stage is stress, and the wave velocity increases linearly with the increasing stress. This research result is in line with the research results of Meng Zhaoping [44], Zhou Feng [14], and Li Qiong [45]. In addition, Dong Shouhua et al. [46] focused on the measurement of the ultrasonic wave velocity of anthracite samples under triaxial stress. The research found that the confining pressure increase was accompanied by the increase in ultrasonic wave velocity, and the wave velocity increased slowly and became stable when it exceeded the critical value of confining pressure. Sun Xiaoyuan [47] found that the P-wave velocity of briquettes was positively correlated with its pressure; Zhang Long [48] and others found a logarithmic relationship between ultrasonic LWV and stress. Yu Hongyan [49] and others found that effective stress increase resulted in obviously increased P-wave and S-wave velocities of fractured shale. Cao Anye et al. [50] found that the P-wave velocity was exponentially positively correlated with the applied stress through laboratory experiments. The research results of these scholars show that the change of wave velocity is related to stress, but the influence of stress on wave velocity is different, which shows that there are different variation rules of wave velocity in different stress change stages. Therefore, according to the stress–strain characteristics of coal, the stress loading process is divided into the elastic stage and plastic stage. In these two stages, the variation law of wave velocity is studied respectively, which can reflect the response characteristics of the ultrasonic wave more truly.

Equation (10) suggests that the primary factor that impacts the change of coal wave velocity in the plastic deformation stage is strain. As the strain is on a rise, the wave velocity exhibits an exponential decrease, consistent with the research results of Zhao Mingjie [51]. Zhao Mingjie established an exponential relationship between rock strength and ultrasonic velocity based on damage mechanics. In addition, the results of Yang Sen [52] and Li Jian [53] show that the LWV decreases with the development of coal and rock damage. Zhang Zhibo et al. [54] also found through experimental research that the wave velocity decreases with the increase of stress, and the important factor affecting the change of UTV is the change of microcrack structure. This shows that the wave velocity of coal decreases in the propagation stage of micro-cracks, and the reason for the decrease of wave velocity can be caused by damage or stress increase. In this paper, the reason for the decrease of wave velocity is summarized as strain, which is due to the fact that the increase of coal damage and the propagation of micro-cracks are two expressions of the same result in the plastic deformation stage. Using strain can describe the change of wave velocity more realistically, and it is also easy to quantify the change of wave velocity.

According to Equations (7) and (10), the main influencing factors of wave velocity change of coal in the elastic deformation stage and plastic deformation stage are different. However, further analysis by combining Equations (7) and (10) unveils that porosity is still the fundamental factor influencing the wave velocity change of coal in the process of stress loading. Moreover, a good mathematical relationship exists between the change in coal porosity and the coal stress in the elastic deformation stage, confirming the stress as the primary factor that impacts wave velocity change in the elastic deformation stage. However, a relationship can be hardly established between the coal porosity change and stress in the plastic deformation stage, while a strain can establish a good relationship with the coal porosity change. Hence, the primary factor influencing coal wave velocity change in the plastic deformation stage is strain. This is the influence mechanism of the ultrasonic wave velocity change during the stress loading process. Equations (7) and (10) reveal the variations of ultrasonic velocity during coal stress loading more deeply and comprehensively, with better accuracy and applicability.

4.2. Analysis of Influencing Factors of Ultrasonic Propagation Energy Attenuation

Equations (17) and (18) describe the change in coal quality factor in the elastic deformation stage and plastic deformation stage, respectively. The calculated value of the quality factor is compared with the experimental value in Table 2 with the X-direction longitudinal wave as an example to verify the reliability of Equations (17) and (18). Meanwhile, the changing trend of the ratio of the quality factor in the process of stress loading is obtained (Figures 10 and 11).

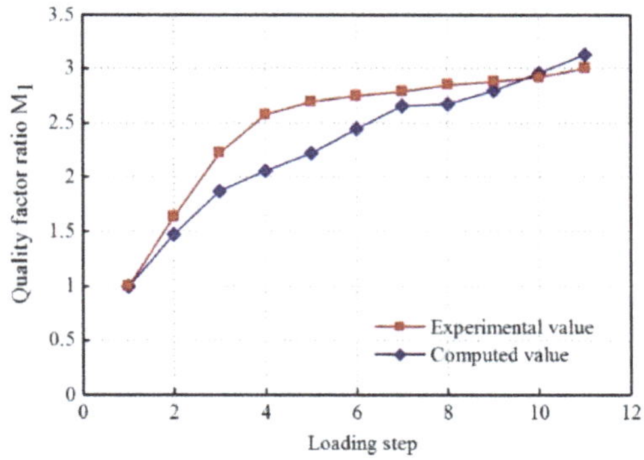

Figure 10. Experimental and the calculated values of quality factor ratio in the elastic deformation stage.

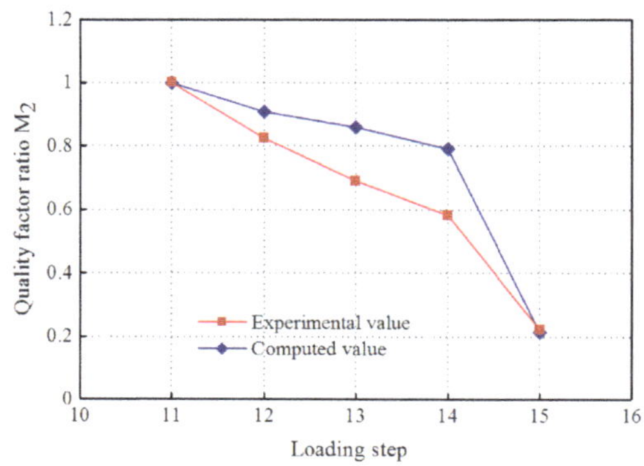

Figure 11. Experimental and the calculated values of quality factor ratio in the plastic deformation stage.

As observed in Figure 10, the ratio of the coal quality factor increases in the form of a power function with the increase in stress in the elastic deformation stage. At the end of the elastic stage, the coal quality factor ratio is about three times the initial time, and the calculated value meets the experimental value, reflecting the changing trend of the quality factor. Equation (17) reveals that the quality factor ratio shows relevance to the porosity and coal wave velocity. In the elastic deformation stage of stress loading, the wave velocity of coal increases, together with the decrease in porosity.

In Figure 11, the coal quality factor ratio decreases as the parabola forms with increasing loading steps during the plastic deformation phase. The coal quality factor ratio at the time of rupture is about 20% of that at the beginning of the plastic deformation phase. The calculated value conforms to the experimental value, implying the changing trend of the quality factor. Equation (18) illuminates that the quality factor ratio of coal is correlated with the coal porosity and wave velocity. In the plastic deformation stage of stress loading, the coal wave velocity decreases, and the porosity increases. Consequently, the quality factor ratio of coal decreases in the form of the parabola function.

Equation (17) suggests that the change of coal quality factor in the elastic deformation stage of coal during stress loading is related to stress. The reason is presented as follows. The porosity decreases with the increase in stress, and there is an acceptable mathematical relationship between the change in coal porosity and coal stress. This implies that the main factor influencing the coal quality factor in the elastic deformation stage is stress. Equation (18) demonstrates that the change in the coal quality factor in the plastic deformation stage is derived from strain. The reason is that the change in stress in the plastic deformation stage is difficult to reflect the change in porosity, and a good mathematical relationship can be established between strain and coal porosity. It is revealed that the main factor affecting the coal quality factor in the plastic deformation stage is strain.

As discovered by combining Equations (15)–(18), coal quality factor Q is inversely proportional to coal porosity φ_4, and the fundamental factor affecting the coal quality factor is coal porosity. The change in coal porosity in the two stress loading stages are expressed by stress and strain, respectively. This is the influence mechanism of energy attenuation during ultrasonic propagation during stress loading.

Zhao Qiufang [55] and Zhang Pingsong [56] reported that the attenuation coefficient of the coal seam presented a logarithmical association with the porosity. Liang Yanxia et al. [57]. revealed that the quality factor presented a negative linear relationship with the porosity, while the coal quality factor (Q) is inversely proportional to the coal porosity (φ_4) in our study. In other words, different researchers have different research results on the coal quality factor–coal porosity relationship. However, the general trend is consistent. Specifically, larger coal porosity indicates a smaller coal quality factor. Generally, the model established in the study based on the ultrasonic wave propagation theory covers the whole process of coal loading and has been verified by the experimental results with high accuracy and reliability.

Since the influence mechanism of ultrasonic dynamic parameters of coal is relatively complex, the relationship model between the ultrasonic attenuation coefficient and stress under loading conditions established in this paper is not perfect. The research results only reflect the variation law of the coal quality factor somewhat. Therefore, the attenuation law of coal ultrasonic should be further investigated in the future.

4.3. Relationship between the Stress–Strain State of the Coal and Ultrasonic Response Characteristics during Stress Loading

The relationship between the coal stress–strain state and ultrasonic response characteristics (ultrasonic velocity and quality factor) can be obtained (Figure 12) by analyzing the influencing factors of coal ultrasonic propagation during stress loading. Figure 12 illustrates that the stress–strain state of coal during stress loading falls into the elastic deformation stage and the plastic deformation stage. In the elastic deformation stage, the wave velocity and coal quality factor exhibit an increasing trend. They increase faster in the pore compaction stage, but increase more slowly in the linear deformation stage. In the plastic deformation stage, the two parameters decrease. They decrease slowly in the microcrack growth stage, but decrease rapidly in the coal destruction stage. Thus, the change in ultrasonic velocity and attenuation of energy have a significant corresponding relevance to the coal stress state, laying a theoretical foundation for the inversion of the stress state and its change trend through the change and attenuation of ultrasonic velocity.

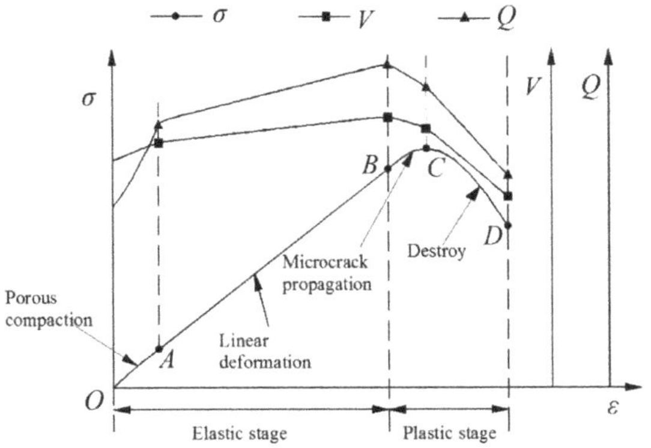

Figure 12. Relationship between the stress–strain state of the coal and the ultrasonic response characteristics.

The following is an application scenario. The working face is an area that possesses a high coal and gas outburst risk in coal mine production, and the high risk is obviously associated with the coal stress state. During the production process of the working face, the ultrasonic characteristics in the coal in front of the working face can be continuously monitored by drilling. The stress state and its change trend of coal are inversed by using the change and attenuation of ultrasonic velocity following the previous research results of this paper. When the ultrasonic speed slows down, the quality factor is reduced, and the stress has not reached the peak stress. Thus, the safety production status of the working face shall be analyzed, and necessary safety protection measures shall be taken to prevent the occurrence of coal mine safety production accidents.

Although the research conclusion of this paper is obtained under the condition of coal with a primary structure, it is also applicable to coal with other structure types. The difference is that the stress and strain process for the primary structure coal is significant during stress loading, and the process is gentle for the fragmented coal and mylonitic coal. As a result, the change in ultrasonic speed and the attenuation of energy are relatively weak, leading to the increased difficulty and requirements of identification. The research in this area will be strengthened in the future. Generally, the method proposed in the paper has a relatively broad application prospect and provides a new idea for inferring the stress state and change trend of coal in the working face, contributing to a positive significance for predicting and preventing the occurrence of mine disasters. Due to the limitation of experimental conditions, the number of coal samples and the range of coal samples selected in this paper are small, which have certain limitations. The ultrasonic response characteristics of loaded coal analyzed in the experiment are only general statistical laws. It is necessary to carry out experimental analysis on more coal samples in order to reduce accidental errors and obtain more accurate general ultrasonic response laws. This is also the direction that the author needs to work hard towards in the future.

5. Conclusions

(1) The coal stress-loading process falls into the elastic deformation stage and the plastic deformation stage. In the process of stress loading, the ultrasonic velocity and the quality factor of the coal sample have the same variation law. In the elastic deformation stage, the ultrasonic velocity and quality factor of the coal sample increase as coal axial strain elevates; in the plastic deformation stage, coal samples' axial strain continues to increase, while the ultrasonic velocity and quality factor significantly decline.

(2) In the process of stress loading, the ultrasonic velocity and quality factor of the coal sample demonstrate strong anisotropy, and are smaller in the Z direction than those in the X and Y directions. The ultrasonic velocity and quality factor of the coal sample in the X and Y directions are similar and present the same variation law.
(3) Coal porosity is the fundamental factor influencing the change in the coal wave velocity and coal quality factor. In the elastic deformation stage, the coal stress increases, and the porosity decreases; the coal wave velocity and coal quality factor both increase, and the coal attenuation coefficient decreases. It is revealed that the primary factor that impacts the coal wave velocity and coal quality factor is stress. In the stage of plastic deformation, the porosity increases as the coal strain is on a rise, the coal wave velocity and coal quality factor decreases, and the coal attenuation coefficient increases, suggesting that the main factor affecting coal wave velocity and coal quality factor was strain.
(4) The wave velocity and coal quality factor rapidly increase in the pore compaction stage and slowly increase in the linear deformation stage. The wave velocity and coal quality factor slowly decrease in the stage of microcrack propagation, while the wave velocity and coal quality factor rapidly decrease in the stage of coal destruction. The stress–strain state of coal corresponds well to the ultrasonic response characteristics during the loading process.

Author Contributions: Conceptualization, G.X.; methodology, G.X.; validation, G.X. and C.W.; formal analysis, T.M.; investigation, Y.W.; writing—original draft preparation, G.X. and Y.W.; writing—review and editing, G.X. and T.M.; visualization, T.M.; supervision, H.J.; project administration, G.X.; funding acquisition, G.X. The published version of the manuscript has been read by and obtained the agreement of all authors. All authors have read and agreed to the published version of the manuscript.

Funding: This work was financially supported by the National Key Research and Development Program (2018YFC0807805), Natural Science Basic Research Program of Shaanxi (2019JM-072).

Institutional Review Board Statement: Not applicable.

Informed Consent Statement: Not applicable.

Data Availability Statement: Not applicable.

Acknowledgments: Many thanks to all the contributions and support given by the authors in preparing the writing of this article. Besides, thanks to the coal sample preparation provided by Shanxi Xinyuan Coal Co., Ltd. of Yangmei Group for this study.

Conflicts of Interest: The authors declare no conflict of interest.

References

1. Cheng, L.; Wang, Y.; Zhao, H. Research status and prospect of acoustic characteristics of coal and rock. *Adv. Geophys.* **2013**, *28*, 452–461. [CrossRef]
2. Karimaei, M.; Dabbaghi, F.; Dehestani, M.; Rashidi, M. Estimating Compressive Strength of Concrete Containing Untreated Coal Waste Aggregates Using Ultrasonic Pulse Velocity. *Materials* **2021**, *14*, 647. [CrossRef] [PubMed]
3. Wang, B.; Hao, J.; Liu, S.; Zhou, F.; Zhang, Z.; Zhang, H.; Sun, H. Experimental Study on the Effect of Gas Pressure on Ultrasonic Velocity and Anisotropy of Anthracite. *Geofluids* **2019**, *2019*, 3183816. [CrossRef]
4. Kim, W.; Jeong, K.; Lee, T.; Park, S. A Study on Correlation between Ultrasonic Pulse Velocity Method and Coarse Aggregate for Estimating Residual Modulus of Elasticity of Concrete Exposed to High Temperatures. *Appl. Sci.* **2022**, *12*, 6644. [CrossRef]
5. Krzesinska, M. Averaged Structural Units in Bituminous Coals Studied by Means of Ultrasonic Wave Velocity Measurements. *Energy Fuels* **2001**, *15*, 930–935. [CrossRef]
6. Kong, X.; He, D.; Liu, X.; Wang, E.; Li, S.; Liu, T.; Ji, P.; Deng, D.; Yang, S. Strain characteristics and energy dissipation laws of gas-bearing coal during impact fracture process. *Energy* **2022**, *242*, 123028. [CrossRef]
7. Solís-Carcalo, R.; Moreno, E. Evaluation of concrete made with crushed limestone aggregate based on ultrasonic pulse velocity. *Constr. Build. Mater.* **2008**, *22*, 1225–1231. [CrossRef]
8. Zhao, M. Research status and prospect of rock acoustic characteristics. *J. Chongqing Jiaotong Univ.* **2000**, *19*, 81–83. [CrossRef]
9. Shea, V.; Hanson, D.; Dai, S. Determination of different stages of loading and failure of coal samples by elastic wave velocity and attenuation. *Adv. Geophys.* **1990**, *5*, 94–102.

10. Meng, Z.; Zhang, J. Joachim Tiedemann: Relationship between rock physical and mechanical parameters of coal measures and acoustic velocity. *J. Geophys.* **2006**, *49*, 1505–1510. [CrossRef]
11. Zhai, X. *Study on Ultrasonic Characteristics of Rock under Uniaxial Loading*; Chengdu University of Technology: Chengdu, China, 2008.
12. Zhao, M.; Wu, D. Ultrasonic classification and strength prediction of engineering rock mass. *J. Rock Mech. Eng.* **2000**, *19*, 89–92.
13. Guo, D.; Han, D.; Feng, Z. Experimental study on wave velocity characteristics of tectonic coal under confining pressure. *Coal Sci. Technol.* **1998**, *4*, 22–24. [CrossRef]
14. Zhou, F. Experiment on the influence of cracks on the ultrasonic velocity of coal and rock-taking -Carboniferous coal seam in Qinshui Basin as an example. *Coalf. Geol. Explor.* **2012**, *40*, 71–74. [CrossRef]
15. Tong, J.; Yang, D.; Li, Z.; Ren, H.; Wang, Z. Experimental analysis of coal seam wave velocity under different confining pressures. *Coal Mine Saf.* **2017**, *48*, 49–52. [CrossRef]
16. Yan, L. Study on the wave velocity characteristics of coal and rock in Yangzhuang Coal Mine and its relationship with strength. *Coal Sci. Technol.* **2006**, *34*, 57–60. [CrossRef]
17. Wu, J.; Jiang, Z.; Fan, C.; Lin, F. Study on wave velocity measurement of coal seam tensile strength. *J. Geotech. Eng.* **2005**, *27*, 999–1003. [CrossRef]
18. Li, Q.; He, J.; Chen, J. Simultaneous ultrasonic experiment of dynamic and static elastic parameters of coal under formation pressure conditions in Qinshui Basin. *Chin. J. Geophys.* **2017**, *60*, 2897–2903. [CrossRef]
19. Adam, L.; Van Wijk, K.; Otheim, T.; Batzle, M. Changes in elastic wave velocity and rock microstructure due to basalt-CO_2-water reactions. *J. Geophys. Res. Solid Earth* **2013**, *118*, 4039–4047. [CrossRef]
20. Engelder, T.; Plumb, R. Changes in situ ultrasonic properties of rock on strain relaxation. *Int. J. Rock Mech. Min. Sci.* **1984**, *21*, 75–82. [CrossRef]
21. Nur, A. Effects of access on velocity anisotropy in rocks with cracks. *Rock Mech. Eng. Geol.* **1971**, *76*, 2002–2034. [CrossRef]
22. Zheng, G.; Zhao, X.; Liu, J.; Li, Y. Experimental study on change in acoustic wave velocity when rock is loading. *J. Northeast. Univ.* **2009**, *30*, 1197–1200. [CrossRef]
23. Liu, X.; Wang, X.; Wang, E.; Liu, Z.; Xu, X. Study on ultrasonic response to mechanical structure of coal under loading and unloading condition. *Shock Vib.* **2017**, *2017*, 7643451. [CrossRef]
24. Chen, Z. *Stress Strain Ultrasonic Response Characteristics of Coal during Loading*; Henan Polytechnic University: Jiaozuo, China, 2018.
25. Close, J. Natural fracture in coal. In *Hydrocarbons from Coal*; Law, B.E., Rice, D.D., Eds.; AAPG: Tulsa, OK, USA, 1993; Volume 38, pp. 119–132.
26. Walker, L. Desities, porposites and surface area of coal macerals as measured by their interaction with gases, vapours and liquids. *Fuel* **1988**, *67*, 1615–1623. [CrossRef]
27. Wang, Y. Basic Research on the Deformation and Rupture Microwave Radiation Law of Loaded Coal Body and Its Mechanism. Ph.D. Thesis, China University of Mining and Technology, Beijing, China, 2008.
28. Wang, X.; Asem, P.; Chen, H.; Labuz, J. Microcracking in tensile fracture of a brittle rock. *Eng. Fract. Mech.* **2021**, *251*, 1–16. [CrossRef]
29. Li, X.; Nie, B.; Yang, C.; Mao, Y. Experimental study on the variation law of sound wave velocity during coal loading. *Coal Mine Saf.* **2016**, *47*, 3–16. [CrossRef]
30. Lu, P.; Shen, Z.; Zhu, G.; Fang, E. Permeability characterization and experimental study in the whole process of stress and strain of rock samples. *J. Univ. Sci. Technol. China* **2002**, *32*, 678–684. [CrossRef]
31. Tao, Y.; Xu, J.; Cheng, M.; Li, S.; Peng, S. Theoretical analysis and experimental study on permeability of coal containing gas. *J. Rock Mech. Eng.* **2009**, *28*, 3363–3370. [CrossRef]
32. Yang, T.; Tang, C.; Zhu, W.; Feng, Q. Coupling analysis of seepage and stress in rock fracture process. *J. Geotech. Eng.* **2001**, *23*, 489–493. [CrossRef]
33. Yang, T.; Xu, T.; Liu, J.; Tang, S.; Yu, Q.; Shi, B. Stress-damage-seepage coupling model and its application in deep coal seam gas pressure relief practice. *Chin. J. Rock Mech. Eng.* **2005**, *24*, 2900–2905.
34. Kong, X.; Wang, E.; Li, S.; Lin, H.; Xiao, P.; Zhang, K. Fractals and Chaos Characteristics of Acoustic Emission Energy about Gas-Bearing Coal During Loaded Failure. *Fractals* **2019**, *27*, 5. [CrossRef]
35. Liu, C.; Jing, H.; Fang, L.; Xu, M. Theoretical study on acoustic attenuation model of gas pipeline leakage. *Vib. Shock* **2018**, *37*, 109–114. [CrossRef]
36. Urick, R. The absorption of sound in suspensions of irregular particles. *J. Acoust. Soc. Am.* **1948**, *20*, 283. [CrossRef]
37. O'Connell, R.J.; Budiansky, R. Seismic velocities in dry and saturated cracked solids. *J. Geophys. Res.* **1974**, *79*, 5412–5425.
38. He, Y.; Ma, X. *Modern Physics Experiment*; Tsinghua University Publishing House: Beijing, China, 2003; pp. 198–206.
39. Wu, X. *Study on Acoustic Velocity and Its Related Properties in Petroleum Fluid*; Institute of Geophysics, Chinese Academy of Sciences: Beijing, China, 2000.
40. RaiC, S.; Hanson, K. Transverse-wave velocity anisotropy in sedimentary rock: A laboratory study. *Geophysics* **1988**, *53*, 800–806. [CrossRef]
41. *GB/T19222-2003*; Coal and Rock Sampling Methods. China Standards Publishing House: Beijing, China, 2003.
42. *GB/T30050-2013*; Coal Structure Classification. China Standards Publishing House: Beijing, China, 2013.
43. Yale, D. Recent advances in rock physics. *Geophysics* **1985**, *50*, 2480–2491. [CrossRef]

44. Meng, Z.; Liu, C.; He, X. Experimental analysis on acoustic velocity of coal measures rocks and its influencing factors. *J. Min. Saf. Eng.* **2008**, *25*, 390–394. [CrossRef]
45. Li, Q.; He, J.; Li, C. Study on the relationship between compressional and shear wave velocities and formation pressure of coalbed methane reservoirs in Qinshui Basin. *Geophys. Geochem. Calc. Technol.* **2013**, *35*, 382–385. [CrossRef]
46. Dong, S.; Wu, H.; Li, D.; Huang, Y. Experimental study of ultrasonic velocity and anisotropy in coal samples. *J. Seism. Explor.* **2016**, *25*, 131–146.
47. Sun, X.; Chen, G.; Li, J.; Xu, X.; Fu, S.; Xie, J.; Liang, L. Propagation characteristics of ultrasonic P-wave velocity in artificial jointed coal briquettes. *J. Geophys. Eng.* **2020**, *5*, 827–837. [CrossRef]
48. Zhang, L.; Zhang, Z.; Zhang, R.; Gao, M.; Xie, J. The Ultrasonic P-Wave Velocity-Stress Relationship and Energy Evolution of Sandstone under Uniaxial Loading-Unloading Conditions. *Hindawi Ltd.* **2021**, *2021*, 9921716. [CrossRef]
49. Yu, H.; Zhang, Y.; Lebedev, M.; Wang, Z.; Li, X.; Squelch, A.; Verrall, M.; Lalauer, S. X-ray micro-computed tomography and ultrasonic velocity analysis of fractured shale as a function of effective stress. *Mar. Pet. Geol.* **2019**, *110*, 472–482. [CrossRef]
50. Cao, A.; Dou, L.; Cai, W.; Gong, S.; Liu, S.; Jing, G. Case study of seismic hazard assessment in underground coal mining using passive tomography. *Int. J. Rock Mech. Min. Sci.* **2015**, *78*, 1–9. [CrossRef]
51. Zhao, M. *Study on Acoustic Characteristics of Fractured Rock Mass under Loading*; Chongqing Jiaotong University: Chongqing, China, 1998.
52. Yang, S.; Zhang, N.; Feng, X.; Kan, J.; Pan, D.; Qian, Q. Experimental Investigation of Sandstone under Cyclic Loading: Damage Assessment Using Ultrasonic Wave Velocities and Changes in Elastic Modulus. *Shock Vib.* **2018**, *2018 Pt 3*, 1–13. [CrossRef]
53. Li, J.; Wang, H.; Zhang, Q. Progressive damage and fracture of biaxially-confined anisotropic coal under repeated impact loads. *Int. J. Rock Mech. Min. Sci.* **2022**, *149*, 4979. [CrossRef]
54. Zhang, Z.; Wang, E.; Liu, X.; Zhang, Y.; Li, S.; Khan, M.; Gao, Y. Anisotropic characteristics of ultrasonic transmission velocities and stress inversion during uniaxial compression process. *J. Appl. Geophys.* **2021**, *186*, 104274. [CrossRef]
55. Zhao, Q. *Test and Research on Seismic Wave Parameters of Coal Seams*; Anhui University of Science and Technology: Huainan, China, 2005.
56. Zhang, P.; Liu, S.; Zhao, Q.; Guo, L. Analysis of attenuation characteristics and quality factors of coal seams in Huainan Mining Area. *Coal Sci. Technol.* **2006**, *5*, 83–85. [CrossRef]
57. Liang, Y. *Experimental Study on Ultrasonic and Pore Change Characteristics of Loaded Coal Containing Water*; Henan Polytechnic University: Jiaozuo, China, 2020.

sustainability

MDPI

Article

Application and Optimization of the Parameters of the High-Level Boreholes in Lateral High Drainage Roadway

Haiqing Shuang [1,2], Weitao Meng [1], Yulong Zhai [1,3,*], Peng Xiao [1,2], Yu Shi [1,2] and Yu Tian [1]

1 School of Safety Science and Engineering, Xi'an University of Science and Technology (XUST), No. 58, Yanta Mid. Rd., Xi'an 710054, China
2 Key Laboratory of Western Mine Exploitation and Hazard Prevention of the Ministry of Education, Xi'an University of Science and Technology (XUST), No. 58, Yanta Mid. Rd., Xi'an 710054, China
3 Department of Civil, Geological and Mining Engineering, École Polytechnique de Montréal, Montréal, QC H3C 3A7, Canada
* Correspondence: yulong.zhai@polymtl.com; Tel.: +86-139-9299-8589

Abstract: The key parameters of high-level boreholes in high drainage roadways affect the gas treatment effect of the working face directly. Therefore, the layout parameters of high-level boreholes in the lateral high drainage roadway (LHDR) are determined and optimized as necessary. Based on the LHDR layout on the 2-603 working face of the Liyazhuang coal mine, the key technological requirements on high-level borehole parameters were analyzed and the distribution characteristics of the gas volume fraction in the coal roof were studied. The gas migration law in the mined-out areas was obtained and the layout locations of high-level boreholes were determined finally. The research demonstrates that the high-level boreholes lag the 2-603 working face distance and the position of the final borehole (the position of the final borehole in this paper refers to the distance between the final borehole and the roof) influence the stability of boreholes and the gas extraction effect. The distribution of the gas volume fraction from the intake airway to the return airway can be divided into the stable stage, slow growth stage, and fast growth stage. Influenced by the flow field in the mined-out areas, the mean volume fraction of the borehole-extracted gas has no obvious relationship with the gas volume fraction at the upper corner. According to the final optimization, the high-level borehole is determined as having a 15 m lag behind the working face and the position of the final borehole is 44 m away from the coal seam roof. These have been applied successfully in engineering practice.

Keywords: lateral high drainage roadway; high-level boreholes; gas extraction; numerical simulation; flow field

Citation: Shuang, H.; Meng, W.; Zhai, Y.; Xiao, P.; Shi, Y.; Tian, Y. Application and Optimization of the Parameters of the High-Level Boreholes in Lateral High Drainage Roadway. *Sustainability* **2022**, *14*, 16908. https://doi.org/10.3390/su142416908

Academic Editor: Tongqiang Xia

Received: 14 September 2022
Accepted: 26 November 2022
Published: 16 December 2022

1. Introduction

Coal bed methane (CBM) is a valuable resource and the disasters (e.g., gas explosions and coal-gas outbursts) affect the safety production in the process of coal mining [1–3]. The overlying rock of the working face will migrate and break after mining. The overlying strata will experience a significant number of fractures, providing pathways for gas storage and flow [4,5]. The gas reserve law in coal mines in China shows the feature of three high (high plasticity structure of coal seam, high gas adsorption ability of coal seam, and high gas content of coal seam) and three low (low gas pressure of coal seam, low gas saturation of coal seam, and low rate of gas permeability of coal seam), which causes the effect of pre-mining gas extraction to be ineffective. Many scholars combine the development characteristics of mining-induced fissures with the migration law of pressure relief gas [6–8] and use high drainage roadways and high-level boreholes to extract pressure relief gas [9–12].

Among them, Li et al. [13] adopted the technology of setting super large diameter (0.3 m) boreholes on the roof to ensure the safety production of high gas and outburst mines. Zhang et al. [14] selected the boreholes parameters reasonably through the observation and trajectory measurement of gas drainage boreholes. Zhou et al. [15] conducted the theoretical analysis on the evolution law of mining-induced overlying strata fracture. The technique of gas extraction in different enrichment areas using different level boreholes was proposed. Shang et al. [16] proposed high directional long borehole drainage technology to control the gas concentration in the upper corner under the safety index of the mined-out areas. Li et al. [17] determined the key parameters of the (LHDR) by analyzing the mining stress. Yu et al. [18] studied the distribution and evolution characteristics of the stress and displacement field in the surrounding rock of the roadway and arranged the floor gas drainage roadway to control gas. Hu et al. [19] proposed a technique of a large-diameter blind shaft joined to the high-level gas drainage roadways to control gas. Many scholars have performed a detailed study on the gas extraction methods in the mined-out areas, among which the high-level boreholes and high drainage roadways are the most frequently employed. Fan [20] and Zhang [21] have also studied the key problems of arranging drilling holes for gas extraction.

Some scholars studied the gas extraction method (e.g., high-level boreholes extraction and high drainage roadway extraction) in the mine-out areas after mining [22,23]. Among them, high-level boreholes extraction is an efficient gas extraction method to control gas over the limit. However, the relationship between the lag working face distance and the boreholes stability is not clear. The influence law of the different layers of the final location of the high-level boreholes on the gas flow field migration is not clear. To study the law of gas flow field migration in the mined-out areas affected by the distance between high-level boreholes lagging working face and the position of the final borehole, $UDEC^{2D}$ was used to simulate the overburden fissures zone after mining and separation zone in the overlying strata of mined-out areas. The distance between high-level boreholes lagging working face and the final borehole position was determined. FLUENT was used to analyze the thermal characteristics of gas volume distribution in the coal seam roof and the distribution characteristics of the coal seam roof gas affected by mine-out areas flow field under the influence of drainage boreholes were obtained, to determine the drainage effect of different layers of high-level boreholes on the coal seam roof. The simulated high-level boreholes layout parameters were used to control gas over the limit in the upper corner of the Liyazhuang coal mine.

To prevent gas over-runs in the Liyazhuang coal mine, the pressure relief gas extraction technology (i.e., two adjacent working faces share the external staggered high drainage roadway) is proposed: vertical (vertical 2-603 working face) and lateral (vertical 2-605 working face) high drainage roadways are placed at the roof along the return airway on the 2-603 working face. In the early service stage of high extraction roadways, pressurized gas extraction technology was applied to the 2-603 working face through high-level boreholes. In the late service stage of high drainage roadways, pressurized gas extraction technology was used to the 2-605 working face through high drainage roadways directly. Based on the development characteristics of mining-induced overburden fissures and the reasonable position of high drainage roadways on the working face, the key position parameters of the final high-level borehole are analyzed to realize high-efficiency pressurized gas extraction in the overburden fissure zone of the 2-603 working face in the early service stage of LHDR.

2. Layout of LHDR

2.1. Determination of Key Parameters

The LHDR layout is proposed based on geological conditions and gas control measures in the Liyazhuang coal mine (Figure 1). Vertical and LHDR are placed at the roof along the return airway on the 2-603 working face. In the early service stage of high drainage roadways, pressurized gas extraction technology is applied to the 2-603 working face through high-level boreholes. In the late service stage of high drainage roadways, pressurized gas extraction technology is used to the 2-605 working face through high drainage roadways directly.

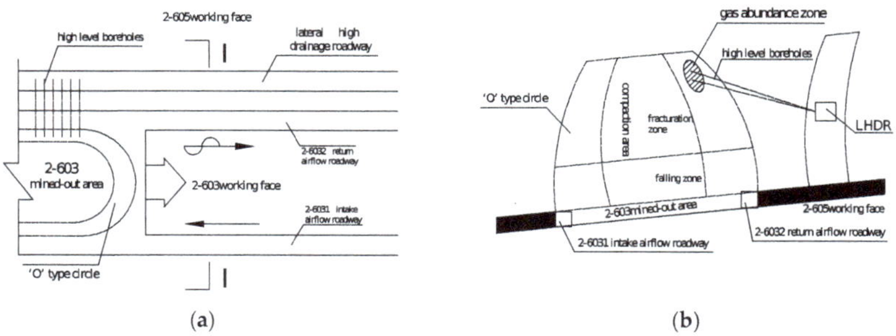

Figure 1. Surrounding rock structure mechanical model of high-level boreholes: (**a**) planar graph; (**b**) I-I section drawing.

The final borehole position directly affects the drainage effect of pressure relief gas in overlying rock mining fissures of the 2-603 working faces in the early stage. Because the high-level boreholes are arranged in the outer staggered high drainage roadway, which is affected by the mining of 2-603 working faces in the early stage, the layer position of the high drainage roadway directly affects the layout parameters of high-level boreholes. The position of the LHDR will influence the surrounding rock stress environment in the roadway directly. Therefore, the technological key of the LHDR layout is to determine parameters for the LHDR and the high-level borehole. Based on previous research results [17], the position parameters of LHDR are determined as follows: roof strike high drainage roadways are arranged along the 2-603 working face, which is 25 m away from the 2-603 working face horizontally and 25 m away from the 2# coal roof vertically. The position parameters of the high-level boreholes are analyzed in the following text.

2.2. Requirements for High-Level Boreholes Parameters in LHDR

From the layout of the LHDR, the high-level boreholes are mainly for pressurized gas extraction in mining-induced overburden fissures on the 2-603 working face. The position parameters of high-level boreholes are the key factors to early extraction pressure relief gas efficiency in the layout of the LHDR. Some requirements on the parameters of the high-level boreholes are provided below:

(1) Overburden mining-induced fissures determine the migration pathways and storage spaces for pressure relief gas directly. Therefore, it is necessary to master the distribution characteristics of overburden mining-induced fissures along the 2-603 working face. The position of final boreholes must be placed at the pressure relief gas enrichment region with consideration to pressure relief gas characteristics.

(2) Three horizontal areas (i.e., coal wall support area, separation area, and re-compaction area) will be formed along the advancing direction of the working face after mining. At the same time, three vertical zones (i.e., bending sagging zone, fracture zone, and caving zone) will be formed from the upper to the lower parts of the overburden areas [24,25], as shown in Figure 2.

From Figure 2, one sees that when the gas extraction boreholes are in the coal wall support area, the supporting stress for advanced mining of the working face will influence the borehole stability and cause the collapse or deformation of boreholes. When boreholes are in the separation area and re-compaction area, the mining-induced overburden fissures will become the gas migration pathways with pressure relief. The arrangement of boreholes in the re-compaction area slightly influences the pressure relief gas flow field on the working face and is easy to cause an exceeding gas concentration at the upper corner. The final high-level boreholes must be placed at the intersection of the fissure zone and separation area.

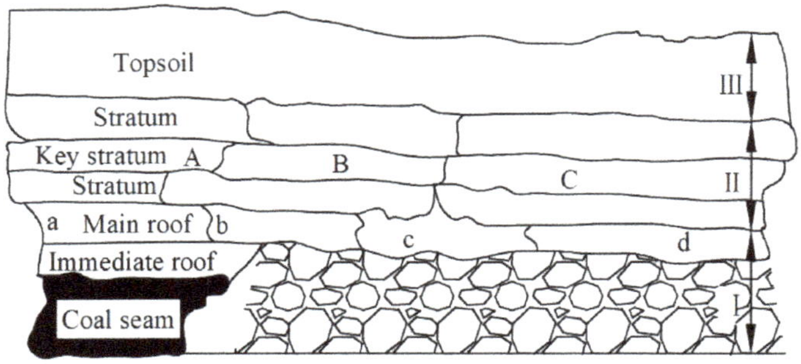

Figure 2. The horizontal zones and vertical zones after working face mining. (A—coal wall support area (a–b); B—separation area (b–c); C—re-compaction area (c–d); I—caved zone; II—fractured zone; III—bending zone).

(3) When high-level boreholes are behind the working face, the LHDR is disturbed by mining the 2-603 working face and the loosing zone of surrounding rocks close to the 2-603 working face increases accordingly. Therefore, the sealing length and strength of the final high-level boreholes must be increased to ensure negative pressure extraction and efficiency.

(4) The high-level boreholes are arranged in the common area of the fractured zone and the separation zone. It is necessary to ensure the stability of the high-level boreholes to achieve efficient gas extraction. The high-level boreholes should not be in the peak stress area [26]. It reduces the influence of mining stress on the stability of the boreholes and ensures the gas extraction effect of the boreholes.

To sum up, when high-level boreholes are behind the working face, the final boreholes must be in the separation area of the fracture zone, and the sealing strength of the high-level boreholes must be increased. At the same time, the mining stress will cause the boreholes to collapse when the working face continues to advance, so it is necessary to strengthen the sealing strength of high-level boreholes and the high-level boreholes should not be in the peak stress area. As a result, the high-level boreholes lag the working face distance as well as the position of the final borehole must be optimized to improve the boreholes stability and achieve a better extraction effect.

3. Distribution Characteristics of Mining-Induced Overburden Fracture Zone and Separation Area

3.1. Overburden Fissure Distribution Range

Figure 3 shows the distribution of the mining-induced fractures after mining. The mining-induced fissures distribution area of the overlying strata at the end of the working face is within 62° of the uphill mining angle. The fissures in the vertical direction are concentrated in two main areas: (1) the first area: 13–25 m away from the floor at the height, 12 m away from the stope boundary, and 65 m in width, and (2) the second area: 38.6–50 m away from the floor at the height, 28 m away from the stope boundary, and 50 m in width. The first place is closer to the working face, the air leakage is more serious. Since they are affected by the ascending characteristics of the gas, the extraction boreholes should be positioned above the second zone.

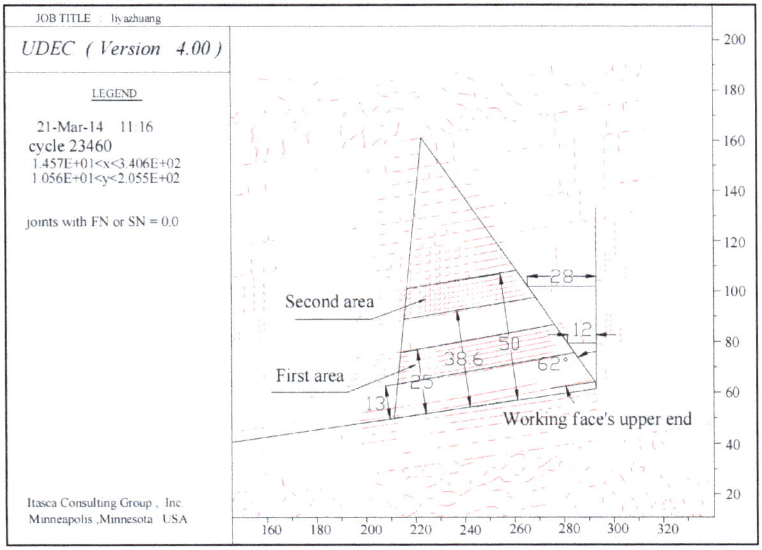

Figure 3. Distribution characteristics of overburden fissures after mining.

3.2. Distribution Characteristics of Separation Area

A numerical model was constructed according to the geological conditions of the 2-603 working face. The model parameters are shown in [17,27,28]. The distribution characteristics and extent of the mining overburden fissures after excavation of the 2-603 working face were simulated. The first pressurization and periodic pressurization occur with the advancing of the working face. The distribution characteristics of mining-induced overburden fissures when the working face advances by 95 m are shown in Figure 4.

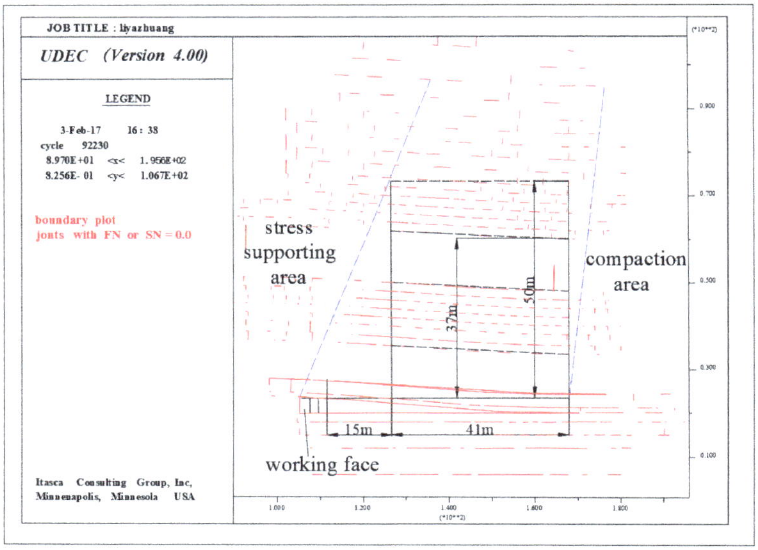

Figure 4. Distribution characteristics of mining-induced overburden fissures.

Figure 4 shows that the stress supporting area, separation area, and compaction area are developed on the working face after mining. There are few mining-induced overburden fissures in the stress supporting area and the compaction area, which limit pressure relief gas storage and migration. Since two layers of thick sandstones play a key role in controlling the overburden activities, the mining-induced fissures are mainly distributed in two areas, which are consistent with the distribution characteristics of mining-induced fissures on inclined overburden. The main storage pathways of pressure relief gas are mainly 15–41 m away from the working face horizontally and 37–50 m away from the roof vertically. Therefore, boreholes are set 15–41 m away from the working face and the final borehole is 37–50 m away from the roof.

4. Parameters of High-Level Boreholes Layout

4.1. Theoretical Basis of the Numerical Model

Gas flow in the mining-induced fracture zone is very complicated and can be simplified as a seepage in porous media, including the laminar region, transition region, and turbulence region. Generally, there is large air leakage in small mined-out areas range close to the working face, but air leakages in other regions are similar with seepage of a small Renault coefficient [29]. Hence, gas migration in the mining-induced fracture zone conforms to the continuity equation, momentum equation, and mass conservation equation of fluid flow in porous media [30,31].

(1) Equations of laminar flow and turbulence flow

Laminar flow and turbulence flow of pressure relief gas in the mining-induced overburden fissures on the working face using borehole extraction were designed. The $RNGk-\varepsilon$ model with high and low Reynolds numbers was applicable.

① Equation of turbulent kinetic energy

$$\frac{\partial}{\partial t}(\rho k)+\frac{\partial}{\partial x_i}(pku_i)=\frac{\partial}{\partial x}\left[a_k(u+u_l)\frac{\partial k}{\partial x_j}\right]+G_k+G_b-\rho\varepsilon-Y_M+S_K \tag{1}$$

where t is time (s); ρ is air density (kg/m^3); k is turbulent kinetic energy (m^2/s^2); u_i is the temporal average velocity (m/s); a_k is the number of turbulence flow; u is the dynamic viscosity of the fluid (Pa · s); u_l is turbulence viscosity (Pa · s); $G_b=-g_i\frac{u_t}{\rho pr_t}\frac{\partial\rho}{\partial x_i}$ (g_i is the acceleration of gravity (m/s^2) and pr_t is *Prandtl* number of turbulence); $Y_M=2\rho\varepsilon M_t$ (M_t is Mach number of turbulence, $M_t=\sqrt{k/a^2}$ and α is the sound velocity (m/s)); ε is dissipation rating (m^2/s^3); $G_K=u_l\left(\frac{\partial u_i}{\partial x_i}+\frac{\partial u_j}{\partial x_i}\right)\frac{\partial u_i}{\partial x_j}$; and S_K is the source item (kg/(m · s^3)).

② Diffusion equation of the dissipation rating of turbulent kinetic energy (ε)

$$\frac{\partial}{\partial_t}(\rho\varepsilon)+\frac{\partial}{\partial x_i}(\rho\varepsilon u_i)=\frac{\partial}{\partial x_i}\left[a_\varepsilon(u+u_t)\frac{\partial\varepsilon}{\partial x_i}\right]+C_{1\varepsilon}\frac{\varepsilon}{k}(G_k+C_vG_b)-C_{2\varepsilon}\rho\frac{\varepsilon^2}{k}-R_\varepsilon+S_\varepsilon \tag{2}$$

where a_ε is the *Prandtl* number of turbulence; $C_{1\varepsilon}$, $C_{2\varepsilon}$ and C_v are dimensionless constants of the model; $R_\varepsilon=\frac{0.0845\rho\eta^3\left(1-\frac{\eta}{4.377}\right)}{1+0.012\eta^3}\cdot\frac{\varepsilon^3}{k}$; kg/(m · s^4); $\eta=\left(2E_{ij}\cdot E_{ij}\right)^{1/2}\frac{k}{\varepsilon}$; $E_{ij}=\frac{1}{2}\left(\frac{\partial u_i}{\partial x_j}+\frac{\partial u_j}{\partial x_i}\right)$, s^{-1}; and S_ε is the source item (kg/(m · s^4)).

Moreover, the model equation of effective velocity and eddy modification gram mark was involved to adapt to the influences of the eddy on low Reynolds number, near-wall flow, and modified turbulence flow in the laminar flow:

$$d\left(\frac{\rho^2k}{\sqrt{u_t\varepsilon}}\right)=1.72\frac{\hat{v}}{\sqrt{\hat{v}^3-1+C_v}}d\hat{v} \tag{3}$$

where $\hat{v} = (u + u_t)/u$; u_t differs for different liquid states. Turbulence flow and eddy are determined according to Equation (4).

$$u_l = 0.0845\rho k^2/\varepsilon u_t = u_{t0} f(a_s, \Omega, k/\varepsilon) \tag{4}$$

where u_{t_0} is a non-revised value of viscosity (Pa · s); a_s is constant; Ω is the estimated value.

(2) Mass conservation equation of gas

In the mining-induced fracture zone, gas transmission in the air must meet obey the mass conservation law.

$$\frac{\partial}{\partial t}(\rho c_g) + \frac{\partial}{\partial x_i}(\rho c_g u_i) = -\frac{\partial}{\partial x_i}(J_g) + S_g \tag{5}$$

where c_g is the volume fraction of gas, m^3/m^3; J_g is gas diffusion flux $(kg/(m^2 \cdot s))$; laminar flow and turbulence flow are determined by Equation (6). S_g is the additional productivity of the gas source item $(\frac{kg}{(m^3 \cdot s)})$.

$$J_g = -D\rho\frac{\partial}{\partial x_i}(c_g), J_g = -\left(D\rho + \frac{u_t}{Sc_t}\right)\frac{\partial}{\partial x_i}(c_g) \tag{6}$$

where D is the diffusion coefficient of gas (m^2/s) and Sc_t is Schmidt number of turbulence flow.

(3) Equation of continuity

According to the mass conservation law of gas in mining-induced fissures, the equation of continuity of gas migration is gained:

$$\frac{\partial \rho}{\partial t} + \frac{\partial}{\partial x_i}(\rho u_i) = S_g \tag{7}$$

where S_g is the gas source item $(kg/(m^3 \cdot s))$.

(4) Momentum conservation equation

The momentum conservation equation of porous medium on the direction i in the inertial coordinate system is:

$$\frac{\partial}{\partial_t}(\rho u_i) + \frac{\partial}{\partial x_j}(\rho u_i u_j) = \frac{\partial \tau_{ij}}{\partial x_j} - \frac{\partial p}{\partial x_i} + \rho g_i + \sum_{j=1}^{3} D_{ij}(\mu + \mu_i)q_j + \sum_{j=1}^{3} C_{ij}\frac{1}{2}\rho|q_j|q_j \tag{8}$$

where P is the pore pressure (Pa); q_j is the seepage velocity (m/s); D_{ij} and C_{ij} are specified matrixes; τ_{ij} is the stress tensor (Pa); $\tau_{ij} = (\mu + \mu_i)\left[\left(\frac{\partial u_i}{\partial x_j} + \frac{\partial u_j}{\partial x_i}\right) - \frac{2}{3}\frac{\partial u_i}{\partial x_i}\delta_{ij}\right]$ and δ_{ij} is the Kronecker mark.

Equations (1), (2), (5), (7) and (8) form the mathematical model of gas migration in the mining-induced fracture zone. If variables, time-varying terms, convective terms, and diffusion terms in all equations are expressed in standard forms, the general forms could be gained (Equation (9)). In the application, only the solver of Equation (9) is programmed, thus enabling to solve gas migration in the mined-out areas under different boundary conditions.

$$\frac{\partial}{\partial_t}(\rho\phi) + div(\rho\phi u) = div(\Gamma grad\phi) + S \tag{9}$$

where ϕ is the universal variable; Γ is the generalized diffusion coefficient and S is the generalized source item.

4.2. Parameter Setting

The permeability of the coal seam and overlying rock is the main factor that controls gas outburst on the working face. The mining-induced stress distribution influences the permeability of both the exploitation bed and the adjacent bed. The permeability is determined by the fissure development in the strata in front of the working face as well as the stress relief behind the working face. The stress distribution caused by mining influenced the permeability of both mining and adjacent layers. Based on analysis of stress distribution law in the goaf and previous experiences on FLUENT simulation study, permeability distribution in the goaf was determined. Permeability changes in different regions were 10^{-4}–10^{-6} m^2 and the maximum viscosity permeability in the goaf was about 10^{-10} m^2 [17,32].

Other parameters of the model are: intake airway is 5 × 3 m; the net fracture surface is 15 m^2; return airway is 4.8 × 3.8 m; the net fracture surface is 18.24 m^2. The inclined length, width, and height of the mining working face are 200 m, 5 m, and 3.5 m, respectively. The dip angle is 8°. The strike length and height of goaf are 250 m and 100 m, respectively. The model uses a local grid subdivision and the grid quantity of the whole model is about 1.12 million. The reference pressure of the model is set at 101.325 KPa. The simulated calculating pressure is the relative pressure value. The inlet of the intake airway is set free and the outlet of the return airway is set as a pressure outlet. The negative pressure ventilation is employed. The extraction borehole is pressure out and the negative pressure for extraction is 30 KPa.

4.3. Scheme Settings

To verify the reasonability of the established model, the working face open-off cut when no gas extraction in the mined-out areas and the distribution characteristics of gas volume fraction in the return airway were analyzed (Figure 5).

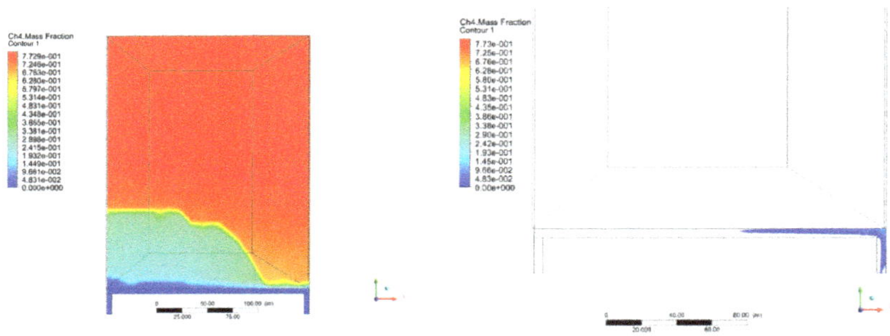

(**a**) (**b**)

Figure 5. Gas distribution characteristics on the working face without extraction: (**a**) Contour of gas volume fraction at the roof; (**b**) Regions with gas volume fraction>1% on the working face.

Distribution characteristics of gas volume fraction (>1%) at the upper corner of the working face under the ventilation conditions are shown in Figure 5b. This region is mainly in the position that is 180–200 m away from the intake airway. The gas volume fraction at the joint between the working face and the return airway reaches 1.45–2.10% and the gas volume fraction on the working face exceeds the limit. Furthermore, practices have proved that increasing the ventilation quantity alone is difficult to solve excessive gas volume fraction at the upper corner of the working face. On one hand, increasing the ventilation quantity will discharge the gas into the working face and roadways. On the other hand, it will intensify air leakage in the goaf and more gases in the goaf will rush into the upper corner of the working face. Therefore, borehole gas extraction in the goaf must be considered when increasing the ventilation quantity fails to improve gas extraction. Combined with practical distribution characteristics of gas volume fraction on the working face, the established model is proved reasonable.

Combined with the distribution characteristics of mining-induced overburden fissures on the working face in Section 2.1, the periodic weighting step on the working face is about 15 m and the angle of mining-induced fissures at end of the 2-603 working face is 62°.

High-level boreholes are placed behind the working face and the experimental scheme is presented in Table 1. According to the line pressure for extraction in LHDR, 20 high-level boreholes were set at an interval of 2 m.

Table 1. Experimental Scheme of the different positions of the final boreholes.

Distance from Working Face/m	35 m away from Coal Seam Roof	45 m away from Coal Seam Roof	55 m away from Coal Seam Roof
10	Scheme1	Scheme2	Scheme3
15	Scheme4	Scheme5	Scheme6
20	Scheme7	Scheme8	Scheme9

4.4. Analysis of Simulation Results

Influenced by floating diffusion, pressure relief gas mostly concentrates in the top of the fissure region. Distribution characteristics of gas volume fraction close to the roof can represent the pressurized gas extraction effect of high-level boreholes to some extent. In the following text, reasonable parameters of high-level boreholes are analyzed from the distribution characteristics of gas volume fraction on the roof, the distribution law of gas volume fraction on the working face, and the average extracted gas volume fraction of high-level boreholes.

1. Distribution characteristics of gas volume fraction on the roof;

Figure 6 shows that the distribution characteristics of gas volume fraction on the roof vary for high-level boreholes located at different positions. The region with low gas volume fraction close to the working face is A area (1.30% > gas volume fraction > 0) and the region with high gas volume fraction close to the working face is B area (gas volume fraction > 1.30%) (Figure 6a).

When boreholes are 10 m away from the 2-603 working face, the A area increases significantly with an increase in the height of the final borehole. B area expands to the return airway gradually, indicating the flow field close to the roof is beneficial for reducing gas discharge on the working face when the final borehole is 55 m away from the working face. When boreholes are 15 m away from the working face, the area of A decreases gradually with the increase in height of the final borehole, while the area of B remains the same. In addition, the B area gradually moves away from the return airway and develops to the deep part of the mining area on the side of the intake airway; when the boreholes lag behind the working face by 20 m, the area of the A area gradually decreases with the rise of the final hole position of the high drilling hole, and the B area gradually moves away from the return airway and moves to the deep part of the mining area on the side of the itay.

2. Distribution characteristics of gas volume fraction on the roof;

The distribution law of gas volume fraction on the side of the coal seam roof will change under the influence of the mined-out areas flow field after extraction, which is reflected by the distribution characteristics of gas volume fraction on the coal seam roof. The distribution law is the macroscopic display of the flow field in the goaf. The distribution characteristics of gas volume fraction on the working face are closely related to this distribution law. To reflect the distribution law of gas volume fraction on the working face intuitively, the monitoring line of gas volume fraction is set between the intake airway and return airway, which is 3.0 m away from the bottom board of the coal bed. Distribution laws of gas volume fraction on the working face under different schemes are presented in Figure 7.

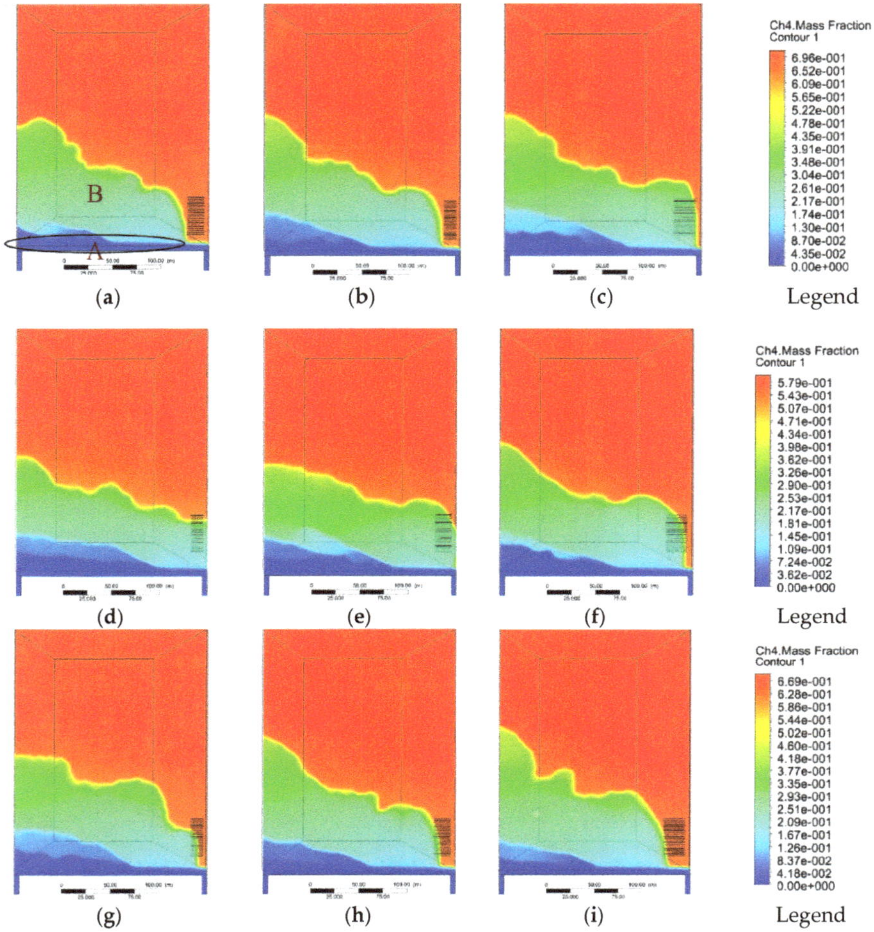

Figure 6. Distribution characteristics of gas volume fraction in different schemes: (**a**) scheme1; (**b**) scheme2; (**c**) scheme3; (**d**) scheme4; (**e**) scheme5; (**f**) scheme6; (**g**) scheme7; (**h**) scheme8; (**i**) scheme9.

In Figure 7, the gas volume fraction at the upper corner under extraction schemes decreases significantly compare with non-extraction schemes. The maximum gas volume fraction at the upper corner is 2.2% for no extraction scheme. For the borehole extraction, Schemes 3, 6, and 9 are over 1% gas volume fraction at the upper corner and the maximum gas volume fraction is 1.5%. This reflects that the borehole extraction can change the flow field for gas migration in the goaf and change distribution characteristics of gas volume fraction on the working face effectively.

The gas volume fraction on the working face is positively correlated with the distance increase from the position of the final borehole to the inlet airway. The growth of gas volume fraction is divided into three stages: the stable stage, slow growth stage, and fast growth stage. The variation laws of gas volume fraction on the working face under non-extraction and borehole extraction are different. Under the non-extraction technique, three stages are 0–26 m, 26–180 m, and 180–200 m in the intake airway, and the corresponding variation ranges are 0.0–0.012%, 0.012–1.00%, and 1.00–2.10%, respectively. Under borehole extraction technique, three stages are 0–110 m, 110–180 m, and 180–200 m, and the corresponding variation ranges are 0–0.10%, 0.10–0.41%, and 0.41–1.49%, respectively.

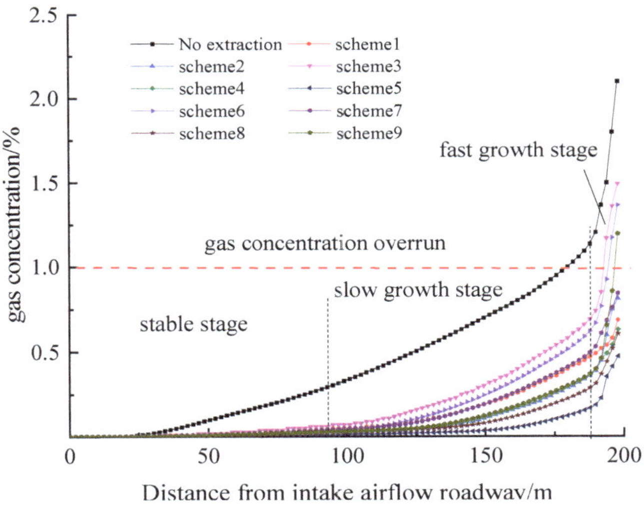

Figure 7. Distribution of gas volume fraction in mining working face.

Figure 7 shows that position parameters of boreholes influence the distribution law of gas volume fraction on the working face directly. According to schemes 1, 4, and 7, boreholes at the lower part of the fracture zone are conducive to gas extraction when the final borehole is 35 m away from the roof. The mining-induced overburden fissures are the main pathways for migration of pressure relief gas rather than the concentration area of pressure relief gas. The maximum gas volume fraction at the upper corner is 0.68%, 0.63%, and 0.84%, respectively. For schemes 2, 5, and 8, boreholes at top of the fracture zone can effectively extract pressure relief gas accumulated in this area when the final borehole is 45 m away from the roof. The maximum gas volume fraction at the upper corner is 0.81%, 0.47%, and 0.61%, respectively. It can be known from schemes 3, 6, and 9 that the borehole negative pressure cannot guide the gas flow field effectively when boreholes are 55 m away from the roof and the pressure relief gas volume fraction at the lower part rather than migrating to the 55 m, thus resulting in excessive gas volume fraction on the working face. The maximum gas volume fraction at the upper corner is 1.49%, 1.37%, and 1.20%, respectively.

3. Average drainage gas volume fraction of high-level boreholes and gas volume fraction at the upper corner.

The gas volume fraction of high-level boreholes influences the late use and extraction effect of gas directly. The gas volume fraction at the upper corner of the working face can reflect the gas control efficiency on the working face directly. Hence, position parameters of high-level boreholes for gas extraction must be determined by combining distribution characteristics of gas volume fraction on the roof and the distribution law of gas volume fraction on the working face. The gas volume fraction of high-level boreholes and gas volume fraction at the upper corner of the working face is shown in Figure 8.

Figure 8 shows that when the position of the final borehole is at different positions, the average drainage gas volume fraction of the high-level boreholes is different from that at the upper corner of the working face. The gas volume fractions at the upper corner in schemes 3, 6, and 9 exceed the limit (the safety concentration is below 1% at the upper corner), while those in schemes 2 and 7 are close to the safety warning line (0.81% and 0.84%, respectively), which are neglected in this paper. Only schemes 1, 4, 5, and 8 are feasible. The corresponding gas volume fraction at the upper corner is 0.69%, 0.63%, 0.47% and 0.61%, and the average drainage gas volume fraction of boreholes are 66%, 60%, 64%, and 55%, respectively. Therefore, the maximum variations of gas volume fraction at the upper corner and the average drainage gas volume fraction of boreholes in these 4 schemes are 0.22% and 11.00%, respectively. The gas volume fraction of borehole extraction changes slightly. Combined with Figure 6, scheme 1 achieves the highest gas volume fraction at the same position 120 m behind the intake airway, followed by scheme 4, scheme 8, and scheme5 successively. This reveals that scheme 5 is more reasonable compared to the rest schemes.

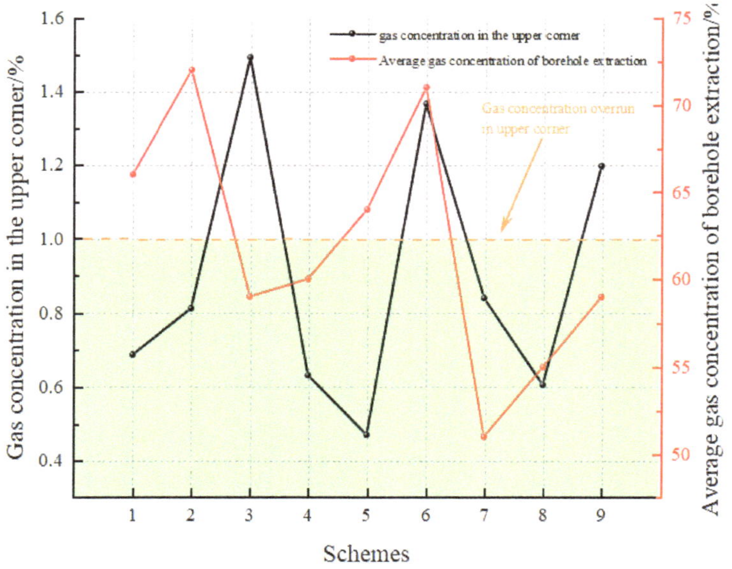

Figure 8. Average gas concentration of drainage borehole and gas concentration in the upper corner.

To sum up, gas extraction effect of high high-level boreholes represented by scheme 5 is best. In other words, when boreholes are 15 m away from the working face and the final boreholes are 45 m away from the coal roof, the average drainage gas concentration of boreholes is 64.00% and the gas volume fraction at the upper corner of the working face is within the limit, meeting the gas control requirements of the working face.

5. Engineering Application

5.1. Layout Parameters of Test Boreholes and Effect Analysis

1. Layout parameters of test boreholes:

To determine the reasonable position of the final borehole, the final borehole will be 15.0 m away from the working face and six boreholes are placed in the sector pattern close to the 115# borehole. The borehole diameter is 0.113 m and the interval of boreholes is set to 5.0 m. The construction position of boreholes are 1.0 m from roadway floor. The detailed borehole parameters and fissures distribution are shown in Figure 9.

The high-level boreholes are placed behind the working face. Based on the distribution characteristics of the mining-induced overburden fissures, the boreholes are connected with the mining-induced overburden fissures. Each borehole is sealed immediately after finishing the construction. Self-plugging pocket sealing technology (it uses polyurethane/polyurea composite material) is applied and the sealing length is determined as 10 m [26,33]. At this moment, high-level boreholes are mainly distributed in the separation area. The LHDR and borehole walls are affected by mining-induced stress on the working face slightly. The polyurethane/polyurea composites material is adequate to meet the sealing requirements. However, the distance from the external obstruction of the metal casing pipe to the internal mouth must be no less than 0.5 m to avoid air leakage caused by the connection between boreholes and the surrounding rock loose zone in the LHDR.

The orifice plate flowmeter and tailrace were connected outside the metal casing pipe. The buried pipelines were connected with the branching units on the gas drainage pipe through the air outlet of the tailrace.

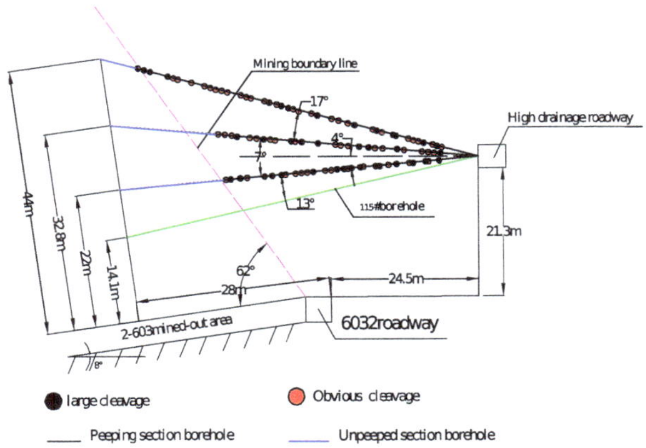

Figure 9. Specific parameters of test boreholes.

2. Sealing technology and key parameters of test boreholes:

To improve the connectivity and sealing quality of the high-level boreholes, the ponding and rock debris in the boreholes must be cleaned before the sealing. Due to timely sealing, the water on the borehole wall or mining-induced overburden fissures have not been eliminated in time, thus requiring irregular water drainage during the borehole extraction and protecting the high-level borehole efficient gas extraction of the pressurized gas.

3. Analysis of test borehole extraction effect:

The monitoring results of the gas volume fraction of six testing boreholes are shown in Figure 10. The effective gas extraction days of the boreholes 1-1 and 1-2 are 42 days and 39 days, respectively, with the corresponding average gas volume fractions of 52.4% and 43.0%, respectively. There are 12 and 13 days with gas drainage concentrations higher than 80%. The effective gas extraction days of the boreholes 2-1 and 2-2 are 23 days and 36 days and the corresponding average gas volume fraction is 12.6% and 17.8%, respectively. The gas drainage concentration mainly ranges between 10% and 30%, lasting for 17 days and 25 days. The effective gas extraction days of the boreholes 3-1 and 3-2 are 26 days and 32 days and the corresponding average gas volume fraction is 11.0% and 8.0%, respectively. The gas drainage concentration mainly ranges between 10% and 30%, lasting for 17 days and 21 days. To sum up, the position parameters of boreholes 1-1 and 1-2 are reasonable, which contribute to the large gas drainage concentration of boreholes and the long duration. They are capable of meeting the requirements for gas extraction in high-level boreholes.

4. Stability analysis of test borehole:

To analyze the borehole fissure morphology, the theoretical design borehole trajectory is expressed in two dimensions, the distance from the working face to the drilling field is represented by the horizontal axis, and the distance from the boreholes to the coal roof is represented by the vertical axis; the graphics represent the development of borehole fracture. The original fracture is represented by ⊗ and the new fracture is represented by ◑, where the bubble size is the crack size. Among them, the bubble size of 0.20 represents small fracture, 0.35 represents soft rock, 0.45 represents obvious fracture, 0.60 represents small-scale hole cutting, 0.65 represents small-scale hole collapse, 0.75 represents large-scale hole collapse, 0.80 represents complete hole cutting, and 1.00 represents borehole bottom. The fracture morphology of each borehole is shown in Figure 11.

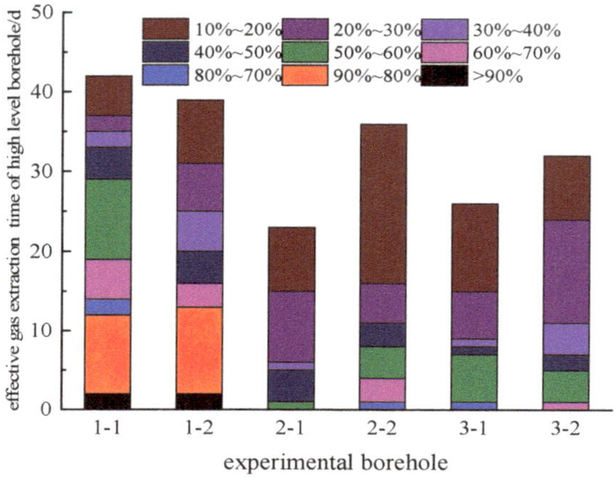

Figure 10. Test boreholes extraction gas volume fraction characteristics.

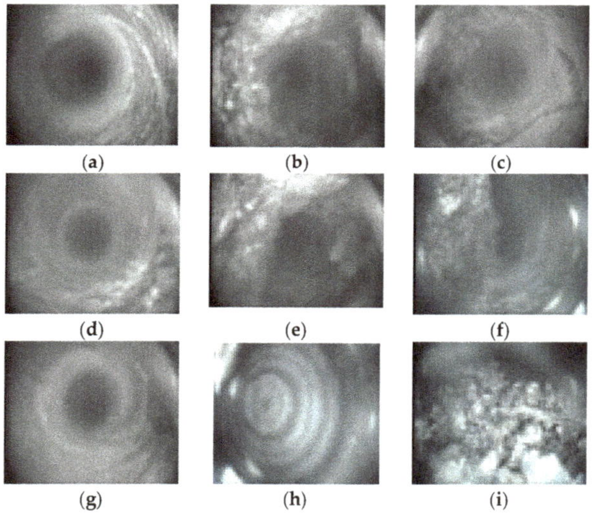

Figure 11. Classification of borehole observation fissures: (**a**) Small fissure; (**b**) Soft rock; (**c**) Obvious fissure; (**d**) Small-scale hole cutting; (**e**) Small-scale hole collapse; (**f**) Large-scale hole collapse; (**g**) Complete hole cutting; (**h**) Borehole bottom; (**i**) Complete hole collapse.

According to the analysis of the gas volume fraction characteristics of the test boreholes, it is known that the drainage effect of boreholes 1-1 and 1-2 are the best. According to the fracture evolution characteristics of the above two boreholes, the above two boreholes are analyzed in stages when the working face is advanced by 15 m and 30 m.

(1) Working face advancing 15 m:

Figure 12 shows that there is a slight hole collapse phenomenon in the borehole between 12 m and 18 m. The fracture in the other borehole section is unevenly released and the borehole forming quality is poor. There is no fracture development and obvious deformation and the bottom of the hole is complete because the borehole has not received the influence of mining pressure at this time. Because the 2-603 working face is far away from the drilling sites, the end of the borehole has not yet fallen the square of the mined-out area. Combined with the borehole observation, it can be seen that the microfracture in the hole section is mainly caused by the damage in the borehole forming process, borehole disturbance, water immersion, and the influence of pre-mining abutment pressure.

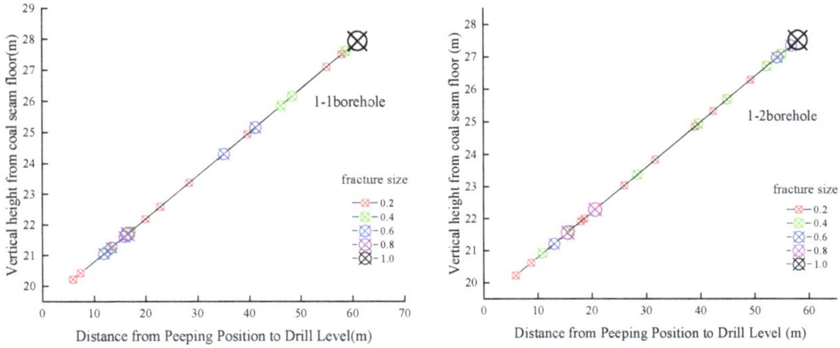

(**a**) (**b**)

Figure 12. Distribution characteristics of borehole fissures when the working face is advanced 15 m: (**a**) Distribution characteristics of fissures in No.1-1 borehole; (**b**) Distribution characteristics of fissures in No.1-2 borehole.

(2) Working face advancing 30 m:

Figure 13 shows that the final high-level borehole position is directly above the mined-out area at this time and the fracture development of each borehole has changed significantly. The fractures of the No. 1-1 borehole are dense at 24–36 m, they are obvious fractures and hole collapse, and there are cuts at the end of the borehole, which indicates that the end of the borehole is in the fracture zone; the situation of the No.1-2 borehole is similar to that of borehole 1-1. The results show that the high-level borehole stability is enough to ensure the extraction effect.

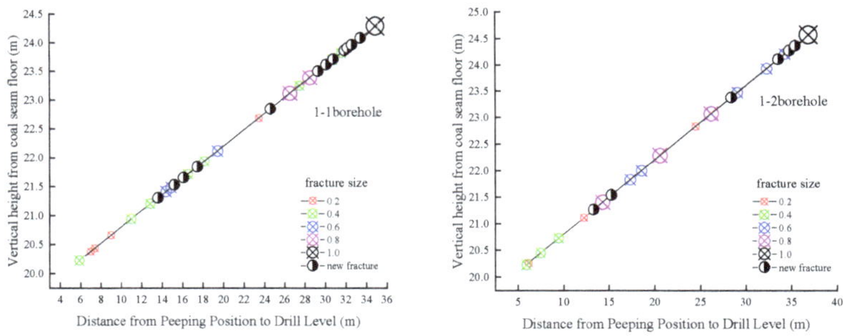

(**a**) (**b**)

Figure 13. Distribution characteristics of borehole fractures when the working face is advanced 30 m: (**a**) Distribution characteristics of fissures in the No.1-1 borehole; (**b**) Distribution characteristics of fissures in No.1-2 borehole.

5.2. Late Engineering Applications

1. Borehole layout parameters:

The extraction boreholes are arranged 15 m behind the 2-603 working face with a spacing of 1.8 m and an orifice of 1.0 m from the LHDR bottom board. The distance is about 17–29 m from the bottom of the high drainage roadway to the working face. The extraction drilling angle should be adjusted in real time to ensure the final borehole is 44 m above the roof of the 2-603 working face and the projection length of the borehole in the goaf is at least 28 m.

2. Layout parameters of test boreholes:

During the mining at the 2-603 working face, the gas volume fractions of the branch pipelines, LHDR boreholes, and the upper corner of the working face are monitored daily. The results show that the extraction branch is connected with from 15 to 20 boreholes, the extraction time can reach 20–40 days, and the gas volume fraction varies between 10% and 65%. The average flow rate of the pure gas is 22.3 m^3/min. The gas concentrations varied from 0.50–0.95% for the production shift and 0.47–0.89% for the maintenance shifts, respectively.

6. Conclusions

1. Key technological requirements for the high-level borehole layout are proposed based on the early service of the high-level boreholes in the layout of the two adjacent working faces sharing the LHDR. Specifically, the high-level boreholes are placed behind the working face and the position of the final borehole is in the overburden fracture zone and the separation area. The sealing strength of the gas extraction boreholes must be enhanced.
2. Gas migration in the mining-induced fracture zone is composed of equations of laminar flow and turbulence flow, continuity equation, momentum equation, and mass conservation equation. The FLUENT numerical simulation model is established according to the distribution law of the mining-induced overburden fissures on the 2-603 working face.
3. The layout parameters of high-level boreholes influence the gas distribution characteristics in the mined-out area directly, thus resulting in different gas migration characteristics on the roof, different gas volume fraction distribution patterns on the working face, and different average drainage gas volume fractions of high-level boreholes in nine schemes. According to the test results on the gas drainage concentration of boreholes, it concludes that the high-level boreholes must be 15.0 m away from the 2-603 working face and the final borehole must be 45 m away from the roof.
4. In engineering practice, the high-level boreholes are behind 15 m away from the working face and 15–20 boreholes are placed (enhancing sealing strength) at an interval of 1.8 m. The final borehole is located at 44.0 m away from the coal roof. According to the application effects, the average flow rate of pure gas is 22.3 m^3/min, which controls the gas volume fraction at the upper corner within the limit and thereby ensures safety and high-efficiency mine of the working face.

Author Contributions: Conceptualization, H.S.; software, H.S. and W.M.; data curation, W.M.; formal analysis, P.X., H.S. and Y.Z.; funding acquisition, Y.S.; investigation Y.T. and W.M.; writing—original draft, W.M.; writing—review and editing, Y.T., H.S. and Y.Z. All authors have read and agreed to the published version of the manuscript.

Funding: This research was funded by the National Natural Science Foundation of China (no. 51904238, no. 11802231, no. 51774235), China Postdoctoral Science Foundation (2019M663937XB), Shaanxi Natural Science Youth Foundation (no. 2019JQ-337), and Special Scientific Research Plan of Shaanxi Provincial Department of Education (no. 19JK0534).

Institutional Review Board Statement: Not applicable.

Informed Consent Statement: Not applicable.

Data Availability Statement: Not applicable.

Acknowledgments: This work is supported by the National Natural Science Fund Project (no. 51904238, no. 11802231, no. 51774235), China Postdoctoral Science Fund (2019M663937XB), Shaanxi Natural Science Youth Fund (no. 2019JQ-337), Special Scientific Research Plan of Shaanxi Provincial Department of Education (no. 19JK0534).

Conflicts of Interest: The authors declare no conflict of interest.

References

1. Li, C.; Pang, X.; Ma, X.; Wang, E.; Hu, T.; Wu, Z. Hydrocarbon generation and expulsion characteristics of the Lower Cambrian Qiongzhusi shale in the Sichuan Basin, Central China: Implications for conventional and unconventional natural gas resource potential. *J. Petrol. Sci. Eng.* **2021**, *204*, 108610. [CrossRef]
2. Wang, K.; Du, F. Coal-gas compound dynamic disasters in China: A review. *Process. Saf. Environ.* **2020**, *133*, 1–17. [CrossRef]
3. Yan, G.; Gui, F.; Nieto, A. A comparative study of gas explosion occurrences and causes in China and the United States. *Int. J. Min. Reclam. Environ.* **2016**, *30*, 269–278.
4. Peng, G.Y.; Gao, M.Z.; Xie, J.; Liu, Q.; Deng, G.D.; Li, C.; He, Z.Q. Field test research on gas migration law of mining coal and rock. *Therm. Sci.* **2019**, *23*, 1591–1597. [CrossRef]
5. Gao, M.Z.; Ai, T.; Qiu, Z.Q.; Zhang, Z.T.; Xie, J. Analysis of gas migration patterns in fractured coal rocks under actual mining conditions. *Therm. Sci.* **2017**, *21*, S275–S284. [CrossRef]
6. Guo, H.; Todhunter, C.; Qu, Q.D.; Qin, Z.Y. Longwall horizontal gas drainage through goaf pressure control. *Int. J. Coal Geol.* **2015**, *150*, 276–286. [CrossRef]
7. Liu, H.; Shu, Z.; Shi, Y.; Wang, X.; Xiao, X.; Lin, J. Gas Migration Patterns with Different Borehole Sizes in Underground Coal Seams: Numerical Simulations and Field Observations. *Minerals* **2021**, *11*, 1254. [CrossRef]
8. Yan, Z.; Wang, Y.; Fan, J.; Huang, Y.; He, Y. Study on Key Parameters of Directional Long Borehole Layout in High-Gas Working Face. *Shock Vib.* **2021**, *2021*, 5579967. [CrossRef]
9. Junxiang, Z.; Bo, L.; Yuning, S. Dynamic leakage mechanism of gas drainage borehole and engineering application. *Int. J. Min. Sci. Technol.* **2018**, *28*, 505–512. [CrossRef]
10. Kong, S.; Cheng, Y.; Ren, T.; Liu, H. A sequential approach to control gas for the extraction of multi-gassy coal seams from traditional gas well drainage to mining-induced stress relief. *Appl. Energy* **2014**, *131*, 67–78. [CrossRef]
11. Sun, H.; Zhao, X.; Li, R.; Jin, H.; Sun, D. Emission reduction technology and application research of surface borehole methane drainage in coal mining-influenced region. *Environ. Earth Sci.* **2017**, *76*, 336. [CrossRef]
12. Yang, Y.; Han, P.; Zhao, Z.; Chen, W. The Method of Determining Layer in Bottom Drainage Roadway Taking Account of the Influence of Drilling Angle on Gas Extraction Effect. *Sustainability* **2022**, *14*, 5449. [CrossRef]
13. Li, H.; Liu, Y.; Wang, W.; Liu, M.; Ma, J.; Guo, X.; Guo, H. The integrated drainage technique of directional high-level borehole of super large diameter on roof replacing roof extraction roadway: A case study of the underground Zhaozhuang Coal Mine. *Energy Rep.* **2020**, *6*, 2651–2666. [CrossRef]
14. Zhao, E.B.; Li, K.N.; Yang, X.; Deng, N. Speculum Observation and Trajectory Measurement in Gas Extraction Drilling: A Case Study of Changling Coal Mine. *Geofluids* **2021**, *2021*, 5545067. [CrossRef]
15. Aitao, Z.; Kai, W. A new gas extraction technique for high-gas multi-seam mining: A case study in Yangquan Coalfield, China. *Environ. Earth Sci.* **2018**, *77*, 150. [CrossRef]
16. Shang, Y.Q.; Wu, G.Y.; Liu, Q.Z.; Kong, D.Z.; Li, Q. The Drainage Horizon Determination of High Directional Long Borehole and Gas Control Effect Analysis. *Adv. Civ. Eng.* **2021**, *2021*, 3370170. [CrossRef]
17. Li, S.; Shuang, H.; Wang, H. Determining the rational layout parameters of the lateral high drainage roadway serving for two adjacent working faces. *Int. J. Min. Sci. Technol.* **2016**, *26*, 795–801. [CrossRef]
18. Zhao, P.X.; Zhuo, R.S.; Li, S.G.; Lin, H.F.; Shu, C.M.; Laiwang, B.; Jia, Y.Y.; Suo, L. Fractal characteristics of gas migration channels at different mining heights. *Fuel* **2020**, *271*, 117479. [CrossRef]
19. Hu, G.; Xu, J.; Ren, T.; Gu, C.; Qin, W.; Wang, Z. Adjacent seam pressure-relief gas drainage technique based on ground movement for initial mining phase of longwall face. *Int. J. Rock Mech. Min.* **2015**, *77*, 237–245. [CrossRef]
20. Fan, C.J.; Li, S.; Zhang, H.H.; Yang, Z.H. Rational Boreholes Arrangement of Gas Extraction from Unloaded Coal Seam. *Adv. Civ. Eng.* **2018**, *2018*, 1501860. [CrossRef]
21. Zhang, X.B.; Gao, J.L.; Jia, G.N.; Zhang, J.W. Study on the influence mechanism of air leakage on gas extraction in extraction boreholes. *Energy Explor. Exploit.* **2022**, *40*, 1344–1359. [CrossRef]
22. Xiang, Z.Z.; Si, G.Y.; Wang, Y.H.; Belle, B.; Webb, D. Goaf gas drainage and its impact on coal oxidation behaviour: A conceptual model. *Int. J. Coal Geol.* **2021**, *248*, 103878. [CrossRef]
23. Li, S.; Shuang, H.; Wang, H.; Song, K.; Liu, L. Extraction of Pressurized Gas in Low Air-Conductivity Coal Seam Using Drainage Roadway. *Sustainability* **2017**, *9*, 223. [CrossRef]
24. Palchik, V. Formation of fractured zones in overburden due to longwall mining. *Environ. Geol.* **2003**, *44*, 28–38. [CrossRef]
25. Qu, Q.; Xu, J.; Wu, R.; Qin, W.; Hu, G. Three-zone characterisation of coupled strata and gas behaviour in multi-seam mining. *Int. J. Rock Mech. Min. Sci.* **2015**, *78*, 91–98. [CrossRef]
26. Feng, X.; Ding, Z.; Ju, Y.; Zhang, Q.; Ali, M. "Double Peak" of Dynamic Strengths and Acoustic Emission Responses of Coal Masses Under Dynamic Loading. *Nat. Resour. Res.* **2022**, *31*, 1705–1720. [CrossRef]
27. Gong, Y.; Guo, G.; Zhang, G.; Guo, K.; Du, Q.; Wang, L. A Vertical Joint Spacing Calculation Method for UDEC Modeling of Large-Scale Strata and Its Influence on Mining-Induced Surface Subsidence. *Sustainability* **2021**, *13*, 13313. [CrossRef]
28. Gao, F.; Stead, D.; Coggan, J. Evaluation of coal longwall caving characteristics using an innovative UDEC Trigon approach. *Comput. Geotech.* **2014**, *55*, 448–460. [CrossRef]
29. Sedghi-Asl, M.; Rahimi, H.; Salehi, R. Non-Darcy flow of water through a packed column test. *Transp. Porous Media* **2014**, *101*, 215–227. [CrossRef]

30. Martínez-Moreno, E.; Garcia-Ros, G.; Alhama, I. A different approach to the network method: Continuity equation in flow through porous media under retaining structures. *Eng. Comput.* **2020**, *37*, 3269–3291. [CrossRef]
31. Teng, T.; Xue, Y.; Zhang, C. Modeling and simulation on heat-injection enhanced coal seam gas recovery with experimentally validated non-Darcy gas flow. *J. Petrol. Sci. Eng.* **2019**, *177*, 734–744. [CrossRef]
32. Cao, J.; Li, W. Numerical simulation of gas migration into mining-induced fracture network in the goaf. *Int. J. Min. Sci. Technol.* **2017**, *27*, 681–685. [CrossRef]
33. Zheng, C.; Kizil, M.; Chen, Z.; Aminossadati, S. Effects of coal damage on permeability and gas drainage performance. *Int. J. Min. Sci. Technol.* **2017**, *27*, 783–786. [CrossRef]

sustainability

MDPI

Article

Numerical Simulation Study on Mechanical Characteristics and Width Optimization of Narrow Coal Pillar in Gob-Side Coal Seam Tunnel

Pengxiang Zhao [1,2], Wenjin Zhang [1,2,*], Shugang Li [1,2], Zechen Chang [1,2], Yajie Lu [1,2], Congying Cao [1,2], Yu Shi [1,2], Yongyong Jia [3], Fang Lou [3], Zongyong Wei [1,2] and Jun Liu [4]

1 College of Safety Science and Engineering, Xi'an University of Science and Technology, Xi'an 710054, China
2 Key Laboratory of Western Mine Exploitation and Hazard Prevention Ministry of Education, Xi'an University of Science and Technology, Xi'an 710054, China
3 Xinjiang Coal Science Research Institute, Urumqi 830091, China
4 State Key Laboratory of Gas Disaster Monitoring and Emergency Technology, Chongqing Research Institute, China Coal Technology and Engineering Group, Chongqing 400037, China
* Correspondence: 20220226150@stu.xust.edu.cn

Abstract: To investigate the influence of coal pillar width on the stress variation of narrow coal pillar (NCP) in the gob-side tunnel in an inclined thick coal seam, theoretical analysis, numerical modeling, and field monitoring are performed to determine the optimal width of the narrow coal pillars in inclined coal seams. The mechanical characteristics of the NCP for varying widths were investigated. Furthermore, vertical and horizontal stress were calculated for various widths of the NCP. The results revealed that with the rise in the width, the vertical stress initially increased dramatically and then stabilized, whereas the mean horizontal stress increased gradually. The mathematical relation between stress and NCP widths was represented by two fitting equations. The evolution process of the plastic zone in the NCP under various widths and the damage form of various widths were obtained; that is, when the width was small, the position of the roadway near the shoulder corner of NCP was inclined to the top of NCP. The field monitoring data revealed that the optimum NCP width was 4 m. This NCP width could stabilize the roadway and improve the loss prevention of the NCP at the gob-side tunnel of similar mines.

Keywords: narrow coal pillar; inclined thick coal seam; optimal width; damage form; loss prevention

Citation: Zhao, P.; Zhang, W.; Li, S.; Chang, Z.; Lu, Y.; Cao, C.; Shi, Y.; Jia, Y.; Lou, F.; Wei, Z.; et al. Numerical Simulation Study on Mechanical Characteristics and Width Optimization of Narrow Coal Pillar in Gob-Side Coal Seam Tunnel. *Sustainability* **2022**, *14*, 16014. https://doi.org/10.3390/su142316014

Academic Editor: Chaolin Zhang

Received: 14 October 2022
Accepted: 17 November 2022
Published: 30 November 2022

1. Introduction

Coal energy occupies the main position in the Chinese energy structure. With the increase of mining years, the reserves of horizontal coal seams and gently inclined coal seams are dramatically reduced. At present, the focus of mining in China is gradually shifting to inclined coal seams [1,2]. The main reasons for the low recovery rate of coal resources are the limitation of mining technology and the unreasonable width of the protective coal pillar, which causes great waste of coal resources. In addition, compared with the gently inclined coal seam, the inclined coal seam is often accompanied by complex geological conditions, which is easy to cause problems such as roof stress concentration, rock burst, serious roadway deformation, and a more serious waste of coal resources [3–5]. In order to solve these problems, domestic and foreign scholars have carried out a lot of research on the theory and technology of narrow coal pillar retention to improve the resource recovery rate.

In view of the characteristics of the NCP size, position, stress change, and the corresponding roadway support technology, researchers conducted in-depth and extensive research using various methods [6–9]. Wu et al. [10] studied the characteristics of deformation and stress distribution of the NCP under advanced abutment pressure, and the results showed an acceleration in the deep zone of the NCP is larger than that in the shallow zone. Wang et al. [11] found that with the decrease in the distance between the monitoring region and the working area, the shape of the vertical stress core area of the NCP changed from an ellipse to a rectangle. Yang et al. [12] performed a conversion of the double-yield constitutive model, and the results revealed that increasing the NCP size could result in a gradual transformation of the peak stress from the solid coal side to the NCP side. Bai et al. [13] further established the roof destruction criteria and studied the roof bending deformation behavior model. Qi et al. [14] provided a method to support the steadiness of the roadway by using the roof fracture and collapse filling effect, which successfully solved the large deformation problem of the roadway in the process of excavation and panel retreat. Moreover, Some scholars [15–17] established models of fracture mechanics and elastic-plastic mechanics to study the stress distribution of strip pillars and the displacement and stress distribution of small pillars in a horizontal coal seam, which provided a theoretical foundation for the NCP stability and control technology of adjoining rock.

In recent years, with the development of DEM models and computational resources, DEMs have been used by scholars to study the crack/damage behavior of the coal pillar [18–20]. Zhu et al. [21] used PFC to study the deformation characteristics, failure behavior, and stress distribution of the combined support column (CSP) of the residual coal pillar and filling body, and evaluated its stability. Fu et al. [22] revealed the influence of axial load width and direct roof thickness on the failure mechanism of a notched roof in chamber-pillar mining through experimental tests and numerical simulation. Sarfarazi et al. [23] studied the influence of non-persistent joints and holes on the failure behavior of rock pillars under the uniaxial compression test by using the discrete element method, which provides a reference and method for studying the failure mode strength of NCP in gob-side entry tunnel. Fan et al. [24] and Wang et al. [25] studied the NCP widths of the gob-side tunnel in a horizontal coal seam and established a numerical model for revealing the damage characteristic of the NCP. Fan et al. [26] investigated the weak adjoining rock roadways under substantial mining deformation and established a creep model of anchored adjoining rocks. Zhang et al. [27] combined the support control technique and non-symmetrical control technique for evaluating the gob-side tunnel.

In summary, the current research primarily concentrates on the stability of adjoining rocks and the determination of NCP width along the empty lane driving in horizontal coal seams and gently inclined coal seam. However, the research on the stability and stress features of NCP along the empty lane driving in inclined thick coal seam has not received much attention. Therefore, this paper takes the (4–5) 06 work-face of Liuhuanggou coal mine in Xinjiang as the research object, analyzes the stress and plastic zone variation law under different NCP sizes in inclined thick coal layer, determines the rational NCP width and the key supporting parameters for the gob-side tunnel in an inclined thick coal seam, and verifies the experimental results by field engineering practice, so as to provide guidance for greatly reducing the waste of coal resources and effectively avoiding the disasters caused by in-situ stress, as well as realize the underground safety production and the material and energy saving.

2. Engineering Background

The (4–5) 06 working area of the Liuhuanggou coal mine in Xinjiang province is principally mined in the 4–5 # coal seam in the middle of the lower section of the Xishanyao Formation. The burial depth of the 4–5 # coal seam is 300–600 m, while the burial depth of the gob-side tunnel is 490 m. The mean thickness is 6.15 m, and the dip angle of the coal seam is 24–26°. The direct roof is dominated by thick carbonaceous mudstone (1.35 m) and thick fine sandstone (2.89 m). The basic roof is dominated by siltstone (17.39 m). The rock strata are greyish-white blocks and fine parallel joints, and the upper and middle parts are silt, fine sandstone and mudstone. The floor belongs to argillaceous siltstone (1.85 m). (4–5) 06 working area is 20 m away from the boundary of the minefield, while the east is the protective coal pillar of the auxiliary inclined shaft, the north is the solid coal without mining, and the south is (4–5) 04 gob.

As shown in Figure 1, (4–5) 06 working area inclined length is 180 m, and the strike length is 1850 m. The working area takes once mining full height retreat stoping method. The spatial representation of the gob-side tunnel arrangement in the (4–5) 06 working area is illustrated in Figure 2. The RMT–150C rock mechanics test (Figure 3) was performed on coal and rock to determine the strength of the experimental work face and the lithology parameters of the top and bottom coal seam 4 (Table 1).

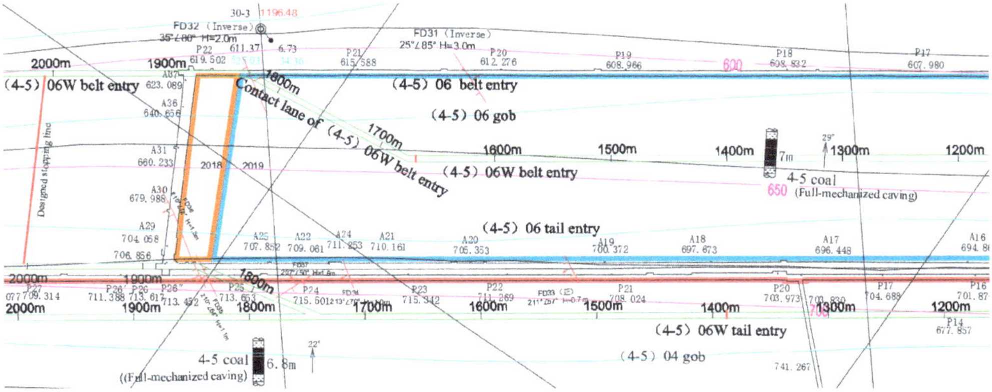

Figure 1. (4–5) 06 working face layout diagram.

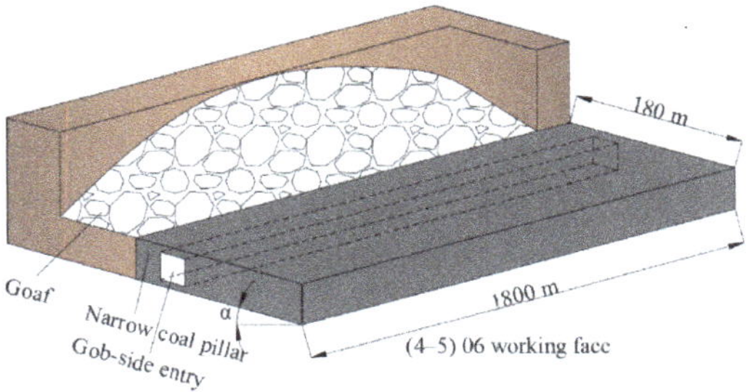

Figure 2. Schematic of the spatial position of the gob-side tunnel.

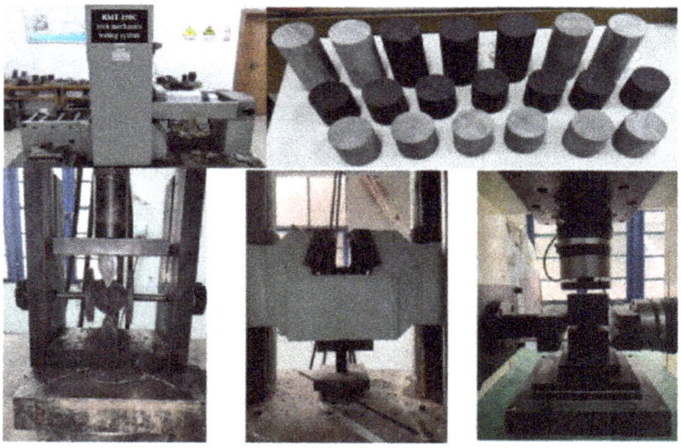

Figure 3. The RMT–150C rock mechanics test.

Table 1. 4–5# Roof and floor rock parameters.

Position	Lithology	γ (kN/m^3)	E (MPa)	Poisson (p)	C_0 (MPa)	Dilatancy Angle (°)	Internal Friction Angle (°)	Tensile Strength (MPa)
Roof	Mudstone	20.8	20,019	0.19	0.93	8	31	0.67
	Middle sandstone	26.6	50,430	0.28	2.27	10	31	0.85
	Fine sandstone	26.2	43,020	0.26	1.93	10	31	0.56
	Siltstone	20.0	34,739	0.25	1.3	12	35	0.87
Coal seam	Coal	14.6	14,142	0.27	0.72	8	20	0.53
Bottom plate	Coarse sandstone	26.4	56,767	0.27	1.38	8	34	0.34
	Sandy mudstone	26.4	56,767	0.27	1.38	8	34	0.34

3. Experimental

3.1. Analysis of Stress Distribution in Narrow Coal Pillars

The coal seam dip angle and mining thickness are large, and the falling rock refuse and residual floating coal accumulate on the side of the NCP, resulting in lateral pressure on the NCP. Excavating the roadway changes the coal and rock layers, and the state of stress equilibrium will be changed due to both lateral and vertical pressures being acted on the tunnel. To obtain the optimal NCP width (x in Figure 4), the following assumptions are made:

- The coal body is a homogeneous continuum.
- The coal strength is the highest in the case of the plain strain.
- The coal body damaged by shear effects is in accordance with the Mohr–Coulomb criterion [28].

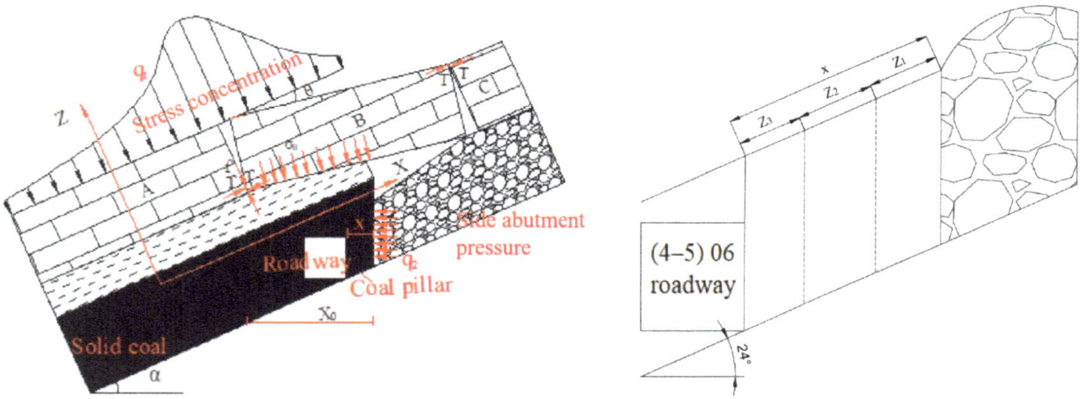

Figure 4. Schematic of the rock structure adjoining the NCP.

To analyze the stress of adjoining rock of the NCP, a mechanical model of the gob-side tunnel in a thick layer is established, as expressed in Figure 4. In Figure 4, q_1 denotes normal pressure on the top of a coal body, MPa; σ_B annotates the lateral pinching pressure of the key block B, MPa; q_2 represents the lateral pressure of the overburden rock on the coal wall in gob area, MPa; θ annotates the rotation angle of the key rock B, °; α expresses the inclination angle of coal layer, °; X_0 denotes the distance from the point of the maximum vertical pressure to the coal wall of the gob, m; x indicates the NCP width, m.

Equation (1) can be obtained by solving the stress balance equation at the yield zone interface as below:

$$\begin{cases} \frac{\partial \sigma_x}{\partial x} + \frac{\partial \tau_{xy}}{\partial y} + X = 0 \ (x \text{ direction}) \\ \frac{\partial \sigma_y}{\partial y} + \frac{\partial \tau_{xy}}{\partial x} + Y = 0 \ (y \text{ direction}) \\ \tau_{xy} = -(C_0 + \sigma_y \tan \varphi) \end{cases} \tag{1}$$

where X and Y represent the volumetric force of the coal in the x and y directions in the limit equilibrium zone, MPa; C_0 denotes the cohesion force at the interface between the coal layer, roof, and floor, MPa; σ_x refers to the horizontal stress in the limit equilibrium zone, MPa; σ_y refers to the vertical stress in the limit equilibrium zone, MPa; τ_{xy} expresses the shear stress at the interface between roof and floor, MPa; φ indicates the angle of internal friction between coal layer and rock mass of roof and floor, °.

The solution to the stress point at any point in the limit equilibrium zone is expressed as below [29]:

$$\begin{cases} \tau_{xy} = -\left\{ \left[\frac{1}{A}(P_x + \gamma x_0 \sin \alpha) + \frac{2C_0 - m\gamma \sin \alpha}{2 \tan \varphi} \right] \cdot \right. \\ e^{\frac{mA\gamma \cos \alpha - 2\tan^2 \varphi}{2A} + \frac{2\tan \varphi}{mA} + (\frac{2\tan^2 \varphi}{mA} - \gamma \cos \alpha)\gamma} \tan \varphi + C_0 \\ \sigma_y = \left[\frac{1}{A}(P_x + \gamma x_0 \sin \alpha) + \frac{2C_0 - m\gamma \sin \alpha}{2 \tan \varphi} \right] \cdot \\ e^{\frac{mA\gamma \cos \alpha - 2\tan^2 \varphi}{2A} + \frac{2\tan \varphi}{mA} + (\frac{2\tan^2 \varphi}{mA} - \gamma \cos \alpha)\gamma} \end{cases} \tag{2}$$

where A represents the constant of lateral pressure; P_x denotes the support resistance of the coal, MPa; γ annotates the mean bulk density of the overburden, kN/m^3; m refers to the thickness of coal layer, m; α denotes the inclination angle of coal layer, °.

Considering the effect of the coal seam inclination angle and the gob area, the horizontal distance X_0 is expressed as follows:

$$X_0 = \frac{mA}{2\tan \varphi} \ln \left[\frac{A(K\gamma H \cos \alpha \tan \varphi + m\gamma \sin \alpha)}{A(2C_0 + m\gamma \sin \alpha) + 2P_x \tan \varphi} \right] \tag{3}$$

where K indicates the lateral bearing pressure concentration factor.

Based on elastic-plastic limit equilibrium theory, the optimal NCP width is expressed as follows [30]:

$$x = (Z_3 + Z_2 + Z_1) \tag{4}$$

$$Z_1 = \frac{mA}{2\tan \varphi} \ln \left[\frac{k\gamma H + \frac{C_0}{\tan \varphi}}{\frac{C_0}{\tan \varphi} + \frac{P_x}{A}} \right] \tag{5}$$

where Z_1 denotes the width of the plastic region in the NCP; Z_2 denotes the side bolt's effective length in the NCP along the empty roadway: 1.2 m. Considering the stability of the NCP, Z_3 denotes the coal pillars stability coefficient considering the large thickness of the coal seam (assumed to be 0.2 (Z_1 + Z_2)); m annotates the thickness of the coal layer: 6.15 m; H refers to the depth of tunnel: 490 m; A represents the constant of lateral pressure: 1.2; K indicates the lateral bearing pressure concentration factor: 2.5; φ denotes the angle of internal friction between the coal layer and rock mass of roof and floor: 12°; C_0 represents the cohesion force: 6 MPa; γ expresses the mean bulk density of the overburden: 20 kN/m^3; P_x refers to the support resistance of coal: 0.3 MPa; and k denotes the safety factor: 1.15–1.45. The parameters are substituted in Equations (3)–(5) to obtain x (4.5 m) and X_0 (10 m).

3.2. Establishment of the Numerical Model

(1) Model establishment

According to the layout of (4–5) 06 working face, a three-dimensional excavation numerical model is established by using ABAQUS4.0 numerical calculation software to analyze the stability of coal pillar and gob-side entry and the variation characteristics of surrounding rock stress under different pillar widths.

The model is established as shown in Figure 5. The model dimension (length × width × height) is determined to be 200 m × 250 m × 180 m, and the dip angle of the coal seam is 25°. In order to eliminate the influence of the boundary effect on the tunneling process of the gob-side tunnel, a 50 m protection boundary is set at both ends of the coal seam strike. The simulated mining length is 100 m, and the simulated dimension (width × height) of (4–5) 06 gob-side tunnel is 4.5 m × 4 m. The initial ground stress was balanced through the initial analysis step calculation.

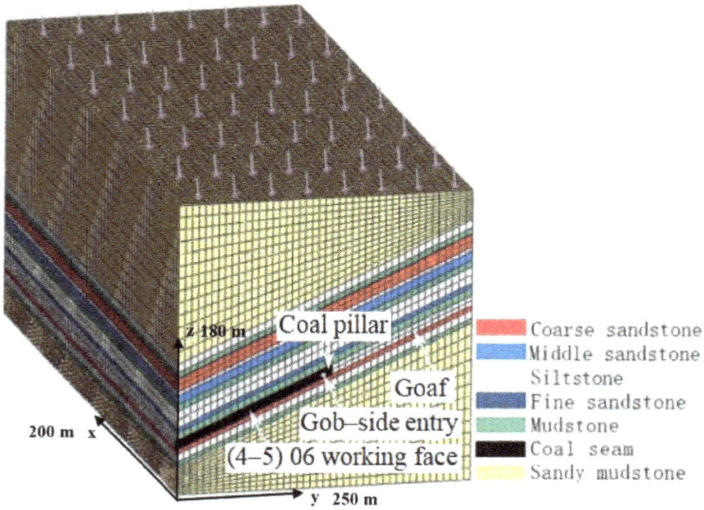

Figure 5. Three-dimensional numerical model.

(2) Mesh subdivision and boundary condition setting

Model boundary conditions are calculated by determining the overburden weight. Based on the actual mining depth of 290 m, a vertical pressure of 10^6 MPa is applied on the top of the model. The bottom and the four sides were considered to be fixed ends, and the initial ground stress was balanced through the initial analysis step calculation. Finally, the model is meshed, in which the method of over-densification is used to make the mesh density around the coal seam reach 1.

(3) Calculation method and mining scheme

The material failure in the model satisfies the Mohr–Coulomb criterion, and the asymmetric solver method is used for the convergence of the calculation results. The simulation calculation refers to the borehole histogram of the Liuhuanggou coal mine and some laboratory test results, as shown in Table 1.

The numerical analysis of the model was performed in two steps. The first step was gob excavation, and the second step was gob-side tunnel excavation. According to the theoretical width of NPC and the distance from the peak abutment pressure of NPC to the coal wall in goaf, the design scheme of the coal pillar width of 3, 4, 5, 6, 7, 8, 9, 10, 15, 20 m was selected to study the stress variation law of coal pillars with different widths.

4. Results and Discussion

4.1. Vertical Stress Distribution

NCP widths of 3, 4, 5, 6, 7, 8, 9, 10, 15, and 20 m were investigated at a yield zone width of X_0 and the NCP width of x (Figure 4). The vertical stress values formed under various NCP widths during roadway excavation are illustrated in Figure 6.

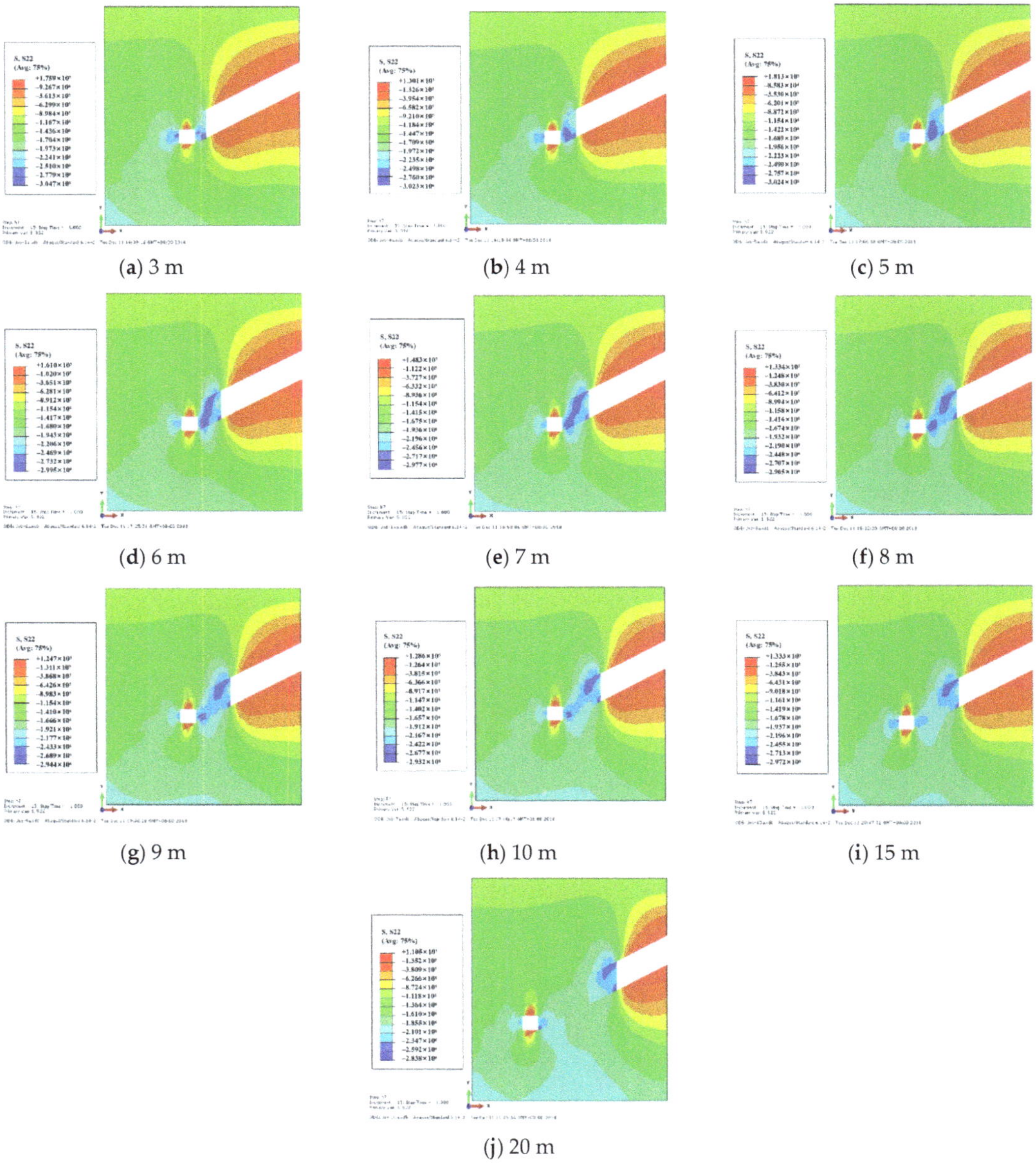

Figure 6. Vertical stress distribution of the NCP of various sizes.

Excavation of the gob-side tunnel changed the stress of the rock masses around the gob area, and a new stress balance state was reached.

The results of numerical analysis, that is, the relation between vertical stress and NCP widths, are illustrated in Figure 7.

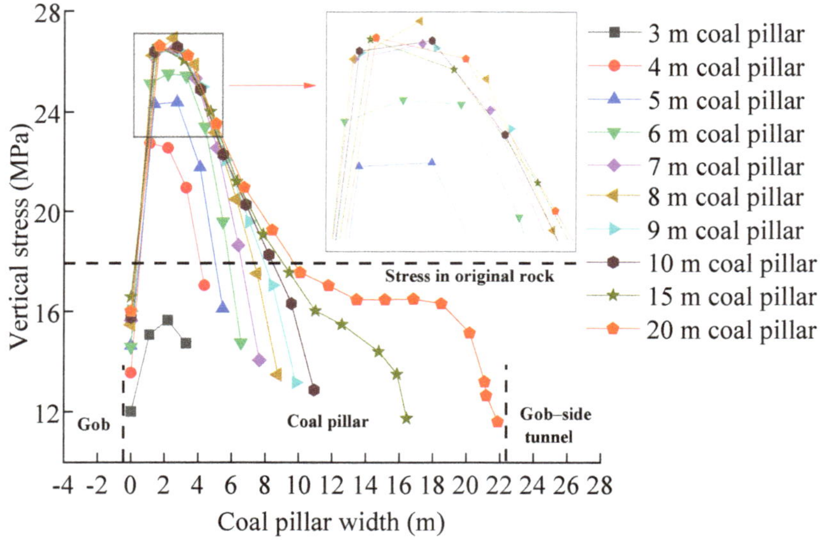

Figure 7. Vertical stress distribution curves of the NCP of various widths.

When the NCP width was 3 m, the vertical peak stress was only 15.6 MPa, which was explicitly lower than the primal rock stress. The NCP promptly collapsed under the load because of the absence of a stable bearing area. When the width was 4 m, the stress reached its peak point of 22.8 MPa, and an effective bearing area was formed. The vertical stress was focused at the center of the NCP and was symmetrically distributed. When the width was between 5 and 10 m, the vertical maximum stress increased with the rise in the width. The peak stress position was gradually offset from the center position in the NCP. When the width was 8 m, it reached a maximum of 27 MPa and subsequently stabilized. When the width was between 15 and 20 m, the vertical stress gradually exhibited an asymmetric distribution, and the vertical stress shifted toward the adjacent gob.

4.2. Horizontal Stress Distribution

During the excavation of the roadway, because of the influence of the weight of the overburden and the coal seam pitch, the magnitude, and direction of the force generated by the horizontal force on the NCP changed. The horizontal stress distribution curves on the right side of the NCP under various widths of the gob-side tunnel were determined through numerical calculations (Figure 8).

The following observations were made regarding the horizontal stress distributions for various widths of the NCP:

- When the width of the NCP increased, the force at both ends of the NCP became more uniform. Therefore, the mean horizontal stress increased. In the NCP, the stress at the bottom was greater than the stress at the top, and the horizontal stress at the two ends was greater than intermediate horizontal stress.
- The horizontal stress of the NCP first decreased from the bottom to the top and then increased. In the NCP, from the bottom up to one-third of the NCP, horizontal stress alleviated linearly; from one-third to two-thirds length of the NCP, horizontal stress remained unchanged. From two-thirds up to the top, the horizontal stress tended to enhance linearly.

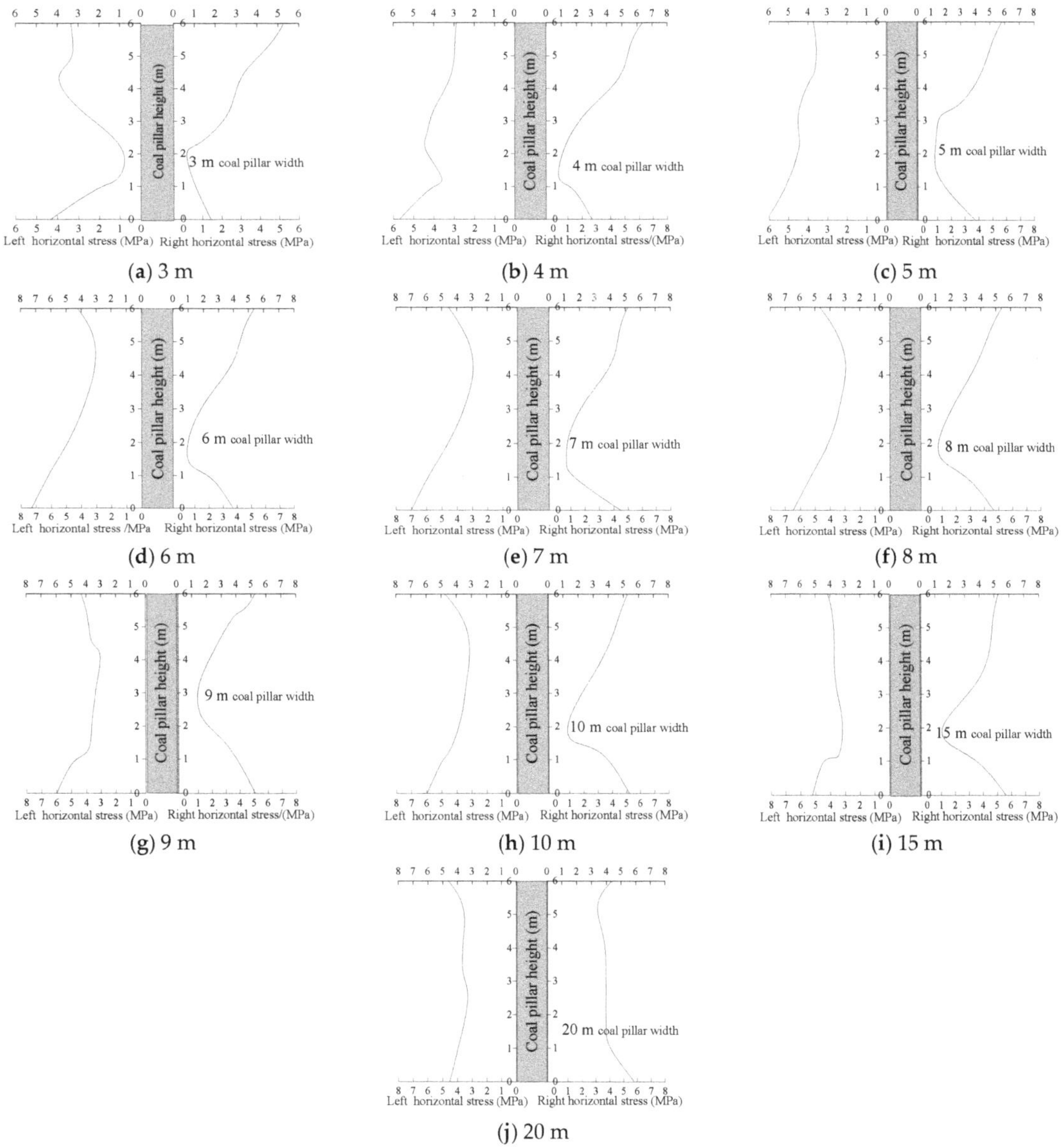

Figure 8. Horizontal stress distribution curves with the NCP widths of 3, 4, 5, 6, 7, 8, 9, 10, 15, and 20 m.

4.3. Vertical Displacement of the Roof of Adjoining Rock

To obtain the deformation characteristics of adjoining rock of NCP, two rows of displacement measuring lines were arranged at the roof of the NCP model and 5 m above it, and 10 displacement measuring points were set in each row. The vertical offset of the roof on the adjoining rock is displayed in Figure 9. The deformation of the adjoining rock varied with the NCP widths. Therefore, the NCP width was critical for controlling the stress and deformation of adjoining rock.

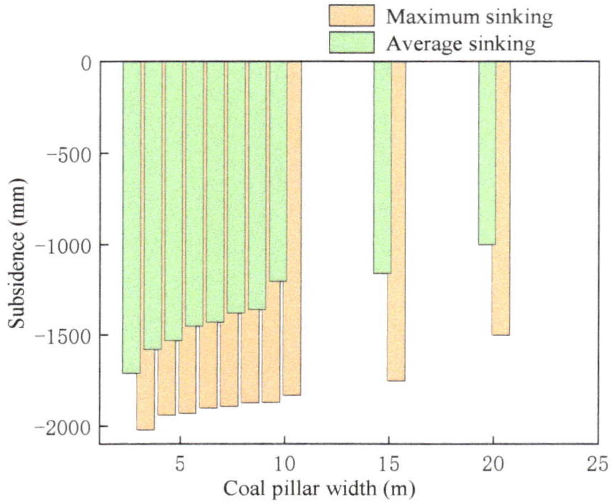

Figure 9. Vertical displacement of the roof of the adjoining rock with respect to the coal pillar width.

When the width of the NCP increased, subsidence occurred for the roof plate. The maximum subsidence occurred when the width was 3 m, but the supporting conditions of the roadway were limited. When the width was 4–10 m, the roadway subsidence varied between 1200 and 1600 mm, which was in the controllable range. Even though the width of the NCP was between 15 and 20 m, the NCP was subjected to high vertical stress (Figure 7), and the subsidence of the roof was not salient.

4.4. Deformation Distribution of Narrow Coal Pillars in the Plastic Zone

The gob-side tunnel destroyed the lateral stress equilibrium state of the gob, and the NCP was affected by the vertical stress and horizontal stress of the upper strata and the key rock block B (Figure 4). The stress-focused area was formed at the shoulder angle of the roadway, which made the coal pillars easy to break. In the inclined thick coal seam, the plastic zone in the NCP increased with its width, and under the combined action of shear and tensile, it was continuously distributed from the inclination to both sides of the NCP. Therefore, it was necessary to accurately grasp the failure scope and distribution characteristics of the coal pillar's plastic zone so as to provide a foundation for the selection of critical parameters of roadway support.

Through the numerical calculation of stress distribution characteristics in NCP and adjoining rock during gob-side tunnel driving, the distribution characteristics of the plastic zone in NCP with various widths were obtained, as demonstrated in Figure 10. The graph showed that when the NCP width was 3 m, the upper end of the NCP was seriously damaged. The plastic zone formed by the NCP near the upper end of the gob and the shoulder angle of the roadway runs through the whole NCP, which made the NCP have no stable bearing area. When the NCP width was 4 m, the stress of the plastic zone tended to transfer to the bottom plate, which reduced the stress of adjoining rock and the deformation of NCP. When the NCP width was 5 m, the plastic zone near the gob expanded, indicating that the stress intensity of coal pillars increased. When the NCP width was 6 m, due to the gradual decrease of the plastic zone in the center of the NCP, the stress of the NCP transferred to the two sides, and the roadway support pressure was increased. When the NCP width was 7 m, the plastic zone appeared completely discontinuous area, and the two sides were still dominated by shear failure. When the NCP width was 8–20 m, with the width increased, the range of plastic zone near the NCP side of the roadway shoulder angle gradually shifted to the lower end of the roadway, and the stress transferred to the roof and floor roadway. The deformation of the adjoining rock was mainly a shear and tensile failure, and the adjoining rock was in a high-stress state.

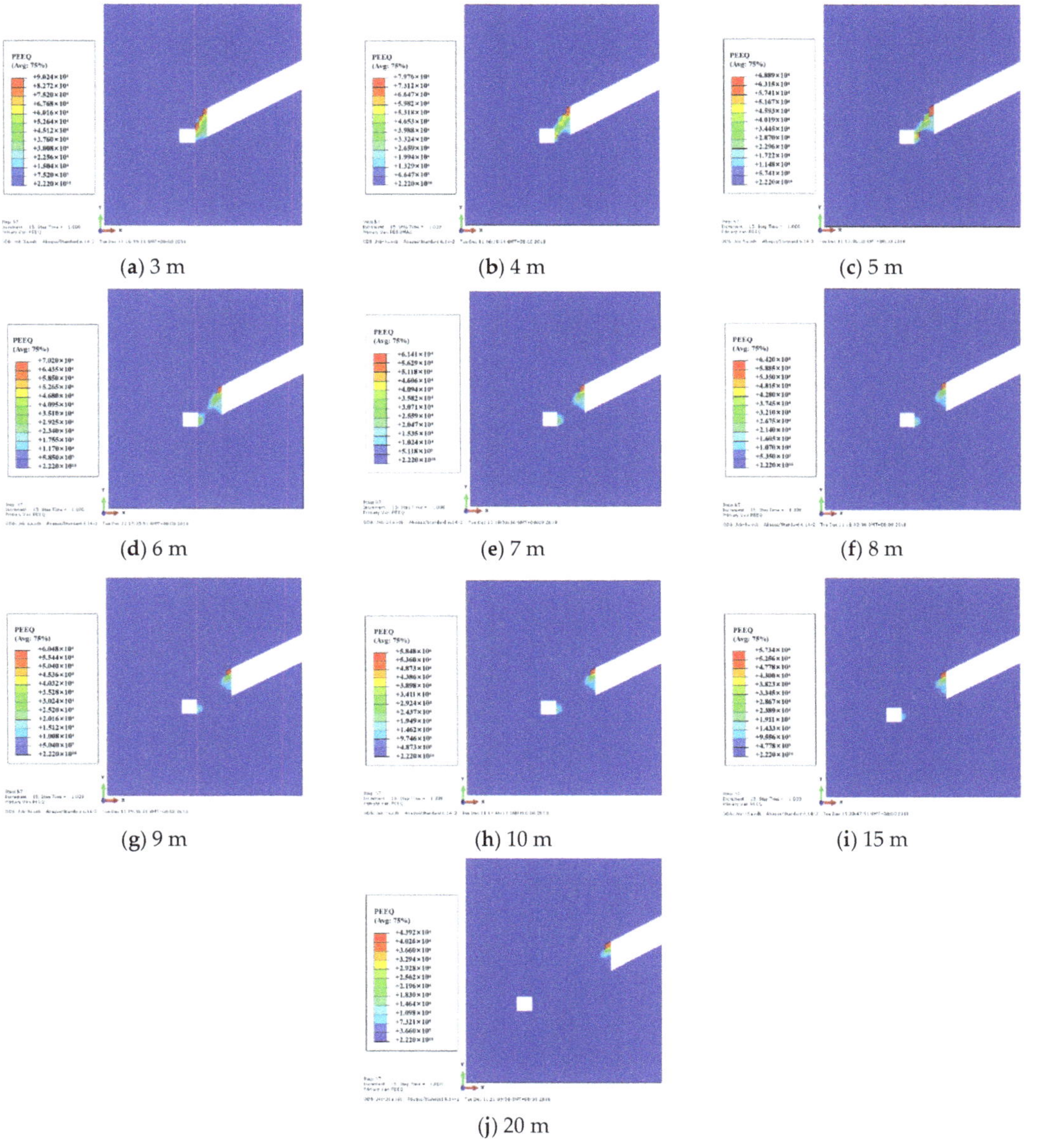

Figure 10. Cloud distribution of the plastic zone with the NCP widths of 3, 4, 5, 6, 7, 8, 9, 10, 15, and 20 m.

Therefore, considering the resource recovery rate and roadway support pressure, the NCP width of 4 m was the most reasonable.

5. Coal Pillars Stress and Deformation Analysis

5.1. Vertical Stress of the Coal Pillars for Various Widths

The NCP width had an evident effect on the variation law of NCP vertical stress. To analyze the association between the NCP widths and the variation of NCP vertical stress, the variation association between the peak value of vertical stress and the size of NCP was plotted in Figure 11.

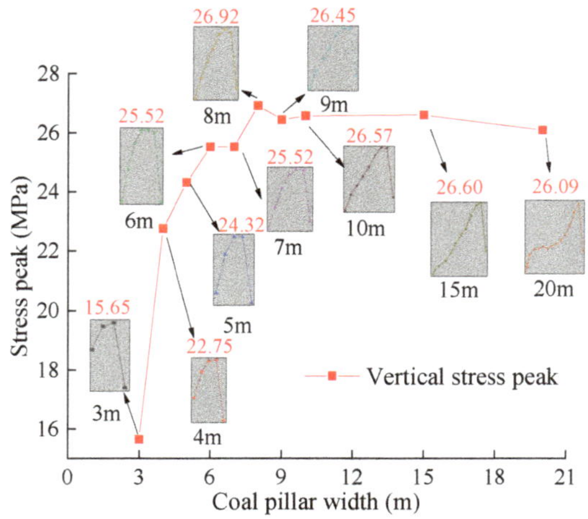

Figure 11. Vertical stress peak change versus the NCP width.

The interrelation between the peak value of NCP vertical stress and the widths in Figure 11 was fitted to obtain the relationship shown in Equation (6), and the fitting degree R^2 was 0.986. The peak vertical stress increased when the width increased. Apparently, the increase in the width was not favorable for roadway support during mining activities. According to the previous analysis, 3 m NCP had no load-carrying capacity. When the NCP width was greater than or equal to 3 m, its bearing capacity was unknown. Thus, the optimal economic width was 4 m.

$$\sigma_v = 26.524 - \frac{24.99}{1 + e^{\frac{x_i - 1.64}{1.34}}} \quad (6)$$

where σ_v denotes the vertical peak stress of the NCP, x_i expresses the width of various NCPs (3, 4, 5, 6, 7, 8, 9, 10, 15, and 20 m).

5.2. Horizontal Stress of Coal Pillars for Varying Widths

The horizontal stress of the NCP for various NCP widths was calculated from previous numerical analysis. The horizontal stress distribution for various NCP widths is displayed in Figure 12.

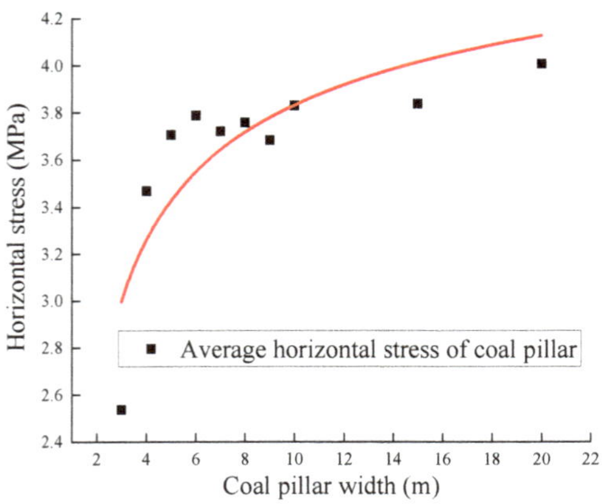

Figure 12. Horizontal stress distribution with the NCP widths of 3, 4, 5, 6, 7, 8, 9, 10, 15, and 20 m.

The interrelation between the mean horizontal stress and the widths in Figure 12 was fitted to obtain the connection shown in Equation (7), and the fitting degree R^2 was 0.971. The equation revealed that the horizontal stress increased logarithmically when the width increased. As mentioned, a 3 m wide NCP was the starting point and basic design for analysis. Thus, 3-m width can not be selected as the optimal width. As depicted in Figure 10, when the NCP width was 4 m, the mean horizontal stress was the least, and the force had a limited effect on the NCP during mining activities. Therefore, the 4-m wide NCP was selected to evaluate if it was a suitable option for the support of roadway mining by using the following expression:

$$\sigma_{\mathrm{H}} = 1.131 \ln(12.2884 \ln x_{\mathrm{i}}) \quad (7)$$

where σ_{H} is the horizontal peak stress of the NCP; x_{i} is the width of different NCPs (3, 4, 5, 6, 7, 8, 9, 10, 15, and 20 m).

6. Case Study

6.1. Determining the Width of the Narrow Coal Pillars

Figures 11 and 12 indicate that 4 m NCP width is the ideal width because the mean vertical stress and horizontal stress are minima. Furthermore, when the width was 4 m, its distribution range of the plastic zone and its overall stability exhibited superior properties to that of other widths. Therefore, a 4 m NCP width was selected to design the gob-side tunnel in the (4–5) 06 work face of the Liuhuanggou coal mine in Xinjiang province.

6.2. Anchor Rod and Cable Stress Monitoring

The monitoring of the anchoring force of the anchor rod and cable is shown in Figure 13. Due to the low bearing capacity of the NCP, the anchor rod of the solid coal side and the roof anchor cable were generally subjected to high stress. The pressure of the overlying strata was mainly borne by the solid coal side. The load of the anchor rod varied between 110–200 kN, and the load of the anchor cable varied between 200–230 kN. The specific parameters of the KMG 500 anchor rod of the solid coal side are shown in Table 2. When the anchor rod with a diameter of 20 mm was used, some of the anchor rods were broken, while the anchor cable with a diameter of 22 mm and a strength level of 1860 had an ultimate load of 448.6 kN and still possessed a large bearing space.

Table 2. Selected parameters for anchors and anchor cables.

Selection	Type	Distance between Anchors (mm)	Diameter (mm)	Length (mm)	Resin Roll Anchoring
Anchors	KMG500	800	22	2200	MSCK2370
Anchor cables	KMG800	1600	22	7300	MSCK2335

In summary, the anchor rod and cable of the roof and both sides in the gob-side tunnel born greater pressure, and some of the anchor rods at the solid coal side corner were broken, while the anchor cable still had a large bearing space. In the geological structure and watering section, the adjoining rock of the roadway was seriously deformed, and it was necessary to strengthen the support in this section and improve the bearing performance of the corner anchorage.

6.3. Observation

The deformation and stress distribution of the adjoining rock with the NCP on the (4–5) 06 work face were measured on-site. The cross-point method was used to construct a hole 28 mm in diameter and 400 mm in length along the vertical direction of the roof and floor as well as the horizontal direction of both sides. The wooden piles 29 mm in diameter and 400 mm in length were inserted into the hole. The curved measuring nails were installed at the roof, and the flat head measuring nails were installed at the floor. The arrangement of measuring points is listed in Table 3. The time-varying deformation of the roof and two-side are recorded, as demonstrated in Figures 14 and 15.

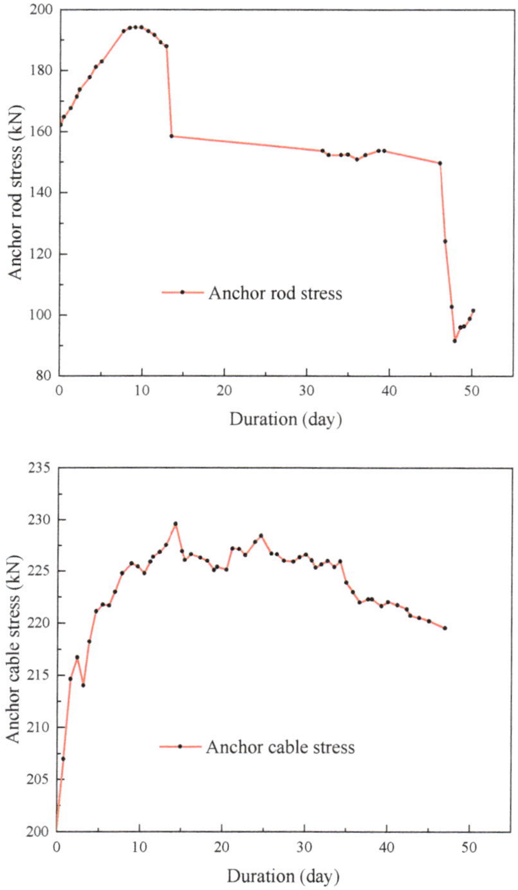

Figure 13. Anchor rod and cable stress monitoring of rock surrounding the roadway versus duration.

Table 3. Measuring point design.

Measuring Point	1	2	3	4	5	6
Distance from heading face (m)	200	400	600	800	1000	1200

As demonstrated in Figures 14 and 15, when the NCP width was 4 m, during roadway excavation, the maximum subsidence of the roadway roof was 85 mm, and the maximum displacement on both sides of the roadway was 105 mm. During the stable period of roadway, the roof subsidence and the two-side displacement increased slightly. The maximum roof subsidence was 91.3 mm, and the maximum two-side displacement was 112.3 mm. After that, the roof subsidence and the two-side displacement remained basically unchanged, indicating that 4 m pillars of roadway protection could meet the requirements of the gob-side tunnel in an inclined coal seam.

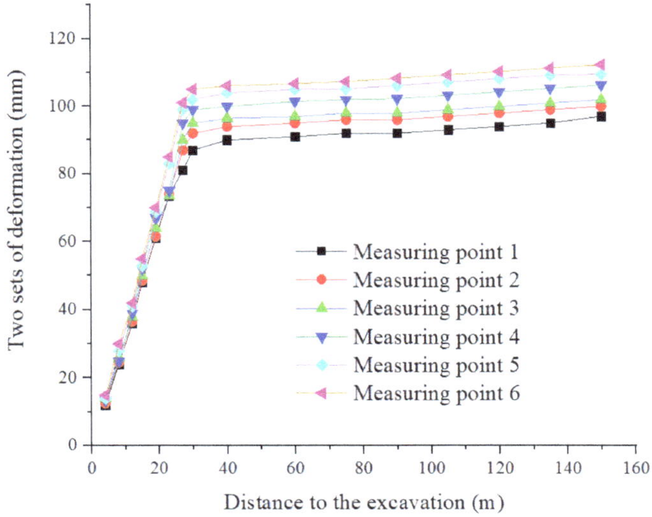

Figure 14. Deformation curves for six measuring points.

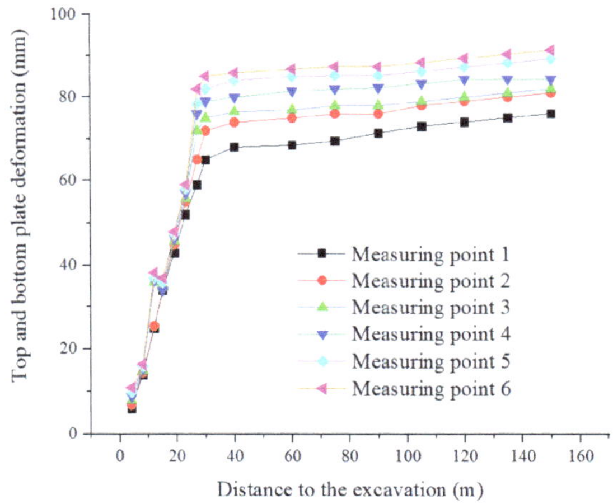

Figure 15. Top and bottom plate deformation curve for six measuring points.

7. Conclusions

The three primary conclusions of this study are made as below:

1. In the NCP of an inclined coal seam, the vertical stress increases first and then decreases, and the horizontal stress decreases first and then increases from bottom to top. With the rising of the NCP width, the peak values of vertical stress and horizontal stress increase in logarithmic function form. The quantitative relationship between pillar size and peak stress is obtained by fitting.
2. Through the numerical simulation method, the relationship between the NCP width in an inclined coal seam and the stress distribution, the plastic zone distribution, and the peak stress are calculated, respectively. Combined with the measured values of the deformation and stress distribution of the adjoining rock, the optimal dimension mathematical model of the NCP is verified. Therefore, the optimal width of the NCP in the (4–5) 06 fully mechanized caving face is 4 m.

3. The NCP of an inclined coal seam is in the form of oblique cutting failure, and it is necessary to strengthen the support of the roadway shoulder corner. The anchor rod and cable support system designed according to the theoretical and numerical calculation results can provide higher support strength and stiffness, which can effectively control the NCP from slipping to the inside of the roadway, prevent the serious deformation of NCP, meet the support requirements of adjoining rock of gob-side tunnel in (4–5) 06 working face, and ensure the overall stability of the roadway.

Author Contributions: Conceptualization, P.Z.; methodology, S.L.; data curation, W.Z. and Z.C.; writing—original draft preparation, W.Z.; investigation, Y.L. and C.C.; writing—review and editing, Y.S.; resources, Y.J. and F.L.; funding acquisition, Z.W. and J.L. All authors have read and agreed to the published version of the manuscript.

Funding: This research was funded by National Natural Science Foundation Grants of China (51974237, 51704228), the Basic Research on National Natural Science Foundation key project of China (51734007), and the Innovation guidance project of Chongqing Research Institute, China Coal Technology and Engineering Group in 2019 (2019YBXM46).

Institutional Review Board Statement: Not applicable.

Informed Consent Statement: Not applicable.

Data Availability Statement: Data are available on request due to privacy restrictions.

Conflicts of Interest: The authors declare no conflict of interest.

References

1. Yuan, L. Research progress of mining response and disaster prevention and control in deep coal mine. *J. China Coal Soc.* **2021**, *46*, 716–725. (In Chinese)
2. Xie, H.P.; Wang, J.H.; Wang, G.F.; Ren, H.W.; Liu, J.Z.; Ge, S.R.; Zhou, H.W.; Wu, G.; Ren, S. New ideas of coal revolution and layout of coal science and technology development. *J. China Coal Soc.* **2018**, *43*, 1187–1197. (In Chinese)
3. Wang, J.; Jiang, J.Q.; Li, G.B.; Hu, H. Exploration and numerical analysis of failure characteristic of coal pillar under great mining height longwall influence. *Geotech. Geol. Eng.* **2016**, *34*, 689–702. [CrossRef]
4. Liu, S.F.; Wan, Z.J.; Zhang, Y.; Lu, S.F.; Ta, X.P.; Wu, Z.P. Research on evaluation and control technology of coal pillar stability based on the fracture digitization method. *Measurement* **2020**, *158*, 107713. [CrossRef]
5. Dai, J.J.; Shan, P.F.; Zhou, Q. Study on intelligent identification method of coal pillar stability in fully mechanized caving face of thick coal seam. *Energies* **2020**, *13*, 305. [CrossRef]
6. Wu, H.; Wang, X.K.; Wang, W.J.; Peng, G.; Zhang, Z.Z. Deformation characteristics and mechanism of deep subsize coal pillar of the tilted stratum. *Energy Sci. Eng.* **2020**, *8*, 544–561. [CrossRef]
7. Qi, J.D.; Chai, J.; Zhang, J.Z. Numerical simulation on stress distribution of the inclined coal seam under different pillar width. *Adv. Mater. Res.* **2014**, *977*, 374–377. [CrossRef]
8. Jaiswal, A.; Shrivastva, B.K. Numerical simulation of coal pillar strength. *Int. J. Rock Mech. Min. Sci.* **2008**, *46*, 779–788. [CrossRef]
9. Zhang, Z.Z.; Deng, M.; Wang, X.Y.; Yu, W.J.; Zhang, F.; Dao, V.D. Field and numerical investigations on the lower coal seam entry failure analysis under the remnant pillar. *Eng. Fail. Anal.* **2020**, *115*, 104638. [CrossRef]
10. Wu, H.; Zhang, N.; Wang, W.J.; Zhao, Y.; Cao, P. Characteristics of deformation and stress distribution of small coal pillars under leading abutment pressure. *Int. J. Min. Sci. Technol.* **2015**, *25*, 921–926. [CrossRef]
11. Wang, B.N.; Dang, F.N.; Gu, S.C.; Huang, R.B.; Miao, Y.P.; Chao, W. Method for determining the width of protective coal pillar in the pre-driven longwall recovery room considering main roof failure form. *Int. J. Rock. Mech. Min. Sci.* **2020**, *130*, 104340. [CrossRef]
12. Yang, R.S.; Zhu, Y.; Li, Y.L.; Li, W.Y.; Lin, H. Coal pillar size design and surrounding rock control techniques in deep longwall entry. *Arabian J. Geosci.* **2020**, *13*, 453. [CrossRef]
13. Bai, J.-B.; Shen, W.-L.; Guo, G.-L.; Wang, X.-Y.; Yu, Y. Roof deformation, failure characteristics, and preventive techniques of gob-side entry driving heading adjacent to the advancing working face. *Rock Mech. Rock Eng.* **2015**, *48*, 2447–2458. [CrossRef]
14. Qi, F.Z.; Ma, Z.G.; Yang, D.W.; Li, N.; Li, B.; Wang, Z.L.; Ma, W.X. Stability control mechanism of high-stress roadway surrounding rock by roof fracturing and rock mass filling. *Adv. Civil Eng.* **2021**, *2021*, 317–334. [CrossRef]
15. Zhao, Y.-M.; Feng, X.-T.; Jiang, Q.; Han, Y.; Zhou, Y.-Y.; Guo, H.-G.; Kou, Y.-Y.; Shi, Y.-E. Large deformation control of deep roadways in fractured hard rock based on cracking-restraint method. *Rock Mech. Rock Eng.* **2021**, *54*, 2559–2580. [CrossRef]
16. Wu, X.-Y.; Jiang, L.-S.; Xu, X.-G.; Guo, T.; Zhang, P.-P.; Huang, W.-P. Numerical analysis of deformation and failure characteristics of deep roadway surrounding rock under static-dynamic coupling stress. *J. Cent. South Univ.* **2021**, *28*, 543–555. [CrossRef]

17. Li, Z.L.; Wang, L.G.; Lu, Y.L.; Li, W.S.; Wang, K. Effect of principal stress rotation on the stability of a roadway constructed in half-coal-rock stratum and its control technology. *Arabian J. Geosci.* **2021**, *14*, 292–303. [CrossRef]
18. Xu, B.; Li, Y.; Wang, S.; Lu, B. Study on the strength characteristics and failure characteristics of the composite load-bearing structure in the cemented filling field. *Constr. Build. Mater.* **2022**, *330*, 127242. [CrossRef]
19. Ji, S.T.; Karlovšek, J. Calibration and uniqueness analysis of microparameters for DEM cohesive granular material. *Int. J. Min. Sci. Technol.* **2022**, *32*, 121–136. [CrossRef]
20. Ji, S.T.; Karlovšek, J. Optimized differential evolution algorithm for solving DEM material calibration problem. *Eng. Comput.* **2022**, 1–16. [CrossRef]
21. Zhu, X.J.; Guo, G.L.; Liu, H.; Peng, X.; Yang, X.Y. Stability analysis of the composite support pillar in backfill-strip mining using particle flow simulation method. *Environ. Earth Sci.* **2022**, *81*, 124. [CrossRef]
22. Fu, J.W.; Haeri, H.; Sarfarazi, V.; Babanouri, N.; Rezaei, A.; Manesh, M.O.; Bahrami, R.; Marji, M.F. Effects of axial loading width and immediate roof thickness on the failure mechanism of a notched roof in room and pillar mining: Experimental test and numerical simulation. *Rock Mech. Rock Eng.* **2022**, 1–27. [CrossRef]
23. Sarfarazi, V.; Abharian, S.; Ghorbani, A. Physical test and PFC modelling of rock pillar failure containing two neighboring joints and one hole. *Smart Struct. Syst.* **2021**, *27*, 123–137.
24. Fan, N.; Wang, J.R.; Zhang, B. Reasonable width of segment pillar of fully-mechanized caving face in inclined extra-thick coal seam. *Geotech. Geol. Eng.* **2020**, *38*, 4189–4200. [CrossRef]
25. Wang, Q.; Gao, H.K.; Jiang, B.; Li, S.; He, M.; Wang, D.; Lu, W.; Qin, Q.; Gao, S.; Yu, H.C. Research on reasonable coal pillar width of roadway driven along goaf in deep mine. *Arabian J. Geosci.* **2017**, *10*, 466–483. [CrossRef]
26. Fan, L.; Wang, W.J.; Yuan, C.; Peng, W.Q. Research on large deformation mechanism of deep roadway with dynamic pressure. *Energy Sci. Eng.* **2020**, *8*, 3348–3364. [CrossRef]
27. Zhang, Y.; Wan, Z.J.; Li, F.C.; Zhou, C.B.; Zhang, B.; Feng, G.; Chengtan, Z. Stability of coal pillar in gob-side entry driving under unstable overlying strata and its coupling support control technique. *Int. J. Min. Sci. Technol.* **2013**, *23*, 193–199. [CrossRef]
28. Wojciechowski, W. A note on the differences between Drucker-Prager and Mohr-Coulomb shear strength criteria. *Stud. Geotech. Mech.* **2018**, *40*, 163–169. [CrossRef]
29. Loboda, V.; Gerge, I.; Khodanen, T.; Mykhail, O. A crack between orthotropic materials with a shear yield zone at the crack tip. *Math. Prob. Eng.* **2019**, *2019*, 9723089. [CrossRef]
30. Zhang, K.X.; Jiang, Y.D.; Zhang, Z.B. Determination of reasonable width of narrow coal pillars along goaf in large coal pillars. *J. Min. Saf. Eng.* **2014**, *31*, 255–262. (In Chinese)

sustainability

MDPI

Article

Experimental Study on the Effect of Gas Adsorption and Desorption on Ultrasonic Velocity and Elastic Mechanical Parameters of Coal

Gang Xu [1,*], Jiawei Liu [1], Yunlong Wang [1], Hongwei Jin [1] and Chaofeng Wang [2]

[1] College of Safety Science and Engineering, Xi'an University of Science and Technology, Xi'an 710054, China
[2] Xin'an Coal Mine, Henan Dayou Energy Co., Ltd., Luoyang 471842, China
* Correspondence: xugang25193@xust.edu.cn

Abstract: The rapid and accurate identification of the physical characteristics of coal by means of ultrasonic detection is of great significance to ensure safe mining of coal and efficient development of coal seam methane. In this paper, the ultrasonic velocity testing experiments of coal during gas adsorption and desorption were carried out, utilizing a low frequency petrophysical measurement device with primary and fractured coal as the research objects. The variations in the elastic mechanical parameters and ultrasonic velocity of coal samples were analyzed to elucidate the influence mechanism that gas adsorption and desorption have on them. During gas adsorption and desorption, the longitudinal wave velocity of the primary structure coal varies from 1990 m/s to 2200 m/s, and the transverse wave velocity varies from 1075 m/s to 1160 m/s, while the longitudinal wave velocity of the fractured structure coal varies from 1540 m/s to 1950 m/s, and the transverse wave velocity varies from 800 m/s to 1000 m/s. The elastic modulus and wave velocities, in both directions of the primary structural coal, were higher than those of the fractured structural coal. In comparison to the fractured structural coal, the main structural coal had a lower Poisson's ratio. In addition, the spread of the elastic mechanical parameters and wave velocities, in both the longitudinal and transverse directions, was more pronounced in the fracture–structured coal than in the primary–structured coal. During gas adsorption and desorption, the speed of the coal's longitudinal waves increased, and then decreased, due to the combined effect of gas adsorption expansion and pore gas pressure compression matrix effect. For this experiment, the maximum longitudinal wave velocity of the coal occurred at a gas pressure of 1.5 MPa. Primary structural coal has a longitudinal wave speed of 2103 m/s, whereas fragmented structural coal has a speed of 1925 m/s. The variation in the shear wave velocity of the coal is controlled only by the gas adsorption expansion effects. The shear wave velocity increases during gas adsorption and decreases during gas desorption. With the change of gas pressure, the longitudinal wave velocity can increase by 23.34%, and the shear wave velocity can increase by 17.97%. Coal undergoes changes to both its Poisson's ratio and elastic modulus as a result of gas adsorption and desorption; these modifications are analogous to the velocity of longitudinal and shear waves, respectively.

Keywords: gas adsorption; gas desorption; ultrasonic velocity; elastic mechanical parameters; coal; porosity

Citation: Xu, G.; Liu, J.; Wang, Y.; Jin, H.; Wang, C. Experimental Study on the Effect of Gas Adsorption and Desorption on Ultrasonic Velocity and Elastic Mechanical Parameters of Coal. *Sustainability* **2022**, *14*, 15055. https://doi.org/10.3390/su142215055

Academic Editors: Xiangguo Kong, Dexing Li and Xiaoran Wang

Received: 13 October 2022
Accepted: 10 November 2022
Published: 14 November 2022

1. Introduction

Coal is one of the important basic energy sources, which provides a solid guarantee for the sustainable and healthy development of society and economy [1–4]. In particular, China is characterized by a coal–based energy resource endowment, which concludes that coal will continue to be China's primary energy source for the foreseeable future [5,6]. Gas, also known as coalbed methane, is a gaseous medium that is produced with the formation of coal and exists mainly in the coal seam, in an adsorbed state. Gas is not only an energy, but is also the largest source of disaster in coal mining. This is closely related to its degree of enrichment. The control and utilization of gas is closely related to its enrichment level, and the enrichment degree and its identification method have become one of the important technical problems to be solved in coal resources development [7,8].

The ultrasonic detection method is a geophysical exploration technique, which has the characteristics of good directionality, strong penetration, concentrated acoustic energy, long propagation distance and being difficult to attenuate, especially for anisotropic and biphasic media [9,10]. Gas primarily exists in the pores and fractures of coal seams in the adsorption state, and gas−bearing coals are both an anisotropic and biphasic media; it is therefore practical to use ultrasonic detection methods to identify the degree of gas enrichment in coal seams [11–14].

Since 1956, the year Biot first put forward his hypothesis of elastic wave propagation in saturated porous medium [15], the saturated fluid porous media model has been developed rapidly. However, the presence of solid medium coal, fluid phase gas and adsorbed phase gas in gas−bearing coals renders them more complicated than conventional saturated fluid porous media. The response characteristics of ultrasonic waves in gas−bearing coal have been studied intensively and extensively by domestic and foreign scientists through field tests, laboratory tests and theoretical analysis [16–20].

In their field test, Shengdong Liu and Qiufang Zhao [21,22] obtained a good negative linear correlation between gas content and inherent principal frequency of coal seam through an in−situ test of coal. Wenlin Liu [23] recognized that adsorption of methane could reduce the longitudinal wave velocity by 10–16%. Zhijun Wang and Shengdong Liu [24–26] concluded from field tests that both the longitudinal and shear wave velocities of seismic activity are reduced when the gas concentration of the coal is high. The coal seam quality factor Q was thought to be inversely linked to gas concentration, whereas the attenuation coefficient was found to be positively related to gas pressure. Pingsong Zhang [27] considered that gas content is logarithmically related to the attenuation coefficient of coal seam and linearly related to the quality factor through field acoustic detection tests, and concluded that coal seams with a quality factor less than 1 and an attenuation coefficient greater than 0.005 are high gas outburst seam.

In laboratory tests, Lixin Jiang [28] believed that saturation had no effect on shear waves in saturated porous media but had a significant effect on longitudinal waves. Jialin Hao [29] predicted that velocity of coal samples' longitudinal and transverse waves initially increases and then decrease with increasing gas pressure. Hongyi Yu [30] thought that the wave velocity of the nitrogen−containing gas continues to increase during the gas injection process. The elastic wave propagation law for gas−bearing coals is theoretically analyzed.

The author [31] concluded that the elastic wave velocity of the non−adsorbed phase model decreases slightly with increasing gas pressure, which is similar to the elastic wave velocity of a single−phase medium coal. The elastic wave velocity of the adsorbed phase model decreases significantly with increasing gas pressure, showing obvious characteristics of a two−phase medium. The results of existing studies indicate that there is a relationship between the degree of gas enrichment in coal seams and the wave velocity of coal. However, due to the limitations of the test environment and test conditions, although gas enrichment is thought to increase coal's wave speed, researchers have not reached a consensus on whether this is really the case or not.

Therefore, during the adsorption and desorption of gases in a laboratory setting, the ultrasonic response properties of a coal mass are experimentally investigated in great detail. The examination of coal's wave speed and elastic mechanical characteristics, in relation to gas pressure variations, will not only provide a basis for identifying the physical properties of coal by ultrasonic detection methods, but will also provide important guidance for the utilization of gas in mines and disaster prevention and control.

2. Materials and Methods

2.1. Experimental Equipment

The apparatus used in the ultrasonic detection experiments under gas adsorption and desorption conditions is a low frequency rock physical measurement device. Figure 1 displays the operating principle of the device. The device is primarily made up of a system for controlling the perimeter pressure, a system for controlling the pore pressure, a system for testing core samples, and a system for collecting data.

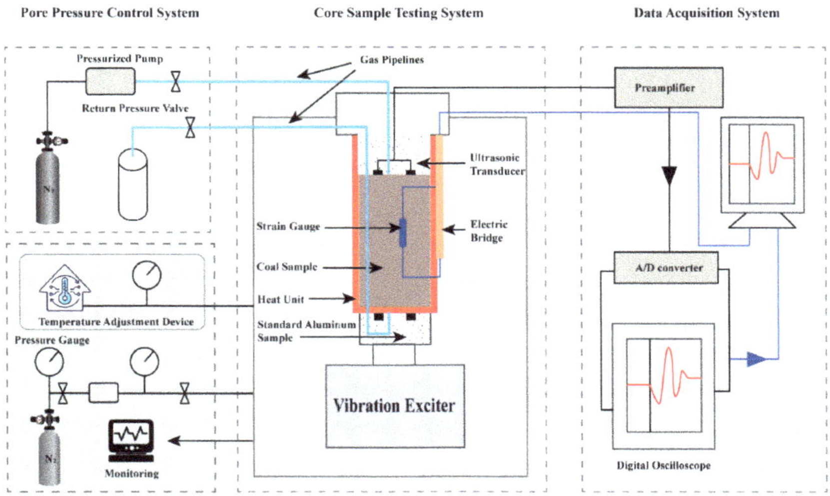

Figure 1. Principle of operation of the low frequency petrophysical measurement device.

(1) Perimeter pressure control system.

This technique used a supply of nitrogen to regulate the core sample's perimeter pressure in order to maintain the coal's pressure within a reasonable range, and the temperature adjustment device was added to the experimental system. The heating unit is controlled by the hot and cold temperature cycle regulation system in the temperature adjustment device to ensure that the experimental temperature of the coal sample is kept constant. The system was equipped with temperature and pressure monitors to allow temperature and pressure values to be read at any time. The main components included a nitrogen cylinder, booster pump, pressure indicator, pressure reducing valve, hot and cold temperature circulator, temperature indicator and heating unit.

(2) Pore pressure control system.

The system used a high precision pressure control pump to achieve pore pressure control and fluid displacement through a fluid rejection reactor. The high precision pressure control pump maintains a constant pressure and rate. It also allows the remote control and monitoring of pump details, including parameters such as flow rate, pressure, cylinder volume and cumulative pumping volume via flow sensors. The main components of the system are a high precision digital hydraulic pump, pressure transducer, return valve, fluid supply system and fluid displacement reactor. For this experiment, we decided to use nitrogen, rather than gas, due to safety aspects, as well as taking into consideration that coal has an equivalent adsorption effect on nitrogen to meet the experimental requirements.

(3) Core sample testing system.

This system is the core part of the low frequency petrophysical measurement device and is physically illustrated in Figure 2. The system allows ultrasonic measurements of coal−bearing gas to be achieved by transmitting and receiving ultrasonic waves through an ultrasonic transducer. The frequency range for ultrasonic velocity measurements is 1–5 MHz. The axial pressure measurements can be achieved by applying pressure to the test coal sample via a servo vibrator. Pore fluid tubes can be used to achieve preheated aeration of the coal sample to simulate the variation in coal at various gas pressures. The system's essential elements include an ultrasonic transducer, servo vibrator, semiconductor strain gauge, pore fluid tube, standard aluminum sample and epoxy resin plate.

Figure 2. Core sample test system.

Nine layers of perforated tinfoil were chosen as the coupling agent, which solves the coupling problem without affecting the implementation of the gas injection process and effectively increases the experiment's accuracy.

(4) Data collection method.

The system mainly realizes high−precision data acquisition, which can record the strain and ultrasonic velocity changes of coal during the experiment. The main components of the system include control software, signal generator, 12−channel high−precision differential amplifier, servo vibration amplifier, pulse generator receiver, digital oscilloscope, digital acquisition board to place the computer and analogue digital input board.

In addition, auxiliary tools such as balances and vernier calipers are required for the experiments.

2.2. *Principle of Measurement*

The pulse transmission method is used to measure the ultrasonic velocities in low−frequency rock physics measurement systems [32,33], where ultrasonic shear wave velocities are measured primarily by picking up the first arriving shear wave using polarization cracking and then calculating its vector position. Due to the experimental conditions, the time difference between the two arrivals at the receiver decreases when the ultrasonic shear wave propagates in an anisotropic thin layer, where the fast and slow shear waves propagate in the medium in orthogonal polarization. The detector can only identify the point in time at which the shear wave is received by the fast wave, whereas identifying the point in time at which the shear wave arrives by the slow wave is difficult. Therefore, considering the influence of practical conditions, only the speed of the shear wave is studied in this experiment.

This experiment adopted a pulse transmission ultrasonic system, and the acoustic wave velocity can be read directly by the computer. The test principle is:

$$\begin{cases} V_p = L/(t_p - t_0) \\ V_s = L/(t_s - t_0) \end{cases} \tag{1}$$

where V_p is the speed of an ultrasonic longitudinal wave (m/s), V_s is the longitudinal wave speed of ultrasound (m/s), L is the transmitting and receiving transducer spacing (m), t_p is the duration of a longitudinal wave (s), t_s is the duration of the transverse wave (s), t_0 is the zero delay of the instrument system (s).

2.3. Coal Sample Collection and Processing

Coal samples were collected in accordance with the requirements of "Geotechnical inquiry and testing—Sampling techniques and groundwater measurements—Part 1: Technical concepts for the sampling of soil, rock, and groundwater". They were then prepared, in accordance with international standards, into cylinders that were 50 mm in diameter and 100 mm in length [34]. The raw coal collected for the experiments was mainly primary structural coal and fractured structural coal. Considering the different orientations of the coal seam cracks, and in particular, how coal seam anisotropy affects ultrasonic transmission, care should be taken to mark the orientation, tendency and vertical lamination direction of the coal samples extracted from the coal seams when collecting them in the field.

The sampling site Is in 31,004 working faces of Shanxi Xinyuan Coal Industry Co. The coal quality of this working face is medium−ash, low−sulfur, high−quality lean coal, mainly light coal, with endogenous fissures, containing 1 to 2 layers of muddy gangue, with a thickness of 0.01 to 0.04 m and an average of 0.02 m. The 3# coal seam mined at this workface is stable, with a simple structure and a solidity factor of 0.51 to 0.82. The coal sample collection requirements for this experiment were met.

Coal samples should be taken from structurally well−preserved bulk coal, placed in a pre−prepared woven bag, and marked with the type of structure and the direction in which the coal sample was collected. After collecting the primary samples, they are brought to the lab, where they are drilled into 50 mm−diameter and 100 mm−high cylindrical samples. The coal cores were ground to a flatness of less than 0.02% to allow better contact between the ultrasonic sensors and the cores.

To facilitate experimental statistics and analysis, the numbering of the coal samples was specified as follows: the direction of the parallel face cuttings was designated as the *X* direction; the direction of the vertical face cuttings was designated as the *Y* direction; and the direction of the vertical layer was designated as the *Z* direction (Figure 3). The drilling of this experimental coal sample was carried out in the *X* direction, and a total of two experimental coal samples were processed. According to the criteria for the classification of coal body structure types, primary structure coal is type I coal and fracture structure coal is type II coal. Therefore, these two coal samples can be named as IX1 and IIX1, respectively, and Table 1 displays the coal sample's specification parameters.

Table 1. Statistical table of processed coal sample parameters.

Coal Sample Number	Type of Coal Structure	Length (mm)	Diameter (mm)	Fissure Development	$R_{0,max}$	V_{daf} (%)	Weight Capacity ($t \cdot m^{-3}$)
IX1	Primary structural coal	99.9	49.9	Clearly stratified, no obvious fractures	1.86	15.7%	1.43
IIX1	Fractured structural coal	99.9	50.1	Generally clear lamination, a few fissures	1.97	16.8%	1.39

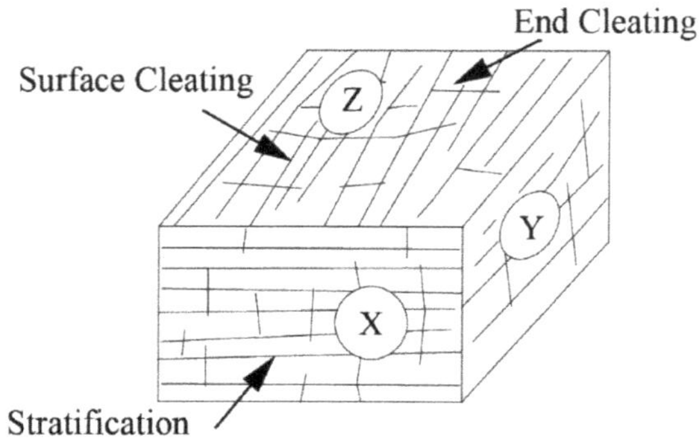

Figure 3. Schematic diagram of coal cuttings and laminations.

2.4. Experimental Protocol and Experimental Steps

To obtain the response characteristics of the ultrasonic velocity of the coal samples during gas adsorption and desorption, the following experimental scheme was developed, according to the function of the experimental setup.

Experimental conditions: to ensure accurate experimental findings, the temperature was maintained at room temperature, approximately 25 °C, throughout the experiments. Taking into account the coal sample's low compressive strength as a whole, the axial and circumferential pressures at the beginning of the experiment were set to 3 MPa to ensure better and sufficient contact between the probe and the coal column. According to the purpose of the experiment, the experimental process was divided into three stages:

- Evacuation stage (I)
- Gas injection and adsorption stage (II)
- Pressure reduction and desorption stage (III)

Evacuation stage (I): Maintaining the axial pressure 3 MPa and the surrounding pressure 3 MPa, the coal sample was evacuated to vacuum (0.01 MPa) and the ultrasonic properties of the coal sample were tested.

Gas injection and adsorption stage (II): The technique of concurrently raising the axial pressure was used to guarantee that the coal sample's effective stress remained constant, surrounding pressure and gas pressure was used in the gas injection and adsorption stages. The coal samples were left to attain the adsorption equilibrium condition for six hours, at which point their ultrasonic properties were assessed. At this point, the injected gas pressure was 0.1 MPa, and the axial and surrounding pressures were 3.1 MPa. The ambient pressure and axial pressure were successively 3.5 MPa, 4.0 MPa, 4.5 MPa, 5.0 MPa, and 5.5 MPa when the pressure of the methane gas being injected rose to 0.5 MPa, 1.0 MPa, 1.5 MPa, 2.0 MPa, and 2.5 MPa. Subsequently, the samples were allowed to reach equilibrium by waiting for 6 h at each pressure state before testing the ultrasonic performance of the coal samples.

Desorption stage (III): The desorption stage is the antithesis of the gas injection and adsorption stages. For the desorption stage, a simultaneous reduction in shaft pressure, surrounding pressure, and gas pressure is conducted. This was achieved by reducing the shaft pressure, surrounding pressure and methane pressure in the gripper by 0.5 MPa, every 2 h, until the test was completed, at which point the methane pressure in the gripper was 0.1 MPa. The ultrasonic characteristics of the samples were tested every 2 h during the decompression process.

3. Results and Discussion

3.1. Results from Experiments and Determination of Elastic Mechanical Characteristics

The coal sample's longitudinal wave velocity V_p and transverse wave velocity V_s during the experiments can be obtained according to the calculation method adopted by Yun Wang [35]. The dynamic−elastic mechanical parameters Poisson's ratio (ν) and elastic modulus (E) may be determined from the known density ρ with V_p and V_s. Tables 2 and 3 show the findings of the evaluation of the coal sample's elastic mechanical characteristics and wave velocity.

Table 2. Results of tests on coal sample IX1's elastic mechanical characteristics and wave velocity.

Test No.	Shaft Pressure (MPa)	Surrounding Pressure (MPa)	Absolute Pressure (MPa)	Measurement Time (h)	Dissemination Time ($\times 10^{-4}$ s)	Longitudinal Wave Velocity (m s^{-1})	Dissemination Time ($\times 10^{-4}$ s)	Transverse Wave Velocity (m s^{-1})	Poisson's Ratio ν	Elastic Modulus E (GPa)
1	3.0	3.0	0.10	-	0.501	1995	0.924	1081	0.292	4.319
2	3.0	3.0	0.01	-	0.492	2031	0.914	1093	0.296	4.429
3	3.1	3.1	0.10	6	0.486	2056	0.906	1103	0.298	4.516
4	3.5	3.5	0.50	12	0.481	2075	0.900	1110	0.300	4.579
5	4.0	4.0	1.00	18	0.478	2092	0.895	1116	0.301	4.635
6	4.5	4.5	1.50	24	0.475	2103	0.891	1121	0.302	4.678
7	5.0	5.0	2.00	30	0.476	2100	0.886	1127	0.298	4.714
8	5.5	5.5	2.50	36	0.477	2095	0.883	1132	0.294	4.742
9	5.0	5.0	2.00	38	0.474	2106	0.884	1130	0.298	4.740
10	4.5	4.5	1.50	40	0.470	2125	0.888	1125	0.305	4.725
11	4.0	4.0	1.00	42	0.473	2110	0.893	1119	0.304	4.671
12	3.5	3.5	0.50	44	0.476	2098	0.896	1115	0.303	4.634
13	3.1	3.1	1.10	46	0.480	2081	0.900	1110	0.301	4.585

Table 3. Test results of wave velocity and elastic mechanical parameters of coal sample IIX1.

Test No.	Shaft Pressure (MPa)	Surrounding Pressure (MPa)	Absolute Pressure (MPa)	Measurement Time (h)	Dissemination Time ($\times 10^{-4}$ s)	Longitudinal Wave Velocity (m s^{-1})	Dissemination Time ($\times 10^{-4}$ s)	Transverse Wave Velocity (m s^{-1})	Poisson's Ratio ν	Elastic Modulus E (GPa)
1	3.0	3.0	0.10	-	0.617	1618	1.181	846	0.312	2.610
2	3.0	3.0	0.01	-	0.580	1722	1.120	892	0.317	2.912
3	3.1	3.1	0.10	6	0.559	1786	1.089	917	0.321	3.088
4	3.5	3.5	0.50	12	0.544	1836	1.066	937	0.324	3.231
5	4.0	4.0	1.00	18	0.531	1882	1.046	955	0.327	3.364
6	4.5	4.5	1.50	24	0.519	1925	1.030	970	0.330	3.478
7	5.0	5.0	2.00	30	0.522	1915	1.014	985	0.320	3.561
8	5.5	5.5	2.50	36	0.525	1902	1.001	998	0.310	3.627
9	5.0	5.0	2.00	38	0.519	1926	1.009	990	0.320	3.598
10	4.5	4.5	1.50	40	0.513	1948	1.019	980	0.331	3.553
11	4.0	4.0	1.00	42	0.523	1910	1.035	965	0.329	3.440
12	3.5	3.5	0.50	44	0.533	1873	1.050	951	0.326	3.335
13	3.1	3.1	1.10	46	0.549	1821	1.072	932	0.323	3.194

3.2. Changes in the Ultrasonic Velocity of Coal Samples during Gas Adsorption and Desorption

According to Tables 2 and 3, it is possible to make the longitudinal and transverse wave velocity variation law of the coal sample during the test, as shown in Figures 4 and 5.

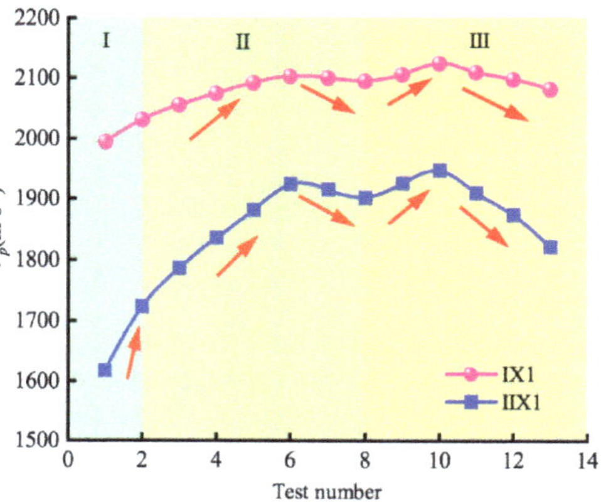

Figure 4. Each sample longitudinal wave velocity varies during the experiment.

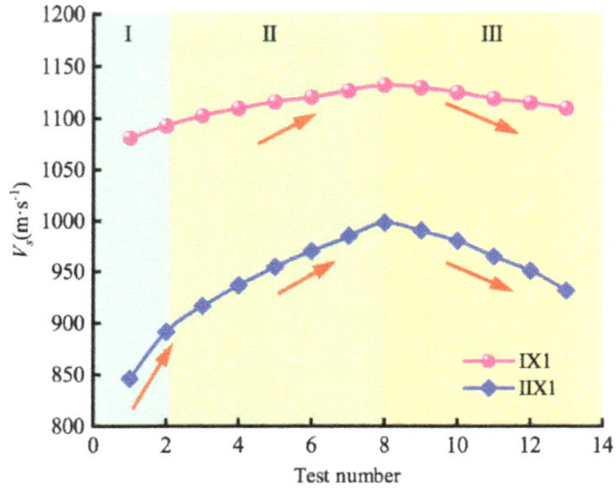

Figure 5. Variations in the coal sample's transverse wave velocity during the experiment.

From Figures 4 and 5, it is evident that the coal samples' longitudinal and transverse wave velocities slightly increase throughout the evacuation stage.

When the gas pressure was increased during the gas adsorption stage, from 0 MPa to 2.5 MPa, the transverse wave velocity tended to increase along with the increase in gas pressure, whereas the longitudinal wave velocity exhibited a pattern of increasing and then decreasing with the increase in gas pressure. The longitudinal wave velocity decreased as the gas pressure rose from 1.5 MPa to 2.5 MPa.

In contrast to the longitudinal wave velocity, which exhibited a pattern of growing and subsequently decreasing with the reduction in gas pressure, the transverse wave velocity showed a declining tendency, as the gas pressure dropped from 2.5 MPa to 0 MPa. The longitudinal wave velocity dropped along with the reduction in gas pressure, as it went from 1.5 MPa to 0 MPa.

Compared with the gas injection stage, the longitudinal and transverse wave velocities of the desorption stage were larger than those of the adsorption process at the same gas pressure.

In addition, the variance in the longitudinal velocity of the basic structural coal varied from 1990 m/s to 2200 m/s, as shown by Figures 4 and 5, and the variation in transverse velocity ranged from 1075 m/s to 1160 m/s, while the variation in longitudinal velocity of fractured structural coal ranged from 1540 m/s to 1950 m/s and the variation in transverse velocity ranged from 800 m/s to 1000 m/s. In all stages of the experiment, the variation in longitudinal and transverse velocities of the primary structural coal was larger than that of the fractured structural coal. At all stages of the experiment, the main structural coal has higher longitudinal and transverse wave velocities than the fractured structural coal. At all stages of the experiment, the changes of longitudinal and transverse wave velocities of the fractured structural coal were more obvious than those of the primary structural coal.

3.3. Changes in the Elastic Mechanical Parameters of Coal Samples during Gas Adsorption and Desorption

Based on Tables 2 and 3, the variation patterns of Poisson's ratio and the elastic modulus of the coal samples during the test period can be derived. The elastic modulus and Poisson's ratio of the coal samples both slightly rose throughout the evacuation stage, as illustrated in Figures 6 and 7. The elastic modulus exhibited an increasing trend with the increase in gas pressure in the gas adsorption stage, when the gas pressure rose from 0 MPa to 2.5 MPa, whereas the Poisson's ratio displayed a variation law of rising and then falling with the rise in gas pressure. The Poisson's ratio declined when gas pressure rose from 1.5 MPa to 2.5 MPa.

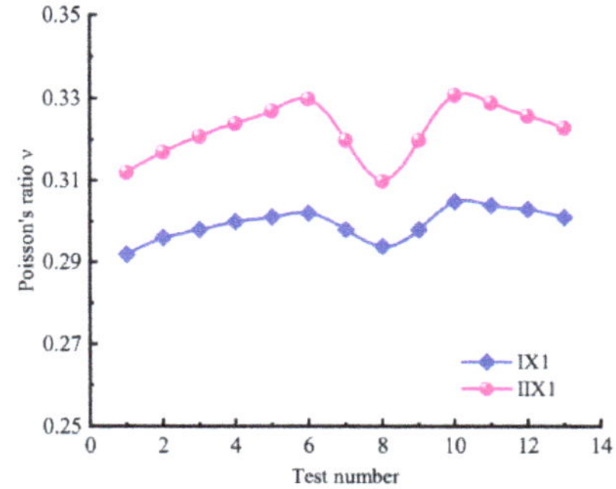

Figure 6. Coal sample Poisson's ratio ν variation pattern during the course of the experiment.

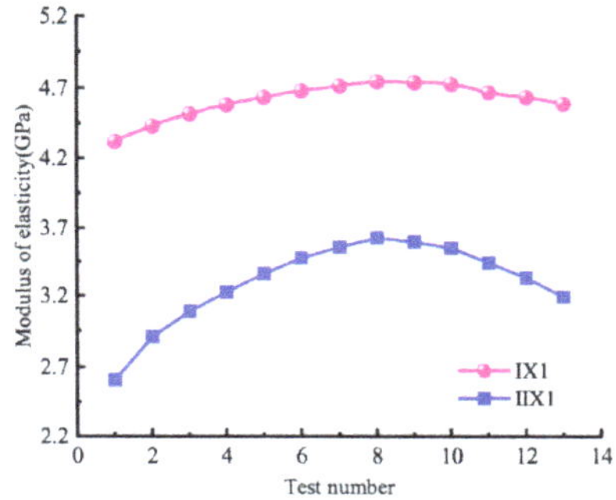

Figure 7. Variations in the coal sample's elastic modulus E during the course of the experiment.

The Poisson's ratio exhibited a rise and, subsequently, a reduction with the decrease in gas pressure, when the gas pressure went from 2.5 MPa to 0 MPa, but the elastic modulus showed a declining trend with the decrease in gas pressure. The Poisson's ratio rises when gas pressure falls from 1.5 MPa to 0 MPa, as seen in the graph.

Additionally, Figures 6 and 7 show that the elastic modulus E of the primary structural coal is greater than that of the fractured structural coal, whereas Figure 8 shows that the fractured structural coal's Poisson's ratio v is greater than that of the primary structural coal.

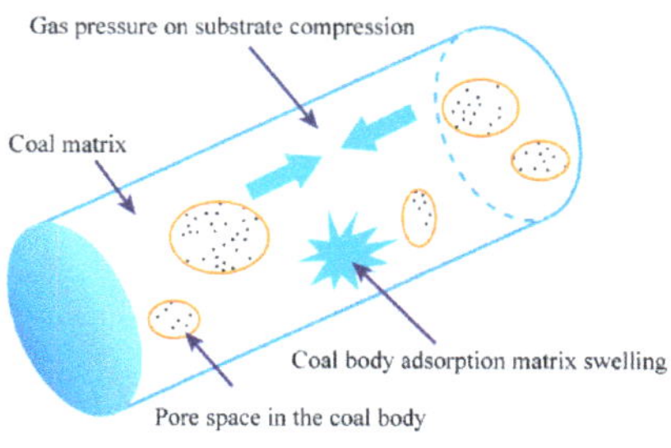

Figure 8. Schematic diagram of gas action inside coal body.

3.4. The Mechanism of How Gas Adsorption and Desorption Affect the Coal's Ultrasonic Velocity

Ultrasound is an elastic wave that is generally considered to have the greatest speed when propagating in solids and the least speed when propagating in liquids. It is generally believed that ultrasonic waves propagate with the greatest velocity in solids, followed by liquids, and the least in gases. Transverse waves can go through solids but only in liquids and gases, while longitudinal waves can move through all three.

Coal is a dual–structured pore medium consisting of matrix pores and pores [36,37]. Both free gas and adsorbed gas are present in the gas–bearing coal that results from gas adsorption; the free gas is present as a gas in the matrix pores and in the pores of the coal, while the adsorbed gas is bound to the solid surface of the coal [38]. The matrix pores, and pores of the coal that contain gas will be examined in this research to determine how gas adsorption and desorption affect coal's ultrasonic velocity when both free and adsorbed gases are present.

The following basic assumptions have been made to facilitate the study:

- The gas–bearing coal's pores and cracks were investigated, and their porosity was quantified.
- Gas containing coal underwent linear elastic minor deformation.
- It was a homogenous continuous isotropic medium that included coal.
- Gas containing coal exhibits a stress–strain relationship that abides with the generalized Hooke's law.

The deformation of coal that contains gas can be classified into two types: structural deformation and intrinsic deformation [39]. Generally, the structural deformation of gas–bearing coal is negligible because it is in the three–way ground stress field and is governed by the geological conditions of the surrounding coal seam. In addition, the experimental results in this work show that the structural deformation of the coal containing gas was constant when the axial pressure, ambient pressure, and gas pressure all increased at the same time. As illustrated in Figure 8, the ontogenetic deformation of gas–bearing coal is caused by the adsorption and desorption of gas, and the parameters influencing this deformation include gas pressure and adsorption.

The strain in the coal, due to the expansion of the adsorbed gas, is [40]:

$$\varepsilon_{BX} = \frac{2a\rho RT(1-2v)}{3EV_m(1-\varphi_0)}\ln(1+bp) \tag{2}$$

where ε_{BX} is the strain caused by the expansion of adsorbed gas, ρ is the apparent density of coal (kg/m^3), v is the Poisson's ratio, E is the elastic modulus (MPa), a is the ultimate adsorption capacity at a given temperature (m^3/t), V_m = 22.4×10^{-3} m^3/mol, V_m is the molar volume of gas, b represents half of the pressure corresponding to the pressure at which the ultimate adsorption capacity is reached (MPa^{-1}), R is the universal gas constant, T is the temperature (K), p is the gas pressure (MPa), R = 8.3143 J/(mol·K), φ_0 is the porosity.

Following a shift in the pore gas pressure, coal matrix compression experiences a linear strain [41]:

$$\varepsilon_{BP} = \frac{C_S}{3}p \tag{3}$$

where ε_{BP} is the linear strain, a result of the coal matrix mass's compression caused by pore gas pressure, and C_S is the coal matrix mass compression factor.

The coal's longitudinal and transverse wave velocities were computed using the following formulas, taking into account the effects of gas adsorption and gas pressure:

$$v_z = \frac{L}{\frac{L-L(1-\varphi_0)(1+\varepsilon_{BX}+\varepsilon_{BP})}{v_k}+\frac{L(1-\varphi_0)(1+\varepsilon_{BX}+\varepsilon_{BP})}{v_{zs}}} \tag{4}$$

$$v_h = v_{hs}(1-\varphi_0)(1+\varepsilon_{BX}) \tag{5}$$

where v_{zs} represents ultrasonic longitudinal velocity in the matrix (m/s), L represents length of the coal unit (m), v_k represents ultrasonic velocity in the pore gas (m/s), v_z represents ultrasonic longitudinal velocity in the coal unit (m/s), v_h represents ultrasonic transverse velocity in the coal unit (m/s), v_{hs} represents ultrasonic transverse velocity in the matrix (m/s).

As gas adsorption and desorption affect the same mechanism of wave velocity variation in primary and cracked structural coals, the ultrasonic velocity variation during gas adsorption is analyzed below as an example for coal sample IX1, and the basic parameters of coal sample IIX1 are represented in Table 4.

Table 4. Coal sample IX1's elastic mechanical characteristics and gas seepage parameters.

$\rho/(\mathrm{t{\cdot}m^{-3}})$	E/MPa	ν	$a/(\mathrm{m^3{\cdot}t^{-1}})$	$b/(\mathrm{MPa^{-1}})$	φ_0	T/K	$C_S/(\mathrm{MPa^{-1}})$	$V_p/(\mathrm{m{\cdot}s^{-1}})$	$V_s/(\mathrm{m{\cdot}s^{-1}})$
1.43	4400	0.296	35.2	0.71	0.06	298	0.016	2031	1093

The calculation regarding the speed of propagation of sound waves in gas gases is based on the equation [42]:

$$v_k = \sqrt{\frac{rRT}{\mathrm{M}}} \tag{6}$$

where r is the proportion of a gas's constant−pressure heat capacity to its constant−temperature heat capacity, for methane 1.33, M is the gas's molecular weight, for methane 16×10^{-3} (kg·mol^{-1}), and T is the absolute temperature (K).

According to Equation (6), the ultrasonic velocity in the gas v_k = 453.94 m/s. According to Equations (4) and (5), simultaneously, the coal matrix velocity v_{zs} = 2607.9 m/s and v_{hs} = 1162.8 m/s.

By substituting the combination of Equations (4) and (5) into the elastic dynamics and gas percolation parameters of different coal samples, the calculated values of the longitudinal and transverse wave velocities of coal containing gas at different test stages could be obtained, and the comparison between the experimental values is shown in Figures 9 and 10.

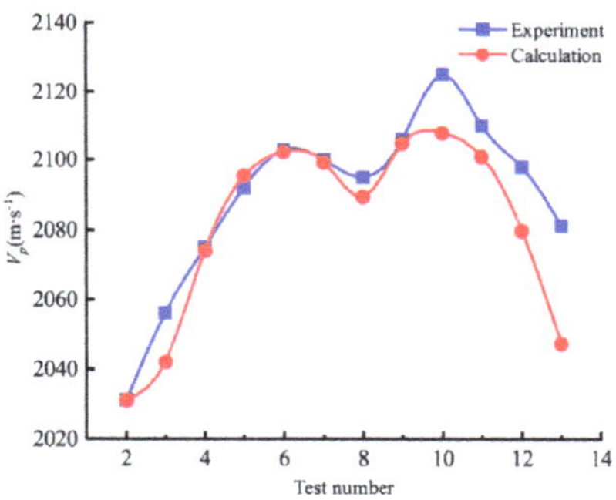

Figure 9. Variation of calculated and experimental values of longitudinal wave velocity of coal sample during the experiment.

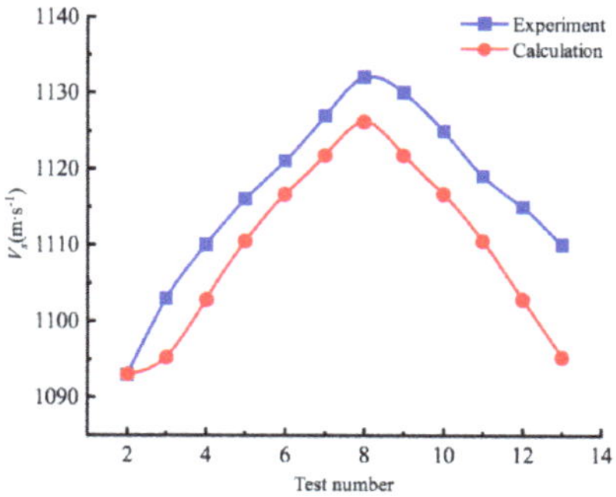

Figure 10. Variation of calculated and experimental values of transverse wave velocity of coal sample during the experiment.

Figures 9 and 10 show that there is a general agreement between the estimated and experimental values for the coal's longitudinal and transverse wave velocities during gas adsorption and desorption, as well as for their magnitudes.

There is a crucial gas pressure value (1.5 MPa) that produces the greatest longitudinal wave velocity of coal during the jet adsorption stage, where the longitudinal wave velocity of coal exhibits a trend of growing and subsequently dropping. As the gas pressure rises during the jet adsorption step, the coal's transverse wave velocity also rises. The coal's longitudinal wave velocity, similarly, has a tendency to rise and then fall throughout the stage of decreasing desorption. The greatest longitudinal wave velocity of coal is caused by a critical gas pressure of 1.5 MPa. The coal's transverse wave velocity drops when the gas pressure drops during the downward desorption stage.

Additionally, during the descending desorption stage, the longitudinal wave velocity of coal, measured experimentally, is a little bit higher than the longitudinal wave velocity estimated. The coal's longitudinal wave velocity in the descending desorption stage was higher, under the same gas pressure circumstances, than it was in the gas injection and adsorption phases.

The above results indicate that it is feasible to analyze the variation in longitudinal and transverse wave velocities of coal during gas adsorption and desorption by using factors such as the adsorption and expansion of gas on the coal matrix and the compression of the coal matrix by pore gas pressure. The mechanism of coal's longitudinal and transverse wave velocity fluctuation will be further explained in the section that follows, from the viewpoint of how coal's porosity is affected by gas adsorption and desorption.

The results show that the porosity of coal is the main factor affecting the longitudinal wave velocity of coal, which will change during gas adsorption and desorption due to the influence of gas adsorption and desorption, resulting in the coal's longitudinal wave velocity changing.

Figure 11 shows the variation of strain in coal with gas pressure. The compression substrate is subjected to equal−sized stresses. The gas adsorption expansion strain was larger than the strain on the gas compression substrate when the gas pressure was less than the critical gas pressure p_c, and vice versa when the gas pressure was higher than the critical gas pressure p_c.

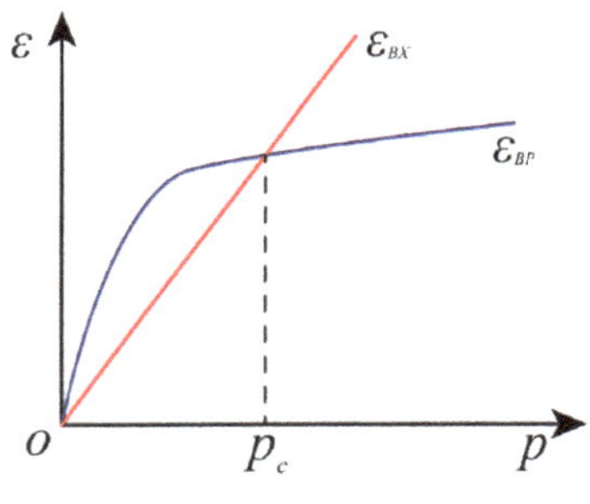

Figure 11. Variation of strain in coal with gas pressure.

Figure 12 shows the variation in the porosity of the coal with gas pressure, and combining Figures 11 and 12, it is clear that the coal's porosity suffered as a result of the gas adsorption expansion strain, whereas the gas compression matrix strain exhibited a positive effect on the porosity of coal. When the gas pressure dropped below the critical gas pressure p_c, the gas adsorption expansion strain was greater than the gas compression matrix strain, thus showing a continuous decrease in coal porosity. When the gas pressure dropped below the critical gas pressure p_c, the gas adsorption expansion strain was less than the gas compression matrix strain, thus showing a continuous increase in coal porosity. Therefore, as gas pressure rises, coal porosity exhibits a pattern of falling and then rising.

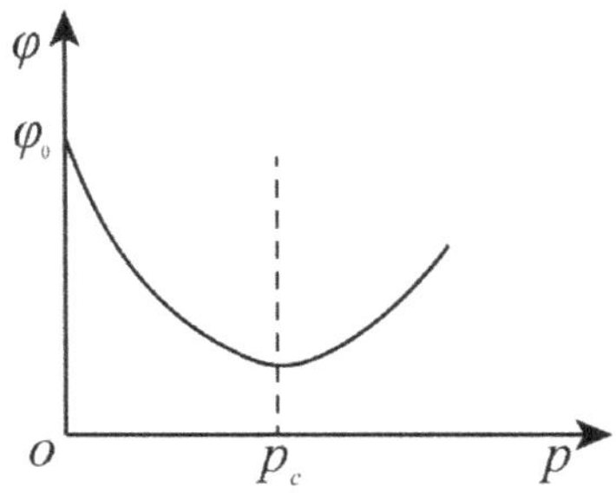

Figure 12. Variation of porosity in coal with gas pressure.

It is generally believed that there is a negative correlation between the coal's porosity and longitudinal velocities. Whenever there is a drop in gas pressure, to below the critical gas pressure (1.5 MPa), the negative effect of gas adsorption and expansion strain is greater than the positive effect of the gas compression matrix strain, and the porosity of coal de−crumples, leading to an increase in the longitudinal velocity of coal. Additionally, if the gas pressure exceeds the critical gas pressure (1.5 MPa), the negative effect of gas adsorption and the expansion strain is smaller than the positive effect of the gas compression matrix strain, and the porosity of the coal increases, leading to the decrease in the longitudinal velocity of the coal. If the gas pressure is higher than the critical gas pressure (1.5 MPa), the negative effect of the gas adsorption expansion strain is smaller than the positive effect of the gas compression matrix, at which point the porosity of the coal body increases, leading to a reduction in the coal body's longitudinal velocity. In the decompression desorption stage, the positive effect of the gas compression matrix strain decreases more significantly as the gas pressure decreases, if the gas pressure is higher than the critical gas pressure (1.5 MPa), at which point the porosity of the coal body decreases, leading to an increase in the longitudinal velocity of the coal body.

The negative impact of the gas adsorption expansion strain is lessened when the gas pressure is below the critical gas pressure (1.5 MPa). At this point, the coal body's porosity rises, increasing the coal body's longitudinal velocity. The negative effects of the gas adsorption expansion strain become less pronounced, and the coal body's porosity rises when the gas pressure is below the critical gas pressure (1.5 MPa), which results in a drop in the longitudinal wave velocity.

The transverse wave can only travel through solids; therefore, the impact of gas adsorption and expansion is the sole factor affecting the transverse wave velocity of coal bodies. The coal body's adsorption and expansion strain rise with the gas pressure, increasing the transverse wave velocity. The coal body's adsorption and expansion strain reduce with a fall in gas pressure, and as a result, so does the transverse wave velocity. The coal body's transverse wave velocity is highest when the gas pressure is 2.5 MPa.

3.5. Mechanisms through Which Variations in the Elastic Mechanical Properties of Coal Masses Are Influenced by Gas Adsorption and Desorption

The elastic mechanical parameters of the coal body are an external reflection of the physical properties of the coal body. Research shows that gas adsorption and desorption can affect the mechanical properties of a coal body. Moreover, gas adsorption and desorption are the result of the combined effect of gas adsorption expansion and gas compression matrix effect. Therefore, gas adsorption expansion and gas compression matrix effect are the root causes of the changes to the elastic mechanical parameters of coal body during gas adsorption and desorption.

During gas injection, if the gas pressure is below the critical gas pressure (1.5 MPa), the gas adsorption expansion effect of the coal skeleton is greater than the gas compression matrix effect, and the overall coal body shows rheological characteristics due to the gas adsorption expansion, and therefore, the Poisson's ratio increases. The critical gas pressure is exceeded when the gas pressure is high, the gas compression matrix effect dominates, and the overall coal body shows hardening due to the gas compression matrix effect. The critical gas pressure is exceeded when the gas pressure is high, the gas compression matrix effect is dominant, and the overall coal body shows hardening due to the gas compression matrix effect. In the desorption process, the critical gas pressure is exceeded when the gas pressure is high, the gas compression matrix effect decreases more blatantly, and the Poisson's ratio increases again. If the gas pressure drops more and falls below the critical gas pressure (1.5 MPa), the gas adsorption expansion effect drops more blatantly, so the Poisson's ratio diminishes again. Meanwhile, the theoretical formula of Poisson's ratio is accurately represented by $v = \left(V_p^2 - 2V_s^2\right)/\left[2\left(V_p^2 - V_s^2\right)\right]$. Poisson's ratio is positively correlated with the longitudinal and transverse wave velocity ratios. The longer the ratio of longitudinal to transverse wave speed, the bigger the Poisson's ratio, which is inversely proportional to the ratio of wave speed. The same variation rule applies to both the Poisson's ratio and the longitudinal wave speed.

A measurement of an object's resistance to elastic deformation is its elastic modulus. The expansion rate of the coal skeleton during the process of gas adsorption rises with an increase in gas pressure, which is represented as a rise in the elastic modulus of the coal body. As the gas pressure drops during the pressure desorption process, the coal skeleton's rate of gas adsorption expands less quickly, which is stated as a reduction in the elastic modulus of the coal body. Meanwhile, the theoretical equation of elastic modulus can also illustrate the above conclusion. From $\mu = \rho V_s^2$, it is known that the larger the transverse wave velocity is, the larger the shear modulus of the coal sample is; subsequently, according to $E = 2\mu(1 + v)$, the elastic modulus is positively related to the shear modulus, so the elastic modulus is in line with the trend of the transverse wave velocity.

4. Conclusions

(1) During the gas adsorption and desorption processes, the longitudinal and transverse wave velocities of the primary structural coal are greater than those of the fractured structural coal, and the change amplitude of the longitudinal and transverse wave velocities of the fractured structural coal is more noticeable than that of the primary structural coal.

(2) The coal body's longitudinal wave velocity can only increase beyond a critical gas pressure threshold, which in this experiment is 1.5 MPa. As a result of the gas adsorption and desorption processes, the coal's longitudinal wave velocity initially rises and subsequently falls. The transverse wave velocity and gas pressure have a strong positive association.

(3) The combined effects of the gas adsorption expansion effect and the gas compression matrix impact result in the change in the longitudinal wave velocity of the coal body.

(4) The change in the transverse wave velocity is compatible with the changing law of gas pressure. The variation in the transverse wave velocity of coal is exclusively regulated by the gas adsorption expansion effect.

(5) During the gas adsorption and desorption process, the coal's elastic modulus and Poisson's ratio both change at the same rates as the longitudinal and transverse wave velocities.

Author Contributions: Conceptualization, G.X.; methodology, G.X.; validation, G.X., C.W. and J.L.; formal analysis, J.L. and Y.W.; investigation, Y.W.; writing—original draft preparation, G.X. and Y.W.; writing—review and editing, G.X. and J.L.; visualization, J.L.; supervision, H.J.; project administration, G.X.; funding acquisition, G.X. All authors have read and agreed to the published version of the manuscript.

Funding: This research was funded by the National Key Research and Development Program of China, No. 2018YFC0807805; the Natural Science Basic Research Program of Shaanxi, No. 2019JM-072.

Institutional Review Board Statement: Not applicable.

Informed Consent Statement: Not applicable.

Data Availability Statement: Not applicable.

Acknowledgments: We appreciate all of the writers' assistance and efforts in helping to write this article.

Conflicts of Interest: The authors state that they have no conflict of interests.

References

1. Liu, P.; Liu, A.; Zhong, F.; Jiang, Y.; Li, J. Pore/fracture structure and gas permeability alterations induced by ultrasound treatment in coal and its application to enhanced coalbed methane recovery. *J. Pet. Sci. Eng.* **2021**, *205*, 108862. [CrossRef]
2. Wang, X.; Wang, E.; Liu, X.; Zhou, X. Failure mechanism of fractured rock and associated acoustic behaviors under different loading rates. *Eng. Fract. Mech.* **2021**, *247*, 107674. [CrossRef]
3. Kong, X.; He, D.; Liu, X.; Wang, E.; Li, S.; Liu, T.; Ji, P.; Deng, D.; Yang, S. Strain characteristics and energy dissipation laws of gas-bearing coal during impact fracture process. *Energy* **2022**, *242*, 123028. [CrossRef]
4. Liu, X.; Jia, T.; Wei, J.; Wei, G.; Yan, J.; Wu, C. Modeling and evolution characteristics of coal reservoir energy during gas and water production. *J. Nat. Gas Sci. Eng.* **2021**, *96*, 104329. [CrossRef]
5. Zhou, A.; Du, C.A.; Wang, K.; Hu, J.; Fan, X. Experimental research on the law of the deformation and damage characteristics of raw coal/briquette adsorption-instantaneous pressure relief. *Fuel* **2022**, *308*, 122062. [CrossRef]
6. Yuan, L. Strategies of high efficiency recovery and energy saving for coal resources in China. *J. Southwest Pet. Univ. Sci. Technol. Ed.* **2018**, *20*, 3–12. [CrossRef]
7. Liu, X.; Zhang, F.; Ma, G.; Su, X. Analysis and applied study on key factors of coal and rock crack network reconstruction. *Coal Sci. Technol.* **2017**, *45*. 85–89+169. [CrossRef]
8. Wang, X.; Asem, P.; Hu, C.; Labuz, J.F. Microcracking in tensile fracture of a brittle rock. *Eng. Fract. Mech.* **2021**, *251*, 107789. [CrossRef]
9. Cheng, L.; Wang, Y.; Zhang, Y.; Zhao, H. The present situation and prospect of the acoustic properties research in coal. *Prog. Geophys.* **2013**, *28*, 452–461. [CrossRef]
10. Shen, X.; Xiong, Q.; Shi, W.; Liang, S.; Shi, X.; Wang, K. A New Algorithm for Reconstructing Two-Dimensional Temperature Distribution by Ultrasonic Thermometry. *Math. Probl. Eng.* **2015**, *2015*, 916741. [CrossRef]
11. Karacan, C.Ö. Swelling-induced volumetric strains internal to a stressed coal associated with CO_2 sorption. *Int. J. Coal Geol.* **2007**, *72*, 209–220. [CrossRef]
12. Li, H.; Lin, B.; Yang, W.; Zheng, C.; Hong, Y.; Gao, Y.; Liu, T.; Wu, S. Experimental study on the petrophysical variation of different rank coals with microwave treatment. *Int. J. Coal Geol.* **2016**, *154–155*, 82–91. [CrossRef]
13. Liu, P.; Fan, L.; Fan, J.; Zhong, F. Effect of water content on the induced alteration of pore morphology and gas sorption/diffusion kinetics in coal with ultrasound treatment. *Fuel* **2021**, *306*, 121752. [CrossRef]
14. Wang, L.L.; Vandamme, M.; Pereira, J.M.; Dangla, P.; Espinoza, N. Permeability changes in coal seams: The role of anisotropy. *Int. J. Coal Geol.* **2018**, *199*, 52–64. [CrossRef]
15. Biot, M.A. Theory of propagation of elastic waves in a fluid-saturated porous solid. I. Low-frequency range. *J. Acoust. Soc. Am.* **1956**, *28*, 168–178. [CrossRef]
16. Busch, A.; Gensterblum, Y.; Krooss, B.M.; Siemons, N. Investigation of high-pressure selective adsorption/desorption behaviour of CO2 and CH4 on coals: An experimental study. *Int. J. Coal Geol.* **2006**, *66*, 53–68. [CrossRef]
17. Liu, P.; Qin, Y.; Liu, S.; Hao, Y. Non-linear gas desorption and transport behavior in coal matrix: Experiments and numerical modeling. *Fuel* **2018**, *214*, 1–13. [CrossRef]
18. Mao, Y.; Chen, Y.; Bu, X.; Xie, G. Effects of 20 kHz ultrasound on coal flotation: The roles of cavitation and acoustic radiation force. *Fuel* **2019**, *256*, 115938. [CrossRef]
19. Pan, Z.; Connell, L.D. A theoretical model for gas adsorption-induced coal swelling. *Int. J. Coal Geol.* **2007**, *69*, 243–252. [CrossRef]
20. Wang, G.X.; Wei, X.R.; Wang, K.; Massarotto, P.; Rudolph, V. Sorption-induced swelling/shrinkage and permeability of coal under stressed adsorption/desorption conditions. *Int. J. Coal Geol.* **2010**, *83*, 46–54. [CrossRef]
21. Liu, S.; Zhao, Q.; Zhang, P.; Guo, L. Test and research on relationship between seam gas features and vibration wave parameters. *Coal Sci. Technol.* **2005**, *33*, 33–36. [CrossRef]
22. Zhao, Q.; Hou, Y.; Liu, S. The pilot study on characteristics of the seismic wave spectrum and gas content of coal seam. *J. Henan Polytech. Univ.* **2008**, *27*, 615–618. [CrossRef]
23. Liu, W. Geophysical response characteristics of coalbed methane. *Lithol. Reserv.* **2009**, *21*, 113–115. [CrossRef]
24. Liu, S.; Zhao, Q.; Zhang, P.; Guo, L. Study on Gas Content and the Attenuation Characteristics of Coal Seams of Huainan Xinzhuangzi Mine. *J. Anhui Univ. Sci. Technol. Nat. Sci.* **2005**, *25*, 1–4. [CrossRef]
25. Wang, Z.; Liu, S.; Lu, T.; Zhang, P.; Zhao, L. Experimental study on the relationships of coal gas and seismic attributes. *Coal Geol. Explor.* **2011**, *39*, 63–65. [CrossRef]
26. Wang, Z.; Liu, S.; Zhang, X. Correlation analysis of apparent seismic wave speeds and gas parameters in front of the coal roadway. *Prog. Geophys.* **2012**, *27*, 349–354. [CrossRef]
27. Zhang, P.; Liu, S.; Qiufang, Z.; Guo, L. Factor analysis on seam reduced features and quality in Huainan Mining Area. *Coal Sci. Technol.* **2006**, *34*, 83–85. [CrossRef]

28. Jiang, L.; Shi, X. Relation between wave velocity in sandstone and fluid content in porous medium under high frequency condition. *Oil Geophys. Prospect.* **1998**, *33*, 355–362. [CrossRef]
29. Hao, J. Experimental research on ultrasonic elastic characteristics of outburst coal. Master's Thesis, China University of Mining and Technology, Xuzhou, China, 2019.
30. Yu, H. Ultrasonographic Features and Mechanism of Anisotropic about the Coal Containing Nitrogen (Water). Master's Thesis, Henan Polytechnic University, Jiaozuo, China, 2016.
31. Xu, G.; Deng, X.; Zhang, K.; Liu, W. Research on propagation law of waves in coal seam containing gas. *Saf. Coal Mines* **2009**, *40*, 1–4. [CrossRef]
32. Xu, X. Research on the Ultrasonic Characteristics of Coal Under Normal Temperature and Pressure Conditions. Master's Thesis, Henan Polytechnic University, Jiaozuo, China, 2013.
33. Zhao, Y.; Zhang, Y.; Wang, S. Experimental Study on Ultrasonic Anisotropic Characteristics of Nitrogen-containing Coal Bodies. *J. Southwest Pet. Univ. Sci. Technol. Ed.* **2018**, *40*, 83–90. [CrossRef]
34. *Geotechnical Investigation and Testing—Sampling Methods and Groundwater Measurements—Part 1: Technical Principles for the Sampling of Soil, Rock and Groundwater*; International Organization for Standardization: Geneva, Switzerland, 2021.
35. Wang, Y.; Xu, X.; Zhang, Y. Characteristics of P-wave and S-wave velocities and their relationships with density of six metamorphic kinds of coals. *Chin. J. Geophys.* **2012**, *55*, 3754–3761. [CrossRef]
36. Law, B.E.; Rice, D.D. *Hydrocarbons from Coal*; American Association of Petroleum Geologists: Tulsa, OK, USA, 1993.
37. Kong, X.; Wang, E.; Li, S.; Lin, H.; Xiao, P.; Zhang, K. Fractals and chaos characteristics of acoustic emission energy about gas-bearing coal during loaded failure. *Fractals* **2019**, *27*, 1950072. [CrossRef]
38. Jia, L.; Yu, Y.; Li, Z.-P.; Qin, S.-N.; Guo, J.-R.; Zhang, Y.-Q.; Wang, J.-C.; Zhang, J.-C.; Fan, B.-G.; Jin, Y. Study on the Hg0 removal characteristics and synergistic mechanism of iron-based modified biochar doped with multiple metals. *Bioresour. Technol.* **2021**, *332*, 125086. [CrossRef] [PubMed]
39. Lu, P.; Sheng, Z.; Zhu, G.; Fang, E. The effective stress and mechanical deformation and damage characteristics of gas-filled coal. *J. Univ. Sci. Technol. China* **2001**, *31*, 686–693. [CrossRef]
40. Wu, S.; Zhao, W. Analysis of effective stress in adsorbed methane-coal system. *Chin. J. Rock Mech. Eng.* **2005**, *24*, 1674–1678. [CrossRef]
41. Li, P.; Kong, X.; Lu, D. Mathematical modeling of flow in saturated porous media on account of fluid-structure coupling effect. *Chin. J. Hydrodyn.* **2003**, *18*, 419–426. [CrossRef]
42. Han, X.; Xu, D.; Guo, J.; Yang, L.; Wang, Z.; Zhang, D.; Luo, X.; Nie, J.; Li, H.; Jiang, J. Rock physics modelling for acoustic velocities of sandstone considering effects of cementation and compaction. *Chin. J. Geophys.* **2018**, *61*, 5044–5051. [CrossRef]

sustainability

Article

Response and Application of Full-Space Numerical Simulation Based on Finite Element Method for Transient Electromagnetic Advanced Detection of Mine Water

Changfang Guo [1], Tingjiang Tan [2,*], Liuzhu Ma [3], Shuai Chang [4], Yiding Chen [5] and Ke Zhao [6]

1 Artificial Intelligence Research Institute, China University of Mining and Technology, Xuzhou 221116, China
2 School of Safety Engineering, China University of Mining and Technology, Xuzhou 221116, China
3 Xi'an Research Institute, China Coal Technology and Engineering Group Corp, Xi'an 710077, China
4 School of Mines, China University of Mining and Technology, Xuzhou 221116, China
5 Xuzhou Trident Information Technology Co., Ltd., Xuzhou 221116, China
6 Department of Physics, Changzhi University, Changzhi 046000, China
* Correspondence: tb20120019b4@cumt.edu.cn

Abstract: The issue of water hazards has led to the restriction of safe and efficient coal mine production in China. The transient electromagnetic method (TEM) is one of the most effective means of detecting the hidden dangers of water hazards in coal mines. However, the current understanding of the whole-space transient electromagnetic response of mine water is only on the general law due to the late start of the forward research. Therefore, this paper established multiple sets of simulation models in the whole area in order to study the rules and factors of transient electromagnetic responses. Subsequently, these laws are used to explain the detection data of TEM in the field. According to the simulation results, the electric properties, distance, and size had the greatest influence on the transient electromagnetic response of regular anomalous geological bodies, while the electromagnetic field projection area also had an impact on irregular ones. Furthermore, field application demonstrated that the response law and TEM's affecting factors are acceptable for directing the detection of transient electromagnetic in coal mines. This research can advance the TEM's data processing and interpretation technology and offer a theoretical basis for detailed investigation.

Keywords: transient electromagnetic method; numerical modeling; transient electromagnetic response; factor; advanced detection

Citation: Guo, C.; Tan, T.; Ma, L.; Chang, S.; Chen, Y.; Zhao, K. Response and Application of Full-Space Numerical Simulation Based on Finite Element Method for Transient Electromagnetic Advanced Detection of Mine Water. *Sustainability* **2022**, *14*, 15024. https://doi.org/10.3390/su142215024

Academic Editors: Xiangguo Kong, Dexing Li and Xiaoran Wang

Received: 1 October 2022
Accepted: 10 November 2022
Published: 14 November 2022

1. Introduction

As the slow decrease and depletion of extractable resources in the shallow strata, the development to deep geological resources has become an inevitable strategy for many countries around the world [1–3]. Nevertheless, as the mining depth increases, the hydrogeological conditions become more and more complex, such as geological faults, the old cellar water accumulation and collapse columns [4–6]. Due to the high crustal stress, high surface temperature and high fluid pressure in the deep part of the rock body characteristics change, coupled with the stratum itself fracture fragmentation zone development and mining-induced water conductivity fragmentation zone conducts different water sources into the mine, that the mine occurs sudden water disaster [7,8]. Water disaster prevention and control are based on the detection of these water-bearing aberrant bodies [9]. The mine transient electromagnetic method (MTEM) has been employed in tunnel excavation advanced detection, working face water-rich region detection, and other fields [10,11] because of its benefits, which include being sensitive to low-resistance body, large anomaly intensity close to the target object, and convenient construction. The three-dimensional forward research of the MTEM is able to not only direct data interpretation in practical work and increase the accuracy of interpretation, but it can also provide the groundwork for three-dimensional inversion, which is still in the research stage, and play a significant role in advancing this method's development.

TEM is a time-domain detection method based on the electromagnetic induction principle, which first emits a step pulse magnetic field, called a primary field, to the ground through an ungrounded loop. Subsequently, the time-varying variation of the induced secondary field in the underground medium is measured during the pause of the primary field. Finally, by processing, analyzing, and interpreting the generated secondary field information, detecting various geological details is achieved [12–14] (see Figure 1).

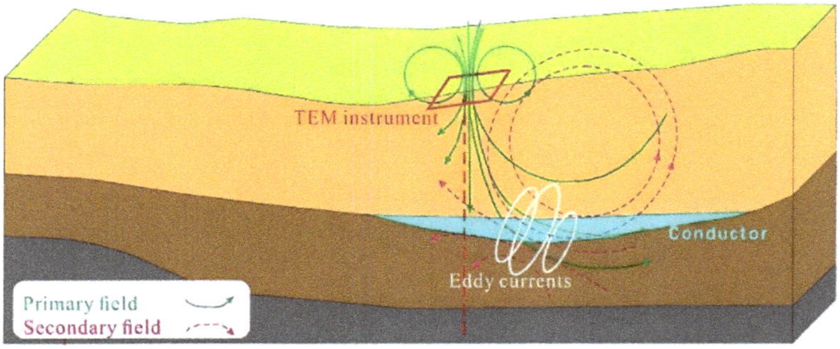

Figure 1. Working principle of TEM.

Transient electromagnetic forward modeling primarily includes the time domain finite difference method (FDTD), finite volume method (FVM), integral equation method (IEM) and finite element method (FEM). The full-space transient electromagnetic response of mined-out water was numerically modelled by Chang et al. [15], utilizing the three-dimensional FDTD. Sun et al. studied the Crank-Nicolson FDTD 3D forward modeling for the transient electromagnetic transition [16]. A 3D forward modeling of loop-source transient electromagnetic response in biaxial anisotropic formation is proposed through the use of the mimetic finite volume method (MFVM) [17]. Liu et al. implemented 3D forward modeling and analysis of the loop-source transient electromagnetic method based on the FVM for an arbitrarily anisotropic medium [18]. The response properties of the highway-borehole transient electromagnetic field under the circumstance of tunnel full space were studied by Chen et al., using the IEM to create a numerical model [19]. The FEM has the advantages of strong application and high precision, and it overcomes the structure-related singularity problem when compared to other methods [17]. Based on earlier research, Zhang et al. added absorption boundary conditions to quicken the three-dimensional finite element forward evolution of the transient electromagnetic technique [20]. The issues of lengthy computation times and high memory requirements have been significantly reduced in recent years due to the quick development of computer hardware, and the finite element method is now frequently used in three-dimensional numerical simulations of transient electromagnetic phenomena [21–24]. Unfortunately, as surface TEM is the main focus of current finite element research, there is little numerical simulation of mine transient electromagnetic with multiple turns and a small loop source. In particular, the study on the transient electromagnetic response characteristics of various water-bearing geological bodies widely present in mines is not ideal, such as the precise law of TEM response and the factors influencing the propagation of transient electromagnetic field. Therefore, it is of great significance to study the factors and rules of the transient electromagnetic response by the TDFEM.

Against the above background, this paper established multiple sets of various numerical models based on the TDFEM to study the transient electromagnetic response law and its influencing factors in detail. Finally, the interpretation of field data was successfully guided by these findings. The main research contents of the following paper are as follows: the second section introduces the theoretical basis of numerical simulation, the third section studies the response characteristics of different geological water-bearing anomalies, the fourth section studies the factors affecting the transient electromagnetic response, and the fifth section is the field application.

2. Method

2.1. Time-Domain Electromagnetic Field Equation

The finite element method is a numerical method for solving the approximate solution of boundary value problems of partial differential equations. The benefits of using the finite element method are that it offers great freedom in the selection of discretization, both in the elements that may be used to discretize space and the basic functions. Kuo et al. first used the two-dimensional nodal FEM to simulate the transient electromagnetic response of the low resistivity orebody under the low resistivity layer [25]. Following this, scholars have conducted a lot of research on the three-dimensional time domain finite element [26–28]. For the electromagnetic field problem, the control equation is primarily the Helmholtz equation.

$$\nabla \times \boldsymbol{H} = \boldsymbol{J} + \boldsymbol{J}_s \tag{1}$$

$$\nabla \times \boldsymbol{E} = -\frac{\partial \boldsymbol{B}}{\partial t} \tag{2}$$

$$\nabla \cdot \boldsymbol{D} = 0 \tag{3}$$

$$\nabla \cdot \boldsymbol{B} = 0 \tag{4}$$

where ∇ is the Hamiltonian operator, $\boldsymbol{H}$ is the magnetic field intensity, $\boldsymbol{E}$ is the electric field intensity, $\boldsymbol{D}$ is the electric displacement vector, $\boldsymbol{B}$ is the magnetic induction intensity, t is time, $\boldsymbol{J_s}$ is the field source current density, $\boldsymbol{J}$ is the conduction current density excited by the field source.

The wave equation of electromagnetic field can be obtained by transformation, as follows:

$$\nabla^2 \boldsymbol{E} + k^2 \boldsymbol{E} = 0 \tag{5}$$

$$\nabla^2 \boldsymbol{H} + k^2 \boldsymbol{H} = 0 \tag{6}$$

where k is the propagation constant.

2.2. Time Discretization

The implicit Euler method, which can be regarded as a backward differentiation formula, is utilized to discretize the time domain in this study [29]. The calculation principle is as follows:

$$\frac{\mathrm{d}x}{\mathrm{d}t} = F(x, t) \tag{7}$$

Assuming that the value of the dependent variable x_i is known at an initial value t_i, then a Taylor approximation is used to relate the value of x at $t = t_{i+1}$, namely $x(t_{i+1})$ with $h = t_{i+1} - t_i$. However, in contrast to the explicit Euler method, the Taylor series around the point $x(t_{i+1})$ was used, that is:

$$x_i = x(t_{i+1}) - \left.\frac{\mathrm{d}x}{\mathrm{d}t}\right|_{t_{i+1}} (h) + \mathcal{O}\left(h^2\right) \tag{8}$$

Substituting the differential equation into the above equation yields:

$$x_i = x(t_{i+1}) - F(x_{i+1}, t_{i+1})(h) + \mathcal{O}\left(h^2\right) \tag{9}$$

Therefore, as an approximation, an estimate for $x(t_{i+1})$ can be taken as x_{i+1}, as follows:

$$x(t_{i+1}) \approx x_{i+1} = x_i + F(x_{i+1}, t_{i+1})(h) \tag{10}$$

Using this estimate, the local truncation error is thus proportional to the square of the step size, with the constant of proportionality related to the second derivative of x, which is the first derivative of the given initial-value problem.

2.3. Solving Sparse Equation

The finite element method inevitably reverts to the solution of huge sparse equations when utilized to calculate complex electromagnetic field issues. The electromagnetic field problem is solved in this study using the MUMPS parallel solver, a package for solving linear equations in the form of $Ax = b$ [30].

2.4. Verification of Algorithm Accuracy

To verify the accuracy of the finite unit algorithm, the geoelectrical model proposed in this paper is based on a 3D sphere model, due to the good symmetry of a sphere. The sphere model has a radius of 180 m, and the 18 m thick spherical shell at its outermost edge is set as the infinite-element domain. The procedure is as follows: A circular coil with a radius of 1 m and a cross-sectional radius of 25 mm is placed at the center of the sphere, with a coil emission current waveform of 2 A, a shutdown time of 20 μs, a transmit backline of 60, and a receiving line of 40. Using non-uniform meshing, it is divided into 114,834 tetrahedron meshes. The analogue time starts at 0 s, and the current starts to shut down at 0.5 ms, with a shutdown time of 20 μs. The 1.5 ms analogue time ends when the current is switched off. The method of local-refinement time-step sectioning is applied, where a time step of 2 μs is used in the current shutdown process and in the range of 20 s before and after, and a time step of 20 μs is used for the rest of the dura-tion. This can also ensure the calculation accuracy and significantly improve the computational efficiency.

Figure 2a,b compares the numerical values obtained from the full-space model and the analytical solution. The compared parameter is the induced electromotive force at the center of the coil. The full-space numerical simulation results agree the analytical solution in each period, and the relative error in the induced electromotive force is below 4.5%.

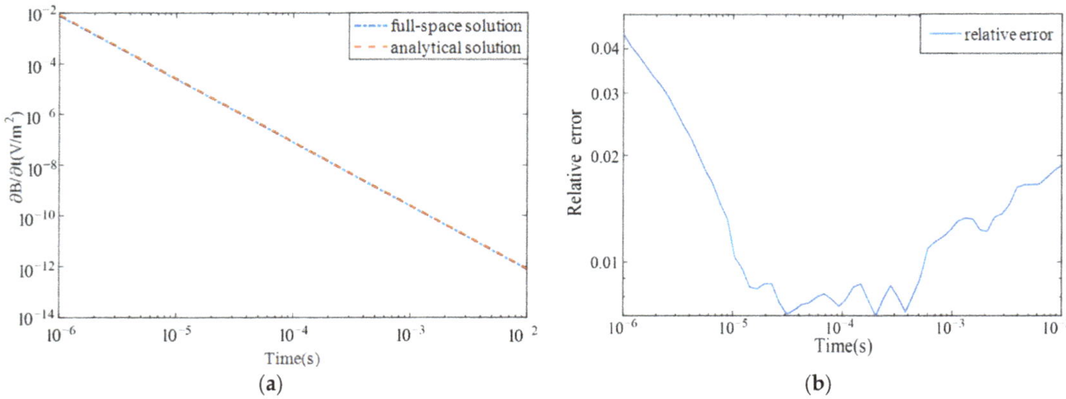

Figure 2. Comparison of numerical and analytical solutions. (**a**) Comparison of solutions. (**b**) Relative error.

3. Simulation of Water-Containing Bodies in Front of the Tunnel

This section develops several geoelectric models, based on the full-space model, for regular water-bearing bodies, water-bearing faults, sinking columns, and other water-bearing bodies in front of the roadway in order to study the transient electromagnetic reaction and compare the effect of the shape on its response.

3.1. Spheres, Cubes, and Cylinders

A layered geoelectrical model is established in this paper, which is divided into three layers, representing the roof, coal seam, and floor. The specific geometry and geoelectrical parameters are shown in Table 1. The tunnel's length and cross-sectional measurements are set at 100 m and 4 m × 4 m, respectively. The interior area of the laneway is set as air, the resistivity of which is 1×10^5 Ω·m. The peripheral surrounding rock resistivity of the laminated geoelectrical model is set to 100 Ω·m. A wire device serves as the excitation source, the coil is taken as a circular coil with a radius of 1 m, the emission coil is 60, the receiving coil is 40, the coil is placed on the digger head, the firing direction is in front of the digger head in order to be close to the real situation, and the coil and the digging head have a 25 mm pore. An abnormal geologic body is placed in front of the digger head, with a resistivity of 1 Ω·m; its parameters are listed in Table 2. The coil emission current is a square wave with a magnitude of 2 A, and the shutdown time is 20 s. Figure 3 shows the specific model and meshing results. Figure 4 demonstrates the electromotive attenuation law curve of three different shape models.

Table 1. Geoelectric parameters of the model.

	Toward the Length (x)/m	Tending Length (y)/m	Thickness (z)/m	Apparent Resistivity (Ω·m)
Clad rock	200	150	80	100
Coal seam	200	150	4	50
Floor	200	150	60	100

Table 2. Geoelectric parameters of the abnormal geologic body.

	Radius (Edge Length)/m	Height/m	Volume/m^3
Sphere	30	/	113,097.3
Cube	48	/	110,592
Cylinder	26	52	110,433.3

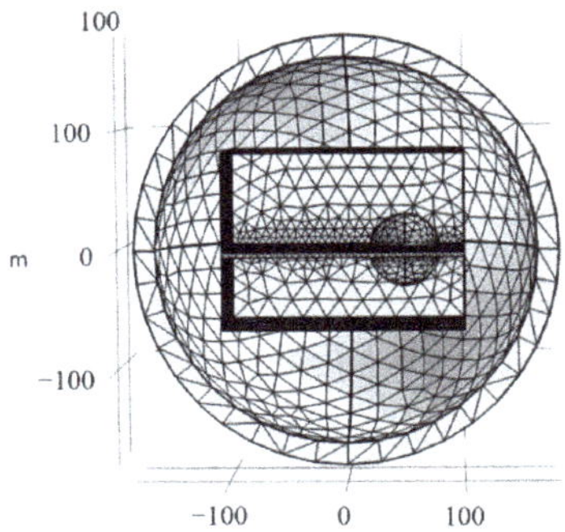

Figure 3. Overall structure of the model and meshing results.

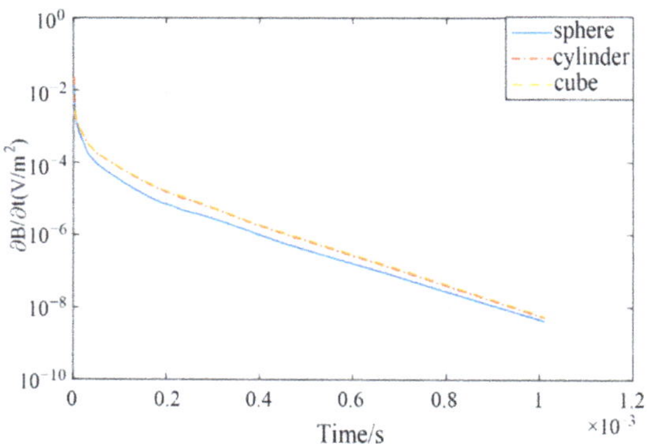

Figure 4. Electromotive attenuation law obtained using three full-space models.

3.2. Aquifer Faults

To study the transient electromagnetic response of the aquifer fault in front of the digger head (see Figure 5), based on the previous location model, the abnormal object is replaced by a fault with dimensions of 10 m × 50 m × 50 m and an apparent resistivity of 1 Ω·m.

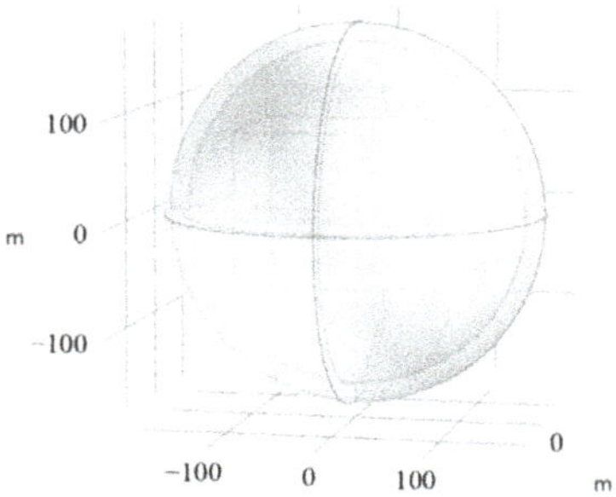

Figure 5. Water-bearing fault model placed in front of the digging head.

As the normal direction of the coil is toward the center of the fault, this section extracts the electromagnetic field *Bx* on its horizontal surface to plot the isoline for analysis, as shown in Figure 6. The induced electromagnetic field is formed in the fault when the current is shut off for 1 μs, and the peak value of *Bx* is close to 3×10^{-10} T. The *Bx* contours at the two short sides of the fault are dense and concave in the early stages. Over time, the electromagnetic field gradually spreads to the central area of the fault and decays continuously. The electromagnetic field becomes concentrated in the fault's center after 50 μs, and the *Bx* contour map progressively becomes steady and starts to degrade.

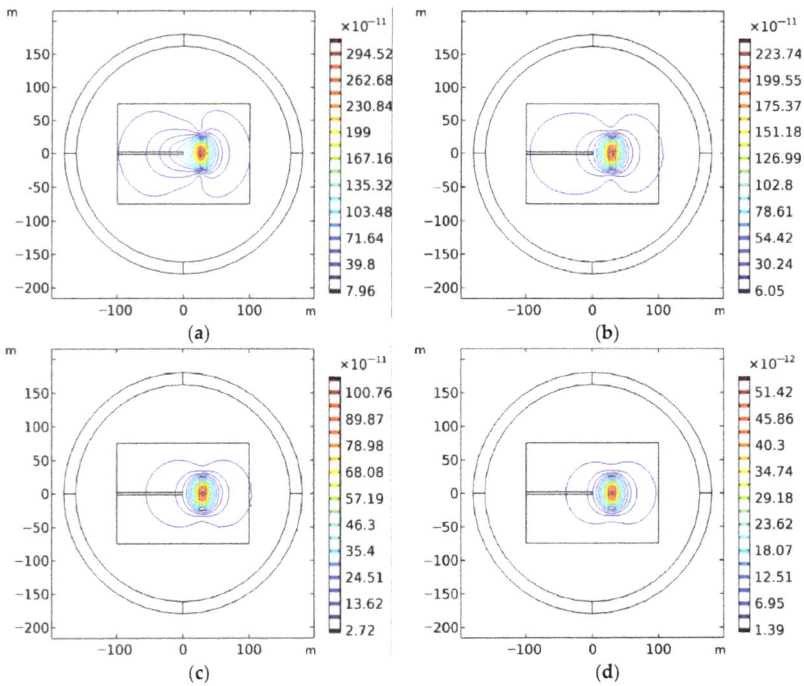

Figure 6. *Cont.*

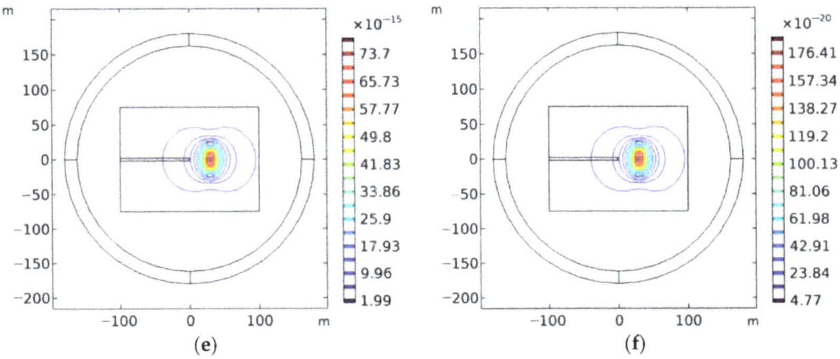

Figure 6. Transient electromagnetic field distributions on the horizontal plane of the fault model after switching off the current. (**a**) $t = 1$ μs. (**b**) $t = 10$ μs. (**c**) $t = 50$ μs. (**d**) $t = 200$ μs. (**e**) $t = 500$ μs. (**f**) $t = 1$ ms.

4. Factors Affecting the Law of Transient Electromagnetic Response

The research in Section 3 demonstrates that the numerical simulation in this work complies with the fundamental rule of the transient electromagnetic field, which states that once the current is cut off, the electromagnetic field reaches a maximum value in the geological anomaly body and then progressively spreads to the center of the anomaly body, whilst constantly degrading.

4.1. Electrical Characteristics

To investigate the impact of electrical characteristics on the transient electromagnetic response law, models with cylinder resistivity values of 0.1, 0.5, 1, 2, 5, and 10 Ω·m were created based on the cylinder model in Section 3.1. The computation's findings are displayed in Figure 7. The coil picks up the response of the adjacent strata at first, therefore the curve's attenuation speed is higher early on. After the cylinder generates the electromagnetic field and is received by the coil, the curve shows a significant difference: the attenuation rate of the 0.1 Ω·m model is the slowest, and then, from slow to fast, is 0.5, 1, 2, 5, 10 Ω·m. In particular, the 10 Ω·m model is attenuated to a less convergent stage after receiving an induced electromotive force of 1.47×10^{-14} V/m at the center of the coil, at 710 s, which is seven orders of magnitude lower than that of the 0.1 Ω·m model.

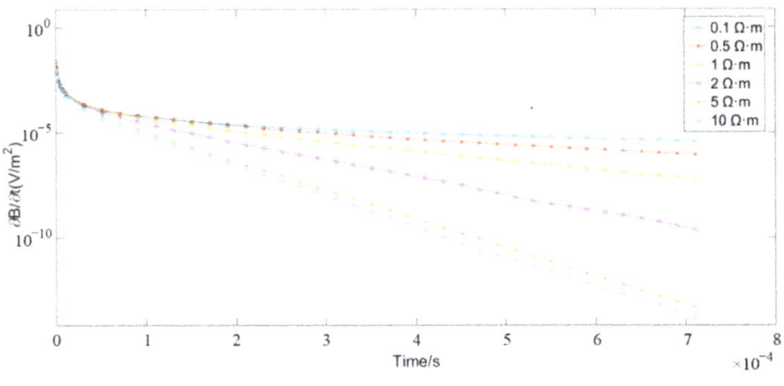

Figure 7. The induced electromotive force of different electrical anomalies.

4.2. Distance

To study the effect of distance between the abnormal geologic body and the coil on the transient electromagnetic response, this paper establishes the same forward model as is presented in Section 3.1. The distances between the cylinder and the coil are set to 5, 10, 20, 30, and 50 m. Figure 8, showing the forward simulation results, indicates that the difference in the distances between the abnormal object and the multiple small return-line sources determines the difference in the intensity of the transient electromagnetic field; however, it has no effect on the attenuation law of the transient electromagnetic field.

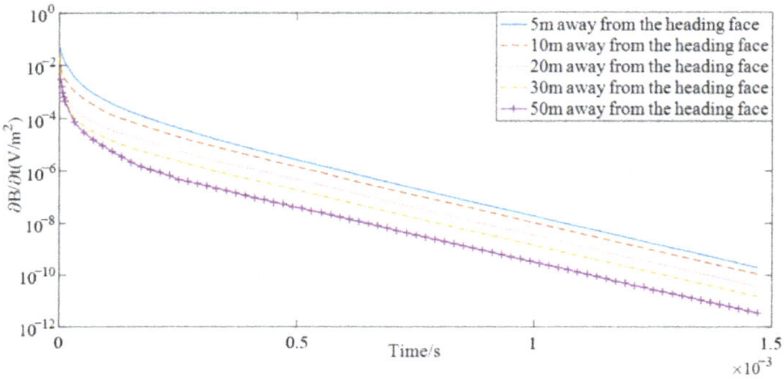

Figure 8. The induced electromotive of abnormal bodies placed at different distances.

4.3. Size

To study the influence of the abnormal geologic body size on the transient electromagnetic response at different times, six different columnar drop column models were established and divided into small, medium and large groups, according to the model size. In each set of models, the round table model has a larger volume than the cylindrical model, and in particular, the volume of the first group of small round tables is smaller than that of the small cylinder (see Table 3).

Table 3. Size parameters of the abnormal bodies.

Exception Body Number	Upper Surface Radius (m)	Lower Surface Radius (m)	Height (m)	Volume (m^3)	Shape
a	5	10	30	5498	Small round table
b	10	10	30	9424.8	Small cylinder
c	10	20	50	36,651	Medium round table
d	15	15	50	35,343	Medium cylinder
e	15	30	100	164,933	Large round table
f	20	20	100	125,663	Large cylinder

Figure 9 shows the attenuation curve of the induced electromotive force at the center of the coil in each model, after the current is switched off. The attenuation curve is consistent with the grouping law.

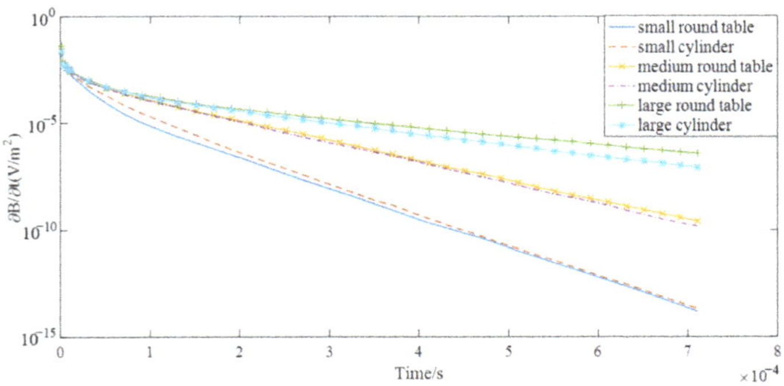

Figure 9. Induced electromotive force attenuation curves for anomalies of different sizes.

4.4. Projected Area

Based on the above findings, we speculated that the attenuation speed of the induced electromagnetic field is influenced by the difference between the contact surface of a low resistivity anomaly and an electromagnetic field. The concept of "electromagnetic field projection area" is formally proposed in this paper. After the current is switched off, an inductive electromagnetic field is generated in the abnormal geologic body, and the projection on the abnormal coil plane at the peak center of the inductive electromagnetic field at a certain moment is called the electromagnetic field projection area at that moment. It can also be understood as a cross-section, at which the abnormal geologic body is parallel to the coil plane. It is of great significance to use this concept to analyze the law of electromagnetic attenuation.

Based on the findings of Section 3.2, this section creates models with normal plane inclination angles of −60°, −30°, 0°, 30°, and 60° for study in order to confirm the accuracy of this supposition. The size and electrical properties of the fault are consistent with the model in Section 3.2. The model is displayed in Figure 10. The distance between the various models and coils is shown to be 0° > −30° = 30° > −60° = 60° in descending order.

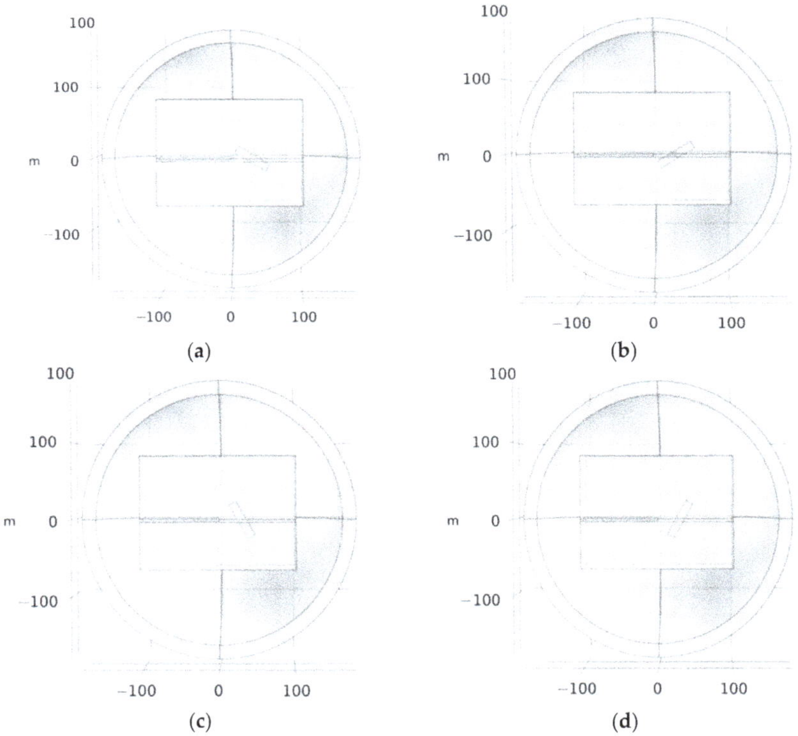

Figure 10. Different dip fault models. (**a**) Deg. = 60°. (**b**) Deg. = −60°. (**c**) Deg. = 30°. (**d**) Deg. =−30°.

Figure 11a–c shows the induced electromotive attenuation curve at the center of the coil after each model is closed. The process of inductive electromagnetic attenuation of each model can be divided into two stages: At the early stage after the shutdown, depending on the nearest distance of each model to the coil, the order of the induced electromotive force is 60° = 60° > 30° = 30° > 0°; at the later stage, as the 0° model has the largest cross-sectional area and the slowest decay rate in comparison to the other models, its induced electromotive force eventually exceeds that of the −30° and 30° models. The late-induced electromotive force from large to small follows the order: −60° = 60° > 0° −30° = 30°. Moreover, both the symmetrical −60° and 60° models and the −30° and 30° models have the same attenuation rules. Similarly, the −60° (60°) model and −30° (30°) model show a similar attenuation pattern due to the small difference in the projection area of their electromagnetic fields.

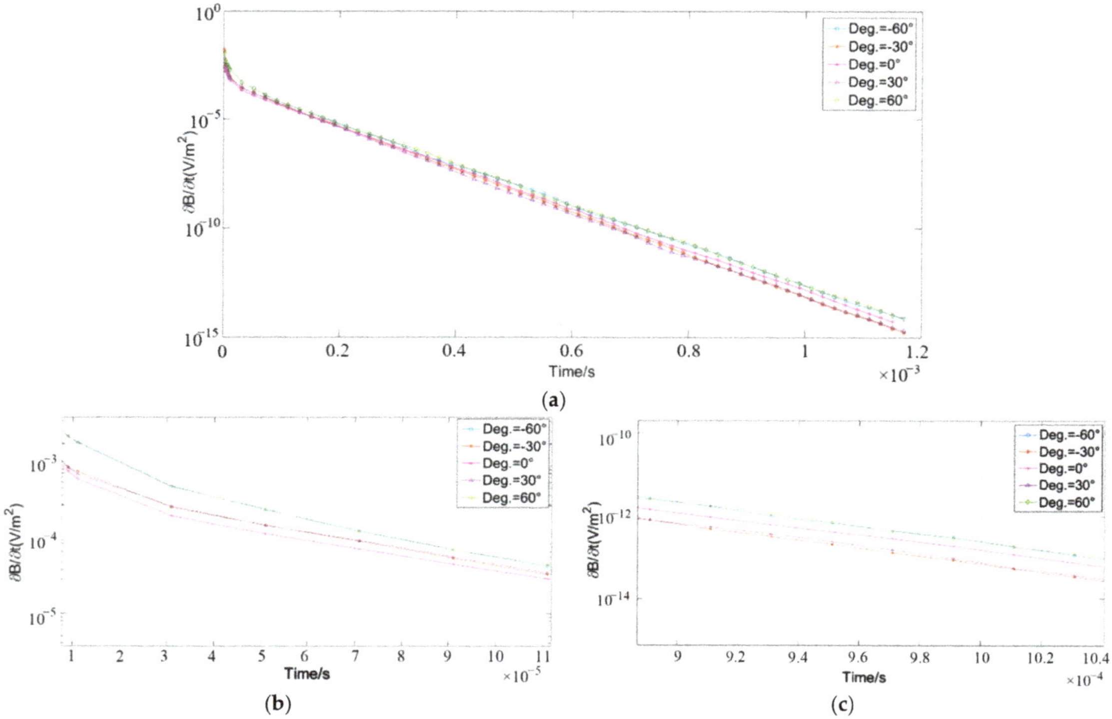

Figure 11. Induced electromotive force attenuation curves of different dip angle collapse columns. (**a**) Overall attenuation law. (**b**) Law of early attenuation (0–100 µs). (**c**) Law of late attenuation (900 µs–1 ms).

To examine the aforementioned theory further, based on the cube, sphere, and cylinder models in Section 3.1, two cone models with large differences in the projection area of the early and late electromagnetic fields are added, as shown in the following illustration. The conical inclination in Figure 12a is 90° and that in Figure 12b is −90° (see Table 4).

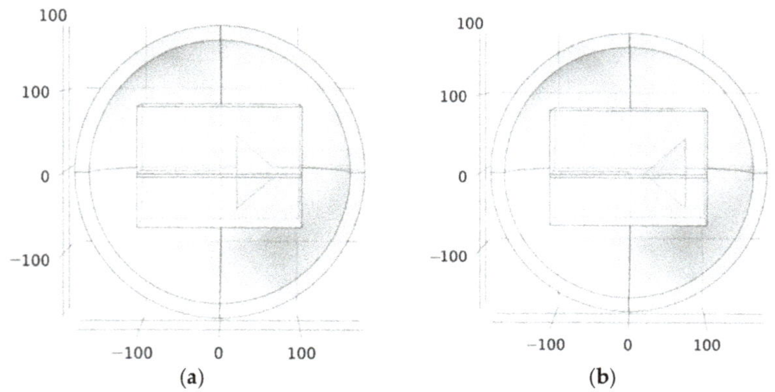

Figure 12. Models with cones of different angles. (**a**) Deg. = 90°. (**b**) Deg. = −90°.

Table 4. Geometric parameters of the four models.

	Edge Length/Radius (m)	Height (m)	Magnetic Field Projection Area (m^2)	Volume (m^3)
Sphere	30	/	Minimum value	113,097.3
Cylinder	26	52	2704	110,433.3
Cube	48	/	2304	110,592
Cone	45	52	Minimum value	110,269.9

Figure 13a–c shows a comparison of the induced electromotive attenuation law at the center of the coil, after each model is closed, and after the current is just switched off, the coil has not received the transient electromagnetic field from the conical model; the curve attenuation law is the same. In 3 μs after shutdown, the induced electromotive force amplitude on the coil is the same because the distance between each model and the coil is also the same. Thereafter, the induced electromotive potential of each model shows a large difference in terms of the attenuation speed. The 90° conical model's attenuation speed is lowest because it has the biggest electromagnetic field projection area at the start of the inductive electromagnetic field. Cubes and cylinders follow the same law. Early on, the sphere's electromagnetic field projection area is modest, which causes a higher attenuation speed. Later, when its electromagnetic field projection area grows and surpasses that of the cube and cylinder models (the greatest volume), its attenuation speed lowers. The decay rate of the −90° conical model is the fastest, and the decay rate decreases with the passage of time. When the induced electromagnetic field propagates to its geometric center of gravity, the decay rate is stable and consistent with the 90° conical model.

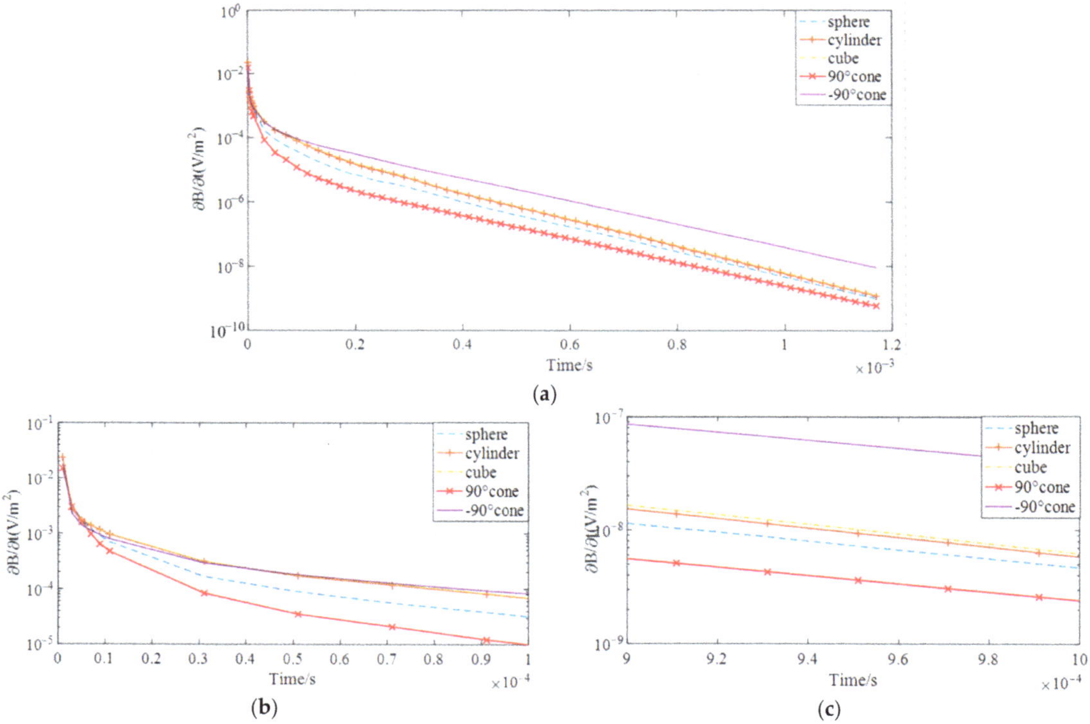

Figure 13. Induced electromotive attenuation curves for different projection area models. (**a**) Overall attenuation law. (**b**) Law of early attenuation (0–100 μs). (**c**) Law of late attenuation (900 μs–1 ms).

5. Field Application

The mine, which predominantly mines the Carboniferous Permian 6th coal seam, uses the inclined shaft development method and was designed to produce 1000 Mt per year. Coarse sandstone aquifer is overlaid on top of the mine's main coal seam's roof, and Ordovician limestone aquifer is buried beneath it. The underlying Ordovician ash confined aquifer places pressure on the 6th coal in the area, and the Ordovician ash water level in this coal mine is +869.7 m. Limestone water may inrush to mining sites through the vertical guide water structure and weak parts of the water-barrier layer when it meets faults and collapse columns as a result of the unequal water wealth of the Ordovician limestone aquifer and strong water wealth in karst development areas. There was water flowing from the roof, around 10 m behind the digging head, when underground production was underway on the site.

5.1. Arrangement

TEM is used to identify the water-rich anomalous area between 7.5 m and 107.5 m before the Z53 point at the design and construction location, which is located 7.5 m before the measurement point Z53 of the six main coal transportation lanes. The detecting region is located in the middle of the magenta section, as shown in Figure 14.

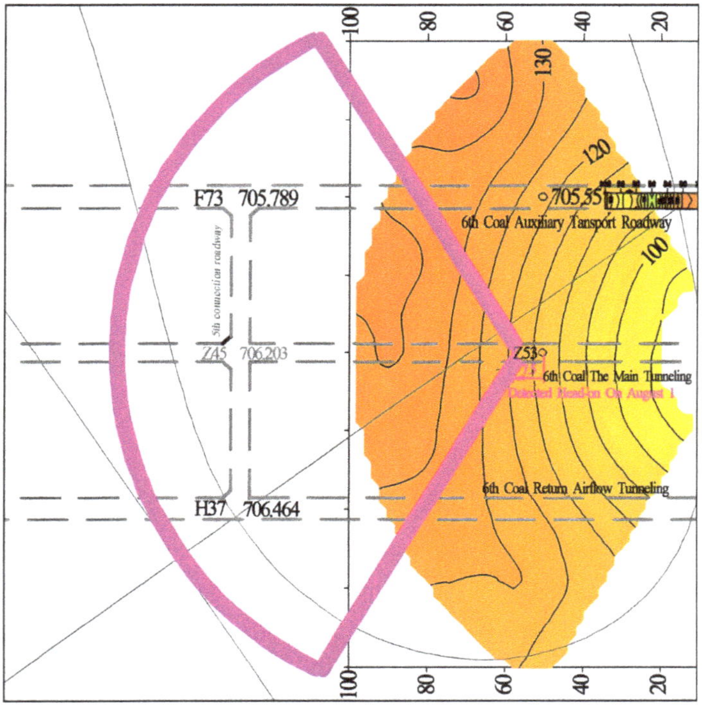

Figure 14. Schematic diagram of the advance detection area.

A multi-turn coil, with a side length of 2–3 m, is typically utilized for transmitting and receiving coils due to the space constraints of underground roadways. According to the results of the previous numerical simulation, the fan-shaped detection method is used to acquire all of the information regarding the front geological body, as illustrated in Figure 15. First, the measuring points are placed at the coal digging head, and the normal directions of the transmitting and receiving antennas are perpendicular to the right side of the roadway for measurement (such as vertical measurement, should be from bottom to top), next, the antenna will then be rotated at an angle (often 15°). When the antenna's normal direction is perpendicular to the digging head, 1–3 measuring sites are placed in accordance with the width of the digging head.

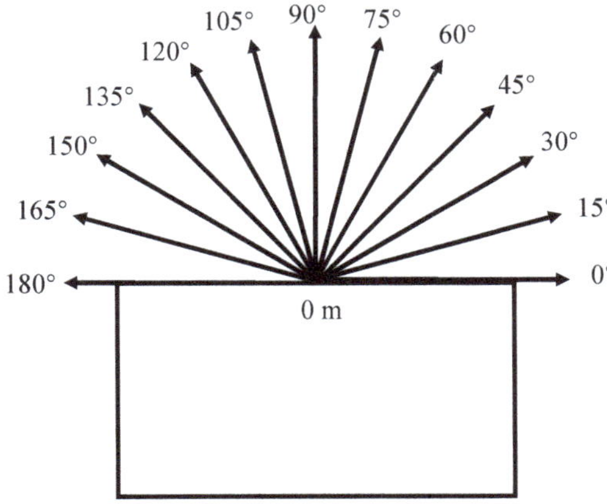

Figure 15. Schematic diagram of the arrangement of advanced detection points.

In the advanced detection of mine TEM, the normal direction of the starting ray frame is computed in accordance with the relative position relationship between coal mining and the aquifer of the roof and floor, as illustrated in Figure 16. This ensures that the rock strata to be examined are within the detection range of the instrument.

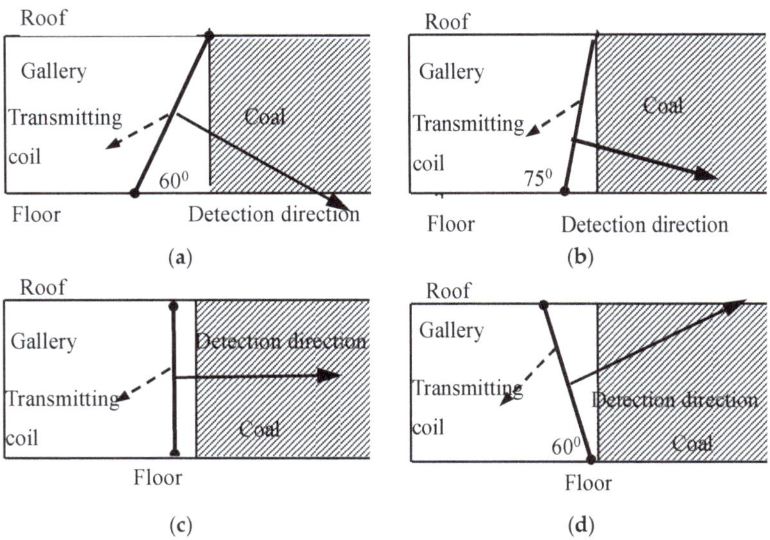

Figure 16. Schematic diagram of advanced detection direction. (**a**) The angle between the detection direction and the floor is 60°; (**b**) The angle between the detection direction and the floor is 75°; (**c**) The angle between the detection direction and the floor is 90°; (**d**) The angle between the detection direction and the floor is 120°.

5.2. Detection Results

Three levels of $\leq \overline{\delta} - \delta_n/3$, $\overline{\delta} - \delta_n/3 \sim \overline{\delta} + \delta_n/3$, and $\geq \overline{\delta} + \delta_n/3$—can be assigned to the apparent resistivity outlier δ, based on the electrical characteristics of this region (where $\overline{\delta}$ is the arithmetic mean value of parameters and δ_n is the standard deviation value of parameters); $\overline{\delta} - \delta_n/3$ and $\overline{\delta} - \delta_n$ are set to abnormal thresholds. Before the resulting map of advanced exploration and interpretation of the digging head is produced, the influence of the road's roof, floor, and geological body behind it should be removed in order to conduct out roadway space correction on the data that has been acquired.

The formation's comprehensive conductivity is better, and its water content is higher when the apparent conductivity value is lower. In the illustration, orange or colorlessness denotes a high resistivity anomaly (a weakly water-bearing or water-free area), green denotes a low resistivity anomaly (a water-rich area), and various intermediate colors denote the zone between low resistivity and high resistivity anomalies.

Figure 17 shows the advance detection results. The area with low apparent resistivity anomalies is represented by the blue-green region, the region with high apparent resistivity values is represented by the yellow region, and the remaining colors denote transitional regions. After processing the detection data in light of the outcomes of the numerical simulation, two clearly visible low-resistance anomaly zones are discovered within 100 m of the tunneling face. The No. 1 anomaly, which has a wide anomaly range and a moderate intensity, is primarily found in the right front of the digging head and extends to the left front of the bedding direction. The No. 2 anomaly is largely centered between 60 and 70 m, in front of the tunneling face on the left, with a modest and weak anomalous range. Additionally, there are secondary anomalies in the pitch angle direction in the detection edge area. It is assumed that the low resistance anomaly in front is the comparatively water-rich section in this area, given the occurrence of dripping from the roof and in the excavation face during detection. Combining the response law from the numerical simulation with the low-resistance anomaly geologic body's volume effect, it is possible that the No. 2 anomaly is created by the same hydrogeological anomaly geologic body that is responsible for the No. 1 anomaly, which stretches from the roof to the coal seam. The results obtained by drilling in the later stage show that the No. 1 anomaly is a low resistivity anomaly caused by the fluctuation of coal seam, and there are water flowing fractures in the No. 2 anomaly area.

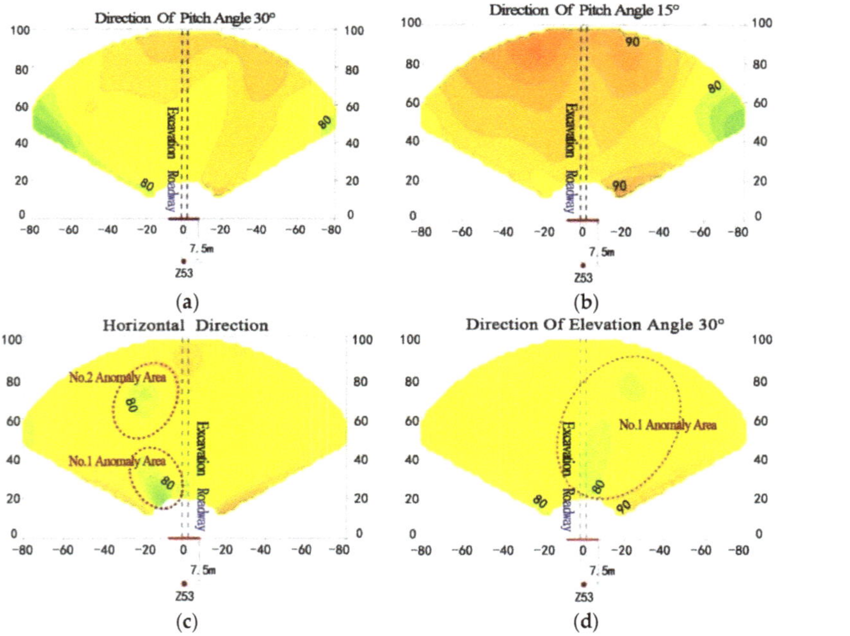

Figure 17. Advance detection results map. (**a**) The angle between the detection direction and the floor is 60°; (**b**) The angle between the detection direction and the floor is 75°; (**c**) The angle between the detection direction and the floor is 90°; (**d**) The angle between the detection direction and the floor is 120°.

6. Discussion

Some notable phenomena and representative laws are discovered through the examination of various response results:

(1) After the current is cut off in the third section, the electromagnetic field will peak in the bad geological body and slowly spread to the center of the fault, before continuing to degrade. This is consistent with the law of full-space transient electromagnetic response in the tunnel studied by Li et al. [31], which also provides feasibility for the subsequent exploration of the electromagnetic response law of the TEM. The cause of this phenomena is that the eddy current will start to decay as soon as the current is cut off; because the surrounding rock's conductivity is so much worse than that of the geological anomaly, the eddy current attenuation rate will be much higher than that of the low resistance body.

(2) The results of the different models, given in Section 4.1, indicate that the lower the apparent resistivity of the abnormal object (the higher the water content), the slower the attenuation of its inductive electromagnetic field. The results of Li and others [31] are also consistent with this claim: the higher the water content, the stronger the electromagnetic response signal and the slower the attenuation of the transient response.

(3) The comparison of the forward results in Section 3.1 and the different distance models in Section 4.2 shows that the distance between the abnormal geologic body and the excitation source controls the intensity of the transient electromagnetic field. The strength of the transient electromagnetic field decreases with distance; nevertheless, distance has no effect on the attenuation law of the water-containing body's transient electromagnetic field.

(4) According to the numerical simulation results in Section 4.3, the magnetic induction intensity is the same in the early stage for each model as the distance between the anomalous body and the transmitting coil is the same. At the later stage, magnetic induction intensity is proportional to volume and is ordered from large to small for the large circular table, large cylinder, circular table, small cylinder, and small circular table.

(5) It is clear from examining the responses in Section 4.4 that the attenuation law of the coil is significantly influenced by the electromagnetic field's projection area on the coil at a moment. Therefore, the idea of "electromagnetic field projection area" is put forth: the larger the electromagnetic field projection area, the slower the transient electromagnetic field decays at the corresponding moment; on the other hand, the smaller the electromagnetic field projection area, the faster the transient electromagnetic field decays at the corresponding moment. Additionally, it can be observed from the results in Sections 4.3 and 4.4 that the attenuation rate of the transient electromagnetic response of the anomalous geologic body with a regular shape has remained consistent throughout the attenuation process, as the electromagnetic field projection area reached its maximum value at the time of the occurrence of the transient electromagnetic field and did not vary in the subsequent electromagnetic field attenuation. The irregular anomalous geologic body's electromagnetic field projection area and transient electromagnetic response attenuation speed have both changed over time.

7. Conclusions

This study built geological models under numerous full-space situations using the time-domain finite unit approach, investigated the key factors impacting the transient electromagnetic response of anomalous bodies and the associated laws. The electrical characteristics and size have the greatest influence on the transient electromagnetic response for regular anomalous geological bodies. The transient electromagnetic response for irregular anomalous bodies is additionally impacted by the projection area of the electromagnetic field because of their unequal volume distribution. Finally, the fan-shaped arrangement of field detection and the processing and interpretation of data are guided by the conclusions drawn from numerical simulation and other geological data, and it is successfully determined that two low-resistance anomalies are developed from the same low-resistance anomaly geologic body.

The findings of this study demonstrate the suitability of the TDFEM for investigating complicated geological anomalies in three dimensions. Therefore, the geological conditions in the actual field are more complex and changeable, and in future research, specific numerical models should be established based on the TDFEM for research to complete the transient electromagnetic response law under three-dimensional full-space. Moreover, the forward evolution performed in this paper, utilizing the benefits of the finite element method, is fundamental to theoretical research, and it will be important to study the inversion under full-space conditions and its specific field application in the future.

Author Contributions: Data curation, Y.C.; Formal analysis, K.Z.; Investigation, L.M.; Software, S.C.; Writing—original draft, C.G.; Writing—review & editing, T.T. All authors have read and agreed to the published version of the manuscript.

Funding: This work was supported by the Natural Science Foundation of Jiangsu Province (No. BK20221130).

Conflicts of Interest: The authors declare no conflict of interest.

References

1. Kong, X.G.; He, D.; Liu, X.F.; Wang, E.Y.; Li, S.G.; Liu, T.; Ji, P.F.; Deng, D.Y.; Yang, S.R. Strain characteristics and energy dissipation laws of gas-bearing coal during impact fracture process. *Energy* **2022**, *242*, 12308. [CrossRef]
2. Wang, C.; Cheng, Y. Role of coal deformation energy in coal and gas outburst: A review. *Fuel* **2023**, *332*, 126019. [CrossRef]
3. Wang, X.; Wang, E.; Liu, X.; Zhou, X. Failure mechanism of fractured rock and associated acoustic behaviors under different loading rates. *Eng. Fract. Mech.* **2021**, *247*, 107674. [CrossRef]
4. Hu, W.; Zhao, C. Evolution of water hazard control technology in China's coal mines. *Mine Water Environ.* **2021**, *40*, 334–344. [CrossRef]
5. Zhang, Y.; Li, F. Prediction of Water Inrush from Coal Seam Floors Based on the Effective Barrier Thickness. *Mine Water Environ.* **2022**, *41*, 168–175. [CrossRef]
6. Liu, S.; Peng, G.; Yin, G. A study on the in-situ stress conditions at the Kailuan mining area in China and their influence on coal mine water inrush. *Arab. J. Geosci.* **2021**, *14*, 2057 . [CrossRef]
7. Kong, X.G.; Wang, E.Y.; Li, S.G.; Lin, H.F.; Xiao, P.; Zhang, K.Z. Fractals and Chaos Characteristics of Acoustic Emission Energy About Gas-Bearing Coal during Loaded Failure. *Fractals* **2019**, *27*, 1950069. [CrossRef]
8. Wang, X.; Asem, P.; Hu, C.; Labuz, J.F. Microcracking in tensile fracture of a brittle rock. *Eng. Fract. Mech.* **2021**, *251*, 107789. [CrossRef]
9. Tan, Y.L.; Liu, X.S.; Ning, J.G.; Lu, Y.W. In Situ Investigations on Failure Evolution of Overlying Strata Induced by Mining Multiple Coal Seams. *Geotech. Test. J.* **2017**, *40*, 244–257. [CrossRef]
10. Wang, F.T.; Tu, S.H.; Zhang, C.; Zhang, Y.W.; Bai, Q.S. Evolution mechanism of water-flowing zones and control technology for longwall mining in shallow coal seams beneath gully topography. *Environ. Earth Sci.* **2016**, *75*, 1309. [CrossRef]
11. Xue, G.Q.; Bai, C.Y.; Yan, S.; Greenhalgh, S.; Li, M.F.; Zhou, N.N. Deep sounding TEM investigation method based on a modified fixed central-loop system. *J. Appl. Geophys.* **2012**, *76*, 23–32. [CrossRef]
12. JiaWe, Z.; GuangDe, G. A geological interpretation method for non-uniform survey. *Int. J. Environ. Sci. Technol.* **2021**, *19*, 9047–9054. [CrossRef]
13. Chang, J.; Yu, J.; Li, J.; Xue, G.; Malekian, R.; Su, B. Diffusion Law of Whole-Space Transient Electromagnetic Field Generated by the Underground Magnetic Source and Its Application. *IEEE Access* **2019**, *7*, 63415–63425. [CrossRef]
14. Yan, S.; Xue, G.Q.; Qiu, W.Z.; Li, H.; Zhong, H.S. Feasibility of central loop TEM method for prospecting multilayer water-filled goaf. *Appl. Geophys.* **2016**, *13*, 587–597. [CrossRef]
15. Chang, J.H.; Yu, J.C.; Liu, Z.X. Three-dimensional numerical modeling of full-space transient electromagnetic responses of water in goaf. *Appl. Geophys.* **2016**, *13*, 539–552. [CrossRef]
16. Sun, H.; Liu, S.; Yang, Y. Crank-Nicolson FDTD 3D forward modeling for the transient electromagnetic field. *Chin. J. Geophys.* **2021**, *64*, 343–354.
17. Li, J.H.; Zhu, Z.Q.; Liu, S.C.; Zeng, S.H. 3D numerical simulation for the transient electromagnetic field excited by the central loop based on the vector finite-element method. *J. Geophys. Eng.* **2011**, *8*, 560–567. [CrossRef]
18. Liu, Y.; Hu, X.; Peng, R.; Yogeshwar, P. 3D forward modeling and analysis of the loop-source transient electromagnetic method based on the finite-volume method for an arbitrarily anisotropic medium. *Chin. J. Geophys.* **2019**, *62*, 1954–1968.
19. Chen, D.; Cheng, J.L.; Wang, A.M. Numerical simulation of drillhole transient electromagnetic response in mine roadway whole space using integral equation method. *Chin. J. Geophys. Chin. Ed.* **2018**, *61*, 4182–4193.

20. Zhang, Y.; Wang, G.; Li, H.; Lian, Y.; Li, W.; Qiu, H.; Mou, Y. Efficient 3D vector finite element modeling for TEM based on absorbing boundary condition. *Chin. J. Geophys.* **2021**, *64*, 1106–1118.
21. Wang, R.; Wang, M.Y.; Di, Q.Y.; Xue, G.Q. 2D FEM modeling on the multi-channel transient electromagnetic method. *Chin. J. Geophys. Chin. Ed.* **2018**, *61*, 5084–5095.
22. Lu, K.L.; Fan, Y.N.; Zhou, J.M.; Li, X.; Li, H.; Fan, K.R. 3D anisotropic TEM modeling with loop source using model reduction method. *J. Geophys. Eng.* **2022**, *19*, 403–417. [CrossRef]
23. Li, X.; Hu, W.; Xue, G. 3D modeling of multi-radiation source semi-airborne transient electromagnetic response. *Chin. J. Geophys.* **2021**, *64*, 716–723.
24. Wang, L.-Y.; Yin, C.-C.; Liu, Y.-H.; Su, Y.; Ren, X.-Y.; Hui, Z.-J.; Zhang, B.; Xiong, B. Three-dimensional forward modeling for the SBTEM method using an unstructured finite-element method. *Appl. Geophys.* **2021**, *18*, 101–116.
25. Kuo, J.T.; Cho, D.-H. Transient time-domain electromagnetics. *Geophysics* **1980**, *45*, 271–291. [CrossRef]
26. Velímský, J.; Martinec, Z. Time-domain, spherical harmonic-finite element approach to transient three-dimensional geomagnetic induction in a spherical heterogeneous Earth. *Geophys. J. Int.* **2005**, *161*, 81–101. [CrossRef]
27. Mur, G. The finite-element modeling of three-dimensional time-domain electromagnetic fields in strongly inhomogeneous media. *IEEE Trans. Magn.* **1992**, *28*, 1130–1133. [CrossRef]
28. Li, J.-H.; Zhu, Z.-Q.; Zeng, S.-H.; Liu, S.-C. Progress of forward computation in transient electromagnetic method. *Prog. Geophys.* **2012**, *4*, 1393–1400.
29. Haber, E. *Computational Methods in Geophysical Electromagnetics*; SIAM: Philadelphia, PA, USA, 2014.
30. Amestoy, P.R.; Guermouche, A.; L'Excellent, J.-Y.; Pralet, S. Hybrid scheduling for the parallel solution of linear systems. *Parallel Comput.* **2006**, *32*, 136–156. [CrossRef]
31. Li, H.; Yu, G.; Zhang, D. Electromagnetic Response Analysis of Tunnel Full Space 3D Forward Numerical Simulation. *Geotech. Geol. Eng.* **2022**, *40*, 273–284. [CrossRef]

sustainability

MDPI

Article

Particle Flow Analysis on Mechanical Characteristics of Rock with Two Pre-Existing Fissures

Zhenzi Yu [1,2], Ang Li [3,*], Bo Zhang [1,2], Hongyue Li [4], Qian Mu [5], Yonggen Zhou [3] and Shuai Gao [3]

1 State Key Laboratory of Coking Coal Exploitation and Comprehensive Utilization, China Pingmei Shenma Group, Pingdingshan 467099, China
2 Institute of Coal Mining and Utilization Pingdingshan Tianan Coal Mining Co., Ltd., Pingdingshan 467099, China
3 School of Architecture and Civil Engineering, Xi'an University of Science and Technology, Xi'an 710054, China
4 No.12 Mine, Pingdingshan Tian'an Coal Mining Co., Ltd., Pingdingshan 467000, China
5 CCTEG Chongqing Research Institute, Jiulongpo, Chongqing 400039, China
* Correspondence: angli@xust.edu.cn

Abstract: Many research results show that under any stress state the rock mass is most likely to crack, swell, bifurcate, and infiltrate from the fissure tip, resulting in rock engineering instability and failure. In order to study the influence of double fissure angles on rock mechanical characteristics, five rock numerical models with different fissure angles were established by numerical simulation software. Uniaxial compression tests were carried out, and the variation characteristics of rock stress, strain, failure, microcrack, and acoustic emission (AE) were recorded. The test results show that: With increases in the fissure angles, the elastic modulus of rock increased, while the peak strength decreased first and then increased. The number of microcracks in rock was greater at 15° and 75° than at other angles. The microcracks in rock were mainly tensile cracks, and relatively few were shear cracks. The angles of microcracks were mostly concentrated between 0 and 180°, most of which were between 60 and 110°. The failure of rock was relatively light when the fissure angle was15° or 75°, but it produced more and smaller fragments, and the failure was the most serious when the fissure angle was 30°. The angles of the fissures affected the maximum number of AE events, the strain values for the initial AE event, and the maximal AE event. This research can provide some reference for disasters caused by rocks with pre-existing fissures.

Keywords: pre-existing fissures; uniaxial compression; microcrack; failure characteristics; acoustic emission

Citation: Yu, Z.; Li, A.; Zhang, B.; Li, H.; Mu, Q.; Zhou, Y.; Gao, S. Particle Flow Analysis on Mechanical Characteristics of Rock with Two Pre-Existing Fissures. *Sustainability* **2022**, *14*, 14862. https://doi.org/10.3390/su142214862

Academic Editors: Xiangguo Kong, Dexing Li and Xiaoran Wang

Received: 21 July 2022
Accepted: 15 October 2022
Published: 10 November 2022

1. Introduction

Rock has become a geological body that must be dealt with in rock engineering, such as mining engineering, tunnel engineering, carbon dioxide geological storage, nuclear waste storage, and geothermal mining [1–5]. As a naturally formed geological body, rock mass is composed of various joint fissures and rock blocks cut by fissures. Many research results show that under any stress state rock mass is most likely to crack, expand, bifurcate, and penetrate from the fissure tips, resulting in instability and failure [6–8]. Therefore, it is of great significance to study the basic mechanical characteristics, failure mechanism, and mechanical model of fissures for the safety and stability of rock engineering.

Fissures include primary fissures and secondary fissures. Primary fissures are formed during diagenesis, and secondary fissures are formed by external forces after rock diagenesis [9]. Domestic scholars have conducted a lot of research on the influence of fissures on rocks, including large-scale in situ tests and laboratory-scale fissure research [6–8]. Among them, most of the fissure research in the laboratory involves taking rock samples with a height/diameter ratio of 2:1, according to the standards of the international society of rock mechanics, and then forming fissures through hydraulic cutting to study their influence on rock strength, deformation, and failure. Wong and Chau [10] studied the influence of fissure angles, bridge angles, and friction coefficients on the crack combination modes and strength of rock specimens with two parallel fissures and determined three typical crack combination modes. Wong and Einstein [11] studied the influence of fissure inclination and rock bridge angles on crack growth under uniaxial loading and found nine types of crack coalescence. Cao et al. [12] prepared rock samples with two fissures and carried out uniaxial compression tests to determine seven types of crack coalescence. Lee and Jean [13] conducted uniaxial compression tests on granite containing two nonparallel fissures (one horizontal and one inclined). The results showed that tensile cracks of granite are always accompanied by the germination of shear cracks. Zhou et al. [14] investigated the types of crack coalescence in specimens with multiple fissures and found another two crack types. These studies on the influence of one/two fissures on the mechanical properties and crack evolution of rock/rock-like material provide a good basis for studying the mechanical properties of rock containing two parallel fissures.

Laboratory test research can usually only obtain the shape of crack development and cannot explain its micromechanism well, so the theoretical and numerical analysis research of fissured rock has gradually developed and expanded. In addition, during numerical simulation, only one geometric feature of the fissure can be changed, while other factors remain unchanged, to study its influence on rock mechanical properties. In laboratory tests, even in the same rock or rock-like standard sample, its homogeneity will be different [15,16]. At present, the numerical methods for studying fissured rock can be summarized into three categories: the continuous medium analysis method, the discontinuous analysis method, and the mixed analysis method. Among them, the discrete element method proposed by Cundall [17] has unique advantages in simulating discontinuous medium materials such as rocks. The nonlinear deformation and crack development characteristics in a jointed rock mass can be simulated more realistically by the discrete element method, and the discrete element software PFC2D has the advantage of counting the direction and number of microcracks in the rock and can better master crack initiation and propagation. Zhang and Wong [18] used the two-dimensional discrete element program PFC2D to simulate the law of crack initiation, propagation, and through failure of a gypsum sample with pre-set fissures under uniaxial load and successfully used the same program to simulate the process of bridging and through failure of multiple pre-set fissures in the gypsum sample. Lin et al. [19] simulated a uniaxial compression test of a double-hole jointed rock mass through discrete element modeling and obtained its strength and failure characteristics. Jin et al. [20] analyzed the influence of pre-set fissures on the crack initiation and failure mode of a specimen under uniaxial compression from the perspective of energy. Based on the particle flow theory, Zhuang [21] studied the law of crack propagation and initiation in a fractured rock mass from the perspective of micromechanics using PFC2D and derived the formula of crack propagation and initiation using macrofracture mechanics and other theories. The above results show that PFC numerical simulation can be used to analyze the fracture behavior of rock or rock-like materials with fissures.

A lot of work has been carried out on the influence of single/double fissures on the mechanical properties and failure modes of rocks. However, the relationship between microcrack evolution and AE event behavior under the influence of various factors (including fissure angle) is still limited and needs further research. Therefore, on the basis of previous studies, this paper considers the simultaneous change in angle of double through fissures and studies its influence on rock strength, deformation, failure, and AE characteristics. At the same time, using the advantages of PFC2D numerical simulation software, we focus on the influence of double through fissures on the change in microcracks and try to explain the influence of fissure angle on rock mechanical characteristics from a micro perspective. The research results can better guide the construction and disaster control of rock engineering with pre-existing fissures.

2. Numerical Model Construction

2.1. Parallel Bonding Model

As a discrete element method software, PFC is suitable for studying the fracture and fracture development of particle aggregates. PFC attempts to explain the mechanical properties and behavior of media from a meso perspective, which has been widely used in rock engineering [17,17,22]. The bonding between particles is damaged by external effects, resulting in the separation of particles to simulate the generation and propagation of cracks in the medium. In the process of simulating particle bonding failure, the PFC program provides two basic particle bonding models: contact bonding and parallel bonding [17,17,22]. The particle models of the two kinds of bonding and their micromechanical behavior are shown in Figure 1. However, the contact bonding model uses point contact and cannot transmit torque, so the parallel bonding model is used more often in rock simulation. The normal stress and tangential stress on parallel bonding are expressed by the following formulas:

$$\begin{aligned} \bar{\sigma} &= \frac{-\bar{F}_i^{n}}{A_2} + \frac{\left|\bar{M}_i^{s}\right|\bar{R}}{I} \\ \bar{\tau} &= \frac{-\bar{F}_i^{s}}{A_2} + \frac{\left|\bar{M}_i^{n}\right|\bar{R}}{J} \end{aligned} \tag{1}$$

where A_2 is the area of the parallel bonding section, J is the polar moment of inertia of the section, and I is the moment of inertia of the section in the direction of rotation along the contact point. $\bar{R}$ is the bonding radius, and $\bar{F}_i^{n}$ and $\bar{F}_i^{s}$ are the normal and tangential components of the force after bonding. $\bar{M}_i^{n}$ and $\bar{M}_i^{s}$ are the normal and tangential components of the bending moment after bonding. When the normal or tangential stress exceeds the corresponding parallel bonding strength, the parallel bonding failure will produce tensile microcracks or shear microcracks, respectively.

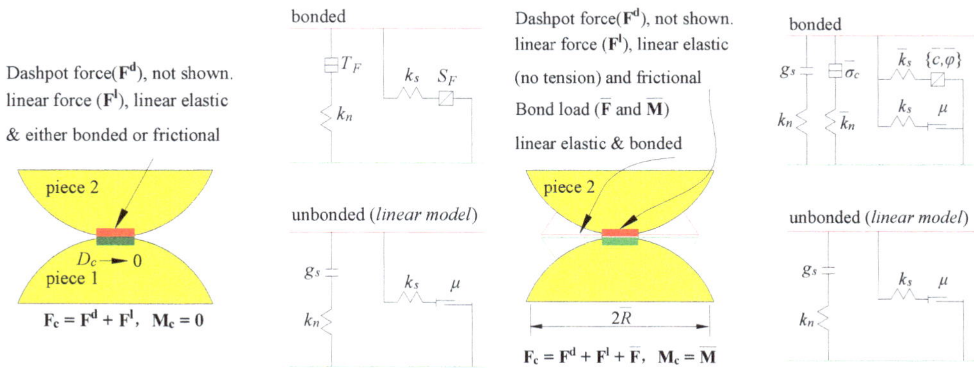

(**a**) (**b**)

Figure 1. Bonded-particle model and its micromechanical behavior in PFC numerical simulation [17,17,22]. (**a**) Contact bond model. (**b**) Parallel bond model.

2.2. Parameters of Numerical Rock Models

In PFC simulation, the macroscopic mechanical properties of a rock model are determined by the microscopic mechanical properties of particles and bonds. However, these microscopic parameters cannot be directly derived from field and laboratory tests. Before numerical simulations are performed, the selection and verification of microscopic parameters is required. Typically, the microscopic parameters of a PFC rock model are calibrated by simulating uniaxial compression experiments. During the calibration process, the microscopic parameters of the particles and bonds are adjusted many times by "trial and error" until these parameters can better reflect the mechanical properties of the real rock [17,17,22]. The load of the numerical model is applied by moving the upper wall at a loading rate of 0.05 mm/s. All conditions of the numerical test are the same as the laboratory test conditions. Through the "trial and error method" of repeated inspection and comparison, the physical and mechanical parameters listed in Table 1 can accurately reflect the macroscopic mechanical properties of real sandstone. The stress–strain curves and failure modes of the PFC model (Figure 2a) are in good agreement with the laboratory results of real sandstones. Based on the uniaxial compression laboratory tests of sandstone, three PFC numerical models were established, as shown in Figure 2b.

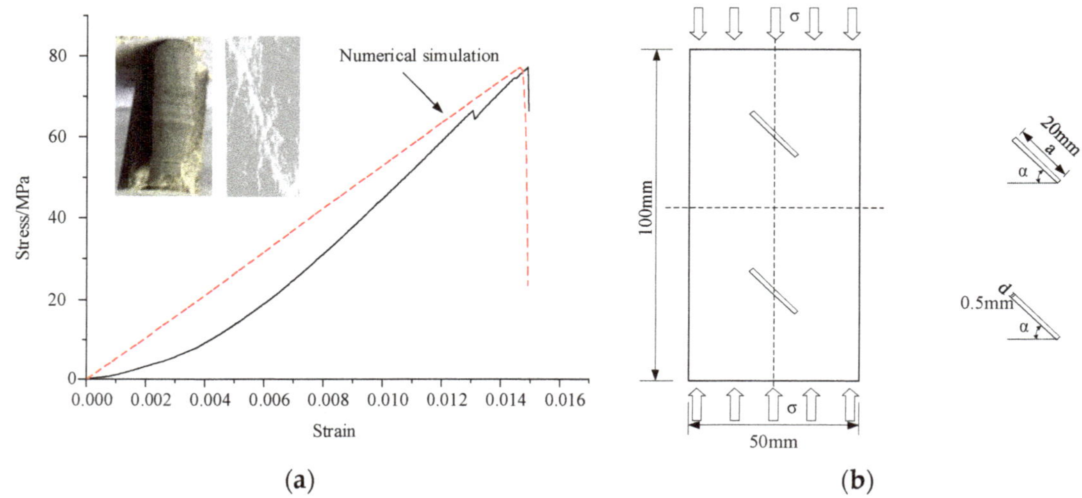

Figure 2. Stress–strain curves, failure modes, and numerical model in uniaxial compression experimental tests and numerical simulations. (**a**) Stress–strain curves and failure modes. (**b**) Numerical model based on PFC.

Table 1. Micromechanical parameters of rock in the numerical simulation.

Parameters	Value	Parameters	Value
Minimum particle size (R_{min})	0.2	Ratio of normal to tan gential bonding contact stiffness ($\bar{K}_n/\bar{K}_s$)	1.5
Ratio of maximum particle size to minimum particle size (R_{max}/R_{min})	1.5	Average and standard deviation of normal bond strength (σ_b)	16
Effective modulus of particles (E_c)	1.8	Mean and standard deviation of cohesive force (c_b)	20
Ratio of the contact stiffness between the normal direction and the tangential bond of particles (K_n/K_s)	1.5	Bond internal friction angle (ϕ)	42
Bond effective modulus ($\bar{E}$)	2.4	Linear friction coefficient of particles ($\bar{\mu}$)	0.5

2.3. Numerical Rock Models with Two Pre-Existing Fissures of Different Angles

In order to study the mechanical characteristics of rock with two pre-existing fissures of different angles, five numerical rock models with the same L (length) and d (width) values were built with different α (angle) values, as shown in Figure 3.

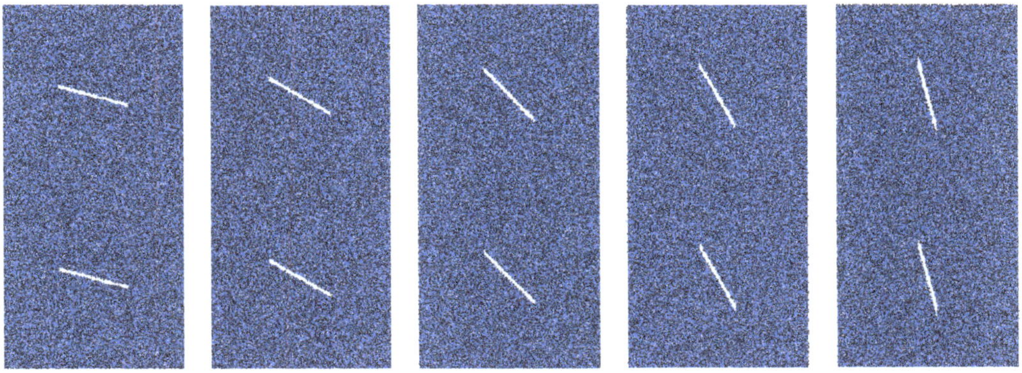

Figure 3. Numerical models and loading conditions of rock samples under uniaxial compression.

2.4. Acoustic Emission Simulation by PFC

Under the influence of external load, when the stress intensity transmitted between particles exceeds the bond strength between particles, bond fracture produces microcracks in rock samples [23]. When the microcracks propagate in a rock sample, the damage energy is rapidly released in the form of sound waves, that is, the phenomenon of acoustic emission (AE) [24,25]. Therefore, AE events can be simulated by calculating the number of particle bond breaks during numerical experiments. Due to the limitation of computing power, the particle size and particle number of PFC2D cannot directly reach the mechanical response level of real macroscopic rocks, but the reflected mechanical laws are helpful for understanding the AE phenomenon of rocks [26–28].

3. Numerical Simulation Results

3.1. Strength and Deformation Characteristics

Figure 4 shows the stress–strain curves, elastic modulus, peak strain, and peak strength of different fissure dip angles. It can be seen from Figure 4a that the variation trends of stress–strain curves of rocks with different dip angles were basically the same. The difference was that with the change in dip angles the changes in the elastic modulus, peak strain, and peak strength were different. With increases in fissure dip angles, the elastic modulus increased from 3.74 GPa at 15° to 4.91 GPa at 75°, an increase of 31.3% (Figure 4b). The peak strength first decreased and then increased with increases in angle, from 45.25 MPa at a 15° angle to 41.66 MPa at a 30° angle, and then increased to 53.57 MPa at 75° (Figure 4c). The peak strain had no fixed law with an increase in angle, but it increased gradually overall (Figure 4d).

3.2. Microcrack Evolution Characteristics

The change trend of the number of cracks in rock at different fissure angles is shown in Figure 5a. The change in crack number and the change in the stress–strain curves are shown in Figure 5b (with limited space, only the case with an angle of 45° is shown). The change in microcracks with strain can be roughly divided into three stages, as shown in Figure 5b. When the stress is small, there are basically no microcracks. Then, microcracks gradually occur with the increase in stress, and a large number of rock fracture microcracks occur at the peak strength of the rock. The number of microcracks at the peak strength is not the most in the whole process, but it is a turning point with a large number of microcracks.

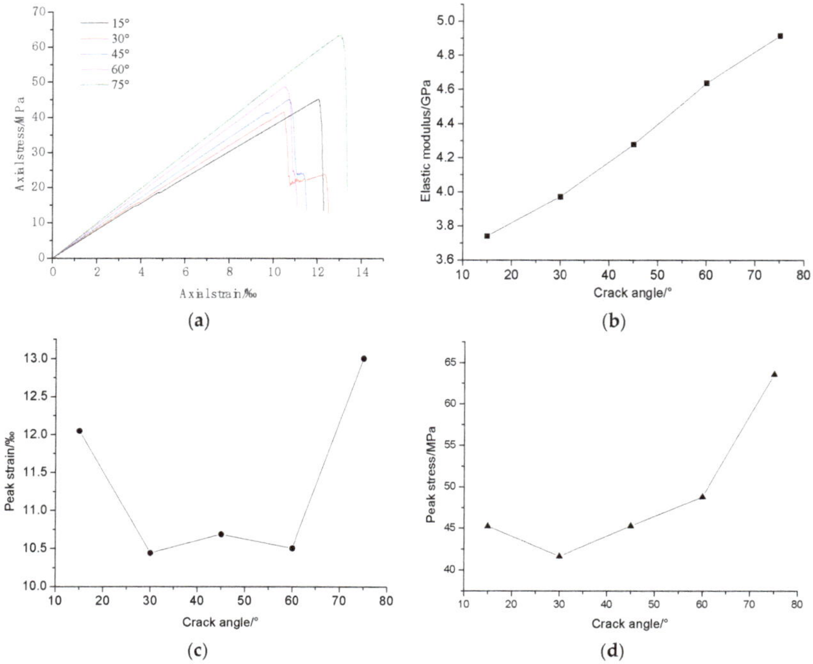

Figure 4. Strength and deformation characteristics with different fissure angles under uniaxial compression. (**a**) Stress–strain curves. (**b**) Elastic modulus. (**c**) Peak strain. (**d**) Peak strength.

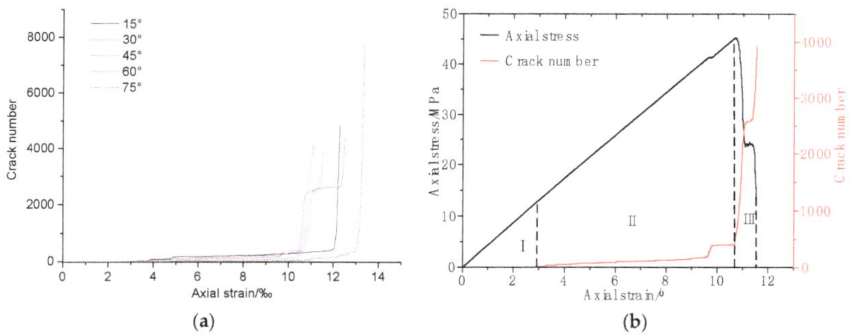

Figure 5. Microcrack evolution with axial strain at different fissure angles under uniaxial compression. (**a**) Microcrack evolution with axial strain. (**b**) Microcrack evolution at 45°.

In the process of uniaxial compression, the total number of microcracks and the number of cracks at the peak strength are shown in Figure 6. The total number of microcracks decreased first and then increased with increases in the fissure angles. The minimum number was 3926 when the angle was 45°. Compared with 4832 cracks at the angle of 15°, the number of cracks decreased by 18.8%, and when the angle increased to 75°, the number of cracks was 7851, which was an increase of 100% compared with 45°. However, the number of microcracks at the peak strength had no fixed law with the change in angle.

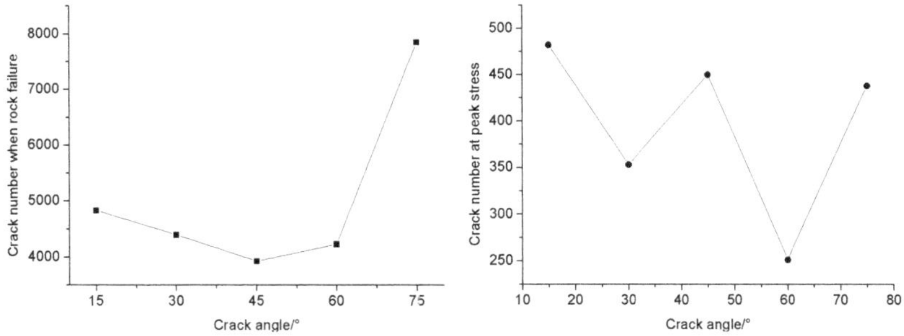

Figure 6. The total number of microcracks and the number of cracks at the peak strength under uniaxial compression.

In order to further study the evolution characteristics of microcracks. Figure 7 shows rose diagrams of rock microcracks at different angles of fissures. The angles of microcrack generation were concentrated between 0 and180°, most of which were between 60 and 110°. With increases in the angle, the 90° microcracks decreased first and then increased, from 830 at the fissure angle of 30° to 693 at the angle of 45° and then to 1419 at the angle of 75°. That is, when the fissure angle was 15° or 75°, more 90° microcracks were generated.

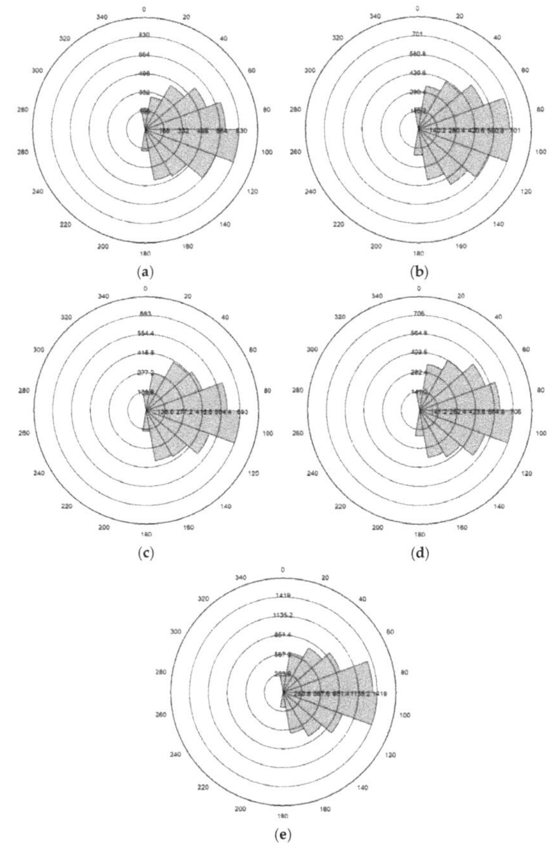

Figure 7. Rose diagrams of rock microcracks with different angles of pre-existing fissures under uniaxial compression: (**a**) 15°, (**b**) 30°, (**c**) 45°, (**d**) 60°, (**e**) 75°.

Figures 8 and 9, respectively, show the distribution of tensile cracks and shear cracks during rock failure at different dip angles. It can be seen that the microcracks of rock during uniaxial compression were mainly tensile cracks. Tensile cracks did not occur only at the fissure tip. At the distance of the fissure tip, tensile crack lines also occurred. Shear cracks mostly occurred at the fissure tip. In addition, both tensile and shear cracks increased with increases in angle.

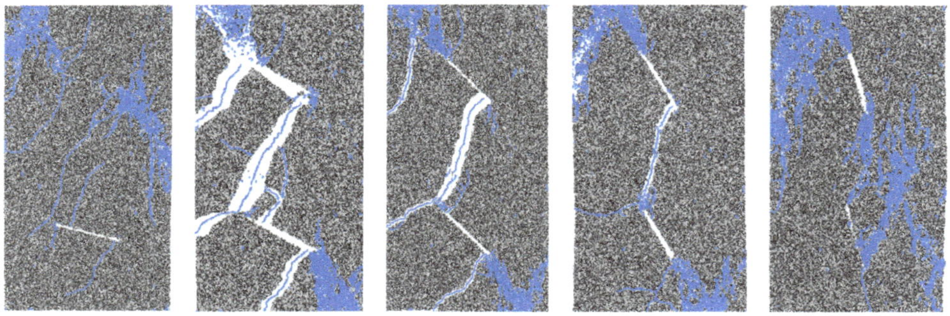

(**a**) (**b**) (**c**) (**d**) (**e**)

Figure 8. The distribution of tensile cracks during rock failure with different fissure angles under uniaxial compression: (**a**) 15°, (**b**) 30°, (**c**) 45°, (**d**) 60°, (**e**) 75°.

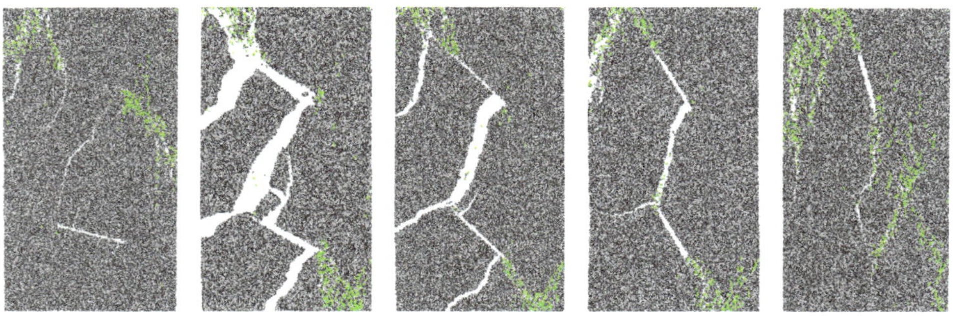

(**a**) (**b**) (**c**) (**d**) (**e**)

Figure 9. The distribution of shear cracks during rock failure with different fissure angles under uniaxial compression: (**a**) 15°, (**b**) 30°, (**c**) 45°, (**d**) 60°, (**e**) 75°.

3.3. Failure Modes

The fissure dip angle has a great influence on the failure modes of rock. As can be seen from Figure 10, when the angles are 15° and 75° the rock fracture was relatively light, but the failure produced more fragments. It can also be explained from Figure 5 that when the angle was 15° or 75° there were more microcracks in the rock than at other angles as well as more cracks and more energy dissipation. Therefore, during rock failure, although the rock fracture was not violent, the fragments produced were relatively small and greater in number. When the angle was 30°, the rock was broken most seriously. The fracture not only ran between the two fissures but also ran through the whole rock due to the failure of the cracks. The failure modes of 45° and 60° were basically the same as those at 30°, but the degree of fracture was smaller than at 30°.

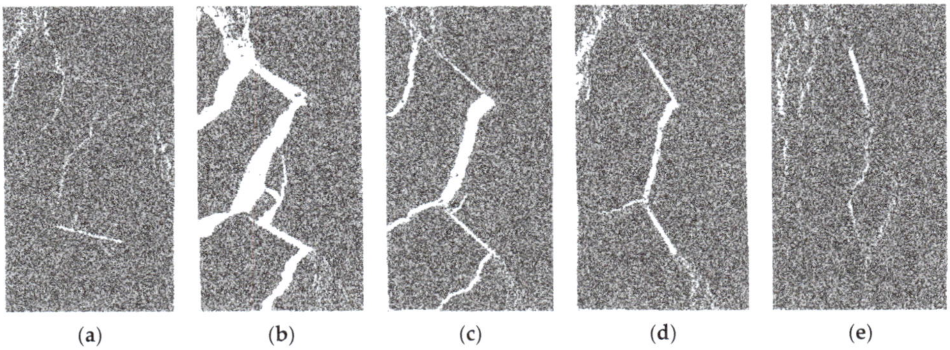

Figure 10. The failure modes of rock with different fissure angles under uniaxial compression: (**a**) 15°, (**b**) 30°, (**c**) 45°, (**d**) 60°, (**e**) 75°.

3.4. Acoustic Emission Characteristics

Figure 11 shows the curves of the stress–strain AE events of rocks with different α values. It can be seen from the figures that the evolution characteristics of the AE events were closely related to the stress–strain relationship. Before the peak intensity of the stress–strain curve, the number of AE events was very small. When the stress–strain behavior was in the failure stage, the number of AE events peaked and declined rapidly, which means that the rock with two pre-existing fissures was severely damaged in this stage.

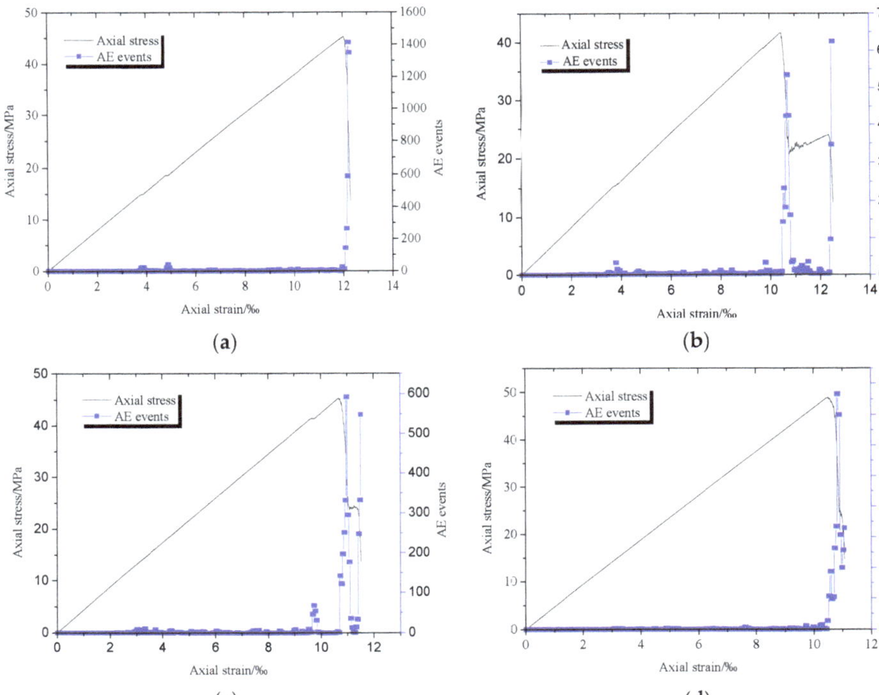

Figure 11. *Cont.*

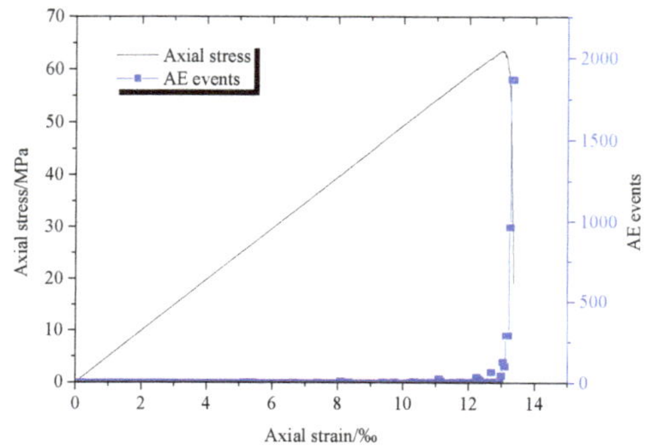

(e)

Figure 11. Stress–strain AE event curves of numerical rocks with different angles under uniaxial compression: (**a**) 15°, (**b**) 30°, (**c**) 45°, (**d**) 60°, (**e**) 75°.

The effects of different α values on the AE characteristics of rocks are described as follows.

(1) The angle of the fissures affected the maximum number of AE events. As the rock fracture angle increased from 15° to 75°, the maximum AE events decreased first and then increased. When the angle increased from 15° to 45°, the maximum number of AE events decreased from 1453 to 602. The maximum number of AE events decreased by 58.6%. When the angle increased from 45° to 75°, the maximum number of AE events increased from 602 to 1896. The maximum number of AE events increased by 68.2. The evolution characteristics of the maximum number of AE events with different α values were similar to the changes in the peak strength of rocks with two pre-existing fissures.
(2) It affected the strain values for the initial AE event and the maximal AE event. As α increased from 15° to 75°, the strain values of the initial AE event were 3.84‰, 2.46‰, 2.23‰, 2.24‰, and 7.37‰ and the strain values of the maximum AE event were 12.12‰, 12.41‰, 10.93‰, 10.86‰, and 11.51‰.
(3) It affected the strain range for severe AE events near the peak strength. The strain range of severe AE events near the peak strength increased first and then decreased with an increase in α. When the angle was 30°, the strain influence range was the largest: 10.51 to 12.42. At the angles of 15° and 75°, the strain influence range was the smallest, and the strain rapidly decreased and disappeared when the peak strength failure was reached.

4. Discussion

In this paper, the influence of fissure angles on rock strength, deformation, failure, microcrack evolution, and AE characteristics was studied by using the discrete element simulation software PFC. The elastic modulus increased with an increase in the fissure angle. The peak strength decreased first and then increased. As the rock fracture angle increased from 15° to 75°, the maximum AE events decreased first and then increased. These change characteristics were basically consistent with previous studies [5,10,14]. The difference is that the rose diagrams of the microcrack distributions were obtained in this paper. They can accurately describe the failure directions of microcracks and the number of microcracks in that direction. In the control of engineering the fracture and failure of rock with pre-existing fissures, they can be controlled according to the propagation direction of a large number of microcracks. In this way, the waste of manpower and material resources caused by the control technology, such as grouting or bolting due to a lack of a clear control direction, can be avoided.

However, the research of this paper also has some limitations. In numerical simulation, rock is regarded as an isotropic material. In addition, when other geometric dimensions of the fissures and loading conditions change, the strength, deformation, microcrack evolution, and other characteristics of the rock change. In addition, the information to employ the rose diagrams of microcrack distribution effectively, such as an accuracy analysis, is not provided. These will be the focus of our next research work.

5. Conclusions

(1) With increases in double fissure angles, the elastic modulus of rock increased gradually. The peak strength of rock decreased first and then increased. The peak strain had no fixed law with the increase in angle, but it increased gradually overall.

(2) There were more microcracks when the angle was 15° or 75°. The most microcracks were produced when the fissure angle was 75°, and the least microcracks were produced when the fissure angle was 45°. No matter the fissure angles, the microcracks of rock under uniaxial compression were mainly tensile cracks, with relatively few shear cracks. The angles of the microcracks were concentrated between 0 and 180°, the majority of which were between 60 and 110°.

(3) When the angle was 15° or 75°, the fracture degree of rock was relatively light, but the fragments produced by failure were greater in number and smaller. When the angle was 30°, the fracture degree was the most serious, with fewer and larger fragments.

(4) The angles of the fissures affected the maximum number of AE events, the strain values for the initial AE event, and the maximal AE event. As the rock fracture angle increased from 15° to 75°, the maximum AE events decreased first and then increased. As α increased from 15° to 75°, the strain values of the initial AE event were 3.84‰, 2.46‰, 2.23‰, 2.24‰, and 7.37‰ and the strain values of the maximum AE event were 12.12‰, 12.41‰, 10.93‰, 10.86‰, and 11.51‰.

This study did not consider the heterogeneity of the rock and other geometric characteristics of fissures. However, these conclusions have certain reference significance for the disaster control of rock engineering with fissures. The information to employ the rose diagrams of microcrack distribution effectively, such as an accuracy analysis, is not provided. In the next step, we will conduct more in-depth research on these limitations.

Author Contributions: Formal analysis, S.G.; Funding acquisition, B.Z.; Investigation, Q.M.; Methodology, Y.Z.; Software, H.L.; Writing—original draft, Z.Y.; Writing—review & editing, A.L. All authors have read and agreed to the published version of the manuscript.

Funding: This research was funded by the National Natural Science Foundation of China (No. 51874229) and the Natural Science Foundation of Shaanxi Province (2020JZ-52).

Informed Consent Statement: Not applicable.

Data Availability Statement: Not applicable.

Acknowledgments: The authors would like to gratefully acknowledge the reviewers that provided helpful comments and insightful suggestions on a draft of the manuscript.

Conflicts of Interest: The authors declare no conflict of interest.

References

1. Zhou, H.; Meng, F.; Zhang, C.; Lu, J.; Xu, R. Experimental study on effect of joints with different locations and sizes on rock failure. *Chin. J. Rock Mech. Eng.* **2015**, *34*, 3018–3028.
2. Ling, T.; Zhang, L.; Liu, H.; Gu, D.; Zhang, S. Omws method for weak signal processing of gpr and its application in the identification of concrete microcracks. *J. Environ. Eng. Geophys.* **2019**, *24*, 317–325. [CrossRef]
3. Öner, E.; Şabano, B.Ş.; Yaylacı, E.U.; Adıyaman, G.; Yaylacı, M.; Birinci, A. On the plane receding contact between two functionally graded layers using computational, finite element and artificial neural network methods. *J. Appl. Math. Mech.* **2022**, *102*, e202100287. [CrossRef]
4. Zhang, Y.; Liu, X.; Man, K. A semi-analytical analysis on rock thermal stress for nuclear waste repository near field. *Chin. J. Undergr. Space Eng.* **2018**, *14*, 741–747.
5. Bai, S.W.; Ren, W.Z.; Feng, D.X.; Lu, J.Q. Failure mechanism and strength properties of rock mass containing close intermittent joints under plane stress condition. *Chin. J. Rock Mech. Eng.* **1999**, *18*, 635–640.

6. Murat, Y. Simulate of edge and an internal crack problem and estimation of stress intensity factor through finite element method. *Adv. Nano Res.* **2022**, *12*, 405–414.
7. Li, D.; Wang, E.; Li, N.; Kong, X.; Wang, X. Research on the coal characteristics of macro-crack dip angles under uniaxial compression. *Chin. J. Rock Mech. Eng.* **2017**, *36*, 3206–3213.
8. Yin, Q.; Jing, H.; Su, H.; Zhao, H. Strength degradation and loading rate effect of sandstone containing a longitudinal fissure. *J. Min. Saf. Eng.* **2016**, *33*, 128–133.
9. Lin, Q.B.; Cao, P.; Wen, G.P.; Meng, J.J.; Cao, R.H.; Zhao, Z.Y. Crack coalescence in rock-like specimens with two dissimilar layers an pre-existing double parallel joints under uniaxial compression. *Int. J. Rock Mech. Min. Sci.* **2021**, *139*, 104621. [CrossRef]
10. Wong, R.H.C.; Chau, K.T. Crack coalescence in a rock-like material containing two cracks. *Int. J. Rock Mech. Min. Sci.* **1998**, *35*, 147–164. [CrossRef]
11. Wong, L.N.Y.; Einstein, H.H. Crack coalescence in molded gypsum and Carrara marble: Part 1. Macroscopic observations and interpretation. *Rock Mech. Rock Eng.* **2009**, *42*, 475–511. [CrossRef]
12. Cao, P.; Liu, T.; Pu, C.; Lin, H. Crack propagation and coalescence of brittle rock-like specimens with pre-existing cracks in compression. *Eng. Geol.* **2015**, *187*, 113–121. [CrossRef]
13. Lee, H.; Jeon, S. An experimental and numerical study of fracture coalescence in pre-cracked specimens under uniaxial compression. *Int. J. Solids Struct.* **2011**, *48*, 979–999. [CrossRef]
14. Zhou, X.; Cheng, H.; Feng, Y. An experimental study of crack coalescence behaviour in rock-like materials containing multiple flaws under uniaxial compression. *Rock Mech. Rock Eng.* **2014**, *47*, 1961–1986. [CrossRef]
15. You, M.Q. Effects of confining pressure on strength scattering of rock specimen. *Chin. J. Rock Mech. Eng.* **2014**, *33*, 929–937.
16. Ma, Q.; Tan, Y.L.; Liu, X.S.; Gu, Q.H.; Li, X.B. Effect of coal thicknesses on energy evolution characteristics of roof rock-coal-floor rock sandwich composite structure and its damage constitutive model. *Compos. Part B Eng.* **2020**, *198*, 108086. [CrossRef]
17. Cundall, P.A.; Strack, O.D.L. A discrete numerical model for granular assemblies. *Geotechnique* **1979**, *29*, 47–65. [CrossRef]
18. Zhang, X.; Wong, L.N.Y. Cracking processes in rock-like material containing a single flaw under uniaxial compression: A numerical study based on parallel bonded-particle model approach. *Rock Mech. Rock Eng.* **2011**, *45*, 711–737. [CrossRef]
19. Lin, Q.B.; Cao, P.; Meng, J.J.; Cao, R.H.; Zhao, Z.Y. Strength and failure characteristics of jointed rock mass with double circular holes under uniaxial compression. *Theor. Appl. Fract. Mech.* **2020**, *109*, 102692. [CrossRef]
20. Jin, J.; Cao, P.; Chen, Y.; Pu, C.Z.; Mao, D.W.; Fan, X. Influence of single flaw on the failure process and energy mechanics of rock-like material. *Comput. Geotech.* **2017**, *86*, 150–162. [CrossRef]
21. Zhuang, N. *Propagation Rule of Crack Rock Mass and Its Mathematics Model Research under the Seepage-Stress Coupling*; Tongji University: Shanghai, China, 2006.
22. Potyondy, D.O.; Cundall, P.A. A bonded-particle model for rock. *Int. J. Rock. Mech. Min. Sci.* **2004**, *41*, 1329–1364. [CrossRef]
23. Wu, S.C.; Zhou, Y.; Gao, B. Study of unloading tests of rock burst and pfc3d numerical simulation. *Chin. J. Rock Mech. Eng.* **2010**, *29*, 4082–4088.
24. Wang, L. Effect of initial flaw on tensile properties of rock. *Coal Sci. Technol.* **2020**, *48*, 66–70.
25. Hazzard, J.F.; Young, R.P.; Maxwell, S.C. Micromechanical modeling of cracking and failure in brittle rocks. *J. Geophys. Res.* **2000**, *105*, 16683–16697. [CrossRef]
26. Wen, Z.; Wang, X.; Chen, L.; Lin, G.; Zhang, H. Size effect on acoustic emission characteristics of coal-rock damage evolution. *Adv. Mater. Sci. Eng.* **2017**, *2017*, 3472485. [CrossRef]
27. Liu, W.; Yin, Z.; Yuan, A.; Li, C. Particle-discrete-element method based research on acoustic emission characteristics and energy evolution laws of surrounding rock in different lithologic roadway. *J. Min. Saf.* **2017**, *34*, 363–370. (In Chinese)
28. Zhang, Q.; Wang, X.; Tian, L.G.; Huang, D.M. Analysis of mechanical and acoustic emission characteristics of rock materials with double-hole defects based on particle flow code. *Shock. Vib.* **2018**, *2018*, 7065029. [CrossRef]

sustainability

MDPI

Article

Characteristics of Mining-Induced Slope Movement and Ground Behavior under Gully Landforms

Shaojie Ma [1] and Yueming Kang [2,*]

[1] Chongqing Vocational Institute of Engineering, Chongqing 402260, China
[2] China Coal Technology Engineering Group Chongqing Research Institute, Chongqing 400039, China
* Correspondence: kangyueming01@163.com

Abstract: Exploring the phenomenon of surface cracks and the abnormal phenomenon of ground behavior during coal mining under gully landforms, laboratory physical similarity simulation models were established to study the movement characteristics and ground behavior laws of working faces of different mining methods under gully landforms. The results indicate that in cases when a roadway is situated below the top of the slope, the corresponding deformation of the surrounding rock of the roadway is aggravated by the mining stress. Furthermore, when a roadway is located below the base of the gully, it is less affected by mining; thus, it could remain relatively stable. The ground behavior laws of working faces in gully geomorphology areas are associated with the position of the working face and the direction of working: when mining towards the gully, the ground behavior appeared relatively gentle and the surface slipped towards the gully; when mining away from the gully, the ground behavior appeared intense, the pressure was sudden and short, and with the increase in overburden thickness, the interval distance periodically decreased. When a working face passes through a gully, dumping of the hydraulic support should be prevented in the section of mining towards the gully; in sections of mining away from gully, the support strength should be strengthened to prevent the support and other equipment from being crushed. In actual mining, mining towards a gully should be adopted as much as possible in the stoping of the working face. In this way, the ground behavior is gentle, the interval distance periodical weight is longer, and the advance abutment pressure is small.

Keywords: coal mine; gully landform; movement mechanism; ground behavior; mining method valley landform; simulation of similar materials

Citation: Ma, S.; Kang, Y. Characteristics of Mining-Induced Slope Movement and Ground Behavior under Gully Landforms. *Sustainability* **2022**, *14*, 13941. https://doi.org/10.3390/su142113941

Academic Editor: Marc A. Rosen

Received: 22 August 2022
Accepted: 24 October 2022
Published: 26 October 2022

1. Introduction

With the reduction in coal resources in the east, China's coal mining has gradually shifted to the west. Taking Shanxi Province as an example, coalfields are distributed in 40% of the land in the region. Due to the large number and thickness of coal seams, as well as the excellent coal quality [1], Shanxi Province has attracted worldwide attention and has become an energy base for national economic development [2,3]. Most of the coal fields in Shanxi are medium-thick coal seams with shallow burial and good mining conditions. With the advancement of mining technology and equipment, most coal mines utilize fully mechanized caving. Many scholars have conducted in-depth research on the characteristics of overlying strata movement, surface motion characteristics, and ore pressure behavior under fully mechanized caving mining conditions [4,5].

Some researchers have analyzed the coal seam overlying strata movement and strata behavior regularity of coal seam under fully mechanized caving mining conditions by numerical simulation and similar material simulation. For example, the deformation and failure of surrounding rocks and stress transfer at different roof thicknesses was analyzed using three-dimensional modeling software through a combination of numerical modeling and analogue modeling [6,7]. A macro stress shell was developed for high-stress bunches in the surrounding rocks [8,9], and a physical model analyzed the influence of barrier pillars on overburden failure at the working face [10]. Patterns of overburden movement at a working face with a large mining height were determined using numerical simulations [11,12], and the deformation and failure mechanisms of front top coal were used to propose the caving number [13]. Furthermore, a dynamic balance arch structure was formed using the roof and surrounding rock in the working face [14] and the relationship of roof fall and rib spalling with support parameters. The structure of an immediate roof studied in mechanical equilibrium revealed the formation mechanism [15]; some researchers conducted the on-site monitoring of mine pressure on a fully mechanized caving face [16,17] and analyzed the mine pressure data and the laws of overlying strata under different conditions [18,19].

In mining-induced surface deformation movements [20,21], there were several notable studies related to the surface slope movement induced by underground mining [22,23], conducted in two scenarios [24,25]: underground mining in a mountain area, and the simultaneous operations of underground mining and surface mining [26,27].Due to surface subsidence caused by underground mining, it is possible to induce landslides under certain geological [28,29], topographical, and natural conditions. These are referred to as mining landslides [30,31]. The reasons for the occurrence of mining landslides are as follows: surface cracks in the goaf, mountain landforms, geological structures, atmospheric precipitation or snow melting, work surface layout, etc.

Most contemporary studies have focused on the influence of underground mining on surface movement, overburden movement, and mine pressure characteristics [32]. However, there are few studies on the influence of geomorphic features of surface undulations on underground mining [33]. It is generally believed that it is easy to mine in medium- and shallow-buried coal fields; thus, the influences of surface fluctuations on mine pressure are ignored [34]. Some data and experiments have shown that the ore pressure under uneven landforms appears to be sensitive and more intense under certain conditions, which makes the mine pressure appear more prominent [35].

Coalfields in western China exhibit a small depth of occurrence and diverse topography. Compared with southwest coalfields, the surface vegetation of northwest coalfields is sparse, the soil and water conservation ability is weak, the surface water erosion is serious, the gullies are vertical and horizontal, the slope body changes greatly, the terrain is fragmented, and the relative height difference is more than 100 m. The typical landform of the area is shown in Figure 1. The influence of landforms on underground mining is clear and sometimes even plays a leading role. The phenomena of sudden, strong ground pressure during the mining of some working faces, large-scale caving due to overburdening, high opening rates of safety valves of hydraulic supports, long-term instability of goafs, and severe pressure in a certain section of roadway are difficult to explain with conventional rock pressure theory.

Figure 1. A typical gully landform.

Here, we studied the influence of surface mountain occurrence on the mining surface pressure of a working face and analyzed the rock movement characteristics and other mining pressure parameters of the working face under the gully landform.

2. Background of Case Study

The Wangjialing Minefield is located in the southwestern part of the Hening Mining Area in Hedong Coalfield, Shanxi Province. It is characterized by clear geomorphology in the western part of China, situated in the southern foothills of the Lvliang Mountains. It is a highly eroded high and medium mountainous area with complex topography and gullies. The bedrock is exposed on the hillside, and the small slopes in the beam and the reclamation area are steep. The slope angle is between 25° and 40°, and most of them are "V" shaped valleys. Most of the strata in the area are covered by the Quaternary loess layer. The mine field ranges northeast to southwest, 10~22 km, tends to the northwest, is 5~10 km wide, and has an area of approximately 194 km^2. Among them, the intensive area is 92.4 km^2, and the detailed investigation area is 101.62 km^2.

The 20,109 working face is located in the west of the central roadway of mining area I, the south side is adjacent to the goaf area of the 20,107 working face, and the north side is the village protecting the coal pillar. The working face is arranged in the east–west direction and the return airway adopts gob-side entry driving, establishing an 8 m coal pillar roadway in the goaf of the 20,107 working face. The total length of the roadway is 1466 m. The roadway layout is shown in Figure 2.

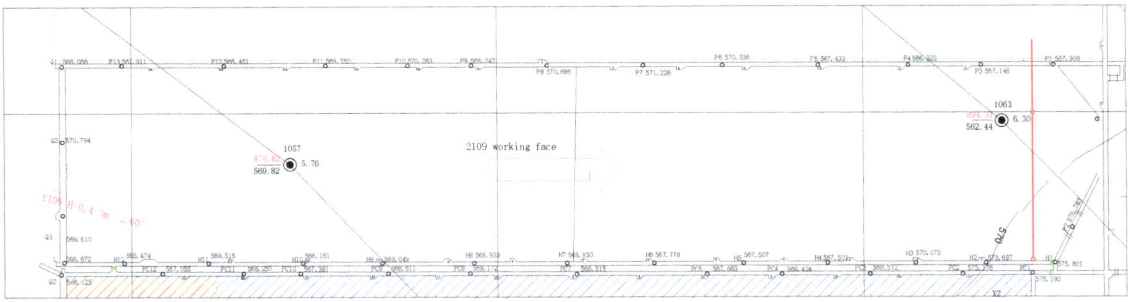

Figure 2. Layout plan of the working face.

The coal seam of the 20,109 working face is the 2# coal seam in the middle and lower part of the Shanxi Formation. There are continental lakes deposited, and the coal seam is stable, with an average thickness of 6.02 m. The coal is fragmented, black, and powdery. The angle of inclination is generally less than 10°. The roof of the 2# coal seam is generally gray-black mudstone, siltstone, etc. The thickness is generally approximately 2 m, containing plant leaf fossils. The bottom is composed of black sand mudstone, fine sandstone, and rich fossilized root fossils, with a thickness of 3~6 m.

3. Model Design and Production of Similar Simulation

According to the mining geological conditions of the 20,109 working face, two sets of physically similar material simulation tests were carried out: the first model was completed on 20 January 2018, simulating the north–south strike section, studying the roadway layout and coal pillar stability; the second model was completed on 23 March, where the east–west section was simulated. The laws of mining pressure of mining towards the gully or away from the gully were studied. The model size (length × width × height) was 5000 mm × 300 mm × 1700 mm, and the scale of the geometric model was 1:200; model materials are detailed in Table 1. After the model was dried to meet the test strength requirements, the mining test was carried out, and corresponding observations and records were performed.

Table 1. Proportioning table of simulation materials.

Thickness/m	Rock Layer	Layer Thickness (m)	Sand	Carbonate (kg)	Gypsum (kg)
7	Medium sandstone	2.00	12.80	0.96	2.24
		1.50	9.60	0.72	1.68
5	Mudstone	2.50	16.66	1.66	1.66
12	Sandy mudstone	2.50	17.14	0.86	2.00
		2.00	13.71	0.68	1.60
		2.00	13.71	0.68	1.60
		2.00	13.71	0.68	1.60
7	Mudstone	2.00	13.33	1.33	1.33
6	Fine sandstone	2.00	12.00	2.00	2.00
		2.00	12.00	2.00	2.00
		2.00	12.00	2.00	2.00
4	Mudstone	2.00	13.33	1.33	1.33
6.3	Coal	2.00	13.71	0.68	1.60
		1.50	10.28	0.52	1.20
4	Medium sandstone	2.00	12.80	0.96	2.24
15	Mudstone	2.50	16.66	1.66	1.66
		2.50	16.66	1.66	1.66
		2.50	16.66	1.66	1.66
18	Sandy mudstone	3.00	20.57	1.28	2.40
		2.00	13.71	0.68	1.60
		2.00	13.71	0.68	1.60

The model used sand as the aggregate, with gypsum and calcium carbonate as the bonding materials. According to the on-hole drilling histogram and the physical and mechanical parameters of the coal rock measured in the laboratory, the proportions of ingredients were determined, and the proportion of the materials required for laying each rock (coal) layer was calculated.

When the model was made, the materials required for the preparation of the coal seam were weighed, stirred with water to a suitable degree, layered, and vibrated; then, mica powder was used to simulate the contact surface of the rock layer. After the model was completed, it was cured for 15 to 20 days. After the model reached the required strength for the test, the front surface of the model was painted white, and a black ink bullet line was drawn with a grid of 10 cm × 10 cm to facilitate observations of the deformation of the model rock mass and destruction characteristics. The experimental model is shown in Figure 3.

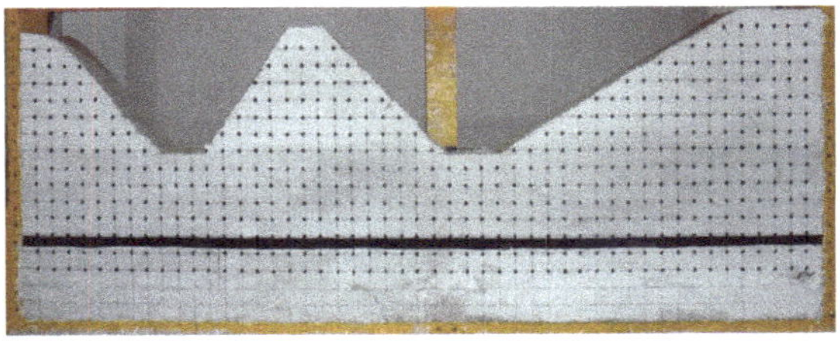

(a)

(b)

Figure 3. Original appearance of the test model. (**a**) Model 1. (**b**) Model 2.

In order to analyze the variation law of vertical stress (advanced abutment pressure), vertical displacement, and horizontal displacement of roof overburdening during the mining of the working face, a pressure sensor was embedded in the model. The measuring point arrangement and instruments are shown in Figure 4.

(a)

(b)

Figure 4. Test systems of two models. (**a**) Embedded pressure sensor. (**b**) Static resistance strain gauge.

4. Failure Characteristics of the Surrounding Rock of the Roadway under Different Positions

4.1. The Roadway Located below the Top of the Slope

When the roadway was located below the top of the slope, it was analyzed according to the geological mechanics surface structure formation principle. The area was affected by horizontal tectonic stress, and it was more difficult to support both sides of the roadway along the mountain range and below the top of the slope. After successfully finishing the mining, the horizontal tectonic stress was relieved, and a crack above the roadway developed on both sides, which led to an increase in the bearing capacity of the coal pillars and increased stress concentration, as shown in Figures 5 and 6.

Figure 5. Roadway located below the top of the slope.

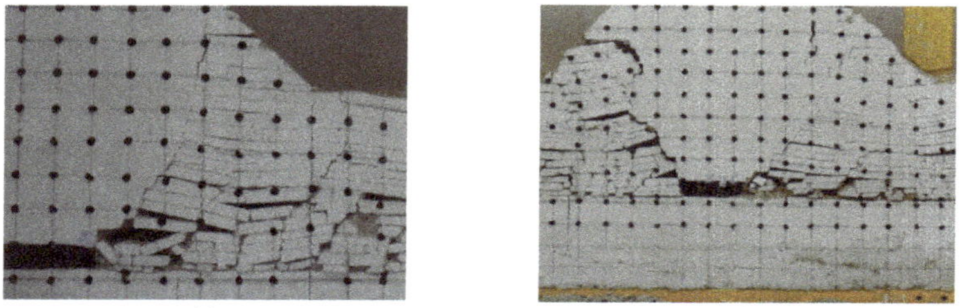

(a) (b)

Figure 6. Overburden damage of roadway. (**a**) Cracks develop to the surface. (**b**) Roadway damaged by extrusion.

After completing the work, the main roof above the working face collapsed and tended to be stable. Cracks developed from the bottom to the top on both sides, the surface exhibited tensile fracture, and the rock layer above the roadway was "diamond shaped". Deformation and failure occurred at both sides of the actual mining roadway affected by mining stress. At this time, the pressure of coal pillars on both sides of the roadway increased.

4.2. The Roadway Located below the Slope Surface

When the roadway was located below the slope surface, the roadway was less affected by the horizontal structure. Due to the asymmetry of the overburden structure above it, the overburden rock near the slope bottom of the roadway was damaged by tensile fracture after finishing the mining work, and maintaining the roof support will be more difficult, as can be seen from Figures 7 and 8.

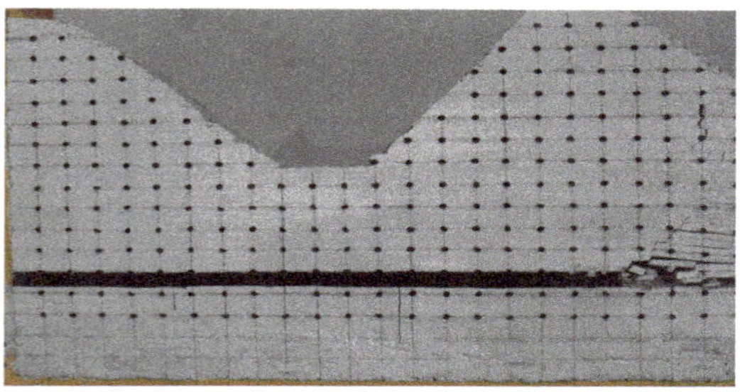

Figure 7. Roadway located below the slope surface.

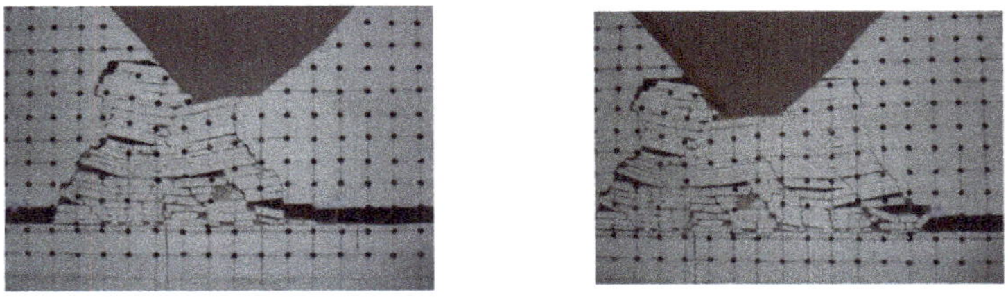

(**a**) (**b**)

Figure 8. Overburden damage of the roadway. (**a**) Overburden rock collapse. (**b**) Asymmetric damage.

After finishing mining, collapse fissures of the overlying strata developed from bottom to top and to the side of the slope bottom, and the surface exhibited tensile fracture. The roadway below the slope surface showed obvious asymmetric deformation and damage, both sides of the top of the roadway were particularly obvious, and the stability of the two sides was relatively good.

4.3. The Roadway Located below the Bottom of Slope

When the gateway was located below the surface of the slope, according to the formation principle of geomechanical surface structures, the two sides of the roadway section along the direction of the valley and near the bottom of the valley were in a tension state, affected by the horizontal tectonic stress. The overburden weight above was smaller, and excavation of the gateway was relatively easy. Overburdening rock above the roadway maintained good stability after stoping in the working face, as shown in Figures 9 and 10.

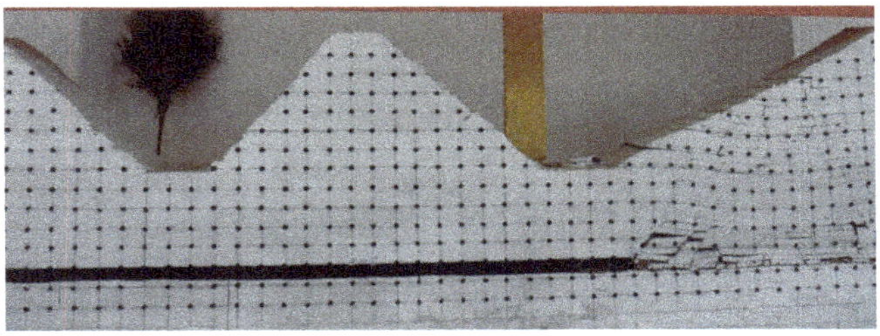

Figure 9. Roadway located below the bottom of slope.

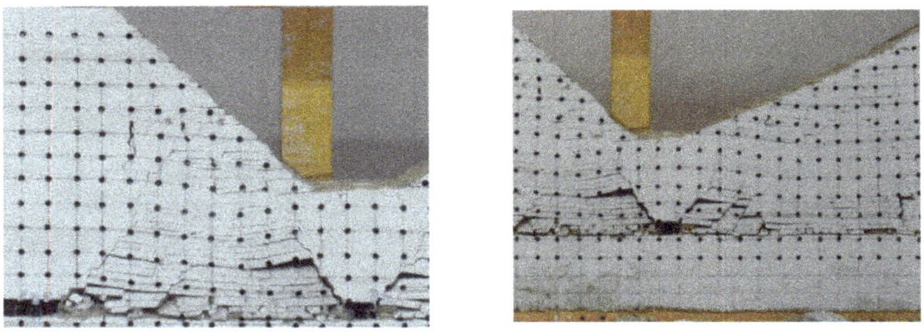

(**a**) (**b**)

Figure 10. Overburden damage of the roadway. (**a**) Slope damage. (**b**) Roadway roof sinking damage.

After finishing the simulation, the overburden fractures were developed from the bottom to the top of the slope, and the strata above the roadway was presented in the form of an "inverted triangle". Due to the support of the fallen strata on both sides, the surrounding rock of the actual mining roadway bore less stress, and the surrounding rock could maintain stability.

5. Laws of Ground Behavior in Different Directions of Working

(1) Mining towards the gully.

The laws of ground behavior appeared to be different for the different directions of working. First, mining towards the gully was carried out; according to the simulation calculation theory, each excavation was 5 cm, i.e., the actual distance was 10 m.

When the working face stoped to the bottom of the slope body, the height of the overlying rock gradually decreased, and the movement and collapse laws of overburden were different from those of the conventional surface. When the working face advanced to 300 m, as shown in Figures 11 and 12, a tensile crack was formed near the top of the slope, and the tensile crack developed downward with mining of the working face. When the working face advanced to 340 m, as shown in Figure 12l, longitudinal fissures formed due to overburden collapse, and surface tension cracks ran through and converged. During the mining at this stage, the slope surface exhibited subsidence towards the bottom of the gully, and slips occurred at the interface of different strata, as shown in Figure 12m,n.

From the above tests, the following characteristics of the overburden movement were determined:

1. The right side of the slope was goaf; therefore, the restriction degree of the slope body decreased, especially in the right horizontal direction, which led to a tendency of the overlying strata sliding to the bottom right of the gully, in addition to downward displacement in the process of stoping. Therefore, when the pulling force towards the right reached a certain degree, a relatively obvious tensile crack was generated at the top of the slope, and the crack developed downward along with working face mining, and finally, the surface tension cracks connected with longitudinal fissures formed by overburden collapse, causing overlying strata with weakened constraints to slip towards the bottom of the gully.
2. Due to the influence of actual mining, the shear strength of the weak surface of overburden gradually became smaller. The slope of the gully was near the goaf; therefore, resistance to additional horizontal stress overburden depended only on the friction between the layers. When the additional horizontal stress was greater than the shear resistance between strata, the strata of the slope body would slip horizontally towards the bottom of the gully to release the additional horizontal stress caused by actual mining.
3. Laws of advanced abutment pressure and displacement of slope

Figure 11. Schematic diagram of mining towards the gully.

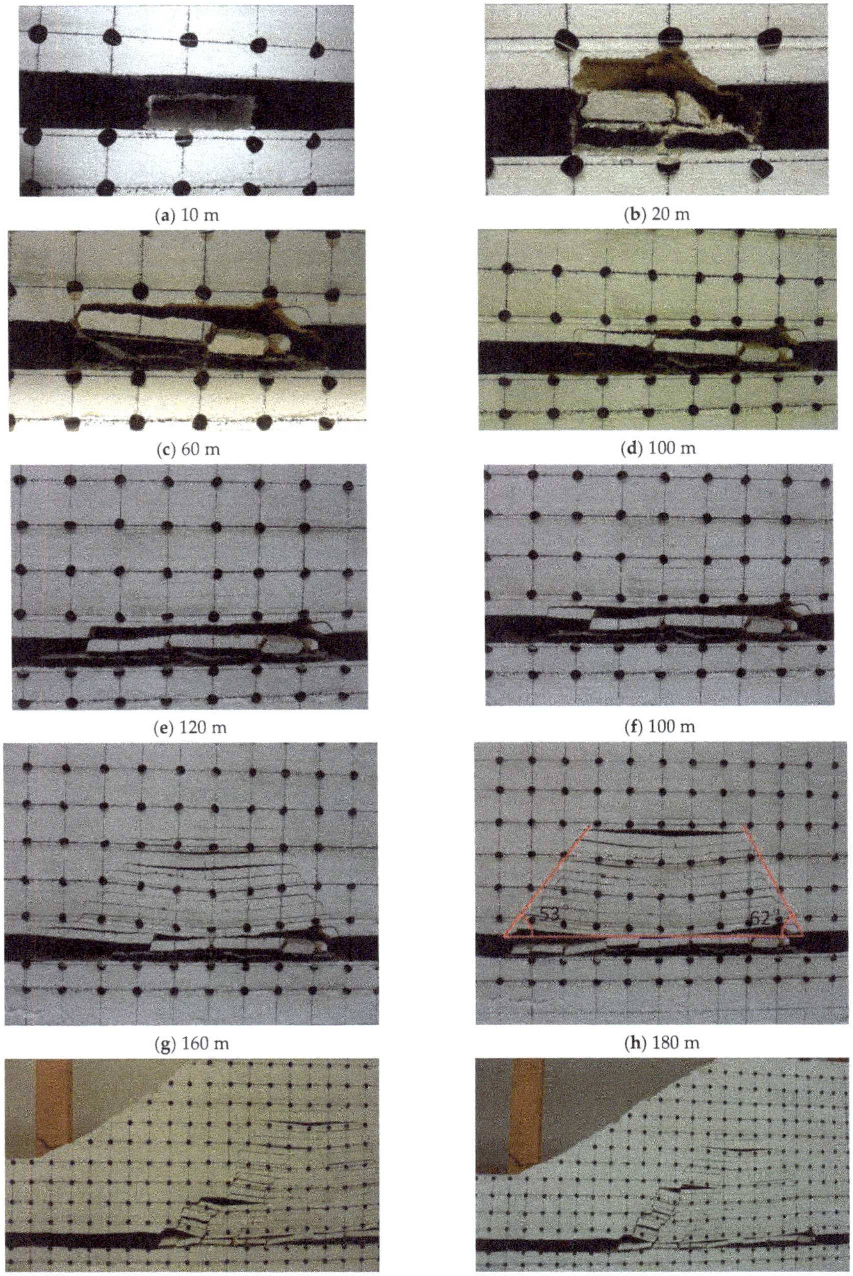

(**a**) 10 m (**b**) 20 m

(**c**) 60 m (**d**) 100 m

(**e**) 120 m (**f**) 100 m

(**g**) 160 m (**h**) 180 m

(**i**) 220 m (**j**) 250 m

Figure 12. *Cont.*

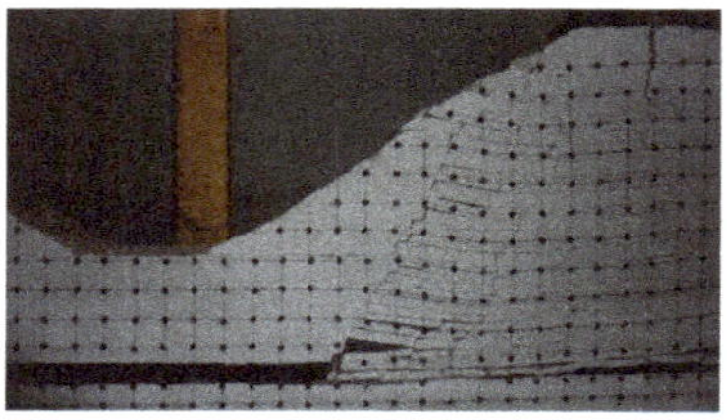

(**k**) 300 m

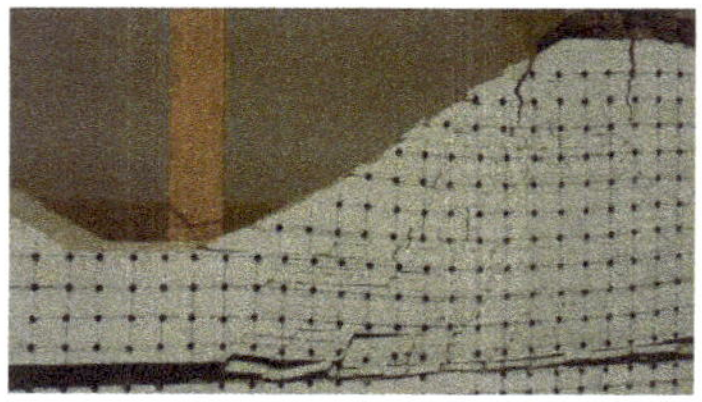

(**l**) 340 m

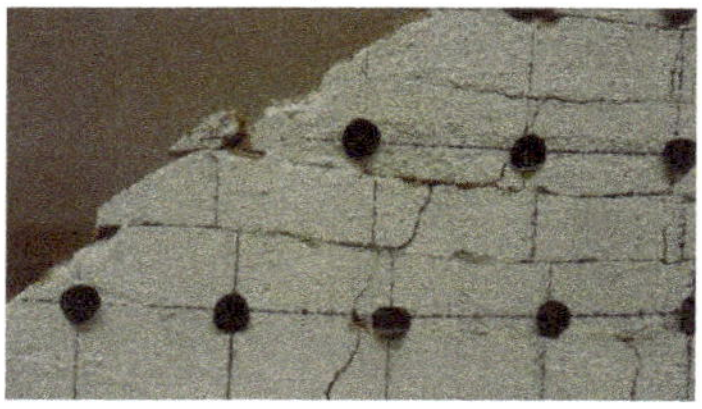

(**m**) 340 m

(**n**) 340 m

Figure 12. Stages of mining towards the gully.

Figure 13 shows the advanced abutment pressure variation in the roof of the working face. It can be seen from the curve on the graph that variations in the 1# and 2# pressure sensors buried in the non-slope section are similar to those of the conventional surface shallow coal seam. It exhibits obvious periodicity, and the influence range of advance abutment pressure is 45 m. The maximum stress concentration factor of the advance abutment pressure is approximately 2.1.

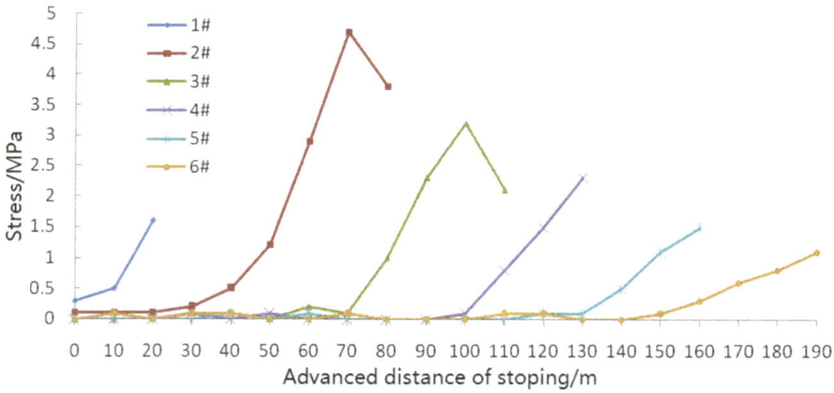

Figure 13. Vertical stress curves of measuring points.

The 3#, 4#, 5#, and 6# pressure sensors were located in the downhill section. It can be seen from the sensor pressure curve that the distribution law of advance abutment pressure during the mining of working face is different from that of the conventional surface working face. The specific performance indicates that the influence range of advance abutment pressure increased, and the concentration factor of advance abutment pressure decreased gradually, from 1.8 to 1.3, and nearly to 1.0 when approaching the bottom of the gully, but the roof could easily be cut off directly, causing dynamic pressure damage. After the occurrence of tensile cracks on the top of slope, the overlying strata lost the tension restraint on the left side due to the tensile cracks, and the slope had a tendency of turning over towards the bottom of the gully until the longitudinal fissures formed by overburden collapse and surface tension cracks ran through and converged.

(2) Mining away from the gully.

Figures 14 and 15 show the response to mining away from the gully. The characteristics of the overlying strata were obviously different from those of mining towards the gully. It can be seen from the graph that with actual mining of the working face, the basic roof did not produce separation and stratified collapse, as in mining towards the gully; instead, almost-vertical tensile cracks were created directly. The cracks developed from the surface of the slope to the coal wall of the working face. The rock cut by the tensile crack turned over to the goaf and formed a multi-block articulated structure at the coal wall, but the articulated structure was unstable and temporary. With the actual mining of the working face, the articulated structure slid and lost stability, inducing dynamic pressure.

With the gradual increase in collapsed rock mass, due to the swelling property of broken rock blocks, the width of the tensile crack became smaller and smaller. With the bending and subsidence of the unfractured overburden on the right side, the fault blocks slowly came into contact with the complete overburden, forming an occlusion zone, which restricted the formation of tensile cracks to a certain extent. It can be seen from Figure 15h that the working face entered the non-slope section; at this point, no tensile cracks were generated, and the movement and collapse characteristics of overburden rock were similar to those of conventional surface coal seam mining.

It can be seen from the curve in Figure 16 that the advance abutment pressure increased with the increase in the length of the cantilever until the formation of the tensile cracks reached the maximum; then, the advance abutment pressure rapidly decreased after the multi-block articulated structure was turned over to the bottom of the gully. As the working surface continued to advance, the new cantilever was formed, the advance abutment pressure was gradually increased, it reached the maximum before the formation of the next tensile cracks, and then it rapidly reduced after the collapse and reversal of the multi-block. Due to the increase in the overburden height, the geometric volume of the fracture blocks increased, as did the peak value of the advance abutment pressure. However, due to the reduction in horizontal dimensions of the multi-block, the periodical weight interval distance and duration decreased, i.e., when mining away from the gully, the roof behavior appeared intense, the peak value grew larger and larger, and the duration was shorter and shorter.

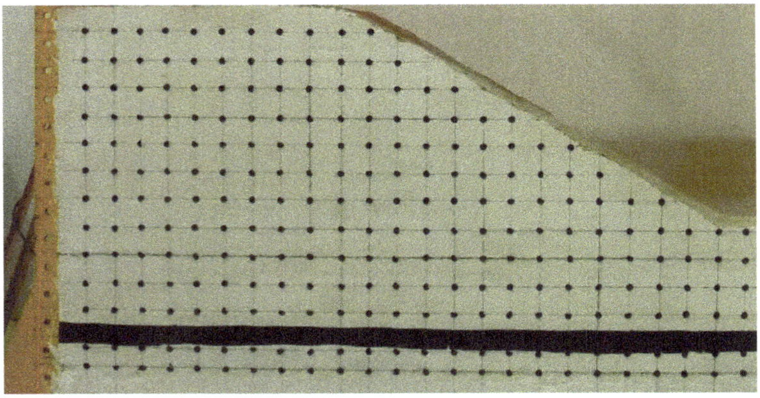

Figure 14. Schematic diagram of the back groove section.

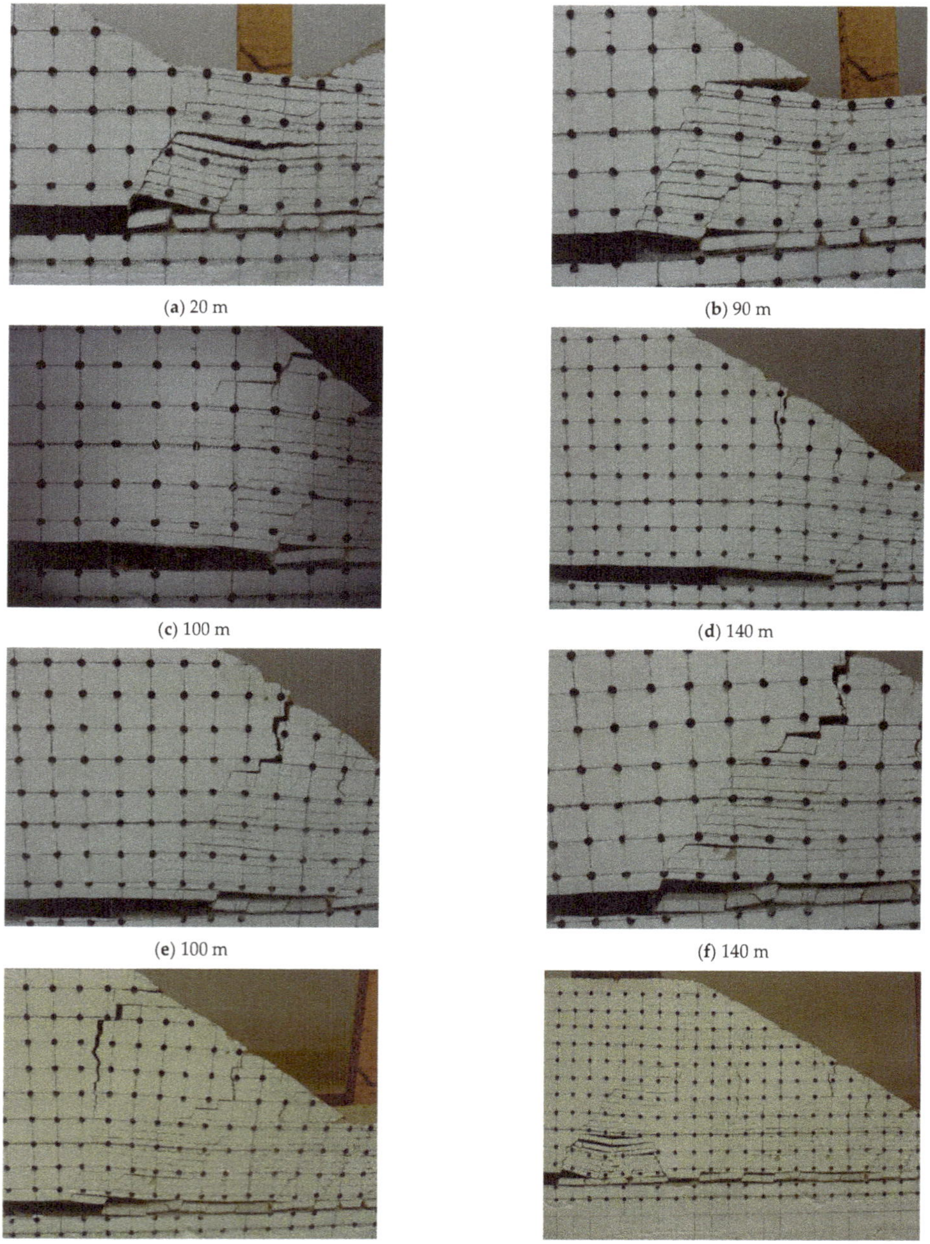

(**a**) 20 m
(**b**) 90 m
(**c**) 100 m
(**d**) 140 m
(**e**) 100 m
(**f**) 140 m
(**g**) 240 m
(**h**) 380 m

Figure 15. Stages of mining away from the gully.

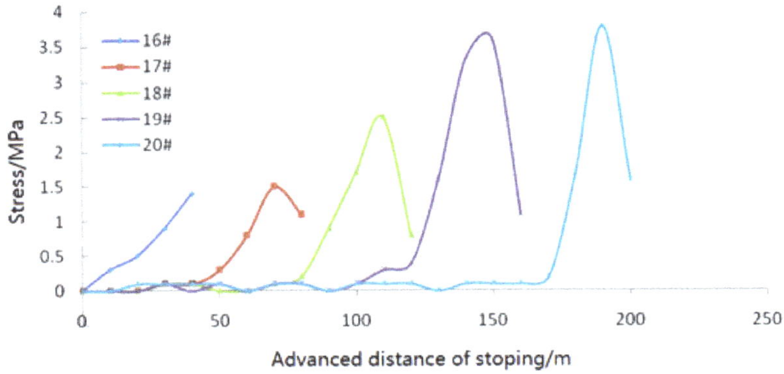

Figure 16. Vertical stress curves of measuring points.

6. Comprehensive Analysis of Test Results

From Figure 17, it can be seen that when the actual mining roadway was located in different positions with different geomorphological features, the stability of the surrounding rock and the degree of disturbance damage were different:

1. When the roadway was located below the top of the slope, the two sides of the roadway were vulnerable to compression damage during excavation; after mining, the deformation of rock surrounding the roadway was aggravated by mining stress, and the abutment pressure of the coal body on the two sides of the roadway increased;
2. When the roadway was located below the bottom of the gully, the stability of rock surrounding the roadway was good during the excavation; after mining, the roadway was less affected by mining and could remain stable;
3. When the roadway was located below the slope surface, the failure of surrounding rock exhibited significant asymmetry under the influence of mining stress.

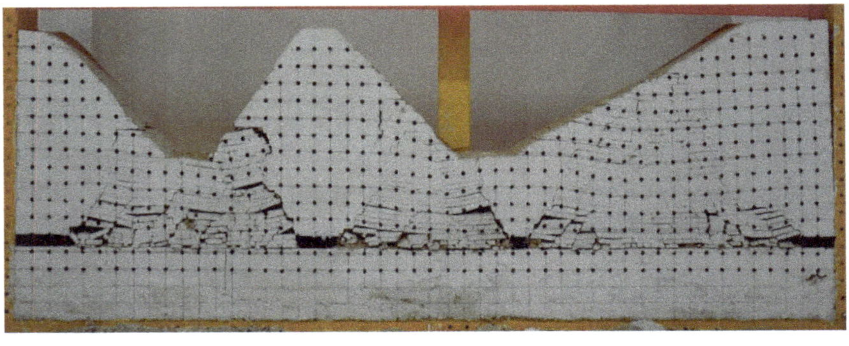

Figure 17. Overview of the model after excavation.

The model 2 test reflected the difference between mining towards or away from gully, as shown in Figures 18 and 19.

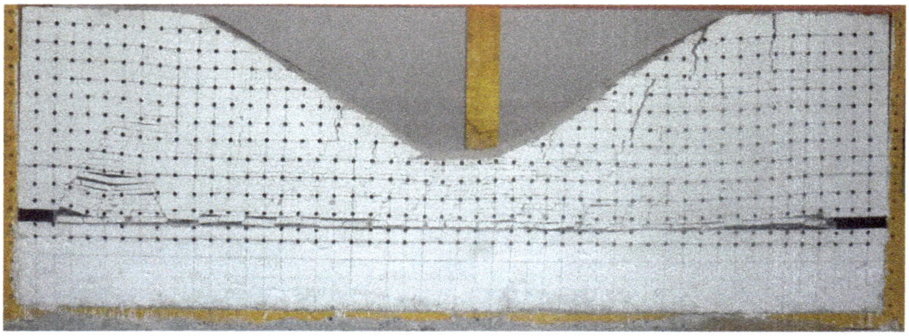

Figure 18. Overview after the excavation of model two.

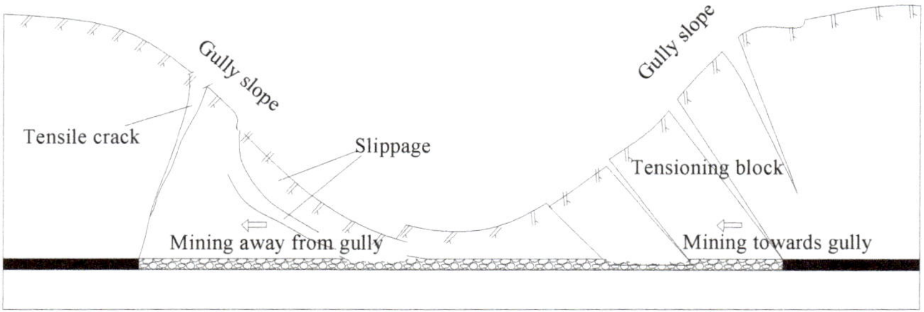

Figure 19. Schematic diagram of overburden failure in coal seam mining under a gully slope.

1. When mining towards the gully, the ground behavior of the working face was different from that of the conventional surface mining section. The concrete manifestation showed that the ground behavior was gentle, the interval distance periodical weight was longer, and the advance abutment pressure was small, i.e., the concentration factor of the advance abutment pressure was small.
2. When mining away from the gully, the ground behavior laws were very different from those of mining towards the gully; the ground behavior appeared intense. With the increase in overburden thickness, the interval distance periodical weight became shorter and shorter, and the roof behavior appeared intense. Dynamic pressure was easily induced, with a high concentration factor of advance abutment pressure.

7. Conclusions

1. When a roadway is located below the top of the slope, the deformation of rock surrounding the roadway is aggravated by mining stress; when a roadway is located below the bottom of a gully, it is less affected by mining and could remain stable. Therefore, the actual mining roadway should be situated in the area below the bottom of the gully as much as possible.
2. The ground behavior laws of the working face in the gully geomorphology area are related to the position of the working face and the direction of working. When mining towards a gully, the ground behavior is gentle, the interval distance periodical weight is longer, and the surface slips towards the gully; when mining away from a gully, the ground behavior appears intense, and the pressure is sudden and short with an increase in overburden thickness and decrease in interval distance periodical weight.
3. When the working face passes through a gully, dumping of the hydraulic support should be prevented in sections with mining towards the gully; in sections of mining away from a gully, the supports should be strengthened to prevent the support and other equipment from being crushed.

4. In actual mining, mining towards the gully should be adopted as much as possible in the stoping of the working face. In this way, the ground behavior is gentle, the interval distance periodical weight is longer, and the advance abutment pressure is small.

Author Contributions: Data curation, S.M.; Funding acquisition, Y.K.; Methodology, S.M.; Writing—original draft, S.M.; Writing—review & editing, Y.K. All authors have read and agreed to the published version of the manuscript.

Funding: This research was funded by the projects "Development of all-weather displacement monitoring device under complex terrain conditions" (2021YBXM31) and "Research on general survey and control technology and equipment of coal mine disaster causing factors" (2018-JIZHUANGSI-ANQUANSHENGCHANYANJIU-01).

Institutional Review Board Statement: Not applicable.

Informed Consent Statement: Not applicable.

Data Availability Statement: The data used to support the findings of this study are available from the corresponding author upon request.

Conflicts of Interest: The authors declare that they have no conflict of interest.

References

1. Zhang, D.S.; Fan, G.W.; Wang, X.F. Characteristics and stability of slope movement response to underground mining of shallow coal seams away from gullies. *Int. J. Min. Sci. Technol.* **2012**, *22*, 47–50. [CrossRef]
2. Zhang, Z.Q.; Xu, J.L.; Zhu, W.B.; Zhenjun, S. Simulation research on the influence of eroded primary key strata on dynamic strata pressure of shallow coal seams in gully terrain. *Int. J. Min. Sci. Technol.* **2012**, *22*, 51–55. [CrossRef]
3. Si, G.; Jamnikar, S.; Lazar, J.; Shi, J.Q.; Durucan, S.; Korre, A.; Zavšek, S. Monitoring and modelling of gas dynamics in multi-level longwall top coal caving of ultra-thick coal seams, part I: Borehole measurements and a conceptual model for gas emission zones. *Int. J. Coal Geol.* **2015**, *144*, 98–110. [CrossRef]
4. Xie, Y.S.; Zhao, Y.S. Numerical simulation of the top coal caving process using the discrete element method. *Int. J. Rock Mech. Min. Sci.* **2009**, *46*, 983–991. [CrossRef]
5. Vakili, A.; Hebblewhite, B.K. A new cavability assessment criterion for Longwall Top Coal Caving. *Int. J. Rock Mech. Min. Sci.* **2010**, *47*, 1317–1329. [CrossRef]
6. Wang, J.C.; Yang, S.L.; Li, Y.; Wei, L.K.; Liu, H.H. Caving mechanisms of loose top-coal in longwall top-coal caving mining method. *Int. J. Rock Mech. Min. Sci.* **2014**, *71*, 160–170. [CrossRef]
7. Wang, J.C.; Zhang, J.W.; Li, Z.L. A new research system for caving mechanism analysis and its application to sublevel top-coal caving mining. *Int. J. Rock Mech. Min. Sci.* **2016**, *88*, 273–285. [CrossRef]
8. Xie, G.X.; Chang, J.C.; Yang, K. Investigation on displacement field characteristics of tunnels surrounding rock and coal seam at FMTC face. *J. Coal Sci. Eng.* **2006**, *12*, 1–5.
9. Xie, G.X.; Chang, J.C.; Yang, K. Investigations into stress shell characteristics of surrounding rock in fully mechanized top-coal caving face. *Int. J. Rock Mech. Min. Sci.* **2009**, *46*, 172–181. [CrossRef]
10. Li, L.; Liu, Z.G.; Chen, P.P.; Liu, G. Study on fully-mechanized top coal caving mining under water body with shallow overburden and thin bedrock. *Adv. Mater. Res.* **2013**, *734–737*, 644–649. [CrossRef]
11. Zhu, Y.J.; Peng, G. Similar material simulation research on movement law of roof over-lying strata in stope of fully mechanized caving face with large mining height. *J. Coal Sci. Eng.* **2010**, *16*, 6–10. [CrossRef]
12. Liu, J.; Chen, S.L.; Wang, H.J.; Li, Y.C.; Geng, X.W. The migration law of overlay rock and coal in deeply inclined coal seam with fully mechanized top coal caving. *J. Environ. Biol.* **2015**, *36*, 821–827.
13. Alehossein, H.; Poulsen, B.A. Stress analysis of longwall top coal caving. *Int. J. Rock Mech. Min. Sci.* **2010**, *47*, 30–41. [CrossRef]
14. Shen, J.; Meng, D.; Wei, L. Study on the structural system of roof in fully mechanized top coal caving. *Appl. Mech. Mater.* **2011**, *90–93*, 2041–2044. [CrossRef]
15. He, F.L.; Wang, X.M.; Zhang, D.Q.; He, S.S. Study on parameters of support for control of roof fall and rib spalling in large fully mechanized top coal caving end face. *Adv. Mater. Res.* **2013**, *616–618*, 421–425. [CrossRef]
16. Du, F.; Bai, H.B. Mechanical model for immediate roof structure of thin bedrock in fully-mechanized sublevel caving face. *Electron. J. Geotech. Eng.* **2012**, *17*, 3075–3088.
17. Yin, G.; Li, X.; Guo, W. Photo-eiastic experimental and field measurement study of ground pressure of surrounding rock of large dip angle working coal face. *Chin. J. Rock Mech. Eng.* **2010**, *29*, 3336–3343.
18. Zhang, J.; Xu, G.; Ma, M.; Wang, H.; Qi, F. Strata behavior law at fully mechanized caving face in thick coal seam with large dip angle Safety in Coal Mines. *Saf. Coal Mines* **2014**, *45*, 181–183.
19. Wu, X.Z.; Jiang, Y.J.; Guan, Z.C.; Wang, G. Estimating the support effect of the energy-absorbing rock bolt based on the mechanical work transfer ability. *Int. J. Rock Mech. Min. Sci.* **2018**, *103*, 168–178. [CrossRef]

20. Gu, S.; Chen, P.; Wang, J.; Wang, H. Field measurement study of mine strata pressure behavior law of coal mining face under goaf. *Coal Mine Eng.* **2013**, *9*, 64–67.
21. Lin, S.Z. A modelling research into slope deformation activities associated with underground mining. *J. Xi'an Coll. Geol.* **1989**, *11*, 70–79+112.
22. Singh, V.K. Slope stability study for optimum design of an opencast project. *J. Sci. Ind. Res.* **2006**, *65*, 47–56.
23. Sun, L.; Zhang, Y.; Wang, Y.; Liu, Q. Study on the Reoxidation Characteristics of Soaked and Air-Dried Coal. *J. Energy Resour. Technol.* **2019**, *141*, 022203. [CrossRef]
24. Erginal, A.E.; Türkeş, M.; Ertek, T.A. Geomorphological investigation of the excavation-induced Dundar landslide, Bursa, Turkey. *Geogr. Ann. Ser. A Phys. Geogr.* **2008**, *90*, 109–123. [CrossRef]
25. Kang, J.R. Analysis of effect of fissures caused by underground mining on ground movement and deformation. *Chin. J. Rock Mech. Eng.* **2008**, *27*, 59–64.
26. Singh, R.; Mandal, P.K.; Singh, A.K.; Kumar, R.; Maiti, J.; Ghosh, A.K. Upshot of strata movement during underground mining of a thick coal seam below hilly terrain. *Int. J. Rock Mech. Min. Sci.* **2008**, *45*, 29–46. [CrossRef]
27. Fan, K.G.; Liu, G.L.; Xiao, T.Q. Study on overlying strata movement and structure features aroused by mountainous shallow buried coal seam mining. In Proceedings of the 2nd International Conference on Mine Hazards Prevention and Control, Qingdao, China, 15–17 October 2010; Volume 12, pp. 76–82.
28. Yuan, T. Research on Mining Subsidence Laws in Shaanxi Loess Gully Region. Master's Thesis, Xi'an University of Science and Technology, Xi'an, China, 2011.
29. Ji, H.; Yu, X.Y. Analysis and study on formation mechanism of subsidence disaster in gully region of shallow depth seam. *Coal Eng.* **2012**, *6*, 69–71.
30. Li, W.X.; Mei, S.H.; Zai, S.H.; Zhao, S.; Liang, X. Fuzzy models for analysis of rock mass displacements due to underground mining in mountainous areas. *Int. J. Rock Mech. Min. Sci.* **2006**, *43*, 503–511. [CrossRef]
31. He, M.C.; Feng, J.L.; Sun, X.M. Stability evaluation and optimal excavated design of rock slope at Antaibao open pit coal mine, China. *Int. J. Rock Mech. Min. Sci.* **2008**, *45*, 289–302. [CrossRef]
32. Yu, L.; Ignatov, Y.; Ivannikov, A.; Khotchenkov, E.; Krasnoshtanov, D. Common features in the manifestation of natural and induced geodynamic events in the eastern regions of Russia and China. *IOP Conf. Ser. Earth Environ. Sci.* **2019**, *324*, 012004. [CrossRef]
33. Rybak, J.; Khayrutdinov, M.M.; Kuziev, D.A.; Kongar-Syuryun, C.B.; Babyr, N.V. Prediction of the geomechanical state of the rock mass when mining salt deposits with stowing. *J. Min. Inst.* **2022**, *253*, 61–70. [CrossRef]
34. Khayrutdinov, A.M.; Kongar-Syuryun, C.B.; Kowalik, T.; Tyulyaeva, Y.S. Stress-strain behavior control in rock mass using different-stregth backfill. *Min. Inf. Anal. Bull.* **2020**, *2020*, 42–55. [CrossRef]
35. Rybak, J.; Gorbatyuk, S.M.; Kongar-Syuryun, C.B.; Khairutdinov, A.M.; Tyulyaeva, Y.S.; Makarov, P.S. Utilization of Mineral Waste: A Method for Expanding the Mineral Resource Base of a Mining and Smelting Company. *Metallurgist* **2021**, *64*, 851–861. [CrossRef]

sustainability

MDPI

Article

Patterns of Influence of Parallel Rock Fractures on the Mechanical Properties of the Rock–Coal Combined Body

Yakang Li, Jiangwei Liu * and Qian Yu

College of Energy and Mining Engineering, Shandong University of Science and Technology, Qianwangang Road 579, Qingdao 266590, China
* Correspondence: jiangweiliu@cumt.edu.cn; Tel.: +86-0532-86057948

Abstract: Hydraulic fracturing of a roof can attenuate the strong mineral pressure and stress transfer by creating a series of parallel artificial fractures in it, and different forms of hydraulic parallel fractures have different effects on the attenuation effect of the coal rock system. In this regard, this study investigated the influence law of different forms of parallel pre-cracks on the mechanical properties of a rock–coal combined body through PFC^{2D} numerical simulation, and the following conclusions were obtained. Parallel pre-cracks significantly affected the mechanical properties of the rock–coal combined body. The stress–strain curve of the coal–rock assemblage containing Parallel pre-cracks changes significantly following reductions in peak strength, peak strain, elastic modulus, and crack initiation stress. The closer the angle θ between the single set of parallel pre-cracks and the horizontal is to 30°, the longer the length, L, and the lower the peak strength of the specimen, the peak strain, the modulus of elasticity, and the crack initiation stress. Macroscopic damage patterns are given for a rock–coal combined body containing single sets of parallel pre-cracks of different parameters, with coal sample damage, coal–rock sample damage and rock sample damage; a rock–coal combined body containing parallel pre-cracks is divided into three modes of fracture initiation when pressurized. These are the cracking of the coal sample, the simultaneous cracking of the parallel pre-crack tips in coal and rock samples, and the cracking of the parallel pre-crack tips in rock samples.

Keywords: pre-cracks; rock–coal combined body; hydraulic fracturing; crack initiation

Citation: Li, Y.; Liu, J.; Yu, Q. Patterns of Influence of Parallel Rock Fractures on the Mechanical Properties of the Rock–Coal Combined Body. *Sustainability* **2022**, *14*, 13555. https://doi.org/10.3390/su142013555

Academic Editors: Xiangguo Kong, Dexing Li and Xiaoran Wang

Received: 9 September 2022
Accepted: 17 October 2022
Published: 20 October 2022

1. Introduction

The deformation, transport, and breakage patterns of the roof plate during the coal mining face retrieval process can result in a mineral pressure phenomena. When the top plate is harder, the fracture of the hard top plate can easily cause strong dynamic pressure to appear [1–5]. The application of the hydraulic fracturing method can create artificial cracks in the rock formation, attenuate the strength of the rock formation, reduce the degree of stress concentration, ease the mine pressure, and reduce the possibility of the sudden release of energy, which is of great significance for the management of strong dynamic pressure roadways and impact ground pressure control [6–9]. Studies have shown that the spatial morphology of hydraulic fractures, including their length, direction, and density, significantly influences the attenuation of roof rocks [10,11]. Conventional experiments with a single fracture mostly simulate hydraulic fractures or primary rock fractures. However, in engineering practice, most of the hydraulic fracturing boreholes are arranged parallel to each other in the direction of the roadway. Furthermore, the fracturing method mostly employs segmented retreating fracturing, which produces nearly parallel fractures in the same stress environment, generally resulting in one or some parallel fractures [12–14]. The study of the mechanical laws of roof pre-cracking on the coal–rock assemblage, seeking to reveal the mechanism of damage to the coal–rock body under load, is of great significance for the application of hydraulic fracturing technology in the pre-fracturing of coal–rock seams to increase the penetration, pressure relief from the roof, and the recovery of released top coal. Therefore, there is an urgent need to carry out physical experimental studies of parallel fractures in the roof on a rock–coal combined body.

Local and international scholars have conducted extensive research on the strength and damage characteristics of "coal–rock" structures. There are currently two main approaches to the study of coal–rock structural bodies. The first is the collection and processing of coal and rock samples into standard coal–rock combination specimens for uniaxial or triaxial testing in the laboratory. The second entails borrowing numerical simulation tools [15]. Chen et al. [16] assessed the effect of rock-to-coal height ratio on the uniaxial compressive strength, macroscopic damage cracking stress, and elastic modulus of the "roof–column" structural body, and that of the loading rate on the uniaxial compression damage of the combined coal rock, by numerical simulation. Guo et al. [17] analyzed the effects of different inclination angles of a rock–coal combined body on its overall deformation and damage, obtained its strength and damage mechanisms under uniaxial and triaxial compression conditions, and analyzed the effects of intersection inclination angles on its overall deformation and damage. Zuo et al. [18] analyzed the differences in the mechanical properties and impact propensities of different coal–rock assemblages. Zhang et al. [19] investigated the influence of the combination method on the mechanical characteristics and damage mechanisms of a rock–coal combined body. Bai et al. [20] sought to investigate the progressive failure characteristics and mechanisms of different kinds of sandwiched coal–rock specimens. Zhao and Xia et al. [21–23] investigated the mechanical behavior and damage characteristics of 3D-printed irregular columnar jointed rock masses under uniaxial, biaxial, and true triaxial actions.

Reyes et al. [24] studied the damage of two cracked rock specimens under uniaxial compression conditions, and Zhang et al. [25] used two-dimensional discrete element software to simulate the crack expansion process in single- and double-fractured rock samples, considering the size of the specimens and influencing factors such as the loading rate. Qian and Li [26] simulated the crack extension process and characteristics of single fractured rocks under bidirectional compression, and analyzed the influence of the surrounding pressure on the crack extension of the rocks. Yin et al. [27] used PFC fine-scale granular flow software to simulate uniaxial compression tests on single-nodular granite with different nodal lengths and inclination angles, in order to investigate the characteristics of crack initiation stresses and damage stresses, and to analyze the effects of nodal length and inclination angle on crack initiation stresses and damage stresses. Liu et al. [12] analyzed the effect of a single fracture in the roof on the uniaxial compression mechanical properties and damage characteristics of combined coal–rocks. Zhang [28] analyzed the effect of parallel double-cracked disc specimens on the cleavage damage process. Yang et al. [29–32] analyzed the law of the influence of the dip angle of coplanar double fissures on the crack evolution process, and studied the deformation and strength characteristics of interrupted prefabricated fractured rocks in different surroundings.

Previous research has focused on the mechanical properties of intact combined coal bodies and on combined coal bodies with fractures in parts of the coal seam. Fewer studies have been conducted on combined coal–rock bodies with rock sections containing cracks. Based on this, this paper simulates the uniaxial compression test of a coal–rock assemblage with parallel cracks in the rock body through PFC^{2D} software, and analyzes the influence law of the angles and lengths of artificial cracks on the mechanical properties of the assembled coal–rock.

2. Project Background

As shown in Figure 1, in engineering practice, most hydraulic fracturing boreholes are arranged parallel to each other in the direction of the roadway. The fracturing method mostly employs segmented retreating fracturing, which produces nearly parallel fractures in the same stress environment, generally resulting in multi-parallel fractures.

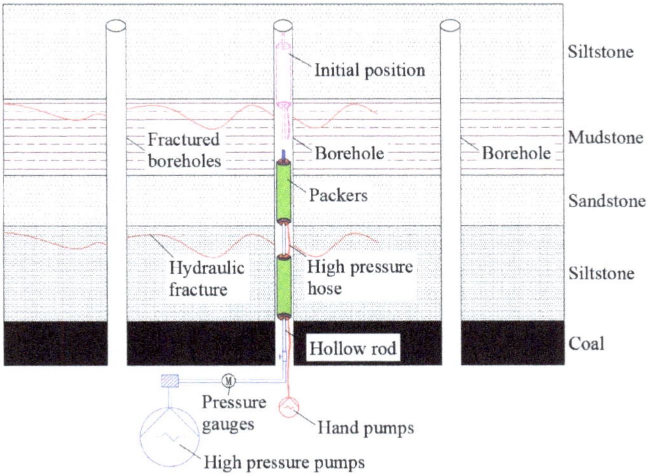

Figure 1. Schematic diagram of hydraulic fracturing in a coal mine.

As the roof of the working face is influenced by the weak coal seam underneath, the roof and coal seam should be considered jointly in the study of the damage evolution process of multi-parallel hydraulic fractures during loading, and the mechanical properties of the rock–coal combined body containing parallel hydraulic fractures, which can be used for the further analysis of the modification mechanism of the stress field.

3. Numerical Models

3.1. Fine Mechanical Parameters of Coal and Rock

There are two main types of cohesive models in the PFC granular flow program: the parallel cohesive model can transfer forces and bending moments, and is therefore commonly used to simulate dense materials such as coal and rock [27,33–35].

Figure 2, and Tables 1 and 2, show that the results of the simulated specimens have small errors compared to those of the test specimens. Moreover, the final fine-scale parameters of the coal–rock can be used to study the effect of a single group of parallel pre-cracks on the mechanical properties of the combined coal rock.

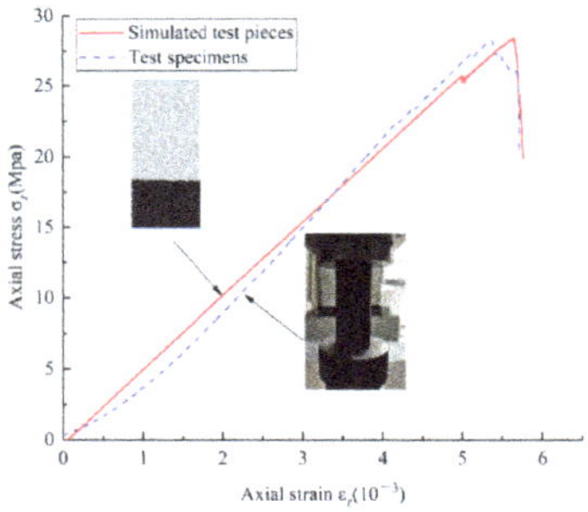

Figure 2. Comparison of stress–strain curves for simulated and test specimens.

Table 1. Comparison between results of simulated and test specimens.

	Peak Strength (MPa)	Peak Strain	Modulus of Elasticity
Simulated test pieces	28.438	0.00565	4.595
Test specimens	28.352	0.00537	4.666
Error	0.086	0.00028	−0.071

Table 2. Fine view parameters of coal and rock [33]. Note: Adapted from Liangfu Xie (2020).

Parameters	Rock	Coal	Parameters	Rock	Coal
Minimum particle size (mm)	0.2		Parallel bond elastic modulus (Gpa)	12	4
Particle size ratio	1.5		Parallel bond normal strength (Mpa)	45	15
Density (kg/m^3)	2533	1800	Parallel bond tangential strength (Mpa)	45	15
Contact modulus of the particle (Gpa)	12	4	Parallel bond normal stiffnes/tangential stiffnes	2.5	
Parallel bond radius multiplier	1		Normal stiffnes/tangential stiffnes	2.5	
Coefficient of friction			0.577		

Notably, the geometry of the prefabricated fractures simulated by the deleted particle approach depends on the microscopic material properties of the rock, and there is no need to recalibrate the microscopic parameters of the prefabricated fractures. Therefore, in this study, the results of physical experiments on the intact rock–coal combined body are used to determine microscopic parameters [36–39].

3.2. *PFC^{2D} Modeling*

The model of the combined coal–rock is shown in Figure 3, with the rock in the upper part and the coal in the lower part. The specimen was 50 mm in diameter and 100 mm in height. According to the geological conditions of the Linhuan coal mine of Huaibei Mining Company Limited, the thickness of the rock layer of the roof is 6.1 m and the thickness of the coal layer is 3.1 m. Therefore, the heights of the rock and coal parts of the specimen can be obtained according to an equiproportional calculation—the heights of the rock and coal were 67 mm and 33 mm, respectively. The model contained 15,129 rock particles and 7457 coal particles, all with particle radii ranging between 0.2 and 0.3 mm.

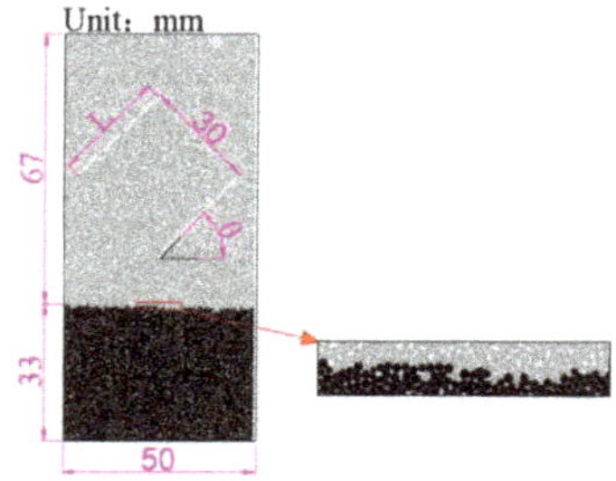

Figure 3. Numerical model of a coal–rock assemblage containing parallel pre-cracks.

Based on the available research results, the coal–rock interface was established, as shown in Figure 3. The friction factor at the coal–rock interface was taken as 0.1, the parallel bond normal and tangential stiffnesses were both 0, and the parallel bond normal and tangential strengths were both 0 [33,38].

3.3. *Numerical Test Conditions*

A parallel fracture was prefabricated in the rock sample of the rock–coal combined body. As shown in Figure 4, the widths of the hydraulic fractures in the field ranged from 0.1 to 1.5 mm; therefore, the width of this prefabricated fracture was determined to be 0.5 mm. Displacement loading was used, and the loading rate was set at 0.01 mm/s until the specimen broke. The specimen numbers and parameters were as follows:

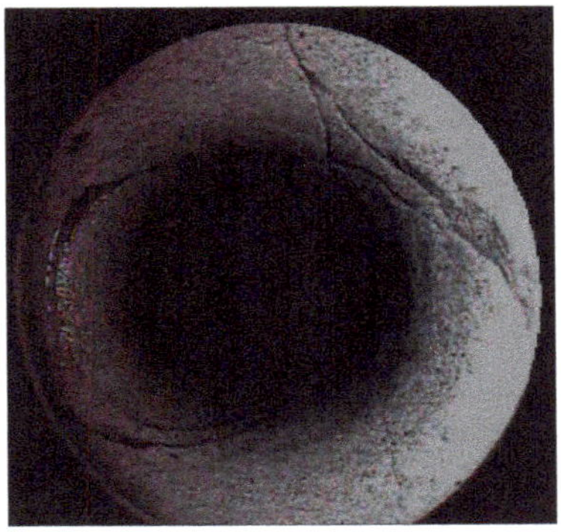

Figure 4. Hydraulic fracture patterns [40].

(1) An intact coal–rock assemblage without parallel pre-cracks was set as the comparison specimen, as shown in Table 3 (1);

(2) The length L was set as 30 mm and the crack spacing D as 30 mm, and a series of experiments with different crack angles θ were conducted. The angles of θ were 0°, 15°, 30°, 45°, 60°, 75°, and 90°, as shown in Table 3 (2–8);

(3) The crack angle θ was set to 45° and the crack spacing D to 30 mm, and a series of experiments was carried out at different crack lengths L: 10, 15, 20, 25, 30, 35, and 40 mm, as shown in Table 3 (5, 9–14).

Table 3. Specimen parameters.

Specimen Number	Parallel Pre-Cracks Angle θ (°)	Parallel Pre-Cracks Length L (mm)
1	The intact rock–coal combined body	
2	0	30
3	15	30
4	30	30
5	45	30
6	60	30
7	75	30
8	90	30
9	45	10
10	45	15
11	45	20
12	45	25
13	45	35
14	45	40

4. Patterns of Influence of Parallel Rock Fractures on the Mechanical Properties of the Rock–Coal Combined Body

4.1. Effect of Parallel Pre-Cracks on the Stress–Strain Characteristics of the Rock–Coal Combined Body

As seen in Figure 5a, the slope of the stress–strain curve is the smallest for a parallel pre-crack angle θ of 30°, and the peak strength and strain are significantly lower. Subsequently, the slope of the stress–strain curve, peak strength, and peak strain decreased for θ = 15° and 45°, respectively. For a θ of 90° and 75°, the curves were similar to those of the intact specimens. The closer the angle of the parallel pre-cracks was to 30°, the lower the peak strength of the specimen, and the slower the slope of the stress–strain curve. From Figure 5a, it can be seen that the parallel pre-cracks with different angles had less influence on the line–elastic stage and compression–density stage of the specimen, and the stages with the most significant influence are the plastic stage and the damage stage, which is mainly because, in the line–elastic stage and compression–density stage, we mainly focus on the elastic change of the specimen, which is related to the material properties of the specimen itself, and has little influence on the angle of the cracks. As the pre-crack is compressed to produce the crack, the expansion changes the plastic structure of the specimen before destruction, such that the specimen is influenced by the angle of the parallel pre-crack in the plastic stage and the destruction stage. From Figure 5b, the stress–strain curve slows down as the length L of the parallel pre-cracks increases. The stress–strain curve is slowest at an L of 40 mm, with the lowest peak strength and peak strain. The stress–strain curve at L = 10 mm was close to that of the intact specimen, and the peak strength was higher than that of the intact specimen. As can be seen from Figure 5b, with a change in the length of the parallel pre-cracks, the stages with the most significant influence are mainly the line–elastic stage and the damage stage, which is mainly because the line–elastic stage and the pressure–density stage are mainly dominated by the elastic change of the specimen, which is related to the material properties of the specimen itself, and has little to do with the length of the cracks. As the pre-crack is compressed to produce a crack expansion that changes the plastic structure of the specimen before destruction, the specimen is affected by the lengths of the parallel pre-cracks in the plastic stage and the destruction stage.

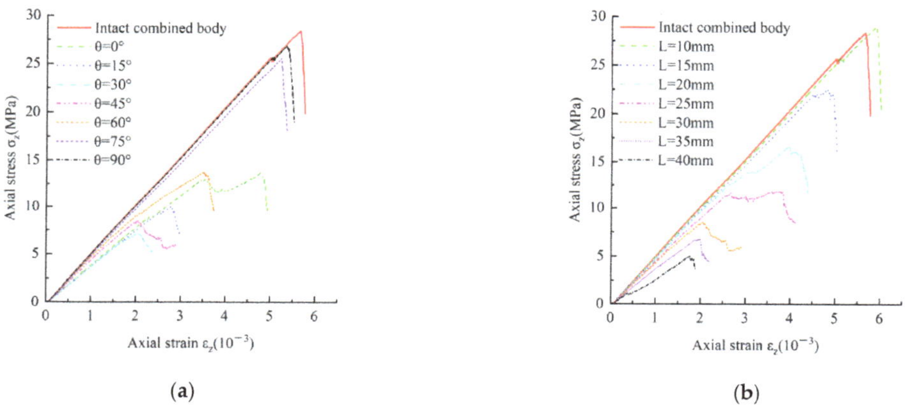

Figure 5. Stress–strain curves for the rock–coal combined body with parallel pre-cracks in different forms. (**a**) Stress–strain curves for different parallel pre-cracking angles. (**b**) Stress–strain curves for different parallel pre-crack lengths.

Figure 5 shows that the strain intervals corresponding to the compression–density and linear–elastic stages of the specimens with different parallel pre-crack parameters did not differ significantly, but the difference was greater after the elastic–plastic stage. The closer the parallel pre-crack angle θ is to 30°, the greater the length L of the parallel pre-cracks, and the greater the change in the stress–strain curve relative to the intact specimen.

4.2. Effect of Parallel Pre-Cracks on the Strength and Deformation Characteristics of the Rock–Coal Combined Body

As shown in Figure 6a, the peak strength decreased and then increased as the angle of the parallel pre-cracks increased. The peak strength of the rock–coal combined body is the smallest for a parallel pre-crack angle θ of 30°, which is 74.71% lower than that of the intact specimen. This was followed by a reduction of 69.96% at a θ of 45° compared to the intact specimen. As shown in Figure 6b, the peak strain first decreased and then increased as the angle of the parallel pre-cracks increased. At an angle θ of 45°, the peak strain in the rock–coal combined body was the smallest, with a reduction of 63.89% compared to the intact specimen. This was followed by a reduction of 63.72% at a θ of 30° compared to the intact specimen. As shown in Figure 6c, the modulus of elasticity first decreased and then increased as the angle of the parallel pre-cracks increased. The smallest modulus of elasticity of the rock–coal combined body was seen at an angle θ of 30°, with a decrease of 39.82% compared to the intact specimen. This was followed by a 34.77% drop at a θ of 15° compared with the intact specimen. As shown in Figure 6, the peak strength, peak strain, and modulus of elasticity of the rock–coal combined body containing parallel pre-cracks at different angles were significantly reduced relative to the intact specimens. The closer the parallel pre-crack angle θ is to 30°, the lower the peak strength, peak strain, and modulus of elasticity of the rock–coal combined body.

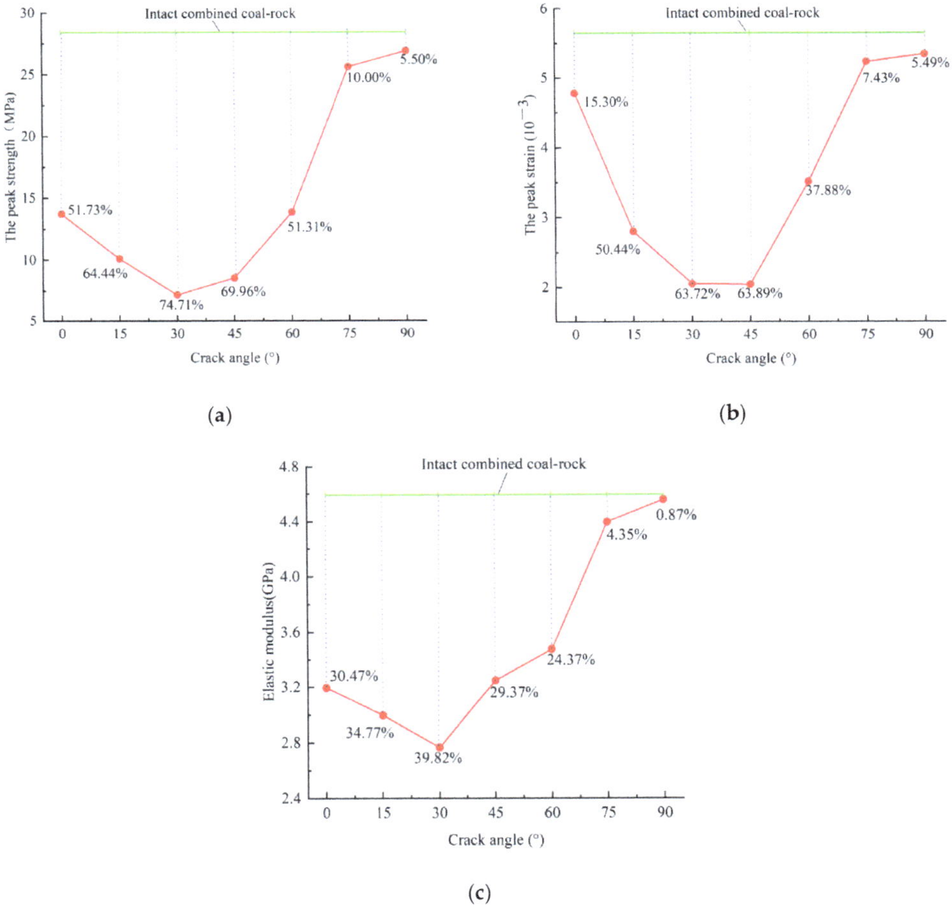

Figure 6. Effects of parallel pre-cracks at different angles on the strength of the rock–coal combined body. (**a**) Effect of parallel pre-cracks at different angles on axial stress. (**b**) Effect of parallel pre-cracks at different angles on axial strain. (**c**) Effect of parallel pre-cracks at different angles on elastic modulus.

As shown in Figure 7a, the peak strength decreased as the length L of the parallel pre-cracks increased. When the L was 40 mm, the peak strength decreased by 82.34% relative to that of the intact specimen. As shown in Figure 7b, the peak strain decreased as the length L of the parallel pre-cracks increased. When the L was 40 mm, the peak strain decreased by 68.67% relative to that of the intact specimen. As shown in Figure 7c, the modulus of elasticity decreased as the length L of the parallel pre-cracks increased. When the L was 40 mm, the peak strain decreased by 53.95% relative to that of the intact specimen. As shown in Figure 7, the peak strength, peak strain, and modulus of elasticity of the rock–coal combined body were significantly reduced by the different lengths of parallel pre-cracks compared to the intact specimens. The peak strength, peak strain, and modulus of elasticity of the rock–coal combined body decreased as the length of the parallel pre-cracks continued to increase.

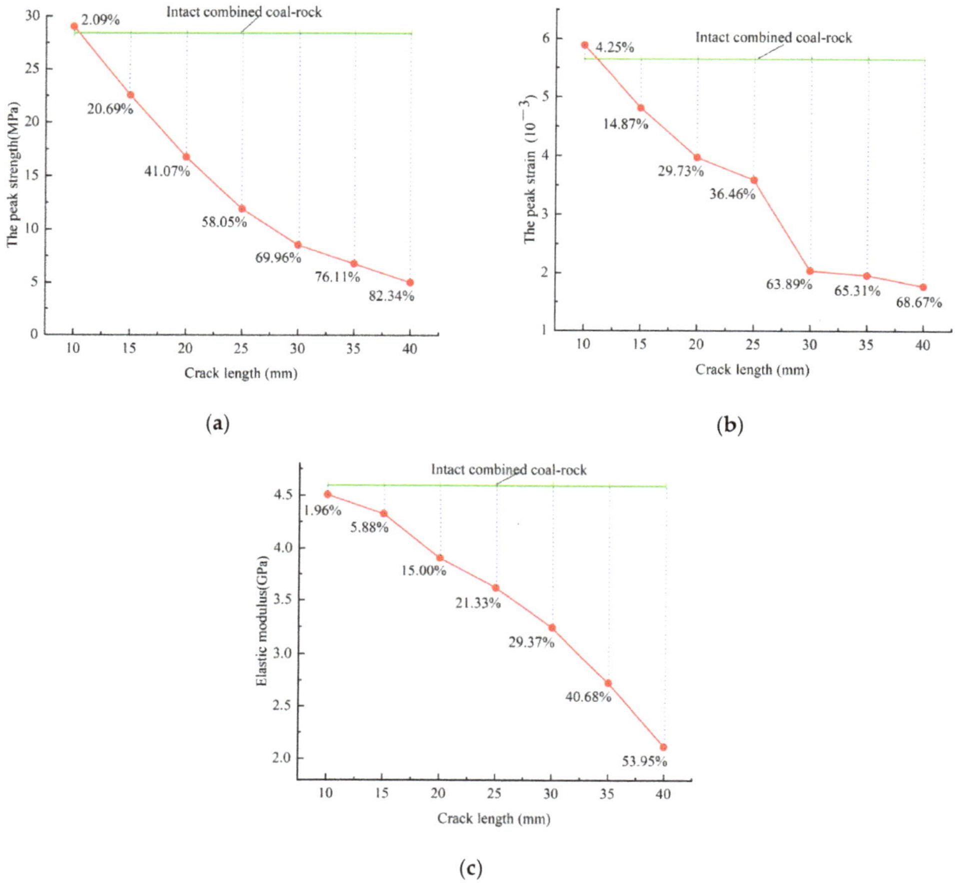

Figure 7. Effect of parallel pre-cracks of different lengths on the strength of the rock–coal combined body. (**a**) Effect of parallel pre-cracks of different lengths on the axial stress. (**b**) Effect of parallel pre-cracks of different lengths on axial strain. (**c**) Effect of parallel pre-cracks of different lengths on elastic modulus.

4.3. Macroscopic Damage Characteristics of the Rock–Coal Combined Body by Parallel Pre-Cracks

In the following descriptions, the direction along the parallel pre-cracks is regarded as tangential, the direction perpendicular to the parallel pre-cracks as vertical, and the direction between the tangential and vertical as oblique. The final damage pattern of the rock–coal combined body with different parallel pre-cracks strains up to 0.01 is shown in Figures 8 and 9.

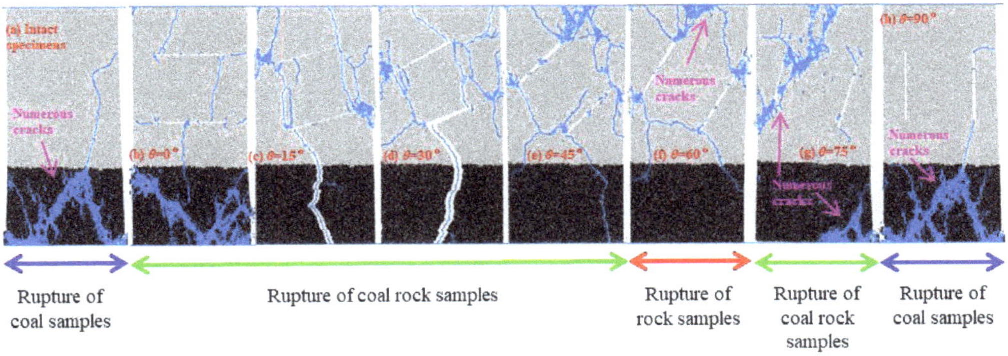

Figure 8. Macroscopic damage characteristics of parallel pre-cracks at different angles.

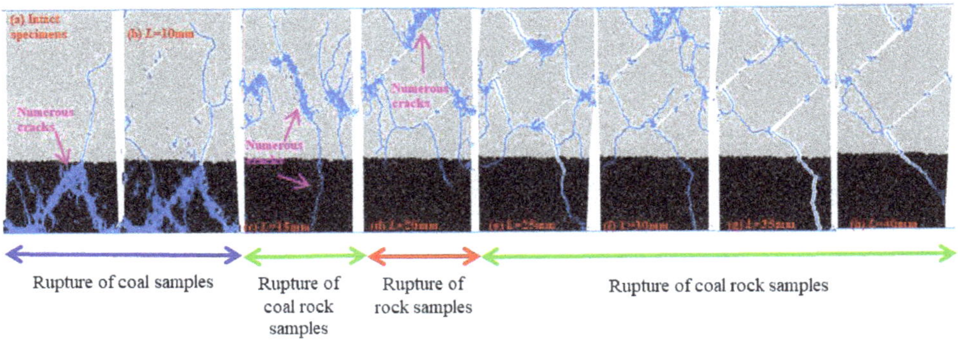

Figure 9. Macroscopic damage characteristics of parallel pre-cracks of different lengths.

Figure 8 shows that there are three macroscopic damage characteristics of the rock–coal combined body with different angles of parallel pre-cracks: coal rupture, rock and coal rupture, and rock rupture. As shown in Figure 8a,h, for intact specimens and $\theta = 90°$, the coal samples in the rock–coal combined body exhibited more intensive cracking, the rock samples produced less cracking, and the damage occurred mainly in the coal samples. Figure 8b–e,g show that for $\theta = 0°$, $\theta = 15°$, $\theta = 30°$, $\theta = 45°$, and $\theta = 75°$, damage occurred in both the coal and rock samples, with more cracks around parallel pre-cracks and a reduced degree of damage in the coal samples relative to the intact specimens. At $\theta = 15°$, the crack extended in the vertical direction, with the top end extending to the top of the specimen and the bottom end extending to the bottom of the specimen. At $\theta = 30°$, parallel pre-cracks were observed along the vertical direction, and developed a major crack that destroyed the coal sample in the middle. It developed tangentially toward the boundary of the specimen and extended to the rock boundary, destroying the rock sample. At $\theta = 45°$, parallel pre-cracks penetrated along the oblique direction, forming multiple oblique cracks that fractured the rock sample. Simultaneously, a number of seam networks were formed around the two parallel cracks, with certain cracks developing in the boundary to break the rock sample. At $\theta = 75°$, an oblique crack extended along the direction of the parallel pre-cracks and produced a network of many seams. Destruction occurred at the bottom of the coal sample. As shown in Figure 8f, at $\theta = 60°$, damage occurred mainly in the rock sample, with oblique damage to the rock and only a few small cracks in the coal body.

Therefore, the rock–coal combined body containing parallel pre-cracks at different angles exhibits three macroscopic damage patterns: coal sample rupture, coal–rock sample rupture, and rock sample rupture. As the parallel pre-crack angle θ increases, the damage pattern cycles from coal sample rupture to both coal and rock sample rupture, then to rock sample rupture, and back to coal sample rupture.

As shown in Figure 9, there are three macroscopic damage characteristics of the rock–coal combined body with different lengths of parallel pre-cracks: coal rupture, rock and coal rupture, and rock rupture. Figure 9a,b show that the form of damage of the rock–coal combined body for a parallel pre-crack length L of 10 mm is nearly identical to that of the intact specimen, occurring mainly in the coal sample, and a small amount of cracking occurred at the tip of the crack. As shown in Figure 9c, numerous seam networks were produced between the parallel pre-cracks and at the tips, and both the coal and rock samples ruptured, with the degree of rupture in the coal samples being significantly lower than that in the intact specimens. Figure 9d shows that the tips of the parallel pre-cracks penetrated each other, and the tips of the parallel pre-cracks produced a large network of seams, with damage occurring mainly in the rock samples. As shown in Figure 9e–h, the coal rock samples all fractured. Figure 9e,f show that the form of damage was essentially the same, except for the location whereat the seam network was created. Damage occurred preferentially in the rock samples and to a lesser extent in the coal samples. As shown in Figure 9g, a small seam network was produced, and the parallel pre-crack tips penetrated each other to form several main cracked damage specimens. As shown in Figure 9h, a small seam network was produced, and the tips under the parallel pre-cracks penetrated each other, preferentially breaking the rock sample in a tangential direction.

As shown in Figure 9, as the length L of the parallel pre-cracks increased, the rock–coal combined body, with different lengths of parallel pre-cracks, ruptured from only coal to both rock and coal, then to rock only, and finally cycled back to both coal and rock samples breaking.

4.4. *Effect of Parallel Pre-Cracks on the Fracture Initiation Pattern of the Rock–Coal Combined Body*

As shown in Figure 10, the crack initiation stress of the rock–coal combined body containing parallel pre-cracks was reduced relative to that of the intact specimen. As the parallel crack angle θ increased, the cracking stress in the rock–coal combined body with parallel pre-cracks decreased and then increased relative to the intact specimen. The greatest change in cracking stress was observed at a θ of 30°, with a reduction of 72.15%. As shown in Figure 11, the parallel pre-cracks' initiation patterns at different angles can be divided into three types. The first type is shown in Figure 11a,h, with crack initiation in the coal sample. The second type is shown in Figure 11b,g, where the fracture started from both the tips of the parallel pre-cracks and the coal sample. The third type is shown in Figure 11c–f, with fracturing from the tip of the parallel pre-crack in the rock sample. As the angle θ of the parallel pre-cracks increased, the rock–coal combined body cycled from coal to coal–rock simultaneously, then to rock-only initiation, and then to cracking from coal. When cracking from the tip of the parallel pre-crack, the direction of crack development gradually changed from vertical to tangential as the θ increased.

As shown in Figure 12, the crack initiation pressure of the rock–coal combined body with parallel pre-cracks was reduced compared to that of the intact specimen. As the parallel pre-crack length L increased, the crack initiation stress decreased relative to the intact specimen. When the L was 40 mm, the crack initiation stress was minimized, with a reduction of 84.16% compared to the intact specimen. As shown in Figure 13, the crack initiation patterns of the specimens can be classified into three types. The first type is shown in Figure 13a, where the crack started from the lower end of the coal sample. The second type is shown in Figure 13b, where the crack started from both the tips of the parallel pre cracks and the sample. The third type is shown in Figure 13c–h, where the crack started at the tips of the parallel pre-cracks.

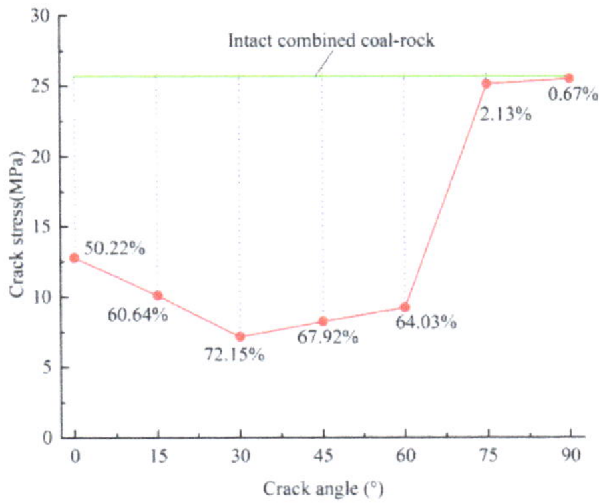

Figure 10. Crack stresses of parallel pre-cracks at different angles.

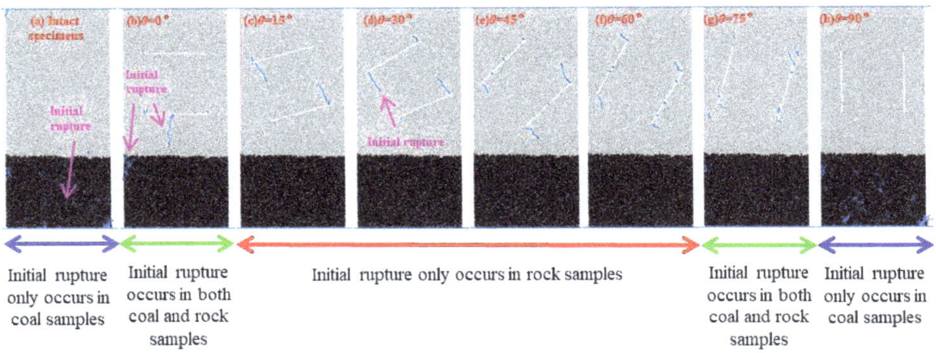

Figure 11. Rock–coal combined body initiation pattern with parallel pre-cracks at different angles.

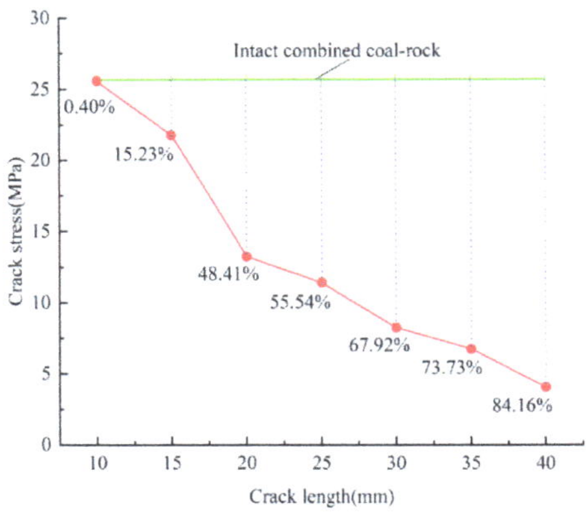

Figure 12. Crack stresses of parallel pre-cracks of different lengths.

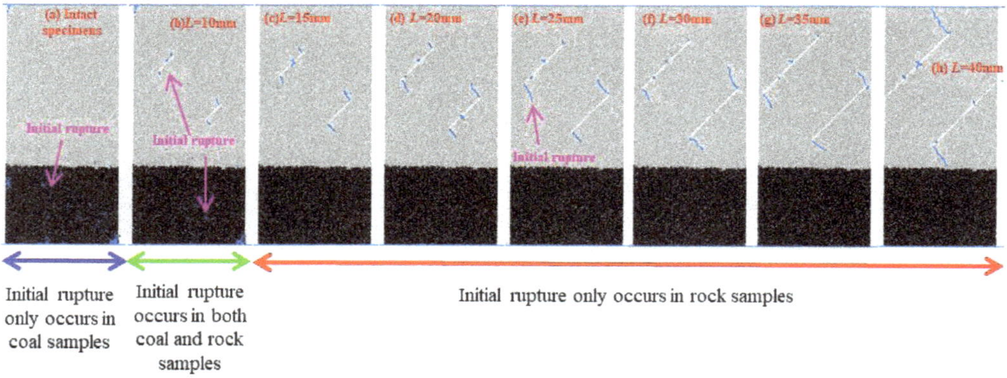

Figure 13. Rock–coal combined body initiation pattern with parallel pre-cracks at different lengths.

In summary, the rock–coal combined body with parallel pre-cracks exhibited three forms of crack initiation: crack initiation in the coal sample, simultaneous crack initiation at the tips of the parallel pre-cracks in the coal and rock samples, and crack initiation at the tips of the parallel pre-cracks in the rock sample. The highest crack initiation stress is required for the cracking of coal samples; the lowest crack initiation stress is required only for crack initiation at the tips of parallel pre-cracks in rock samples. The closer the angle of the parallel pre-cracks is to 30° and the longer the length, the lower the crack initiation stress. This study on the effects of parallel pre-cracks on fracture initiation pattern can be used to determine the parameters of hydraulic fractures in rock seams in practical engineering, to control the fracture initiation pattern and damage pattern of rock seams and lower coal seams, etc. For example, the effect on fracturing resulting from controlling the angle and length of the hydraulic fracture tip can make the rock seam rupture under pressure to absorb energy, so as to achieve the purpose of protecting the coal seam below.

4.5. Effect of Parallel Pre-Cracks on Acoustic Emission Characteristics of the Rock–Coal Combined Body

From Figure 14, we see that the total number of acoustic emission events was highest at a parallel pre-crack angle of $\theta = 90°$ and lowest at $\theta = 15°$. The overall change in the total number of acoustic emission events took the form of a gradual decrease and then a gradual increase as the angle θ of the parallel pre-cracks changed. The rate of change of the acoustic emission phenomenon curve was the slowest at θ of 30°, followed by 15°. This is because the acoustic emissions from intact specimens were mainly concentrated in the coal sample area, with the total number of acoustic emission events becoming smaller and then larger as the parallel pre-crack angle increased to 45°, before the acoustic emissions from the coal sample became progressively smaller and the acoustic emissions from the rock sample gradually increased. The number of acoustic emission events decreased in the rock samples and increased in the coal samples as the angle of parallel pre-cracks increased from 45° to 90°, with the total number of acoustic emission events first decreasing and then increasing. When θ was 90°, the total number of acoustic emission events increased relative to that of the intact specimen. This is due to the fact that the coal specimens had almost identical fracture patterns, while the initiation of fracture tips in the rock samples increased the total number of acoustic emission specimens.

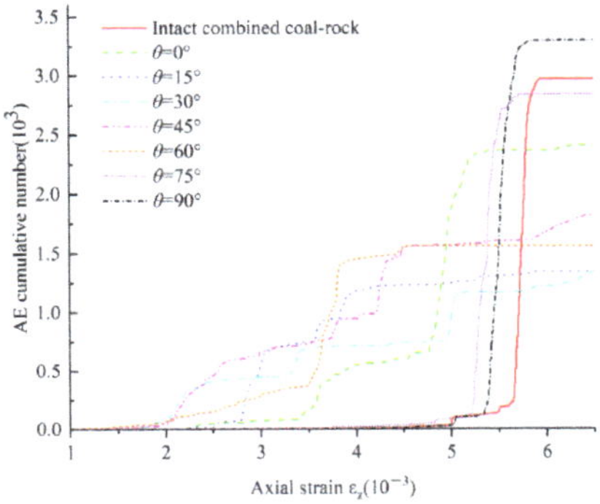

Figure 14. Effects of parallel pre-cracks at different angles on acoustic emission characteristics.

As shown in Figure 15, the total number of acoustic emission events was higher than that in the intact specimen when L was 10 mm, and both produced acoustic emissions at almost the same time. However, owing to the effects of parallel pre-cracks, the number of cracks produced in the rock sample was almost the same; however, owing to the effects of parallel pre-cracks, the number of cracks produced in the rock sample increased for a length L of 10 mm compared to the intact specimen. As the parallel pre-crack length L increased, the overall total number of acoustic emission events gradually decreased.

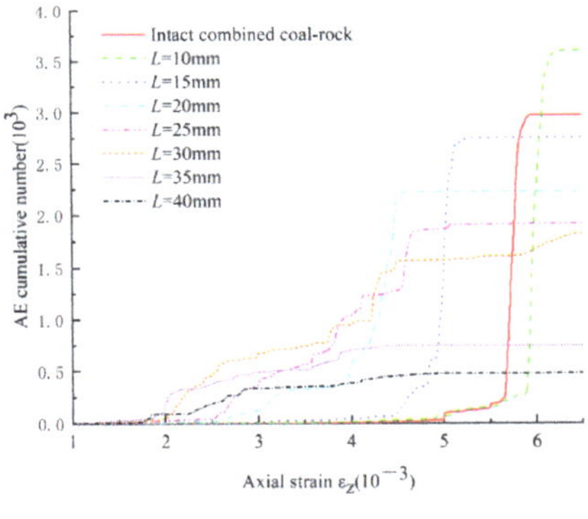

Figure 15. Effects of parallel pre-cracks at different lengths on acoustic emission characteristics.

Combined with the analysis of the macroscopic damage characteristics of the rock–coal combined body, the macroscopic damage characteristics of specimens 8 (θ = 90°) and 9 (L = 10 mm) indicate that only the coal sample ruptured, and the total number of acoustic emission events was higher than that of the intact specimen. As the parameters of the parallel pre-cracks changed, the total number of acoustic emission events also changed significantly, as the macroscopic damage characteristics changed from only the rock samples breaking to both coal and rock samples breaking. This suggests that parallel pre-cracks contribute to the generation of acoustic emissions under the same conditions.

The acoustic emission characteristics of specimens with three different macroscopic damage modes were selected separately for comparison. The macroscopic damage manifested in the form of coal sample damage, and the acoustic emission characteristics of specimen 1 are shown in Figure 16a; the macroscopic damage manifested in the form of both coal and rock damage, and the acoustic emission characteristics of parallel pre-crack specimen 7 containing an angle θ of 75° and a length L of 30 mm are shown in Figure 16b. The acoustic emission characteristics of the parallel pre-crack specimen 11, containing an angle θ of 45° and a length L of 20 mm, are shown in Figure 16c. The four photos in each figure correspond to the linear–elastic stage, the compression-density stage, the elastic–plastic stage, and the damage stage, respectively.

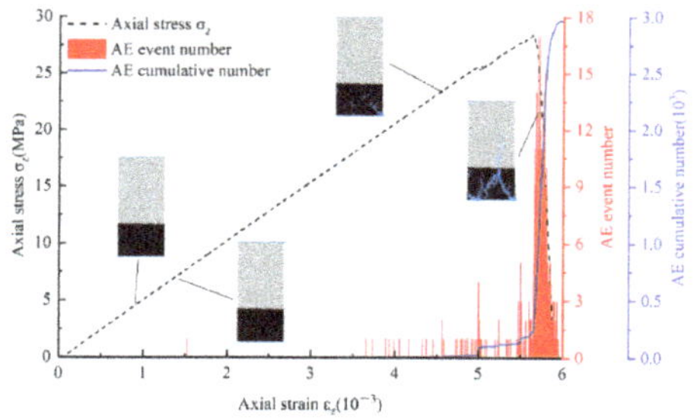

(**a**)

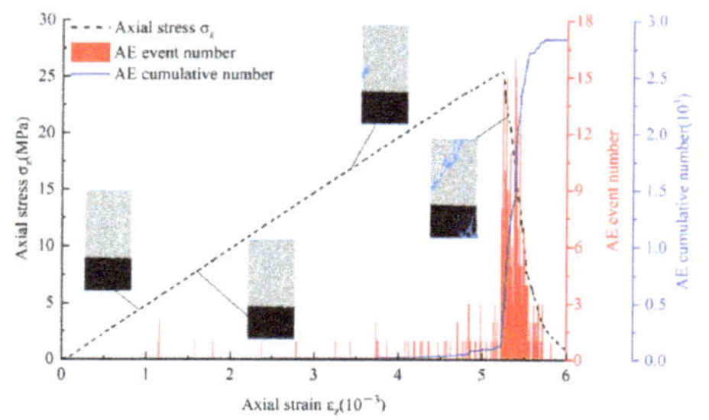

(**b**)

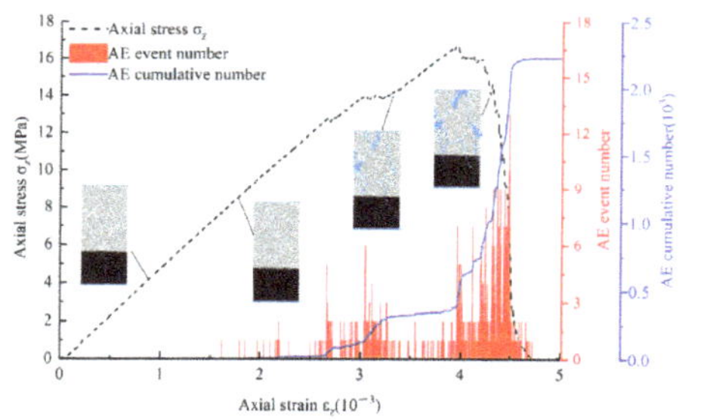

(**c**)

Figure 16. Acoustic emission characteristics of different macroscopic damage forms. (**a**) Acoustic emission characteristics of the macroscopic rupture form of coal rupture (specimen 1). (**b**) Acoustic emission characteristics of the macroscopic fracture form in which all the coal rock samples fractured (specimen 7). (**c**) Acoustic emission characteristics of the macroscopic fracture form where only the fracture of the rock sample occurred (specimen 11).

As shown in Figure 16a, only the specimens damaged by the coal sample showed no acoustic emission in the linear–elastic stage, a small amount of acoustic emission in the compression–density stage, and gradually increasing in the elastic-plastic stage; the greatest increase in the number of acoustic emission events was seen in the specimens after reaching peak intensity, with the number of acoustic emission events concentrated after the peak intensity. As shown in Figure 16b, in the specimens where both rock and coal were damaged, a small acoustic emission phenomenon appeared at the linear–elastic stage and the compression–density phase; the acoustic emission events gradually increased in the elasto-plastic phase, and a large acoustic emission phenomenon appeared after reaching the peak intensity. The acoustic emission events were more evenly distributed compared to the intact specimens. As shown in Figure 16c, in the specimens with damaged rock samples, no acoustic emission events appeared in the linear–elastic stage, and they gradually appeared in the compression–density stage, becoming more prominent in the elastic–plastic stage, and concentrating after reaching the peak intensity. Compared with the intact specimen, the distribution of acoustic emission events was the most dispersed, with a large amount of acoustic emissions before and after the peak strain. As seen in Figure 16, the cumulative number of acoustic emission events for specimens containing parallel pre-cracks decreased compared with those for intact specimens. The highest total number of acoustic emission events was observed with coal sample damage, the second highest total number of acoustic emission events with both rock and coal damage, and the smallest total number of acoustic emission events with rock sample damage. Specimens containing parallel pre-cracks showed significantly more acoustic emissions in the elastic and plastic phases compared to the intact specimens, owing to their cracking at the tips of the parallel pre-cracks before reaching peak strength. After the peak strength was reached, the acoustic emissions of specimens containing parallel pre-cracks slowed down relative to the intact specimens, owing to the partial damage that occurred before the peak strength was reached.

5. Conclusions

(1) Parallel pre-cracks had a significant effect on the mechanical properties of the rock–coal combined body. Compared to the intact specimens, the stress–strain curve of the rock–coal combined body containing parallel pre-cracks changed significantly, with a decrease in peak strength, peak strain, elastic modulus, and cracking stress;

(2) The closer the angle θ is to 30° and the longer the length L of the parallel pre-cracks, the lower the peak strength, peak strain, modulus of elasticity, and crack initiation stress of the specimen;

(3) For a rock–coal combined body with parallel pre-cracks, the macroscopic damage patterns are coal sample damage, coal–rock sample damage, and rock sample damage. The coal samples broke to different degrees with different parallel pre-crack parameters, and some showed no destruction;

(4) The rock–coal combined body with parallel pre-cracks exhibited three forms of crack initiation: crack initiation in the coal sample, simultaneous crack initiation at the tips of the parallel pre-cracks in the coal and rock samples, and crack initiation at the tips of the parallel pre-cracks in the rock sample.

Author Contributions: Conceptualization, Y.L.; methodology, J.L.; software, Y.L. and Q.Y.; validation, J.L.; data curation, Y.L.; writing—original draft preparation, Y.L.; writing—review and editing, J.L.; funding acquisition, J.L. All authors have read and agreed to the published version of the manuscript.

Funding: This work was supported by the Natural Science Foundation of Shandong Province China (grant number: ZR2022ME112, Jiangwei Liu).

Institutional Review Board Statement: Not applicable.

Informed Consent Statement: Not applicable.

Data Availability Statement: The data are available and explained in this article; readers can access the data supporting the conclusions of this study.

Conflicts of Interest: The authors declare no conflict of interest.

References

1. Chen, S.J.; Xia, Z.G.; Guo, W.J.; Shen, B.T. Research status and prospect of mining catastrophic response of rock mass under the influence of fault. *Coal Sci. Technol.* **2018**, *46*, 20–27. [CrossRef]
2. Li, F.X.; Yin, D.W.; Zhu, C.; Wang, F.; Jiang, N.; Zhang, Z. Effects of Kaolin Addition on Mechanical Properties for Cemented Coal Gangue-Fly Ash Backfill under Uniaxial Loading. *Energies* **2021**, *14*, 3693. [CrossRef]

3. Chen, S.J.; Zhang, J.C.; Yin, D.W.; Cheng, X.Z.; Jiang, N. Relative permeability measurem-ent of coal microchannels using advanced microchip technology. *Fuel* **2021**, *312*, 122633. [CrossRef]
4. Li, X.L.; Chen, S.J.; Wang, S.; Zhao, M.; Liu, H. Study on in situ stress distribution law of the deep mine taking linyi Mining area as an example. *Adv. Mater. Sci. Eng.* **2021**, *9*, 5594181. [CrossRef]
5. Deng, X.J.; Li, Y.; Wang, F.; Shi, X.M.; Yang, Y.C.; Xu, X.C.; Huang, Y.L.; Dewit, B. Experimental study on the mechanical properties and consolidation mechanism of microbial grouted backfill. *Int. J. Min. Sci. Technol.* **2022**, *32*, 271–282. [CrossRef]
6. Li, X.L.; Chen, S.J.; Liu, S.M.; Li, Z.H. AE waveform characteristics of rock mass under uniaxial loading based on Hilbert-Huang transform. *J. Cent. South Univ.* **2021**, *28*, 1843–1856. [CrossRef]
7. Liu, J.W.; Wu, N.; Si, G.Y.; Zhao, M.X. Experimental study on mechanical properties and failure behaviour of the pre-cracked coal-rock combination. *Bull Eng. Geol. Environ.* **2017**, *80*, 2307–2321. [CrossRef]
8. Kang, H.P. Seventy years development and prospects of strata control technologies for coal mine roadways in China. *Chin. J. Rock Mech. Eng.* **2021**, *40*, 1–30. [CrossRef]
9. Shi, X.D.; Bai, J.W.; Feng, G.R.; Wang, K.; Cui, B.Q.; Guo, J.; Yang, X.Y.; Song, C. Crack propagation law at the interface of FRP wrapped coal-backfilling composite structure. *Constr. Build. Mater.* **2022**, *344*, 128229. [CrossRef]
10. Dehghan, A.N.; Goshtasbi, K.; Ahangari, K.; Jin, Y. Experimental investigation of hydraulic fracture propagation in fractured blocks. *Bull Eng. Geol. Environ.* **2015**, *74*, 887–895. [CrossRef]
11. Kang, H.P.; Feng, Y.J. Hydraulic fracturing technology and its applications in strata control in underground coal mines. *Coal Sci. Technol.* **2017**, *45*, 1–9. [CrossRef]
12. Liu, J.W. Study on Change Mechanism and Control of Stress Field of Artificial Cracking Coal and Rock Mass. Ph.D. Thesis, China University of Mining, Xuzhou, China, 2021. [CrossRef]
13. Chen, S.J.; Li, Z.Y.; Ren, K.Q.; Feng, F.; Xia, Z.G. Experimental study on development process of reverse fault in coal measures strata and law of stress evolution in hanging wall strata. *J. Min. Saf. Eng.* **2020**, *37*, 366–375. [CrossRef]
14. Liu, C.; Zhang, D.M.; Zhao, H.G.; Li, M.H.; Song, Z.L. Experimental study on hydraulic fracturing properties of elliptical boreholes. *Bull Eng. Geol. Environ.* **2022**, *81*, 18. [CrossRef]
15. Chen, S.J.; Yin, D.W.; Zhang, B.L.; Ma, H.F.; Liu, X.Q. Mechanical characteristics and progressive failure mechanism of roof-coal pillar structure. *Chin. J. Rock Mech. Eng.* **2017**, *36*, 1588–1598. [CrossRef]
16. Chen, S.J.; Yin, D.W.; Jiang, N.; Wang, F.; Guo, W.J. Simulation study on effects of loading rate on uniaxial compression failure of composite rock-coal layer. *Geomech. Eng.* **2019**, *17*, 333–342. [CrossRef]
17. Guo, D.M.; Zuo, J.P.; Zhang, Y.; Yang, R.S. Research on strength and failure mechanism of deep coal-rock combination bodies of different inclined angles. *Rock Soil Mech.* **2011**, *32*, 1333–1339. [CrossRef]
18. Zuo, J.P.; Chen, Y.; Cui, F. Investigation on mechanical properties and rock burst tendency of differernt coal-rock combined bodies. *J. China Univ. Min. Technol.* **2018**, *47*, 81–87. [CrossRef]
19. Zhang, Z.T.; Liu, J.F.; Wang, L.; Liu, Z.J.; Yang, X.C. Effects of combination mode on mechanical properties and failure characteristics of the coal-rock combinations. *J. China Coal Soc.* **2012**, *37*, 841–847. [CrossRef]
20. Bai, J.W.; Feng, G.R.; Wang, Z.H.; Wang, S.Y.; Qi, T.Y.; Wang, P.F. Experimental Investigations on the Progressive Failure Characteristics of a Sandwiched Coal-Rock System Under Uniaxial Compression. *Appl. Sci.* **2019**, *9*, 1195. [CrossRef]
21. Zhao, D.C.; Xia, Y.J.; Zhang, C.Q.; Zhou, H.; Tang, C.A.; Liu, N.; Chen, J.; Wang, P.; Wang, C.l. Laboratory test and numerical simulations for 3D printed irregular columnar jointed rock masses under biaxial compression. *Bull Eng. Geol. Environ.* **2022**, *81*, 124. [CrossRef]
22. Xia, Y.J.; Liu, B.C.; Zhang, C.Q.; Liu, N.; Zhou, H.; Chen, J.; Tang, C.A.; Gao, Y.; Zhao, D.C.; Meng, Q.K. Investigations of mechanical and failure properties of 3D printed columnar jointed rock mass under true triaxial compression with one free face. *Geomech. Geophys. Geo-Energ. Geo-Resour.* **2022**, *8*, 26. [CrossRef]
23. Xia, Y.J.; Zhang, C.Q.; Zhou, H.; Hou, J.; Su, G.S.; Gao, Y.; Liu, N.; Singh, H.K. Mechanical behavior of structurally reconstructed irregular columnar jointed rock mass using 3D printing. *Eng. Geol.* **2020**, *268*, 105509. [CrossRef]
24. Reyes, O.; Einstein, H.H. Failure mechanisms of fractu-red rock: A fracture coalescence model. *Isrm Congr.* **1991**, *V1*, 333–340.
25. Zhang, X.P.; Wong, L.N.Y. Cracking processes in rock-like material containing a single flaw under uniaxial compression: A numerical study based on parallel bonded-particle model approach. *Rock Mech. Rock Eng.* **2012**, *45*, 711–737. [CrossRef]
26. Tang, Q.; Li, Y.A. Particle Flow Simulation on the Influence of Confinement on Crack Propagation in pre-cracked rock. *J. Yangtze River Sci. Res. Inst.* **2015**, *32*, 81–85.
27. Yin, D.W.; Chen, S.J.; Liu, X.Q.; Wang, G.; Ma, H.F.; Shen, B.T. Particle Flow Simulation of Effects of Joint Characteristics on Crack Initiation Stress and Crack Damage Stress of Jointed Granite. *J. Basic Sci. Eng.* **2018**, *26*, 808–820. [CrossRef]
28. Zhang, X.L. Research on Crack Initiation and Propagation Characteristics of Disc Specimens Containing a Parallel Prefabricated Double Cracks. Master's Thesis, Henan Polytechnic University, Jiaozuo, China, 2020. [CrossRef]
29. Yang, S.Q. Crack coalescence behavior of brittle sandstone samples containing two coplanar fissures in the process of deformation failure. *Eng. Fract. Mech.* **2011**, *78*, 3059–3081. [CrossRef]
30. Yang, S.Q.; Wen, S.; Li, L.Q. Experimental study on deformation and strength properties of coarse marble with discontinuous pre-existing cracks under different confining pressures. *Chin. J. Rock Mech. Eng.* **2007**, *26*, 1572–1587.
31. Yang, S.Q.; Dai, Y.H.; Han, L.J.; He, Y.N.; Li, Y.S. Uniaxial compression experimental research on deformation and failure properties of brittle marble specimen with pre-existing fissures. *Chin. J. Rock Mech. Eng.* **2009**, *28*, 2391–2404.

32. Yang, S.Q.; Huang, Y.H.; Wen, S. Experimental study of mechanical behavior of red sandstone with two non-coplanar fissures after high temperature heating. *Chin. J. Rock Mech. Eng.* **2015**, *34*, 440–451. [CrossRef]
33. Xie, L.F.; Zhu, Q.Y.; Qin, Y.J.; Wang, J.H.; Qian, J.G. Study on evolutionary characteristics of toppling deformation of anti-dip bank slope based on energy field. *Sustainability* **2020**, *12*, 7544. [CrossRef]
34. Chen, S.J.; Ge, Y.; Yin, D.W.; Yang, H.S. An experimental study of the uniaxial failure behaviour of rock-coal composite samples with pre-existing cracks in the coal. *Adv. Civ. Eng.* **2019**, *2019*, 8397598. [CrossRef]
35. Zhang, X.P.; Wong, L.N.Y. Crack initiation propagation and coalescence in rock-like material containing two flaws: A numerical study based on bonded-particle model approach. *Rock Mech. Rock Eng.* **2013**, *46*, 1001–1021. [CrossRef]
36. Kong, X.G.; He, D.; Liu, X.F.; Wang, E.Y.; Li, S.G.; Liu, T.; Ji, P.F.; Deng, D.Y.; Yang, S.R. Strain characteristics and energy dissipation laws of gas-bearing coal during impact fracture process. *Energy* **2022**, *242*, 123038. [CrossRef]
37. Shen, J.Y.; Zhan, S.X.; Karakus, M.; Zuo, J.P. Effects of flaw width on cracking behavior of single-flawed rock specimens. *Bull Eng. Geol. Environ.* **2021**, *80*, 1701–1711. [CrossRef]
38. Ma, J.B.; Jiang, N.; Wang, X.J.; Jia, X.D.; Yao, D.H. Numerical study of the strength and characteristics of sandstone samples with combined double hole and double fissure defects. *Sustainability* **2021**, *13*, 7090. [CrossRef]
39. Kong, X.G.; Li, S.G.; Wang, E.Y.; Wang, X.; Zhou, Y.X.; Ji, P.F.; Shuang, H.Q.; Li, S.R.; Wei, Z.Y. Experimental and numerical investigations on dynamic mechanical responses and failure process of gas-bearing coal under impact load. *Soil Dyn. Earthq. Eng.* **2021**, *142*, 106579. [CrossRef]
40. Liu, J.W.; Liu, C.Y.; Yao, Q.L.; Si, G.Y. The position of hydraulic fracturing to initiate vertical fractures in hard hanging roof for stress relief. *Int. J. Rock Mech. Min. Sci.* **2020**, *132*, 104328. [CrossRef]

sustainability

MDPI

Article

Safety and Protection Measures of Underground Non-Coal Mines with Mining Depth over 800 m: A Case Study in Shandong, China

Li Cheng [1], Qinzheng Wu [1], Haotian Li [2], Kexu Chen [1], Chunlong Wang [1], Xingquan Liu [1], Xuelong Li [2,*] and Jingjing Meng [3]

1 Deep Mining Laboratory of Shandong Gold Group Co., Yantai 264000, China
2 College of Energy and Mining Engineering, Shandong University of Science and Technology, Qingdao 266590, China
3 Department of Civil, Environmental and Natural Resources Engineering, Luleå University of Technology, 971 87 Luleå, Sweden
* Correspondence: lixl@sdust.edu.cn

Abstract: With the increase in mining depth, the risk of ground pressure disasters in yellow gold mines is becoming more and more serious. This paper carries out a borehole test for the pressure behavior in a non-coal mining area with a mining depth of more than 800 m in the Jiaodong area. The test results show that under a depth of 1050 m, the increase in the vertical principal stress is the same as the increase in the minimum horizontal principal stress, which is about 3 MPa per 100 m. When the depth increases to 1350 m, the vertical principal stress increases by about 3% per 100 m, and the self-weight stress and the maximum horizontal principal stress maintain a steady growth rate of about 3 MPa per 100 m. In addition, based on the test results, the operation of the ground pressure monitoring system in each mine is investigated. The investigation results show that in some of the roadway and stope mines with depths of more than 800 m, varying degrees of rock mass instability have occurred, and a few mines have had sporadic slight rockbursts, accounting for about 5%. There was a stress concentration area in the lower part of the goaf formed in the early stage of mining, and slight rockburst phenomena such as rock mass ejection have occurred; meanwhile, the area stability for normal production and construction was good, and there was no obvious ground pressure. This paper compares the researched mines horizontally as well as to international high-level mines and puts forward some suggestions, including: carrying out ground pressure investigations and improving the level of intelligence, which would provide countermeasures to balance the safety risks of deep mining, reducing all kinds of safety production accidents and providing a solid basis for risk prevention and supervision.

Keywords: non-coal mine; great mining depth; ground pressure disaster; safety investigation

Citation: Cheng, L.; Wu, Q.; Li, H.; Chen, K.; Wang, C.; Liu, X.; Li, X.; Meng, J. Safety and Protection Measures of Underground Non-Coal Mines with Mining Depth over 800 m: A Case Study in Shandong, China. *Sustainability* **2022**, *14*, 13345. https://doi.org/10.3390/su142013345

Academic Editors: Xiangguo Kong, Dexing Li and Xiaoran Wang

Received: 2 September 2022
Accepted: 28 September 2022
Published: 17 October 2022

1. Introduction

China is one of the primary gold-producing areas in the world. As a critical non-coal mineral resource, gold accounts for a large proportion of the national economy [1,2]. The Jiaodong area of Shandong Province [3], which is the largest gold producing area in China, produces many world-class large gold deposits (more than 100 tons) and many small- and medium-sized gold deposits [4,5], including the Zhaoyuan-Laizhou, Penglai-Qixia, and Muping-Rushan gold deposits. Among them, the gold resources of the Zhaoyuan-Laizhou metallogenic area account for more than 80% of the total reserves of Shandong Province. Sai et al. inferred the southward extension of the Zhaoyuan-Pingdu gold belt under the Quaternary cover area via comprehensive multivariate information, such as geological exploration, geophysical, and geochemical data [6]. Cheng et al. established metallogenic models of altered rock type gold deposits and quartz vein type gold deposits in the fracture zone, which verifies the great prospecting potential in the Jiaodong area [7].

With the increase in mining depth, the risk of ground pressure disasters is becoming more serious. Defining the distribution of the in situ stress field can effectively and accurately measure, prevent, and control ground pressure disasters. At present, in situ stress surveys have been carried out in more than thirty countries globally [8–10], but the application of technical research in China lags. The main task of detecting underground stress fields in deep mines is determining the activity intensity and the mode of ground stress, as well as the direction of the main stress and its fundamental mechanism of changing with depth [11]. Liu et al. obtained the rockburst tendency evaluation results by drilling cores in the Sanshandao gold mine. It was proposed that the corresponding intensity grade would increase with the mining depth [12]. Qin et al. proposed that in situ stress measurements should select measuring points according to rock mass structures and should avoid interference sources, such as stress fields and unstable regions, to ensure the authenticity of the stress data. Their theory played a guiding role in the selection of the Xiling mining area in Sanshandao as the drilling location [13]. Based on the measurements of the in situ stress field of a specific mine in Jiaodong, it was found that the stress in the deep mining area is mainly horizontal tectonic stress [14], and proper support was suggested to be provided in time. Li et al. drilled fifty-three measuring points in the Sanshandao, Xincheng, and Linglong gold mines [15], and they found that the maximum horizontal stress increases with depth. Du et al. improved the traditional coring technique and predicted the potential location of rockbursts and their intensity in future mining activities of the Sanshandao gold mine using a theoretical analysis and numerical simulation [16].

The appearance of ground pressure will have an impact on production safety, so some mines have adopted new support schemes [17–19]. Jinqingding mine of the Jinzhou Mining Group is gradually replacing its rigid support with flexible support [20,21], and advanced grouting is used in local broken areas. In the Sanshandao gold mine, resin bolts are widely used in deep support to replace pipe joint bolts, anchor-shotcreting nets, and piercing belts during excavation [22–24]. These measures effectively reduce the damage caused by ground pressure, but at the same time, their high cost makes it difficult to popularize them in all mines. We aim to find a more reasonable support scheme. In addition, numerical simulation software, such as COMSOL and FLAC3D, can establish mathematical models more intuitively and promote the research of mine safety production [25–28]. Some studies have used the constitutive model or time–frequency transformations to study rock mass and coal-rock assemblage, and an instability prediction method has been established [29–31]. However, the above research does not consider the problem of high temperature, ignores the complex physical and chemical interactions between rock masses, and is not suitable for deeper mining.

Compared with foreign advanced mines, there are some common problems in domestic gold mines, such as a lack of rock mechanics data [32], unsuitable support in some areas [33], and so on. Based on previous studies, this paper found some new problems, such as inadequate mine pressure monitoring, imperfect ventilation systems, and so on. This paper focuses on the present situation and the existing problems of ground pressure, and puts forward some suggestions for preventing ground pressure disasters and ensuring safe productions. In addition, a basic microseismic monitoring and analysis method is determined through in situ stress measurement and analysis, and a long-term construction strategy of the mine is proposed, which provides a feasible scheme for the construction of a green, intelligent mine.

2. Basic Situation of Mines

In regards to their geographical location, the nineteen underground non-coal mines investigated in this study are distributed in Yantai and Weihai. There is one in Muping, Yantai, three in Penglai, one in Longkou, twelve in Zhaoyuan, one in Laizhou, and one in Rushan, Weihai. Sixty percent of the mines are located in Zhaoyuan and fifteen percent in Penglai. The geographical location of each mine is shown in Figure 1. In terms of ownership, six belong to the Shandong Gold Group, one belongs to a local state-owned enterprise, four belong to the Shandong Zhaojin Group, four belong to the Shandong China Mining Group, and the remaining four belong to collective enterprises or private enterprises. Collective city-owned mines accounted for 37%, and provincial state-owned mines accounted for 32%. When classifying the mines by their production and construction stages, one is newly built. Five have completed reconstruction and expansion as well as the completion and acceptance of their safety facilities. Thirteen are in stages of reconstruction and expansion during production, accounting for 68%. Among them, the safety facilities of eleven mines of the project have passed acceptance inspections, and two have not yet had complete acceptance of their safety facilities. In terms of their states of production and construction, twelve are in normal production and construction, and six are in policy shutdown or accident shutdown, accounting for 30%. The situation is shown in Table 1. Pvso means provincial state-owned, Cco means collective city-owned, Tse means township enterprises, Pve means private enterprise, Pdt means production, Cst means construction, Nm means normal construction, Plc std means policy shutdown, and Acd std means accident shutdown.

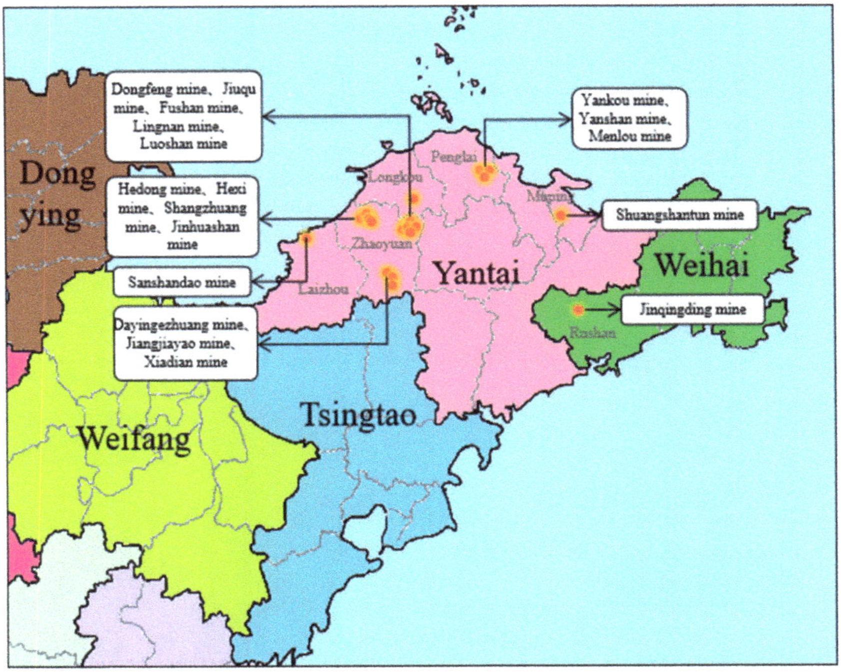

Figure 1. Geographical map of nineteen mines with mining depths of more than 800 m.

Table 1. List of location, ownership, stage, and status of nineteen mines with mining depths of more than 800 m.

No	Abbreviation	Location	Owner-Ship	Stage	Status
1	Yantai Baihen Shuangshantun Mine	Muping	Pve	Cst	Nm
2	Shnajin Jinchuang Yanshan Mine	Penglai	Pvso	Pdt+Cst	Nm
3	Shanjin Jinchuang Yankou Mine	Penglai	Pvso	Pdt+Cst	Nm
4	Penglai Menlou Mine	Penglai	Pve	Pdt+Cst	Acd std

Table 1. *Cont.*

No	Abbreviation	Location	Owner-Ship	Stage	Status
5	Longkou Jintai Damoqujia Mine	Longkou	Pve	Pdt+Cst	Plc std
6	Shanjin Linglong Jiuqu Mine	Zhaoyuan	Pvso	Pdt	Plc std
7	Shanjin Dongfeng Mine	Zhaoyuan	Pvso	Pdt	Nm
8	Zhaojin Xiadian Mine	Zhaoyuan	Cco	Pdt+Cst	Nm
9	Zhaojin Dayingezhuang Mine	Zhaoyuan	Cco	Pdt+Cst	Nm
10	Zhaojin Shangzhuang Mine	Zhaoyuan	Cco	Pdt+Cst	Nm
11	Zhaojin Hedong Mine	Zhaoyuan	Cco	Pdt	Nm
12	Zhongkuang Lingnan Fifth Mine	Zhaoyuan	Cco	Pdt+Cst	Plc std
13	Zhongkuang Luoshan Forth Mine	Zhaoyuan	Cco	Pdt+Cst	Plc std
14	Zhongkuang Fushan Dongfeng Mine	Zhaoyuan	Cco	Pdt	Plc std
15	Zhaoyuan Jiangjiayao Mine	Zhaoyuan	Tse	Pdt+Cst	Nm
16	Zhaoyuan Hexi Mine	Zhaoyuan	Tse	Pdt	Nm
17	ZhaoyuanLingshan Jinhuashan Mine	Zhaoyuan	Tse	Pdt+Cst	Nm
18	Shanjin Sanshandao Mine	Laizhou	Pvso	Pdt+Cst	Nm
19	Shanjin Jinzhou Jinqingding Mine	Rushan	Pvso	Pdt+Cst	Nm

2.1. Geological Condition

The nineteen mines investigated are all gold mines, and all of them are distributed in the Jiaodong area. In terms of geotectonic location, the Jiaodong area is located on the southeastern margin of the North China Craton, bounded by the Tan-Lu fault zone in the west and the northern segment of the Sulu ultra-high-pressure metamorphic belt in the east (Figure 2). It is China's largest gold-producing area, producing several world-class large gold deposits (more than 100 tons) and many small- and medium-sized gold deposits. There are mainly three gold metallogenic areas in the Jiaodong area: Zhaoyuan-Laizhou, Penglai-Qixia, and Muping-Rushan. Most of the investigated mines are located in the Zhaoyuan-Laizhou metallogenic area. The gold reserve in this area accounts for more than 80% of the gold reserves in Shandong Province.

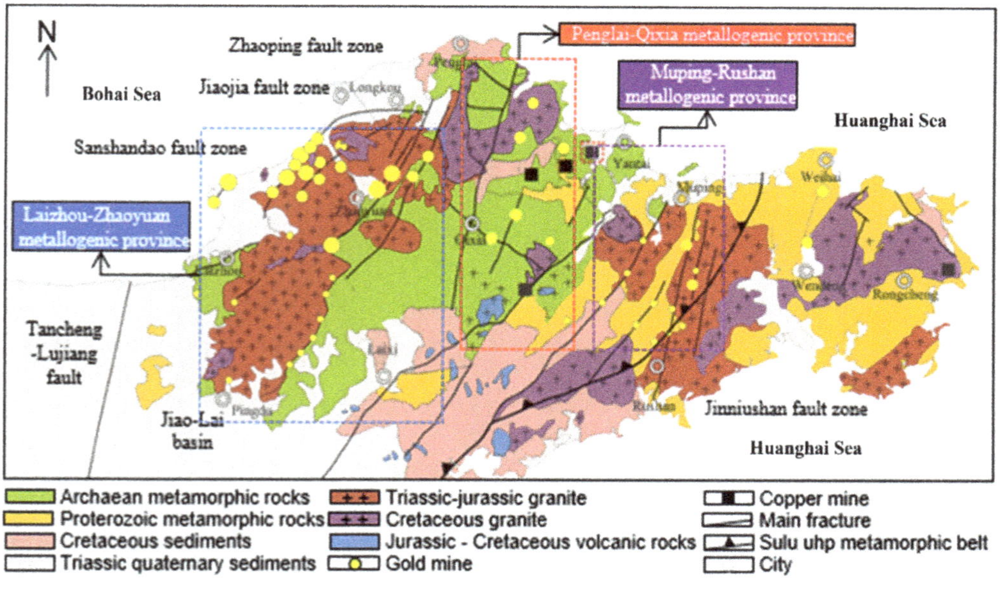

Figure 2. Regional geological map of Jiaodong area.

The strata exposed in the Jiaodong area are mainly Upper Archean Jiaodong Group, Lower Proterozoic Jingshan Group, Fanzishan Group, Upper Proterozoic Penglai Group, Mesozoic Jurassic and Cretaceous, and the Tertiary and Quaternary of Cenozoic. Magmatic rocks are widely distributed in Jiaodong, and intrusive rocks are distributed everywhere, accounting for more than half of the bedrock area (Figure 3). The magmatic rocks in the area can be divided into five types: exquisite granite, Guojialing granite, Aishan granite, Kunyushan granite, and Sanfoshan granite.

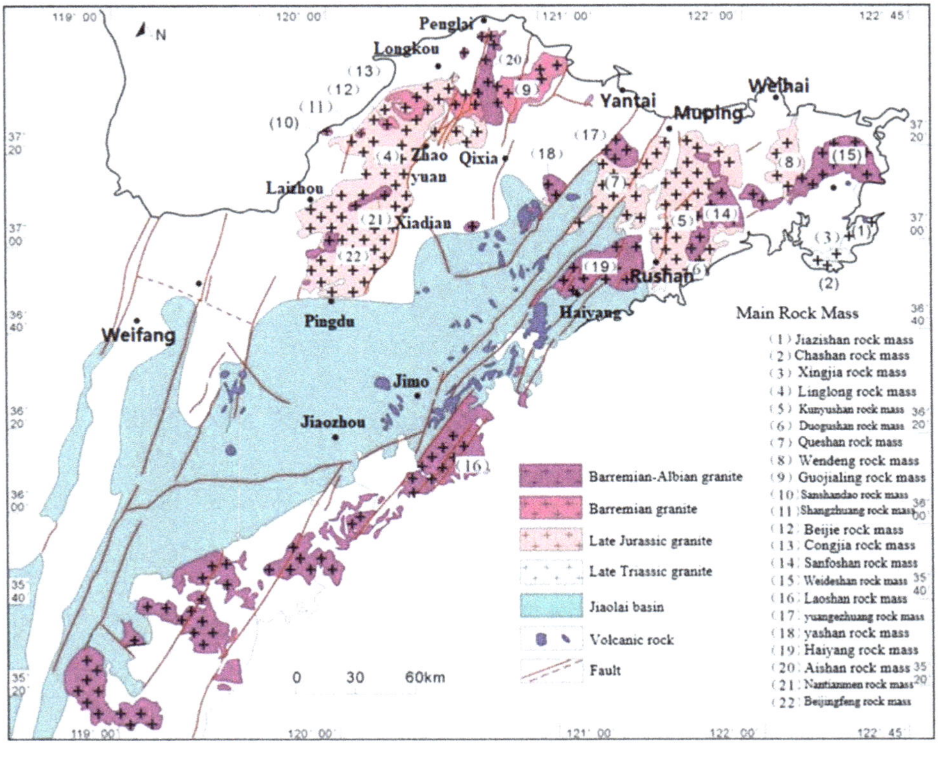

Figure 3. Magmatic rock distribution map in the Jiaodong area.

The development of fault structures characterizes the basic structural framework of Jiaodong. Its main fault zones are along the NE and NNE directions. The secondary fault zones are along the NW and SN directions (Figure 4). There are seven major faults in the region: the Yishu fault (Shandong section of Tanlu fault), Wulian-Jimo-Muping fault, Zhao-Ping fault, Long-Lai fault, San-Cang fault, Mu-Ru fault, and the Peng-Qi fault. Among them, the the Yishu fault and the Wulian-Jimo-Muping fault zone are the first-order structures in the region. In contrast, the Zhao-Ping fault, Long-Lai fault, San-Cang fault, Mu-Ru fault, and Peng-Qi fault are the second-order faults in the region. They are the main ore-controlling structures of gold deposits.

The Dongfeng Mining area, Jiuqu Sub-Mine of Linglong Gold Mine, Fushan Gold Mine, Lingnan Gold Mine, Luoshan Gold Mine, Damoqujia Gold Mine, and the Dayingezhuang, Xiadian, and Jiangjiayao gold deposits are located in the Zhaoping-Ping fault zone. The Hexi mining area, Hedong mining area, Shangzhuang mining area, and Jinhuashan mining area are located in the Long-Lai fault zone. The Xishan sub-deposit of the Sanshandao gold mine is located in the San-Cang fault zone. The Jinqingding mining area and Shuangshan-tun mining area are located in the Mu-Ru fault zone. The Yanshan mining area, Yankou mining area, and Menlou mining area are located in the Peng-Qi fault zone.

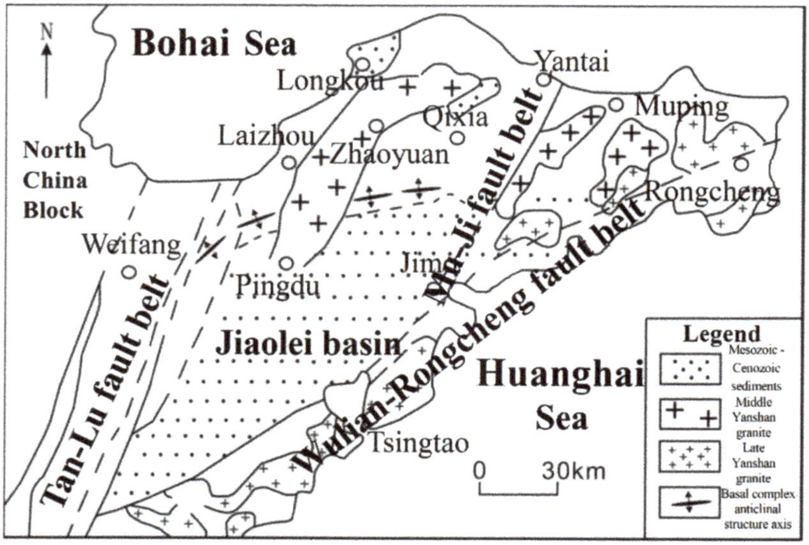

Figure 4. Brief map of tectonic geology in Jiaodong area.

2.2. Production Situation

Among the nineteen mines, there are nine large mines with an annual production scale of more than 150,000 tons, nine medium-sized mines with an annual production capacity of 60,000–150,000 tons, and one small mine with an annual production capacity of fewer than 60,000 tons. These mines have a total designed annual production capacity of 7.9301 million tons.

Nine mines adopt the combined development method of the open shaft and blind shaft. Two mines adopt the combined development method of the open shaft and (auxiliary) ramp. One mine adopts the combined development method of the open shaft, inclined shaft, and blind shaft. One mine adopts the combined development method of drift, open shaft, blind shaft, and blind inclined shaft. There is one mine with a development depth of less than 800 m, twelve mines with depths of 800~1000 m, and six mines with depths of more than 1000 m. The mine with the deepest development depth is the Zhaojin Xiadian gold mine, which is 1565 m. There are eleven mines with the main production horizon above −800 m and one mine with −800~1000 m, and the rest of the mines are in a state of shutdown or have not yet been put into production.

3. Present Situation of Deep Ground Pressure in Non-coal Mines

3.1. Regional Tectonic Stress Level

Related studies have calculated the linear regression of the relationship between the maximum horizontal principal stress and the minimum horizontal principal stress with depth in the northwest of Jiaodong. The results show that when the depth reaches 1000 m, the maximum horizontal principal stress is about 40 MPa. Most of the levels of minimum horizontal principal stress are close to 20 MPa, and the highest can reach 30 MPa, which belongs to the high ground stress areas.

In the Sanshandao Xiling mining area, the coring hole position is determined according to a comprehensive and significant principle. The coring line is arranged in a cross shape with the 88# borehole exploration line as the axis, taking the northeast and southwest directions into account. Finally, five representative prospecting boreholes are identified, as shown in Figures 5 and 6.

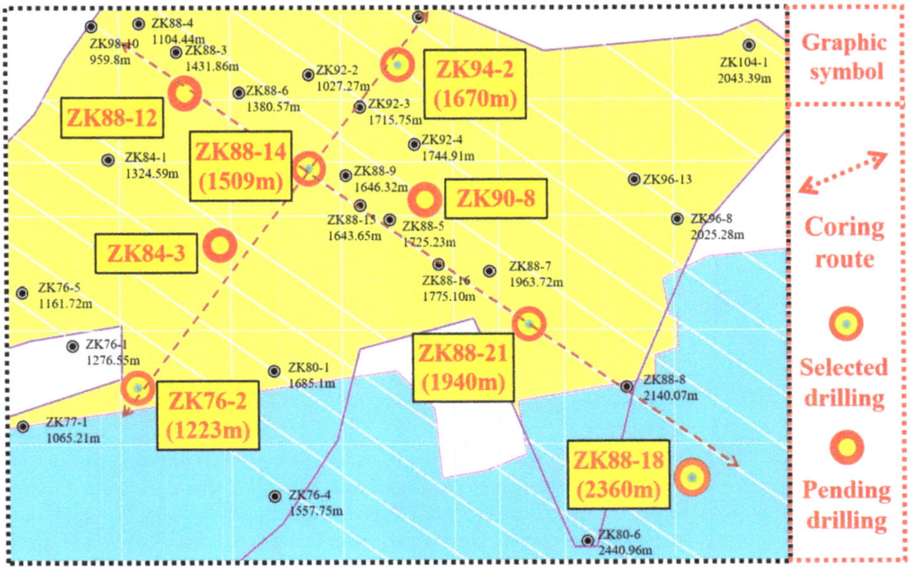

Figure 5. Selection of core drilling location.

Figure 6. Part of the experimental specimens.

In the Xincheng Gold Mine, based on core reorientation technology [34], the in situ stress measurement is carried out by drilling cores with seven different depths of −950 m, −1050 m, −1150 m, −1250 m, −1350 m, −1450 m, and −1550 m. The in situ stress is measured by acoustic emission. The results are shown in Table 2.

Table 2. Stress distribution at different depths, ref. [35].

Measuring Point Depth (m)	Vertical Principal Stress σ_v (MPa)	Self-Weight Stressσ(MPa)	Maximum Horizontal Principal Stress σ_H (MPa)	Minimum Horizontal Principal Stress σ_h (MPa)	Horizontal Maximum Principal Stress Direction θ
950	26.41	26.46	32.78	14.52	N103°34′ E
1050	29.22	29.16	41.83	17.18	N103°42′ E
1150	33.33	31.86	35.37	16.64	N95°43′ E
1250	34.25	34.56	38.95	16.85	N110°13′ E
1350	35.98	37.26	41.23	16.44	N105°26′ E
1450	40.01	39.96	40.63	14.02	N111°20′ E
1550	40.95	42.66	45.84	17.56	N107°33′ E

It is shown that the vertical principal stress, the self-weight stress, and the minimum horizontal principal stress increase the same amount at the depth of 1050 m, which is about 3 MPa. When the depth increases to 1350 m, the vertical principal stress increases by about

3% per 100 m, and the self-weight stress and the maximum horizontal principal stress maintain a steady growth rate of about 3 MPa per 100 m. When the depth increases from 1350 m to 1550 m, due to the influence of geological structure, the vertical principal stress increases rapidly by 14%, while the self-weight stress continues to increase at a relatively stable rate. It is worth noting that due to the existence of geological structures, both the maximum horizontal principal stress and the minimum horizontal principal stress decrease slightly from depths of 1350 m to 1450 m, and then rise slightly to the normal level at the speed of 16% with the increase in depth.

(1) The principal vertical stress increases nearly linearly with the increase in coring depth, which is consistent with the buried depth's dead weight stress (Figure 7).
(2) The in situ stress in the borehole is located mainly in horizontal tectonic stress. In the different depths of the boreholes, the maximum stress is the horizontal maximum principal stress, and the intermediate principal stress is the vertical stress. With the increase in the borehole depth, the dominant role of the horizontal tectonic stress field weakens, and the effect of the self-weight stress field tends to increase, as shown in Figure 8.
(3) The horizontal maximum principal stress increases with the increase in hole depth. It is worth noting that the maximum principal stress shows a sudden increase at −1050 m, reaching 41.83 MPa (higher than the stress value of −1150 m and −1250 m at greater depths). This indicates an unstable local change in the maximum principal stress of this area.
(4) The azimuth of the maximum horizontal principal stress at different depths is consistent; all values are along the NWW~SEE direction, between N95° E and N111° E, as shown in Figure 9. This shows that the location of the new shaft of the Xincheng Gold Mine is consistent with its geological structure and the orientation of the maximum principal stress measured by many scientific institutions above its −950 m elevation.
(5) The maximum horizontal principal stress ratio to the principal vertical stress (lateral pressure coefficient) in the measured area is between 1.01 and 1.43, which is consistent with the distribution law of the lateral pressure coefficient in China [36,37].

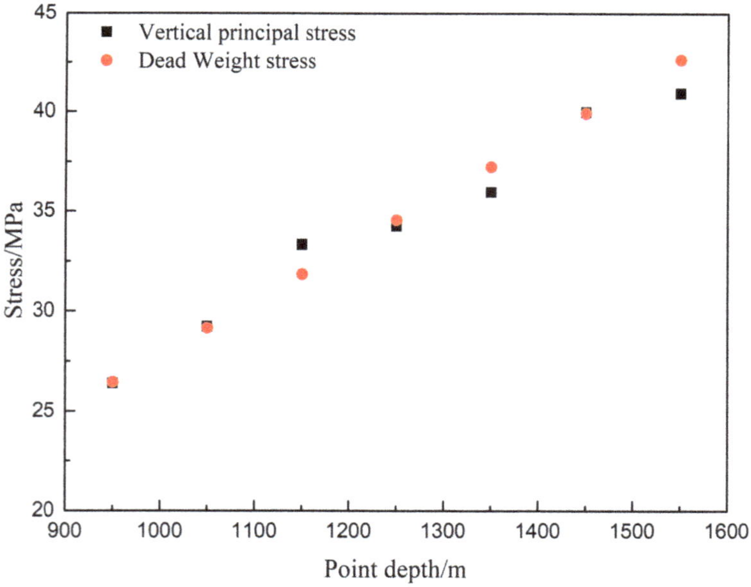

Figure 7. Relationship between in situ vertical stress, self-weight stress, and drilling depth.

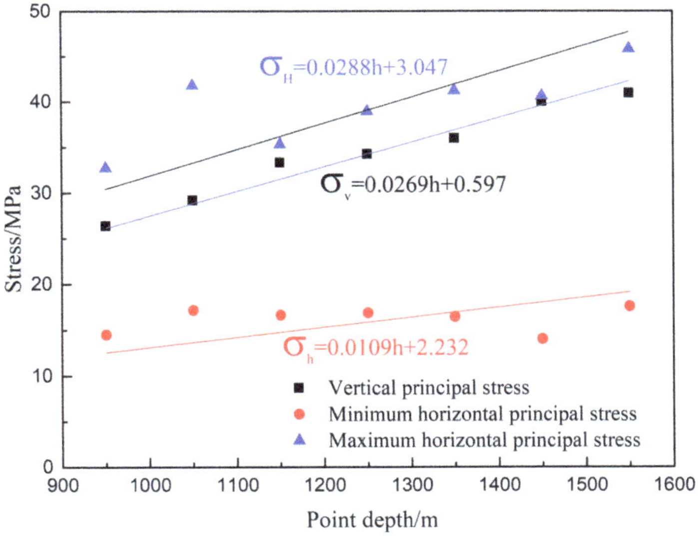

Figure 8. Comparison of stresses in different directions.

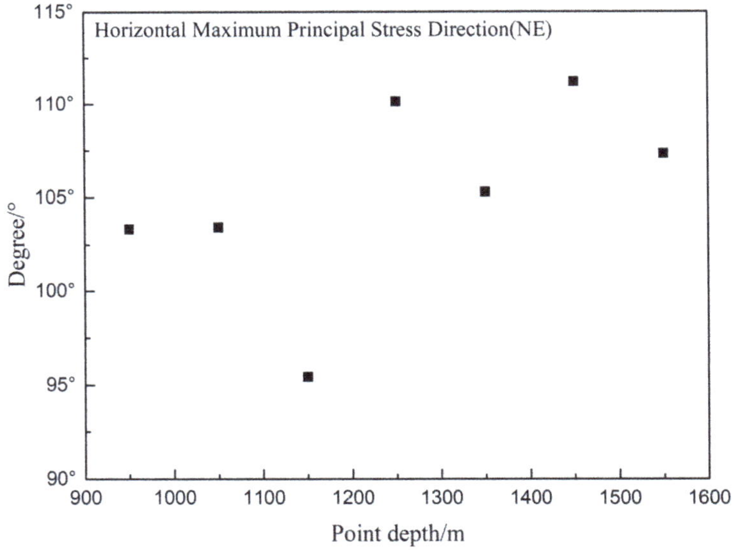

Figure 9. Variation trend of horizontal maximum principal stress direction with depth.

3.2. Deep Ground Pressure Appearance

In the roadways and stope mines with mining depths of more than 800 m, rock mass instability has occurred to varying degrees in some mines, and slight rockburst has occurred in a few mines. In the deep development process, some mines have a failure tendency of roadway surrounding rock. This is because of the complex regional geological structure, broken rock mass, high in situ stress level, or because the roadway axis is not parallel to the direction of horizontal maximum principal stress. Few mines have long mining histories and many pressure concentration areas in the lower part of the goaf have formed at early mining stages. Therefore, slight rockburst phenomena such as rock mass ejection have occurred.

Most of the investigated areas of production and construction in the mines are stable, and there is no obvious ground pressure. There are also slight roof falls, slopes, or rock spalling in some local areas of the mines, as shown in Figure 10.

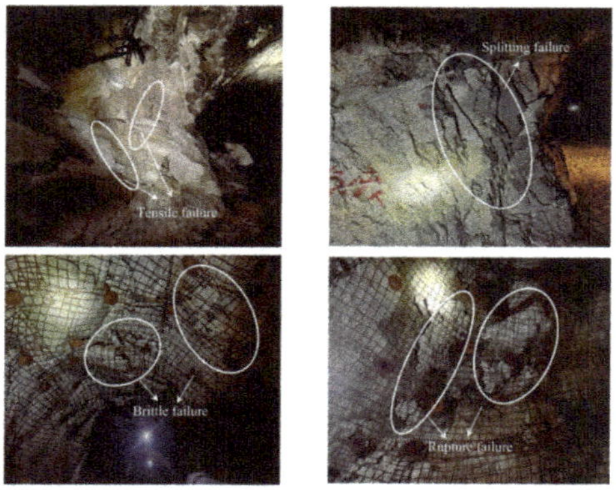

Figure 10. Deep ground pressure appearance.

3.3. Control Measures of Deep Ground Pressure

At present, bolting and shotcreting net flexible support measures are widely used to control rock mass instability in deep non-coal mines in Shandong province. The Jinqingding mining area of the Jinzhou Mining Group is gradually replacing rigid support with flexible support and uses advanced grouting in local broken areas [38]. In the Sanshandao gold mine, resin bolts are widely used in deep support to replace pipe joint bolts, anchor-shotcreting nets, and piercing belts for support when excavating. At the same time, the Jinzhou Mining Group actively cooperates with relevant departments to develop prestressed anchors and constantly optimize its deep support technology [39–41]. Long anchor cables are used to support the stope's roof in the Jiangjiayao gold mine.

3.4. Risk of Rockburst in Deep Mines

Among the nineteen mines investigated, the typical lithology of the upper and lower surrounding rock was amphibolite, monzonitic granite, sericite, granite, potash granite, granulite, and so on. Therefore, the evaluation result of the rockburst tendency based on the drilling core of the Shanjin Sanshandao gold mine is instructive in evaluating the rockburst tendency in the region. Typical monzonitic granite, granite, and sericite have more than moderate rockburst tendencies (Figure 11). With the increase in burial depth, the corresponding intensity grade increases.

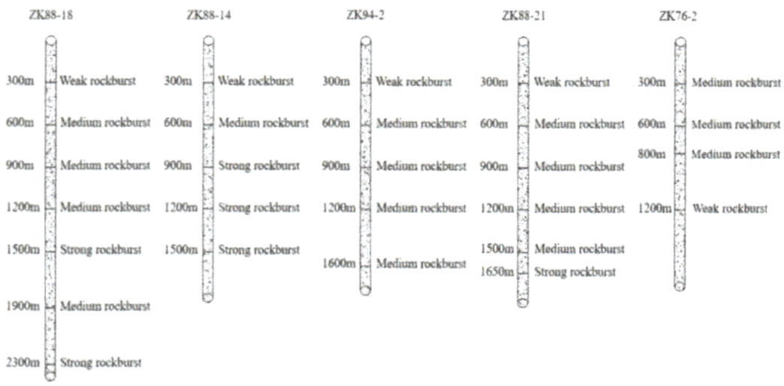

Figure 11. Schematic diagram of rockburst tendency grade of boreholes varying with depth in Xiling mining area of Sanshandao.

4. Problems Existing in the Mines

4.1. The Deep Development System Is Complex

Overall, the deep development system of the investigated mines is more complex, and the development engineering includes open shafts, blind shafts, blind inclined shafts, ramps, and so on. Commonly, there are more than ten shafts in a mine. From the point of view of mine construction, the mining projects are not reasonably planned. Design and construction are often needed due to later expansion and capacity expansion, resulting in the complexity of the development system. As a result, there is a high cost for mine capital construction. Additionally, system maintenance requires a heavy workload. Furthermore, it is difficult to manage multiple lifting systems and there are many hidden dangers in the processes of operation.

4.2. Lack of Rock Mechanics Data in Most Deep Mines

Among the nineteen mines, the Shanjin Linglong Jiuqu mine, Shanjin Dongfeng mining area, Zhaojin Xiadian gold mine, Zhaojin Dayingezhuang gold mine, and Zhongkuang Lingnan No. 5 mining area have carried out preliminary research on deep rock mechanics and accumulated some basic data on rock mechanics. The Sanshandao gold mine has carried out more systematic rock mechanics research, while the other thirteen mines have not yet carried out rock mechanics data collection or related research. The research on in situ stress measurement, rock mass quality classification, and rock mass stability analysis is an essential reference for mining engineering design.

Mine rock mechanics data are the basis of mining design. With the increase in mining depth, engineering designs based on empirical analogies gradually show some problems of unsuitability and low reliability. Two problems have been found in the investigation of this paper. First, most of those working in the mines lack an understanding of the importance of rock mechanics data collection, management, and ideological research work. Second, the obtained basic data of rock mechanics have locality and discontinuity. Usually there are only simple strength test data or partial area data, which lack systematic studies, and the guiding significance for mining engineering design is limited.

4.3. Some Support Engineering in Deep Mining Is Unsuitable and Not in Place

Nineteen mines have adopted the necessary shotcreting or combined anchor supports, shotcreting nets, or steel arch supports in unstable areas, especially in areas with broken structures. The roofs of some unstable areas have adopted long anchor cable supports. However, there are still three problems.

First, the support scheme in some areas is unsuitable. For example, roadways with obvious deformations adopt rigid support under high stress. This often does not have an excellent effect on support, but it increases the cost of the support due to the deformation and failure of the support.

Second, the support is not timely nor in place. For example, when the stope, roadway, and drift are temporarily stable in the deep part, many mines do not take or take insufficient support measures to save support costs.

Third, the support design is unsuitable due to the lack of basic rock mechanics data and rock mass stability classification standards. Moreover, the selection of support parameters lacks references, resulting in excessive support or support space.

4.4. The Ground Pressure Monitoring System Lacks Maintenance, and the Early Warning Ability Is Poor

Among the investigated nineteen mines, thirteen mines have installed ground pressure online monitoring systems, and some mines have also installed several monitoring systems. However, only a few of the mines with the system can operate normally, and the monitoring and early warning systems are poor, so there are the following problems:

① After installation, most online ground pressure monitoring systems are not inspected and maintained. A few mines carry out maintenance after equipment damage, which lacks timeliness, resulting in the loss of a lot of monitoring data and the omission of microseismic events.

② The monitoring data cannot be analyzed and processed in time. Only a few mines entrust technical service institutions for data analysis. There is a severe lag in data analysis, and the ground pressure online systems' monitoring and early warning functions cannot be fully utilized.

③ Transmission formats of monitoring data are various. The online monitoring systems of ground pressure in mines investigated are independently developed by domestic manufacturers, universities, or scientific research institutes, and some are imported from abroad. As a result, the data storage formats are various. The intellectual property rights of imported equipment are strictly protected abroad, so there are some difficulties converting different data formats into a unified standard format, which causes some obstacles to the integrated management of the online monitoring system of underground non-coal mine ground pressure.

④ The ground pressure online monitoring system has a low positioning accuracy and a poor early warning ability for microseismic events. The online microseismic monitoring system of ground pressure has difficulty locating rock fractures accurately. At the same time, the installed online monitoring system of ground pressure only has a real-time monitoring function and cannot reduce noise in real time. The risk grade of ground pressure disasters cannot be judged, so it is challenging to implement ground pressure disaster prediction and early warning.

4.5. The Temperature in the Deep Mine Is High, the Humidity Exceeds the Standard, and the Ventilation System Is Not Perfect

At present, some non-coal mines with a mining depth of over 800 m are suffering from heat damage caused by surrounding rocks and groundwater during mining [40]. Imperfect ventilation systems and lagging ventilation projects result in high local working surface temperatures, excessive humidity, and a poor working environment, which affects the efficiency of work and equipment.

4.6. The Intelligence Level of Deep Mine Is Poor

Except for the Sanshandao gold mine, the intelligence levels of the other mines are relatively poor. Whether it is the construction of the related intelligent data platforms, centralized management, the control platform, underground mining, excavation, loading, or transportation equipment, there is a significant gap in the implementation of intelligent mining.

5. Prevention and Control Countermeasures and Suggestions

Given the above problems and the current situation of each production system, the following suggestions are put forward.

5.1. Establish and Improve the Working Organization and System of Safety Risk Prevention in Deep Mining

To prevent the safety risks, deep mines should establish and improve the work organization and safety risk prevention system, and set up a leading group for safety risk prevention headed by technical personnel and management personnel.

5.2. Comprehensively Consider the Planning and Layout and Optimize the System Layout as a Whole

In mine construction planning, all deep mines should consider the overall layout and optimize the system layout to avoid a large number of new development projects in the process of production succession.

5.3. Start the Work of Mine Rock Mechanics and Establish the Working System of Rock Mechanics

To ensure the safety of deep mining, all deep mines should be equipped with engineers and technicians to collect rock mechanics data, such as investigations on rock mass structural planes, rock sampling, physical and mechanical index tests, in situ stress tests, point load index tests, and rock mass quality classification and stability analyses, etc. A computer-aided system for deep rock mass stability and support should be developed to provide guidance for the deep shaft and roadway, the stope design, and the support design and to improve the rationality and reliability of the design.

5.4. Carry Out Ground Pressure Investigation

It is necessary to investigate the ground pressure of the roadway, stope, and drift with a mining depth of more than 800 m, record the instability and failures of rock mass and support failures, and form investigation files. Ground pressure monitoring should be strengthened for the areas with frequent ground strata behaviors.

5.5. Test and Evaluate the Tendency of Rockburst

Deep mines should entrust scientific institutions to conduct rockburst tendency tests and evaluate the possible rockburst intensity by prospecting borehole cores or sampling in deep mining faces. Pressure relief or energy absorption support methods should be adopted in moderate rockburst areas.

5.6. Strengthen the Maintenance and Management of the Online Monitoring System of Ground Pressure

All deep mines should be equipped with maintenance and management personnel for the online ground pressure monitoring system to timely process monitoring data, identify microseismic events, compile weekly and monthly monitoring data reports, statistically analyze the law of ground pressure activities, and provide early warning information. Thereby maintenance and inspections can ensure the normal operation of the monitoring system.

5.7. Establishment of a Unified Ground Pressure Monitoring and Early Warning Platform

To fully implement the monitoring and early warning function of the ground pressure monitoring system, it is necessary to unify the online monitoring and early warning platform for underground non-coal mining pressure. This can help implement the unified management and integrated display of ground pressure monitoring data. Mines should carry out relevant positioning algorithms and early warning indicators and models to improve ground pressure monitoring, early warning, and supervision.

5.8. Innovate the Concept of Support Methods

Deep mines should establish a computer-aided decision-making system for deep support based on fundamental data, such as the rock's physical and mechanical parameters, in situ stress, joints and fissures, rock quality index, and groundwater state, to implement an intelligent dynamic stability classification of ore and rock. Mine companies should determine a reasonable and practical deep support system. Under deep high stress, the support system should specifically address energy absorption and deformation.

5.9. Strengthen Cooling Measures

Deep mines should optimize the design of deep ventilation systems to meet the ventilation needs under severe conditions, expedite the construction of deep ventilation systems, build complete deep ventilation systems in time, and strengthen the operation and management of deep ventilation systems.

5.10. Improve the Intelligent Level of Deep Mining

Deep mines should continuously strengthen scientific and technological investment, carry out digital and information system construction, and centralize the management and control of various production systems, such as backfilling, lifting, drainage, and ventilation. Mines can popularize and apply advanced large-scale mining, excavation, loading, and transportation equipment and make full use of big data, cloud computing, the Internet of things, 5G communication, and other technologies to improve the intellectual level of deep mining.

6. Conclusions

This manuscript studies the current situation of main safety systems in non-coal mines with a mining depth of more than 800 m in Shandong province and mainly draws the following conclusions:

(1) There are no large-scale ground pressure behaviors in underground non-coal mines with a mining depth of more than 800 m in Shandong province. No destructive ground pressure accidents have occured, indicating they are still in the early stages of ground pressure disasters, and rock mass instability can be controlled by strengthening support. However, with the increase in mining depth, the rockburst risk rises, and the support work becomes difficult, so it is necessary to strengthen scientific study and explore effective control measures.

(2) All the mines with normal production and construction have built ground pressure online monitoring systems. However, few systems can play an effective role in monitoring and early warning. Hence, it is necessary to strengthen the maintenance and management of the ground pressure monitoring system and establish a unified ground pressure monitoring and early warning platform for the whole province. The safety facilities of each mine meet the needs of safe production. However, their maintenance and management in the operation process need to be further strengthened, and the level of intelligence needs to be improved.

(3) The phenomena of high temperatures and high humidity in the deep mines are relatively common, and cooling can be achieved by strengthening ventilation at the current mining depth. Each mine needs to improve its ventilation systems further, strengthen ventilation management, and adopt local cooling if necessary.

(4) The construction of safety facilities in a few of the surveyed gold mines is similar to that of international mines with high development levels, but other mines can only meet the minimum requirements. It is necessary to introduce advanced digital intelligent mine technology, such as ground pressure monitoring and early warning systems, as well as big data platforms for upgrading and transforming mine construction in the future.

Author Contributions: Formal analysis, L.C.; Funding acquisition, X.L. (Xingquan Liu); Investigation, Q.W.; Methodology, C.W. and J.M.; Project administration, K.C.; Resources, X.L. (Xuelong Li); Software, H.L. All authors have read and agreed to the published version of the manuscript.

Funding: This work was financially supported by the National Natural Science Foundation of China (52104204, 52204226), the Natural Science Foundation of Shandong Province (ZR2021QE170), Qingdao Postdoctoral Applied Research Project, and China Postdoctoral Science Foundation Funded Project (2022M711961).

Data Availability Statement: The data that support the findings of this study are available from the corresponding author upon reasonable request.

Conflicts of Interest: The authors declare that there are no conflicts of interest regarding the publication of this paper.

References

1. Chen, W.; Geng, Y.; Hong, J.L.; Dong, H.J.; Cui, X.W.; Sun, M.X.; Zhang, Q. Life cycle assessment of gold production in China. *J. Clean Prod.* **2018**, *179*, 143–150. [CrossRef]
2. Lu, J.; Xie, H.P.; Li, M.H.; Li, C.B.; Gao, M.Z.; Shang, D.L.; Li, J.H. Effect of microwave radiation on mechanical behaviors of tight fine sandstone subjected to true triaxial stress. *Int. J. Rock Mech. Min.* **2022**, *10*, 5063–5078. [CrossRef]

3. Deng, J.; Yang, L.Q.; Li, R.H.; Groves, D.I.; Santosh, M.; Wang, Z.L.; Sai, S.X.; Wang, S.R. Regional structural control on the distribution of world-class gold deposits: An overview from the giant Jiaodong gold province, China. *Geol. J.* **2019**, *54*, 378–391. [CrossRef]
4. Yu, X.F.; Yang, D.P.; Li, D.P.; Shan, W.; Xiong, Y.X.; Chi, N.J.; Liu, P.R.; Yu, L.H. Mineralization characteris-tics and geological significance in 3000 m depth of Jiaojia gold metallogenic belt, Jiaodong Peninsula. *Acta. Petrol. Sin.* **2019**, *35*, 2893–2910.
5. Liu, S.M.; Li, X.L.; Wang, D.K.; Zhang, D.M. Investigations on the mechanism of the microstructural evolution of different coal ranks under liquid nitrogen cold soaking. *Energ. Source Part A.* **2020**, *10*, 1–17. [CrossRef]
6. Sai, S.X.; Qiu, K.F. Ore-forming processes of the Rushan gold deposit, Jiaodong: Fluid immiscibility in-duced by episodic fluid pressure fluctuations. *Acta. Petrol. Sin.* **2020**, *36*, 1547–1566.
7. Cheng, N.N.; Hou, Q.L.; Shi, M.Y.; He, M.; Liu, Q. New insight into the genetic mechanism of shear zone type gold deposits from Muping-Rushan metallogenic belt (Jiaodong Peninsula of eastern China). *Minerals* **2019**, *9*, 775. [CrossRef]
8. Wu, N.; Liang, Z.Z.; Zhang, Z.H.; Li, S.H.; Lang, Y.X. Development and verification of three-dimensional equivalent discrete fracture network modelling based on the finite element method. *Eng. Geol.* **2022**, *306*, 6759–6772. [CrossRef]
9. Wang, S.; Li, X.L.; Qin, Q.Z. Study on surrounding rock control and support stability of Ultra-large height mining face. *Energies* **2022**, *15*, 6811. [CrossRef]
10. Zhou, X.M.; Wang, S.; Li, X.L. Research on theory and technology of floor heave control in semicoal rock roadway: Taking Longhu coal mine in Qitaihe mining area as an Example. *Lithosphere* **2022**, *11*, 988–999. [CrossRef]
11. Liu, G.; Ma, F.S.; Zhao, H.J.; Li, G.; Cao, J.Y.; Guo, J. Study on the fracture distribution law and the influence of discrete fractures on the stability of roadway surrounding rock in the Sanshandao coastal gold mine, China. *Sustainability* **2019**, *11*, 2758. [CrossRef]
12. Qin, Z.C.; Cao, B.; Liu, Y.L.; Li, T. Study on in situ stress measurement and surrounding rock control tech-nology in deep mine. *Geofluids* **2020**, *6*, 8839333.
13. Li, T.; Gong, H.; Xu, G.L. Study on the influence of in situ stress distribution on the stability of roadway surrounding rock. *Adv. Civ. Eng.* **2021**, *21*, 523–534. [CrossRef]
14. Li, P.; Cai, M.F.; Guo, Q.F.; Miao, S.J. In situ stress state of the northwest region of the Jiaodong Peninsula, China from overcoring stress measurements in three gold mines. *Rock Mech. Rock Eng.* **2019**, *52*, 4497–4507. [CrossRef]
15. Kong, X.G.; He, D.; Liu, X.F.; Wang, E.Y.; Li, S.G.; Liu, T.; Ji, P.F.; Deng, D.Y.; Yang, S.R. Strain characteristics and energy dissipation laws of gas-bearing coal during impact fracture process. *Energy* **2022**, *242*, 3028–3040. [CrossRef]
16. Chang, J.C.; He, K.; Yin, Z.Q.; Li, W.F.; Li, S.H.; Pang, D.D. Study on the instability characteristics and bolt support in deep mining roadways based on the surrounding rock stability index: Example of pansan coal mine. *Adv. Civ. Eng.* **2020**, *5*, 8855335. [CrossRef]
17. Shen, W.L.; Shi, G.C.; Wang, Y.; Bai, J.B.; Zhang, R.F.; Wang, X.Y. Tomography of the dynamic stress coeffi-cient for stress wave prediction in sedimentary rock layer under the mining additional stress. *Int. J. Min. Sci. Techno.* **2021**, *31*, 653–663. [CrossRef]
18. Liu, H.Y.; Zhang, B.Y.; Li, X.L. Research on roof damage mechanism and control technology of gob-side entry retaining under close distance gob. *Eng. Fail. Anal.* **2022**, *138*, 6331–6345. [CrossRef]
19. Kong, X.G.; Li, S.G.; Wang, E.Y.; Wang, X.; Zhou, Y.X.; Ji, P.F.; Shuang, H.Q.; Li, S.R.; Wei, Z.Y. Experimental and numerical investigations on dynamic mechanical responses and failure process of gas-bearing coal under impact load. *Soil Dyn. Earthq. Eng.* **2021**, *142*, 6579–6591. [CrossRef]
20. Yokota, Y.; Zhao, Z.Y.; Nie, W.; Date, K.; Iwano, K.; Koizumi, Y.; Okada, Y. Development of a new defor-mation-controlled rock bolt: Numerical modelling and laboratory verification. *Tunn. Undergr. Space Technol.* **2020**, *98*, 3305–3318. [CrossRef]
21. Qiu, L.; Liu, Z.; Wang, E.; He, X.; Feng, J.; Li, B. Early-warning of rock burst in coal mine by low-frequency electromagnetic radiation. *Eng. Geol.* **2020**, *279*, 5755–5765. [CrossRef]
22. Li, X.L.; Chen, S.J.; Wang, S. Study on in situ stress distribution law of the deep mine taking Linyi mining area as an example. *Adv. Mater. Sci. Eng.* **2021**, *9*, 5594181. [CrossRef]
23. Li, X.L.; Cao, Z.Y.; Xu, Y.L. Characteristics and trends of coal mine safety development. *Energy Sources Part A* **2020**, *12*, 1–19. [CrossRef]
24. Fan, C.J.; Yang, L.; Wang, G.; Huang, Q.M.; Fu, X.; Wen, H.O. Investigation on coal skeleton deformation in CO_2 injection enhanced CH_4 drainage from underground coal seam. *Front. Earth Sci.* **2021**, *12*, 6011–6022. [CrossRef]
25. Chen, S.J.; Zhang, J.C.; Yin, D.W.; Cheng, X.Z.; Jiang, N. Relative permeability measurement of coal microchannels using advanced microchip technology. *Fuel* **2022**, *312*, 2633–2646. [CrossRef]
26. Sun, X.M.; Zhao, C.W.; Tao, Z.G.; Kang, H.W.; He, M.C. Failure mechanism and control technology of large deformation for Muzhailing Tunnel in stratified rock masses. *Bull. Eng. Geol. Environ.* **2021**, *80*, 4731–4750. [CrossRef]
27. Fan, C.J.; Wen, H.O.; Li, S.; Bai, G.; Zhou, L.J. Coal seam gas extraction by integrated drillings and punch-ings from floor roadway considering hydraulic-mechanical coupling effect. *Geofluids* **2022**, *10*, 1155–1165.
28. Yin, D.W.; Chen, S.J.; Sun, X.Z. Effects of interface angles on properties of rock-cemented coal gangue-fly ash backfill bi-materials. *Geomech. Eng.* **2021**, *24*, 81–89.
29. Li, X.L.; Chen, S.J.; Zhang, Q.M.; Gao, X.; Feng, F. Research on theory, simulation and measurement of stress behavior under regenerated roof condition. *Geomech. Eng.* **2021**, *26*, 49–61.
30. Li, X.L.; Chen, S.J.; Liu, S.M.; Li, Z.H. AE waveform characteristics of rock mass under uniaxial loading based on hilbert-huang transform. *J. Cent. South Univ.* **2021**, *28*, 1843–1856. [CrossRef]

31. Ghorbani, M.; Shahriar, K.; Sharifzadeh, M.; Masoudi, R. A critical review on the developments of rock support systems in high stress ground conditions. *Int. J. Min. Sci. Technol.* **2020**, *30*, 555–572. [CrossRef]
32. Kong, X.G.; Wang, E.Y.; Li, S.G.; Lin, H.F.; Xiao, P.; Zhang, K.Z. Fractals and chaos characteristics of acoustic emission energy about gas-bearing coal during loaded failure. *Fractals* **2019**, *27*, 195–206. [CrossRef]
33. Ma, C.D.; Li, X.B.; Chen, J.Z.; Zhou, Y.N.; Gao, S. Geological core ground reorientation technology applica-tion on in situ stress measurement of an over-kilometer-deep shaft. *Adv. Civ. Eng.* **2020**, *20*, 8830593.
34. Cheng, L.; Wang, C.L.; Wang, X.; Chen, K.X. Stability analysis of surrounding rock of large section ul-tradeep shaft wall. *Adv. Mater. Sci. Eng.* **2021**, *9*, 4391759.
35. Yang, C.; Zhou, K.P.; Li, Z.C.; Xiong, X.; Lin, Y.; Luo, Z.W. Numerical modeling on the fracturing and energy evolution of large deep underground openings subjected to dynamic disturbance. *Energies* **2020**, *13*, 6102. [CrossRef]
36. Yin, D.W.; Chen, S.J.; Ge, Y. Mechanical properties of rock-coal bi-material samples with different lithologies under uniaxial loading. *Jmr. T. J. Mater. Res. Technol.* **2021**, *10*, 322–338. [CrossRef]
37. Wu, W.D.; Bai, J.B.; Feng, G.R.; Wang, X.Y. Investigation on the mechanism and control methods for roof collapse caused by cable bolt shear rupture. *Eng. Fail. Anal.* **2021**, *130*, 5724–5740. [CrossRef]
38. Fu, M.X.; Liu, S.W.; Jia, H.S.; He, D.Y. Experimental study of an orientation and resin-lifting device for improving the performance of resin-anchored roof bolts. *Rock Mech. Rock Eng.* **2020**, *53*, 211–231. [CrossRef]
39. Liang, W.Z.; Zhao, G.Y.; Wu, H.; Chen, Y. Optimization of mining method in subsea deep gold mines: A case study. *Trans. Nonferrous Met. Soc. China* **2019**, *29*, 2160–2169. [CrossRef]
40. Zhang, C.; Wang, E.; Li, B.; Kong, X.; Xu, J.; Peng, S.; Chen, Y. Laboratory experiments of CO_2-enhanced coalbed methane recovery considering CO_2 sequestration in a coal seam. *Energy* **2022**, *12*, 5473–5488. [CrossRef]
41. Xie, H.P.; Lu, J.; Li, C.B.; Li, M.H.; Gao, M.Z. Experimental study on the mechanical and failure behaviors of deep rock subjected to true triaxial stress: A review. *Int. J. Min. Sci. Technol.* **2022**, *5*, 1–37. [CrossRef]

sustainability MDPI

Article

Acoustic Emission Waveform Characteristics of Red Sandstone Failure under Uniaxial Compression after Thermal Damage

Herui Zhang [1,2] and Weihong Guo [1,3,4,*]

1 School of Urban Geology and Engineering, Hebei GEO University, Shijiazhuang 050031, China
2 School of Resources and Geosciences, China University of Mining and Technology, Xuzhou 221116, China
3 Hebei Technology Innovation Center for Intelligent Development and Control of Underground Built Environment, Hebei GEO University, Shijiazhuang 050031, China
4 Key Laboratory of Intelligent Detection and Equipment for Underground Space of Beijing-Tianjin-Hebei Urban Agglomeration, Ministry of Natural Resources, Hebei GEO University, Shijiazhuang 050031, China
* Correspondence: guo@cumt.edu.cn

Abstract: During the exploitation of deep resources, rocks are often in a high-temperature, high-pressure environment. It is of great significance to study the acoustic emission (AE) characteristics of thermal damaged rock under load to improve the accuracy of monitoring in practical engineering. In this paper, sandstone was heated at different temperatures, before a uniaxial compression test was performed and the AE in the process was monitored. The results show that the strength and AE energy of sandstone decrease gradually with an increase in heating temperature. Through frequency domain analysis of the AE waveform at the time of failure, it was found that the frequency and intensity of AE also showed a downward trend with an increase in temperature. In addition, multifractal theory is introduced to deconstruct the waveform data. The multifractal characteristics of the waveforms decrease with an increase in temperature. It provides new parameters for waveform analysis, which can be combined with frequency analysis as parameters to more accurately identify rock failure in engineering practice. The attenuation of AE of thermally damaged sandstone may be related to an increase in porosity and a decrease in elastic energy release.

Keywords: acoustic emission; frequency domain; multifractal; thermal damage; rock fracture

Citation: Zhang, H.; Guo, W. Acoustic Emission Waveform Characteristics of Red Sandstone Failure under Uniaxial Compression after Thermal Damage. *Sustainability* **2022**, *14*, 13285. https://doi.org/10.3390/su142013285

Academic Editor: Cinzia Buratti

Received: 28 August 2022
Accepted: 13 October 2022
Published: 16 October 2022

1. Introduction

The rock mass in the deep part of the stratum is often in the mechanical environment of high temperature and high pressure, which brings great challenges to the engineering stability of underground development activities such as coal resource mining, tunnel construction, and deep drilling [1–4]. In the process of geothermal energy exploitation, temperature effect and rock damage evaluation are especially important [5,6]. Therefore, the mechanical properties and fracture behavior of rock under high temperatures are of great concern for scholars.

Extensive research shows that high-temperature treatment will damage rock and reduce a series of mechanical parameters such as compressive strength, tensile strength, and elastic modulus [7–9]. Moreover, there may be a temperature threshold of 200 °C for sandstone and granite. Below this threshold, the rock damage can be ignored [10]. The higher the temperature, the more serious the rock damage. Pyrolysis of some minerals leads to an increase in porosity and decrease in P-wave velocity. When the temperature exceeds 600 °C, quartz will undergo phase transformation, resulting in a significant reduction in rock strength [6,11]. Related research involves different rocks such as granite [12,13], sandstone [14,15], shale [5], etc. In addition, many scholars have explored rock characteristics under different temperature levels [16], different cooling methods [17], cyclic thermal shock [11], and other conditions.

The acoustic emission (AE) signal will be generated when the rock fractures. It is a transient elastic wave generated by the rapid release of energy from the local source in the material [7,11,12,18–20]. It is widely used in the field of underground engineering disaster monitoring and material nondestructive monitoring [21,22]. By monitoring and analyzing AE signals, rock damage can be evaluated and rock fracture behavior can be inversed [18]. The analysis methods of AE signals generally include parameter analysis and waveform analysis. Waveform analysis refers to the frequency domain performance of the signal obtained by time-frequency transformation. The fracture scale of rock is generally inversely proportional to the AE frequency. There have been recent reports [6,11,12] on the AE characteristics of heat-damaged rocks, especially the performance of parameters such as count and energy. The AE signal generally has good activity in the crack initiation and propagation stage loading, which is closely related to the stress drop process [19]. Kong et al. [23] also studied the nonlinear characteristics of the time-varying process of the AE signal in the uniaxial compression process of sandstone after high-temperature treatment. However, few studies have paid attention to the attenuation behavior of AE signals of rock during loading after thermal damage, especially the analysis of waveform.

Therefore, standard samples of red sandstone were processed in this paper, and they were heated at 0 °C (no treatment), 300 °C, 600 °C, and 900 °C, respectively. The AE signals were then monitored and analyzed when they were loaded in the unconfined condition. In particular, the waveforms at the time of failure were analyzed by multiple methods. In addition to the common time-frequency transform analysis of the waveform, including fast Fourier transform (FFT) [22] and short-time Fourier transform (STFT) [24], multifractal theory [25,26] was introduced to deeply deconstruct the nonlinear characteristics of the waveform data. The research results are of great significance in improving the accuracy of rock damage monitoring in the process of sustainable development of deep energy engineering.

2. Specimen Preparation and Test System

The rock used for the test was red sandstone, and it was obtained from the Linfen area of ShanXi, China. Its density was approximately 2545.65 kg/m^3. Large blocks with no obvious cracks on the surface were selected for drilling and coring. Further, standard specimens of 50 mm in diameter and 100 mm in height were cut and made for mechanical testing of uniaxial compression. The top- and bottom-end faces were smoothed according to the standards of the International Society of Rock Mechanics (ISRM).

Next, the specimens were divided into four groups, with two replicate experiments performed in each group. One group of specimens (Group S25) was not subjected to heat treatment. We regard the treatment temperature as the room temperature at the time of the experiment, which was 25 °C. The other three groups were heated in a QSH-1200T chamber-type high-temperature furnace. The three groups were heated from room temperature to 300 °C (Group S300), 600 °C (Group S600), and 900 °C (Group S900), respectively, and cooled after two hours of constant temperature. The heating rate and cooling rate were 5 °C/min and 1 °C/min, respectively. The process of heat treatment is shown in Figure 1. All specimens were tested for P-wave velocity. The specimen numbers and test results are shown in Table 1. The wave velocity of the untreated specimens was approximately 3.51 km/s and it gradually decreased as the temperature increased. This shows that the homogeneity of the specimens was satisfactory and the heat treatment was very effective.

After heat treatment, all the specimens were subjected to uniaxial compression mechanical tests. In addition, the AE signals generated by the rock during the loading process were collected synchronously. The press used for the test was the new SANS microcomputer-controlled electro-hydraulic servo pressure testing machine. It can control and collect the axial stress and strain of the specimen with high precision. The AE collector was a 24-channel Micro-II type AE monitoring host of American Physical Acoustics Corporation and the NANO-30 AE probe. The center frequency of the probe is 150 kHz and it can respond well to a wider frequency range. During the experiment, the probe was attached to the surface of the specimen and coupled with a specially formulated reagent. The threshold value, amplification, and acquisition frequency of AE acquisition were set at 45 dB, 40 dB, and 2×10^6/s, respectively. The loading rate of the press was set at 120 N/s.

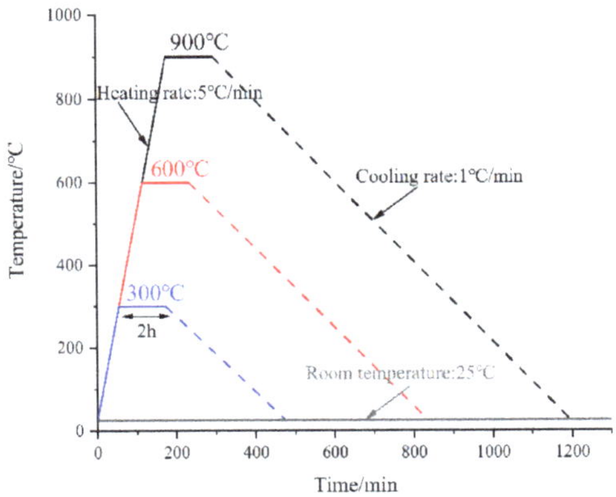

Figure 1. The heating process.

Table 1. Specimen number and test results.

Sample NO.	P-Wave Velocity (km/s)		*USC* (MPa)		*E* (GPa)		Total AE Energy (mV·μs)		Peak AE Energy (mV·μs)	
	Value	Mean	Value	Mean	Value	Mean	Value	Mean	Value	Mean
S25-1	3.52	3.51	62.58	65.74	10.65	11.52	308,982	297,311.5	30,120	31,788
S25-2	3.51		68.89		12.38		285,641		33,456	
S300-1	2.84	2.82	55.83	57.4	8.9	8.52	182,911	168,878.5	25,657	26,657.5
S300-2	2.81		58.98		8.14		154,846		27,658	
S600-1	2.38	2.36	46.03	43.76	7.87	7.5	79,938	87,521	13,093	12,037
S600-2	2.34		41.48		7.12		95,104		10,981	
S900-1	2.03	2.01	38.07	35.32	6.97	6.04	54,006	46,231	9125	8433
S900-2	1.98		32.57		5.1		38,456		7741	

3. Mechanical Properties and AE Response

Figure 2 shows the axial stress–strain curves of a group of specimens after heat treatment at different temperatures. The final failure occurred at the peak stress point after the specimens had undergone the stages of primary pore compacting (where the curves have an upward concave shape), elastic deformation (where the curve is linear), and yielding (where the curve has a downward concave shape before the peak). In Figure 3, the uniaxial compressive strength (*USC*, i.e., peak stress) and elasticity modulus (*E*, i.e., the slope of the curve in the elastic deformation phase) of all specimens are counted. The *USC* and *E* of red sandstone gradually decrease with increasing temperature of heat treatment. The *UCS* and *E* of the samples without heat treatment are approximately 65.74 MPa and 11.52 GPa, respectively. However, the *USC* and *E* of the samples treated at 900 °C are approximately 35.32 MPa and 6.04 GPa, respectively. The reductions in *USC* and *E* reached 46.27% and 47.57%, respectively. All the data are shown in Table 1.

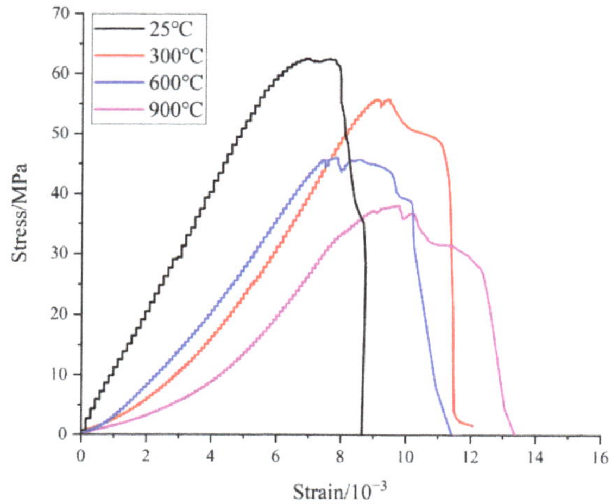

Figure 2. Stress–strain curves.

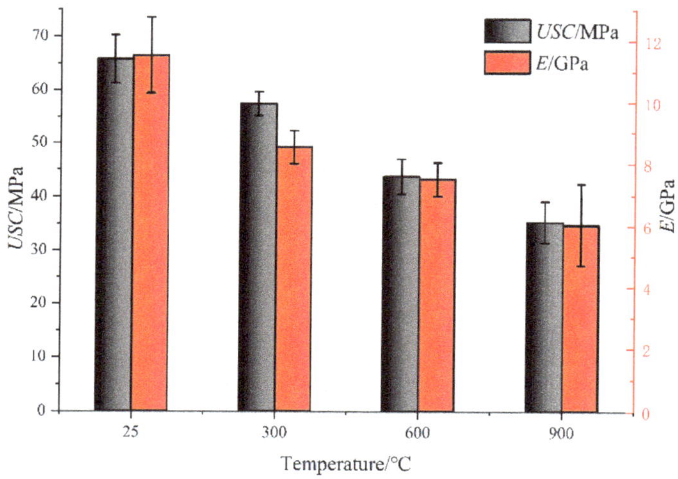

Figure 3. Changes of *UCS* and *E*.

Energy is an important parameter of the AE signal. It refers to the area under the envelope of the waveform signal, which generally reflects the intensity of AE. Figure 4 shows the AE energy of a set of specimens subjected to heat treatment at different temperatures during the loading process. AE energy has a good response to rock fracture. It is kept at a low value in the early stage, but increases sharply in the later stage when the stress fluctuates (with significant fracture). When the stress is loaded to the peak value, the main fracture of the specimen occurs and leads to complete failure. The AE energy also reaches its peak value at this time. With an increase in heat treatment temperature, the level of the AE energy value of the specimens in the whole loading process generally shows a downward trend. This can be clearly seen from the curves of cumulative AE energy. The AE energy accumulation of the untreated sample increases rapidly with an increase in stress. After high-temperature treatment, the increase rate slows down. Figure 5 and Table 1 show the comparison between total AE energy and peak AE energy. After high-temperature treatment at 900 °C, AE energy and peak AE energy decreased by 84.45% and 73.47%, respectively, compared with the untreated samples.

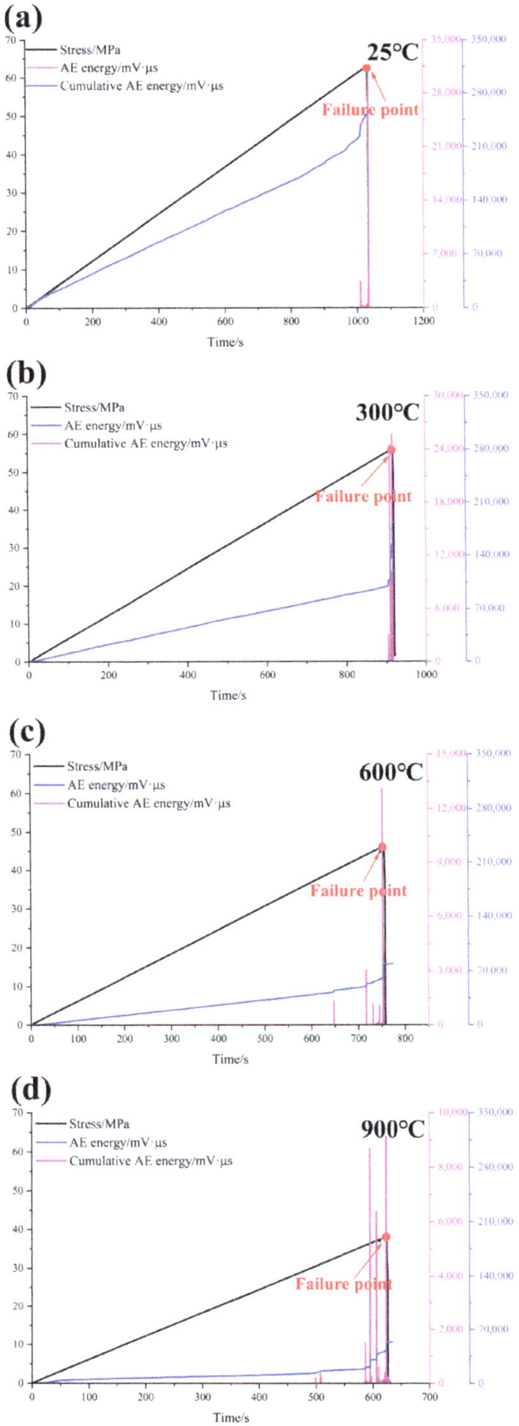

Figure 4. AE energy of the specimens during loading: (**a**) 25 °C, (**b**) 300 °C, (**c**) 600 °C, (**d**) 900 °C.

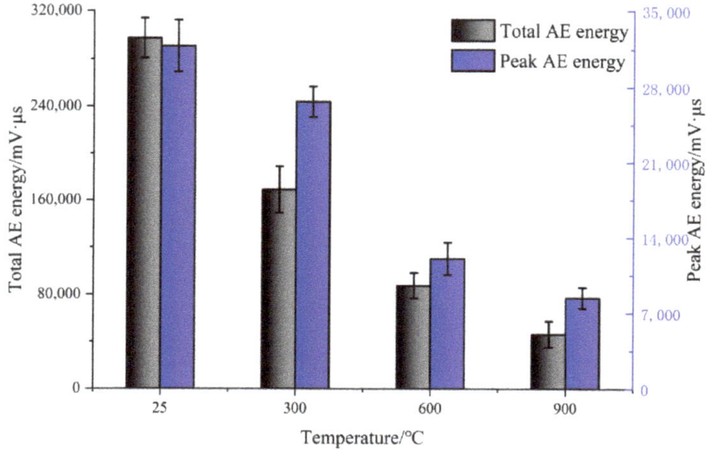

Figure 5. Changes of total AE energy and peak AE energy.

4. Waveform Characteristics at Failure

4.1. Original Waveform and FFT Analysis

The AE waveform at the failure point (peak stress) was extracted and further analyzed. The original waveforms of a group of specimens treated at different temperatures are shown in Figure 6. The AE waveforms all show the shape of increasing at first and then decreasing gradually. With an increase in the treatment temperature, the maximum amplitude of the waveform decreases significantly. In order to further analyze its frequency characteristics, fast Fourier transform (FFT) [22] was performed on all waveforms. The obtained frequency spectrums are shown in Figure 7. Further, the main frequency (frequency of the highest amplitude point in the spectrum) and the amplitude of the main frequency for all specimens in the experiment are counted and shown in Figure 8. As the temperature increases, both the main frequency and its amplitude gradually decrease. From no treatment to the 900 °C treatment, the average value of the main frequency decreases from 24.63 kHz to 18.91 kHz. The average value of the main frequency decreases by 23.22%. Moreover, these frequencies belong to the range of low-frequency AE signals caused by macroscopic large-scale fracture of rock under a laboratory test [6]. The amplitude of the main frequency decreases from 0.12 mv to 0.04 mv, a decrease of 66.67%. The decrease in amplitude is more significant than the main frequency. In addition, it is obvious from the FFT spectrum that the signal has significant subdominant frequency characteristics between 60–70 kHz. With an increase in the heat treatment temperature, the amplitude of the subdominant frequency also gradually decreased.

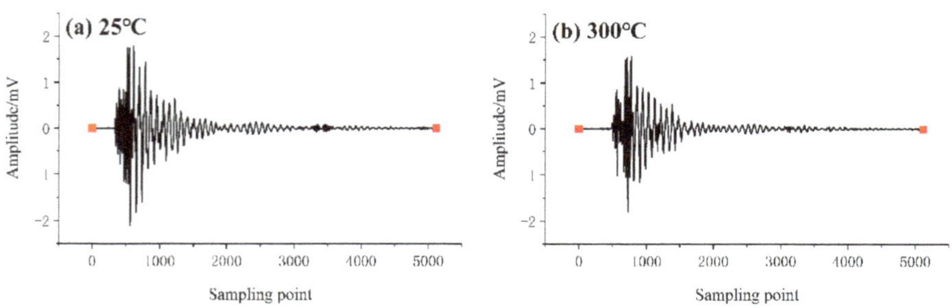

Figure 6. *Cont.*

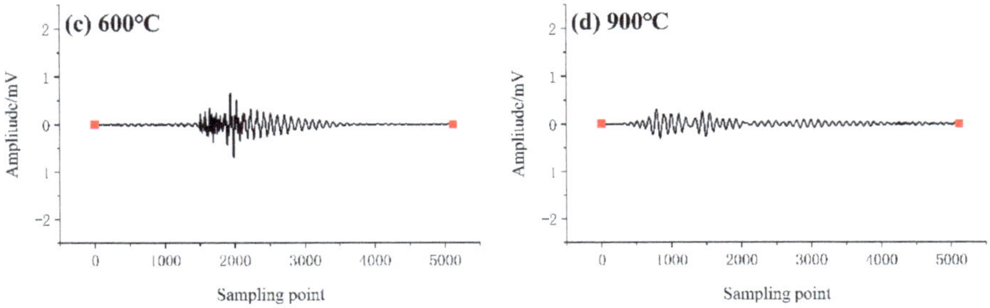

Figure 6. The AE waveform at failure point: (**a**) 25 °C, (**b**) 300 °C, (**c**) 600 °C, (**d**) 900 °C.

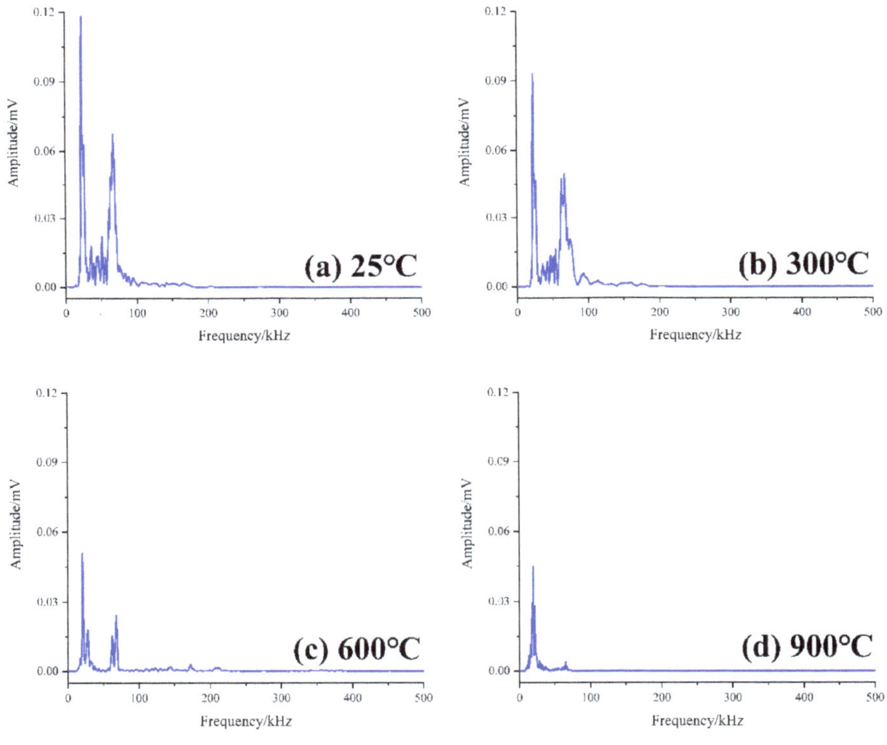

Figure 7. Frequency spectrums after FFT: (**a**) 25 °C, (**b**) 300 °C, (**c**) 600 °C, (**d**) 900 °C.

4.2. STFT Analysis

Compared with the FFT, the short-time Fourier transform (STFT) [24] has the resolution on the time scale and is more advanced. By setting the window function, it can obtain the frequency domain performance of the signal at different times. Figure 9 shows the calculated time-frequency spectrums of a set of specimens treated at different temperatures. Consistent with the trend of the FFT results, the highest energy (color axis in Figure 9) of the signal gradually decreases as the temperature increases. The highest energy value of the untreated specimen is approximately 6, while the highest energy value of the sample treated at 900 °C is approximately 0.9. The frequencies at the highest energy are all below 100 Hz. On the time scale, the signal gradually attenuates after reaching the highest energy, with its frequency and energy range both showing a gradually decreasing trend. For example, in Figure 9a, the signals at approximately the 600th sampling point show good performance in the range of 0–500 Hz. With the increase in time, at the 1500th sampling point, the signal only performs well at 0–200 Hz. Moreover, the attenuation of signal strength is also obvious.

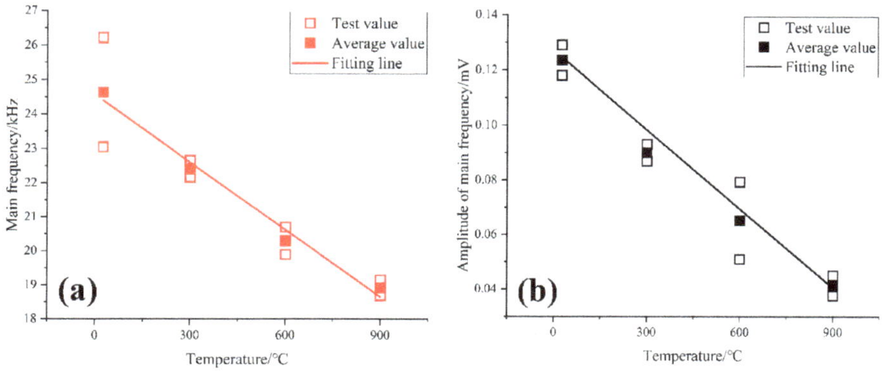

Figure 8. Changes of main frequency and its amplitude: (**a**) Main frequency, (**b**) Amplitude of main frequency.

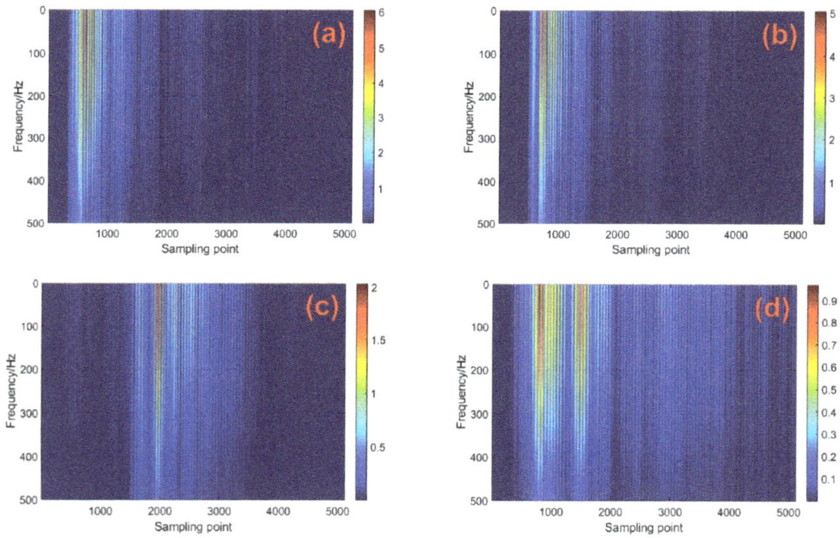

Figure 9. Time-frequency spectrums after STFT: (**a**) 25 °C, (**b**) 300 °C, (**c**) 600 °C, (**d**) 900 °C.

4.3. Multifractal Analysis

Fractal theory is proposed by scientists to describe the existence of irregularities in nature [25,26]. At present, multifractal theory is also used by many scholars to analyze AE parameters in the field of rock mechanics, but it is rarely used for waveform analysis [19,22]. Multifractal analysis can deconstruct data and reflect the characteristics of an uneven distribution of data. Through multifractal analysis of the waveform, new signal characteristic parameters can be obtained, which is beneficial to the monitoring of rock failure. The larger the width $\Delta\alpha = \alpha_{max} - \alpha_{min}$ of the calculated multifractal spectrum $\alpha - f(\alpha)$, the more uneven the data [19]. In addition, $\Delta f = f(\alpha_{max}) - f(\alpha_{min})$ reflects the proportion of large data and small data [22]. If Δf is small, it means that large data prevails; if Δf is large, it means that small data prevails. The multifractal spectrum of waveform data is calculated by the box dimension method [19]. The formulas used for the calculations are listed below, with details provided in the literature [22].

$$X_q(\varepsilon) \equiv \sum P_i(\varepsilon)^q \sim \varepsilon^{\tau(q)} \quad (1)$$

$$\tau(q) = \lim_{\varepsilon \to 0} \frac{lnX_q(\varepsilon)}{ln\varepsilon} \quad (2)$$

$$\alpha = \frac{d(\tau(q))}{dq} = \frac{d}{dq}\left(\lim_{\varepsilon \to 0} \frac{lnX_q(\varepsilon)}{ln\varepsilon}\right) \quad (3)$$

$$f(\alpha) = \alpha q - \tau(q) \quad (4)$$

The data x_i are divided into multiple subsets of length ε. $P_i(\varepsilon)$ is the probability distribution of the subset; $X_q(\varepsilon)$ is the statistical moment; $\tau(q)$ is the quality index; α is the singularity index, which reflects the inhomogeneity of the subset; and $f(\alpha)$ denotes the frequency of the represented subset in the whole subset, i.e., the fractal dimension of the subset.

Figure 10 shows the multifractal spectrums of a set of specimens treated at different temperatures. They all have a right hook shape ($\Delta f > 0$), reflecting the fact that the data are all dominated by small values. Figure 11 tallies and shows the $\Delta\alpha$ and Δf for all specimens. As the temperature increases, the multifractal characteristics of the AE waveform weaken and the multifractal parameters $\Delta\alpha$ and Δf tend to decrease in general. The mean value of α decreases from 2.14 to 0.73, while the mean value of f decreases from 0.45 to 0.08. This reflects that the data distribution of the waveform is becoming increasingly less different, with the large data becoming increasingly more dominant.

In summary, all analyses show that heat treatment causes a weakening of the AE signal of the rock, both in terms of the temporal variation of the energy and the time-frequency domain performance of the waveform at the time of failure. This may be due to various reasons. Heat treatment leads to damage of sandstone and an increase in porosity [27–29], which will aggravate attenuation of the AE signal during propagation. In addition, the strength of sandstone after thermal damage is reduced, while the storage of elastic properties is weakened during loading [6]. Less elastic energy is released, resulting in weaker AE signals [13]. In addition, it is useful to analyze the waveform by multiple means, which is beneficial to the depth inversion of rock fracture and improves the monitoring accuracy. Moreover, the introduction of multifractal theory provides new characteristic parameters for the analysis of AE waveforms.

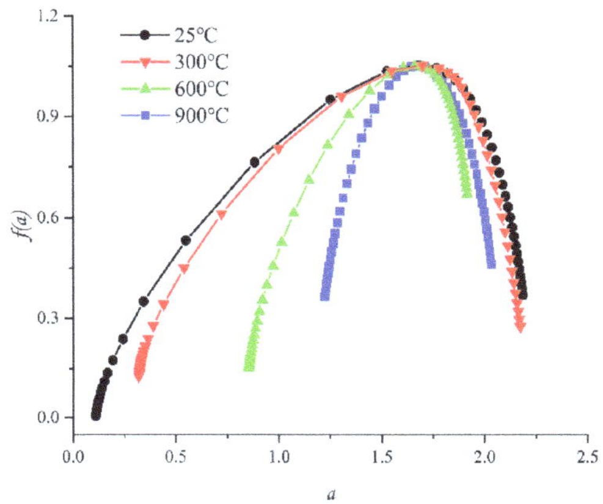

Figure 10. Multifractal spectrums.

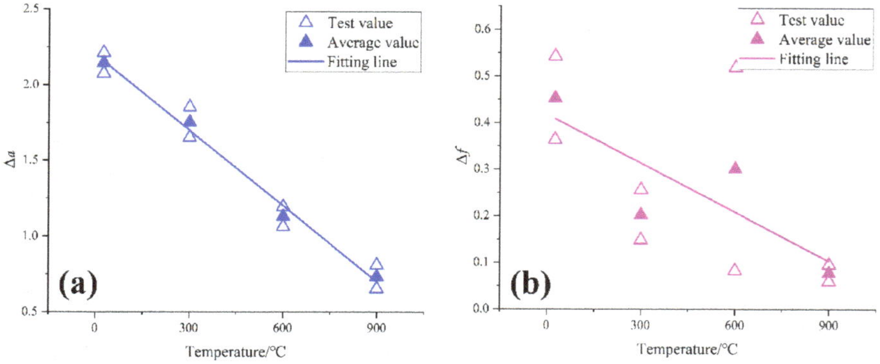

Figure 11. Changes of multifractal parameters: (**a**) $\Delta\alpha$, (**b**) Δf.

5. Conclusions

In this paper, in order to improve the accuracy of acoustic emission monitoring of rock failure after thermal damage, sandstone was heated at different temperatures (25 °C, 300 °C, 600 °C, 900 °C) and then subjected to an unconfined compression test, with the AE monitored. The mechanical properties and AE response of heat-damaged sandstone were analyzed. In particular, frequency domain analysis (including FFT and STFT) and multifractal analysis of the AE waveform at the time of failure were performed. The main conclusions are as follows:

(1) When the heating temperature increases from 25 °C to 900 °C, the uniaxial compressive strength and elastic modulus of sandstone decrease by 46.27% and 47.57%, respectively. Moreover, with an increase in temperature, the AE energy level in the loading process tends to decay. Compared with the untreated samples, the total value and peak value of AE energy of the samples after the 900 °C treatment decrease by 84.45% and 73.47%, respectively.

(2) The FFT analysis shows that with an increase in temperature, the dominant frequency and amplitude of the AE waveform of sandstone failure are decreased. The dominant frequency and amplitude of the samples treated at 900 °C are 23.22% and 66.67% lower, respectively, than those of the untreated samples. The STFT analysis also has basically consistent laws. In addition, STFT shows that the frequency range and intensity are gradually attenuated after the waveform reaches a peak. The attenuation of AE may be related to the decrease in elastic energy release and the increase in porosity.

(3) Multifractal theory was introduced to analyze the waveform of failure. The multifractal characteristic ($\Delta\alpha$ and Δf) of the signal gradually decreases with an increase in temperature, and it provides new parameters for the waveform characteristic analysis, which has an engineering application value and is conducive to improving the accuracy of rock-failure monitoring.

Author Contributions: Conceptualization, W.G.; methodology, H.Z.; data curation, H.Z.; writing—original draft preparation, H.Z.; writing—review and editing, W.G. All authors have read and agreed to the published version of the manuscript.

Funding: This research received no external funding.

Institutional Review Board Statement: Not applicable.

Informed Consent Statement: Not applicable.

Data Availability Statement: Not applicable.

Conflicts of Interest: The authors declare that they have no conflict of interest.

References

1. He, M.C.; Xie, H.P.; Peng, S.P.; Jiang, Y.D. Study on rock mechanics in deep mining engineering. *Chin. J. Rock. Mech. Eng.* **2005**, *24*, 2803–2813.
2. Kong, X.G.; He, D.; Liu, X.F.; Wang, E.Y.; Li, S.G.; Liu, T.; Ji, P.F.; Deng, D.Y.; Yang, S.R. Strain characteristics and energy dissipation laws of gas-bearing coal during impact fracture process. *Energy* **2022**, *242*, 123028. [CrossRef]
3. Kong, X.G.; Li, S.G.; Wang, E.Y.; Wang, X.; Zhou, Y.X.; Ji, P.F.; Shuang, H.Q.; Li, S.R.; Wei, Z.Y. Experimental and numerical investigations on dynamic mechanical responses and failure process of gas-bearing coal under impact load. *Soil Dyn. Earthq. Eng.* **2021**, *142*, 106579. [CrossRef]
4. Kong, X.G.; Li, S.G.; Wang, E.Y.; Ji, P.F.; Wang, X.; Shuang, H.Q.; Zhou, Y.X. Dynamics behaviour of gas-bearing coal subjected to SHPB tests. *Compos. Struct.* **2021**, *256*, 113088. [CrossRef]
5. Han, S.B.; Hu, X.D.; Zhou, F.J.; Qiu, Y.; Li, M.H.; Huang, G.P. Influence of temperature on the interaction of metal veins and hydraulic fractures in shale formations: Experiment and simulation. *J. Pet. Sci. Eng.* **2021**, *200*, 108372. [CrossRef]
6. Kong, B.; Wang, E.Y.; Li, Z.H.; Wang, X.R.; Liu, J.; Li, N. Fracture mechanical behavior of sandstone subjected to high-temperature treatment and its acoustic emission characteristics under uniaxial compression conditions. *Rock Mech. Rock Eng.* **2016**, *49*, 4911–4918. [CrossRef]
7. Giordano, M.; Calabro, A.; Esposito, C.; Damore, A.; Nicolais, L. An acoustic-emission characterization of the failure modes in polymer-composite materials. *Compos. Sci. Technol.* **1998**, *58*, 1923–1928. [CrossRef]
8. Brotons, V.; Tomas, R.; Ivorra, S.; Alarcon, J.C. Temperature influence on the physical and mechanical properties of a porous rock: San Julian's calcarenite. *Eng. Geol.* **2013**, *167*, 117–127. [CrossRef]
9. Ozguven, A.; Ozcelik, Y. Investigation of some property changes of natural building stones exposed to fire and high heat. *Constr. Build. Mater.* **2013**, *38*, 813–821. [CrossRef]
10. Ozguven, A.; Ozcelik, Y. Effects of high temperature on physico-mechanical properties of Turkish natural building stones. *Eng. Geol.* **2014**, *183*, 127–136. [CrossRef]
11. Zhu, D.; Jing, H.W.; Yin, Q.; Han, G.S. Experimental study on the damage of granite by acoustic emission after cyclic heating and cooling with circulating water. *Processes* **2018**, *6*, 101. [CrossRef]
12. Chmel, A.; Shcherbakov, I. Temperature dependence of acoustic emission from impact fractured granites. *Tectonophysics* **2014**, *632*, 218–223. [CrossRef]
13. Chmel, A.; Shcherbakov, I. Microcracking in impact-damaged granites heated up to 600 °C. *J. Geophys. Eng.* **2015**, *12*, 485–491. [CrossRef]
14. Sun, Q.; Lu, C.; Cao, L.W.; Li, W.C.; Geng, J.S.; Zhang, W.Q. Thermal properties of sandstone after treatment at high temperature. *Int. J. Rock. Mech. Min. Sci.* **2016**, *85*, 60–66. [CrossRef]
15. Zuo, J.P.; Xie, H.P.; Zhou, H.W.; Peng, S.P. SEM investigation on thermal cracking behaviour of PingDingShan sandstone at elevated temperatures. *Geophys. J. Int.* **2010**, *181*, 593–603.
16. Hueckel, T.; Peano, A.; Pellegrini, R. A thermo-plastic constitutive law for brittle-plastic behavior of rocks at high temperatures. *Pure. Appl. Geophys.* **1994**, *143*, 483–510. [CrossRef]
17. Han, G.S.; Jing, H.W.; Su, H.J.; Liu, R.C.; Yin, Q.; Wu, J.Y. Effects of thermal shock due to rapid cooling on the mechanical properties of sandstone. *Environ. Earth Sci.* **2019**, *78*, 146. [CrossRef]
18. Wang, X.R.; Wang, E.Y.; Liu, X.F.; Zhou, X. Failure mechanism of fractured rock and associated acoustic behaviors under different loading rates. *Eng. Fract. Mech.* **2021**, *247*, 107674. [CrossRef]
19. Li, H.R.; Qiao, Y.F.; Shen, R.X.; He, M.C.; Cheng, T.; Xiao, Y.M.; Tang, J. Effect of water on mechanical behavior and acoustic emission response of sandstone during loading process: Phenomenon and mechanism. *Eng. Geol.* **2021**, *294*, 106386. [CrossRef]
20. Shen, R.X.; Li, H.R.; Wang, E.Y.; Li, D.X.; Hou, Z.H.; Zhang, X.; Han, X. Mechanical behavior and AE and EMR characteristics of natural and saturated coal samples in the indirect tensile process. *J. Geophys. Eng.* **2019**, *16*, 753–763. [CrossRef]
21. Li, H.R.; Shen, R.X.; Qiao, Y.F.; He, M.C. Acoustic emission signal characteristics and its critical slowing down phenomenon during the loading process of water-bearing sandstone. *J. Appl. Geophys.* **2021**, *194*, 104458. [CrossRef]
22. Li, H.R.; Qiao, Y.F.; Shen, R.X.; He, M.C. Electromagnetic radiation signal monitoring and multi-fractal analysis during uniaxial compression of water-bearing sandstone. *Measurement* **2022**, *196*, 111245. [CrossRef]
23. Kong, B.; Wang, E.Y.; Li, Z.H.; Wang, X.R.; Chen, L.; Kong, X.G. Nonlinear characteristics of acoustic emissions during the deformation and fracture of sandstone subjected to thermal treatment. *Int. J. Rock Mech. Min. Sci.* **2016**, *90*, 43–52. [CrossRef]
24. Masera, D.; Bocca, P.; Grazzini, A. Frequency analysis of acoustic emission signal to monitor damage evolution in masonry structures. *J. Phys. Conf. Ser.* **2011**, *305*, 12134–12143. [CrossRef]
25. Li, D.X.; Wang, E.Y.; Kong, X.G.; Ali, M.; Wang, D.M. Mechanical behaviors and acoustic emission fractal characteristics of coal specimens with a pre-existing flaw of various inclinations under uniaxial compression. *Int. J. Rock Mech. Min. Sci.* **2019**, *116*, 38–51. [CrossRef]
26. Wu, Z.X.; Liu, X.X.; Liang, Z.Z.; You, X.; Yu, M. Experimental study of fractal dimension of AE serials of different rocks under uniaxial compression. *Rock. Soil. Mech.* **2012**, *33*, 3561–3569.
27. Yavuz, H.; Demirdag, S.; Caran, S. Thermal effect on the physical properties of carbonate rocks. *Int. J. Rock. Mech. Min. Sci.* **2010**, *47*, 94–103. [CrossRef]

28. Hajpál, M. Changes in sandstones of historical monuments exposed to fire or high temperature. *Fire. Technol.* **2002**, *38*, 373–382. [CrossRef]
29. Ito, K.; Enoki, M. Acquisition and analysis of continuous acoustic emission waveform for classification of damage sources in ceramic fiber mat. *Mater. Trans.* **2007**, *48*, 1221–1226. [CrossRef]

sustainability

MDPI

Article

A Novel Method for Gas Disaster Prevention during the Construction Period in Coal Penetration Tunnels—A Stepwise Prediction of Gas Concentration Based on the LSTM Method

Penghui Li [1,2], Ke Li [1,2,*], Fang Wang [1,2], Zonglong Zhang [1,2], Shuang Cai [2] and Liang Cheng [1,2,*]

1 China Merchants Chongqing Communications Technology Research & Design Institute Co., Ltd., Chongqing 400067, China
2 School of Civil Engineering, Chongqing Jiaotong University, Chongqing 400074, China
* Correspondence: like1@cmhk.com (K.L.); chengliang1@cmhk.com (L.C.)

Abstract: Aiming at the tunnel gas disaster can produce major safety problems such as combustion, explosion, and coal and gas outbursts. Firstly, a time series consisting of the tunnel gas concentration was used as the entry point for the article, and the gas prediction models based on multiple intelligent computational methods were established for comparison to determine the optimal network prediction model. Then, this study proposed a stepwise prediction method which is based on the optimal network prediction model for gas disaster prevention during the construction period of tunnels at the excavation workface. The length of the input step, output step, and interval step were considered by the method to investigate the effect on the predictive performance of the model. The model was extrapolated by the rolling prediction method, and the adaptive grid search method was used to determine the optimal parameter combination of stepwise prediction. Finally, a stepwise prediction of short-term gas concentration trends was achieved for each construction process at the excavation workface. As a result, the best LSTM network prediction model was preferred with an R^2 value of 0.94 for the fit and MAE and RMSE values of 3.2% and 4.3%. Results based on stepwise predictions showed that single-step prediction is more accurate than multi-step predictions when a reasonable input step size was determined. Moreover, with the length of both the interval step and the output step, the model prediction accuracy showed a decreasing trend. Generally speaking, the single-step continuous and interval prediction of the gas concentration at the excavation workface can be realized by the gas stepwise prediction method, and the gas concentration value can be obtained at any time in the prediction. It can also realize the transformation from single-step point prediction to multi-step trend prediction, and obtain the accurate prediction of gas concentration change trends in the stepwise prediction range (t + 1~t + 5). Therefore, the important security guarantee can be provided by stepwise prediction for subsequent gas disaster safety prevention and efficient tunnel production.

Keywords: tunnel; gas disaster prevention; time series; long short-term memory network; prediction

Citation: Li, P.; Li, K.; Wang, F.; Zhang, Z.; Cai, S.; Cheng, L. A Novel Method for Gas Disaster Prevention during the Construction Period in Coal Penetration Tunnels—A Stepwise Prediction of Gas Concentration Based on the LSTM Method. *Sustainability* **2022**, *14*, 12998. https://doi.org/10.3390/su142012998

Academic Editors: Xiangguo Kong, Dexing Li and Xiaoran Wang

Received: 3 August 2022
Accepted: 1 October 2022
Published: 11 October 2022

1. Introduction

A large number of extra-long tunnels have emerged as China's infrastructure construction gradually covers remote mountainous areas as well as the western plateau zone. Meanwhile, due to the narrow, long, and closed characteristics of the tunnels themselves, it is inevitable that the coal strata need to be crossed during the tunnel construction process, which leads to a large influx of coal seam gas into the tunnel interior [1,2]. On the one hand, high gas concentrations can easily ignite, resulting in a fire at the working face or a gas explosion that could result in a lot of victims. On the other hand, tunnel face building can also result in coal (rock) and gas explosions, which destroy and obstruct the tunnel's interior area and result in asphyxiation [3]. Consequently, it is of great practical importance to adopt accurate and reliable gas prevention and control methods in the process of tunnel construction safety, as well as green energy saving and efficient production, in order to reduce the number of casualties and property damage caused by gas disaster accidents.

The prediction of gas concentration has been an important tool for preventing and controlling the occurrence of gas disasters, and has been widely used in the field of coal mine gas prediction. The comprehensive influence of gas concentrations on existing coal seam gas conditions, the coal mining method, the roadway boring process and other factors [4], and the complex and dynamic nonlinear relationship among the factors [5] make it difficult for general linear prediction methods to meet the prediction accuracy requirements [6]. There are essential differences between coal penetration tunnels and coal mine roadway boring processes based on the characteristics of coal seam gas gushing and the characteristics of the tunnels themselves. As a result, both domestic and international researchers have put forth numerous efficient nonlinear prediction methods by combining the necessary machine learning algorithms, such as gray theory [7,8], support vector machine [9], random forest [10], and the artificial neural network [11,12]. Although the majority of gas tunnels can provide gas monitoring data, they rarely identify all relevant factors affecting gas concentration and the corresponding complete data, which causes nonlinear models based on multiple factors to diverge from reality [13].

Despite this, these approaches are still able to capture data from nonlinear mapping relationships and learn from trends. Due to this, domestic and international researchers have attempted to start from the gas time series itself and use prediction methods such as the autoregressive sliding average (ARMA), support vector machine (SVM), Bayesian network (BN), and matrix decomposition (MF) to mine the series′ effective information in order to achieve the prediction of future information [14,15]. The advantage is that it makes full use of the raw gas monitoring data collected by the field monitoring system and does not require the effort of determining the causal relationships between the factor variables, but simply extrapolates the historical trends identified by the time series model to predict future trends. Furthermore, given that time series impacted by gas concentrations are taken into account, recurrent neural networks (RNNs) have strong processing power when learning nonlinear aspects of sequences, but they also experience issues such gradient disappearance or loss of historical data [16]. The long short-term memory network (LSTM) is a superb RNN version that has the ability to handle problems that are highly associated with time series and effectively overcome the limitations of RNN in dealing with serial data and automatically mining probable data correlations [17].

The LSTM network model has been used in the prediction of gas concentration in tunnels, mainly to build a single-step prediction model for gas concentrations at a particular step in the future, achieving high prediction accuracy [18,19]. However, in the field process construction, the current moment gas concentration may depend on the change pattern of several historical samples for a long time, and the single-step iterative training mode is difficult to capture the temporal characteristics efficiently, which makes the single-step prediction results neither intuitively portray the trend of gas concentration changes in the future period nor provide a reliable reference basis for the field construction. At the same time, multi-step prediction models have been applied more often in wind speed prediction, traffic flow prediction, and stock prediction, while very few studies have been applied in gas tunnels [20,21]. However, the multi-step prediction results can help to both prevent and regulate the incidence of gas incidents and offer a definite reference foundation for process operations and site ventilation efficiency.

Based on the potential disaster risks of gas combustion, explosion, and coal and gas outbursts in coal penetration tunnel construction, a novel stepwise prediction method based on the LSTM network prediction model was proposed in this paper for gas disaster prevention. Single-step and multi-step prediction models were established by the method to consider the effects of the input step length, output step length, and interval step length on the prediction performance of the model. The optimal combination of parameters was determined to achieve a stepwise prediction of the short-term gas concentration trend for each construction process at the excavation workface. For the sustainable growth of tunnel ventilation in terms of energy conservation, effective production, and construction safety, the theoretical foundation and assistance can be offered.

2. Prediction Model Building

2.1. LSTM Network Theory

The long short-term memory (LSTM) network [22,23] is a special type of recurrent neural network. When processing time series data, it can obtain changes by adding input gates, forget gates, and output gates. In addition, the LSTM network can more accurately characterize the sequence data with spatial and temporal correlations, and realize the efficient use of historical data when the model parameters are fixed. This prevents the problem of gradient explosion and disappearance of RNN. The LSTM's structure is depicted in Figure 1 below.

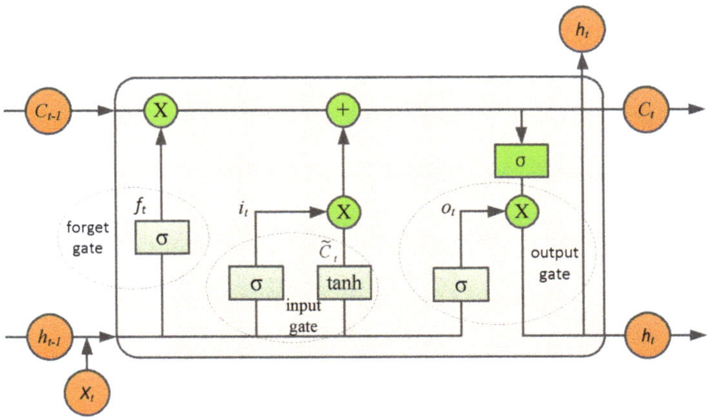

Figure 1. The hidden layer structure of LSTM.

Each of the aforementioned three gate architectures gradually updates the cell state, as shown by the LSTM structural diagram. In the first step, the forget gate determines the information discarded from the cell state c_{t-1} in the previous moment. In the second step, the information that needs to be updated in the old cell state is determined by the input gate, and the generation of alternative information $\tilde{C}_t$ for updating by the tanh layer to determine how much new information needs to be allowed to be added to the cell state. In the third step, the two steps in the second step are combined to update, and a new cell state is obtained C_t. Finally, the σ output gate filters and scales from the new cell state to obtain the output information h_t of the hidden layer [18]. The specific algorithm is as follows [17].

Gating unit:

$$f_t = \sigma\left(W_f \cdot [h_{t-1}, x_t] + b_f\right) \tag{1}$$

$$i_t = \sigma(W_i \cdot [h_{t-1}, x_t] + b_i) \tag{2}$$

$$o_t = \sigma(W_o \cdot [h_{t-1}, x_t] + b_o) \tag{3}$$

Storage unit:

$$\tilde{C}_t = \tanh(W_c \cdot [h_{t-1}, x_t] + b_c) \tag{4}$$

$$C_t = f_t \cdot C_{t-1} + i_t \cdot \tilde{C}_t \tag{5}$$

Output state:

$$h_t = o_t \cdot \tanh(C_t) \tag{6}$$

where f_t, i_t, and o_t are the outputs of the forget gate, input gate, and output gate at time t, respectively; W_f, W_i, and W_o and b_f, b_i,and b_o are the weight matrix and bias of the neural unit forgetting gate, input gate, output gate, and unit information update, respectively; σ is the sigmoid activation function; and tanh is the tangent activation function of hyperbolas.

2.2. Model Construction Ideas

The steps for predicting short-term gas concentrations at the tunnel palm face are as follows:

Step 1: Data source: Extraction of gas data from the database based on online monitoring data from field sensors;

Step 2: Data processing: Pre-processing of missing data values and outliers to obtain sample datasets;

Step 3: Model comparison: Building a gas prediction model based on multiple intelligent computational methods and testing the model errors and comparing them by the stepwise trial and error method to select the best training model for the dataset;

Step 4: Stepwise prediction: Using a combination of single-step and multi-step stepwise prediction means, the extrapolation performance of the model is studied using a rolling prediction method, using a grid search to determine the hyperparameters, while considering the effect of different input steps and interval steps on the model performance, to determine the optimal combination of parameters for stepwise prediction, and finally to achieve the short-term prediction of gas concentrations at a future moment or time period.

2.3. Dataset Preprocessing

In order to prevent the problems of excessive model prediction error and overfitting due to insufficient sample size in model training, this paper refers to the four-step method of the sample size calculation [24], and the minimum sample size was estimated to be about 600 h based on certain empirical parameters. In selecting the sample size of the data, the gas concentration data were taken from the real-time air exhaust gas concentrations collected by the tunnel automatic monitoring system crossing the same coal seam working face, with a sampling time of 30 d in April 2022 and a time node of a 1 h equidistant interval, for a total of 720 > 600 h.

As gas monitoring data are affected by a number of factors such as process cycle operations, i.e., construction process cycle operation, fan capacity adjustment, and sensor interference or failure, there are often missing values and abnormal values. If they are removed, this will not only destroy the time series and data stability of the data themselves, but also lead to low accuracy of model prediction due to the lack of data samples.

Therefore, the average value is used to correct the abnormal value that is the average value of the first, and the last non-abnormal values is used to replace the abnormal value [25], so as to reduce the influence of data anomalies on the performance of the prediction model. In addition, if the data sample itself does not have many missing values, the number of the original gas sequences and the number of the missing values are determined and the missing values are filled in using cubic spline interpolation [26]. The dataset was divided according to the following proportion: 70% was used as a training set for network training, and 30% as a test set for out-of-sample prediction, namely 504 training sets and 216 test sets, as shown in Figure 2.

Aiming at the time series prediction requires a certain sequence length of data input. The innovation of the model prediction method was that the training set was transformed into multiple training sub-samples of n groups of m-dimensional vectors by using a sliding window with a step length of 1 h, and then the rolling prediction was realized by using the previous output results as the input data [27,28]; where m = input step + output step, the input step is the sub-length of the training data sequence, and the output step is the predicted sequence length. The prediction of gas concentration with different steps can be completed by inputting different steps, i.e., the prediction performance of the model with different steps can be studied by updating the input value through continuous sliding prediction by rolling prediction. Step = 1 is single-step prediction; step ≥ 2 is multi-step prediction. The rolling prediction diagram is shown in Figure 3.

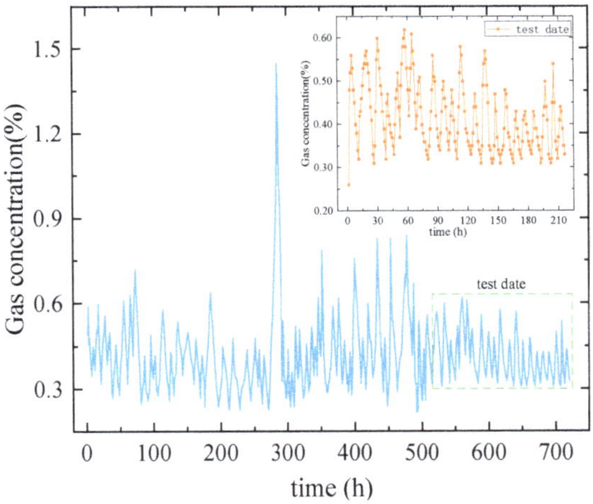

Figure 2. Sample gas data.

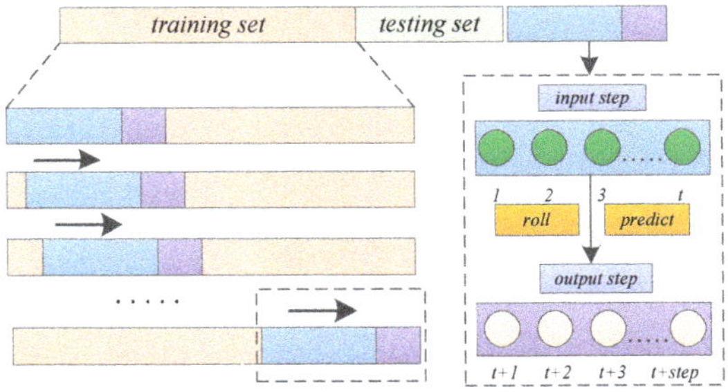

Figure 3. Scroll prediction schematic.

The single step mentioned in this paper represents an hour: 24 h a day is 24 steps. The sample set of 30 days resulted in a total of 720 steps, corresponding to the horizontal coordinates of the original data sample shown in Figure 2. Single-step prediction, also known as one-dimensional one-step prediction, is based on one-dimensional time series, using the previous steps (previous hours) as the input variable interval n steps (interval several hours) to predict the average gas concentration of a future step (a certain hour). Multi-step prediction is also called single-dimensional multi-step prediction, which means that the average gas concentration of a certain step in the future is predicted by using the previous steps as the input variables and interval n steps, where interval 0 steps are multi-step continuous prediction.

2.4. Model Parameter Settings

Super-parameter selection can significantly increase the LSTM network model's predictive accuracy [27], the number of input, output, and interval steps in a batch of training, the random inactivation rate (dropout), the number of training sessions (epoch), the number of layers (L) of the LSTM hidden layer, and the number of hidden layer neurons (rnn unit). The interval between the projected value and the actual value is represented by the interval step. Some super-parameters were trained using the grid search approach to guarantee the accuracy of gas predictions. Stepwise prediction was accomplished using the data sliding window in the code, as shown in Table 1 below.

Table 1. Super-parameter selection.

Serial Number	Hyper Parameter	Grid Search Scope
1	epoch	[50,100,150,200,250,300]
2	dropout	[0.1,0.2,0.3,0.4,0.5,0.6]
3	rnn unit	[4,8,12,16,20,24]
4	output step	[1,2,3,4]
5	interval step	[0,1,2]

The LSTM prediction model is composed of a single hidden layer and a dense full connection layer, and the dropout layer is set up in the LSTM layer to prevent overfitting. The batch size was set to 1, MSE was utilized as the loss function, and the learning rate was set to lr = 0.001 in order to maximize the suggested model structure. The samples were trained using the Adam optimizer. To reduce the discrepancy between the predicted value and the actual value, the connection weights within the network were changed after each iteration.

2.5. Model Evaluation Indicators

Relative error (δ), mean absolute error (MAE), coefficient of determination (R^2), and root mean square error (RMSE) were employed as evaluation indicators in order to objectively assess the model's suitability [19]. The accuracy of the prediction outcomes was inversely correlated with the values of, MAE, and RMSE; the smaller the value, the better the accuracy. The more closely the R^2 number approached 1, the more accurate the forecast. The following were the ways for calculating errors:

$$\delta = \frac{\widehat{y_i} - y_i}{y_i} \tag{7}$$

$$MAE = \frac{1}{n}\sum_{i=1}^{n}\left|y_i - \widehat{y_i}\right| \tag{8}$$

$$RMSE = \sqrt{\frac{1}{n}\sum_{i=1}^{n}\left(y_i - \widehat{y_i}\right)^2} \tag{9}$$

$$R^2 = \left(\frac{\sum_{i=1}^{n}(y_i - \overline{y})\left(\widehat{y_i} - \overline{y}\right)}{\sqrt{\sum_{i=1}^{n}(y_i - \overline{y})^2}\sqrt{\sum_{i=1}^{n}\left(\widehat{y_i} - \overline{y}\right)^2}}\right)^2 \tag{10}$$

where $\widehat{y_i}$ represents the *i*-th predicted value, y_i represents the *i*-th real value, $\overline{y}$ represents the average value of the real data, and n represents the number of predicted points.

3. Result and Discussion

A tunnel with a total length of 6044 m is located in an alpine and high-altitude area. According to the advanced geological prediction, the tunnel under construction crosses a coal stratum, and when the advance geological borehole was being constructed at the ZK37+845 palm face of the left tunnel, the gas concentration was detected as high as 2.5% in the left- and right-side boreholes 2 m from the bottom plate. The geological structure of coal measures would be destroyed during tunnel excavation, and a huge quantity of dangerous gases would be produced readily, according to on-site monitoring and analysis. Gas, a small amount of hydrogen sulfide, and carbon monoxide are the principal hazardous gases.

As known from the engineering geology report, the tunnel follows the left and right holes and has coal strata that are at least 1400 m long. Using the aforementioned engineering background as a foundation, this paper employed a variety of intelligent calculation techniques to create prediction models, train the samples to choose the best model, and then make stepwise predictions of the gas concentration at the palm face based on this model.

3.1. Model Comparison

Multiple intelligent algorithms, including the regression vector machine (SVR), multi-layer perceptron (MLP), gated loop unit (GRU), and long short-term memory network (LSTM), were used to construct prediction models for comparison in order to choose the best prediction model for the sample. The preprocessed data were used as the sample dataset in this study, and the relative error (δ), average absolute error (MAE), decisive coefficient (R^2), and root mean square error (RMSE) were used as the evaluation indices. To compare the performance of the four prediction models, the minimum evaluation index of the model was chosen. According to the test set, various models′ test sample prediction curves were drawn, as seen in Figure 4.

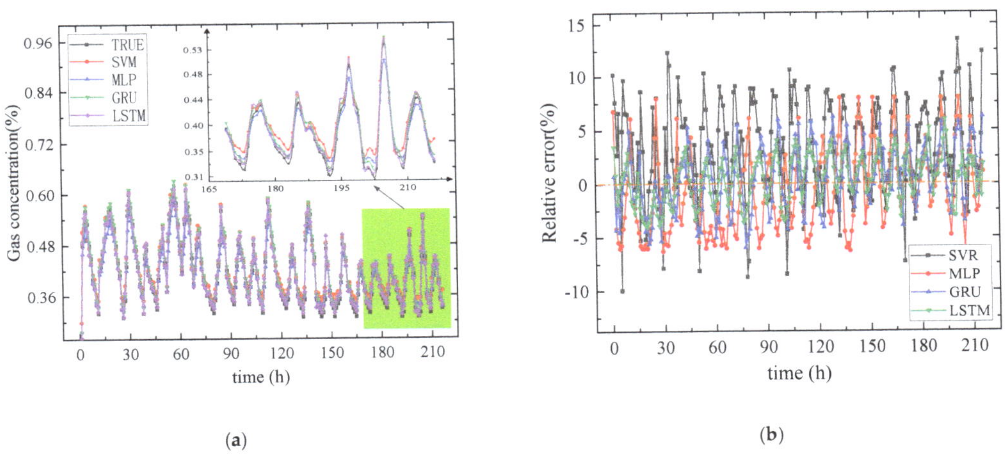

Figure 4. Prediction effect of test samples with different models; (**a**) Comparison of single prediction models; (**b**) Comparison of relative error curves.

Zoomed in on a total of 48 data points from the two days after the test sample, Figure 4a demonstrates that the four prediction models fit in a descending order from LSTM > GRU > MLP > SVR at the crest, trough, extreme value point, and inflection point. The relative errors of SVR and MLP are within 12% and 8%, respectively, as can be shown from Figure 4b. The fluctuation range is wide and the accuracy is low when compared to the other two single prediction models. GRU was within ± 6%, with high accuracy. Nonetheless, LSTM was largely maintained at 3%. The prediction performance at the extreme point was better and the overall error was reduced by two times when compared to GRU, showing that the model was better able to capture the correlation between historical monitoring information.

In order to objectively investigate the performance indices and further assess the model's predictive accuracy, the MAE, RMSE, and R^2 values of the four prediction models were extracted. According to Table 2, the LSTM network model had a fitting degree R^2 of about 0.94, and the MAE and RMSE values were the smallest among the three prediction models, which provided it a greater prediction accuracy and better generalization ability on the test set. Therefore, the preferred LSTM network prediction model was used as the optimal prediction model for this sample to provide some basic support for the subsequent stepwise prediction of gas.

Table 2. Model prediction performance comparison.

Serial Number	Prediction Method	MAE	RMSE	R^2
1	SVM	0.0381	0.0532	0.854
2	MLP	0.0376	0.0468	0.879
3	GRU	0.0355	0.0484	0.913
4	LSTM	0.0318	0.0430	0.942

3.2. Analysis of Single-Step Prediction Results

3.2.1. Different Input Time Steps

As can be seen from the model parameter settings, the predictive performance of the model was determined by a number of hyperparameters together, the most important of which was the determination of the input step for the rolling prediction subsamples [20]. When determining the input step length, it was difficult to capture the long-term dependence of the sample if the step length was too short, and if it was too long it could result in redundant information and reduced learning efficiency [21]. It was therefore important to use a suitable input step length for the prediction of gas concentration trends in order to better suit the prediction needs of the various operational processes in the actual project and to reduce the complexity of the training model. Considering that the duration of each construction process was 2~4 h, the data themselves had the characteristics of randomness, periodicity, volatility, and mutation. In this paper, firstly, based on the preferred LSTM network prediction model, the complete period of single or multiple process change trends was used as the input step length, and the actual capacity of the dataset was combined to set the input step between 1 and 10. Then, considering the influence of the interval step, the single-step prediction model was established by selecting the step length with less loss and error. Three evaluation indexes were used to evaluate the prediction effect of the model under different output step lengths as shown in Figure 5.

The model's predictive performance was influenced by a variety of hyperparameters, the most significant of which was the choice of the input step for the rolling prediction subsamples, as can be seen from the model parameter settings [20]. A little step length made it impossible to capture the sample's long-term reliance, while a big step length might lead to redundant information and decreased learning effectiveness [21]. Therefore, it was important to use appropriate input steps for predicting gas concentration trends in order to better suit the prediction needs of each work process in the actual project and to reduce the complexity of the training model. Considering that the duration of each construction process on site was 2–4 h, this made the data inherently random, periodic, fluctuating, and mutable. In order to better meet the prediction requirements of each work process in the actual project and to lessen the complexity of the training model, it was critical to select the appropriate input step length for predicting gas concentration trends. The data were intrinsically random, periodic, variable, and mutable because each construction process on site lasted between two and four hours.

As seen in Figure 5, when the input step length was 2~8, the model had a better prediction effect under the evaluation index in 3, which also verified that neither the larger the input step length nor the smaller the input step length resulted in the highest prediction accuracy. However, the difference between the MAE and RMSE evaluation index values under different output step lengths was not significant at time t + 1. From the curve fitting degree determination coefficient R^2, when the input step lengths were 2, 4, and 6, the gas prediction data in the past 2, 4, and 6 h could make the model result in the best prediction effect.

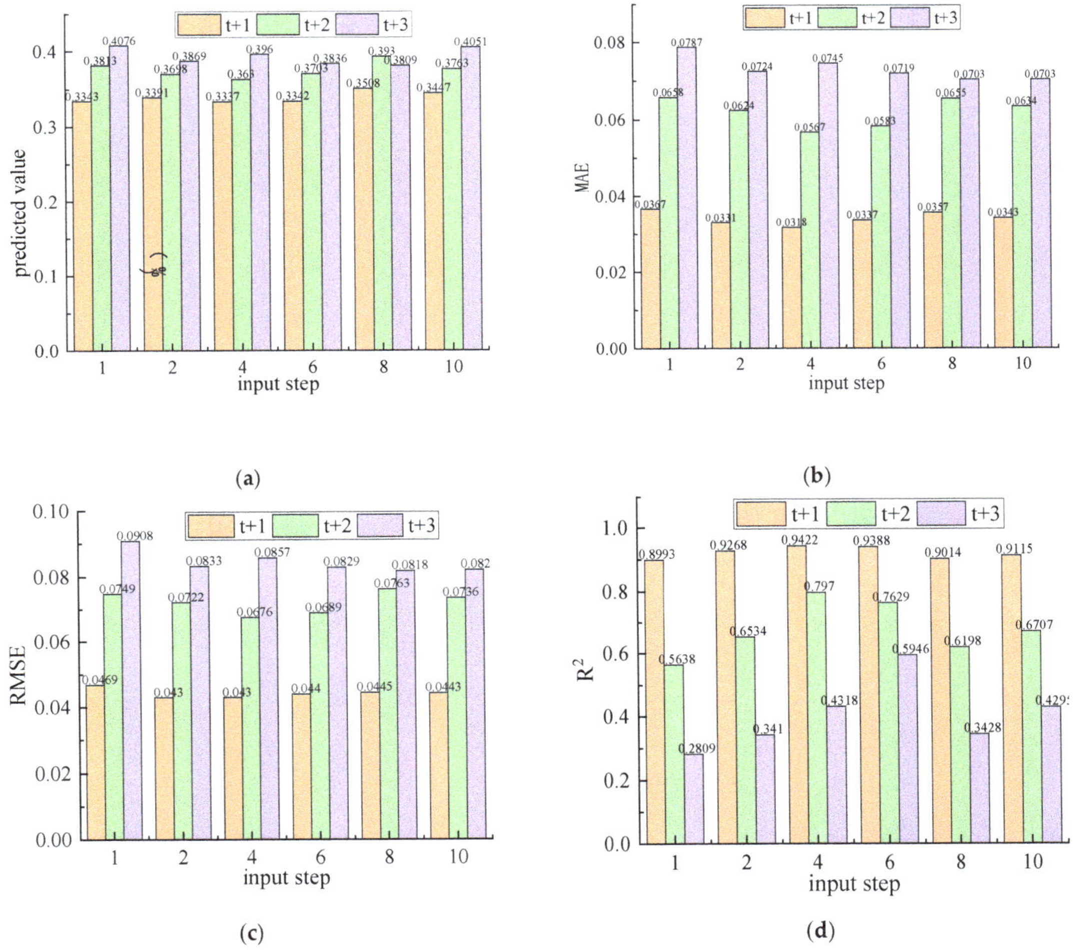

Figure 5. Model prediction results under different output step lengths; (**a**) Predicted gas concentration; (**b**) Mean absolute error (MAE); (**c**) Root mean square error (RMSE); (**d**) Coefficient of determination (R^2).

3.2.2. Analysis of the Effect of Single-Step Prediction

To find out how the interval step affects the accuracy of the gas concentration prediction at a later time, the best time steps for the single-step predictions obtained above were compared and examined. The single-step prediction effect exhibited a discernible decreasing trend with an increase in the interval step, as shown in Figure 6a,c,e from the peak, trough, extreme point, and inflection point of the curve. The relative error of the forecast findings, however, can more accurately and intuitively convey the veracity of the prediction effect. The relative inaccuracy of the single-step prediction effect also tended to diminish to varying degrees as the interval step length was decreased, as shown in Figure 6b,d,f. Meanwhile, the relative error of interval 0 step (interval 0) varied at 5% with excellent accuracy, that of interval 1 step (interval 1) was roughly two to five times bigger, and that of the prediction effect of interval 2 step (interval 2) was considerably higher.

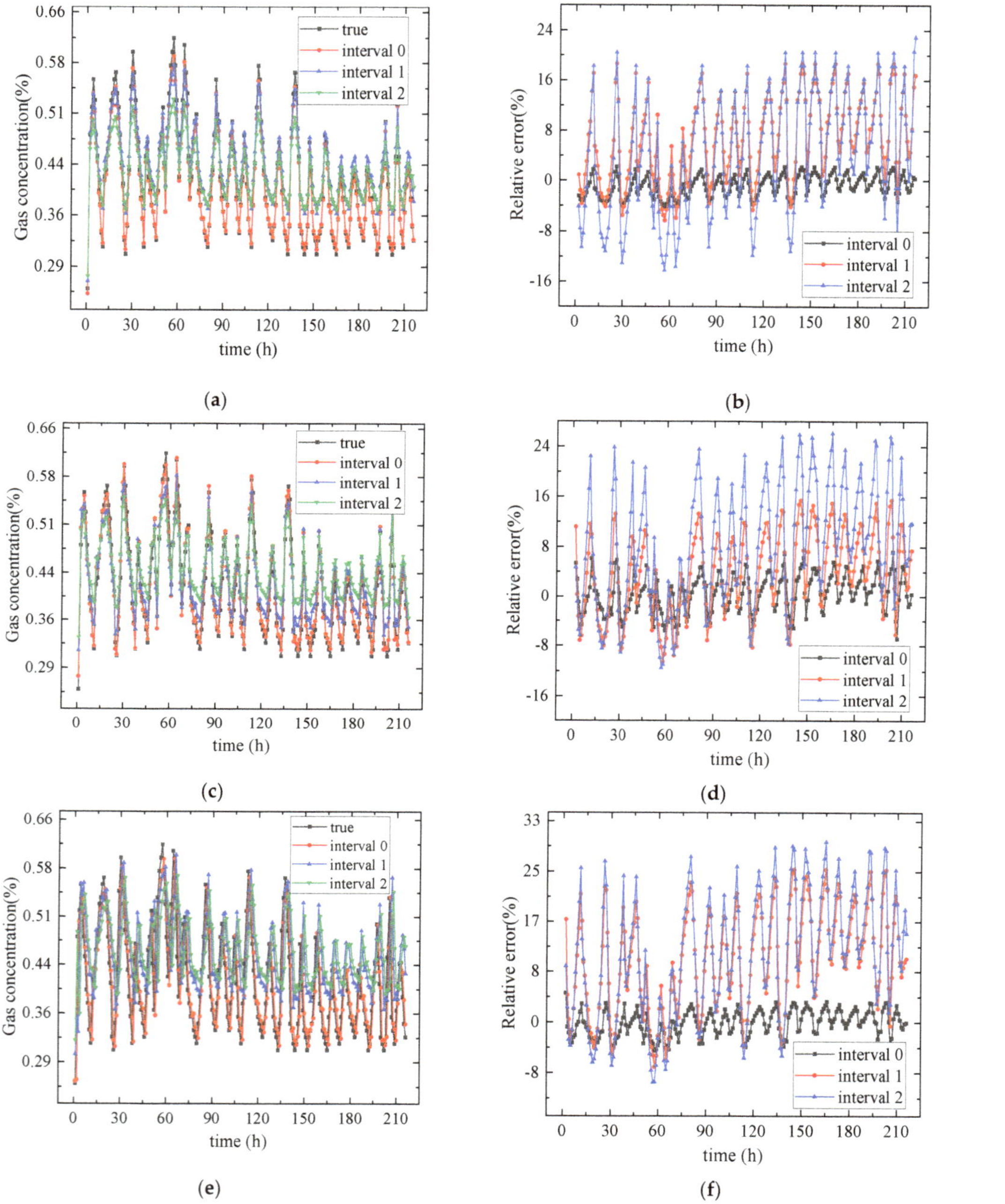

Figure 6. Single-step prediction effect diagram; (**a**) Input step-2; (**b**) The relative error of input step-2; (**c**) Input step-4; (**d**) The relative error of input step-4; (**e**) Input step-6; (**f**) The relative error of input step-6.

The three evaluation indices typically exhibited a phenomenon of continuous deterioration with the increase in the interval step, which results in poor accuracy and reliability of prediction, as can be seen from Table 3. On the other hand, when the input step length was moderate, the prediction effect of the model was better. When the anticipated value for the following 1 h was 0.3337%, the curve fit R^2 value was as high as 0.94 and the MAE and RMSE values were likewise relatively minimum. Although there was only a 0.80 and 0.60 accuracy in predicting the gas value at times t + 2 and t + 3, in the future, accurate, efficient, convenient, and synchronous gas rapid prediction technology can not only realize synchronous monitoring and prevention with the automatic monitoring system on the spot, but also serve as a guide for the following construction process.

Table 3. Comparison of single-step prediction comprehensive evaluation.

Interval Step	Input Step	Output Step	Predicted Value	Evaluating Indicator		
				MAE	RMSE	R^2
interval 0 (t + 1)	2	1	0.3391	0.0331	0.0430	0.9268
	4	**1**	**0.3337**	**0.0318**	**0.0430**	**0.9422**
	6	1	0.3342	0.0337	0.0440	0.9388
interval 1 (t + 2)	2	1	0.3698	0.0624	0.0722	0.6534
	4	**1**	**0.3630**	**0.0567**	**0.0676**	**0.7970**
	6	1	0.3703	0.0583	0.0689	0.7629
interval 2 (t + 3)	2	1	0.3869	0.0724	0.0833	0.3410
	4	1	0.3960	0.0745	0.0857	0.4318
	6	**1**	**0.3836**	**0.0719**	**0.0829**	**0.5946**

3.3. *Analysis of Multi-Step Prediction Results*

In order to study the change trends of gas concentrations in a certain period of time or a certain construction process in the future, the multi-step prediction was compared under different input steps, output steps, and interval steps, and the influence of the multi-step prediction on the prediction performance of the model was further discussed. The interval 0 step was the continuous prediction. According to the analysis of the single-step prediction, the input step lengths 2, 4, and 6 were used to explore the effect of multi-step prediction.

3.3.1. Multi-Step Continuous Prediction Analysis

If the output step length of multi-step continuous prediction was n, there would n data repetitions per rolling, which would increase the test set to n times. As can be seen in Figure 7, the error would accumulate when the rolling prediction method was used for extrapolation. With longer prediction step lengths, the effect of multi-step continuous prediction also exhibited a discernible declining trend. The fit of the curve became worse and the error increased with the increasing prediction step size, resulting in poor model prediction accuracy. As a result, considering the degree of curve fitting, the forecast accuracy revealed that the effects of successive single-step predictions (output step-1) were superior to successive two-step predictions (output step-2) and superior to successive three-step predictions (output step-3) and four-step predictions (output step-4), respectively.

The analysis of Table 4 reveals that the continuous two-step prediction's curve fitting degree R^2 of 0.76 was 18% lower than that of the single-step prediction. However, when compared to the continuous three- and four-step predictions, the MAE and RMSE values were relatively the lowest and the accuracy was highest. Overall, as the number step of prediction increased, the MAE and RMSE values rose, and the model's prediction accuracy rapidly deteriorated. The curve fitting degree deteriorated at the same time that the R^2 index dropped. The accuracy of the continuous prediction of the following three and four steps was less accurate than that of the previous two steps, but the R^2 values of 0.70 and 0.60, respectively, also have some reference relevance.

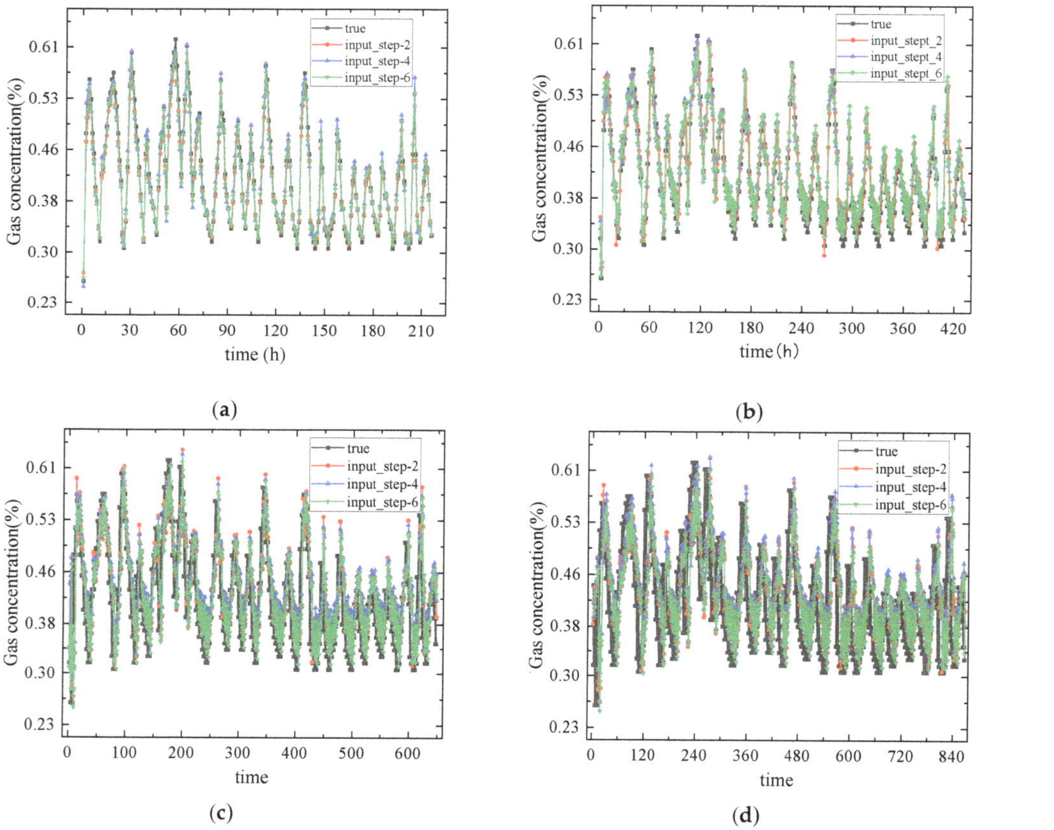

Figure 7. Comparison of multi-step continuous prediction; (**a**) Output Step-1; (**b**) Output step-2; (**c**) Output step-3; (**d**) Output step-4.

Table 4. Comparison of multi-step continuous prediction.

Multi-Step Continuous Prediction		**Evaluating Indicator**		
Output Step	**Input Step**	**MAE**	**RMSE**	**R^2**
	2	0.0460	0.0573	0.7592
output step-2	4	0.0469	0.0578	0.7541
	6	0.0462	0.0584	0.7741
	2	0.0538	0.0670	0.6845
output step-3	4	0.0568	0.0698	0.7102
	6	0.0549	0.0686	0.7272
	2	0.0583	0.0727	0.6016
output step-4	4	0.0617	0.0753	0.6277
	6	0.0605	0.0744	0.5783

3.3.2. Multi-Step Interval Prediction Analysis

The outcomes of the multi-step sequential predictions suggest that the more step in the prediction, the worse the corresponding prediction. Therefore, to increase the output step length of the model training, the number of output steps was chosen to be two for multi-step interval prediction to further explore the prediction performance of the model.

The multi-step prediction effect is shown in Figure 8 a,c,e below to be broadly equivalent to the single-step prediction effect, both showing a clear pattern of a declining curve fit as the interval step increases, with continuous prediction having the highest accuracy. Figure 8b,d,f show that the relative inaccuracy of the multi-step prediction increases along with the lengthening of the interval step. The interval 0 step relative errors varied approximately 14%, with outstanding accuracy; the interval 1 step relative errors were roughly two to three times higher than those for the interval 0 step, and the interval 2 step prediction effect relative errors were significantly higher. As a result, the accuracy of the prediction findings was more influenced by the number of interval steps. According to the degree of curve fitting, the effect of each interval's prediction inaccuracy was as follows: interval 0 > interval 1 > interval 2.

Table 5 shows that interval prediction significantly affected the outcome of the prediction. The accuracy of multi-step interval prediction is equal to that of single-step interval prediction in terms of the three indicators, with the individual evaluation indicators and associated prediction accuracy declining with the increasing interval step size. Although the interval 0 step prediction's R^2 value of 77% was higher than in intervals 1 and 2, the difference in values is insignificant in terms of the number of times the anticipated values will repeat. Therefore, the interval 0 step was used to predict the gas concentration in the early stage of a process in the face of the palm, and the values of the follow-up interval 1 and 2 steps can provide reference for the field.

Table 5. Comparison of multi-step interval predictions.

Multi-Step Interval Prediction			Predicted Value/%		Evaluating Indicator		
Interval Step	Input Step	Output Step	Value 1	Value 2	MAE	RMSE	R^2
interval 0 (t + 1~t + 2)	2	2	0.343	0.357	0.0460	0.0573	0.7592
	4	2	0.344	0.364	0.0469	0.0578	0.7541
	6	2	0.338	0.369	0.0462	0.0584	0.7741
interval 1 (t + 2~t + 3)	2	2	0.368	0.381	0.0652	0.0762	0.6002
	4	2	0.371	0.404	0.0679	0.0797	0.6165
	6	2	0.382	0.410	0.0717	0.0834	0.6135
interval 2 (t + 3~t+4)	2	2	0.380	0.395	0.0732	0.0852	0.2786
	4	2	0.383	0.412	0.0763	0.0887	0.3897
	6	2	0.375	0.413	0.0752	0.0881	0.3761

3.4. Comparison of Stepwise Prediction Effects

3.4.1. Comparison of Single and Multi-Step Prediction Effects

In order to explore the trend of the gas concentration value at a certain point in the future or the change of gas concentration within a certain time period for each construction process at the palm face, the optimal values of the distribution prediction for the three evaluation indicators with different output steps and interval steps were extracted and analyzed separately for comparison, as shown in Tables 6 and 7 below.

Table 6. Comparison of stepwise sequential predictions.

Evaluation Indicators	Output Step-1	Output Step-2	Output Step-3	Output Step-4
MAE	0.032	0.046	0.055	0.062
RMSE	0.043	0.058	0.069	0.075
R^2	0.942	0.774	0.727	0.628

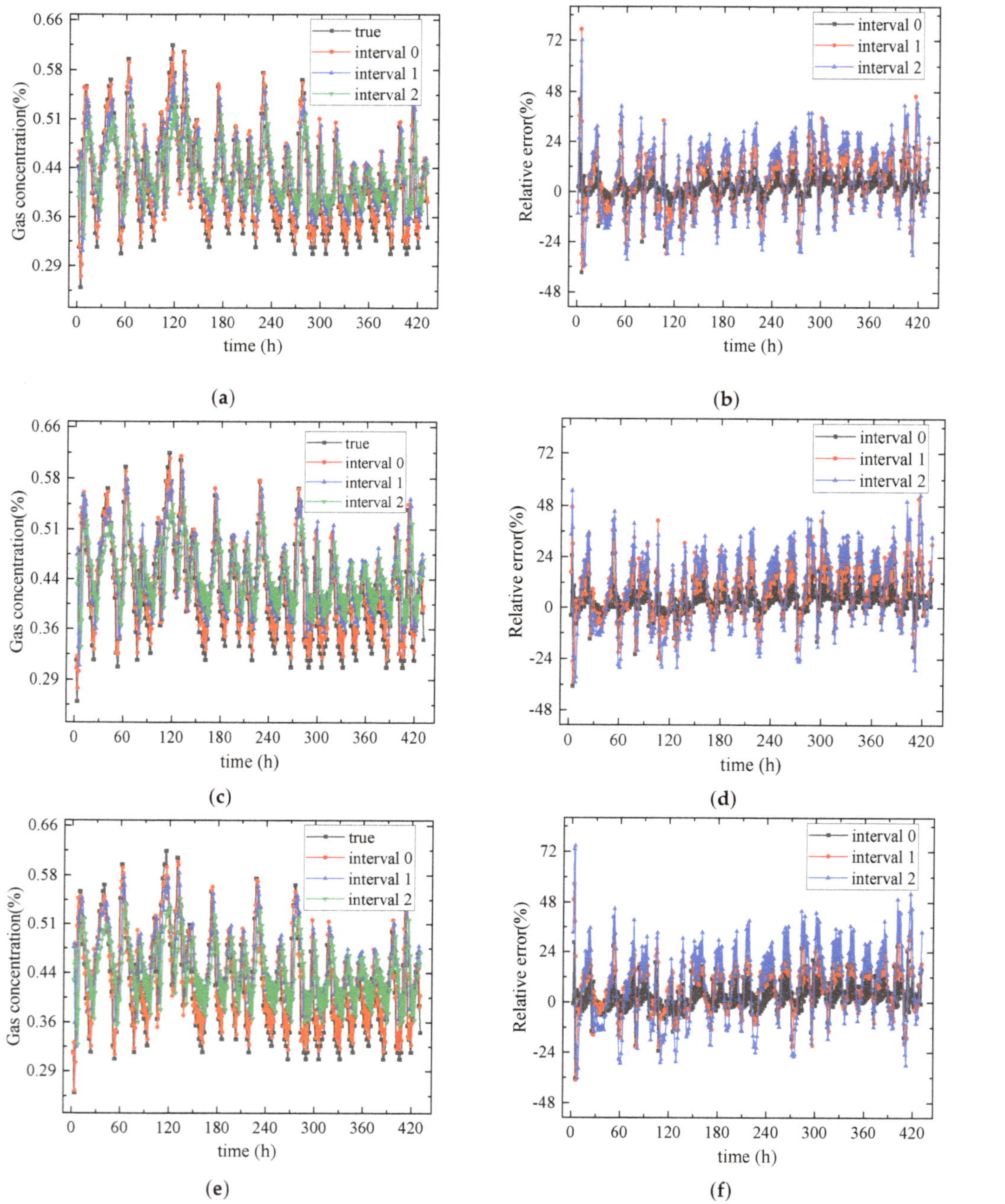

Figure 8. Multi-step interval prediction effect comparison; (**a**) Input step-2; (**b**) The relative error of input step-2; (**c**) Input step-4; (**d**) The relative error of input step-4; (**e**) Input step-6; (**f**) The relative error of input step-6.

Table 7. Comparison of single-step and multi-step predictions.

Stepwise Prediction	Interval 0			Interval 1			Interval 1		
	MAE	RMSE	R^2	MAE	RMSE	R^2	MAE	RMSE	R^2
Single-step prediction	0.032	0.043	0.942	0.057	0.068	0.800	0.072	0.083	0.600
Multi-step prediction	0.046	0.058	0.774	0.068	0.080	0.617	0.076	0.089	0.390
Decay rate	30.40%	25.90%	17.80%	16.20%	15%	22.80%	5.30%	6.70%	35%

Table 6 shows that when the stepwise continuous prediction step length increases, the three evaluation indices continue to worsen, and that there is a general downward trend in the model's prediction performance. Although the accuracy of single-step prediction is higher than that of multi-step prediction, the trend of gas concentration predicted by multi-step prediction was more significant for short-term field construction. The optimal step length of single-step and multi-step were extracted respectively to further explore the effect of stepwise interval prediction on the performance of the model (see Table 7).

Among them, the comparison of stepwise interval prediction as shown in Table 7 are as follows:

(1) As the interval step length increases, the three assessment metrics for both single-step and two-step predictions continue to degrade, which also affects the model's accuracy. For the identical single-step prediction instance, the MAE, RMSE, and R^2 metrics decreased from 0.032, 0.043, and 0.942 to 0.057, 0.068, and 0.80 correspondingly for interval 0 step compared to interval 1 step, a decrease of 43.8%, 36.7%, and 15%, respectively. This is shown by the fact that interval 0 step prediction generalized more accurately and learned the time series data's features more efficiently;

(2) Under different interval steps, single-step prediction had a higher evaluation index than two-step prediction. For instance, the single-step prediction's MAE, RMSE, and R^2 metrics for the interval 0-step scenario are, respectively, 30.4%, 25.9%, and 17.8% lower than the 2-step predictions. This result verifies that the model takes into account the results of the previous prediction step when making rolling prediction, allowing errors to accumulate. The prediction error at each step increased as the length of prediction step increased, thus causing the predicted sequence to deviate from the original sequence. That is, the bigger the length of the output step, the more errors would be accumulated, which in turn leads to a poorer accuracy of the model.

3.4.2. Parameter Selection and Model Optimization

The output step and interval step length will significantly lessen the temporal dependence between data, which will have a significant impact on the accuracy of the prediction results, according to the discussion results of the aforementioned stepwise prediction. The outcomes of continuous versus interval forecasts and single-step versus multi-step predictions are compared in this study, and the best set of parameters is chosen, as shown in Table 8.

Table 8. Stepwise prediction of optimal parameters.

Serial Number	Sliding-Window	Optimal Parameters
1	input step	2, 4, 6
2	interval step	0, 1, 2
3	output step	1, 2, 3

The single-step continuous and interval prediction of gas concentration can be realized by figuring out the best stepwise prediction parameters, and the gas concentration value at any moment within the forecast range may be acquired. In order to achieve high accuracy in the gas concentration change trend in the stepwise prediction range (t + 1~t + 5), it is also capable of converting from single-step to multi-step prediction, as shown in Figure 9.

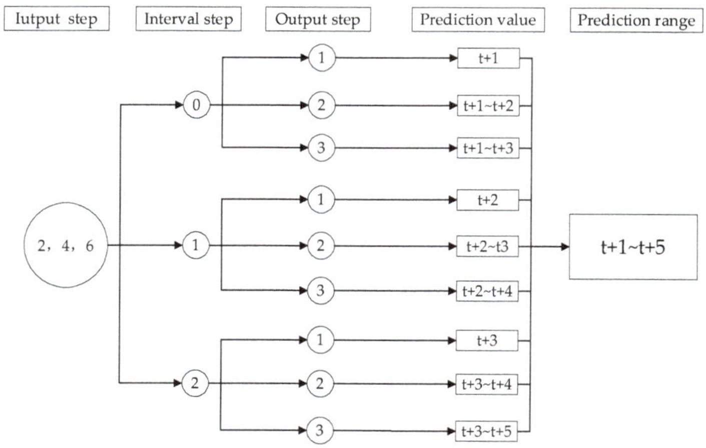

Figure 9. Range of stepwise prediction.

Therefore, the results of stepwise prediction can not only dynamically predict the gas concentration at a high level in the early stage of each construction process, but also provide a reference for mastering the size and trend of gas concentration after the excavation of the tunnel face in advance. At the same time, it can also provide some basic support and time reserve for the site ventilation efficiency and construction progress, and then provide an important guarantee for tunnel ventilation energy saving, efficient production and construction safety.

Additionally, the data samples can be continuously expanded in conjunction with the field automatic monitoring system to fulfill the goal of dynamic updating, which will improve the generalizability of the model. In order to determine whether a more suitable prediction model for the site and a more precise prediction of the trend of short-term gas concentration in the future can be obtained, the prediction method of "adaptive grid search method + real-time updating of data samples" was used, as shown in Figure 10.

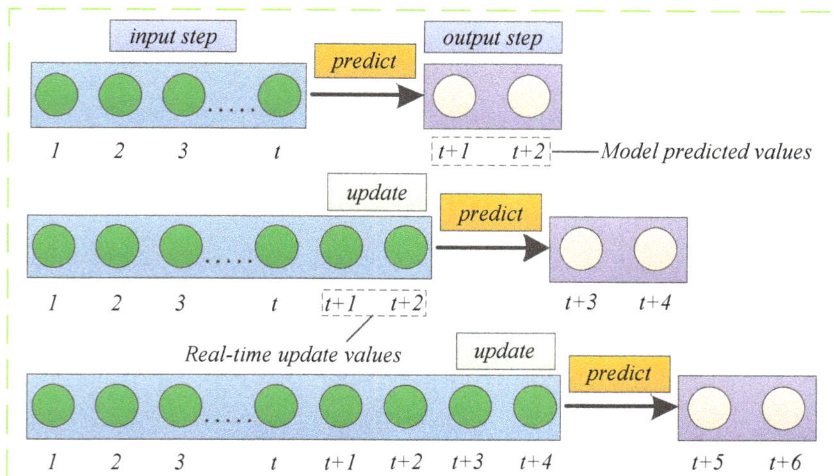

Figure 10. Real-time update of data sample—take 2 steps for example.

4. Conclusions

Aiming at the tunnel gas disaster can produce major safety problems such as combustion, explosion, and coal and gas outbursts. This paper proposed a stepwise prediction method which is based on the optimal network prediction model for gas disaster prevention during the construction period of tunnels at the excavation workface.

(1) The gas prediction models were built separately based on various intelligent computing methods in this paper, and the model errors were trained and compared through a stepwise trial-and-error method. The optimal LSTM network model suitable for this dataset was finally preferred. The rolling prediction method was then used to extrapolate the single- and multi-step prediction models to determine their optimal parameter sets by the adaptive grid search method, thus overcoming the detrimental effect of relying on the empirical selection of hyperparameters on improving the prediction performance of the models and improving their prediction performance;
(2) The model performance is strongly influenced by the length of output step and interval step. The results show that single-step prediction is more accurate than multi-step prediction at different output step lengths, and the model performance decreases as the output step length increases. At different interval step lengths, continuous prediction is more accurate than interval prediction, and the model performance tends to decay significantly with increasing interval step lengths;
(3) The method not only achieves accurate single-step prediction for future time t + 1, but also achieves the transformation from single-step prediction to multi-step trend prediction, and obtains accurate prediction of the trend of gas concentration change within the step-by-step prediction range (t + 1~t + 5). It can play an effective role in controlling and preventing the occurrence of gas disasters and reducing the possibility of casualties caused by gas disasters.

With the application of deep learning theory to gas prediction, optimization of prediction models in conjunction with continuous lengthening of training sample sizes for field monitoring systems and consideration of multiple factors will be the focus of subsequent research.

Author Contributions: P.L. and K.L. proposed the research. F.W. and Z.Z. prepared figures and tables, and interpreted the structural data. S.C. and L.C. developed the main ideas. All co-authors actively contributed to the manuscript with comments, ideas, and suggestions. All authors have read and agreed to the published version of the manuscript.

Funding: This study was financially supported by the National Natural Science Foundation of China (NSFC) (Grant No. 51804055), Chongqing Natural Science Foundation (Distinguished Youth Fund) project (cstc2021jcyj-jqX0012).

Data Availability Statement: The datasets used and/or analyzed during the current study are available from the corresponding author on reasonable request.

Acknowledgments: This work was supported by the National Natural Science Foundation of China (NSFC) (Grant No. 51804055) and the Chongqing Natural Science Foundation (Distinguished Youth Fund) project (cstc2021jcyj-jqX0012). The authors also thank the editor and anonymous reviewers for their useful advice.

Conflicts of Interest: The authors declare no potential conflict of interest with respect to the research, authorship, and/or publication of this paper.

References

1. Zhao, L.T.; Zhao, Y.; Su, P.D.; Quan, F.; Zhu, Q.W.; Zhao, C.J. Study on the construction safety management of toxic and harmful gases in non-coal strata tunnel. *Mod. Tunn. Technol.* **2020**, *57*, 4.
2. Li, K.; Li, P.L.; Cai, S.; Cheng, L.; Yao, X.F. Research on Tunnel Harmful Gas Control Technology Based on Fuzzy Control. *Tunn. Constr.* **2022**, *42* (Suppl. S1), 120–127.
3. Kang, X.B.; Xu, M.; Ding, R. Discussion on Danger Assessment of Gas Disaster in Tunnel. *J. Railw. Eng. Soc.* **2010**, *38*, 2.

4. Xie, H. Research review of the state key research development program of China: Deep rock mechanics and mining theory. *J. China Coal Soc.* **2019**, *44*, 1283–1305.
5. Wu, F.; Huo, Y.; Gao, J. Coal mine gas emission prediction method based on random forest regression. *Ind. Mine Autom.* **2021**, *47*, 102–107.
6. Wang, L.; Liu, Y.; Liu, Z.Z.; Qi, J.Y. Research on prediction model for gas emission based on IABC-LSSVM. *Transducer Microsyst. Technol.* **2022**, *41*, 34–38.
7. Shi, S.; Wu, A. Application of GM (1,1) and line regression for predicting amount of mine gas emission in coal mine. *J. China Coal Soc.* **2008**, *4*, 415–418.
8. Wei, C.; Li, Y.; Sun, J.; Mi, H.; Li, J. Gas emission rate prediction in coal mine by grey and separated resources prediction method. *J. Min. Saf. Eng.* **2013**, *30*, 628–632.
9. Huang, W.; Tong, M.; Ren, Z. Nonlinear Combination Forecast of Gas Emission Amount Based on SVM. *J. China Univ. Min. Technol.* **2009**, *38*, 234–239.
10. Hu, X.F.; Belle, J.H.; Meng, X. Estimating PM2.5 concentrations in the conterminous United States using the random forest approach. *Environ. Sci. Technol.* **2017**, *51*, 6936–6944. [CrossRef]
11. Kuang, L.; Zhao, W.; Yu, Y. Research on the Prediction Model and Case of Coal and Gas Outburst in Tunnel by Using BP Neural Network. *J. Railw. Eng. Soc.* **2018**, *35*, 56–61.
12. Fu, H.; Xie, S.; Xu, Y.; Cheng, Z. Gas emission dynamic prediction model of coal mine based on ACC-ENN algorithm. *J. China Coal Soc.* **2014**, *39*, 1296–1301.
13. Liu, J.; An, F.; Lin, D.; Go, Z.; Zhang, L. Prediction of gas emission from coalface by intrinsic mode SVM modeling. *Syst. Eng.-Theory Pract.* **2013**, *33*, 505–511.
14. Zhang, Y.; Pan, G.; Chen, B.; Han, J.Y.; Zhao, Y.; Zhang, C.H. Short-term wind speed prediction model based on GA-ANN improved by VMD. *Renew. Energy* **2020**, *156*, 1373–1388. [CrossRef]
15. Yang, H.; Pan, Z.; Bai, W. Review of Time Series Prediction Methods. *Comput. Sci.* **2019**, *46*, 21–28.
16. Zheng, H.; Cheng, Y.; Hu, Y. Air Quality Prediction Based on MLP&ST Model. *J. Appl. Sci. Electron. Inf. Eng.* **2022**, *40*, 302–315.
17. Cheng, Z.; Zhang, X.; Liang, Y. Railway Freight Volume Prediction Based on LSTM Network. *J. China Railw. Soc.* **2020**, *42*, 15–21.
18. Zhang, Z.; Zhu, Q.; Li, Q.; Liu, Y.; Zhang, E.H.; Zhao, Q.M.; Qing, X.F. Prediction of Mine Gas Concentration in Heading Face Based on Keras Long Short Time Memory Network. *Saf. Environ. Eng.* **2021**, *28*, 61–67.
19. Li, W.; Wang, L.; Wei, C. Application and design of LSTM in coal mine gas prediction and warning system. *J. Xi'an Univ. Sci. Technol.* **2018**, *38*, 1027–1035.
20. Wang, B.W.; Wang, J.X.; Wang, T.Y.; Zhang, Z.Q.; Liu, Y.; Yu, H. An encoder-decoder multi-step traffic flow prediction model based on long short-time memory network. *J. Chongqing Univ.* **2021**, *44*, 71–80.
21. Yang, Y.; Lin, Y. Multi-step forecasting of stock markets based on fuzzy time series model. *Comput. Eng. Appl.* **2014**, *50*, 252–256.
22. Hochreiter, S.; Schmidhuber, J. Long short term memory. *Nat. Comput.* **1995**, *9*, 1735–1780. [CrossRef] [PubMed]
23. Lecun, Y.; Bengio, Y.; Hinton, G. Deep learning. *Nature* **2015**, *521*, 436–444. [CrossRef] [PubMed]
24. Riley, R.D.; Ensor, J.; Snell, K.I.E.; Harrell, F.E.; Martin, G.P.; Reitsma, J.B.; Moons, K.G.M.; Collins, G.; Van Smeden, M. Calculating the sample size required for developing a clinical prediction model. *BMJ Br. Med. J.* **2020**, *368*, m441. [CrossRef]
25. Wang, J.; Li, W. Ultra-short-term forecasting of wind speed based on CEEMD and GWO. *Power Syst. Protect. Control* **2018**, *46*, 69–74.
26. Cheng, Z.; Ma, L.; Zhang, Y. Prediction of spatiotemporal distribution of gas concentration based on LSTM-FC model. *Comput. Eng. Appl.* **2020**, *56*, 258–264.
27. Wang, J.; Li, X.; Zhou, X.D.; Zhang, K. Ultra-short-term wind speed prediction based on VMD-LSTM. *Power Syst. Protect. Control* **2020**, *48*, 45–52.
28. Zhao, H.S.; Wang, K.; Wang, Z.; Liu, B.C.; Peng, Y.H. Temperature prediction of box-type transformer high-voltage bushing based on VMD-MGRU. *Electr. Mach. Control* **2021**, *25*, 18–28.

sustainability

MDPI

Article

The Effects of True Triaxial Loading and Unloading Rates on the Damage Mechanical Properties of Sandstone

Man Wang [1], Weihang Du [2], Yingwei Wang [1], Xinjian Li [1,3], Liming Qiu [4], Beichen Yu [2], Zehua Niu [1,2] and Dongming Zhang [2,*]

1 China Pingmei Shenma Group, State Key Laboratory of Coking Coal Exploitation and Comprehensive Utilization, Pingdingshan 467000, China
2 State Key Laboratory of Coal Mine Disaster Dynamics and Control, Chongqing University, Chongqing 400000, China
3 State Key Laboratory Cultivation Base for Gas Geology and Gas Control, Henan Polytechnic University, Jiaozuo 454000, China
4 Key Laboratory of the Ministry of Education for Efficient Mining and Safety of Metal Mines, University of Science and Technology Beijing, Beijing 100083, China
* Correspondence: zhangdm@cqu.edu.cn

Abstract: Coal is the main energy source in China. In the process of coal resource mining, the surrounding rock of roadways is often in the complex stress environment of "three heights and one disturbance". At the same time, rocks in the stratum are often in a three-way unequal pressure state under the action of geological structure, and conventional rock mechanics tests cannot study the mechanical properties of rocks under actual stress conditions; thus, this is based on the self-developed true triaxial multifunctional fluid–structure coupling test system to study the damage mechanical Properties of Sandstone. The results are shown as follows: With an increase in loading rate, the peak damage D_{cr} of sandstone decreases, but the initial damage D_a increases in the elastic stage, and the brittleness of sandstone weakens. With the increase in the unloading rate, D_{cr} increases, but D_a decreases in the elastic stage, and the sandstone brittleness increases first, then decreases. In addition, the peak maximum principal strain ε_{1max} first decreases rapidly and then slowly; the peak minimum principal strain ε_{3max} increases first, then decreases slowly, and increases slowly; the peak intermediate principal strain ε_{2max} decreases slowly; and the peak volume strain ε_{vmax} increases rapidly first and then slowly with increases in the loading rate. With an increase in the unloading rate, ε_{1max} increases rapidly first, then decreases slowly, then increases rapidly and finally increases slowly; ε_{3max} first decreases slowly, then increases slowly, and finally decreases slowly; and ε_{2max} increases slowly then decreases slowly. ε_{vmax} decreases rapidly first and then increases slowly with increasing loading rate.

Keywords: true triaxial; loading and unloading rate; damage; sandstone brittleness

Citation: Wang, M.; Du, W.; Wang, Y.; Li, X.; Qiu, L.; Yu, B.; Niu, Z.; Zhang, D. The Effects of True Triaxial Loading and Unloading Rates on the Damage Mechanical Properties of Sandstone. *Sustainability* **2022**, *14*, 11899. https://doi.org/10.3390/su141911899

Academic Editors: Xiangguo Kong, Dexing Li and Xiaoran Wang

Received: 16 August 2022
Accepted: 12 September 2022
Published: 21 September 2022

Publisher's Note: MDPI stays neutral with regard to jurisdictional claims in published maps and institutional affiliations.

1. Introduction

Due to the surging exploitation of coal resources in recent decades, shallow energy cannot meet the growing demand of energy in China, which also makes the momentum of deep energy exploration and exploitation soar. At the same time, with the increasing depth of underground space engineering, safety controllability decreases, the stress conditions are more complex, energy mining is more difficult, and costs are higher [1].

With increasing mining depth under complex stress conditions, the roadway surrounding rock is more prone to instability failure in the process of coal excavation, resulting in casualties and economic losses. In addition, with increasing coal mining depth, it is easier to induce coal and gas outburst, rock burst, and other dynamic disasters. At present, the rock samples used in the tests cannot easily reflect the original occurrence environments and stress paths in deep mines; deep field engineering requires long periods of time, and there is a lack of large-scale in situ monitoring research; therefore, the stress paths used in the laboratory tests are not consistent with the field [2]. In deep mines, the thick and hard bottom layer will produce higher mining stress, which will cause coal and rock roadways to produce larger deformation and even rock burst [3].

To better understand the mechanical properties of deep rocks and prevent or reduce the occurrence of these dangerous accidents, scholars at home and abroad have carried out a large number of studies in indoor laboratories, focusing on the mechanical properties of rocks under complex stress conditions. Xie et al. (2021) [4] conducted a conventional three-week test with different depths of ground stress and found significant differences in the physical and mechanical parameters of rocks at different depths. The results showed different brittleness characteristics of rocks with different occurrence depths. For sandstone with depth of 1600 m, the brittleness decreases with the increase in confining pressure on the whole, showing a transition from brittleness to ductility to strain hardening, and the post-peak plasticity gradually increases until it becomes completely plastic after the peak. Li et al. (2022a) [5] defined rock failure parameters by carrying out true triaxial tests and proposed a characterization method that could reflect the proportion of tensile and shear fractures in the rock failure process. Liu et al. (2021) [6] found that, compared with under the unloading path, granite required more energy when it was destroyed under the loading path in the true triaxial test, but it was more dangerous under the unloading path. Yin et al. (2019) [7] conducted a detailed study of the mechanical properties of sandstone under different loading and unloading rates using a true triaxial testing machine and found that tensile cracks were mostly concentrated on the unloading surface. Li et al. (2021a) [8] used a TRW-3000 true triaxial testing machine to carry out loading and unloading tests under different stress paths. Under DP criterion fitting, the cohesion and internal friction angle under loading conditions were higher than those under unloading conditions. Chu et al. (2022) [9] used MRI to analyze the pore and fracture expansion of coal samples after liquid nitrogen freezing and thawing. Quan et al. (2020) [10] used a true triaxial testing machine and a high-speed camera to study the mechanical properties of marble under different unloading rates and found that the failure process was more stable when unloading rate was lower, and the failure mode of marble changed from shear failure to shear tension failure with the reduction in minimum principal stress. Roohollah et al. (2020) [11] studied a wide range of rock properties and compiled a database, and they established a prediction model of rock burst maximum stress and risk index. Chu et al. (2019) [12], through triaxial cyclic loading and unloading tests on coal samples, found that the cumulative residual strain is related to the number of cycles. The more cycles, the greater the cumulative residual strain, but the relative residual strain gradually decreases, then stabilizes and finally rises sharply. Meanwhile, the total energy of a coal sample increases exponentially with increasing deviatoric stress. Zhai et al. (2020) [13] conducted the rock burst test under the condition of single-side airborne true triaxial loading and combined high-speed camera and SEM to study the results; they determined that the main reason for rock burst of different rock types is differences in the internal microscopic structures of rocks and their evolution under the different loading conditions. Danni et al. (2019) [14] found that the initial static stress is the main important factor in dynamic failure through true triaxial dynamic and static loading system research on rock dynamic failure. Su et al. (2016) [15] used a true triaxial rock burst testing machine to study rock samples at different high temperatures. The results showed that 300 °C was a critical point: The peak strength of rock samples changed little when the temperature was less than 300 °C, and the kinetic energy required by rock burst increased significantly when it was greater than 300 °C. Xiang et al. (2009) [16] studied the mechanical behavior of rock with a single structural plane under simulated excavation and support stress path using a true triaxial testing machine. The results showed that stress state, support strength and the parameters of the structural plane influenced the failure mode and support effect of rock with a structural plane under this stress path. Qiu et al. (2022) [17] used Hilbert-H and multiple analysis theory and studied the nonlinear characteristics of EMR and AE in re-coal cracking failure.

The results showed that the EMR and AE of coal cracking failure were related to the coal crack propagation process. Qiu et al. (2020) [18] established that the deformation and fracture of coal rock was caused by the accumulation of discrete fractures in rock samples. The time series obtained by the moving average method had a good correlation with the inner coal rock fractures and had obvious characteristics of coal instability and dynamic disaster precursor. Han et al. (2021) [19] carried out the true triaxial compression test of pre-cracked rock, and the results showed that the failure behavior of the compressive strength of the sample was related to the crack angle. The peak intensity decreased first and then increased with increasing crack priority angle. Li et al. (2019a) [20] used a true triaxial test system combined with CT scanning technology to study the mechanical properties of sandstone under different medium principal stress, and the results showed that the strength of sandstone increased first and then decreased with increases in medium principal stress. Dong et al. (2018) [21] used a true triaxial testing machine to study the mechanical properties of sandstone under biaxial compression and found that the fracture surface of sandstone specimen was parallel to the directions of intermediate principal stress and minimum principal stress, forming a large angle. Li et al. (2021b) [22] used a true triaxial rock burst testing machine to conduct surrounding rock failure tests on samples under pre-static load and dynamic disturbance and found that the threshold of rock burst occurrence and the frequency and failure degree of rock burst increased with the continuous increase of axial pressure. Li et al. (2022b) [23] used a true triaxial testing machine to study sandstone with holes and found that the holes significantly degraded the mechanical parameters of the specimen, and that the specimen entered the plastic yield stage in advance with decreasing peak strength. Fan et al. (2018) [24] studied the unloading failure strength of red sandstone under true triaxial conditions and found that the failure strength of red sandstone under rapid unloading condition decreased in different levels compared with that under loading condition. Wang et al. (2015) [25] studied the deformation and failure characteristics of fractured rock mass around the roadway under true triaxial conditions. The fracture angle had a great influence on the failure characteristics of rock mass; with the increase of the fracture angle, the compaction phenomenon was obvious, the dilatancy phenomenon showed a rising trend and the stress–brittle drop coefficient increased. Lee and Haimson (2011) [26] used a true triaxial testing machine to study granodiorite and found that rock strength increased with increases in intermediate principal stress. Wang et al. (2018) [27] performed conventional and true triaxial tests, and the results showed that in the first three stages of damage, evolution under the conditions of two different time and space distributions of acoustic emission activities was basically the same. Under the condition of CTT, the fracture surface of the test decreased with the increase of the confining pressure. However, under the condition of TTT, it first decreased and then increased with increasing intermediate principal stress. Wang et al. (2022) [28] used a true triaxial test system to study the mechanical properties of red sandstone under four different unloading stress paths and found that the octahedral shear stress was linearly correlated with the average effective stress. Hu et al. (2018) [29] used a true triaxial test system and DEM to study the characteristics and mechanism of rock burst induced by disturbed stress and found that under true triaxial test conditions, the test failure was mainly tensile splitting, and the generation of tensile cracks generally preceded the generation of shear cracks. In addition, DEM simulation results showed that weak dynamic rock burst was the result of tensile and shear failure. Zhao et al. (2021) [30] used a true triaxial test system to study the mechanical properties of sandstone under different loading and unloading rates and found that the bulk strain under a low unloading rate was mainly caused by axial compression and that rock damage was more serious at a high loading rate. Si et al. (2020) [31] conducted true triaxial testing on specimens with round holes of 50 mm diameter and found that the axial stress of hole wall failure increased with the increase of the loading rate, and the rock burst of the hole wall was more serious when the loading rate was lower. Zheng and Feng (2019) [32] used a true triaxial testing machine to study specimens with stress induction and found that with the decrease in the intermediate principal stress, Young's modulus decreased and the bilateral deformation increased. Duan et al. (2017) [33] used DEM to study the failure mechanism of sandstone; the results show that macroscopic response to σ_2 played an important role. With the increase in σ_2, peak stress increased after the first drop, and damaged the angle increment obtained by numerical simulation increased. The influence of σ_2 on Young's modulus, however, increased as σ_3 and σ_2 had less of an effect on the mechanical properties. Ze et al. (2014) [34] used the true triaxial test system to study the effect of intermediate principal stress and found that the intermediate principal stress coefficient had a quadratic function relationship with rock strength. Li et al. (2019b) [35] used DEM to simulate the unloading process of materials with cracks. With the increase in the unloading rate, the more severe the failure was, the more the cracks split. Xiao et al. (2021) [36] carried out laboratory tests and discrete element simulation to study the mechanical properties of sandstone under the unloading condition of maximum principal stress. The results showed that: with the increase in maximum principal stress, the bearing limit of sandstone could be improved, but it was more likely to be destroyed when unloading. Kong et al. (2021) [37] constructed the dynamic constitutive equation of gas-bearing coal under impact load through an SHPB test, which clearly explained the influence of different conditions on the dynamic mechanical properties of coal samples.

Kong et al. (2019) [38] studied the damage evolution mechanism of gas-bearing coal in this process and the formation reasons of acoustic emission signals by carrying out loading tests on gas-bearing coal. Du et al. (2015) [39] revealed the influence of intermediate principal stress on the failure of platen by true triaxial unloading test. Zhao et al. (2014) [40] used the true triaxial strain explosion test system to study the strain explosion process by changing the unloading rate, and the results show that the strain explosion is more likely to occur when the unloading rate is high. Fan et al. (2020) [41] used the true triaxial test system to carry out the unloading test under cyclic load path. The research results show that: with the increase of cyclic load before unloading, the elastic modulus and unloading strength increase first and then decrease, and the failure mode of rock changes from tensile failure to mixed tensile shear failure. Lu et al. (2021) [42] through uniaxial compression, triaxial compression, and true triaxial unloading tests on basalt, the results show that the stress–strain curves under uniaxial and true triaxial compression show strain softening, and the stress–strain curves under triaxial compression show strain hardening. Miao et al. (2011) [43] conducted rock burst tests on granite with the true triaxial testing machine, and the results showed that the debris or irregular massive debris of granite was related to stress conditions and boundary conditions.

The results of the above research have a guiding significance for understanding coal dynamic disasters such as rock instability and rock burst in the process of coal mining, and provide an important reference for studying the mechanical properties under different actual triaxial loading and unloading conditions. In underground engineering construction and mining processes, the surrounding rock stress state is complex, to better meet the "three highs a disturbance" complex stress conditions; this research uses independent research and development of the multifunctional fluid–structure coupling triaxial test system for studying the sandstone under the different loading and unloading rates; the complex stress conditions of the instability of the roadway surrounding rock were examined in order to carry out the research.

2. Test Device and Scheme

2.1. Sample Device and Sample

Based on the self-developed true triaxial multifunctional fluid–structure coupling test system (as shown in Figure 1), mechanical tests of true triaxial sandstone under different loading and unloading rates are carried out. The system can provide the maximum pressure of 6, 6 and 4 MN in three directions to meet requirements of this test [44].

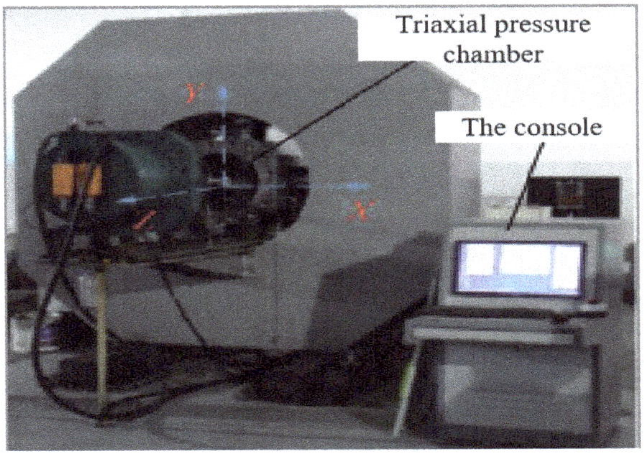

Figure 1. True triaxial fluid–structure coupling test system.

The samples taken in this test were from Mine No. 12 of PingMei ShenMa Groupin China. The sample is a 100 × 100 × 100 mm cubic specimen with the end face flatness within 0.02 mm. Young's modulus E of the sandstone is 10.6 GPA, and Poisson's ratio v of the sandstone is 0.31. The apparent density is 2260 kg/m^3. This indicates that there are no obvious joints and fissures, in line with the standards of the International Society for Rock Mechanics.

2.2. Test Scheme

To better meet conditions of field stress, the loading and unloading methods of this test are σ_3 single-side unloading and σ_1 single-side loading. In this test, two sets of mechanical tests were performed on the sandstone at different loading and unloading rates, named group H and group G. For group H. The loading rate was kept the same and the unloading rate was changed. For group G, the unloading rate was kept the same and the loading rate was changed. The details of the test are shown in Table 1. The test stress path is shown in Figure 2, and proceeds as follows:

Table 1. Test schemes of different loading and unloading rates.

Specimen Number	Loading Rate/(mm•$^{1-}$)	Unloading Rate/(kN•$^{s-}$)	Specimen Number	Loading Rate/(mm•$^{1-}$)	Unloading Rate/(kN•$^{s-}$)
H1		0.2	G1	0.001	
H2		1	G2	0.003	
H3	0.003	2.5	G3	0.005	1
H4		3	G4	0.008	
H5		5	G5	0.012	

First, the stresses in the three directions are added synchronously at40 MPa with a force control of 2 kN/s. The σ_3 remained unchanged, σ_1 and σ_2 continue to be simultaneously loaded to 60 MPa. The σ_2 and σ_3 remained unchanged and σ_1 continued to be loaded to 80 MPa. When the hydrostatic pressure was reached, the σ_1 was loaded in a displacement control way, and the σ_3 was unloaded at a single side in a force control way. When the sandstone sample broke, the test was stopped.

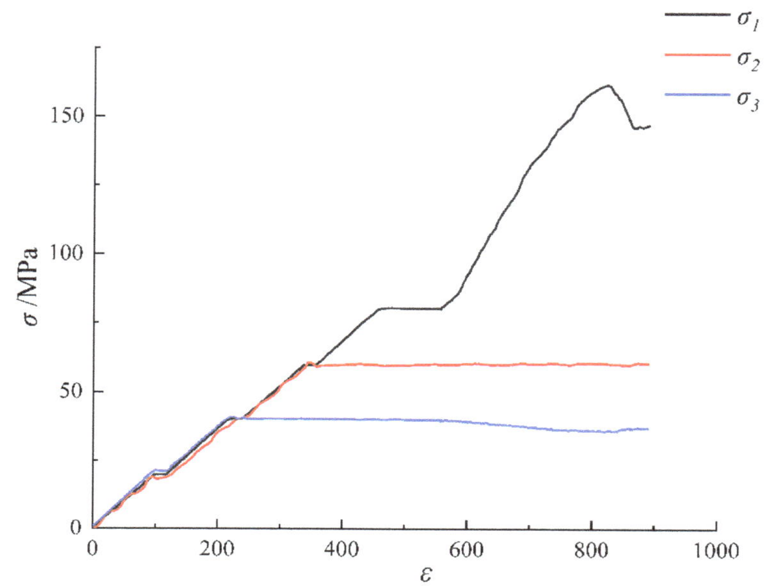

Figure 2. Stress–strain characteristics.

3. Experimental Results and Analysis

3.1. Variation of Peak Strain and Peak Deviant Stress under True Triaxial Loading and Unloading Rates

The peak stress and strain of rock are important indicators to measure the mechanical properties of rock, and important mechanical parameters can be obtained from them. In order to better analyze deformation characteristics of sandstone at different loading and unloading rates, the curves of peak strain-loading (unloading) rate (Figure 3a) and peak deviatoric stress-loading (unloading) rate (Figure 3b) were drawn.

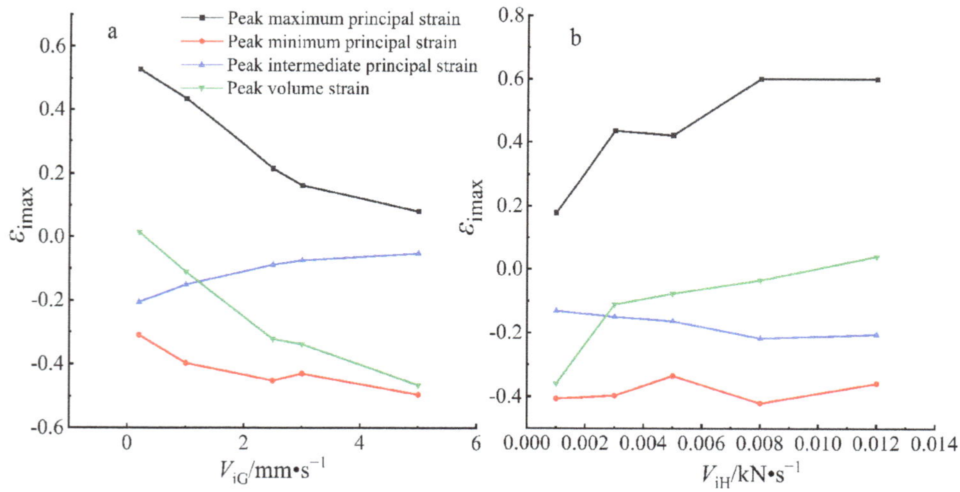

Figure 3. Peak strain-loading (unloading) rate curve: (**a**): different loading rate conditions, (**b**): different unloading rate conditions.

In Figure 3a, ε_{imax} is the peak strain under different loading rates, and Vi_G is the loading rate. It can be found from the figure that the peak maximum principal strain first decreases rapidly and then decreases slowly as the loading rate increases. With the increase in the loading rate, ε_{3max} increases first and then slowly decreases, then slowly increases and then slowly decreases, and ε_{2max} decreases slowly with the increasing loading rate, and ε_{vmax} rapidly first and then slowly with the increase in the loading rate. Meanwhile, it can be found from Figure 3a that the variation of ε_{vmax} is basically consistent with that of ε_{1max}. Vi_H in Figure 3b is the unloading rate. It can be found from Figure 3b that ε_{1max} increases rapidly first and then slowly decreases with the increase in the unloading rate, then increases rapidly and finally slowly increases. With the increase in the unloading rate, ε_{3max} firstly decreases slowly, then increases slowly, and finally decreases slowly. ε_{2max} increases slowly with the increase in the unloading rate. Eventually, it slowly decreased; ε_{vmax} decreases rapidly first, and then slowly with the increase in the loading rate; the variation in ε_{vmax} is similar to that of ε_{1max}.

It can be found from Figure 4a that the peak deviational stress first increases, then decreases rapidly, and finally slowly decreases as the loading rate increases. This is because the increase in loading rate accelerates the fracture rate of the sandstone specimen. Although the loading rate is very high at this time, the time required for the rock to fracture is also reduced, and the increase in the peak deviational stress is greatly reduced when the unloading rate is high. It can be found from Figure 4b that the peak deviationic stress first increases rapidly, then slowly, then rapidly, and finally slowly decreases with the increase in the unloading rate. When the unloading rate is low, the increase in peak deviationic stress is higher.

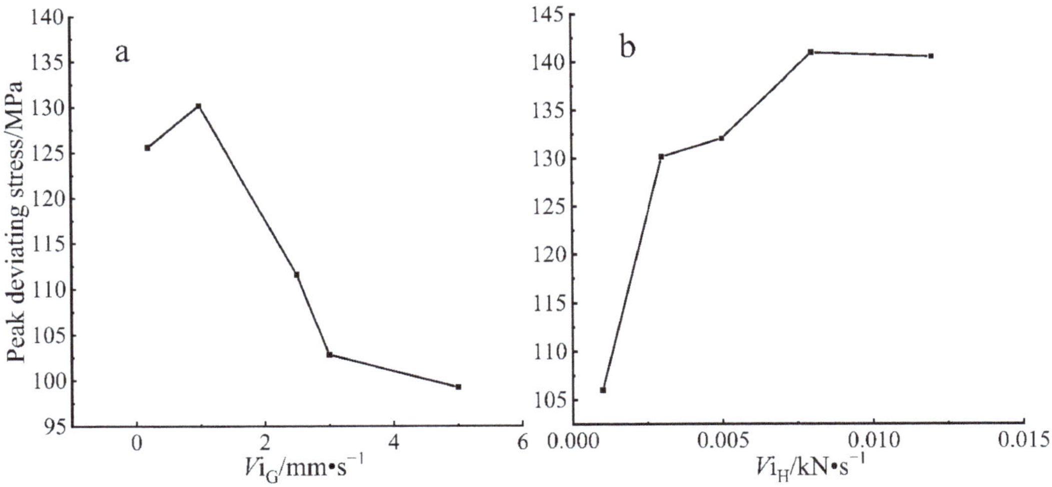

Figure 4. Peak deviatoric stress-loading (unloading) rate curve: (**a**): different loading rate conditions, (**b**): different unloading rate conditions.

To better understand the influence of the increasing loading and unloading rate on peak strain and deviational stress, the slope K of the line between the other four points is plotted against G1 and H1, respectively. The value of K, based on G1 and H1, shows the effect of the increasing loading and unloading rate on peak strain and deviational stress. At the same time, K (displacement generated at unit loading and unloading rate) is an indicator to analyze the influence of increasing loading and unloading rate on rock deformation and strength. Table 2 shows the values of K for different loading rates, and a negative value of K indicating a decreasing displacement. $K_{\varepsilon1}$ represents the response of the increasing loading rate to the displacement in the σ_1 direction. $K_{\varepsilon2}$ represents the response of the increasing loading rate to the displacement in the σ_2 direction. $K_{\varepsilon3}$ represents the response of the increasing loading rate to the displacement in the σ_3 direction. $K_{\varepsilon v}$ represents the response to peak volume strain after increasing the loading rate. K_σ represents the response to deviatoric stress as the loading rate increases.

Table 2. Slope K of the line between the other four points and the base point under the condition of the loading rate.

Slope Number	$K_{\varepsilon 1}$	$K_{\varepsilon 2}$	$K_{\varepsilon 3}$	$K_{\varepsilon V}$	K_σ
G2-G1	−0.1168	0.0683	−0.1096	−0.1576	5.7574
G3-G1	−0.1360	0.0509	−0.0614	−0.1466	−6.1034
G4-G1	−0.1304	0.0466	−0.0428	−0.1263	−8.1351
G5-G1	−0.0932	0.0318	−0.0386	−0.1000	−4.5491

It can be found from Table 2 that $K_{\varepsilon 1}$ decreases first and then increases with the increase in the loading rate, indicating that the response to ε_1 becomes faster and then slowly decreases with the increase in the loading rate, and the response to ε_1 becomes slower with the increase in the loading rate. With the increase in the loading rate, $K_{\varepsilon 2}$ decreases and $K_{\varepsilon 3}$ increases gradually. It shows that the response of ε_2 to the increase in the loading rate is slower and the response of ε_3 to the increase in the loading rate is slower. With the increase in the loading rate, $K\varepsilon_V$ increases gradually, which means that ε_v responds more and more slowly to the increase in the loading rate. In addition, it can be found from the table that the absolute values of $K_{\varepsilon 1}$, $K_{\varepsilon 2}$ and $K_{\varepsilon 3}$ are the largest; thus, the fracture of the sandstone specimen is dominated by the increase in ε_1. It can be found from Table 3 that $K_{\varepsilon 1}$ and $K_{\varepsilon V}$ decrease with the increase in the unloading rate. ε_v and ε_1 respond slowly to the increase in the unloading rate. Although $K_{\varepsilon 2}$ and $K_{\varepsilon 3}$ show no obvious change rule, the change degree is not significant. At the same time, it can be found from the table that the absolute values of $K_{\varepsilon 1}$, $K_{\varepsilon 2}$ and $K_{\varepsilon 3}$ are the largest, which means that under different unloading rates, the increase in ε_1 also leads to the failure of sandstone specimens.

Table 3. Slope K of the line between the other four points and the base point under the condition of unloading rate.

Slope Number	$K_{\varepsilon 1}$	$K_{\varepsilon 2}$	$K_{\varepsilon 3}$	$K_{\varepsilon V}$	K_σ
H2-H1	128.57	9.57	4.88	123.885	11,970.26
H3-H1	60.61	8.1425	18.025	70.49	6503.73
H4-H1	60.59	12.3028	−1.9843	46.2986	4988.28
H5-H1	42.41	6.6945	4.4455	36.3055	3132.19

3.2. Damage Characteristics of Sandstone under True Triaxial Loading and Unloading Rates

In underground space engineering such as tunnels and roadways, damage to sandstone is associated with the conditions of excavation. To explore the damage properties of sandstone under different opening conditions, real triaxial tests were performed at different loading and unloading rates. Under realistic triaxial loading and unloading rates, the extent of damage to the sandstone increases with the increase in time under three-dimensional stress. For the intact sandstone, the initial damage variable $D = 0$. However, the sandstone specimens we took in the laboratory were both macroscopically and microscopically defective. To obtain more accurate damage variables, [45] Qin et al. (2018) optimized the damage state of the traditional constitutive model which was based on Weibull statistical damage mechanics. The damage value and the initial damage coefficient k were obtained for each characteristic point:

$$m = \frac{1}{\ln^{E} - \ln^{E_M}} \tag{1}$$

$$\sigma_{\max} = E\varepsilon_{\max}\left(ke^{-\frac{1}{m}}\right) \tag{2}$$

According to Equation (2), the initial damage coefficient k can be obtained:

$$k = \frac{\sigma_{\max}}{E\varepsilon_{\max}} e^{\ln^{E} - \ln^{E_M}} \tag{3}$$

where m is the mean value of the materials; E is the elastic modulus of rock materials; E_M is the secant modulus of the peak, $E_M = \frac{\sigma_{\max}}{\varepsilon_{\max}}$; $\varepsilon_{\max}$ is the maximum principal strain of the sandstone peak. $\sigma_{\max}$ is the peak maximum principal stress. Table 4 shows the material coefficient m and the damage coefficient k for the sandstone specimens selected for this test.

Table 4. Material coefficient and damage coefficient of the sandstone specimen.

Specimen Number	Damage Coefficient	Degree of Material Mean	Specimen Number	Damage Coefficient	Degree of Material Mean
G1	1.0000	2.0000	H1	0.9999	3.0894
G2	0.9858	2.2216	H2	0.9999	2.2216
G3	1.0000	2.3100	H3	0.9999	2.1465
G4	0.9999	2.8277	H4	0.9999	1.7948
G5	0.8892	3.0415	H5	0.9999	1.6428

Damage values for each characteristic point of the sandstone specimens can be obtained from the damage coefficient, material mean value, peak maximum principal stress and peak maximum principal strain:

$$D_{cr} = 1 - ke^{\frac{-1}{m}}\varepsilon \tag{4}$$

$$D_{th} = 1 - ke^{\frac{-1}{m}(1-D_{cr})^{m}} \tag{5}$$

$$D_{c} = 1 - ke^{\frac{-1}{m}(1-D_{cr})^{2m}} \tag{6}$$

$$D_{b} = 1 - ke^{-\left(\frac{1}{m}\right)^{m}\frac{1}{m}(1-D_{cr})^{m}} \tag{7}$$

$$D_{a} = 1 - e^{(1-D_{cr})^{m}\frac{-D_{cr}^{m}}{m}} \tag{8}$$

where D_{cr} is the peak damage value; D_{th} is the elastic limit point damage value, the end point of the elastic damage; D_b is the damage value at the half peak intensity point; D_a is the initial damage value for linear elasticity. According to Formulas (4)–(8) above, the damage values of each feature point under different loading and unloading rates are calculated as is shown in Tables 5 and 6. Figure 5 shows the distribution of each feature point.

Table 5. Damage values of feature points under different loading rates.

Specimen Number	D_{cr}	D_{th}	D_c	D_b	D_a
G1	0.8636	0.00069	0.0000015	0.000021	0.00044
G2	0.7223	0.0258	0.0015	0.0044	0.0126
G3	0.7358	0.0264	0.0015	0.0052	0.0137
G4	0.6548	0.0793	0.0122	0.0284	0.0379
G5	0.6440	0.1055	0.0330	0.0481	0.0526

Table 6. Damage values of characteristic points under different unloading rates.

Specimen Number	D_{cr}	D_{th}	D_c	D_b	D_a
H1	0.6791	0.0502	0.0053	0.0128	0.0235
H2	0.7223	0.0258	0.0015	0.0044	0.0126
H3	0.8581	0.0038	0.00005	0.00069	0.0033
H4	0.8918	0.0007	0.000002	0.000035	0.00048
H5	0.9477	0.00005	0.0000001	0.000002	0.00004

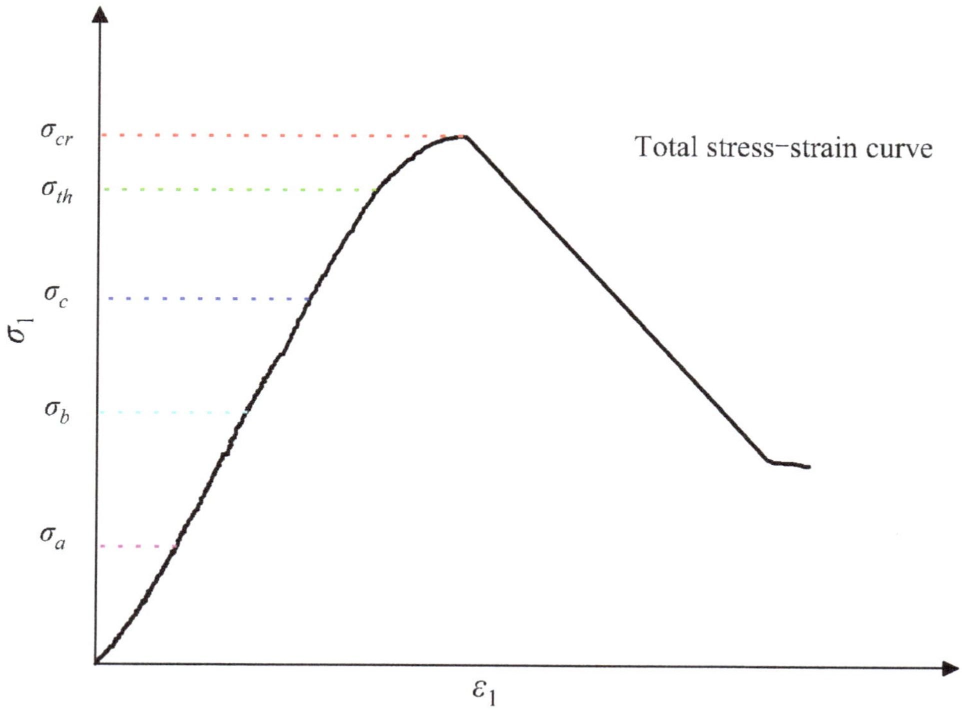

Figure 5. Distribution diagram of each feature point.

As shown in Table 5, under the conditions of different real triaxial loading rates, such as the direction of σ_1 loading and the direction of σ_3 unloading, the sandstone specimen has little damage, with no microcrack occurring before the D_c line elastic damage ending point. The rate of native crack propagation is slow. As the σ_1 direction continues loading and the σ_3 direction continues unloading, the damage value increases significantly at the elastic limit point of D_{th}, but the damage value of sandstone specimen H1 remains small and can be ignored at this point; thus, crack growth and the development of H1 occur after the elastic limit point of D_{th}. The remaining specimens are considered to have begun to expand before D_{th}; thus, it can be found that the loading rate will affect the propagation rate of the fissures in the rock. However, as can be seen from Figure 6 (D is damage value; V loading rate) that shows the loading rate—damage value curve of each characteristic point, the damage value of peak point decreases with the increase in the loading rate; G1 with the lowest loading rate has the largest peak damage value. This is probably because the low loading rate is more conducive to the increase in the number of cracks. At the same time, it can be found from Table 5 that increasing the loading rate when it is low has a larger effect on the peak damage value of sandstone than increasing the loading rate when it is high. In addition, D_{th}, D_c, D_b and D_a all increase with the increase in the loading rate. When the loading rate is low, D_{th}, D_c, D_b and D_a are small. This does not favor the initial damage of the internal structure of sandstone during the elastic stage. Since damage to the rock begins to expand from internal cracks, the macro damage to the sandstone specimen is easier and faster when the unloading rate is higher. It can be seen from Table 5 that the damage of sandstone specimens at different loading rates is mainly in the stage from D_{th} to D_{cr}.

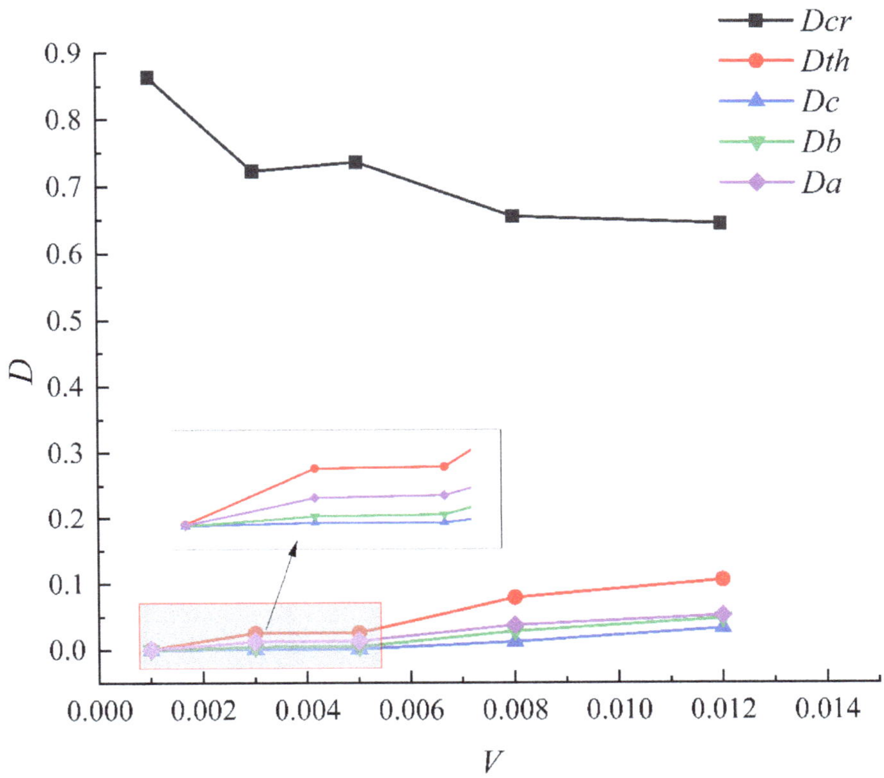

Figure 6. Loading rate–damage value curve of each characteristic point.

It can be found from Table 6, under the conditions of different loading and unloading rates in the true triaxial, with the loading in the σ_1 direction and unloading in the σ_3 direction, that sandstone basically has no damage at the elastic limit point. As the direction of σ_3 continues unloading and the direction of σ_1 continues loading, the only H1 and H2 with a low unloading rate occur an obvious increase before the D_{th} elastic limit point. Meanwhile, from Figure 7 (D is damage value; V1 is the unloading rate) the unloading rate can be seen—the damage value curve of each characteristic point shows that D_{th}, D_c, D_b and D_a all decrease with the increase in the unloading rate. With the increase in the unloading rate, the initial damage rate of the internal structure of sandstone is slower, and the reduction in the unloading rate causes more damage to the sandstone specimen during the initial elastic stage. However, it can be seen from the figure that the peak damage value increases with the increase in the unloading rate; thus, the internal damage of sandstone can be accelerated when the unloading rate is low, but the number of cracks in macroscopic failure will decrease. As the failure of rock begins to expand from the internal cracks, the macro failure of the sandstone specimen is easier and faster when the loading rate is low, but the damage to sandstone will be reduced. It can be seen from Table 6 that the damage of the sandstone specimens with different unloading rates is mainly in the stage from D_{th} to D_{cr}.

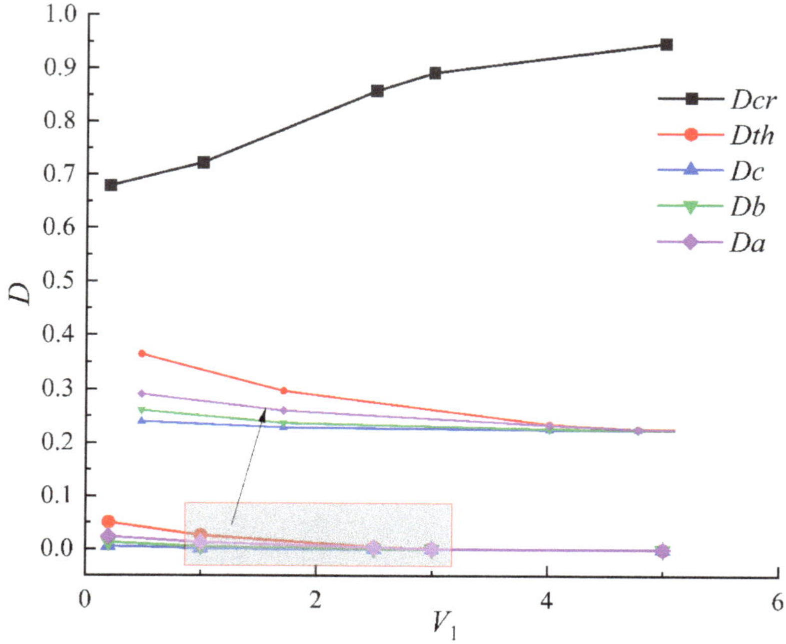

Figure 7. Damage curves of unloading rate and each characteristic point.

Deviational stress is the main factor responsible for the macroscopic failure of the internal structure. To analyze the damage response capacity of deviational stress to rock under different loading and unloading rates, a deviational stress damage compliance Δq is introduced:

$$\Delta q = \frac{D_{cr}}{\sigma_{1-3}} \tag{9}$$

where $\sigma_{1\text{-}3}$ is the deviatoric stress difference between loading and unloading starting point and peak point; D_{cr} is peak damage value, according to Equation (9). The deviational stress calculated is shown in Figures 8 and 9.

It can be found from Figure 8 that as the loading rate increases, the deviational stress damage compliance first decreases and finally increases, and the deviational stress response capacity becomes slower. However, when the loading rate decreases to 0.12 $mm{\cdot}^{1-}$, the response capacity of deviational stress to sandstone increases again, because when the loading rate is too high, the bearing capacity of sandstone decreases.

From Figure 9, it is shown that with the increase of unloading rate of deviator stress damage increase compliance, and the increasing rate is stable, but when the unloading rate is lower, increased compliance deviatoric stress injury after unloading rate change is small; therefore, when the unloading rate increased to a certain value, the deviatoric stress damage compliance of sandstone began to stably increase and the response ability to sandstone damage stably increased.

3.3. Brittleness Characteristics of Rock under True Triaxial Loading and Unloading Rates

3.3.1. Stress Brittle Drop Factor

In traditional materials, materials that produce large deformation but do not crack are ductile materials with good ductility, or, on the contrary, are brittle materials. For rock materials, the key to distinguishing brittle from ductile is the type of rock failure; that is, the form of the rock failure process itself. Nowadays, there are many studies on rock brittleness under unicycle compression or triaxial confining pressure, but there are few studies on rock brittleness under true triaxial condition [46].

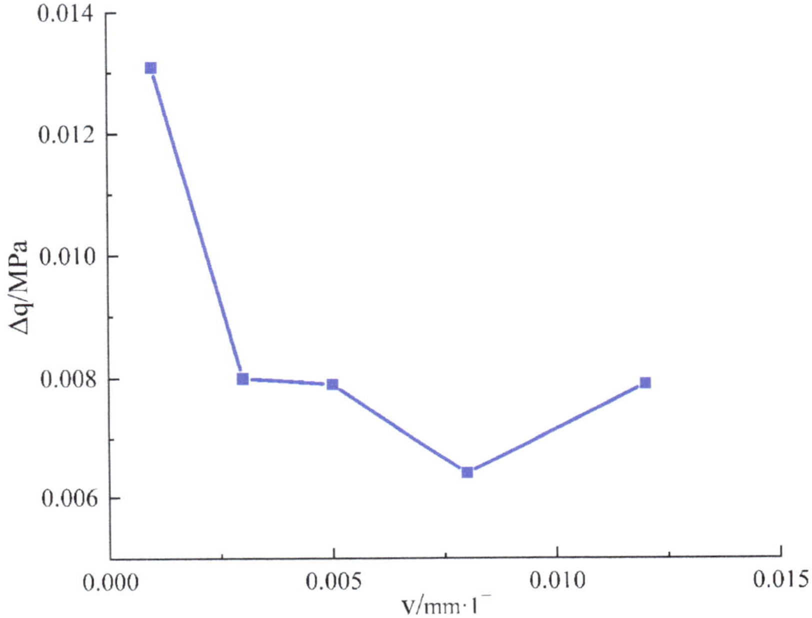

Figure 8. Deviatoric stress damage flexibility–loading rate curve.

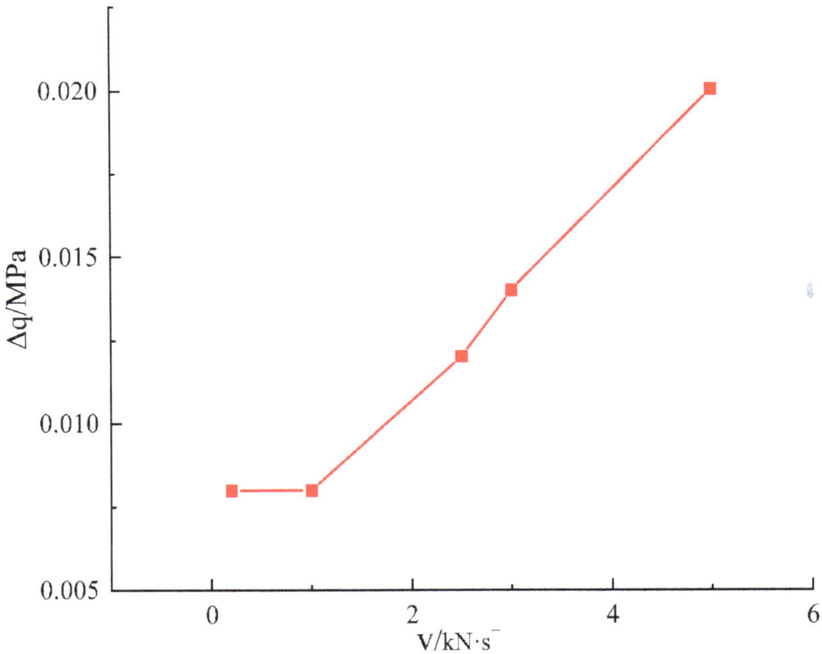

Figure 9. Deviatoric stress damage compliance–unloading rate curve.

Figure 10 shows the stress–strain characteristic curve of specimen H3, calculated to be 47.12%. Based on the stress-brittle drop process of the non-vertical drop model derived from Ge (1997), Shi et al. (2006) [47,48] determined the stress-brittle drop coefficient R in combination with the typical stress–strain curve generalization diagram of rock (Figure 10):

$$R = \frac{b}{a} \tag{10}$$

where a and b are strain-related parameters, $a = \varepsilon_p - \varepsilon_n$, $b = \varepsilon_b - \varepsilon_p$, ε_p is the maximum principal strain of peak strength, ε_b is the maximum principal strain of residual strength, and ε_n is the strain of residual strength corresponding to the initial loading stage. According to Equation (10), the smaller R is, the more serious the stress brittle failure of rock is.

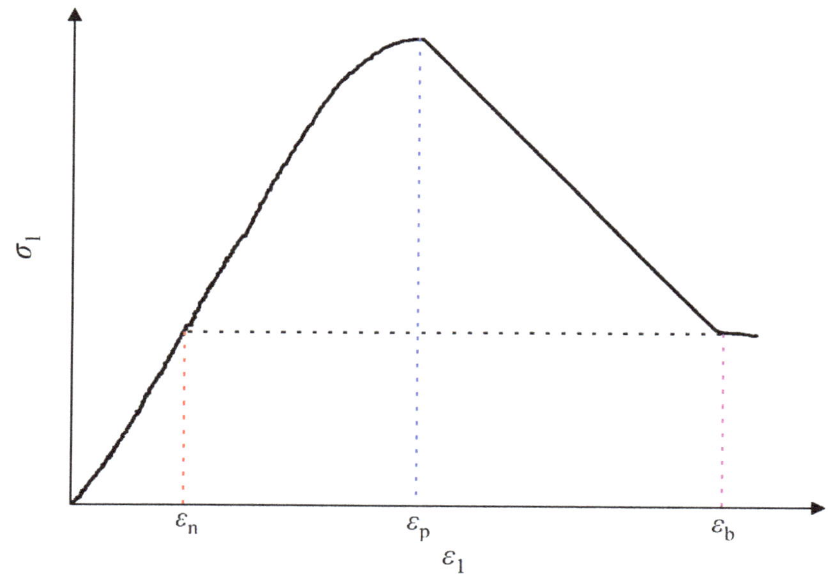

Figure 10. Classical stress–strain curves of brittle rocks.

3.3.2. Variation Characteristics of Specimen Characteristic Parameters under Different Loading Rates

Under the same initial stress path and different loading and unloading rates, the stress brittle drop coefficients and are function of the characteristic paramecium and the loading rates, which can be expressed as follows:

$$R = \frac{\varepsilon_b(v) - \varepsilon_p(v)}{\varepsilon_p(v) - \varepsilon_n(v)} \tag{11}$$

In the experiment, the characteristic parameters at different loading rates are shown in Figure 11. As can be seen from Figure 11, the displacement of the maximum principal strain direction at the peak strength point, the displacement of the residual strength point and the residual strength point all showed an increasing trend for the displacement of the loading section, and the increasing trend was basically the same. The dotted line in the figure fitted curve: $\varepsilon_n = -3714.729v^2 + 0.254 + 91.639v$; $\varepsilon_p = -4594.288v^2 + 0.519 + 94.123v$; $\varepsilon_b = -5046.028v^2 + 0.847 + 97.952v$.

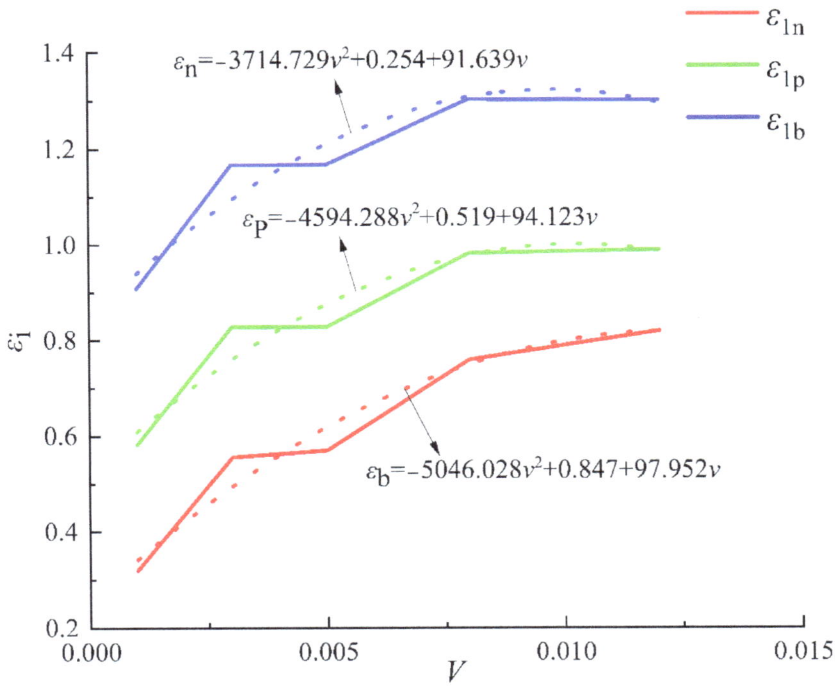

Figure 11. Curves of the relationship between residual strength point, peak point and residual strength point corresponding to the maximum principal strain displacement of loading segment point and loading rate under different loading rates.

Combined results:

$$R(v) = \frac{-451.74v^2 + 0.328 + 3.829v}{-879.559v^2 + 0.265 + 2.484v} \tag{12}$$

Equation (11) is the relation between stress brittle drop coefficient and loading rate at different loading rates. Through Equation (11), the stress brittle drop coefficient at different loading rates can be obtained, as is shown in Figure 12. With the increase in the loading rate, the stress brittle drop coefficient increases, and sandstone brittleness weakens.

3.3.3. Variation Characteristics of Specimen Characteristic Parameters under Different Unloading Rates

Under the same initial stress path and different unloading rates, the stress–brittle drop coefficients and are functions of the characteristic parameters of the specimen, which can be expressed as follows:

$$R = \frac{\varepsilon_b(v) - \varepsilon_p(v)}{\varepsilon_p(v) - \varepsilon_n(v)} \tag{13}$$

In the experiment, the characteristic parameters at different unloading rates are shown in Figure 13. As can be seen from Figure 13, the displacement of the maximum principal strain direction at the peak strength point, the displacement of the residual strength point and the residual strength point all show an increasing trend for the displacement of the loading section, and the increasing trend was basically the same. The dotted line in the figure fitted curve: $\varepsilon_n = 0.02046v^2 + 0.96941 - 0.18168v$; $\varepsilon_p = 0.01959v^2 + 1.21853 - 0.20361v$; $\varepsilon_b = 0.02705v^2 + 1.55467 - 0.25897v$.

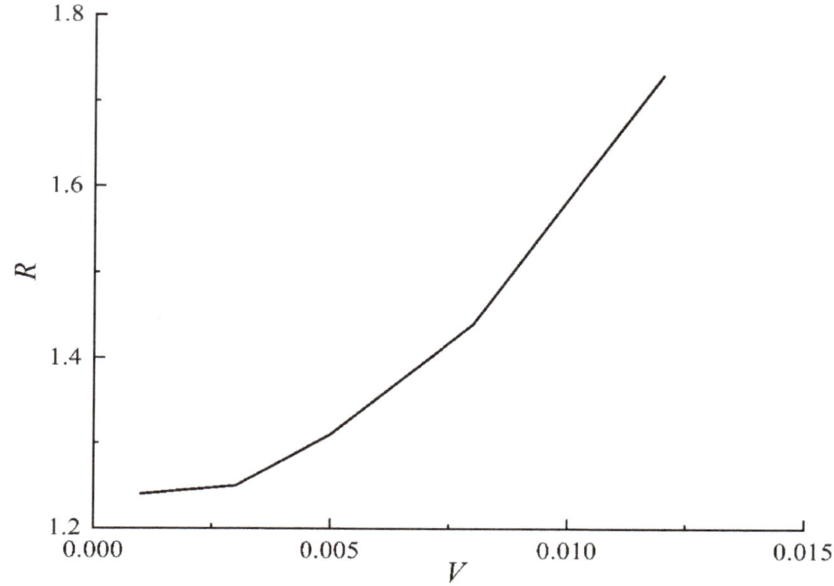

Figure 12. The relation curve between stress brittle drop coefficient and loading rate at loading rate.

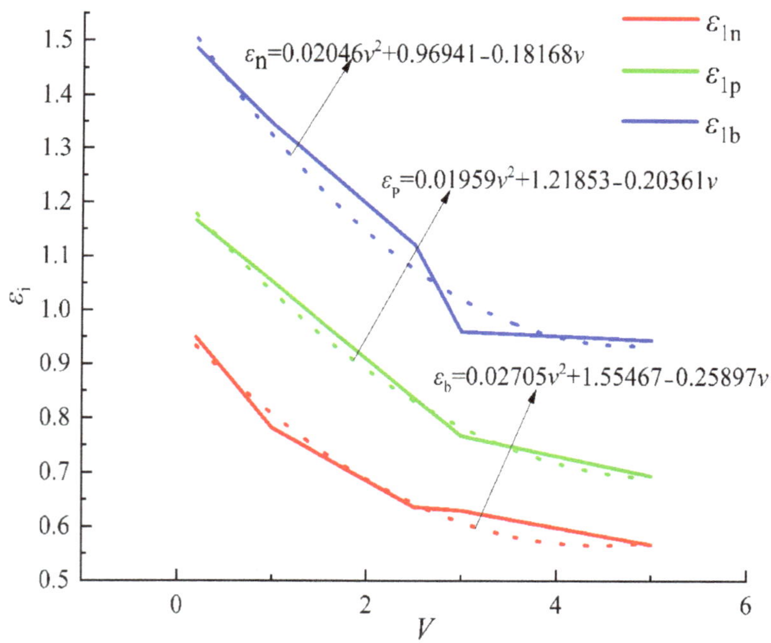

Figure 13. Curves of the relationship between the maximum principal strain displacement and unloading rate of the loading segment corresponding to residual strength point, peak point and residual strength under different unloading rates.

Combined results:

$$R(v) = \frac{0.00746v^2 + 0.33614 - 0.05536}{-0.00087v^2 + 0.24912 - 0.02193v} \quad (14)$$

Equation (14) is the relationship between stress brittle drop coefficient and unloading rate under different unloading rates. According to Equation (14), the stress brittle drop coefficient under different unloading rates can be obtained. As is shown in Figure 14, with the increase in the unloading rates, the stress brittle drop coefficient first decreases and the sandstone brittleness increases, and then the sandstone brittleness decreases. When the unloading rate is low, its brittleness remains relatively stable, and when the unloading rate reaches 5 kN/s, it decreases sharply, which is caused by the vertical drop in the stress–strain curve after the peak, which causes the abnormal increase in $\varepsilon_b - \varepsilon_p$. Therefore, this method is not suitable for rock brittle assessment at high unloading rates.

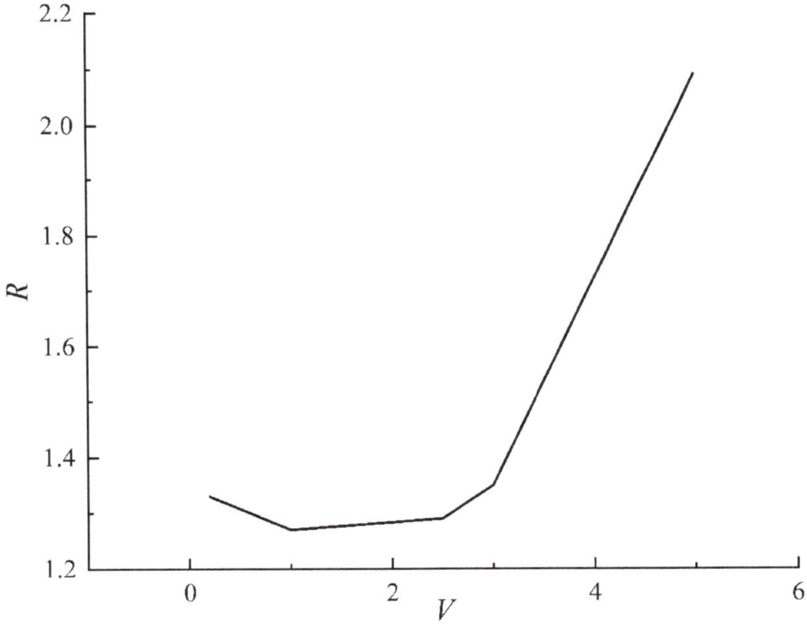

Figure 14. The relation curve between stress brittle drop coefficient and unloading rate at different unloading rate.

4. Discussion

In underground engineering construction, the risk of rock mass instability is often reduced by slowing down the excavation rate and reducing the excavation footage, the essence of which is to adjust the rate of surrounding rock stress loading and unloading caused by excavation, so as to reduce the possibility of rock mass instability and rock burst [49]. During underground excavations, high-stress areas are more prone to failure, and rock mass instabilities and failures are frequent. Under the condition of high-stress single-side unloading, the surrounding rock failure is the composite failure of tensile, splitting and shear [50]. Huang and Huang, 2010, [51] conducted a triaxial unloading test in the laboratory and found that the unloading rate and initial confining pressure have a great influence on the brittleness and tensile fracture characteristics of rock, and this influence is more obvious when unloading at high speed and high initial confining pressure. Whether the roadway is stable is closely related to the supporting conditions and methods. Under the condition of true triaxial single-plane in-flight test, the failure mode of rock samples changes from brittle cracking to dynamic rock burst failure with the increase in the failure rate of support [52].

During coal mining and roadway excavation, the surrounding rock mass subjected to disturbance stress undergoes instability failure, which is different for different degrees of excavation depth and tunneling rate. According to the analysis in Section 3.2, the peak damage value of sandstone decreases with the increase in loading rate for different loading rates. In this case, the likelihood of inducing rock mass instability is lower, and the excavation strength and turning rate can be appropriately increased to ensure rock mass stability during mine excavation and roadway excavation. The damage value of the sandstone increases with the off-loading rate in the case of a two-fold differential off-loading rate. At this point, the likelihood of inducing rock mass instability increases. We can reduce the mining intensity and reduce the running footage of the roadways through the coal to ensure the stability of the rock mass. In addition, when the perturbation stress of the surrounding rock mass is too high, the accumulated strain energy of the empty rock mass is suddenly and violently released, resulting in an explosion-like brittle fracture of the rock mass. Rockfalls can cause large amounts of rock to fall and produce loud sounds and gas waves that can not only destroy mines but also endanger buildings on the surface. According to the analysis in Section 3.3, the brittle failure of sandstone weakens with increasing loading rate for different loading rates. However, the brittle failure of the sandstone is enhanced as the loading rate is increased at different offloading rates. As mentioned above, the offloading rate can be reduced at different offloading rates to ensure the stability of the rock mass. However, according to the analysis in Section 3.3, the brittle failure of sandstone is stronger at low offloading rates. Therefore, in the actual process of mine production and road tunneling, the appropriate excitation intensity and turning rate must be chosen to achieve safe and efficient production.

5. Conclusions

To ensure the safety of underground space excavation and to provide a theoretical basis for its use in laboratory tests, mechanical tests of sandstone real triaxial at different loading and unloading rates were performed based on the self-developed multi-function real triaxial test system. The results are shown as follows:

(1) With the increase in the loading rate, the peak ε_1, ε_3 and ε_v decrease, and the peak ε_2 increases, and the peak deviational stress increases first and then decreases. With the increase in the unloading rate, the peak ε_1 and ε_v increase, and the peak ε_2 decreases first and then increases, and the ε_3 increases first, then decreases and then increases and the peak deviational stress increases.

(2) Under different loading and unloading rates, with the increase in loading and unloading rates, the damage of sandstone specimens is mainly from the online elastic damage end point to the peak point, and the peak damage value decreases and D_{th}, D_c, D_b, D_a increase with the increase in the loading rate. The peak damage values increase and D_{th}, D_c, D_b, D_a decrease with the increase in the unloading rate.

(3) With the increase in the loading rate, the stress brittle drop coefficient of sandstone increases, and the brittle failure weakens. With the increase in the unloading rate, the stress brittle drop coefficient of sandstone decreases first and then increases, and the brittle failure of the rock first becomes stronger first and then weaker. However, the stress brittle drop coefficient appears abnormal at a high unloading rate.

(4) With the increase in the loading rate, $K_{\varepsilon 1}$ first decreases and then increases, and $K_{\varepsilon 2}$ gradually decreases, $K_{\varepsilon 3}$ gradually increases, and $K_{\varepsilon V}$ gradually increases. With the increase in the unloading rate, $K_{\varepsilon 1}$ and $K_{\varepsilon V}$ decrease, and the change in $K_{\varepsilon 2}$ and $K_{\varepsilon 3}$ is not obvious.

Author Contributions: Conceptualization, Z.N. and B.Y.; methodology, B.Y.; formal analysis, L.Q.; investigation, X.L.; resources, Y.W.; data curation, D.Z.; writing—original draft preparation, M.W.; writing—review and editing, W.D.; project administration, D.Z.; funding acquisition, D.Z. All authors have read and agreed to the published version of the manuscript.

Funding: This research was funded by the National Natural Science Foundation of China (51874053, 52064016), the Scientific Research Foundation of State Key Laboratory of Coal Mine Disaster Dynamics and Control (2011DA105287-zd201804), Jiangxi Provincial Thousand Talents Plan Project (jxsq2019102082).

Institutional Review Board Statement: Not applicable.

Informed Consent Statement: Informed consent was obtained from all subjects involved in the study.

Data Availability Statement: The experimental data supporting the conclusions are available from the corresponding author on request.

Acknowledgments: This study was financially supported by the National Natural Science Foundation of China (51874053, 52064016), the Scientific Research Foundation of State Key Laboratory of Coal Mine Disaster Dynamics and Control (2011DA105287-zd201804), Jiangxi Provincial Thousand Talents Plan Project (jxsq2019102082).

Conflicts of Interest: The authors declare that they have no known competing financial interests or personal relationships that could have appeared to influence the work reported in this paper.

References

1. He, M.C.; Xie, H.P.; Peng, S.P.; Jiang, Y.D. Study on rock mechanics in deep mining. *J. Rock Mech. Eng.* **2005**, *16*, 2803–2813.
2. Gao, M.; Ye, S.; Yang, B.; Liu, Y.; Li, J.; Liu, J.; Xie, H. Research progress of deep in-situ rock mechanics. *China Sci. Found.* **2021**, *35*, 895–903.
3. Zhang, Z.; Barla, G. Introduction to the Special Issue "Rock Mechanics Advances in China Coal Mining. *J. Rock Mech. Rock Eng.* **2019**, *52*, 2721–2723. [CrossRef]
4. Xie, H.P.; Bao, L.C.; Zhong, G.M.; Ru, Z.; Feng, G.; Bo, Z.J. Conception and preliminary exploration of deep in-situ rock mechanics. *J. Rock Mech. Eng.* **2021**, *40*, 217–232.
5. Li, B.x.; Yu, S.; Yang, L.; Zhu, W.s.; Xue, Y.g.; Feng, D.; Wang, C.; Chen, Y.j. Multiscale fracture characteristics and failure mechanism quantification method of cracked rock under true triaxial compression. *Eng. Fract. Mech.* **2022**, *262*, 108257. [CrossRef]
6. Liu, J.; Zhang, L.M.; Cong, Y.; Wang, Z.Q. Mechanical characteristics of unloading failure of granite under true triaxial stress path. *Rock Soil Mech.* **2021**, *42*, 2069–2077.
7. Yin, G.Z.; Ma, B.; Liu, C.; Li, M.H.; Lu, J. Effect of loading and unloading rate on mechanical properties and energy characteristics of sandstone under true triaxial stress. *J. Coal* **2019**, *44*, 454–462.
8. Li, J.T.; Liu, S.F.; Zhao, Y.; Guo, Q. True triaxial test and numerical simulation of sandstone under different stress paths. *J. Cent. South Univ. (Nat. Sci.)* **2021a**, *52*, 693–700.
9. Chu, Y.; Zhang, D.; Liu, H.; Wu, X.; Zhai, P.; Sheng, T. Experimental study on mechanical properties, acoustic emission characteristics and energy evolution of coal samples after freezing with liquid nitrogen. *J. Fuel.* **2022**, *321*, 123955. [CrossRef]
10. Quan, J.; Zhang, M.Z.; Yan, F.; Su, G.S.; Feng, X.T.; Xu, D.P.; Feng, G.L. Effect of initial minimum principal stress and unloading rate on the spalling and rockburst of marble: A true triaxial experiment investigation. *Bull. Eng. Geol. Environ.* **2020**, *80*, 1617–1634.
11. Roohollah, S.F.; Abbas, T.; Luis, R.E.S.; Murat, K. Rockburst assessment in deep geotechnical conditions using true-triaxial tests and data-driven approaches. *Int. J. Rock Mech. Min. Sci.* **2020**, *128*, 104279.
12. Chu, Y.; Sun, H.; Zhang, D. Experimental study on evolution in the characteristics of permeability, deformation, and energy of coal containing gas under triaxial cyclic loading-unloading. *Energy Sci. Eng.* **2019**, *7*, 2112–2123. [CrossRef]
13. Zhai, S.B.; Su, G.S.; Yin, S.D.; Zhao, B.; Yan, L. Rockburst characteristics of several hard brittle rocks: A true triaxial experimental study. *J. Rock Mech. Geotech. Eng.* **2020**, *12*, 279–296. [CrossRef]
14. Danni, L.; Su, G.S.; Zhang, G.l. True-Triaxial Experimental Study on Mechanical Behaviours and Acoustic Emission Characteristics of Dynamically Induced Rock Failure. *Rock Mech. Rock Eng.* **2019**, *53*, 1205–1223.
15. Su, G.Y.; Chen, Z.Y.; Yin, Y.; Zhang, X.H.; Mo, J.H. True triaxial test of granite rockburst after high temperature. *J. Geotech. Eng.* **2016**, *38*, 1586–1594.
16. Xiang, T.B.; Feng, X.T.; Chen, B.R.; Quan, J.; Zhang, Q.c. Failure mechanism and true triaxial test of rock samples with single discontinuity under triaxial stress state. *Rock Soil Mech.* **2009**, *30*, 2908–2916.
17. Qiu, L.; Zhu, Y.; Song, D.; He, X.; Wang, W.; Liu, Y.; Xiao, Y.; Wei, M.; Yin, S.; Liu, Q. Study on the Nonlinear Characteristics of EMR and AE during Coal Splitting Tests. *Minerals* **2022**, *12*, 108. [CrossRef]
18. Qiu, L.; Liu, Z.; Wang, E.; Li, B. Early-warning of rock burst in coal mine by low-frequency electromagnetic radiation. *J. Eng. Geol.* **2020**, *279*, 105755. [CrossRef]

19. Han, Y.; Zhao, Y.; Li, J. Experimental and Numerical Study of Strength and Failure Behavior of Precracked Marble under True Triaxial Compression. *J Shock. Vibration.* **2021**, *2021*, 3869045. [CrossRef]
20. Li, W.S.; Wang, L.G.; Lu, Y.L.; Li, Z.L. Experimental study on strength, deformation and failure characteristics of sandstone under true triaxial condition. *J. Min. Saf. Eng.* **2019a**, *36*, 191–197.
21. Dong, L.; Shi, Y.H.; Li, Y.; Wang, J.B.; Fan, P.X. Experimental study on deformation and failure of red sandstone under bidirectional compression. *Prot. Eng.* **2018**, *40*, 17–23.
22. Li, Z.J.; Yang, M.; Cheng, H.J.; Li, Q. Experimental study on true triaxial rockburst of deep rock under different disturbance conditions. *Min. Res. Dev.* **2021b**, *41*, 53–58.
23. Li, Z.Y.; Wu, P.; Jiang, D.Y.; Xiao, F.; Liu, W. True triaxial mechanical properties of sandstone with voids. *J. Undergr. Space Eng.* **2022**, *18*, 120–128.
24. Fan, P.X.; Yin, L.; Tang, Z.Y.; Yang, W.M.; Rong, W.D.; Hua, S.Y. Experimental study on unloading failure strength characteristics of red sandstone. *J. Rock Mech. Eng.* **2018**, *37*, 852–861.
25. Wang, M.; Zhu, Z.M.; Feng, R.Q. Experimental study on deformation and failure of fractured rock mass around roadway under true triaxial loading and unloading condition. *J. Coal* **2015**, *40*, 278–285.
26. Lee, H.; Haimson, B.C. True triaxial strength, deformability, and brittle failure of granodiorite from the San Andreas Fault Observatory at Depth. *Int. J. Rock Mech. Min. Sci.* **2011**, *48*, 1199–1207. [CrossRef]
27. Wang, B.; Zhang, K.I.; Wang, L.G.; Li, W.S.; Meng, X.Y.; Lu, Y.L.; Li, Z.L. Damage Evolution and Failure Behavior of Sandstone under True Triaxial Compression. *Geotech. Test. J.* **2018**, *42*, 1–29.
28. Wang, S.; Wang, L.G.; Tian, J.S.; Fan, H.; Jiang, C.Y.; Ding, K. An Experimental Study on the Effects of True Triaxial Loading and Unloading Stress Paths on the Mechanical Properties of Red Sandstone. *Minerals* **2022**, *12*, 204. [CrossRef]
29. Hu, L.H.; Ma, K.; Liang, X.; Tang, C.N.; Wang, Z.W. Experimental and numerical study on rockburst triggered by tangential weak cyclic dynamic disturbance under true triaxial conditions. *Tunn. Undergr. Space Technol. Inc. Trenchless Technol. Res.* **2018**, *81*, 602–618. [CrossRef]
30. Zhao, H.g.; Song, Z.l.; Zhang, D.m.; Liu, C.; Yu, B. True triaxial experimental study on mechanical characteristics and energy evolution of sandstone under various loading and unloading rates. *Geomech. Geophys. Geo-Energy Geo-Resour.* **2021**, *7*, 22. [CrossRef]
31. Si, X.f.; Huang, L.q.; Gong, F.q.; Liu, X.l.; Li, X.B. Experimental investigation on influence of loading rate on rockburst in deep circular tunnel under true-triaxial stress condition. *J. Cent. South Univ.* **2020**, *27*, 2914–2929. [CrossRef]
32. Zheng, Z.; Feng, X.T. Influence of intermediate principal stress on the mechanical properties of rocks containing fractures. *Géotechnique Lett.* **2019**, *10*, 82–87. [CrossRef]
33. Duan, K.; Kwok, C.Y.; Ma, X. DEM simulations of sandstone under true triaxial compressive tests. *Acta Geotech.* **2017**, *12*, 495–510. [CrossRef]
34. Ze, K.W.; Wei, K.M.; Hu, J.Q.; Fang, Y.L. The Changing Rule of Rock Strength under True Triaxial Condition. *Appl. Mech. Mater.* **2014**, *3013*, 1410–1413.
35. Li, X.B.; Chen, Z.H.; Weng, L.; LI, C.J. Unloading responses of pre-flawed rock specimens under different unloading rates. *Trans. Nonferrous Met. Soc. China* **2019b**, *29*, 1516–1526. [CrossRef]
36. Xiao, F.; Jiang, D.Y.; Wu, F.; Chen, J.; Zhang, J.Z.; Liu, W. Deformation and failure characteristics of sandstone subjected to true-triaxial unloading: An experimental and numerical study. *Fatigue Fract. Eng. Mater. Struct.* **2021**, *44*, 1862–1882. [CrossRef]
37. Kong, X.G.; Li, S.Y.; Wang, E.Y.; Ji, P.X.; Wang, X.; Shuang, H.Q.; Zhou, Y.X. Dynamics behaviour of gas-bearing coal subjected to SHPB tests. *Compos. Struct.* **2021**, *256*, 113088. [CrossRef]
38. Kong, X.G.; Wang, E.Y.; Shuang, L.; Li, H.F.; Xiao, P.; Zhang, K.Z. Fractals and chaos characteristics of acoustic emission energy about gas-bearing coal during loaded failure. *Fractals* **2019**, *27*, 1950072. [CrossRef]
39. Du, K.; Li, X.B.; Li, D.Y.; Weng, L. Failure properties of rocks in true triaxial unloading compressive test. *Trans. Nonferrous Met. Soc. China* **2015**, *25*, 571–581. [CrossRef]
40. Zhao, X.G.; Wang, J.; Cai, M.; Cheng, C.; Ma, L.K.; Su, R.; Zhao, F.; Li, D.J. Influence of Unloading Rate on the Strainburst Characteristics of Beishan Granite Under True-Triaxial Unloading Conditions. *Rock Mech. Rock Eng.* **2014**, *47*, 467–483. [CrossRef]
41. Fan, X.; Jiang, D.y.; Wu, F.; Zou, Q.l.; Chen, J.; Chen, B.; Sun, Z.G. Effects of prior cyclic loading damage on failure characteristics of sandstone under true-triaxial unloading conditions. *Int. J. Rock Mech. Min. Sci.* **2020**, *132*, 104379.
42. Lu, W.; Zhu, Z.; He, Y.; Que, X. Strength Characteristics and Failure Mechanism of a Columnar Jointed Rock Mass Under Uniaxial, Triaxial, and True Triaxial Confinement. *Rock Mech. Rock Eng.* **2021**, *54*, 2425–2439. [CrossRef]
43. Miao, J.l.; Jia, X.N.; Cheng, C. The Failure Characteristics of Granite under True Triaxial Unloading Condition. *Procedia Eng.* **2011**, *26*, 1620–1625.
44. Yin, G.Z.; Li, M.H.; Xu, J.; Wang, Z.W.; Li, W.P.; Li, X.; Song, Z.L.; Deng, B.Z. Development and application of multifunctional true triaxial fluid-structure coupling test system. *J. Rock Mech. Eng.* **2015**, *34*, 2436–2445.
45. Qin, Q.C.; Li, K.G.; Yang, B.W.; Wang, T.; Wei, X.Y.; Guo, W. Damage characteristics analysis of key characteristic points in rock total stress-strain process. *Rock Soil Mech.* **2018**, *39*, 14–24.
46. Wang, S.Z. Brittle-ductility transition and plastic flow networks in rocks. *Prog. Geophys.* **1993**, *8*, 25–37.
47. Ge, X. *Postfailure Behaviour and a Brittle-Plastic Model of Brittle Rock*; Balkema: Rotterdam, The Netherlands, 1997.
48. Shi, G.C.; Ge, X.R.; Lu, Y.D. Experimental study on stress brittle drop coefficient of marble. *J. Rock Mech. Eng.* **2006**, 1625–1631.

49. He, M.; Zhao, F. Experimental analysis of rockburst failure characteristics under different unloading rates. *Rock Soil Mech.* **2014**, *35*, 2737–2747+2793.
50. Zhao, M.; Xu, W.; Liu, Y. Instability mechanism of high stress rock mass under excavation and unloading induced by disturbance. *J. Coal* **2020**, *45*, 936–948.
51. Huang, R.; Huang, D. Under the condition of high ground stress unloading rate on the experimental study of kam screen marble effects of mechanical properties. *J. Rock Mech. Eng.* **2010**, *29*, 21–33.
52. Su, G.; Mo, J.; Chen, Z.; Jiang, J. True triaxial test study on the influence of support failure on rock burst ejection failure. *Rock Soil Mech.* **2017**, *38*, 1243–1250.

sustainability

MDPI

Article

Experimental Study on Strength Weakening of Gypsum Rock with Effect of Long-Term Overlying Strata Pressure

Shuli Wang [1], Zhihe Liu [1,*], Kaizhi Zhang [2], Guangli Zhu [1] and Huaichang Zheng [1]

[1] School of Resources and Environmental Engineering, Shandong University of Technology, Zibo 255000, China
[2] Institute of Mining Engineering, Guizhou Institute of Technology, Guiyang 550003, China
* Correspondence: liuzhihe@sdut.edu.cn

Abstract: A long time lag is the main characteristic of gypsum mined gob collapse disasters. With the coring of a gypsum rock specimen from the supporting pillars in gob, which formed over several years, the strength weakening effect of the gypsum rock with long-term overlying strata pressure is revealed by experimentation. The results show that: uniaxial compression stress–strain curves represent major differences in different lateral depths of the same supporting pillar. With the increase in lateral depth, peak strength increases and the corresponding strain decreases, which becomes more obvious as the age increases. As a function of time, peak strength decreases and the corresponding strain increases in the shallow part of the pillar as the age increases. Peak strength fluctuates in the middle part and increases in the deep part; the corresponding strain fluctuates in the middle and deep parts, but demonstrates the opposite changing law. Finally, the reason for the above law was comprehensively and thoroughly researched and demonstrated. The maximum strength weakening rate of gypsum rock in the shallow part of a supporting pillar of 0.5 m depth was 21.06% in the year 1996. The slow strength weakening effect of gypsum rock with long-term overlying strata pressure is the essential reason why gypsum mined gob collapses occur in subsequent years or even decades.

Keywords: gypsum rock; strength weakening; mine gob collapse; uniaxial compressive strength; elastic and plastic

Citation: Wang, S.; Liu, Z.; Zhang, K.; Zhu, G.; Zheng, H. Experimental Study on Strength Weakening of Gypsum Rock with Effect of Long-Term Overlying Strata Pressure. *Sustainability* **2022**, *14*, 11442. https://doi.org/10.3390/su141811442

Academic Editors: Xiangguo Kong, Dexing Li and Xiaoran Wang

Received: 18 July 2022
Accepted: 29 August 2022
Published: 13 September 2022

1. Introduction

According to incomplete statistics, the amount of underground mined gob in metal–non-metal mines is as high as 432 million cubic meters in China, which formed after other mineral resources had been mined. It has become one of the most serious problems in the mining field, both at present and in the future [1]. Gypsum is an indispensable and important mineral resource in the field of civil and construction engineering. At present, underground gypsum mines mainly use the room-pillar and strip mining methods, which have formed a large area of the hanging mine gob. According to statistics, only in Shandong Province, China is the hanging mine gob of a gypsum mine as high as 2.5 million square meters. Gypsum rock pillar is the main supporting structure in gypsum mined gob, and it can remain stable with long-term loading. After several years or even decades, large areas of hanging mine gob may collapse integrally, induced by damage and the sudden failure of the gypsum rock pillar. For example, the Wanzhuang gypsum mine, located in Hubei Province, China, was put into operation in 1996 and closed in 2012, but a large portion of the mine's gob collapse accident occurred in 2015 [2]. Gypsum mine gob collapse accidents display the significant characteristics of the chaining effect and large scale of the space and a long time lag (20 years or even longer).

Aiming at the gypsum mined gob collapse disaster, many scholars have researched the collapse mechanisms and influential factors. In terms of the collapse mechanism, Xu et al. constructed a plastic supporting system of pillar-beam for the gypsum mined gob and analyzed the collapse mechanism by applying mutation theory [3,4]. Xia et al.

constructed a cusp mutation model of a "pillar-roof and protecting layer" support system and studied the mechanism of structural instability, which was combined with nonlinear dynamics [5,6]. Zheng et al. revealed the long-term stability evolution mechanism and damage mode of gypsum mined gob [7,8]. Extensive works have focused on factors influencing gypsum mined gob's collapse. Li et al. analyzed factors influencing gypsum mined gob's collapse systematically [9]. Aiming at water's effect on the stability of gypsum rock, Liu et al. studied the influence of water on the strength weakening mechanism of gypsum rock, from aspects of the water-filled state and brine immersion through designing different experiments [10–12]. Creep is the typical mechanical property of gypsum rock. Liu et al. researched the creep characteristics of gypsum rock and its stability through uniaxial compression experiments and the step loading test [13–15]. A large number of other scholars have also carried out research work on the collapse hazard of gypsum mined gob based on different methods and perspectives [16,17].

Gypsum rock pillars exist in complex geological environments. During the decades from gypsum mined gob formation to collapse, how does the strength of gypsum rock change under the long-term pressure of the overlying strata and the coupling effect with water and disturbance? The clearing strength and weakening mechanism effects of the gypsum rock with long-term overlying strata pressure are the keys to revealing the collapse mechanism of gypsum mined gob and predicting the collapse's lag time. A few scholars have conducted some studies on time's effect on rock strength. Zhou et al. proposed a model of the temporal evolution of rock strength, based on analyzing a large number of experiment results [18,19]. Liu et al. conducted an experiment and simulation study on the time effect of the fracture extension of deeply buried barite in Jinping, Sichuan Province, China [20]. Jin et al. established a rock weathering model considering time and buried depth [21]. Li et al. established a non-probability reliability prediction model based on interval theory for the time-dependent stability of gypsum rock, considering various factors [22,23].

In summary, revealing the strength evolution law of gypsum rock with long-term loading is the key to predicting and preventing a large percentage of gypsum mined gob collapse accidents. To this end, relatively little research work has been attempted thus far. In the paper on the Luneng gypsum mine in Shandong Province, China, as an engineering background, gypsum rock cores were extracted from different gypsum rock pillars in different types of mined gob, which formed in different years. The uniaxial compressive strength of gypsum rock is used as a reference to reveal the strength evolution law of gypsum rock at different depths in the lateral direction of the gypsum rock pillar and the strength weakening effect of gypsum rock with long-term overlying strata pressure.

2. Coring from the Gypsum Rock Pillar

2.1. Situation of Gypsum Mine Gob

The Luneng gypsum mine is located in Tai'an City, Shandong Province, China and was put into production in 1996, with a production scale of 600,000 t/a. It adopts the shallow-hole and room-pillar mining methods, with a width of 4–5 m for gypsum rock pillar, 4–5 m for room, and 4 m for mining height. Over the past several years, the Luneng gypsum mine has continued mining operations and has formed a large-scale hanging mine gob—more than 750,000 m^2—supported by gypsum rock pillars.

2.2. Coring from Gypsum Rock Pillars

Once the mining area was mined over, the area was closed. Especially for gob that had been mined over for many years, the environment of the mine gob is complex and changeable, and original power and water supply systems are unavailable, which makes the coring work difficult and dangerous. Combined with basic equipment, such as an electric drill, battery and coring sleeve, simple and portable equipment for coring from gypsum rock pillars was used. The coring sites are different gypsum rock pillars of II-1 gypsum layers in different mine gob, which formed in different years. The thickness of the

II-1 gypsum layer is 5.86 m. The mechanical parameters of gypsum rock are as follows: uniaxial compressive strength 24.12 MPa, tensile strength 3.61 MPa, internal friction angle 29.8°, cohesion 2.75 MPa, and Poisson's ratio 0.29.

In order to ensure the accuracy of sampling and comparability of subsequent indoor tests as much as possible, all of the coring sites of gypsum rock pillars were carried out in gob of the relatively stable II-1 gypsum layer, with similar geological and mining technical conditions and the same size. Three to five gypsum rock pillars with high integrity and a width of 5 m were selected for coring in the middle position of mine gob in different years. The coring location was 1.0–1.5 m from the bottom of the gypsum rock pillar, and coring holes were taken at an angle of 45° on both sides of pillar to the oblique top, with a depth of 2.5–3.0 m, so that drill holes on both sides could cross in the middle of the pillar, which ensured that all cores within the lateral range of the gypsum rock pillar could be obtained. Drilled cores were sorted and numbered by year and then processed into standard specimens in the laboratory. The coring process is shown in Figure 1.

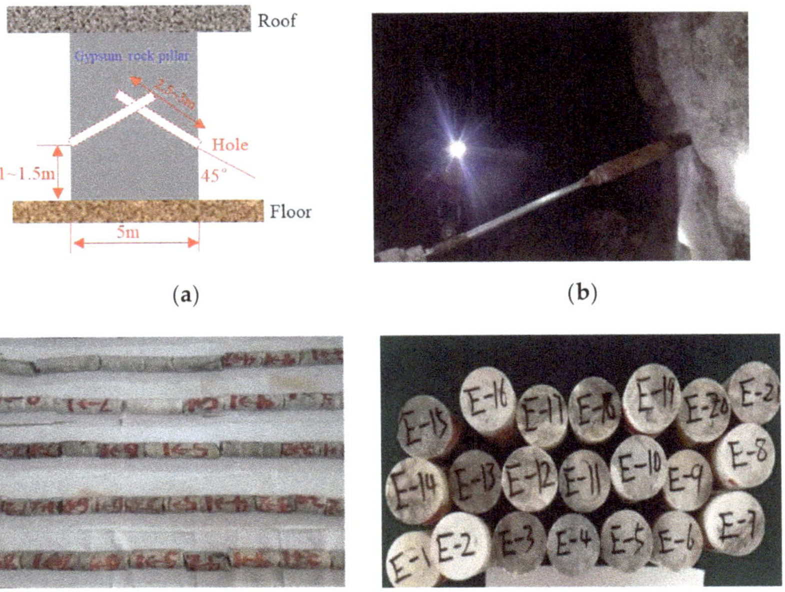

Figure 1. Coring method and gypsum rock specimen. (**a**) Coring parameter, (**b**) scene picture, (**c**) gypsum rock cores, and (**d**) standard gypsum rock specimen.

3. Uniaxial Compression Strength Test of Gypsum Rock

3.1. Experimental Equipment

The experiment mainly uses a Shimadzu AG-X250 rock testing machine, which is mainly used for uniaxial compression, tension, cyclic compression, and other experimental research of rock and concrete, with 0.2 ms interval sampling accuracy. The experimental equipment and process are as shown in Figure 2.

(**a**) (**b**) (**c**)

Figure 2. Experimental equipment and process. (**a**) Shimadzu testing machine, (**b**) process, and (**c**) failure of gypsum rock specimen.

3.2. Uniaxial Compression Strength Curve

In order to study the strength variation law along the transverse of a gypsum rock pillar in different years and the strength weakening effect of gypsum rock with long-term overlying strata pressure, gypsum rock specimens in 0.5 m, 1.5 m, and 2.5 m at the lateral depth of the pillar in different years were selected for a uniaxial compression test. Considering the intermittent and broken features of cores during field coring of gypsum rock pillars in mine gob, gypsum rock test specimens were allowed to fluctuate by 0.1 m at the lateral depth of the pillar, for which the final selected gypsum rock test specimens were 0.5 ± 0.1 m, 1.5 ± 0.1 m, and 2.5 ± 0.1 m at the lateral depth of the pillar. In order to maintain the accuracy of the test, displacement control was selected as the loading mode, with a parameter of 0.01 mm/s. The uniaxial compression stress–strain curves of gypsum rock in different years and lateral depths of the pillar are shown in Figure 3.

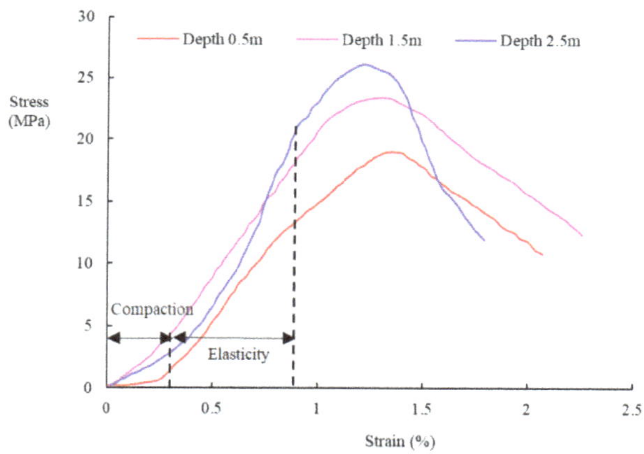

(**a**) In 1996

Figure 3. *Cont.*

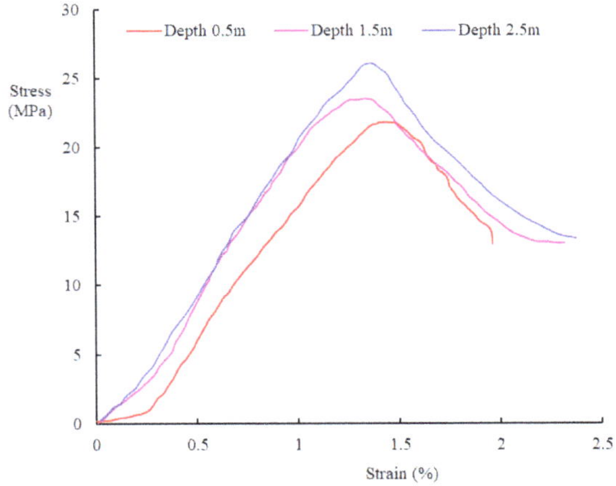

(**b**) In 2000

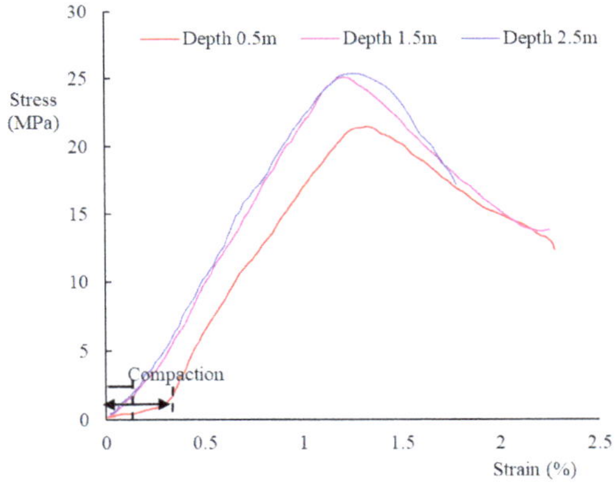

(**c**) In 2008

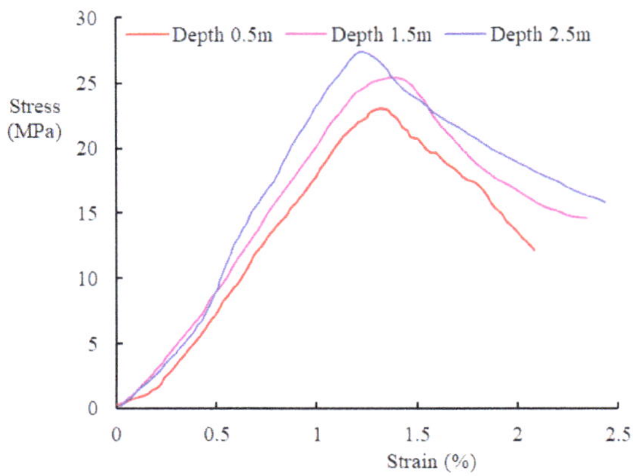

(**d**) In 2013

Figure 3. *Cont.*

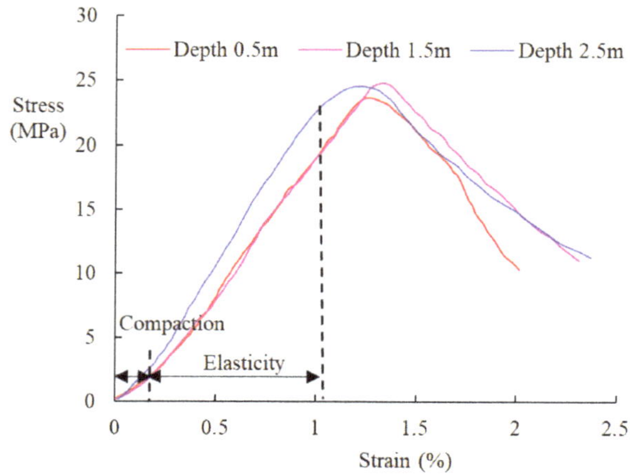

(**e**) In 2017

Figure 3. Uniaxial compression stress–strain curve of gypsum rock in different years. (**a**) Year of 1996, (**b**) year of 2000, (**c**) year of 2008, (**d**) year of 2013, and (**e**) year of 2017.

3.3. Evolution Law of Strength Curve

By comparing the uniaxial compression stress–strain curves of gypsum rock in different years and at different lateral depths of pillars, the following laws can be obtained:

(1) The peak strength of gypsum rock gradually increases with the increase in lateral depth, and the trend is more and more obvious with the increase in age. The difference value of peak strength in different lateral depths gradually increases as the age increases. For example, the maximum difference value of the uniaxial strength of gypsum rock at the lateral depth of the pillar in 1996 was 9.41 MPa, while it was only 1.15 MPa in 2017.

(2) With the lateral depth of the gypsum rock pillar increasing, strain values corresponding to the peak strength of gypsum rock display a decreasing trend, and the difference value of the strain gradually increases as the age increases. For example, the maximum strain difference value corresponding to the peak strength in different lateral depths was 0.3% in 1996, while it was only 0.13% in 2017.

(3) The compaction stage of the uniaxial compression stress–strain curve of gypsum rock gradually increases and the elastic stage decreases as the age increases. For gypsum rocks located in the shallow part of the pillar, such as at 0.5 m, the increase in age was especially obvious. The compaction stage tends to decrease gradually with the increasing lateral depth of the pillar, and it becomes less and less obvious with decreasing age. For example, the compaction stage and stress–strain curves of gypsum rock in different lateral depths of the pillar in 2017 were basically the same.

4. Strength Weakening of Gypsum Rock

4.1. Evolution Law of Peak Strength

The uniaxial compression strength of gypsum rock as a reference parameter can reveal peak strength and strain evolution law of the effect of time in different years, ss shown in Figures 4 and 5.

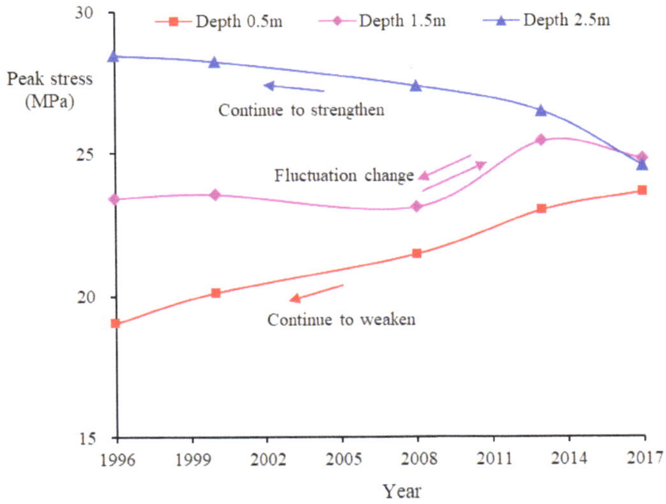

Figure 4. Evolution law of peak strength in different years.

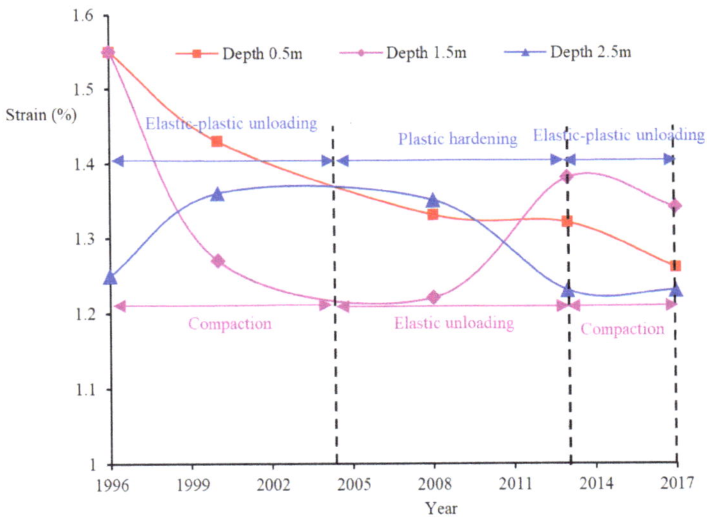

Figure 5. Evolution law of strain in different years.

Based on the changing curve of peak strength and corresponding strain of gypsum rock in different lateral depths of pillars and different years, the following conclusions can be drawn:

(1) Peak strength and strain of gypsum rock in different lateral depths of pillars display different evolutionary laws with increasing age, which indicates that load magnitude and the elastic–plastic state of gypsum rock in different lateral depths are different, with long-term overlying strata pressure [24,25].

(2) In the shallow part (0.5 m) of the gypsum rock pillar, the peak strength of gypsum rock continues to decrease with increasing age, for example, peak strength was 19.04 MPa in 1996, but 23.65 MPa in 2017, and both of them were smaller than the original peak strength of gypsum rock, which was 24.12 MPa, while the corresponding strain to peak strength exhibits an opposite evolutionary law, that is, continues to increase with increasing age.

(3) In the middle of the gypsum rock pillar (1.5 m), the peak strength of gypsum rock fluctuates with increasing age, and is greater or less than the original peak's strength. At a deep location of the gypsum pillar (2.5 m), peak strength of gypsum rock continues to increase with increasing age and greater than original peak strength, which indicates that plastic hardening has occurred in the gypsum rocks [26,27]. The strains corresponding to peak strength in the middle and deep locations both exhibit fluctuating changes, but fluctuating changing patterns are opposite, which indicates that the loading state of the gypsum rock pillar with overlying strata pressure is not fixed.

(4) Gypsum rock pillars are subjected to weathering with long-term loading, and phenomena of surface-to-internal fragmentation and swelling, exfoliation and shedding gradually appear. This intensifies with increasing age, and conical pillars even appear, as shown in Figure 6.

(**a**)

(**b**)

Figure 6. Pictures of gypsum rock pillars in different years. (**a**) Formed in 1996 and (**b**) formed in 2017.

With long-term overlying strata pressure, the most superficial layer of the gypsum rock pillar is gradually broken and dislodged as a function of stress concentration and weathering. The failure of the most superficial layer caused increasing loads at the middle of the gypsum rock pillar, which induced the internal structure of the gypsum rock to be compressed and dense, and strain corresponding to peak strength decreased, while the load at the deep position of the gypsum rock pillar decreases and produces elastic–plastic unloading, inducing strain corresponding to increasing peak strength. When the most superficial layer of the gypsum rock pillar is exfoliated, it leads to stress concentration in the inner layer of the gypsum rock pillar, which induces elastic unloading in the middle position of the gypsum rock pillar and strain corresponding to increasing peak strength. On the contrary, an increasing load in the deep position of the gypsum rock pillar induces plastic hardening, and the peak strength rises and corresponding strain decreases. With long-term overlying strata pressure, the gypsum rock pillar maintains the cycling process mentioned above, which indicates that the broken gypsum rock pillar is formed by layer-by-layer denudation and shedding.

(5) Due to the weathering effect on the surface layer of the gypsum rock pillar, inducing fracture of gypsum rock by extending and crack through causes the quality index RQD of the actual cores taken out at the site to be smaller, which is also the reason why the compaction stage continues to increase and strength continues to decrease in the shallow part of the gypsum rock pillar with increasing age. The shallow part of the gypsum rock pillar causes continuous shedding and is an effective support area of the pillar, which induces a loading state in the middle of the changing gypsum rock pillar. Finally, alternately generating pressure-dense and elastic unloading induces changes in peak strength and strain fluctuations. The deep part of the gypsum rock pillar as the main bearing part continues to creep with a long-term load and plastic hardening occurs, which has been proved by an obvious jamming phenomenon during coring. That leads to continuously increasing peak strength of the gypsum rock and is greater than the original peak strength. The fluctuating change in the loading state of the gypsum rock pillar causes an increasing or decreasing load in the deep part of pillar, which induces elastic–plastic unloading and plastic hardening. This is the reason why the strain corresponds to fluctuating peak strength changes.

4.2. Strength Weakening of Gypsum Rock

To quantitatively evaluate the strength weakening effect of gypsum rock with long-term overlying strata pressure, the strength weakening rate of gypsum rock in different years is assumed to be [28–30]:

$$\eta = \frac{\sigma_c - \sigma_T}{\sigma_c} \times 100\% \tag{1}$$

where σ_c is the original peak strength of gypsum rock, MPa, and σ_T is the peak strength of gypsum rock in different years, MPa.

Based on the peak strength of gypsum rock specimens in different years and at lateral depths, the strength weakening rate curves of gypsum rock were obtained, as shown in Figure 7.

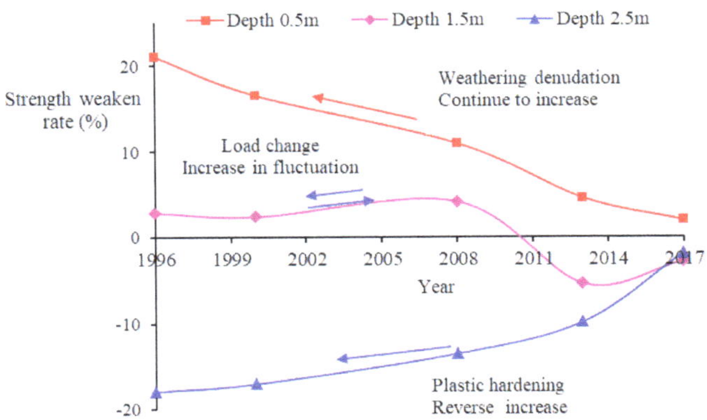

Figure 7. Strength weakening rate of gypsum rock in different years.

The strength weakening rate curves show a similar pattern with peak strength variation. The strength weakening rate of gypsum rock in the shallow part (0.5 m) of the pillar continues to increase with increasing age, to a maximum of 21.06% (1996). The middle part of the pillar (1.5 m) displays a fluctuating increase, while the deep part (2.5 m) exhibits an inverse increase. However, both strength weakening rates are low. The weathering of the gypsum rock pillar in closed gob is an extremely slow process. In other words, peak strength, magnitude of strain change, and strength weakening rate are small and slow. The gypsum rock pillar can maintain long-term stability, which is the essential reason why gypsum mined gob collapses after years or even decades.

5. Conclusions

Aiming at significant characteristics of the lagging gypsum mined gob collapse disaster, the strength weakening effect of gypsum rock with long-term overlying strata pressure was revealed by experiment. The main conclusions are as follows:

(1) There are significant differences in the uniaxial compression stress–strain curve of gypsum rock in different years and at different lateral depths of gypsum rock pillars. The peak strength of gypsum rock increases as the lateral depth of the pillar increases, and the corresponding strains exhibit a decreasing trend. With increasing age, this becomes more obvious. The compaction stage shortens with increasing lateral depth but increases with increasing age. The stress–strain curves in different depths tend to be consistent with decreasing age.

(2) Peak strength and the corresponding strain of gypsum rock are different in different lateral depths and years. In the shallow part of the gypsum rock pillar, peak strength decreases with increasing age, while strain displays the opposite evolutionary law. Peak strength fluctuates in the middle part and continues to increase in the deep part, which becomes greater than the original peak strength. The strain in the middle and deep parts exhibits fluctuation changes, but the fluctuating laws are opposite.

(3) Subjected to long-term weathering, the compaction stage increases and strength decreases in the shallow part of the gypsum rock pillar. The pillar's layer-by-layer shedding causes changes in the pillar's effective supporting area, inducing the loading state of the fluctuating changes in the middle part of the pillar, alternately producing a compaction stage and elastic unloading along with peak strength and strain fluctuating changes. The deep part is the main load bearing part, with the phenomenon of plastic hardening due to continuous creep with a long-term load, inducing a peak strength increase.

(4) The maximum strength weakening rate of gypsum pillar in the shallow part (0.5 m) is 21.06%. Weathering in closed gob is an extremely slow process, and peak strength, strain change, and the strength weakening rate are all small. This is the reason why gypsum mined gob collapse occurs after years or even decades.

Author Contributions: Methodology, S.W.; formal analysis, S.W. and Z.L.; investigation, K.Z.; resources, H.Z.; writing—original draft preparation, S.W. and G.Z.; writing—review and editing, S.W. and Z.L.; supervision, K.Z. All authors have read and agreed to the published version of the manuscript.

Funding: This research was funded by the National Natural Science Foundation of China (Grant No. 51904177), the Natural Science Foundation of Shandong Province (Grant No. ZR2019BEE078), and Guizhou Science and Technology Major Projects (Grant No. [2018]3005).

Institutional Review Board Statement: Not applicable.

Informed Consent Statement: Not applicable.

Data Availability Statement: The data used to support the findings of this study are available from the corresponding author.

Conflicts of Interest: The authors declare no conflict of interest.

References

1. He, R.; Han, Z.; Zhou, Y.; Liu, H.; Liu, Y. Analysis on disaster characteristics and preventive measures of goaf in metal and non-metal mines. *Min. Res. Dev.* **2020**, *40*, 33–38.
2. Luo, G.; Chen, L.; Jia, Q.; Song, S.; Liu, G. Gob area No. 1 subsidence mechanism and development trend prediction in Dangyang gypsum mine, Hubei Province. *Coal Geo. Chin.* **2019**, *31*, 62–65.
3. Xu, X.; Zhou, Y.; Pang, S. Analysis of catastrophic instability of plastic supporting system in old goaf of gypsum mine. *Chin. J. Rock Mech. Eng.* **2018**, *37*, 2548–2555.

4. Zhu, Y.; Huang, X.; Liu, Y.; Liu, Z.; Lan, H.; Tian, W. Nonlinear viscoelastoplastic fatigue model for natural gypsum rock subjected to various cyclic loading conditions. *Int. J. Geomech.* **2021**, *21*, 1004. [CrossRef]
5. Xia, K.; Chen, C.; Liu, X.; Zhou, Y.; Jiang, X. Study of the failure of pillar-roof system in gypsum mines based on catastrophe theory. *Chin. J. Rock Mech. Eng.* **2016**, *35*, 3837–3845.
6. Xia, K.; Chen, C.; Yang, K.; Sun, C.; Liu, X.; Zhou, Y. Influence of relative humidity on the non-linear failure and stability of gypsum mines. *Eur. J. Environ. Civ. Eng.* **2020**, *3*, 1518–1543. [CrossRef]
7. Zheng, H.; Hou, X.; Liu, Z.; Li, B.; Li, H. Disturbance mechanism of rock deformation based on energy accumulation. *Ind. Miner. Process.* **2017**, *46*, 44–49.
8. Caselle, C.; Bonetto, S.; Comina, C. Comparison of laboratory and field electrical resistivity measurements of a gypsum rock for mining prospection applications. *Int. J. Min. Sci. Technol.* **2019**, *29*, 841–849. [CrossRef]
9. Li, M.; Liu, Z.; Miao, Q.; Zheng, H.; Zhang, J. Analysis of major influence factors to stability of mined-out area. *Ind. Miner. Process.* **2013**, *42*, 20–24.
10. Liu, X.; Jiang, X.; Chen, C.; Xia, K.; Zhou, Y. Study of creep characteristics of gypsum rock in natural and saturated state. *Rock Soil Mech.* **2017**, *38*, 277–283. [CrossRef]
11. Ma, H.; Song, Y.; Chen, S.; Yin, D.; Zheng, J.; Shen, F.; Li, X.; Ma, Q. Experimental investigation on the mechanical behavior and damage evolution mechanism of water-immersed gypsum rock. *Rock Mech. Rock Eng.* **2021**, *54*, 4929–4948. [CrossRef]
12. Li, W.; Einstein, H.H. Theoretical and numerical investigation of the cavity evolution in gypsum rock. *Water Resour. Res.* **2017**, *53*, 9988–10001. [CrossRef]
13. Liu, Z.; Zheng, H.; Li, M. Experimental research on creep failure characteristics of gypsum rock based on rock longitudinal wave velocity. *Geotech. Geol. Eng.* **2019**, *37*, 1515–1522. [CrossRef]
14. Wang, S.; Lv, W.; Liu, Z.; Zheng, H. Lagging collapse mechanism of gypsum-mined gob with rock creep. *Geotech. Geol. Eng.* **2022**, *40*, 2489–2499. [CrossRef]
15. Lyu, C.; Liu, J.; Wu, Z.; Liu, H.; Xiao, F.; Zeng, Y. Experimental study on mechanical properties, permeability and acoustic emission characteristics of gypsum rock under THM Coupling. *Rock Mech. Rock Eng.* **2021**, *54*, 5761–5779. [CrossRef]
16. Xu, X.; Cui, X.; Liu, X.; Tang, Q.; Zhang, X.; Sun, Y. Damage analysis of soaking gypsum and safety evaluation of goaf: Based on energy dissipation theory. *Geotech. Geol. Eng.* **2020**, *38*, 6177–6188. [CrossRef]
17. Heidari, M.; Khanlari, G.R.; Torabi Kaveh, M.; Kargarian, S. Predicting the uniaxial compressive and tensile strengths of gypsum rock by point load testing. *Rock Mech. Rock Eng.* **2012**, *45*, 265–273. [CrossRef]
18. Zhou, H.; Yang, Y.; Liu, H. Time-dependent theoretical model of rock strength evolution. *Rock Soil Mech.* **2014**, *35*, 1521–1527.
19. Hogan, J.D.; Boonsue, S.; Spray, J.G.; Rogers, R.J. Micro-scale deformation of gypsum during micro-indentation loading. *Int. J. Rock Mech. Min.* **2012**, *54*, 140–149. [CrossRef]
20. Liu, N.; Zhang, C.; Chu, W. Experimental research on time-dependent behavior of crack propagation in Jinping deep marble. *Rock Soil Mech.* **2012**, *33*, 2437–2443.
21. Jin, J.; Xu, Y.; Li, J.; Wan, L.; Wang, L. Influencing factors analysis on rock mass strength considering time and depth effect. *Chin. J. Under Sp. Eng.* **2016**, *12*, 475–480.
22. Li, X.; Zhang, X.; Zhao, Z.; Liu, G. Non-probabilistic reliability prediction on aging stability of ore pillar in gypsum mine. *Ind. Miner. Process.* **2018**, *47*, 37–41.
23. Wang, J.; Zhang, C.; Zheng, D.; Song, W.; Ji, X. Stability analysis of roof in goaf considering time effect. *J. Min. Strat. Control. Eng.* **2020**, *2*, 013011.
24. Meng, T.; Hu, Y.; Fang, R.; Kok, J.; Fu, Q.; Feng, G. Study of fracture toughness and weakening mechanisms in gypsum interlayers in corrosive environments. *J. Nat. Gas Sci. Eng.* **2015**, *26*, 356–366. [CrossRef]
25. Gómez, S.; Sanchidrián, J.A.; Segarra, P. Near-field vibration from blasting and rock damage prediction with a full-field solution. *Int. J. Rock Mech. Min.* **2020**, *134*, 104357. [CrossRef]
26. Ji, S.; Zhang, J.; Pan, R.; Karlovšek, J. Local acceleration monitoring and its application in physical modelling of underground mining. *Int. J. Rock Mech. Min.* **2020**, *128*, 104282. [CrossRef]
27. Huang, F.; Yan, S.; Wang, X.; Jiang, P.; Zhan, S. Experimental study on infrared radiation characteristics of gneiss under uniaxial compression. *J. Min. Strat. Control. Eng.* **2021**, *3*, 013011.
28. Li, W.; Li, S.; Feng, X.; Li, S.; Yuan, C. Study of post-peak strain softening mechanical properties of rock based on Mohr-Coulomb criterion. *Chin. J. Rock Mech. Eng.* **2011**, *30*, 1460–1466.
29. Ji, S.; Karlovšek, J. Calibration and uniqueness analysis of micro parameters for DEM cohesive granular material. *Int. J. Min. Sci. Techno.* **2022**, *32*, 121–136. [CrossRef]
30. Li, L.; Zhang, X.; Deng, H.; Han, L. Mechanical properties and energy evolution of sandstone subjected to uniaxial compression with different loading rates. *J. Min. Strat. Control. Eng.* **2020**, *2*, 043037.

sustainability

MDPI

Article

Research on Deformation and Failure Control Technology of a Gob-Side Roadway in Close Extra-Thick Coal Seams

Bin Zhao [1,2], Shengquan He [3,*], Xueqiu He [3], Le Gao [3], Zhenlei Li [3], Dazhao Song [3] and Feng Shen [3]

1 School of Safety Engineering, China University of Mining and Technology, Xuzhou 221116, China
2 China Coal Datong Energy Co., Ltd., Datong 037038, China
3 School of Civil and Resources Engineering, University of Science & Technology Beijing, Beijing 100083, China
* Correspondence: shenqhe@163.com

Abstract: Close extra-thick coal seams are subject to the broken overburden of mined coal seams, and the deformation and damage of the roadways is serious, which affects the safe operation of the mine. To reduce the deformation and damage of the roadways, this paper studied the deformation and damage law of the gob-side roadway in close extra-thick coal seams through numerical simulation and field monitoring, compared and analyzed the deformation and damage characteristics of the roadway under different reinforcement support methods, determined the optimal reinforcement support method, and carried out field verification. The obtained results indicated that the deformation and damage of the gob-side roadway showed asymmetric characteristics. The large deformation of the coal body in the deep part of the roadway wall is an important reason for the continuous occurrence of roadway wall heave in the coal pillar. Under the action of unbalanced support pressure, the floor is subject to the coupling effect of horizontal extrusion pressure and vertical stress that cause extrusion mobility floor heave. The horizontal and vertical displacement of the coal pillar side of the roadway under different support methods is much larger than that of the solid coal side. Increasing the anchor cable length and fan-shaped arrangement can improve the support effect. Grouting at the coal pillar side can significantly improve the bearing capacity and stability of the coal pillar. The effect of floor grouting is much better than the anchor cable in controlling the floor heave. The integrated reinforcement method of anchor cable + coal pillar side grouting + floor grouting has the best effect with the least horizontal and vertical deformation. The research results are of great significance for ensuring the stability of similarly endowed roadways.

Keywords: close distance coal seams; roadway support; roadway deformation; grouting reinforcement; control technology

Citation: Zhao, B.; He, S.; He, X.; Gao, L.; Li, Z.; Song, D.; Shen, F. Research on Deformation and Failure Control Technology of a Gob-Side Roadway in Close Extra-Thick Coal Seams. *Sustainability* **2022**, *14*, 11246. https://doi.org/10.3390/su141811246

Academic Editor: Baoqing Li

Received: 23 July 2022
Accepted: 31 August 2022
Published: 8 September 2022

1. Introduction

In order to improve the mining rate of coal resources, reduce unnecessary mining lanes, and solve the problem of mining succession tension, gob-side entry retaining technology has been widely used [1–4]. Close extra-thick coal seams form a special overburden structure due to the mining of the overlying coal seams. Under the action of overburden structure, the deformation and damage of the gob-side roadway are serious. With the increase of mining depth, the degree of damage becomes more and more intense, causing damage to the roadway and equipment.

Related scholars have conducted a lot of research on the stress and deformation damage characteristics of the surrounding rock of gob-side roadways. Wang et al. [5] and Han et al. [6] investigated the structural mechanical causes of key block fracture at the roof of a gob-side roadway and found that bimodal stresses existed on both sides of the gob-side roadway and in the roadway in the low-stress area, and pointed out that the mechanical transfer mechanism of the surrounding rock in the vertical direction was weaker and the surrounding rock was most stable when the roadway was located outside the key block fracture line. Li et al. [7] investigated the spatial and temporal evolution of fractures in the surrounding rock of a gob-side roadway under the pressure generated during mining, and the results showed that coal pillar deformation and rupture were the cause of destabilization and damage of the surrounding rock. Zheng et al. [8] established a mechanical model of the overburden rock of a gob-side roadway and analyzed the large deformation and stress field distribution of the roadway overburden rock, and the results showed that a large deformation of the roadway was caused by the stress field superimposed by the lateral transfer stress formed in the overburden rock formation, the oversupport pressure caused by mining, and the regional ground stress. Slashchov et al. [9] analyzed the minimum principal strain changes of rocks in different working faces and strata, and found that due to the existence of gob, the overburden pressure in part of the goaf decreased, but in the pressure bearing area of the solid coal pillar, the overburden of the main roof of the roadway was reactivated. Some scholars believe that filling the gob area can reduce the movement of the overburden rock layer and decrease the deformation of the gob-side roadway [10,11], but this method becomes more and more difficult to achieve with the increase of mining depth. Guo et al. [12] concluded that the length of the overhanging roof is the main factor affecting the stress distribution and deformation of the rock surrounding the roadway, and with the decrease of the overhanging length, the influence of the movement and load transfer of the overburden rock layer in the mining area on the stability of the tunnel. Liu et al. [13] analyzed the stress characteristics of the direct top of the gob-side roadway, and found that the roof deformation was symmetrically distributed, with the maximum roof deformation occurring in the middle of the roadway, and gradually decreasing with the increase of coal seam stiffness. Feng et al. [14] tested the mechanical properties of rock, and based on the energy analysis results, established a rock damage model by using the relative change of the dissipated energy ratio, and then evaluated the accuracy of the rock damage model with the peak strength and peak strain of rock as evaluation parameters.

Based on the research on the deformation mechanisms and deformation characteristics of the rock surrounding the gob-side roadway, the corresponding control measures are becoming more and more abundant. Studies have shown that the deformation and damage characteristics of roadways also have asymmetric characteristics due to the inhomogeneity of the stress conditions of the rock surrounding the roadways under the action of mining, so the control strategy based on asymmetric, high-prestressed anchor cables has a good effect on controlling the stability of the surrounding rock [15–17]. Peng et al. [18] studied the reasonable coal pillar size and roadway support method for isolated comprehensive mining workings in extra-thick coal seams using stress limit balance theory, the internal stress field distribution law, and the finite element method, and proposed the combined support method of "anchor mesh cable" + "reinforced ladder beam" + "steel belt", which can effectively reduce the displacement. Ren et al. [19] established a mechanical model and numerical simulation model to study the influence of the interaction between the bracket and the roof on the stability of the support of a gob-side roadway, and the results showed that the bracket resistance increased with the increase of width, and the optimal width was 3 m. Zheng et al. [20] derived a formula for calculating the maximum settlement of the roof slab and proposed the fenestration control technology of grating side-entry retaining, and the results showed that with the advancement of the working face, the deformation of the whole process of the gob-side roadway was effectively controlled. Xie et al. [21] and Xie et al. [22] analyzed the damage mechanism and controllable engineering factors causing deformation in the surrounding rock of the gob-side roadway, and proposed an asymmetric support scheme with prestressed anchor cable as the core member, which was verified in engineering practice. Based on the numerical simulation, Begalinov et al. [23] conducted a detailed study on the stress and deformation state of the rock around the transportation roadway, and adopted an adjustable support method according to the rock mass stability to control the deformation of the roadway to the maximum extent. He et al. [24] analyzed the deformation mechanism of the surrounding rock with theoretical and numerical simulations, and found that the high volume ratio of the inelastic zone is the main reason for the severe deformation of the surrounding rock, and proposed a combined support technology with "reinforced anchor cable + W band + π beam + single hydraulic pillar" as the core, which was verified in the test section of the roadway. Yang et al. [25] concluded that coal pillars will gradually lose their bearing capacity under the influence of mining stress, leading to a sudden increase in the overhanging length of the roof and causing abnormal stress concentration in the tunnel surrounding rock, and proposed that strengthening the support density in the stress concentration area could effectively improve the surrounding rock deformation. Ma et al. [26] concluded that one of the reasons for the failure of the adit under the hard roof and soft floor is the insufficient support strength at the end of the gob-side roadway, so they proposed the support concept of "improving the support strength to achieve cutting the roof along the digging side", and the results showed that the anchor net, roof anchor, single hydraulic strut, and roof bolting technology can ensure the support effect. Guo et al. [27] found that after the damage of the main roof in the dynamic pressure zone, its rotary sinking sharply increased the deformation and damage of the tunnel envelope, and proposed a support design consisting of high tension, constant resistance, and the use of large deformation anchors rods + single hydraulic strut + steel beam + U beam. Li et al. [28] studied the mechanical properties of grouted rock from both macroscopic and microscopic aspects by using shear and compression experiments, and found that the elastic modulus of rock mass was greatly improved after grouting. MałKowski et al. [29] studied three different roadway support schemes for six years, and determined the time of secondary equilibrium and the strain of support unit after the rock mass is damaged so as to guide the rationality of support installation under specific stope and geological conditions. Dychkovsky et al. [30] studied the visualization of the formation principle of the stress–strain state in the coal support pressure area and evaluated the rock mass occurrence conditions. The results showed that when the width of the support pressure area was 18 m, the maximum stress was reduced to 70 MPa, which can be used as the basis for identifying the bearing capacity of the mine and the excavation roadway.

In summary, it can be seen that fruitful results have been achieved for the deformation and damage characteristics and support technology of gob-side roadways, but not much research has been conducted for the gob-side roadway support of the overburden coal seams that have been mined under the condition of close extra-thick coal seams. However, in this occurrence condition, the deformation and damage of the roadway are serious; the stability is low, and it is easy to cause damage to personnel, property, and equipment. The existing research results can not meet the requirements of roadway deformation control. Therefore, it is necessary to apply further efforts to improve the stability of the roadway. In this paper, we studied the deformation and damage law of a close extra-thick coal seam gob-side roadway through numerical simulation, compared and analyzed the deformation and damage characteristics of the roadway under different reinforcement support methods, determined the optimal reinforcement support method, and carried out field verification.

2. Background and Method

2.1. Engineering Overview

The 305# coal seam of the Tashan coal mine is one of the few extra-thick coal seams in China, with an average burial depth of 435 m and a variation coefficient of 0.16%, and which is a stable coal seam. The 4.67 m layer above the coal seam 305# is the 102# coal seam of the Carboniferous Taiyuan Group, with an average seam thickness of 3 m, and mining was completed in July 2015. The protected coal pillar between the 10,201 working face and 10,203 working face mining void area of coal seam 102# is 20 m, as shown in Figure 1a. At present, the mining of the 30,503 working face of coal seam 305# has been started by using long-wall comprehensive mining to release the top coal-recovery process, which is typical in mining of extra-thick coal seams under close mining areas, and its gob-side 30,501 working face has been completed. The 30,503 working face has a strike length of 1738.58 m and a tendency length of 193 m. The average thickness of coal seam 305# is 14 m, and between it and coal seam 102# is 4.67 m-thick mudstone seam.

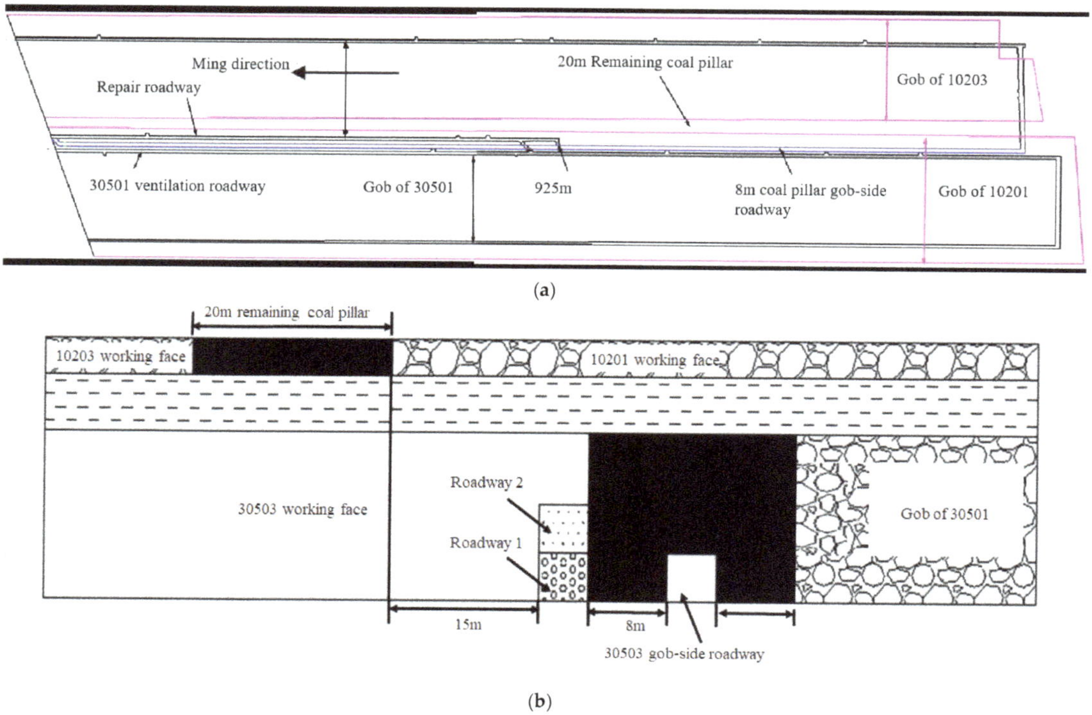

Figure 1. Diagram of the 30,503 working face layout. (**a**) Layout plan. (**b**) Cutaway layout.

The initial design of the 30,501 working face used roadway 1 as the transportation roadway; the horizontal distance of this roadway from the coal pillar left in coal seam 2# was 15 m. When the roadway was dug to 925 m, due to the serious mine pressure, the roadway dig program was changed and roadway 2 was used instead. The current mining is underway for the 30,503 working face of the coal seam 305# seam of the Carboniferous Taiyuan Group. It is planned to repair roadway 3 above roadway 1 as the return airway of 30,503, and in the process of repairing to 925 m, due to the structural damage of roadway 1 and serious roof fall, the working face of 30,503 was rearranged by the method of leaving an 8 m coal pillar in the gob-side entry driving, and the repair roadway was filled with high water material. After the roadway layout of working face 30,503, the horizontal distance between the roadway and the remaining coal pillar of overlying coal seam 102 # is 28 m. The gob-side is a rectangular section along the floor with a height of 4.0 m and a width of 5 m. Under the disturbance of back mining at the working face, the deformation and damage of the gob-side roadway is intense when crossing the filling roadway, which seriously affects the safe back mining at the working face, and the roadway layout is shown in Figure 1. The gob-side roadway is supported by a combination of an anchor rod and an anchor cable. The roadway is supported by left-handed, high-strength, rebarless anchor rods with a roof anchor rod size of Ø22 mm × 2400 mm and a side anchor rods size of Ø20 mm × 2200 mm with a preload torque of no less than 400 N·m. The anchor cable adopts 1 × 19 strands of Ø22 mm × 8300 mm high-strength low-relaxation prestressing strand with an ultimate breaking force of ≥550 kN and a pretensioning force of 200 kN. The roof anchor rods are 6 rows each with inter-row spacing of 1000 mm × 1000 mm, the side anchor rods are 4 rows each with inter-row spacing of 1000 mm × 1000 mm, and the 5 anchor cables are symmetrically arranged along the centerline and both sides of the roadway with inter-row spacing of 1300 mm × 2000 mm as shown in Figure 2. Affected by the mining stress of the working face and the complex mining environment around the gob-side roadway, the deformation and damage of the gob-side roadway under the existing support conditions is still serious. The middle and upper part of the roadway are greater than the lower part, and the coal pillar side is greater than the solid coal side, as shown in Figure 3.

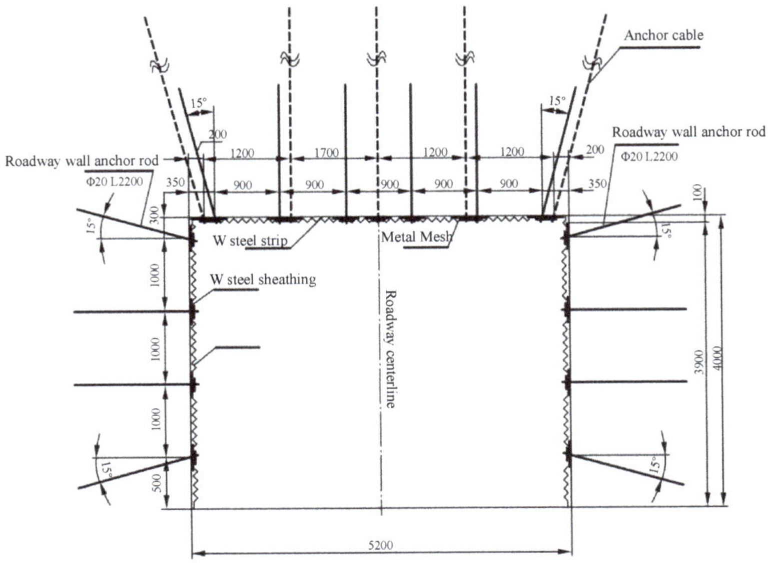

Figure 2. Schematic diagram of section support of the 30,503 gob-side roadway.

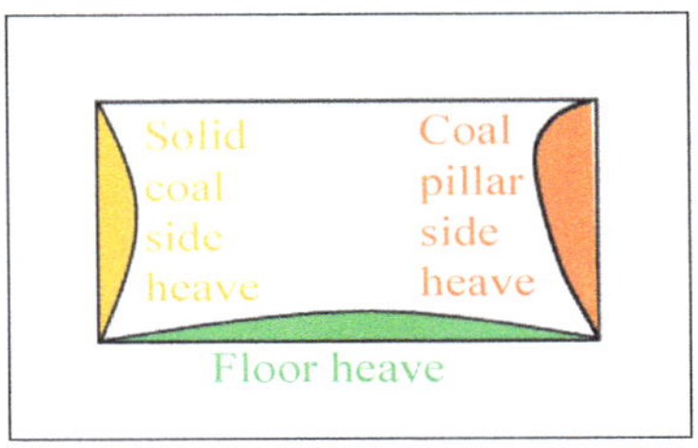

Figure 3. Deformation and damage characteristics of the roadway.

2.2. Model Construction

The FLAC3D numerical model was established with the geological background of the Tashan coal mine, as shown in Figure 4. The coal seam 305# is buried at 435 m, and the horizontal elevation is between +1005~+1019 m. Considering the influence of repeated mining of the coal seam, the model simulated back mining for both coal seams 102# and 305#. Considering the size of the model and the computational efficiency, the top burial depth of the model was set to 335 m. The model size is 640 m × 400 m × 130 m (XYZ) with 530,504 grids. The thickness of coal seam 305# in the model was 14 m, the thickness of immediate roof was 4.6 m, the thickness of coal seam 102# was 3 m, and the width of the remaining coal pillar was 20 m. The width of the 30,503 gob-side roadway was 5 m, the height was 4 m, the west part of the gob-side roadway was 8 m solid coal area for roadway 1 and 3 areas, the east part was 8 m narrow coal pillar for the mining area, the west part was 15–35 m solid coal for the overburden of coal seam 102#'s left coal pillar area, and the rest of the overburden rock thickness is given according to the actual situation of the mine.

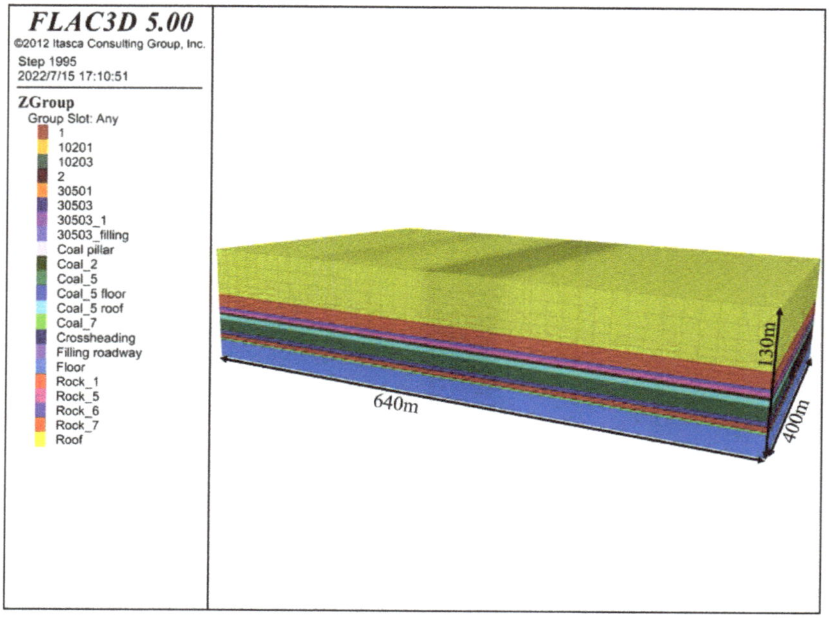

(**a**)

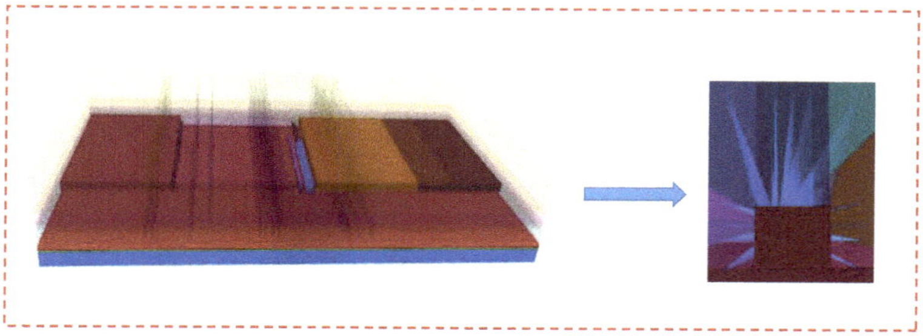

(**b**)

Figure 4. Numerical model. (**a**) Numerical model. (**b**) Model cross-section.

This numerical study selected the Mohr–Coulomb intrinsic structure model to study the deformation and damage control technology of a gob-side roadway in close extra-thick coal seam, and the physical and mechanical parameters used in the numerical model were obtained from the laboratory experimental results of coal rock samples in Tashan mine, as shown in Table 1. The cohesion and tensile strength were measured. The bulk modulus, shear modulus, and internal friction angle were calculated as follows:

$$K = \frac{E}{3(1\text{–}2\mu)} \tag{1}$$

$$G = \frac{E}{2(1+\mu)} \tag{2}$$

$$\varphi = 30\left(1 - \frac{\mu}{1-\mu}\right) + 15 \tag{3}$$

Table 1. Physical and mechanical parameters of the coal rocks.

Rockiness	Thickness (m)	Density (kg/m^3)	Bulk Modulus (GPa)	Shear Modulus (GPa)	Cohesion (MPa)	Internal Friction Angle (°)	Tensile Strength (MPa)
Sandy mudstone	51	2400	5.08	3.5	2.78	32.21	1.32
Clasticite	15	2450	5.49	3.78	2.94	33.15	1.52
Fine Sandstone	3	2470	8.77	6.58	4.77	36.79	2.98
Sandy mudstone	8	2400	5.08	3.5	2.78	32.21	1.32
2# Coal	3	1340	4.93	3.25	2.67	31.22	1.04
Kaolinite	5	2500	7.8	7.63	5.86	32.17	1.57
3~5# coal	14	1340	4.93	3.25	2.67	31.22	1.04
Kaolinite mud	4	2546	6.65	4.33	3.63	35.83	2.38
Siltstone	5	2470	7.81	5.62	4.24	35.93	2.71
7# Coal	2	1340	4.93	3.25	2.67	31.22	1.04
Sandy mudstone	21	2400	5.08	3.5	2.78	32.21	1.32
High water materials	4	1310	1.1	0.7	0.25	31.22	0.5

The parameter values of the abutment structure elements are shown in Table 2.

Table 2. Parameter values of the abutment structure elements.

Type	Bolt Length/mm	Grout Length/mm	Diameter/mm	Tensile Strength/kN	Row Spacing/mm
Anchor rods	8300	2700	20	200	1300 × 1000
Anchor cables	2400	800	22	120	1000 × 1000

2.3. Boundary Condition

The bottom of the model was set as a fixed boundary, the stress boundary conditions were applied around the model, the boundary conditions of the model were set by the results of ground stress measurements, the model vertical load was applied according to the change of burial depth, and the horizontal stress was calculated on the left and right according to the lateral pressure coefficient of this coal bed. Field measurements yielded a vertical load size of 8.25 MPa at the top of the model and a lateral pressure coefficient of 1.2, and then the horizontal stress at different burial depths along the vertical could be calculated by the following equation:

$$\sigma_{xx} = 1.2\gamma H \tag{4}$$

$$\sigma_{yy} = 1.2\gamma H \tag{5}$$

where γ is unit weight, H is the depth below the ground surface.

2.4. Roadway Monitoring Line Layout

Numerical simulations showed that 20 m ahead of the working face is the peak area of mining stress. Therefore, in order to quantitatively reveal the deformation and damage characteristics of the two sides and the floor of the gob-side roadway under the existing support conditions, eight monitoring lines were arranged at the position of 20 m ahead of the working face as shown in Figure 5. The monitoring line A is located at the height of the roof of the roadway. With an interval of 1 m, a total of 4 monitoring lines were arranged. The floor monitoring line starts from the position of 1 m from the floor, with an interval of 1 m, and a total of 4 lines were arranged. The section size of the roadway was 5 m × 4 m. The length of the monitoring line at the solid coal side and the coal pillar side was 8 m, and the length of the floor monitoring line was 6 m. All monitoring lines were arranged parallel to the roadway floor.

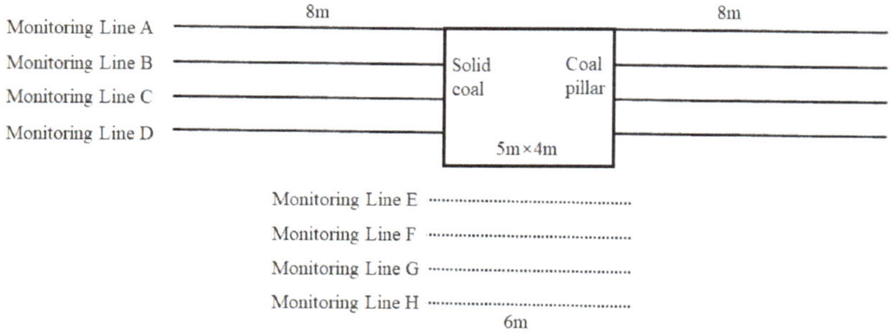

Figure 5. The 30,503 gob-side roadway monitoring line layout.

3. Results and Discussion

3.1. Deformation and Damage Characteristics of the Gob-Side Roadway under Existing Support Conditions

When the working face is mined to 100 m in a direction, the horizontal displacement distribution laws of each monitoring line of the two sides of the roadway 20 m ahead of the working face are shown in Figure 6. Due to the roof sinking, the roof of the roadway is no longer at the height of the monitoring line A, so the data of monitoring line A are not analyzed. It can be seen from Figure 6 that under the influence of mining, the horizontal displacements of the monitoring lines on two sides of the roadway were basically the same. The maximum deformation of the solid coal side was 0.69 m, which occurred 1 m from the roadway wall and then gradually decreased. The horizontal displacement of coal pillar side near the 30,501 gob area showed a continuous increasing trend, and the maximum displacement of the coal pillar side could reach 1.4 m. It can be seen by comparing the displacement of the coal wall of the coal pillar that the deformation at B and C monitoring lines in the middle and upper parts of the roadway were larger than that at D, showing asymmetric deformation characteristics, consistent with the site deformation characteristics. Based on this, it can be seen that the large deformation of the coal body in the deep part is an important reason for the continuous occurrence of roadway wall heave in the coal pillar, and the roadway wall heave in coal pillars cannot be controlled by using only anchor rods and anchor cables for reinforcement support. The deformation within 2~3 m of the solid coal side within the influence area of mining is still large, and the existing 2.2 m anchor rods support method of the roadway side cannot control the roadway wall heave. Improving the bearing capacity of the coal pillar side and using longer anchor cables to support the two sides are feasible measures to improve the stability of the gob-side roadway.

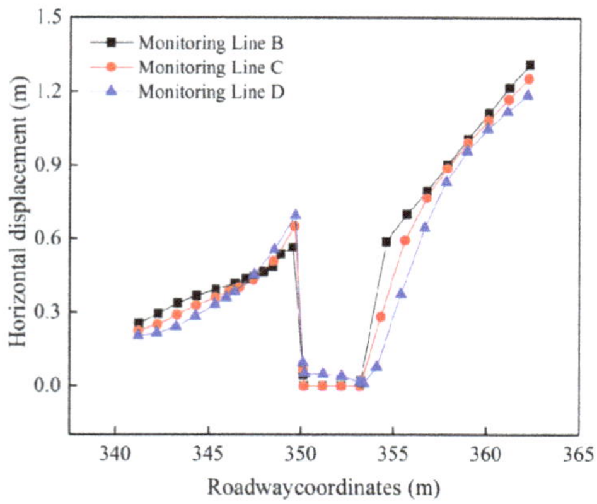

Figure 6. The horizontal displacement distribution law of the two sides of the gob-side roadway at 20 m ahead of the working face.

The vertical displacement of each monitoring line at different distances from the floor is shown in Figure 7. The floor of the roadway was obviously affected by mining. The maximum floor heave of each monitoring line basically occurred in the middle of the roadway, and the maximum displacement at 1 m from the roadway floor was 0.41 m. Under the action of high stress in the two sides of the roadway, the floor heave occurred continuously, and the displacement decreased gradually with the increase of the distance from the floor. The displacement was almost zero at the position of 4 m from the floor, less affected by mining and static load stress. For the 30,503 gob-side roadway, the vertical stress generated after the mining-induced overburden high-level and low-level main roof breakage acted on the gob-side rock surrounding the roadway, and the concentrated stress was transmitted to the floor through the two sides of the roadway. The floor was extruded into the roadway by the coupling of horizontal extrusion pressure and vertical stresses under the action of unbalanced support pressure, and the extruded fluid floor heave occurred [31].

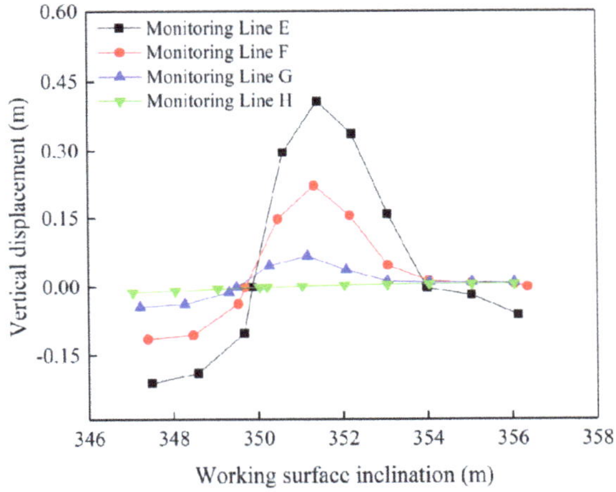

Figure 7. The vertical displacement distribution of the floor 20 m ahead of the working face (the monitoring line layout map shows 6 m).

3.2. *Stress and Displacement Distribution Law of the Roadway Wall under the Roadway Wall Reinforcement*

3.2.1. Optimization of the Support Method of the Roadway Wall

Based on the deformation and damage characteristics of the 30,503 gob-side roadway, it is known that the existing anchor rod support depths are shallow and do not achieve effective support, and the coal deformation inside the coal pillar side of the roadway is large. The support of the two sides under the existing support method of the roadway should be optimized to ensure the safety of the subsequent working face retrieval. Therefore, long anchor rods or anchor cables and coal pillar side grouting were considered for reinforcement support. Considering the cost of support, the existing support materials of the mine were used for reinforcement support, including 2.2 m and 2.4 m anchor rods and 8.3 m anchor cables. With the existing support method, the length of the two sides anchor rods is 2.2 m and the spacing is 1.0 m. The effective control of the roadway can no longer be achieved by using the existing anchor rods, and the number of anchor rods may aggravate the shallow coal fragmentation by reinforcement, which further leads to the increase of deformation. Therefore, the existing 8.3 m anchor cables were divided into two 4.15 m-long anchor cables for reinforcement support of the two sides. Under the influence of mining, the middle part of the solid coal side heaves, and the middle and upper part of the coal pillar side heaves seriously. Therefore, the roadway sides reinforcement method was carried out for these parts, and the anchor cable reinforcement design includes the following four options: (1) three 4.15 m anchor cables were used to support the two sides, which were arranged in the middle of the existing two rows of anchor rods and arranged in 212 three flowers with the existing anchor rods. The distance between rows was 1.0 m, 2.0 m, the first row was 0.8 m from the top of the roadway wall, and the third row was 1.2 m from the floor, as shown in Figure 8a. (2) Three anchor cables were used; the inter-row spacing was 1.3 m, 2.0 m. The first and third rows were 0.8 m from the roof and floor, they were reinforced at 15° to the roadway wall, and the middle anchor cable was perpendicular to the roadway wall, as shown in Figure 8b. (3) Four anchor cables were used, with inter-row distance of 1.1 m and 2.0 m. The first row was 0.3 m from the roof and 15° from the roadway wall and the remaining three rows were arranged vertically on the roadway wall, with inter-row distances of 1.2 m and 2.0 m, as shown in Figure 8c. (4) Four anchor cables were used, with inter-row spacing of 1.2 m, 2.0 m. The first row was 0.3 m from the roof and 15° from the roadway wall, the second and third rows were perpendicular to the roadway wall, and the fourth row was 0.9 m from the floor and 15° from the roadway wall, as shown in Figure 8d. Different support schemes have different support effects on the roadway due to the different number of anchor cables, the row spacing between anchor bolts, and the angle between the anchor cables and the two sides. Under certain conditions, the greater the support density, the longer the support length, and the more the support angle conforms to the stress distribution, the more obvious the support effect that can be achieved.

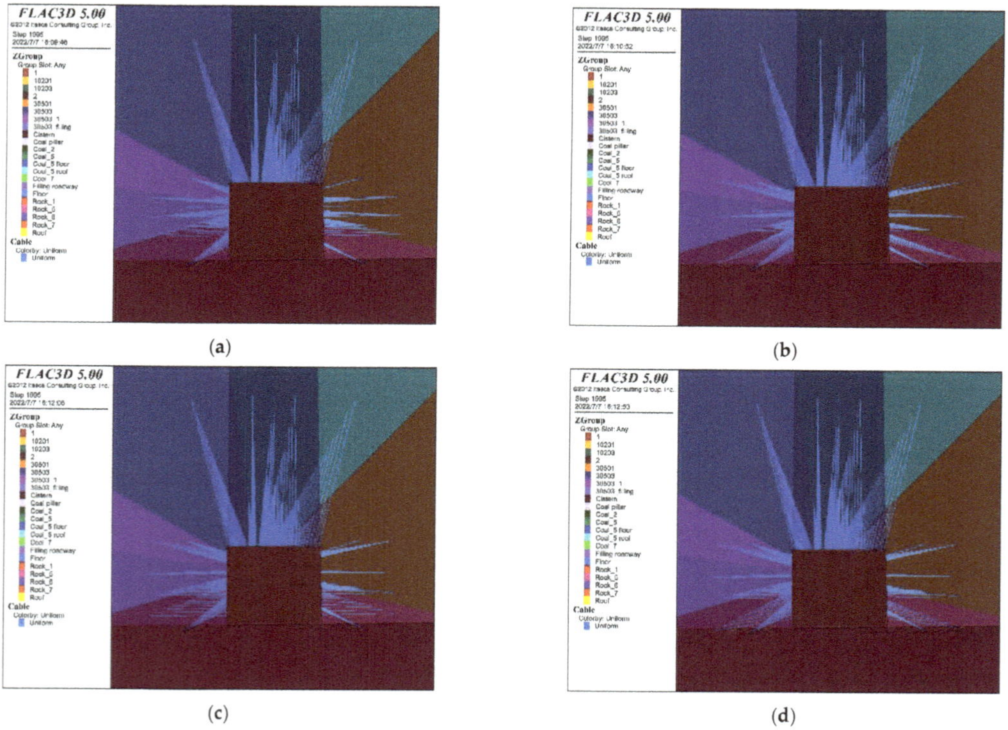

Figure 8. Anchor cable reinforcement method for the two sides of the roadway. (**a**) Reinforcement support method 1. (**b**) Reinforcement support method 2. (**c**) Reinforcement support method 3. (**d**) Reinforcement support method 4.

Based on the results of on-site drilling and numerical simulation analysis, the broken coal pillar is an important cause of wall heave, so grouting reinforcement support was carried out on the side of the coal pillar. The reinforcement method mainly includes four methods with the grouting depths of 2 m, 3 m, 4 m, and 5 m at the side of the coal pillar. The research shows that when the coal is grouted, the stiffness and shear strength of the fracture surface of the coal and rock mass will be greatly improved, thus improving the strength and bearing capacity of the coal and rock mass [32–34]. In this paper, we simulated the effect after grouting by increasing the strength and stiffness of the coal body within the grouting range when simulating the grouting support on the coal pillar side. In the model, the strength and stiffness and friction of the coal body increased by 1.2 times compared with the parameters in Table 1. The diagram of grouting range is shown in Figure 9.

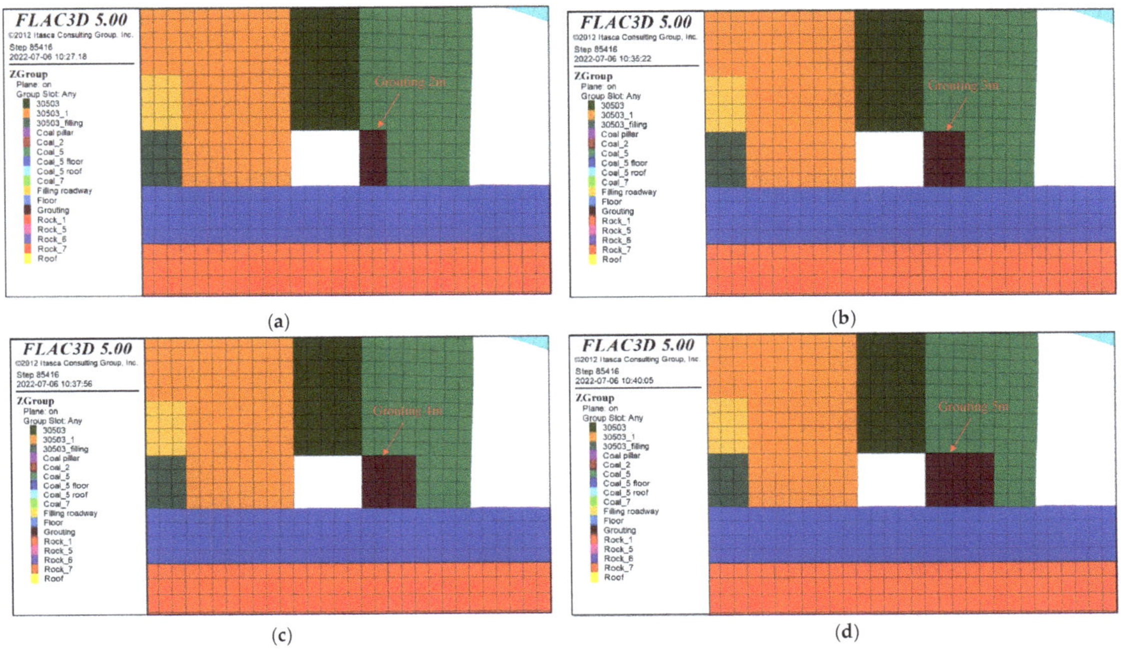

Figure 9. Coal pillar side grouting reinforcement method. (**a**) Grouting 2 m. (**b**) Grouting 3 m. (**c**) Grouting 4 m. (**d**) Grouting 5 m.

3.2.2. Deformation and Damage Characteristics of the Roadway Wall

When the working face was mined to 100 m strike, the horizontal displacement distribution law of each monitoring line on the two sides of the roadway 20 m ahead of the working face under different reinforcement support methods is shown in Figure 10. The horizontal displacement variation law of each monitoring line under different support methods was basically the same, and the horizontal displacement of the roadway wall on the coal pillar side was much greater than that on the solid coal side. Compared with the horizontal deformation of the two sides of the gob-side roadway under the original support, each reinforcement support method reduced the deformation of the two sides of the roadway and played a role in enhancing the support capacity. Among them, the first reinforcement method had the worst effect, the second reinforcement and third reinforcement had few differences, and the fourth reinforcement method had the best control effect. Additionally, the effect of the more broken coal pillar side was more obvious under the anchor cable-reinforced support, compared with the original support deformation reduced by more than 20%, as shown in Tables 3 and 4.

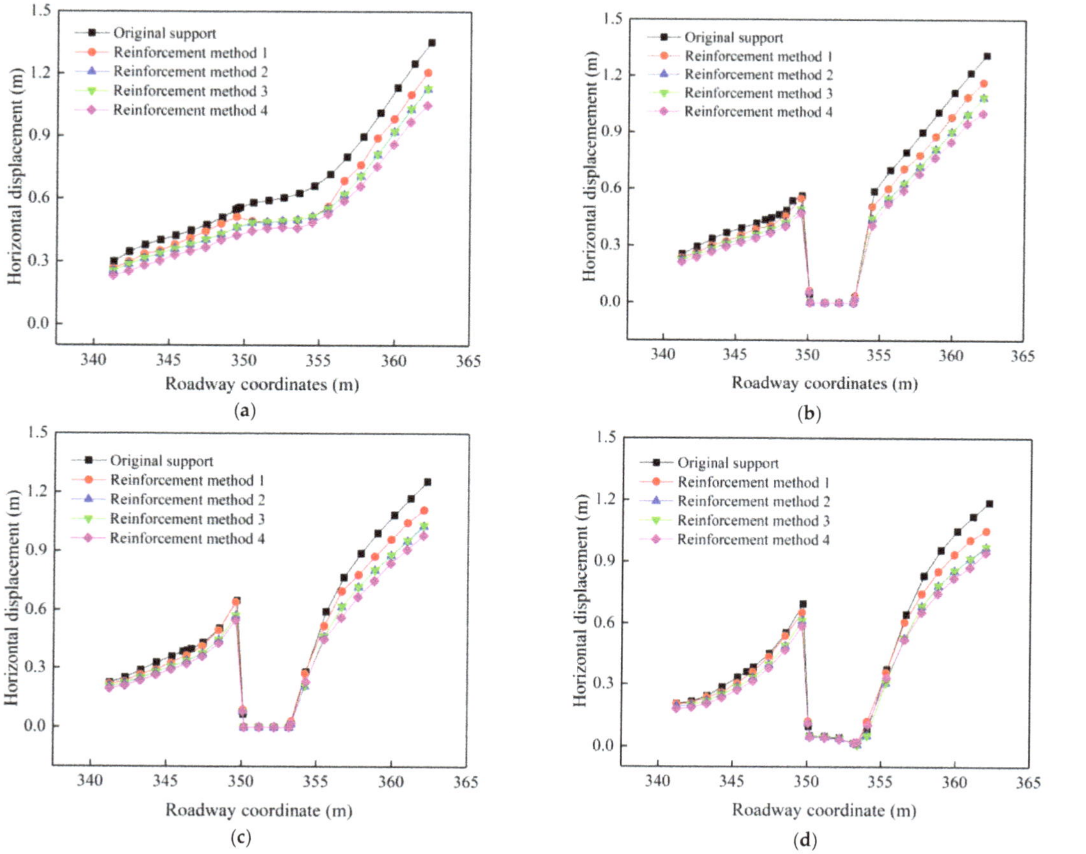

Figure 10. Horizontal displacement of the two sides of the roadway under the anchor cable reinforcement method. (**a**) Monitoring Line A. (**b**) Monitoring Line B. (**c**) Monitoring Line C. (**d**) Monitoring Line D.

Table 3. Percentage reduction in deformation under solid coal side anchor cable reinforcement.

Reinforcement Method/ Location	Method 1	Method 2	Method 3	Method 4
Monitoring Line A	5.00%	14.80%	14.80%	22.00%
Monitoring Line B	5.00%	11.00%	11.00%	16.00%
Monitoring Line C	11.00%	14.00%	14.00%	18.70%
Monitoring Line D	5.00%	16.00%	16.00%	20.00%

Table 4. Percentage reduction of deformation under coal pillar side anchor cable reinforcement.

Reinforcement Method/ Location	Method 1	Method 2	Method 3	Method 4
Monitoring Line A	11.80%	19.80%	19.8%	27.70%
Monitoring Line B	15.50%	22.20%	22.2%	27.00%
Monitoring Line C	11.30%	18.20%	18.2%	26.00%
Monitoring Line D	10.80%	18.00%	18.0%	22.00%

The vertical displacement distribution on the two sides of the roadway under different anchor cable reinforcement methods is shown in Figure 11. It can be seen from Figure 11 that the vertical displacement control on the two sides of the roadway was obvious with different reinforcement methods, especially on the coal pillar side, which indicates that the use of anchor cable reinforcement at this stage can improve the stability of two sides. Comparing the four reinforcement methods, we can see that the fourth reinforcement method had the best effect, followed by the second and third, and the first one had the worst effect. In order to determine the reinforcement effect of anchor cable reinforcement support on the shallow part of solid coal and the deep part of the coal pillar, the deformation at 1 m in the shallow part of the solid coal side and 4 m in the middle of coal pillar side under different reinforcement methods were compared with the deformation under the original support method, and the results are shown in Tables 5 and 6. The deformation of the coal pillar side's vertical pressure under the influence of mining was more serious, the effect of using reinforcement anchor cable was obvious, the first one could reduce the deformation by about 10%, and the best effect was achieved by using the fourth reinforcement method, which could reduce the deformation by more than 20% compared with the original one.

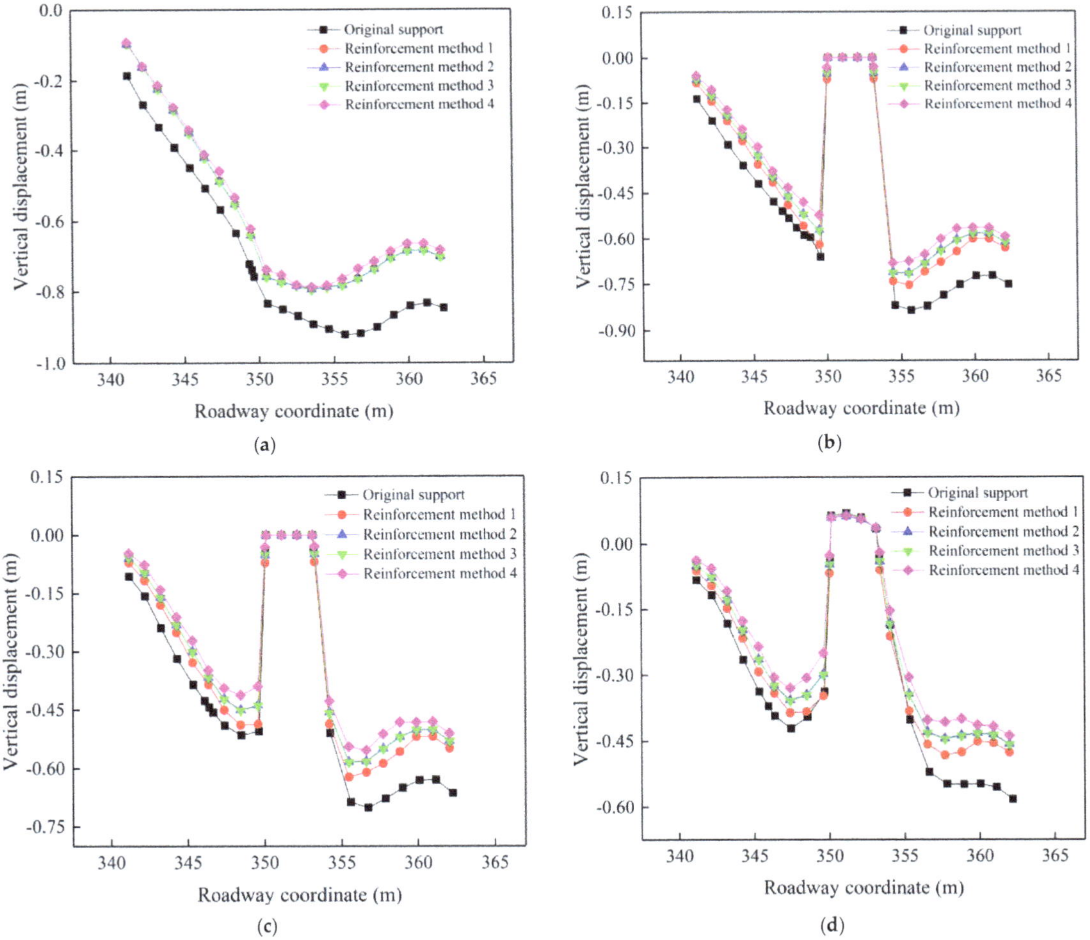

Figure 11. Vertical displacement of the two sides of the roadway under the anchor cable reinforcement method. (**a**) Monitoring Line A. (**b**) Monitoring Line B. (**c**) Monitoring Line C. (**d**) Monitoring Line D.

Table 5. Percentage reduction in deformation under solid coal side anchor cable reinforcement.

Reinforcement Method/ Location	Method 1	Method 2	Method 3	Method 4
Monitoring Line A	14.00%	15.60%	15.60%	17.00%
Monitoring Line B	6.00%	13.00%	13.00%	21.00%
Monitoring Line C	4.00%	11.70%	11.70%	27.50%
Monitoring Line D	4.800%	12.20%	12.20%	26.00%

Table 6. Percentage reduction of deformation under coal pillar side anchor cable reinforcement.

Reinforcement Method/ Location	Method 1	Method 2	Method 3	Method 4
Monitoring Line A	10.00%	20.00%	20.00	21.30%
Monitoring Line B	13.00%	17.90%	17.90%	23.00%
Monitoring Line C	12.00%	19.40%	19.40%	25.00%
Monitoring Line D	14.20%	21.40%	21.40%	28.00%

All reinforcement methods play a role in reducing the deformation of two sides of the roadway, mainly because the 4.15 m long anchor cable can stabilize the broken coal body in the deeper part of the sides compared with the existing 2.2 m-long anchor rods, improving the range of the stability area of the roadway side. The main reason why method 4 is more effective than other support methods is that its fan-shaped arrangement provides multi-angle support to two sides of the roadway and covers a wider area than the vertical roadway.

The coal pillar side is severely broken due to multiple mining disturbances. In order to determine the reasonable grouting depth of the coal pillar side of the roadway wall, vertical stress data were extracted from each monitoring line position of the two sides of the roadway under different grouting depths, as shown in Figure 12. The vertical stress distribution on the solid coal side was basically unchanged compared to the original support method, and the peak stress on the coal pillar side was significantly increased compared with that without grouting. Additionally, this shows that grouting the coal pillar can effectively improve the bearing capacity and stability of the coal pillar. Comparing the stress distribution on the side of the coal pillar when the grouting depth was 2 m, 3 m, 4 m, and 5 m, we can see that the peak stress still appeared at 2 m. It shows that the low-level main roof broke at the position 2 m above the coal pillar, resulting in the formation of a stress concentration zone on the coal pillar. When 2 m grouting was used, the peak stress at each monitoring point on the coal pillar basically reached 23~26 MPa. With the increase of grouting depth, the stress distribution gradually changed to bimodal distribution, with the peak stress near the roadway wall gradually decreasing and the peak stress near the gob area gradually increasing, indicating that the integrity of the coal pillar is continuously improving. The vertical stress distribution in the ungrouted area on the side of the gob is basically the same as under the original support method.

To further determine the reasonable depth of coal pillar side grouting, the horizontal deformation of two sides after different grouting depths and anchor cable support are shown in Figure 13. Although only the shallow part of the coal pillar side was grouted, it still could significantly reduce the deformation of two sides of the roadway, indicating that improving the bearing capacity and integrity of the coal pillar is an effective measure to ensure the stability of the roadway [35]. The grouting at the coal pillar side of the roadway wall reduced the deformation of the solid coal side, mainly because the grouting improves the bearing capacity of the coal pillar to the overburden. Additionally, the coal pillar side can bear more pressure transmitted by the overburden rock layer and share the stress of the solid coal side. The improved integrity of the coal pillar side also improved the ability to resist the deformation and damage of the roadway under the influence of mining. The deformation of the roadway decreased and the performance of coal pillar side was more obvious. Comparing the effect of the anchor cable support and grouting in the roadway, we can see that the effect of grouting was better than anchor cable support. Comparing the effect of different grouting depths, it was found that the deformation reduction on the coal pillar side was not obvious as the grouting depth increased, and the difference in deformation between 2 m and 5 m after grouting was less than 0.05 m. Considering economic and other factors, therefore, 2 m grouting can be considered for the coal pillar side in the reinforcement process.

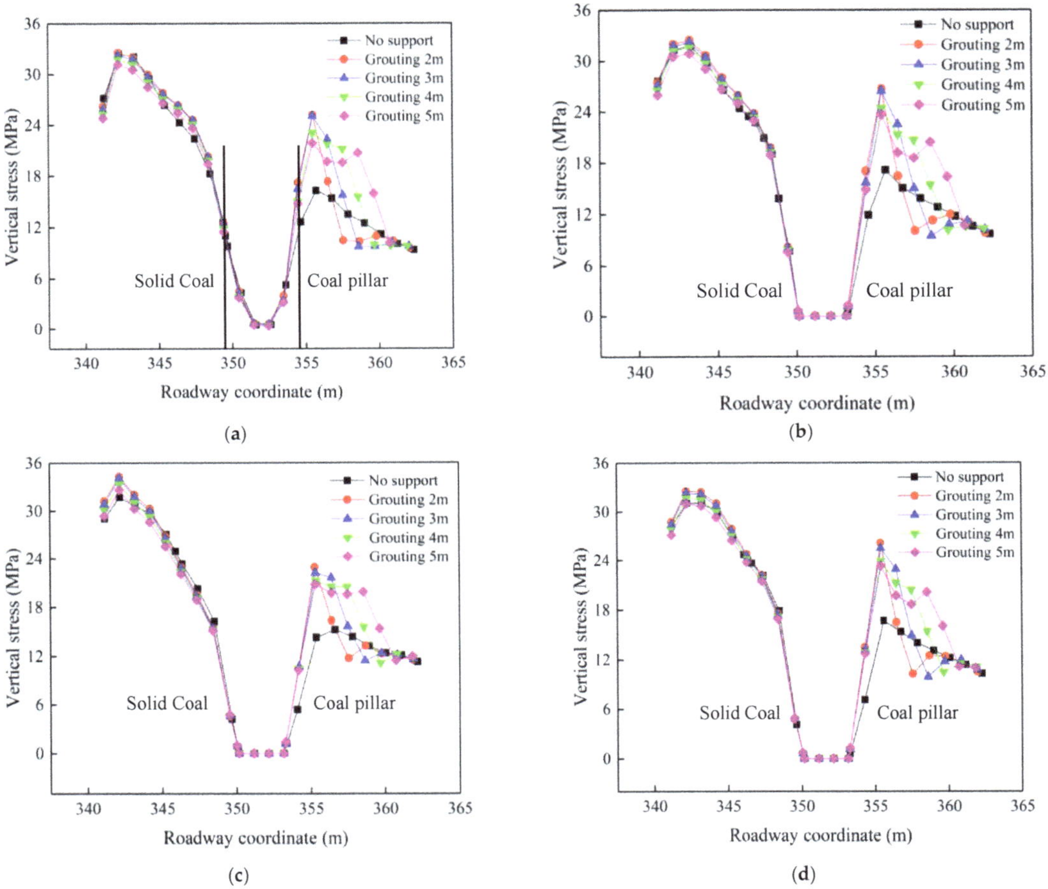

Figure 12. Vertical stress distribution in the two sides of the roadway at different grouting depths. (**a**) Monitoring Line A. (**b**) Monitoring Line B. (**c**) Monitoring Line C. (**d**) Monitoring Line D.

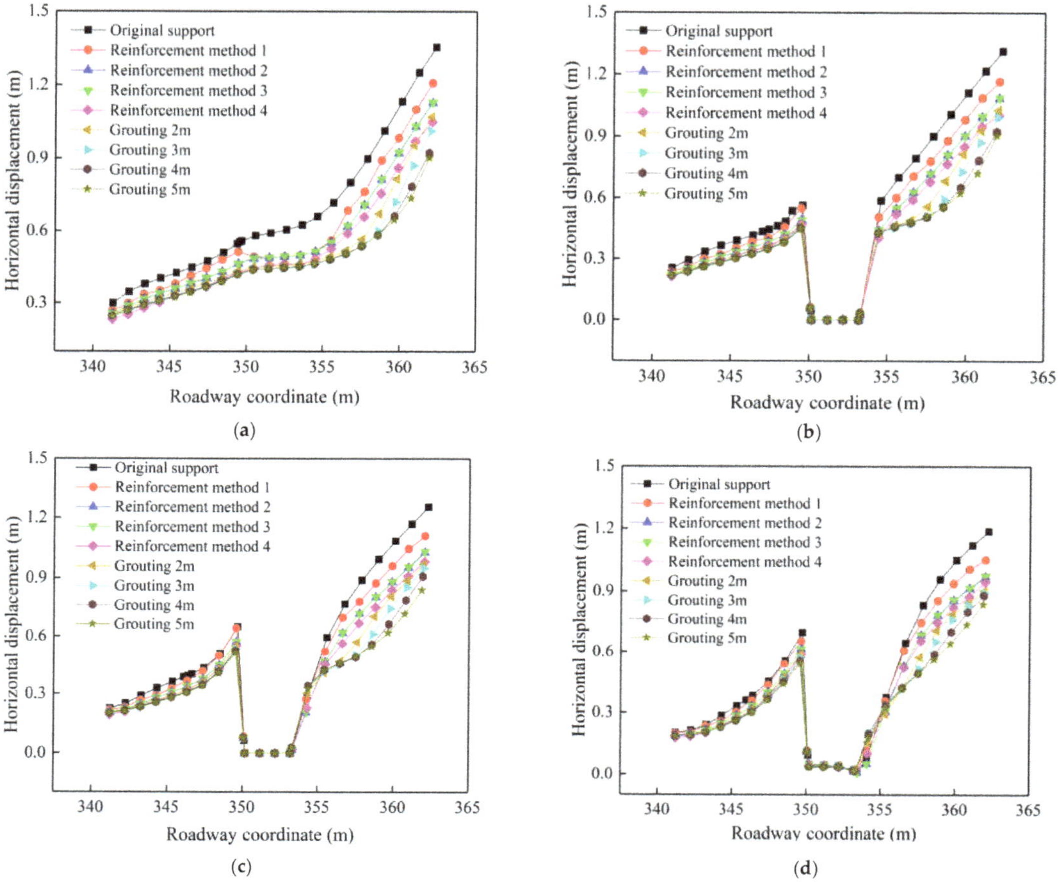

Figure 13. Horizontal displacement of the two sides of the roadway under different reinforcement methods of the roadway. (**a**) Monitoring Line A. (**b**) Monitoring Line B. (**c**) Monitoring Line C. (**d**) Monitoring Line D.

3.3. Stress and Displacement Distribution Pattern of the Roadway Wall under the Floor Reinforcement

3.3.1. Optimization of the Roadway Floor Support Method

The floor reinforcement support of the roadway includes floor anchor cable support and grouting. The grouting method of the floor is the same as that of the coal pillar side. The reinforcement support of the floor anchor cable includes the following three methods. (1) Five 4.15 m anchor cables were used; the first and last two were 0.7 m from the roadway wall, and the inter-row distance was 0.9 m × 2.0 m; (2) four anchor cables were used to support the vertical downward direction 1 m from the roadway wall, and the inter-row distance was 1.0 m × 2.0 m; (3) five anchor cables were used to support the floor of the roadway, with the first and last distance from the two sides being 0.3 m and the inter-row distance being 1.2 m × 2.0 m, corresponding to the way the roof of the roadway was supported. The anchor cable reinforcement method is shown in Figure 14.

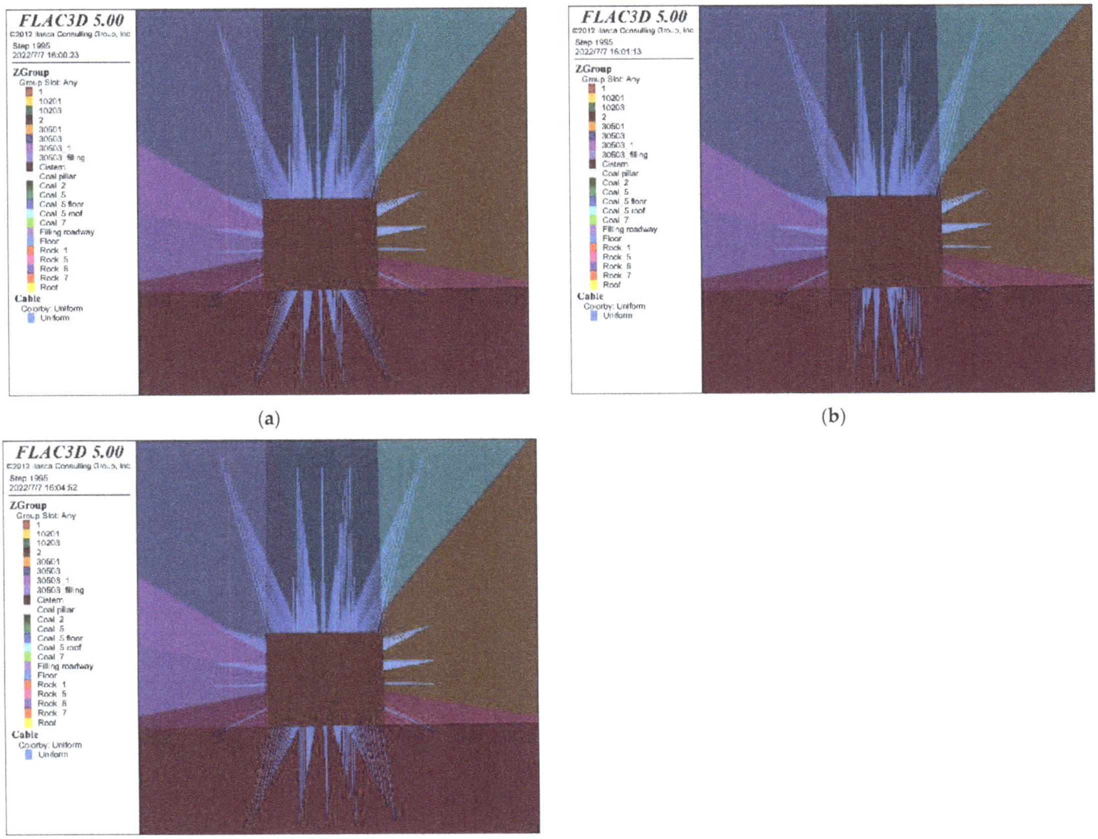

Figure 14. Anchor cable reinforcement method at the floor of the roadway. (**a**) Reinforcement method 1. (**b**) Reinforcement method 2. (**c**) Reinforcement method 3.

3.3.2. Deformation and Damage Characteristics of the Roadway Floor

When the working face was mined to 100 m strike, the vertical deformation distribution of the roadway floor under different anchor cable and grouting reinforcement methods is shown in Figure 15. Under the action of high stress on two sides of the roadway, the floor of the roadway had a continuous floor heave, the deformation damage of the floor was not reduced by increasing the anchor cable only, and the use of floor grouting obviously reduced the amount of floor heave of the roadway. The floor heave in the middle of the roadway floor under unsupported conditions was close to 0.5 m, and the anchor cable support was basically unchanged, indicating that the floor has been very broken under the influence of mining. The rock bearing capacity can not be improved by anchor cable support alone, and the maximum value of floor heave after grouting is 0.2 m, which is 60–70% lower than that before grouting. At 2 m below the floor of the roadway, the maximum value of floor deformation under the unsupported condition was 0.38 m, which decreased to 0.06 m after grouting. After the floor grouting, the sinking amount of the two sides of the roadway is also significantly reduced. This is because after the floor bearing capacity is improved, the force of the two sides cannot make the roadway more deform more, resulting in heaving and the coal body of the two sides will not sink.

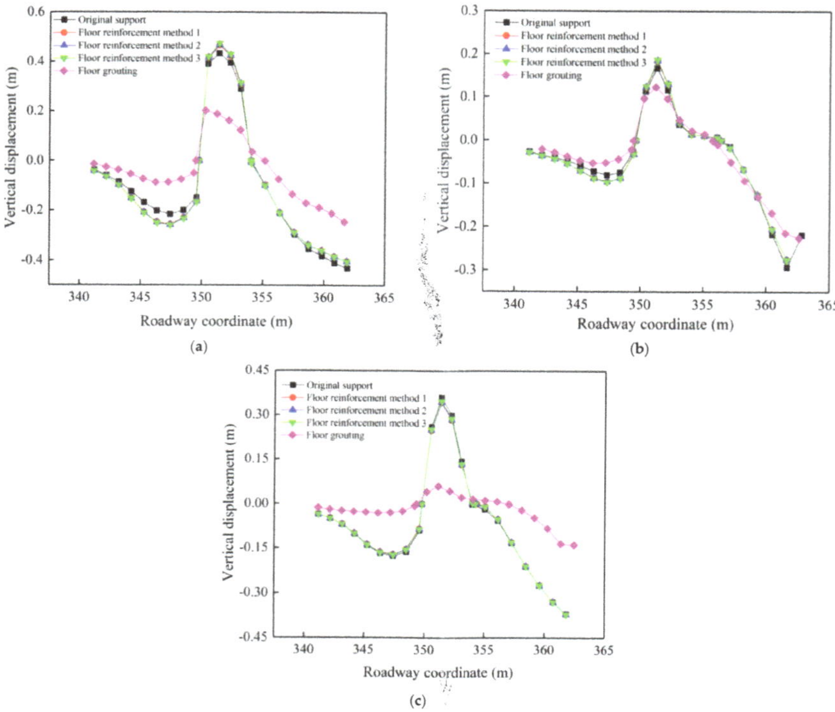

Figure 15. Vertical displacement of the floor under the floor reinforcement method of the roadway. (**a**) Floor. (**b**) 1 m below the floor. (**c**) 2 m below the floor.

In order to determine the optimal integrated reinforcement method for the roadway, a comprehensive analysis of the above reinforcement methods was performed. The deformation of the two sides of the roadway under different reinforcement methods are shown in Figures 16 and 17. The support effect of the anchor cable + the grouting support method was better than that of the simple anchor cable reinforcement support, but there was no obvious difference with the grouting reinforcement method. The comprehensive reinforcement method of the anchor cable + coal pillar side grouting + floor grouting at the side of the roadway had the best effect, with the lowest horizontal and vertical deformation, the reduction of vertical deformation was especially more obvious. The increase of vertical stress caused by mining is the direct cause of roadway deformation and damage. The floor heave of the roadway will further aggravate the increase of deformation on two sides. Based on this, it can be seen that when the strength of the floor is increased, on the one hand, it can bear more vertical stress caused by mining, so that the vertical stress can be transferred to a deeper depth. On the other hand, it can resist a larger floor heave in the floor.

The deformation of the roadway floor under different reinforcement methods is shown in Figure 18. From Figure 18, it can be seen that the use of floor grouting was still the most effective measure to control the floor heave. This is because the anchor cable and grouting reinforcement support increased the bearing capacity of the roadway wall. The stress concentration of the roadway wall increased, and the greater vertical stress was transferred to the floor, which increased the floor heave of the roadway.

Comparing only grouting and joint reinforcement support of the roadway floor, it can be seen that the deformation of the floor heave at depth increased to about 0.1 m after supporting the two sides, while the shallow part was basically unchanged. It can be seen that the joint anchor cable + grouting was the most effective measure to ensure the safety of the roadway, as the deformation and damage of the roadway wall and floor under different reinforcement methods were combined.

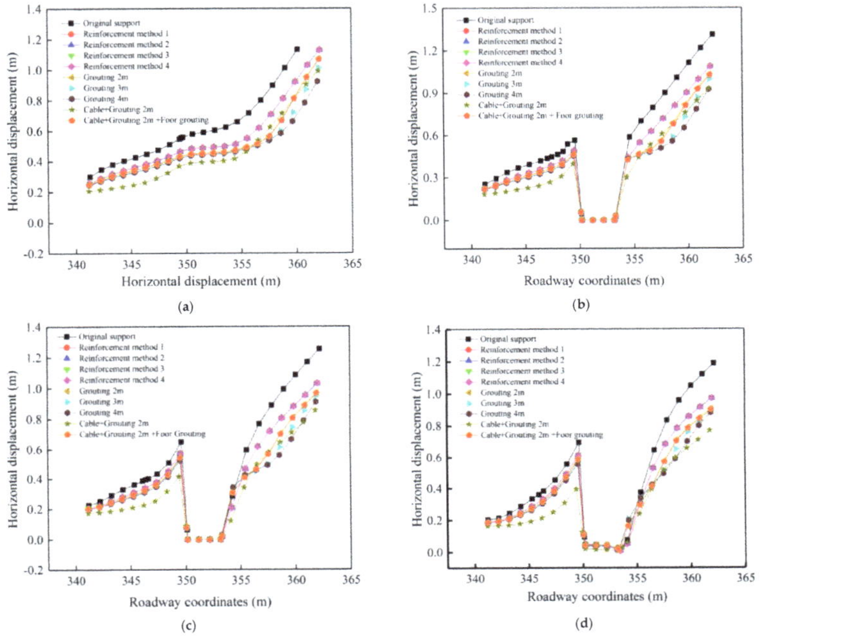

Figure 16. Horizontal displacement of the two sides of the roadway under different reinforcement methods. (**a**) Monitoring Line A. (**b**) Monitoring Line B. (**c**) Monitoring Line C. (**d**) Monitoring Line D.

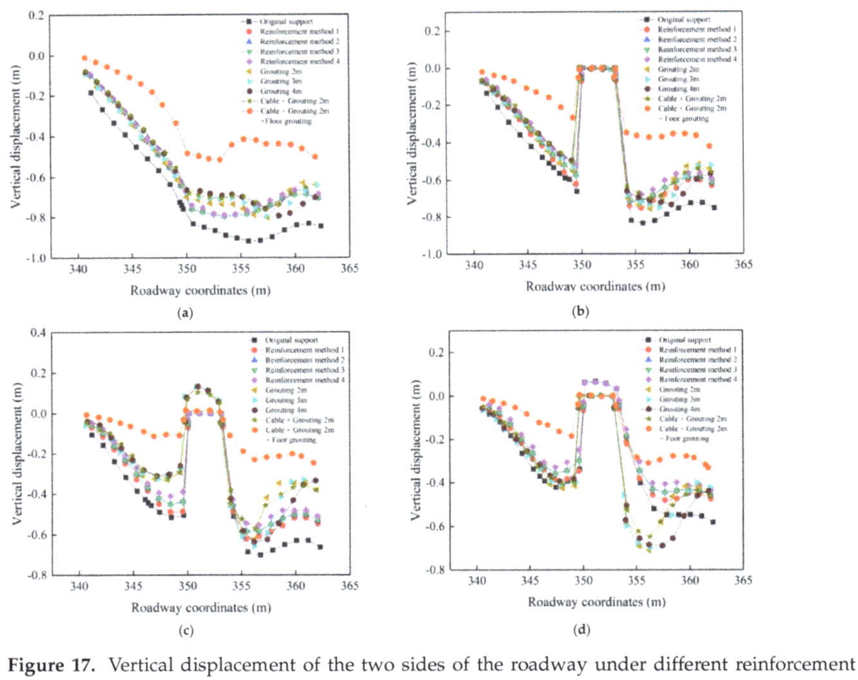

Figure 17. Vertical displacement of the two sides of the roadway under different reinforcement methods. (**a**) Monitoring Line A. (**b**) Monitoring Line B. (**c**) Monitoring Line C. (**d**) Monitoring Line D.

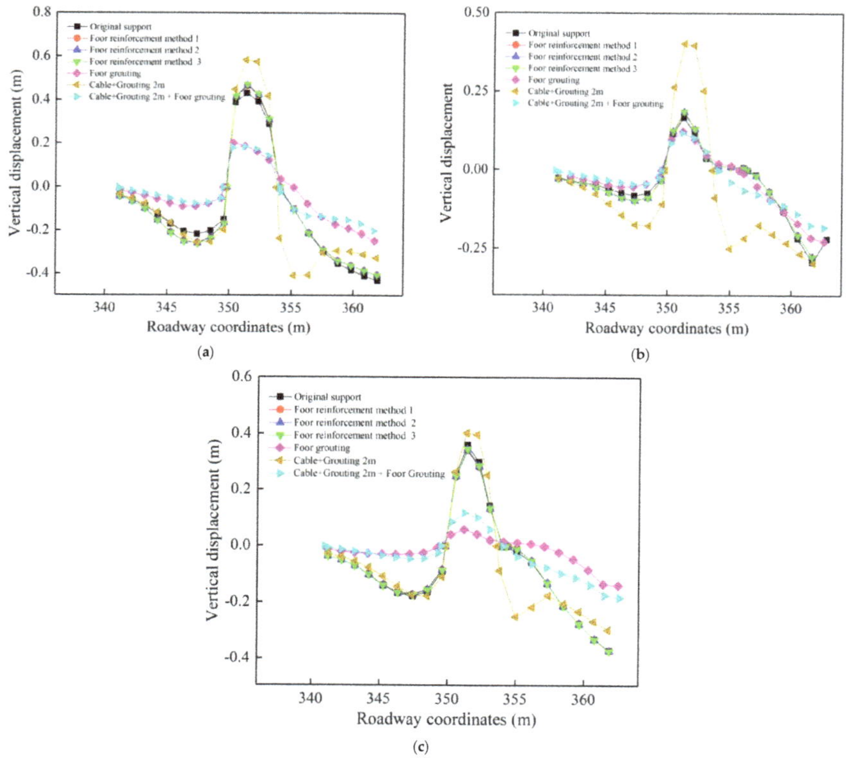

Figure 18. Vertical displacement of the floor under different floor reinforcement methods in the roadway. (**a**) Floor. (**b**) 1 m below the floor. (**c**) 2 m below the floor.

3.4. Field Application and Validation

In order to determine the effect of the joint anchor cable + grouting joint reinforcement support method to reduce roadway deformation and damage after application in the field, a laser rangefinder was used to monitor the roadway deformation within 250 m of the roadway ahead of the working face during the back mining of the 30,503 working face. The displacements of the roof and floor of the roadway and the two sides are shown in Figure 19. From Figure 19, it can be seen that affected by mining, the rapid deformation stage of the roadway was mainly concentrated within 50 m ahead of the working face, and the roadway deformation was large.

Near the working face, the deformation of the two sides was still about 600 mm, and the distortion of the ceiling and floor was still almost 350 mm. However, the deformation gradually decreased and stabilized with the increase of the distance ahead of the working face. This shows that after using the anchor cable + grouting support reinforcement, the deformation pattern of each monitoring point of the roadway was kept consistent and the deformation was better controlled. Compared with the deformation data of two sides of the roadway and the roof and floor without reinforcement support (Figure 20), the overall deformation of two sides was reduced by 50% and the roof and floor were reduced by about 40%. In the field observation, it is found that after the combined support of anchor cable and grouting, the floor heave and roadway wall heave still exists, but compared with the previous support, they are better controlled, which means that the existing support methods can better ensure the stability of the roadway during the recovery period.

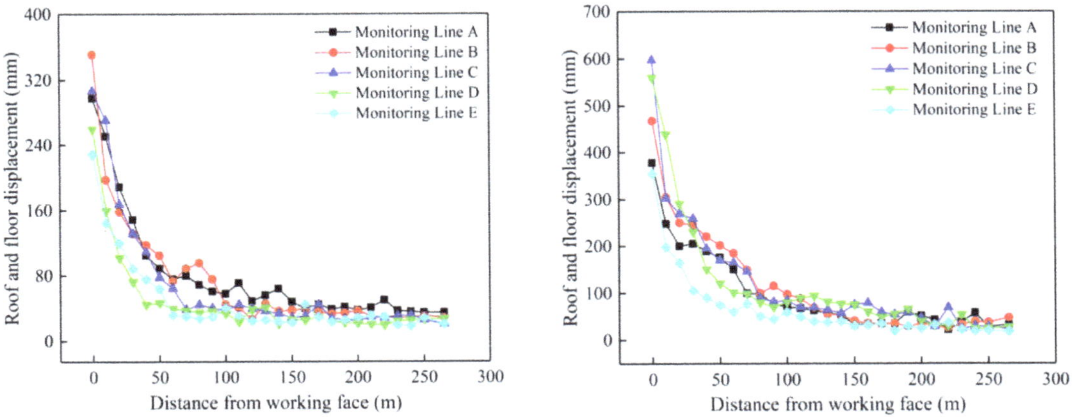

(**a**) (**b**)

Figure 19. Deformation data after roadway reinforcement. (**a**) Roof–floor displacement. (**b**) Two sides displacement.

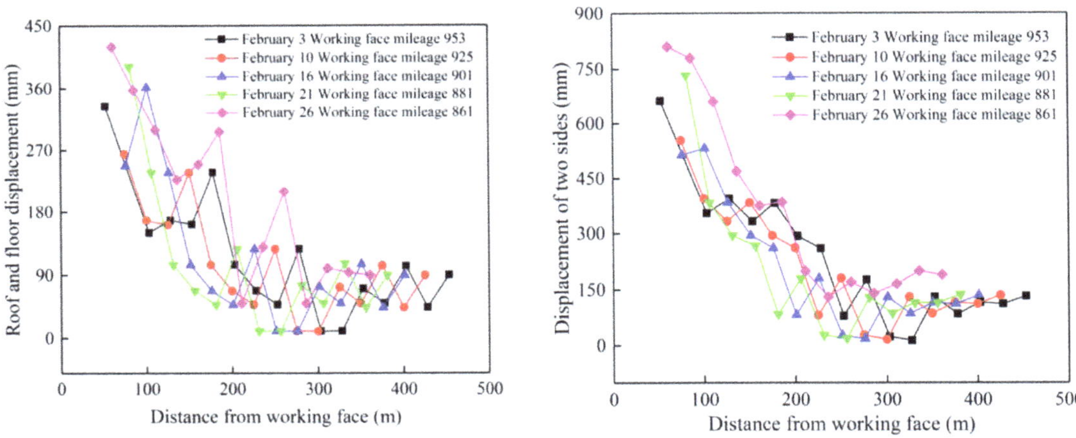

(**a**) (**b**)

Figure 20. Deformation of the gob-side roadway wall. (**a**) Deformation of roof and floor. (**b**) Deformation of two sides.

4. Conclusions

4.1. Main Contributions

The close extra-thick coal seams are subject to the broken overburden of mined coal seams, and the deformation and damage of the roadway is serious, which affects the safe operation of the mine. In this paper, we studied the deformation and damage law of a gob-side roadway in close extra-thick coal seam by numerical simulation and field monitoring and compared and analyzed the deformation and damage characteristics of the roadway under different reinforcement support methods. Finally, the optimal reinforcement support method was determined and we verified it on site. The main contributions of the study are listed below. The gob-side roadway in close extra-thick coal seam has asymmetric deformation damage characteristics; the coal pillar side deformation was greater than that of the solid coal side. The large deformation of the coal body in the deep part of the roadway wall is an important reason for the continuous appearance of roadway wall heave in the coal pillar.

The horizontal and vertical displacement of the coal pillar side under various support methods was significantly different than that of the solid coal side. Each reinforcement support method reduces the deformation of the two sides of the roadway. Among them, increasing the length of anchor cable and arranging it in a fan shape can improve the support effect remarkably. The bearing capacity and stability can be greatly increased by grouting at the coal pillar side. Grouting to strengthen the integrity of the floor is the most effective way to avoid and control floor heave since its effect on regulating floor heave is significantly better than that of anchor cables. Therefore, the comprehensive reinforcement method of anchor cable + coal pillar side grouting + floor grouting at the roadway wall has the best effect, especially the vertical deformation.

4.2. Limitations and Prospects

The limitation of this paper is that the optimized support results for the gob-side roadway in the close extra-thick coal seam obtained in this study are based on the horizontal coal seam, but whether it is applicable to other inclined coal seams needs further research and verification. Additionally, based on the understanding of the limitations of this paper, in the future, the author will conduct large-scale physical similarity simulation to further study the instability characteristics of the gob-side roadway in close extra-thick coal seam, and provide guidance for roadway instability control under this condition.

Author Contributions: B.Z.: Writing—original draft, investigation, conceptualization. S.H.: Validation, conceptualization, methodology, funding acquisition. X.H.: Validation, writing—review and editing. L.G.: Methodology, conceptualization. Z.L.: Investigation, formal analysis. D.S.: Writing—review and editing, formal analysis. F.S.: Methodology, formal analysis. All authors have read and agreed to the published version of the manuscript.

Funding: This work was financially supported by the Postdoctoral Research Foundation of China (Project No. 2021M700371). We thank anonymous reviewers for their comments and suggestions to improve the manuscripts.

Institutional Review Board Statement: Not applicable.

Informed Consent Statement: Not applicable.

Data Availability Statement: Not applicable.

Acknowledgments: The authors confirm that there are no conflicts of interest associated with this publication. The authors gratefully thank the anonymous reviewers for their constructive comments on improving the presentation. All authors have agreed to the listing of authors.

Conflicts of Interest: The authors declare no conflict interests.

References

1. Yang, H.Y.; Cao, S.G.; Wang, S.Q.; Fan, Y.C.; Wang, S.; Chen, X.Z. Adaptation assessment of gob-side entry retaining based on geological factors. *Eng. Geol.* **2016**, *209*, 143–151. [CrossRef]
2. Li, X.L.; Chen, S.J.; Wang, S.; Zhao, M.; Liu, H.; Gong, F. Study on in situ stress distribution law of the deep mine taking Linyi Mining area as an example. *Adv. Mater. Sci. Eng.* **2021**, *2021*, 5594181. [CrossRef]
3. Tan, Y.L.; Ma, Q.; Zhao, Z.H.; Gu, Q.H.; Fan, D.Y.; Song, S.L.; Huang, D.M. Cooperative bearing behaviors of roadside support and surrounding rocks along gob-side. *Geomech. Eng.* **2019**, *18*, 439–448. [CrossRef]
4. Wang, Q.; He, M.C.; Yang, J.; Gao, H.K.; Jiang, B.; Yu, H.C. Study of a no-pillar mining technique with automatically formed gob-side entry retaining for longwall mining in coal mines. *Int. J. Rock Mech. Min. Sci.* **2018**, *110*, 1–8. [CrossRef]
5. Wang, X.; Lu, M.Y.; Gao, Y.H.; Luo, W.B.; Liu, W.G. Structural Mechanical Characteristics and Instability Law of Roof Key Block Breaking in Gob-Side Roadway. *Adv. Civ. Eng.* **2020**, *2020*, 6682303. [CrossRef]
6. Han, C.L.; Zhang, N.; Li, B.Y.; Si, G.Y.; Zheng, X.G. Pressure relief and structure stability mechanism of hard roof for gob-side entry retaining. *J. Cent. South Univ.* **2015**, *22*, 4445–4455. [CrossRef]
7. Li, Y.H.; Zhang, Q.; Lin, Z.B.; Wang, X.D. Spatiotemporal evolution rule of rocks fracture surrounding gob-side roadway with model experiments. *Int. J. Min. Sci. Technol.* **2016**, *26*, 895–902. [CrossRef]
8. Zheng, J.W.; Ju, W.J.; Sun, X.D.; Li, Z.W. Large Deformation Mechanics of Gob-Side Roadway and Its Controlling Methods in Deep Coal Mining: A Case Study. *Adv. Civ. Eng.* **2020**, *2020*, 8887088. [CrossRef]

9. Slashchov, I.; Shevchenko, V.; Kurinnyi, V.; Slashchova, O.; Yalanskyi, O. Forecast of potentially dangerous rock pressure manifestations in the mine roadways by using information technology and radiometric control methods. *Min. Miner. Deposits* **2019**, *13*, 9–17. [CrossRef]
10. Sun, Q.; Zhang, J.X.; Huang, Y.L.; Yin, W. Failure mechanism and deformation characteristics of gob-side entry retaining in solid backfill mining: A case study. *Nat. Resour. Res.* **2020**, *29*, 2513–2527. [CrossRef]
11. Li, C.; Wang, P.; Wu, Y. Study on Movement Laws of Rockmass Structure and its Control in Gob-Side Roadway Retaining. *Conf. Ser. Earth Environ. Sci.* **2018**, *170*, 022017. [CrossRef]
12. Guo, P.F.; Zhang, X.H.; Peng, Y.Y.; He, M.C.; Ma, C.R.; Sun, D.J. Research on Deformation Characteristic and Stability Control of Surrounding Rock During Gob-Side Entry Retaining. *Geotech. Geol. Eng.* **2020**, *38*, 2887–2902. [CrossRef]
13. Liu, X.; Hua, X.Z.; Yang, P.; Huang, Z.G. A study of the mechanical structure of the direct roof during the whole process of non-pillar gob-side entry retaining by roof cutting. *Energy Explor. Exploit.* **2020**, *38*, 1706–1724. [CrossRef]
14. Feng, Q.; Jin, J.; Zhang, S.; Liu, W.W.; Yang, X.X.; Li, W.T. Study on a Damage Model and Uniaxial Compression Simulation Method of Frozen–Thawed Rock. *Rock Mech. Rock Eng.* **2022**, *55*, 187–211. [CrossRef]
15. Wang, Y.; Li, Y.M.; Yu, Z.L.; Zhang, H.; Ding, D. Research on the Surrounding Rock Control Technology of Gob-Side Entry with a Narrow Coal Pillar Reserved in a Fully Mechanized Caving Face with Large Mining Height. *Geotech. Geol. Eng.* **2022**, *40*, 285–300. [CrossRef]
16. Xu, X.H.; He, F.L.; Li, X.B.; He, W.R. Research on mechanism and control of asymmetric deformation of gob side coal roadway with fully mechanized caving mining. *Eng. Fail. Anal.* **2021**, *120*, 105097. [CrossRef]
17. Li, G.J.; Wang, X.Y.; Bai, J.B.; Wu, B.W.; Wu, W.D. Research on the failure mechanism and control technology of surrounding rock in gob-side entry driving under unstable overlying strata. *Eng. Fail. Anal.* **2022**, *138*, 106361. [CrossRef]
18. Peng, D.U.; Min, T.U. Study on Coal Pillar Width and Control Technology on Gob-side Roadway of Isolated Fully Mechanized Working Face in Ultra-thick Coal Seam. *Min. Saf. Environ. Prot.* **2019**, *46*, 16–20. [CrossRef]
19. Ren, Z.; Li, Y. Study on Influencing Factors of Support Stability About Gob-Side Roadway in Fully Mechanized Caving Face. *Geotech. Geol. Eng.* **2020**, *38*, 1579–1589. [CrossRef]
20. Zheng, W.X.; Duan, H.Q. Discussion on stability analysis and support technology of surrounding rock of gob-side entry retaining. *J. Vibroeng.* **2019**, *21*, 1058–1068. [CrossRef]
21. Xie, S.R.; Wang, E.; Chen, D.D.; Sun, Y.H.; Cheng, Q.; Ji, C.W.; Yan, Z.Q.; Xiao, H.B. Failure analysis and control mechanism of gob-side entry retention with a 1.7-m flexible-formwork concrete wall: A case study. *Eng. Fail. Anal.* **2020**, *117*, 104816. [CrossRef]
22. Xie, F.X. Control of Gob-Side Roadway with Large Mining Height in Inclined Thick Coal Seam: A Case Study. *Shock Vib.* **2021**, *10*, 6687244. [CrossRef]
23. Begalinov, A.; Almenov, T.; Zhanakova, R.; Bektur, B. Analysis of the stress deformed state of rocks around the haulage roadway of the Beskempir field (Kazakhstan). *Min. Miner. Deposits* **2020**, *14*, 28–36. [CrossRef]
24. He, F.L.; Xu, X.H.; Qin, B.B.; Li, L.; Lv, K.; Li, X.B. Study on deformation mechanism and control technology of surrounding rock during reuse of gob side entry retaining by roof pre-splitting. *Eng. Fail. Anal.* **2022**, *137*, 106271. [CrossRef]
25. Yang, X.J.; Hou, L.; Xue, H.J.; Yuan, D.; Cao, J.D.; Han, Z.J.; Guo, Y.B. Pressure Distribution and Deformation Control of Gob-Side Entry Retaining Formed by Roof Cutting Influenced by Abandoned Roadways. *Geotech. Geol. Eng.* **2021**, *39*, 2533–2545. [CrossRef]
26. Ma, Q.; Tan, Y.L.; Zhao, Z.H.; Xu, Q.; Wang, J.; Ding, K. Roadside support schemes numerical simulation and field monitoring of gob-side entry retaining in soft floor and hard roof. *Arab. J. Geosci.* **2018**, *11*, 563. [CrossRef]
27. Guo, P.F.; Yuan, Y.D.; Ye, K.K.; Sun, D.J. Fracturing mechanisms and deformation characteristics of rock surrounding the gate during gob-side entry retention through roof pre-fracturing. *Int. J. Rock. Mech. Min.* **2021**, *148*, 104927. [CrossRef]
28. Li, W.S.; Jiang, B.Y.; Gu, S.T.; Yang, X.X.; Shaikh, F.U.A. Experimental study on the shear behaviour of grout-infilled specimens andmicromechanical properties of grout-rock interface. *J. Cent. South Univ.* **2022**, *29*, 1686–1700. [CrossRef]
29. Małkowski, P.; Niedbalski, Z.; Majcherczyk, T.; Bednarek, Ł. Underground monitoring as the best way of roadways support design validation in a long time period. *Min. Miner. Deposits* **2020**, *14*, 1–14. [CrossRef]
30. Dychkovskyi, R.; Shavarskyi, I.; Saik, P.; Lozynskyi, V.; Falshtynskyi, V.; Cabana, E. Research into stress-strain state of the rock mass condition in the process of the operation of double-unit longwalls. *Min. Miner. Deposits* **2020**, *14*, 85–94. [CrossRef]
31. Sun, Y. Analysis of coal mine roadway bottom drum problem. *Silicon Val.* **2012**, *5*, 163–164.
32. Wang, R.F.; Zhang, S.; Shen, W.L. Study on the control of stress concentration in the surrounding rock of deep roadway by slurry injection reinforcement. *Coal Technol.* **2016**, *35*, 65–68.
33. Xing, M. *Research on the Mechanism and Process of Grouting Reinforcement of Fractured Back Mining Roadway Surrounding Rock*; Taiyuan University of Technology: Taiyuan, China, 2015.
34. Jia, S.C. Application of slurry injection reinforcement technology in small coal column dynamic pressure roadway. *Coal* **2007**, *5*, 9–13.
35. Li, W.W. Study on the optimization of reasonable width of coal pillars along the air-excavated roadway in dynamic pressure roadway. *Energy Energy Conserv.* **2019**, *8*, 145–147.

sustainability

MDPI

Article

Research on Gob-Side Entry Retaining Mining of Fully Mechanized Working Face in Steeply Inclined Coal Seam: A Case in Xinqiang Coal Mine

Xuming Zhou [1,2], Haotian Li [2,*], Xuelong Li [1,2,*], Jianwei Wang [3], Jingjing Meng [4], Mingze Li [2] and Chengwei Mei [2]

1 Mine Disaster Prevention and Control-Ministry of State Key Laboratory Breeding Base, Shandong University of Science and Technology, Qingdao 266590, China
2 College of Energy and Mining Engineering, Shandong University of Science and Technology, Qingdao 266590, China
3 Zhengmei Group, Engineering Technology Research Institute, Zhengzhou 450000, China
4 Department of Civil, Environmental and Natural Resources Engineering, Luleå University of Technology, 97187 Luleå, Sweden
* Correspondence: 202183010042@sdust.edu.cn (H.L.); lixlcumt@126.com (X.L.)

Abstract: As a kind of non-coal pillar roadway support technique, gob-side entry retaining is of great significance to improve the production efficiency of a fully mechanized working face. However, the construction of the roadway is often subject to the surrounding rock conditions, the application is mainly concentrated in the nearly horizontal and gently inclined coal seam conditions, and the application in the steeply inclined coal seam conditions is relatively less. This paper is based on the gob-side entry retaining roadway construction of the 58# upper right 3# working face in the fifth district of Xinqiang Coal Mine, and describes the investigation in which we measured the advanced abutment stress, mining stress, and roof stress and analyzed the moving rule of roof. On this basis, in this work, we determined the filling parameters and process and investigated the filling effect from the perspective of the deformation of the filling body and the surrounding rock. The results show that the influence range of the advanced abutment stress in the working face is about 20~25 m, the stress in the upper part is intense, and stress in the middle and lower parts are relaxed. The setting load, the cycle-end resistance, and the time-weighted mean resistance at the upper end of working face along the direction of length are the largest, followed by the middle part, and the lower end is the minimum. When exploiting the steep inclined coal seam, the upper part of the working face is more active than the lower part, and the damaging range of overlaying strata is mainly in the upper part of the goaf. With this research, we established the filling mining process in steeply inclined coal seams and determined the relevant parameters. The gangue cement mortar filling can ensure the deformation of the filling body, the surrounding rock of the roadway is small in the process of roadway retention, and the stress of the filling body is also small, which ensure the successful retention of the roadway. This study verifies the possibility of repair-less exploitation and provides a reference for the popularization and application of the gob-side entry retaining technique in steep inclined coal seam.

Keywords: steeply inclined coal seam; gob-side entry retaining; roadside filling; surrounding rock control

Citation: Zhou, X.; Li, H.; Li, X.; Wang, J.; Meng, J.; Li, M.; Mei, C. Research on Gob-Side Entry Retaining Mining of Fully Mechanized Working Face in Steeply Inclined Coal Seam: A Case in Xinqiang Coal Mine. *Sustainability* **2022**, *14*, 10330. https://doi.org/10.3390/su141610330

Academic Editors: Xiangguo Kong, Dexing Li and Xiaoran Wang

Received: 27 July 2022
Accepted: 15 August 2022
Published: 19 August 2022

1. Introduction

China has rich coal resources, accounting for about 40% of the world's production. For a long time to come, coal will be an irreplaceable dominant energy resource [1,2]. However, coal belongs to non-renewable energy. It has become an urgent problem to select an advanced mining technology to reduce the loss of coal resources [3]. As a kind of non-coal pillar roadway support technique, gob-side entry retaining can reasonably develop existing coal resources and improve the recovery rate of coal resources [4,5]. Its research and promotion are significant to the development of the coal industry.

Although the gob-side entry retaining technique has been widely used worldwide, its application effect is often subject to the surrounding rock conditions of roadway retention, especially the stability of the roof [6–8], and the high cost also restricts its further promotion. To solve these problems, scholars have conducted a lot of research on the strata behavior, applicable conditions, reasonable support forms and new support materials. Through theoretical analysis and numerical simulation, Zhang et al. analyzed the roof cutting stress relief mechanism and the temporal and spatial evolution law of surrounding rock stress, and verified it in the field [9,10]. Bai et al. [11–13] used site investigation data, numerical simulation, and theoretical calculation to establish the dynamic mechanical model of the roof, and studied the roof bending deformation mode of the retained roadway. Fan et al. [14,15] used grouting cable to control roof cutting, built a surrounding rock support model composed of coal wall, support, and goaf gangue, and verified it in the working face. To analyze the subsidence law of roof after gob-side entry retaining in close coal seam, Liu et al. [16,17] calculated the stress distribution of the roof after mining, verified it by numerical simulation, and analyzed the failure process of the roadway roof under closed goaf, which provided guidance for gob-side entry retaining design with similar geological conditions. Chang et al. [18–20] combined numerical simulation and theoretical calculation to study the surrounding rock control effect of high-water fillings with different widths. Kumar et al. [21–23] put forward a construction method for roadway retention by using bag material and verified the feasibility using an indoor mechanical test based on the characteristics of thin coal seam mining. Li et al. [24,25] studied the key parameters of high-water filling material support when the retaining roadway in high-gas thin coal seams and studied the ground stress law by using roof displacement and filling body deformation.

In these studies, the stress state of surrounding rock and the interaction between support and surrounding rock were comprehensively analyzed. On this basis, scholars also carried out various explorations of gob-side entry retaining technology under different geological conditions and coal seam occurrence conditions and accumulated rich experience and lots of application cases. Qin et al. [26,27] used FLAC3D numerical simulation software to analyze the stress distribution law of gob-side entry retaining in thin coal seams and found that the horizontal stress gradually decreased with the advancement of the working face. He et al. [28–31] conducted directional fracturing of the roof with constant resistance and large deformation anchor cables and studied the reasonable support scheme under thick coal seam mining with numerical simulation and field observation. Guo et al. [32–34] studied the construction scheme of gob-side entry retaining technology under thick and hard roof conditions, proposed the pre-splitting blasting method, and conducted experiments on the site. Qi et al. [35–37] proposed a new scheme of gob-side entry retaining technology in three-soft coal seams based on the mechanical and numerical calculation model, including roadway expansion and large section roadway support, which provided theoretical guidance for construction under similar geological conditions. Based on site investigation and numerical analysis, Das et al. [38–40] analyzed the failure mechanism of the soft roof of a roadway in gently inclined coal seams and determined the bolt support parameters after working face advancement.

The above research mainly focuses on the geological conditions of the working face in the near-horizontal and gently inclined coal seam, which further fills the research gaps in the law of strata behavior, roadway support methods, and roadside filling techniques in gob-side entry retaining, and has been successfully applied onsite. However, there are still many problems in the application of gob-side entry retaining technology in medium-thick or thick coal seams with complex occurrence conditions, especially in large dip angle coal seams [41]. Some steeply inclined coal seams have not achieved the expected results in application, which limited the further application of the gob-side entry retaining technique.

The steeply inclined fully mechanized mining face in Xinqiang Coal Mine is rare in China. The research on the gob-side entry retaining technique in the working face needs to solve key problems, such as reasonable retention of the filling body, filling system and method, and stability of the filling body [42–44]. Related research can not only provide guarantees for safe and efficient production under the condition of steeply inclined coal seam, but also enrich and develop the gob-side entry retaining technology. Based on the gob-side entry retaining of the 58# upper right 3# working face in the fifth district of Xinqiang Coal Mine, we measured the advanced abutment stress and roof weighting, and analyzed the roof activity law. On this basis, we determined the filling parameters and process of the steeply inclined working face and investigated the filling effect from the perspective of the stress and deformation of the filling body and the surrounding rock. The theoretical and practical research in this paper can solve the problem of mine production and provide a demonstration for the popularization and application of the gob-side entry retaining technique under special working conditions.

2. Survey and Requirements in the Field

2.1. General Situation of Working Face

Xinqiang Coal Mine belongs to the Qitaihe Mining Bureau, using cross-cut development in a multi-level division of the vertical shaft method; the depth of the two horizons are −50 m and −300 m. There are, altogether, four production districts: 2#, 3#, 5#, and 6#. The location of the mine is shown in Figure 1. The fifth district is located in the eastern part of the mine. There are, altogether, three layers of coal seam, 54#, 58#, and 59#, which are all coking coal and lean coal. The coal seam is roughly east-west strike and south dip. The dip angle is between 27° and 55° with an average of 45°. With the deepening of mining, the dip angle of coal seam gradually increases. The fifth district 58# upper right 3# working face belongs to 2# mining horizon, and the depth is −156.83 m~−190 m, the strike length is 330 m, and the inclination length is 54~60 m. The immediate roof of the coal seam is 12.3 m siltstone, while the upper roof is 2.3 m medium sandstone and the floor is 17.0 m siltstone.

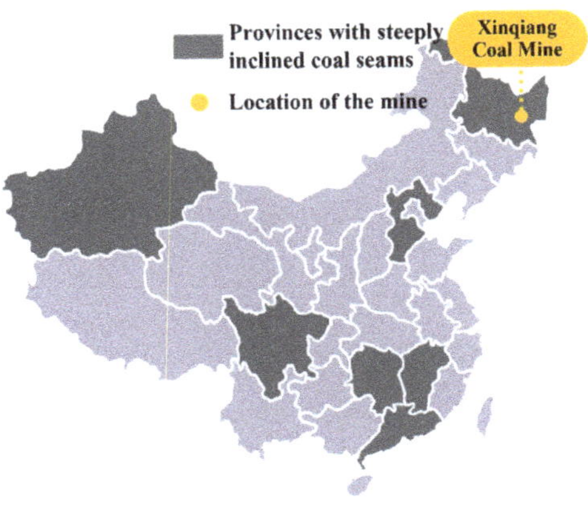

Figure 1. Distribution of steeply inclined coal seams and mine location.

The recoverable reserves of the fifth district 58# upper right 3# working face are 52,000 tons, adopted from the longwall backward mining method. The mining technology used a shearer drum to load coal, an armored face conveyor to transport coal, hydraulic supports to support the roof, and the total caving method to manage the roof. A MG2 × 125 (100)/580 (480) WD drum shearer and a ZY3600/12/28 shield hydraulic support was selected. The upper roadway belongs to 2# working face and the lower roadway is 3#. The gob-side entry retaining roadway located in the 58# upper right 3# working face are shown in Figure 2. The average thickness of coal seam is 2.0 m, the height of the shearer is 1.25~2.58 m, the cutting depth of the drum is 0.8 m, and the effective support height is

1.2~2.8 m. The working face adopts the double drum one-way coal cutting method. The cycle footage is 0.8 m, and the mining height is 2.0 m, which is mined along the coal seam floor. The drum cuts triangular coal in the upward direction after feeds at the upper end, and then cuts coal in the downward direction. After the shearer sweeps the floating coal in the upward direction, the conveyor sequentially moves and supports.

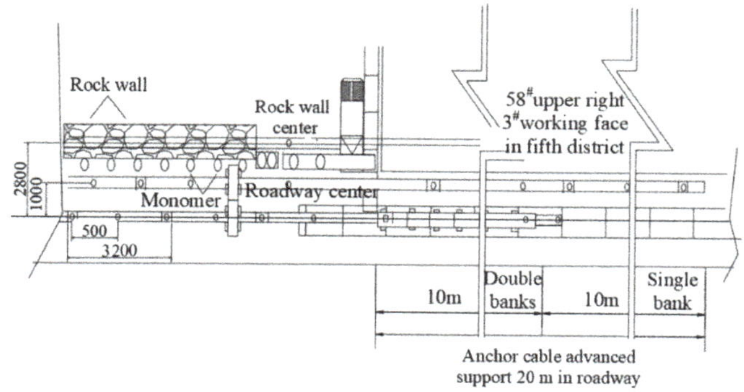

(a)

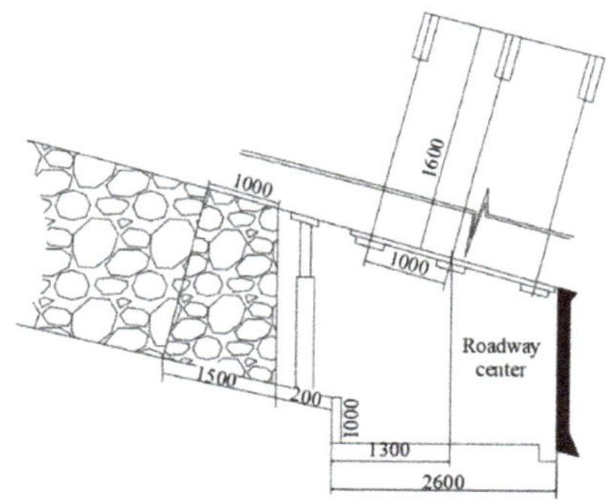

(b)

Figure 2. Views of gob-side entry retaining in 58# upper right 3# working face: (**a**) plan view; (**b**) sectional view.

2.2. Problems and Demands in the Field

Although the gob-side entry retaining technique has been valued and developed in China, it is restricted by the surrounding rock conditions and the stability of the roof is a key factor in construction effects [45]. The 58# upper right 3# working face in the fifth district of Xinqiang Coal Mine adopted the fully mechanized mining method; the advancing speed of the working face was fast, resulting in mining imbalance and a difficult replacement of the working face. To alleviate the production stress caused by mining imbalance, the gob-side entry retaining technology was adopted to retain the transportation roadway of the 3# working face as the return airway of the 4# working faces.

3. Roof Behavior in a Fully Mechanized Working Face in Steeply Inclined Coal Seam

3.1. Law of Advanced Abutment Stress

In the position of 100 m ahead of the working face on the coal side in the lower groove of the 58# upper right 3# working face in the fifth district, we drilled and installed a survey station. There were seven boreholes with depths of 2 m, 5 m, 8 m, 11 m, 14 m, 17 m, and 20 m. The height was about 1.5 m away from the roadway floor and the angle was consistent with the coal seam dip angle (45°). The corresponding length of the stress sensor was installed into the borehole, numbered 1, 2, 3, 4, 5, 6, and 7. Among them, the No. 5 borehole stress sensor has oil leakage due to poor sealing during installation and the stress data has been zero. The other six borehole stress sensors have data, which can be used normally. The layout parameters of drilling and stress sensors are shown in Figure 3.

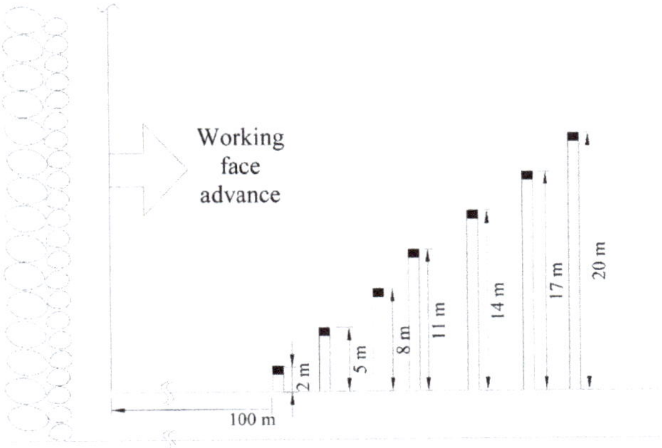

Figure 3. Schematic drawing of stress borehole measuring points.

The advanced abutment stress of the working face increases obviously from the initial stress at the position of about 6 m ahead of the working face, as shown in Figure 4. With the increase of the distance, the stress reaches the maximum at about 10 m from the working face, and the average peak stress is 14 MPa. After reaching the peak value, the stress begins to decrease, but is still higher than the initial stress; the influence continues to the position of 20~25 m from the working face and then restores to the initial stress value gradually. When the measuring point is about 25 m away from the working face, the stress sensor counting begins to change, indicating that the advanced abutment stress of the working face begins to affect the stress. The influence range of the advanced support stress of the working face is about 20~25 m, which is consistent with the empirical value of the end advanced support distance given in the field. Therefore, the end advanced support distance of the working face can meet the requirements of 25 m.

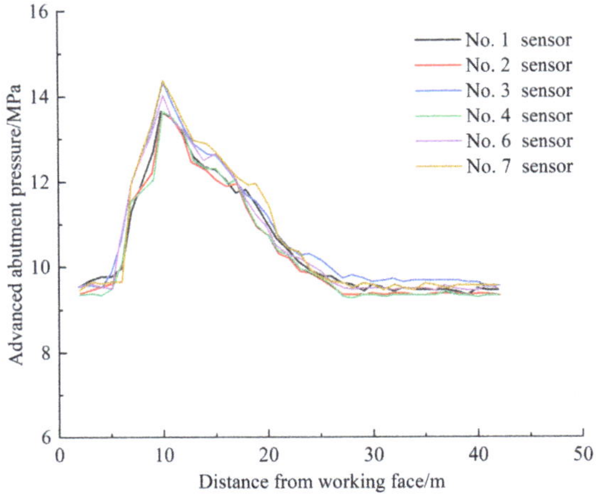

Figure 4. Distribution map of advanced abutment stress in working face.

3.2. Law of Rock Stress Activity

The measuring areas are arranged in the upper, middle, and lower parts along the face length direction of the 58# upper right 3# working face in the fifth district. We selected a support and installed stress gauges in each area to record the time-weighted mean resistance. The working face is a steeply inclined working face and the support quality of the hydraulic support in the working face is closely related to the three supports at the lower end. As a result, the support at the lower end was also selected and the digital stress gauge was installed. There are 60 hydraulic supports in the working face in total, and the stress gauges are installed on the 3#, 20#, and 36# supports to record the support resistance and the strata behavior. The position of the measuring points and the selected support are shown in Figure 5.

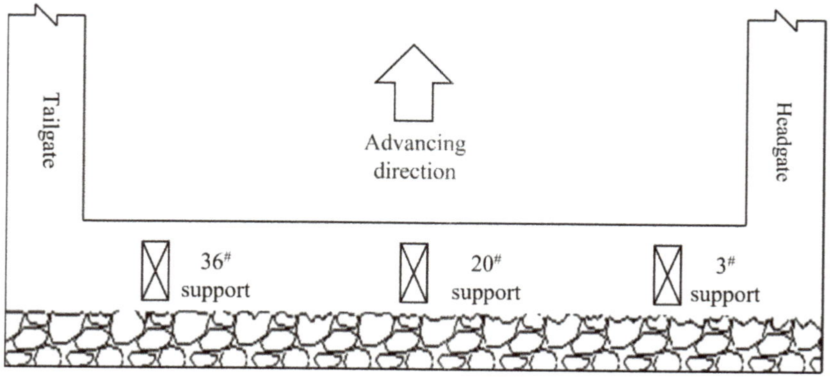

Figure 5. Layout drawing of rock stress measuring point.

The roof weighting data in each part of the working face are shown in Table 1, where K_m represents the end-of-cycle coefficient and K_t represents the time-weighted coefficient. According to Table 1, the 36# support weighted four times, the maximum stress step is 23.3 m, the minimum is 18.6 m, and the average is 20.5 m. The maximum influence range is 2.8 m, the minimum is 2.1 m, and the average is 2.35 m, and the dynamic load coefficients during the weighting period are K_m = 1.29 and K_t = 1.28. Therefore, the stress strength in the upper part of the working surface (36# support) is intense. The 20# support weighted four times, the maximum stress step is 25.6 m, the minimum is 21.6 m, and the average is 23.6 m. The maximum influence range is 3 m, the minimum is 1.2 m, and the average is 2.375 m, and the dynamic load coefficients during the weighting period are K_m = 1.21 and K_t = 1.22. Therefore, the stress strength in central part of working surface (20# support) is mild. The measurement shows that the 3# support weighted three times, the maximum stress step is 34.6 m, the minimum is 30.4 m, and the average is 32.4 m. The maximum influence range is 3.2 m, the minimum is 2.4 m, and the average is 2.8 m, and the dynamic load coefficients during the weighting period are K_m = 1.06 and K_t = 1.04. Therefore, the stress strength in the lower part of the working surface (3# support) is mild. At the same time, it can be seen from Table 1 that the dynamic load coefficients of the upper, middle, and lower roof weighting are 1.28, 1.21, and 1.04, respectively, which are values far less than the dynamic load coefficient of gently inclined coal seam. The average periodic weighting step distance of the upper, middle, and lower parts of the working face is 20.5 m, 23.6 m, and 32.4 m, respectively, which is much larger than that of gently inclined coal seam. The roof stress behavior of coal seam in the working face is mild and the periodic weighting is not obvious.

The roof stress distribution along the face length direction of the steeply inclined fully mechanized working face is different when affected by the mining method, the coal and rock occurrence conditions and support quality, and especially the dip angle of coal seam [46,47]. The stress distribution in the long direction of the working face during the periodic weighting and non-periodic weighting is shown in Figure 6.

Table 1. Roof weighting data in working face.

Support Number	Number of Weighting	Step Distance(m)	Range of Influence (m)	Dynamic Load Coefficient	
				K_m	K_t
36	4	20.5	2.35	1.29	1.28
20	4	23.6	2.37	1.22	1.21
3	3	32.4	2.8	1.06	1.04

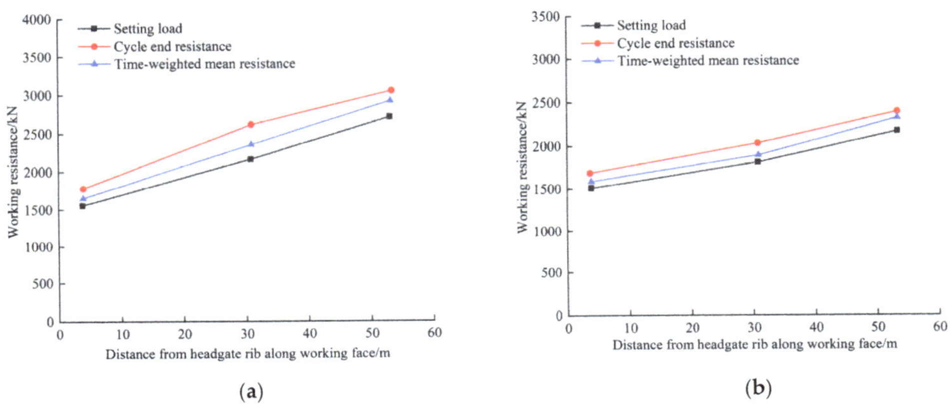

Figure 6. Distribution of roof stress along the length direction of working face: (**a**) weighting period; (**b**) non-weighting period.

It can be seen in Figure 6a that, during the weighting period, the working resistance in the lower end of the working face is the lowest, the setting load P_o is only 1557 kN, the cycle-end resistance P_m is 1782 kN, and the time-weighted mean resistance P_t is 1653 kN. With the increase of distance, the working resistance of the support in the middle of the working face also increases, with an increase of 610 kN, 834 kN, and 707 kN, respectively. With the further increase of the distance from the headgate, the setting load, cycle-end resistance, and time-weighted mean resistance in the end of the working face are the maximum, which are 2713 kN, 3050 kN, and 2922 kN, respectively. During the weighting period, the average setting load of the three supports is 2145 kN, which is 69.39% of the rated setting load (3092 kN). The average cycle-end resistance is 2482 kN, which is 68.94% of the rated working resistance (3600 kN). Therefore, the design makes full use of the rated bearing capacity of the support while ensuring reliability. It is shown in Figure 6b that, during the non-weighting period, the working resistance also shows a trend of increasing with distance. At the lower end of the working face, the setting load, the cycle-end resistance and the time-weighted mean resistance are all minimum values, which are 1506 kN, 1686 kN, and 1589 kN, respectively. With the increase of the distance from the headgate, the working resistance increases slightly, reaching 1811 kN, 2032 kN, and 1894 kN in the middle of the working face. The maximum values of setting load, cycle-end resistance, and time-weighted mean resistance appear at the upper end of the working face, and the resistance values are 2170 kN, 2391 kN, and 2322 kN, respectively, during the same stress period.

3.3. Characteristics of Roof Weighting

The gravity effect of overlying strata in steeply inclined coal seams is different from that in nearly horizontal and gently inclined coal seam, resulting in different roof weighting characteristics.

(1) The roof behavior of steeply inclined coal seam is mild; periodic weighting is not obvious. The weighting strength is significantly lower than that of gently inclined coal seam, while the weighting step distance is significantly larger than that of gently inclined coal seam. The reason is that the component force of the gravity action of overlying strata on the roof of coal seam is much smaller than that of gently inclined coal seam. This also leads to hydraulic support stress in the steeply inclined coal seam working face being small; support performance can meet the requirements, so the support design should mainly consider the support equipment anti-skidding.
(2) The periodic weighting difference of each position in the face length direction of steeply inclined coal seam is large, but the overall weighting difference of the working face is not large, especially the upper and central parts. In comparison, the periodic weighting at the upper end of the working face is stronger than that at the middle of the working face, while the periodic weighting at the middle is stronger than that at the bottom of the working face. The reason is that the dip angle of coal seam is much larger than the natural repose angle of coal and rock mass. When the immediate roof is broken, the weak immediate roof rock mass in the upper part of the goaf rolls down to the lower part under the action of self-weight, forms a state of full enrichment in the top, semi-enrichment in the middle, and hanging in the bottom in the length direction of the working face. Therefore, the periodic weighting step distance and dynamic load coefficient are inconsistent in the length direction of the working face.

4. Gob-Side Entry Retaining and Backfilling Process in Steeply Inclined Coal Seam

4.1. Advantages of the Gob-Side Entry Retaining Technique

The gob-side entry retaining technique has the following advantages for the $58^{\#}$ upper right $3^{\#}$ working face in the fifth district of Xinqiang Coal Mine:

(1) The gob-side roadway is synchronously filled and maintained with the advance of the working face, which ensures the normal transportation of the material, raw coal, and fresh air of the working face. It is also conducive to the safe production of the mine.
(2) The roadways retained along the goaf can be used as the flat roadway in the next working face, which can save the tunneling cost of the roadway and avoid continuous tension and mining imbalance.
(3) The district sublevel coal pillar between working faces is omitted, which greatly improves the recovery rate of coal resources.

4.2. Determination of Filling Parameters

The haulage roadway of the working face is excavated along the coal seam roof. In the process of roadside filling, the retention roadway is strengthened by the support. The roof and floor of the roadway in the goaf side are punched with anchor cables to maintain the stability of the roadside filling body. To prevent the slurry from overflowing into the roadway due to the fluidity of the slurry during the initial grouting of the filling material, the wooden plate is added between the anchor cables and the filling body. The personnel have difficulty entering the goaf side, so the slurry flows naturally in the early stage of backfilling and stops grouting when the size of the filling body is qualified. The physical properties of the slurry determine the natural resting angle of the filling body and the plane section of the filling body is finally trapezoidal, as shown in Figure 7.

According to the geological conditions of the coal seam in the mine, the total stress of overlying strata on the filling zone can be directly calculated according to the multiple rock weight method (estimation method) P_t [48]. The formula is as follows:

$$P_t = ngS\gamma M \cos\alpha \tag{1}$$

The working resistance of filling body P_T is:

$$P_T = \sigma S_t \tag{2}$$

The strength requirements of roadside filling body is:

$$P_T \geq P_t \tag{3}$$

The filling body will not be damaged, the roadway retained along the goaf can be completely preserved [49]. Three formulas stand together and can be derived that:

$$S_t \geq ngS\gamma M \cos\alpha / \sigma \tag{4}$$

In the formula, S_t is the area supported by filling body; n is the rock weight multiple according to the medium stability below the roof consideration, generally taken as 6~8; g is the gravitational acceleration, S is the roof area supported by filling body, which is $S = ab$; a is the length of filling body along the roadway direction, b is the width of the roof supported by the filling body, γ is the density of roof strata, M is the cutting height; α is the dip angle of coal seam, taken as an average of 45°; and σ is the compression of the filling body.

The shape of the filling body is trapezoidal due to the geological conditions of roadside filling in steeply inclined coal seam, so the sum of the upper and lower boundaries of the trapezoid determines that the cross-sectional area of the filling body should not be less than 6 m. Considering that the filling body is incompressible, the filling body supports the overlying strata other than the immediate roof and the main roof. For safety purposes, the designed average width of the filling body is 3 m, as shown in Figure 8.

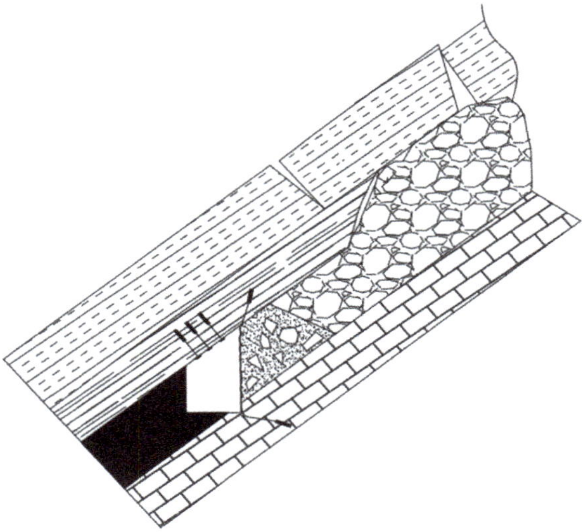

Figure 7. Trapezoid filling body and roof structure.

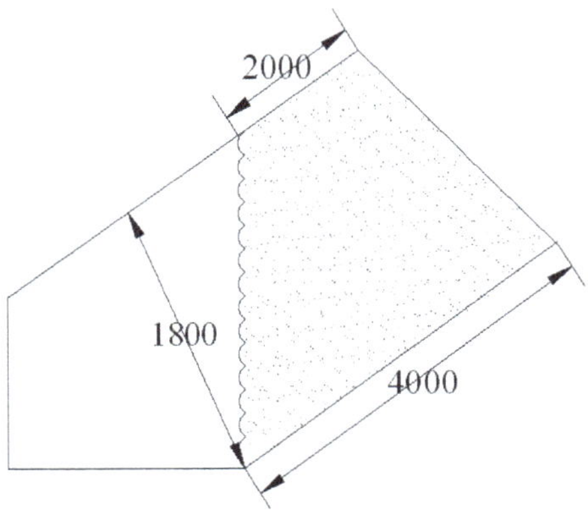

Figure 8. Geometric dimensions of filling body (mm).

4.3. Process of Backfilling

(1) Processing and transportation of filling materials

After drying the sand, stone, and fly ash in the ground workshop, put the materials into the mixer according to the prescribed ratio and mix evenly. Then, load the mixture into the woven bag and transport it to the underground working surface by the mine truck.

(2) Filling and pumping of filling materials

According to the advancing time of the working face, each advancing cycle (0.8 m) of the working face should fill once. The filling pump is used to transport the uniformly stirred material through the pipeline into the filling space isolated by the baffle.

(3) Filling space

Clean the floating coal after the downward cutting of the cutting coal machine in the working face. The support is pushed forward from the bottom to the top, the fourth support that is close to the upper side of the filling body of the working face is retained, and the filling space is formed behind the bent support. The filling bag and the filling partition are laid in the filling space to resist the impact of a small amount of gangue on the goaf side. At the same time, the trapezoidal partition is laid behind the shield beam of the lower end row head frame, the rectangular partition is erected on the roadway wall, and the single pillar is initially blocked. The filling space surrounded by roadway wall, unmoved support (fourth support), bent shield beam, and filling body are shown in Figure 9. Subsequently, the outlet of the filling pipeline enters the filling port of the filling partition and the filling material is injected into the space. At the same time, thick wood plates are placed in advance above the filling bag to buffer the roof subsidence and prevent the roof subsidence of the retaining wall from crushing the wall before it reaches the predetermined strength.

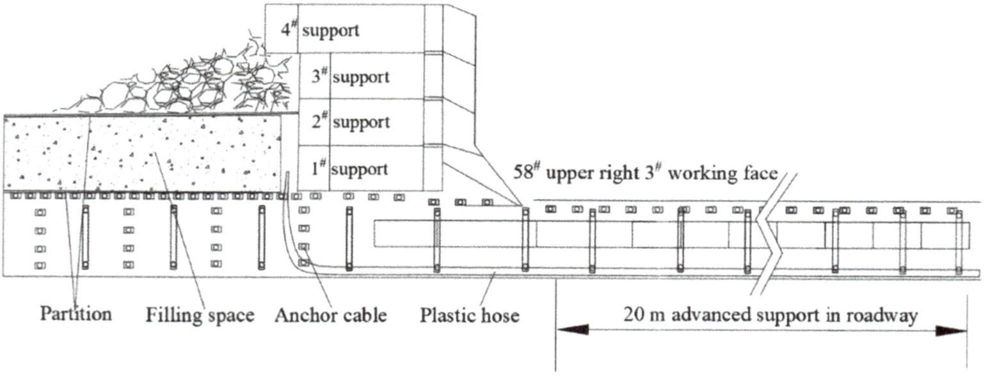

Figure 9. Schematic drawing of filling space.

(4) Complete filling work

At the end of filling, the filling material is no longer filled in the pump, the material in the pipe should be pushed clean with plastic balls, and the full-length pipe should be washed once with water to prevent solidification and blockage of filling materials in the pipe. At the same time, the next filling preparation work should be done. The filling pump is arranged in the transportation roadway. When the working face is pushed over, the filling pump does not need to be moved because it is in the retained roadway, and the filling pipeline only needs to be extended outward.

The sequence and process of gob-side entry retaining in Xinqiang Mine are shown in Figure 10.

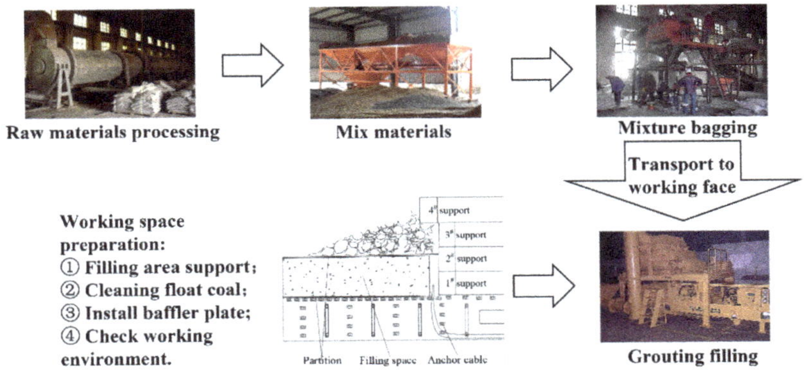

Figure 10. Process of backfilling.

4.4. Filling Effect Monitoring

To examine the filling effect and the strength and deformation characteristics of the roadside filling body, we installed displacement sensors (Figure 11a), borehole stress sensors (Figure 11b), and surrounding rock deformation sensors (Figure 11c) during the process of roadside filling to record experimental data and investigate whether the filling body meets the strength and deformation requirements. The letters A to F in the figures represent sensors in different locations, and the letters with circles represent the corresponding indicators.

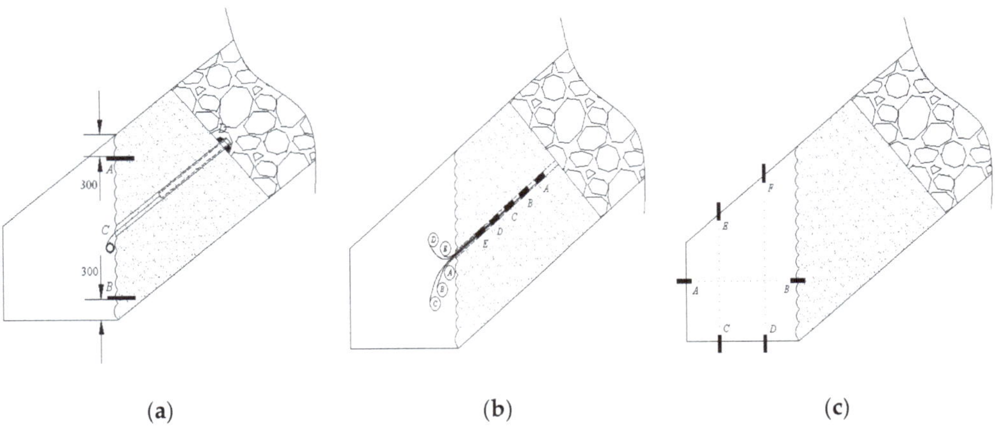

Figure 11. Layout drawing of sensors: (**a**) displacement sensors; (**b**) borehole stress sensors; (**c**) surrounding rock deformation sensors.

5. Investigation of Filling Effect

5.1. Deformation Characteristics of Filling Body

The stress sensors are embedded in the filling body after the roadside filling. The transverse and longitudinal deformation of the filling body obtained by a period of testing is shown in Figure 12.

The deformation of the filling body can be roughly divided into five stages. In the initial construction stage of the filling body, the distance from the working face is generally about 10 m and the deformation of the roadway roof is small. The filling body cannot be actively bear stress and the deformation in two directions is almost zero. In the range of 10~20 m away from the working face, the rock stress begins to appear; the amount and ratio of the roof-to-floor convergence gradually increase and the filling body gradually begins to actively bear stress. Due to the high strength of the cement mortar filling body in the initial solidification stage, the roof subsidence is not large at this stage. In the distance of about 20~40 m from the working face, the overlaying strata of the roof moves violently and the roof-to-floor convergence further increases. The vertical and horizontal deformation of the filling body can reach 57 mm and 31 mm. In the range of 40~80 m away from the working face, the activity of roadway roof strata gradually eases, the deformation of filling body gradually stabilizes, and the longitudinal deformation generally does not exceed 100 mm. At about 80 m away from the working face, the filling body becomes stable and there is basically no vertical and horizontal deformation.

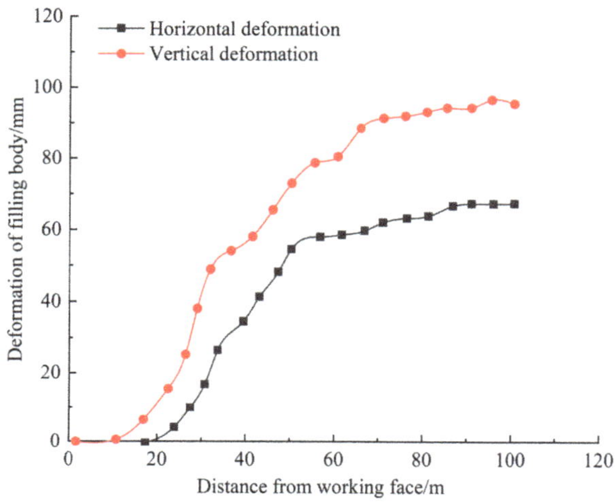

Figure 12. Deformation curves of filling body in different stages.

5.2. Stress Characteristics of the Filling Body

The borehole stress sensors were installed in the roadside backfills in different distances from the working face to measure the loads borne by the backfills at different locations in the backfills, as shown in Figure 13.

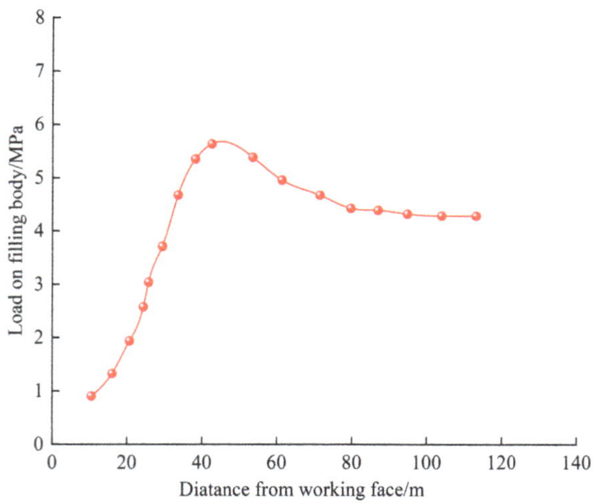

Figure 13. Load variation curve in filling body.

With the increase of the distance from the working face, the trend of the load in the roadside filling body can be roughly divided into three stages. At about 10 m from the working face, the displacement of the roadway roof increases, and the filling body gradually begins to bear a certain load. Subsequently, the load shows an increasing trend with the increase of distance. The fastest increase is in the range of 20~40 m. At about 40 m from the working face, the filling body reaches the ultimate strength, about 6 MPa. In the range of 40~80 m from the working face, the filling body gradually tends toward a plastic yield state under the action of a severe deformation of the roadway roof and the bearing capacity reduces as the load reduces. The deformation of the roadway roof tends to be stable and the load distribution in the roadside filling body tends to be stable and no longer decreases in the range outside of 80 m.

5.3. Deformation Characteristics of Roadway-Surrounding Rock

When the working face advances, the roof and floor of the retained roadway are always in the state of spatial motion. The activity intensity of the roadway roof and floor is related to the distance from the working face, which leads to a different roof convergence. Figure 14a shows the relationship between the displacement of the roof and floor of the retained roadway behind the working face and the distance from the working face. Figure 14b shows the relationship between the displacement of the two sides of the reserved roadway and the distance from the working face.

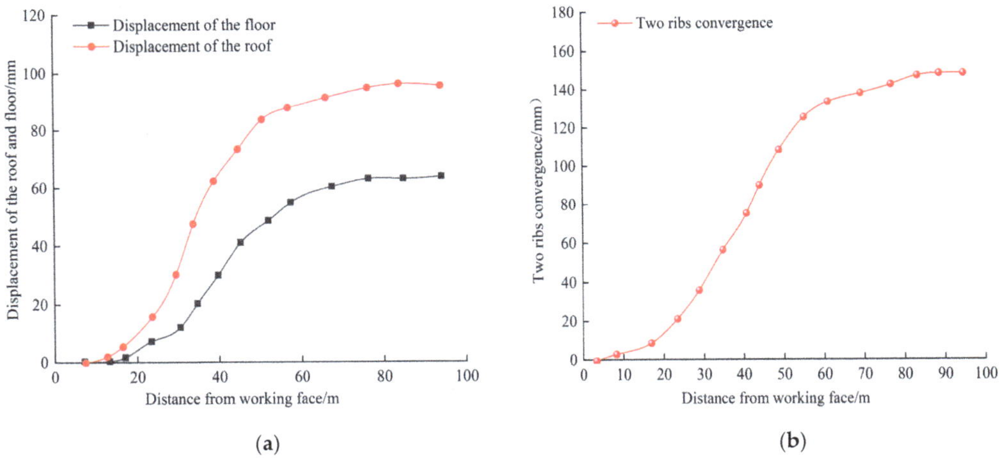

Figure 14. Variation curve of roadway-surrounding rock: (**a**) roof-to-floor convergence; (**b**) two ribs convergence.

It can be seen from the measurement results that the displacement of the roof-to-floor convergence and two ribs of the retained roadway slightly changes within the range of 10~20 m from the working face. In the range of 20~50 m from the working face, the roof-to-floor convergence and the two ribs' displacement increase rapidly, and the displacement is approximately linear with the length of the working face. At about 50 m from the working face, the sum of the two ribs' displacement reaches about 116 mm. In the range of 50~80 m from the working face, due to the stable motion state of the surrounding rock, the displacement of the two ribs of the roadway numerically increases, but the increase rate is less than the previous stage. At about 80 m away from the working face, the surrounding rock movement becomes stable and the roadside filling body bears a stable load. The roof-to-floor convergence is less than 100 mm, and the displacement of the two ribs is not more than 140 mm.

6. Conclusions

To study the gob-side entry retaining technique in steeply inclined coal seam, we measured the advanced abutment stress, mining stress, and roof stress, and analyzed the moving rule of roof based on the gob-side entry retaining roadway construction of the 58# upper right 3# working face in the fifth district of Xinqiang Coal Mine. On this basis, in this work, we determined the filling parameters and process of the steeply inclined working face and investigated the filling effect from the perspective of the stress and deformation of the filling body and the surrounding rock. The results show that:

(1) The influence range of the advanced abutment stress of the working face is about 20~25 m, which is consistent with the empirical value of the advanced support distance in the field. The stress intensity in the upper part of the working face is intense; the stress intensity in the middle and lower part of the working face is mild.
(2) During the period of periodic weighting or not, the setting load, the cycle-end resistance, and the time-weighted mean resistance at the upper end of the working face along the direction of length are the largest, followed by the middle part, and the lower end is the minimum. The increase is greater during the weighting period than during the non-stress period.
(3) The overlaying strata in the upper part of the working face is more active than the lower part when mining steeply inclined coal seam. The range of rock failure is mainly in the upper part of the goaf, the upper roof has a tensile fracture due to sliding instability, and the lower part is unevenly filled by falling gangue. Due to the rotary instability stress on the filled gangue, the floor heave phenomenon is obvious.
(4) The filling mining process of steeply inclined coal seam is established. The multiple estimation method was used to determine and calculate the relevant parameters of trapezoidal backfill and the field observation was carried out. The roadside filling body constructed by gangue cement mortar material has a fast setting speed and strong bearing capacity. After filling, the deformation of the filling body and the roadway-surrounding rock is small and the stress of the filling body is small, indicating that the filling effect is good. This study verifies the possibility of repair-less exploitation and provides reference for mines with the same conditions.

Author Contributions: X.Z., H.L. and X.L. proposed the research. J.W. and J.M. prepared figures and tables, and interpreted the structural data. M.L. and C.M. developed the main ideas. All co-authors actively contributed to the manuscript with comments, ideas, and suggestions. All authors have read and agreed to the published version of the manuscript.

Funding: This study was financially supported by the National Natural Science Foundation of China (52104204), Natural Science Foundation of Shandong Province (ZR2021QE170).

Institutional Review Board Statement: Not applicable.

Informed Consent Statement: Not applicable.

Data Availability Statement: The datasets used and/or analyzed during the current study are available from the corresponding author on reasonable request.

Acknowledgments: This work was supported by the National Natural Science Foundation of China (52104204) and the Natural Science Foundation of Shandong Province (ZR2021QE170). The authors also thank the editor and anonymous reviewers for their useful advice.

Conflicts of Interest: The authors declare no potential conflict of interest with respect to the research, authorship, and/or publication of this paper.

References

1. Mohr, S.H.; Evans, G.M. Forecasting coal production until 2100. *Fuel* **2009**, *88*, 2059–2067. [CrossRef]
2. Bai, X.F.; Ding, H.; Lian, J.J.; Ma, D.; Yang, X.Y.; Sun, N.X.; Xue, W.L.; Chang, Y.J. Coal production in China: Past, present, and future projections. *Int. Geol. Rev.* **2018**, *60*, 535–547. [CrossRef]

3. Wu, R.; He, Q.Y.; Oh, J.; Li, Z.C.; Zhang, C.G. A new gob-side entry layout method for two-entry longwall systems. *Energies* **2018**, *11*, 2084. [CrossRef]
4. Wang, Q.; He, M.C.; Yang, J.; Gao, H.K.; Jiang, B.; Yu, H.C. Study of a no-pillar mining technique with automatically formed gob-side entry retaining for longwall mining in coal mines. *Int. J. Rock. Mech. Min.* **2018**, *110*, 1–8. [CrossRef]
5. Ma, Z.M.; Wang, J.; He, M.C.; Gao, Y.B.; Hu, J.Z.; Wang, Q. Key technologies and application test of an innovative noncoal pillar mining approach: A case study. *Energies* **2018**, *11*, 2853. [CrossRef]
6. Kan, J.G.; Zhang, N.; Wu, J.K.; Wu, H. Effect of main roofs fracture position on the secondary gob-side entry retaining stability. *Disaster Adv.* **2013**, *6*, 189–199.
7. Zhou, X.M.; Wang, S.; Li, X.L. Research on theory and technology of floor heave control in semi-coal rock roadway: Taking Longhu coal mine in Qitaihe mining area as an example. *Lithosphere* **2022**, *11*, 3810988. [CrossRef]
8. Luan, H.J.; Jiang, Y.J.; Zhou, L.J.; Lin, H.L. Stability control and quick retaining technology of gob-side entry: A case study. *Adv. Civ. Eng.* **2018**, *2018*, 7357320. [CrossRef]
9. Zhang, Y.; Xu, H.C.; Song, P.; Sun, X.M.; He, M.C.; Guo, Z.B. Stress evolution law of surrounding rock with gob-side entry retaining by roof cutting and stress release in composite roof. *Adv. Mater. Sci. Eng.* **2020**, *2020*, 1961680. [CrossRef]
10. Islavath, S.R.; Deb, D.; Kumar, H. Numerical analysis of a longwall mining cycle and development of a composite longwall index. *Int. J. Rock. Mech. Min.* **2016**, *89*, 43–54. [CrossRef]
11. Bai, J.B.; Shen, W.L.; Guo, G.L.; Wang, X.Y.; Yu, Y. Roof deformation, failure characteristics, and preventive techniques of gob-side entry driving heading adjacent to the advancing working face. *Rock. Mech. Rock. Eng.* **2015**, *48*, 2447–2458. [CrossRef]
12. Guo, P.F.; Zhao, Y.X.; Yuan, Y.D.; Ye, K.K.; Zhang, H.J.; Gaol, Q. Gob-side entry stability analysis by global-finite and local-discrete modeling approach. *Geomech. Eng.* **2021**, *27*, 391–404.
13. Liu, S.M.; Li, X.L.; Wang, D.K.; Zhang, D.M. Investigations on the mechanism of the microstructural evolution of different coal ranks under liquid nitrogen cold soaking. *Energ. Source Part A* **2020**, *10*, 1–17. [CrossRef]
14. Fan, D.Y.; Liu, X.S.; Tan, Y.L.; Yan, L.; Song, S.L.; Ning, J.G. An innovative approach for gob-side entry retaining in deep coal mines: A case study. *Energy Sci. Eng.* **2019**, *7*, 2321–2335. [CrossRef]
15. Kumar, R.; Mishra, A.K.; Singh, A.K.; Ram, S.; Singh, R. Depillaring of total thickness of a thick coal seam in single lift using cable bolts: A case study. *Int. J. Min. Sci. Technol.* **2016**, *26*, 223–233. [CrossRef]
16. Liu, H.Y.; Zhang, B.Y.; Li, X.L.; Liu, C.W.; Wang, C.; Wang, F.; Chen, D.Y. Research on roof damage mechanism and control technology of gob-side entry retaining under close distance gob. *Eng. Fail. Anal.* **2022**, *138*, 106331. [CrossRef]
17. Li, X.L.; Chen, S.J.; Wang, S.; Zhao, M.; Liu, H. Study on in situ stress distribution law of the deep mine taking Linyi Mining area as an example. *Adv. Mater. Sci. Eng.* **2021**, *2021*, 5594181. [CrossRef]
18. Li, T.; Chen, G.B.; Qin, Z.C.; Li, Q.H.; Cao, B.; Liu, Y.L. The gob-side entry retaining with the high-water filling material in Xin'an Coal Mine. *Geomech. Eng.* **2020**, *22*, 541–552.
19. Xie, S.R.; Wu, X.Y.; Chen, D.D.; Sun, Y.H.; Zeng, J.C.; Zhang, Q.; Ji, C.W.; Cheng, Q.; Wang, E.; Ren, Y.X. Automatic roadway backfilling of caving gangue for cutting roofs by combined support on gob-side entry retaining with no-pillars: A case study. *Adv. Mater. Sci. Eng.* **2019**, *2019*, 8736103. [CrossRef]
20. Chang, Q.L.; Tang, W.J.; Xu, Y.; Zhou, H.Q. Research on the width of filling body in gob-side entry retaining with high-water materials. *Int. J. Min. Sci. Technol.* **2018**, *28*, 519–524. [CrossRef]
21. Du, Z.W.; Chen, S.J.; Ma, J.B.; Guo, Z.P.; Yin, D.W. Gob-side entry retaining involving bag filling material for support wall construction. *Sustainability* **2020**, *12*, 6353. [CrossRef]
22. Luan, H.J.; Jiang, Y.J.; Lin, H.L.; Wang, Y.H. A new thin seam backfill mining technology and its application. *Energies* **2017**, *10*, 2023. [CrossRef]
23. Li, F.X.; Yin, D.W.; Zhu, C.; Wang, F.; Jiang, N.; Zhang, Z. Effects of Kaolin addition on mechanical properties for cemented coal gangue-fly ash backfill under uniaxial loading. *Energies* **2021**, *14*, 3693. [CrossRef]
24. Li, H.; Zu, H.D.; Zhang, K.L.; Qian, J.F. Study on filling support design and ground stress monitoring scheme for gob-side entry retention by roof cutting and stress relief in high-gas thin coal seam. *Int. J. Environ. Res. Public Health* **2022**, *19*, 3913. [CrossRef] [PubMed]
25. Ma, Z.Q.; Zhang, D.Y.; Cao, Y.Q.; Yang, W.; Xu, B. Study of key technology of gob-side entry retention in a high gas outburst coal seam in the Karst Mountain Area. *Energies* **2022**, *15*, 4161. [CrossRef]
26. Qin, T.; Ren, K.; Jiang, C.; Duan, Y.W.; Liu, Z.; Wang, L. Distribution law of mining stress of the gob-side entry retaining in deep mining thin coal seam. *Adv. Civ. Eng.* **2021**, *2021*, 5589948. [CrossRef]
27. Tian, Z.J.; Zhang, Z.Z.; Deng, M.; Yan, S.; Bai, J.B. Gob-side entry retained with soft roof, floor, and seam in thin coal seams: A case study. *Sustainability* **2020**, *12*, 1197. [CrossRef]
28. He, M.C.; Gao, Y.B.; Yang, J.; Gong, W.L. An innovative approach for gob-side entry retaining in thick coal seam longwall mining. *Energies* **2017**, *10*, 1785. [CrossRef]
29. Li, X.L.; Chen, S.J.; Liu, S.M.; Li, Z.H. AE waveform characteristics of rock mass under uniaxial loading based on Hilbert-Huang transform. *J. Cent. South Univ.* **2021**, *28*, 1843–1856. [CrossRef]
30. Hu, J.Z.; He, M.C.; Wang, J.; Ma, Z.M.; Wang, Y.J.; Zhang, X.Y. Key parameters of roof cutting of gob-side entry retaining in a deep inclined thick coal seam with hard roof. *Energies* **2019**, *12*, 934. [CrossRef]

31. Elik, A.; Zelik, Y. Investigation of the efficiency of longwall top coal caving method applied by forming a face in horizontal thickness of the seam in steeply inclined thick coal seams by using a physical model. *Int. J. Rock. Mech. Min.* **2021**, *148*, 104917.
32. Liu, X.Y.; He, M.C.; Wang, J.; Ma, Z.M. Research on non-pillar coal mining for thick and hard conglomerate roof. *Energies* **2021**, *14*, 299. [CrossRef]
33. Guo, Z.C.; Li, S.G. Research and application of key technology of gob-side entry retaining in working face with hard roof. *Fresenius Environ. Bull.* **2022**, *31*, 5967–5972.
34. Qiu, L.M.; Zhu, Y.; Song, D.Z.; He, X.Q.; Wang, W.X.; Liu, Y.; Xiao, Y.Z.; Wei, M.H.; Yin, S.; Liu, Q. Study on the nonlinear characteristics of EMR and AE during coal splitting tests. *Minerals* **2022**, *12*, 108. [CrossRef]
35. Shen, J.C.; Zhang, Y. Theory and application of gob-side entry retaining in thick three-soft coal seam. *Geofluids* **2021**, *2021*, 6157174. [CrossRef]
36. Qi, F.Z.; Ma, Z.G.; Li, N.; Li, B.; Wang, Z.L.; Ma, W.X. Numerical analysis of the width design of a protective pillar in high-stress roadway: A case study. *Adv. Civ. Eng.* **2021**, *2021*, 5533364. [CrossRef]
37. Qiu, L.M.; Liu, Z.T.; Wang, E.Y.; He, X.Q.; Feng, J.J.; Li, B.L. Early-warning of rock burst in coal mine by low-frequency electromagnetic radiation. *Eng. Geol.* **2020**, *279*, 105755. [CrossRef]
38. Kong, X.G.; He, D.; Liu, X.F.; Wang, E.Y.; Li, S.G.; Liu, T.; Ji, P.F.; Deng, D.Y.; Yang, S.R. Strain characteristics and energy dissipation laws of gas-bearing coal during impact fracture process. *Energy* **2022**, *242*, 123028. [CrossRef]
39. Sun, Y.T.; Bi, R.Y.; Sun, J.B.; Zhang, J.F.; Taherdangkoo, R.; Huang, J.D.; Li, G.C. Stability of roadway along hard roof goaf by stress relief technique in deep mines: A theoretical, numerical and field study. *Geomech. Geophys. Geo-Energy Geo-Resour.* **2022**, *8*, 45. [CrossRef]
40. Das, A.J.; Mandal, P.K.; Bhattacharjee, R.; Tiwari, S.; Kushwaha, A.; Roy, L. Evaluation of stability of underground workings for exploitation of an inclined coal seam by the ubiquitous joint model. *Int. J. Rock. Mech. Min.* **2017**, *93*, 101–114. [CrossRef]
41. Sun, X.; Zhao, C.; Li, G.; Zhang, B.; Wang, J.; Cai, F. Physical model experiment and numerical analysis on innovative gob-side entry retaining with thick and hard roofs. *Arab. J. Geosci.* **2020**, *13*, 1245. [CrossRef]
42. Wang, Y.L.; Tang, J.X.; Dai, Z.Y.; Yi, T.; Li, X.Y. Flexible roadway protection technology in medium-thickness coal seam with large dip angle. *Energy Sources Part A-Recovery Util. Environ. Eff.* **2019**, *41*, 3085–3102. [CrossRef]
43. Lv, W.Y.; Guo, K.; Wu, Y.P.; Tan, Y.; Ding, K.; Li, B. Compression characteristics of local filling gangue in steeply dipping coal seam. *Energy Explor. Exploit.* **2022**, *40*, 1131–1150. [CrossRef]
44. Cao, W.X.; Liu, H.L.; Hang, Y.J.; Wang, H.Z.; Li, G.D. Similarity simulation on the movement characteristics of surrounding rock and floor stress distribution for large-dip coal seam. *Sensors* **2022**, *22*, 2761. [CrossRef]
45. Kong, X.G.; Li, S.G.; Wang, E.Y.; Wang, X.; Zhou, Y.X.; Ji, P.F.; Shuang, H.Q.; Li, S.R.; Wei, Z.Y. Experimental and numerical investigations on dynamic mechanical responses and failure process of gas-bearing coal under impact load. *Soil Dyn. Earthq. Eng.* **2021**, *142*, 106579. [CrossRef]
46. He, S.Q.; Song, D.Z.; He, X.Q.; Chen, J.Q.; Ren, T.; Li, Z.L.; Qiu, L.M. Coupled mechanism of compression and prying-induced rock burst in steeply inclined coal seams and principles for its prevention. *Tunn. Undergr. Sp. Technol.* **2020**, *98*, 103327. [CrossRef]
47. Wang, H.W.; Wu, Y.P.; Liu, M.F.; Jiao, J.Q.; Luo, S.H. Roof-breaking mechanism and stress-evolution characteristics in partial backfill mining of steeply inclined seams. *Geomat. Nat. Hazards Risk* **2020**, *11*, 2006–2035. [CrossRef]
48. Jiang, L.L.; Yang, Z.Q.; Li, G.W. Research on the reasonable coal pillar width and surrounding rock supporting optimization of gob-side entry under inclined seam condition. *Adv. Civ. Eng.* **2021**, *2021*, 7145821. [CrossRef]
49. Cui, F.; Dong, S.; Lai, X.P.; Chen, J.Q.; Jia, C.; Zhang, T.H. Study on the fracture law of inclined hard roof and surrounding rock control of mining roadway in longwall mining face. *Energies* **2020**, *13*, 5344. [CrossRef]

sustainability

MDPI

Article

Spatiotemporal Evolution Characteristics of Apparent Resistivity and Its Response Correlation with Acoustic Emission of Coal under Multi-Step Loading

Xinyu Wang [1], Guoqing Zhu [1], Deqiang Cheng [2], Bin Miao [3,*], Fanbao Chen [1] and He Tian [1]

1 School of Safety Engineering, China University of Mining and Technology, Xuzhou 221116, China
2 School of Information and Control Engineering, China University of Mining and Technology, Xuzhou 221116, China
3 School of Resources and Geosciences, China University of Mining and Technology, Xuzhou 221116, China
* Correspondence: zymb@cumt.edu.cn

Abstract: To understand the early warning signs of damage during the coal mass deformation process, an integrated monitoring procedure was designed that combines the apparent resistivity (AR) and acoustic emission (AE) of coal damage under multi-step loading. The spatiotemporal response characteristics of AR and the time-varying evolution of AE were studied and the varying correlation between the two was discussed. Additionally, the macro–microscopic mechanics of the AR response during the coal deformation process was explored. The results show that the AR and AE signals corresponded well with the applied load during the coal deformation process. In the early loading stage, variations in AR and AE signals were not apparent. As the applied load increased, the high-resistance area of AR increased and the AE signals became active. The local variation characteristics of AR could be used to indirectly invert the internal structure of the coal samples. The electrical variation in the loaded coal was mainly controlled by the conductive surface of cracks. The acoustic and resistivity methods can strongly complement the spatial and temporal dimensions of early warning systems for disasters. The AE technique can continuously monitor a test area for abnormal occurrences in the engineering site, and AR tomography images that are obtained can be used to locate inversions in the source coal in order to take pre-emptive action before disaster occurs. This research can provide new ideas for monitoring and early warning systems for coal and rock dynamic disasters.

Keywords: multi-step load; acoustic emission; resistivity method; coal fracture; spatiotemporal evolution; early disaster warning

Citation: Wang, X.; Zhu, G.; Cheng, D.; Miao, B.; Chen, F.; Tian, H. Spatiotemporal Evolution Characteristics of Apparent Resistivity and Its Response Correlation with Acoustic Emission of Coal under Multi-Step Loading. *Sustainability* **2022**, *14*, 10061. https://doi.org/10.3390/su141610061

Academic Editors: Xiangguo Kong, Dexing Li and Xiaoran Wang

Received: 16 July 2022
Accepted: 4 August 2022
Published: 14 August 2022

1. Introduction

With the increasing intensity and depth of coal mining in China, coal–rock dynamic disasters, such as rock burst, roof fall, and coal and gas outburst, are becoming more serious [1]. A coal–rock dynamic disaster is a complex process in which internal microfractures develop and multiply under external load perturbation until visible damage is produced [2–4]. Therefore, studying the evolution characteristics of coal mass deformation and rupture in different loading stages and accurately capturing the early warning signs of damage could provide a theoretical and methodological basis for disaster prediction and prevention at engineering sites. At the same time, the process of coal mass rupture produces a variety of physical effects, such as changes in AE, electrical signals, and other physical information [5–7]. In addition, joint monitoring and multi-source information analysis by various means could help us more thoroughly understand the early warning signs of coal mass deformation and rupture and solve the limitation that single sources of information cannot accurately identify disaster sources.

AE is a phenomenon in which tiny cracks that exist inside a material close under external loading and energy is released in the form of elastic waves when new cracks develop, expand, and penetrate [8]. Recently, many scholars have conducted extensive research on AE signals during the coal mass loading process. Jia et al. revealed the spatial and temporal evolution characteristics of coal damage during this process by studying the physical behavior and AE signals of coal at different depths under triaxial compression conditions [9]. Li et al. adopted the Hilbert–Huang transform (HHT) method to examine the detailed structural characteristics of coal masses that were damaged at different loading stages by analyzing AE waveform characteristics [10]. Dou et al. conducted uniaxial compression tests on sandstone samples that involved pre-cracking at different dip angles and investigated the effect of the crack dip angle on the fracture mechanism using AE and scanning electron microscopy (SEM) [11]. Zhang et al. analyzed the AE signals of coal samples under true triaxial loading and unloading during the process of deformation and failure and found that the multifractal characteristics of AE were able to reflect the degree of stress and damage in the coal samples [12]. Feng et al. studied the fracture characteristics of coal mass under dynamic loading and the AE signal response and found that the dynamic intensity was bimodal. In addition, the AE peak counts and energy increased linearly with the impact velocity but decreased with increased axial static load [13]. Kong et al. studied the variance and autocorrelation coefficients of AE signal sequences by introducing critical moderation theory [14]. Li et al. conducted uniaxial compression experiments on sandstone samples with different water content and collected AE signals. They found that the water content significantly affected the mechanical behavior and AE signals of the sandstone samples [15]. Through AE monitoring, Wang et al. conducted quantitative inversion of the source mechanism of rock fracture and obtained the spatiotemporal evolution law of the source parameters, such as microcracking type, fracture size, azimuth orientation, dissipated energy, etc. [16]. The passive and real-time AE monitoring technique has irreplaceable advantages in coal and rock fracture monitoring.

Resistivity is one of the important physical parameters of coal mass, and there have been many in-depth studies on the variation law of the resistivity of coal under different conditions. Jiang et al. studied the variation in the resistivity of quartz sandstone and limestone under uniaxial loading conditions and introduced Maxwell's conductivity formula to analyze those variations during experiments [17]. Li et al. derived and tested an analytical expression for the damage variables of rock samples based on resistivity characterization and obtained the damage evolution equation for typical rock samples. They also proposed discrimination criteria and pre-damage characteristics for the damage state of rock samples [18]. Jia et al. studied the variation in resistivity response during the process of crack development, expansion, and merging in saturated green sandstone under uniaxial loading and proposed a method for fracture potential prediction based on resistivity anisotropy [19]. Liu et al. investigated the AR of coal samples during the process of uniaxial loading. The correspondence between AR and stress was also analyzed to reveal the mechanism of AR response to coal damage [20]. Xu et al. conducted Brazilian splitting tests on two groups of vertical and parallel limestone samples and obtained the resistivity law for two loading modes. Their test results proved the evolution law for resistivity damage variables [21]. The above results all show that the resistivity method can effectively reflect the characteristics of deformation and rupture during coal loading, which is one of the main methods used for active geophysical exploration in mines.

The above research shows that the study of resistivity change in loaded coal mass has mainly been limited to the temporal variation characteristics of resistivity, and the spatial variation characteristics during the loading process have been less studied. In addition, it is difficult to monitor electrical signals throughout the whole process of coal mass loading in real time due to the limitation of sampling frequency of the measuring instrument. Moreover, while the AE technique can monitor the test area in real time and is sensitive to the moment of coal mass rupture, compared with the active detection technique, the main disadvantage of AE for dynamic disaster prediction is that the advance time is not long enough. In other words, when the AE signal is detected, the damage or fracture has already occurred, while an active detection technique such as resistivity can predict the risk in advance, before the disaster occurs, and it is difficult to reflect the internal evolution state of the coal mass during the loading process in all aspects. The two observation methods, AE and resistivity, both have their advantages, but the correlation between them in the spatial and temporal dimensions of coal fracture warning still needs to be further clarified.

This paper integrates the DC resistivity method and the AE technique to test the AR and AE signal response laws during the multi-step coal mass loading process. The focus is on analyzing the spatial and temporal evolution characteristics of the AR of the coal mass to reveal the change mechanism from the microscopic and macroscopic perspectives. By exploring the spatial and temporal correlation of the AR-AE response to the coal damage process, a new idea for dynamic disaster warning based on the integration of passive monitoring and active detection is proposed. The research results provide an important theoretical foundation and experimental basis for further clarifying the evolution of coal rupture damage and enriching the technical system of dynamic disaster monitoring, prevention, and control.

2. Experimental Preparation and System

2.1. Sample Preparation

The coal samples were taken from the 11061-machine tunnel of the Fangshan coal mine in Yuzhou City, Henan Province, China. The coal samples were processed by a method recommended by the International Society for Rock Mechanics and Rock Engineering (ISRM). Standard specimens with a diameter of 50 mm and height of 100 mm were used, and non-parallelism and non-perpendicularity were controlled to within ±0.05 mm. A total of 18 coal samples were used, numbered sequentially from M-1 to M-18. To facilitate the analysis, the basic parameters of the tested coal mass were analyzed industrially, and the results are shown in Table 1. The tested coal was relatively soft, with the following properties: density $\rho = 980\ kg/m^3$, Poisson's ratio $\nu = 0.26$, Young's modulus E = 1.2 GPa, and uniaxial compressive strength (UCS) = 4.2 MPa. The *p*-wave and S-wave velocity were 1180 and 670 m/s, respectively.

Table 1. Basic parameters of coal samples.

Parameters	Moisture (M_{ad}/%)	Ash Content (A_d/%)	Volatile Fraction (V_{daf}/%)
Value	1.15	7.32	12.20

2.2. Experimental System

The experimental coal mass AR–AE system mainly included four parts: loading system, shielding system, AE acquisition system, and electrical parameter acquisition system, as shown in Figure 1.

The experimental loading system consisted of a SANS-Y4306 electro-hydraulic servo press platform. The platform was mainly composed of a servo press, a DCS loading control box, and a PowerTest V 3.3 control program, with maximum load capacity of 3000 kN and data acquisition frequency up to 20 Hz. Three loading methods could be realized: force control, displacement control, and constant load control.

The experimental shielding system consisted of an AFGP-II high-efficiency shielding room, with a shielding range of 50 MHz–1 GHz ≥ 110 dB, 300 kHz ≥ 110 dB, 100 kHz ≥ 100 dB, and 14 kHz ≥ 80 dB to meet the shielding requirements of the experiment. In order to avoid the conduction of excitation current to the press during electrical detection, insulating paper was placed at the upper and lower ends of the coal mass prior to the experiment to avoid direct contact with the press and ensure the accuracy of the test results.

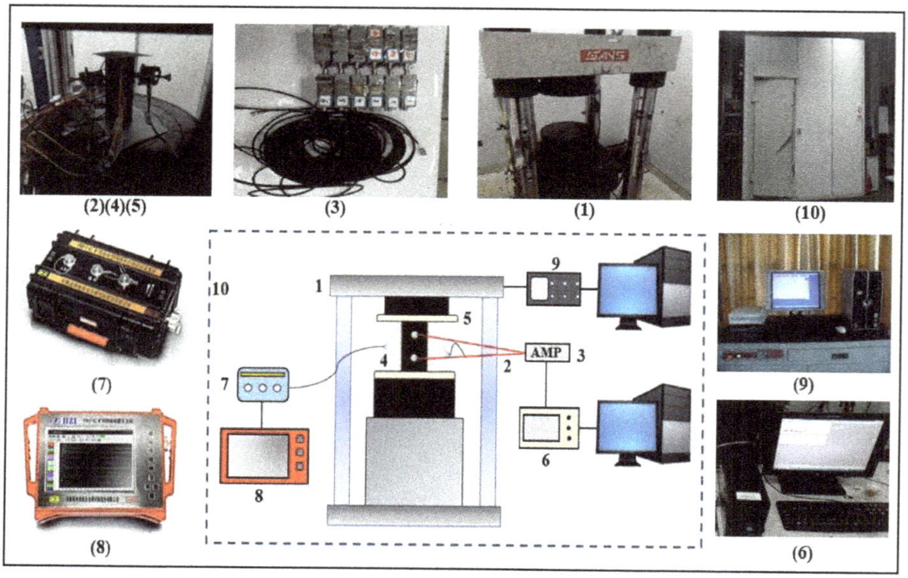

Figure 1. Experimental system: (1) press head; (2) AE probe; (3) preamplifier; (4) electrodes; (5) insulating paper; (6) AE data acquisition system; (7) acquisition base station; (8) electrical method host; (9) press control system; (10) shielding room.

The AE acquisition system was a PAC-II system from Physical Acoustics, USA. The system consists of an A/D converter module, pre- and gain amplifiers, and waveform processing module, which can realize efficient acoustic signal acquisition using multiple channels (up to 32) by computer. In order to ensure that there is no interference between channels, a preamplifier is connected to each channel through a shielded line. Two Nano-30 miniature AE sensors (diameter = 8 mm; height = 8 mm) with a resonant response at 300 kHz were used in the experiments to measure the time-varying parameters (i.e., AE counts) during the deformation process of loaded coal specimens. The AE sensors are placed in the special fixture, and the fixture chuck is glued to the surface of the sample by hot-melt adhesive.

The electrical parameter acquisition system consists of a Huizhou Institute YBD-11Z type network parallel electrical method host, acquisition base station, and patch electrodes. The signal acquisition of the instrument can reach up to 64 channels, the measuring voltage accuracy is ±0.3 mV, and the sampling interval can be set from 1 to 20 ms. The pure copper patch electrode is 7 mm long, 4 mm wide, and 0.5 mm thick, and is fully coupled with the sample by conductive epoxy adhesive and connected to the acquisition base station by silver-plated shielding wire.

2.3. Experimental Schemes and Procedures

In this study, uniaxial multi-step loading was used, the loading control method was force-controlled, the rate was set to 20 n/s, and the step load was maintained for 30 s with increments of 1 kN. The AE acquisition parameters were assigned as sampling frequency of 20 MHz and amplification of 40 dB. The electrodes were arranged in a two-dimensional symmetric arrangement, as shown in Figure 2. The electrical parameters were collected using the AM method with a power supply time of 0.1 s and a sampling interval of 10 ms. The mechanical parameters, AE signals, and electrical parameters of the coal sample were collected simultaneously during the whole loading process.

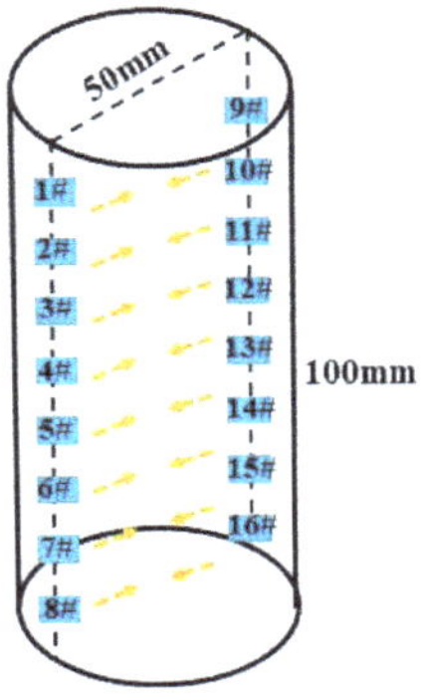

Figure 2. Electrode arrangement.

After the acquisition, the data processing was mainly done by the parallel electro fabrication instrument with WBD 3.0 analysis software. First, the data were imported into the software and preprocessed through the main interface to reject distortion points and bad data. Then, the coordinates were set according to the actual position of the electrode in the 3D space, and the space type and device type were entered. Next, the measurement line was segmented, and the appropriate azimuth and inclination angles were set to form a symmetric profile. Finally, the AR values were calculated and plotted into an AR tomography image.

3. Analysis of Experimental Results

3.1. Analysis of Variation Characteristics of AR and AE

According to the characteristics of the time–load curve, the AR of each sample during the dead load and the unloaded moment was extracted for calculation, and then drawn in an AR tomogram and marked. Tomograms with similar AR data were marked with the same color scale, and the AE counts of the whole loading process were extracted for comparative analysis. Due to space limitations, two representative samples, M-3 and M-11, were selected for description, and the analysis of sample M-3 is emphasized.

The temporal characteristics of AE counts during the multi-step loading of sample M-3 and the changing pattern of tomography images of AR evolution at different stages are shown in Figures 3 and 4. It can be seen that native microscopic cracks in the coal mass gradually closed to form new conductive channels under the action of axial stress at the early stage of loading, resulting in the development of a low-resistance point-like potential in the original strip-like central high-resistance imaging area with increasing load. However, due to the low stress level, the overall structure of the sample did not undergo extensive pose adjustment, so the chromatic aberration of the tomography imaging map changed slowly and the low-resistance region still dominated. The AE counts did not produce significant changes; only discrete signals were generated by the influence of microscopic crack closure in the late stage of this phase, which were maintained at a low level.

At the late stage of loading, the AR tomography image started to show divergent behavior; the left and lower resistance values had complete convergence and produced a significantly increased resistance value, while the central high-resistance area continued to spread and produced signs of penetration in the lower part. At this time, the microscopic fracture development inside the coal mass moved from disorderly to orderly and the expansion speed accelerated sharply, forming a macroscopic main rupture and irreversible damage by converging with and penetrating the primary fracture. When the current flowed through the rupture area, conductivity decreased and the high-resistance imaging area filled the interior of the profile compared with the previous stages. At this stage, the AE signal intensity increased significantly, the counts increased sharply and intensely, and a sudden increase occurred near the stress peak while reaching the instantaneous maximum.

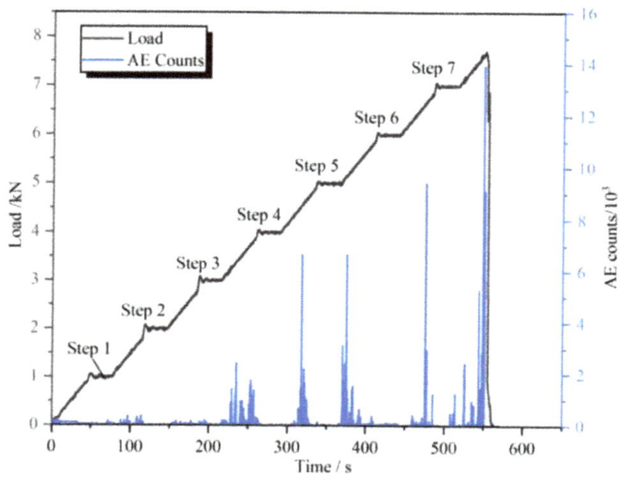

Figure 3. Relationship curves of time–load and AE rate for sample M-3.

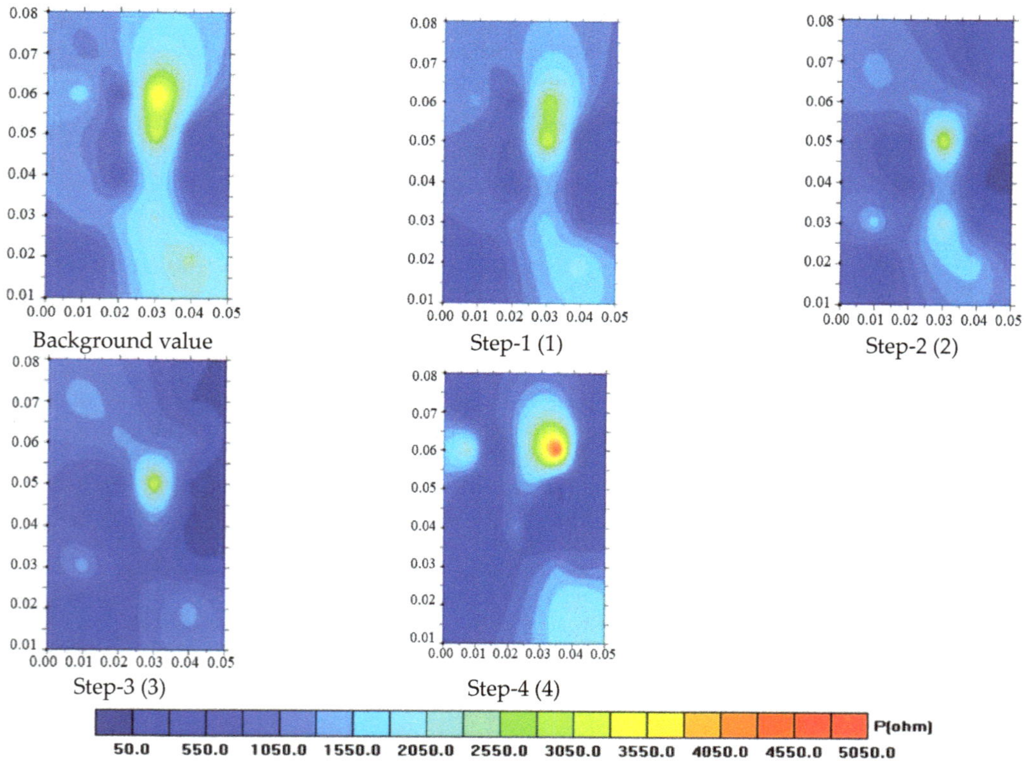

Figure 4. *Cont.*

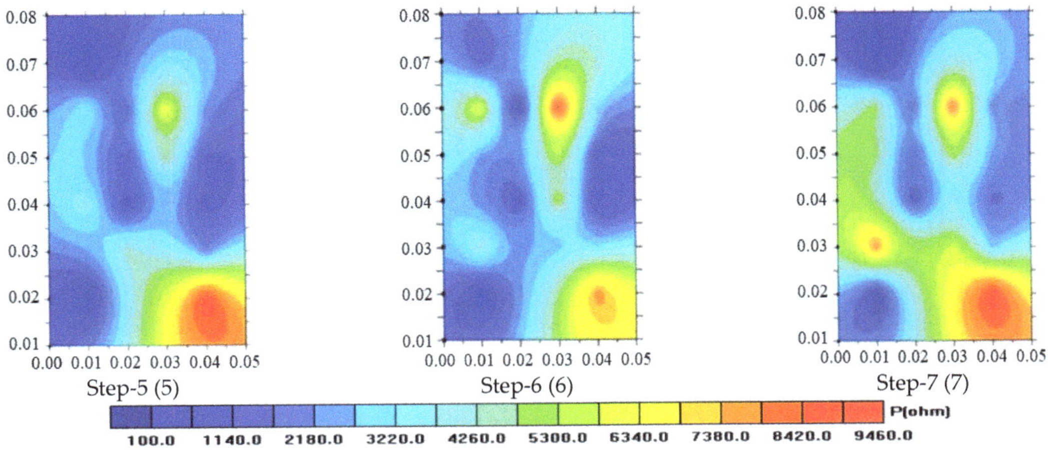

Figure 4. Tomography images of AR evolution of sample M-3 at different dead load stages.

The temporal characteristics of AE counts during the multi-step loading of sample M-11 and the changing pattern of tomography images of AR evolution at different stages are shown in Figures 5 and 6. It can be seen that the AR and AE counts of sample M-11 had similar evolution laws compared with sample M-3. For AR, in the initial loading stage, the high-resistance region showed a decreasing trend due to the effect of primary fracture closure. With the gradually increasing load, AR gradually increased under the influence of fracture development and confluence inside the coal sample. At the same time, the relatively high-resistance area continued to expand with increasing load-bearing degree in the middle and late stages of the loading process, producing a magnitude change. In addition, the AE response and stress change also maintained good consistency. When the coal load increased one level, the active degree of AE counts increased, but in the dead load stage, the AE counts were very weak and a quiet period appeared. Altogether, the overall experience can be roughly divided into three stages: stable silence area, intermittent pulse area, and rapid increase area.

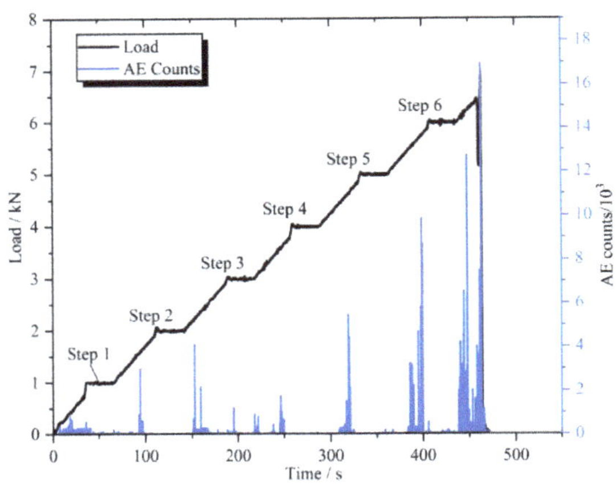

Figure 5. Relationship curves of time with stress and AE rate for sample M-11.

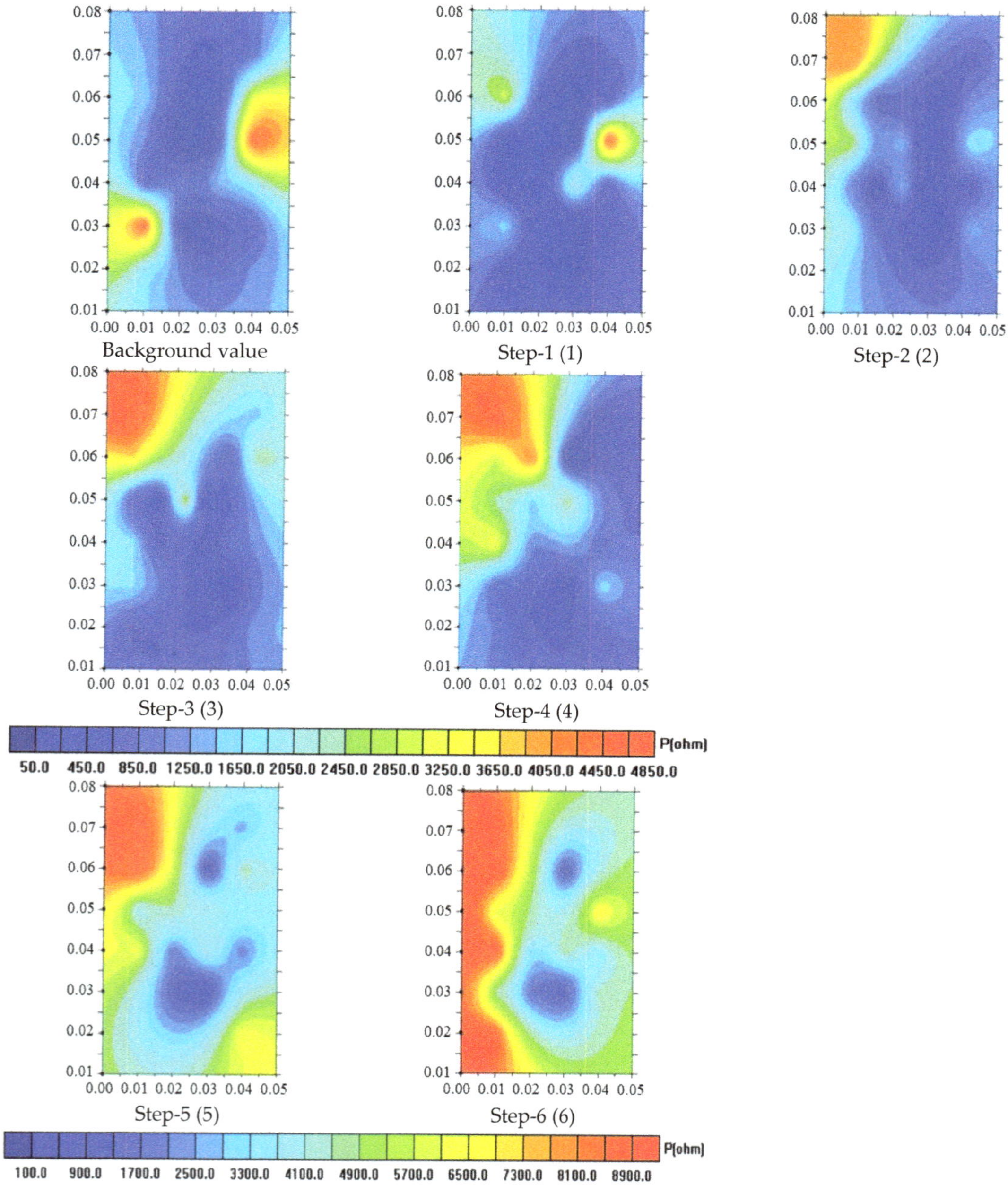

Figure 6. Tomography images of AR evolution of sample M-11 at different dead load stages.

3.2. Analysis of Time Series Variation Characteristics of AR

In order to gain further insight into the time-varying law of local AR during the deformation process of loaded coal samples, some key points with high resistance and obvious AR variation were selected within the tomography image, marked as A–F in

Figures 7 and 8. Then, the time-varying characteristics of AR response at these key points were extracted, and the results are shown in Figures 9 and 10.

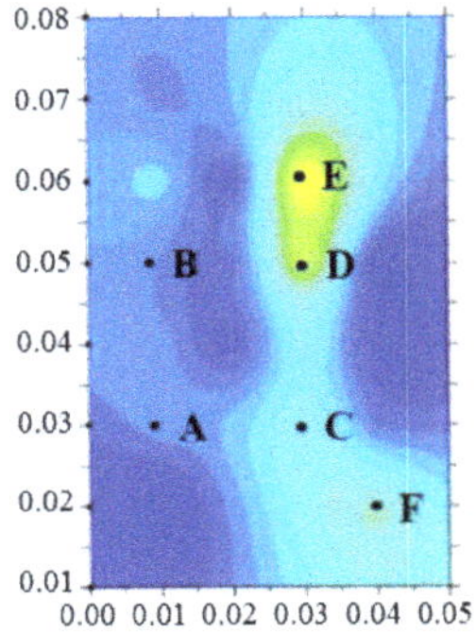

Figure 7. Position of each point of sample M-3.

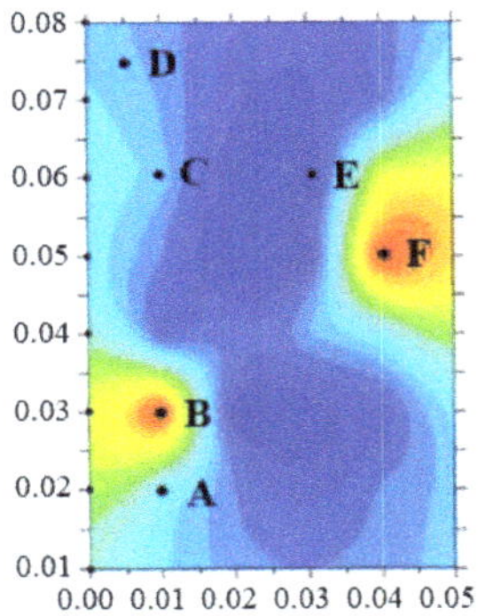

Figure 8. Position of each point of sample M-11.

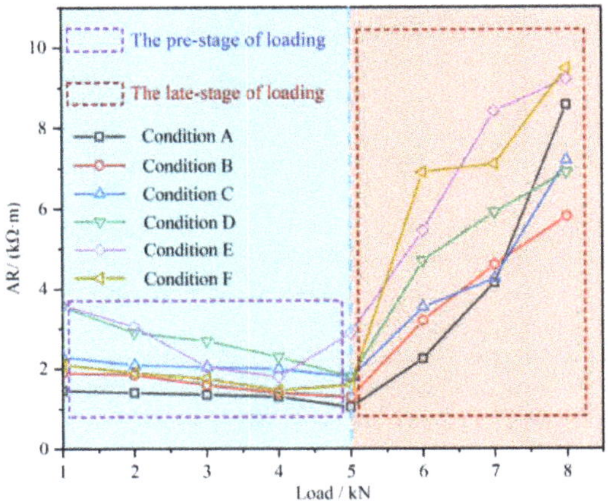

Figure 9. Relationship between load and AR of sample M-3.

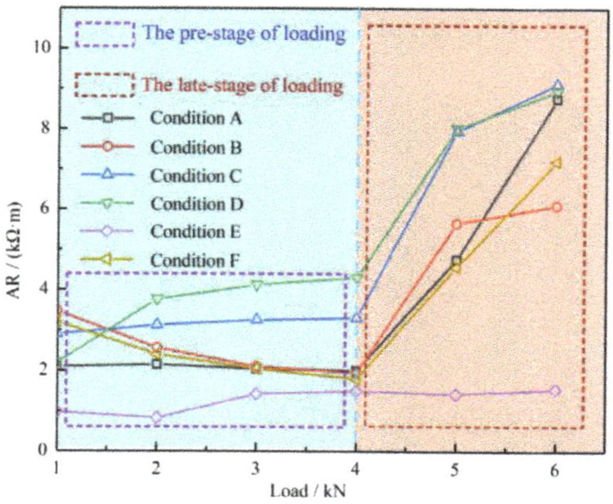

Figure 10. Relationship between load and AR of sample M-11.

As can be seen from Figures 8 and 10, although there are differences in the temporal evolution patterns of the AR high-resistance points of different coal samples, the overall change trends have reasonable similarity. On the whole, there are three main forms of high-resistivity point evolution:

(1) It develops from the low-resistivity point. In the initial loading stage, the change of AR resistance at the corresponding position is relatively stable and maintained at a low level. When loaded to a certain moment, its resistance value increases, indicating that the externally applied elastic strain energy accumulates in the weak unit surface inside the coal mass with increased load, forming a local stress concentration area. When the load exceeds its strength, the strain energy will be released outward, producing local rupture and gradually spreading with the increased stress.

(2) It evolves from the original high-resistance point. In the initial stage, the random distribution of the primary fractures inside the coal sample gradually closes with increased load, resulting in a slowly decreasing trend of AR at the corresponding location. However, when the load reaches a certain level, the resistance value of this part suddenly increases, producing a magnitude change, indicating that the primary fissures or defects that were closed start to redevelop under the influence of high stress, or new fissures are produced.

(3) Part of the original high-resistance point decreases to a lower level and no longer changes. This is caused by the non-homogeneity of the coal mass, in turn causing the bearing capacity of different parts of the coal sample to be strong or weak with the same load, which may lead to no further development of internal primary fissures or defects after closure.

From the AR tomography image and load curve, it can be seen that changes in the high-resistance points constitute the main controlling factor in the change in overall AR value. Furthermore, most of the high-resistance points show a changing trend of slowly decreasing, then sharply increasing with increased stress. The overall AR value should show the same rule, which indicates that the evolution of high-resistance points is mainly controlled by fracture propagation in the coal mass. In addition, it can be seen in the AR tomography image that the location of the final penetration damage of the coal sample coincides with the high-resistance area by comparing the physical maps of M-3 and M-11 after destruction (Figures 11 and 12). Therefore, the time and degree of fracture development at the corresponding location within the coal mass can be inferred from the variation pattern of AR at each point.

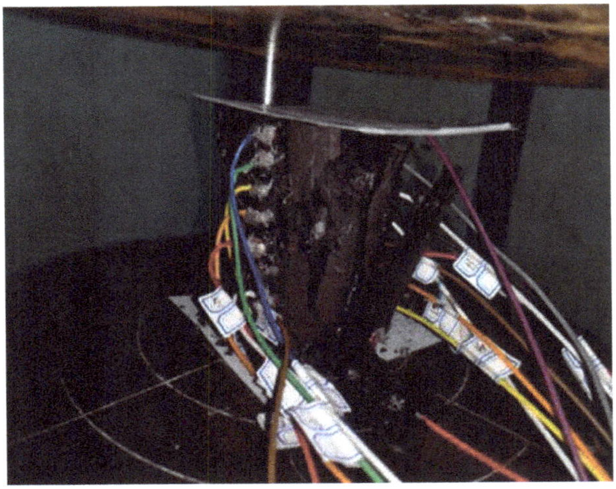

Figure 11. Photograph of sample M-3 after damage.

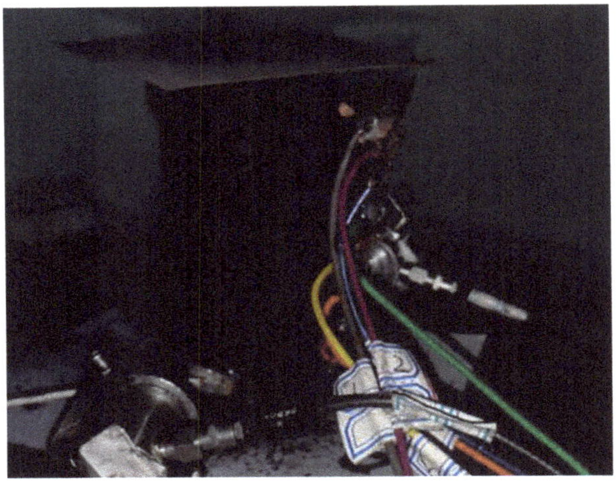

Figure 12. Photograph of sample M-11 after damage.

4. Discussion

4.1. Macro–Microscopic Mechanism Analysis of Resistivity Change of Loaded Coal Mass

A coal mass is an anisotropic non-homogeneous crystalline formation that develops under long-term complex conditions. The natural biochemical structure and composition and degree of water saturation have a significant influence on the original conductivity of coal mass and the type of conductivity (Figure 13) [22,23]. According to solid–dielectric theory, the electrical types of solid medium can be divided into two categories by the type of conductive carriers and the migration mode. One type is electron conductivity, generated by the directional movement of intermolecular electron clouds, and the other type is ion conductivity, caused by the migration of positive and negative ions in the filter-diffusion electric field in minerals and solutions [24]. The conductive properties of coal mass consist of both, but the electrical response of different aspects of coal differ in their sensitivity to the two conductivities in different incubation environments.

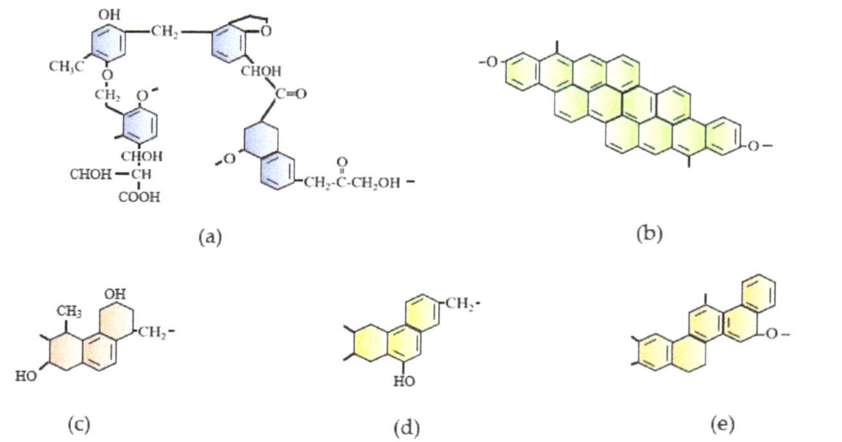

Figure 13. Structure model unit model of coal in different metamorphic grade: (**a**) lignite; (**b**) anthracite; (**c**) sub-bituminous coal; (**d**) high-volatility bituminous coal; (**e**) low-volatility bituminous coal.

For example, in the case of coal with low coalification, the internal aromatic nuclei of its molecules are filled with more interspersed polar oxygen-containing functional groups, which are arranged in a random and disordered cross-linked pattern through side chains and bridge bonds under the influence of van der Waals forces, resulting in weak electrical conductivity, dominated by the ionic conductivity type. For coal with high coalification, the internal functional groups account for a lower percentage. Due to the reduced cross-linking density, the polymerization capacity of the aromatic layer is improved and the internal molecular structure is more regular [25]. This arrangement also enhances the electron transport path and activity range, resulting in stronger electron conductivity than coal mass with low coalification. Therefore, the coal mass, a kind of solid dielectric, will interfere with the directional migration of internal carriers, thus generating current and breaking the equilibrium of the initial electric field in the native medium when carrying a certain load.

Take the ion-conducting coal mass shown in Figure 14, and assume that certain ions in the gap at the lowest potential energy level in the absence of external interference for thermal vibration will be in a semi-stable state, since the surrounding ions have a binding effect on them. If the ion wants to participate in the conductivity, it must break this binding effect. When the energy obtained by the external disturbance is greater than the binding energy, the ions will overcome the barrier and jump to another gap, and the diffusion process is accompanied by the generation of ion current. Here, it is assumed that ions with a vibrational frequency of v at the lowest potential energy in semi-stable position A and B undergo thermal vibration (between the two points can be regarded as equipotential distribution), and the energy generated by the vibration obeys Boltzmann's distribution law. The ion is defined as n_0. The binding potential barrier of adjacent ions is μ_0 [26]. Based on the above conditions, the number of ions that comfortably leap to the semi-stable position of another gap along a certain direction, overcoming the potential barrier per unit time, can be derived as:

$$n = \frac{n_0}{6} \quad (1)$$

The ion potential barrier in the gap does not remain in equilibrium under the action of electric field E. Ions migrate more easily if they conform to the direction of the electric field, and less easily if they don't conform to the direction of the electric field. Therefore, the remaining number during the directional migration of ions along the direction of the electric field in unit time is:

$$\Delta n = n_{A \to B} - n_{B \to A} = \frac{n_0}{6} v e^{-\frac{u_0}{kT}} \left(e^{\frac{\Delta u}{kT}} - e^{-\frac{\Delta u}{kT}} \right) \quad (2)$$

In the presence of a weak electric field, it is:

$$\Delta u = \frac{1}{2} \delta q E << kT \quad (3)$$

$$e^{\pm \frac{\Delta u}{kT}} \approx 1 \pm \frac{\delta q E}{2kT} \quad (4)$$

This leads to the following formula:

$$\Delta n = \frac{n_0 q \delta v}{6kT} e^{\frac{u_0}{kT}} \quad (5)$$

where q is the ionic charge and δ is the average leap distance.

The macroscopic average drift rate v of the ions along the electric field direction is:

$$v = \frac{q \delta^2 v}{6kT} e^{-\frac{u_0}{kT}} E \quad (6)$$

Combining the above equations yields ion mobility μ as:

$$\mu = \frac{q \delta^2 v}{6kT} e^{-\frac{u_0}{kT}} \quad (7)$$

The electrical conductivity γ of the dielectric is:

$$\gamma = \frac{n_0 q^2 \delta^2 v}{6kT} e^{-\frac{u_0}{kT}} \quad (8)$$

For electron conductivity, the expressions for mobility and conductivity were still derived through the ion conductivity expressions. The difference is that δ and n_0 in the electron conductivity expressions represent electron charge and average electronic transition distance, respectively. According to the analysis of the above equation, it can be seen that if the coal mass has not been damaged, the increased stress will cause compression of the space, restricting the free movement of ions between coal molecules, leading to decreased mobility and conductivity of ions, thus hindering the ion leap activity and resulting in increased resistivity. The performance of electrons is the opposite: the electron cloud links more closely before the coal molecules are disturbed by stress, which expands the range of electron transport. This makes the electron leap in the gap easier, so the mobility and conductivity of electrons increases, resulting in a decreasing trend of resistivity with increasing stress. In summary, the two conductivity mechanisms that contribute to the conductivity and resistivity changes of the coal mass under load are opposed. Therefore, to investigate the resistivity changes of coal mass under load, it is necessary to consider the weights of electron and ion conductivity in the conductivity of the coal mass, and to perform individual analyses for different coal qualities. In addition, based on the characteristics of the basic parameters of the tested coal samples and the variation law of AR with increasing stress, it can be noted that the conductivity of the coal mass with the sampling location is mainly dominated by electronic conductivity.

On the other hand, many studies have found that coal resistivity dominated by both electronic and ionic conductivity showed a significant increase after a certain time point when the coal mass was loaded at a high level [17–21]. This is because the high stress level intensifies the instability of the coal mass structure, the strain energy accumulated in the internal weak unit face in the early stage starts to release, and internal fissures develop continuously and form a two-dimensional connection trend in the three-dimensional space. Due to the change in the conductive structure, the conductive medium in the space conducts current both in series and parallel with the fissure development direction. At this time, several conductive mechanisms play a role in the time series successively or simultaneously, and produce different behaviors near the fracture plane. Under the coupling effect, the electrical effect will evolve into mutual promotion and exclusion. The AR evolution tomography image and local AR variation of the samples show that the AR exhibits obvious anisotropic characteristics at the high-stress stage, and the high-resistance region produces a magnitude change relative to the background value. Comparing the morphology of the samples after peak damage reveals that the high-resistance region of all samples is generally consistent with the crack orientation after destabilization.

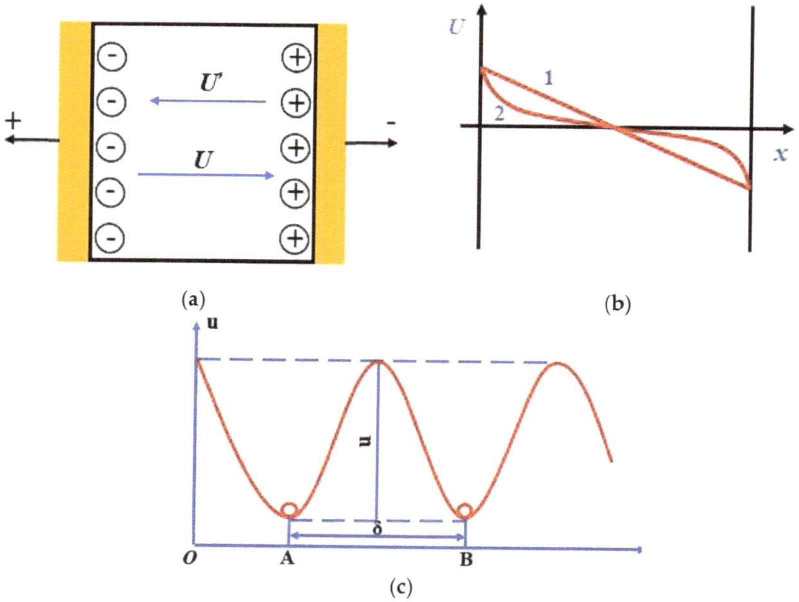

Figure 14. Electric field distribution caused by space charge of medium: (**a**) original potential distribution; (**b**) potential distribution after distortion; (**c**) ionic potential energy in solid medium.

The resistivity response of the coal mass during loading results from the interaction of multiple conductive mechanisms and is influenced by external factors such as electric field strength and test direction. In addition to its properties, the resistivity change of the coal mass during loading is still mainly contributed by the development of macroscopic fracture-conductive surfaces, which cannot hinder the overall resistivity trend even under the disturbance of the applied electric field.

4.2. Exploration of Joint AR-AE Response Method for Coal Rock Fracture Warning

Many scholars have studied the AE technique, and some of their results have been applied to the field of monitoring and forecasting coal dynamic disasters [27–32]. However, based on the feedback from experiments and engineering sites, there are still many bottlenecks hindering early warning by the AE technique. On the one hand, due to the complex geological conditions of an underground coal mine, the elastic stress wave generated by a coal mass during rupture will be irregularly attenuated by its anisotropy and non-homogeneity in the coal seam, which leads to a less strong final signal received by the AE instrument compared with the active source signal noise suppression. In addition, the AE signal is not received far enough ahead of time, i.e., the damage has already occurred when the signal is received. On the other hand, although the resistivity method has a strong ability to distinguish the geological information of the detection area, it is limited by the construction conditions and sampling frequency, and cannot monitor the target area in real time as the AE technique does. Based on the abovementioned factors and the measured experimental results, the characteristics of AE and AR tomography during the process of loading the coal sample were further studied qualitatively. Additionally, passive monitoring–active detection, i.e., the integrated AR–AE technique, for early warning of coal or rock failure disaster was proposed.

The high AE counts and fluctuation degree were used to determine the loaded stage of the coal mass. From the previous experimental results, it can be seen that AE has the characteristics of low amplitude, small amplitude, and continuous fluctuation in the low load-bearing stage. In the middle load stage, the AE signal is relatively active, presenting irregular discrete signal characteristics and a pulse peak higher than 5000. In the high-stress stage, the AE counts show high amplitude, burst-type behavior, and more intensity. In addition, the characteristics of AE counts in the constant load stage are apparent, and all of them have a certain window. It is known from the previous study that sample M–3 still had a small number of AE signals even in the steady pressure stage when it was finally close to destruction, which indicates that coal mass can be subjected to a high degree of load, and the maintenance of stress still causes coal mass fracture expansion and nucleation. The AE signals of sample M-11 had a short calm period before the stress reached the peak. All of the above characteristics can be captured as precursor signals of coal mass destabilization.

The spatiotemporal evolution characteristics of AR tomography were used to identify the internal fracture sources of coal mass. Due to the heterogeneity of coal mass, the load-bearing capacity of different internal parts varies even under the same loaded conditions, so the fracture evolution characteristics also differ, as well as the internal conductive structure evolution. However, it can be seen from the AR tomography images and temporal variation characteristics that the temporal variation characteristics of the loaded coal mass can reflect the local variation and loading state to a certain extent. Combined with the resistivity variation mechanism of the loaded coal mass, the conductive type of coal mass can be obtained, while the spatial variation characteristics can invert the internal stress concentration area, fracture location, and development state. A comprehensive analysis of both can invert the location of the rupture source, which can provide a basis for determining early warning and prevention areas at engineering sites.

The above results show that both acoustic and electrical signals during the loading of the coal mass are controlled by the mechanical behavior. Meanwhile, the strength of the AE signal not only reflects the internal rupture level of the coal mass, but also has a highly positive correlation with the degree of AR change. Through comparative analysis, it can be found that after the AE signal becomes active and appears with high-value pulses greater than 5000 and less than 8000, the overall resistance value of the AR tomography image will show an order of magnitude change, indicating that the coal mass may have entered the destabilization state, which is defined as a secondary warning response. The AE signal appears to be a calm or small dense signal after exhibiting pulse values greater than 8000. At the same time, the high-resistance area of the AR develops to fill the interior of the tomography image, which indicates that damage is coming, and this is defined as the first-level warning response.

In view of the above factors, AE signal amplitude and variation characteristics can be used to monitor the loading state of coal during the whole process of engineering field application, and phased detection apparent resistance tomography images can be used to locate the source of rupture and make a secondary judgment of the loading state.

5. Conclusions

(1) In the process of multi-step loading, the AR and AE signals of coal samples have good correspondence with the load changes. At the early stage of loading, the primary fracture gradually closes under pressure, the original high-resistivity area and overall

resistance value of the sample gradually decrease, and AE is maintained at a low level. The maximum value is reached near the stress peak.

(2) The AR changes in each region during the loading process of coal samples are anisotropic, and the spatial and temporal evolution characteristics of the internal high-resistance region can indirectly reflect the degree and location of fracture development in coal samples. In addition, areas with small AR in the early loading period are more likely to form stress concentration areas and produce deformation and rupture when the loading degree is higher. Therefore, field applications should be focused on areas with these two kinds of AR variation characteristics.

(3) From the microscopic point of view, the coal mass can be regarded as a solid dielectric with both ionic and electronic conductivity, and its natural biochemical structure and composition determine which conductive property is dominant. If ionic conductivity is dominant, the resistivity of the loaded coal mass will increase with stress before damage occurs, while the opposite is true if electronic conductivity is dominant. From the macroscopic point of view, the resistivity change of the loaded coal mass is influenced by the evolution of its internal structure, and the development of its fissure-conductive surface represents the main contribution to the electrical property change.

(4) Before coal mass rupture, AE and AR signals exist with different precursor characteristics. For application at engineering sites, the AE method as a means of passive monitoring can continuously monitor the test area and obtain the degree of coal mass loading through temporal signal changes. With the resistivity method, as an active means of detection, the temporal and spatial changes of AR characteristics can invert and locate the stress concentration area and rupture source, and the two parts of precursor information can verify and complement each other.

Author Contributions: Conceptualization, X.W. and F.C.; methodology, X.W.; software, H.T.; formal analysis, X.W.; investigation, B.M. and H.T.; data curation, B.M.; writing—original draft preparation, X.W.; writing—review and editing, G.Z., D.C. and B.M.; supervision, F.C. All authors have read and agreed to the published version of the manuscript.

Funding: This research received no external funding.

Institutional Review Board Statement: Not applicable.

Informed Consent Statement: Not applicable.

Data Availability Statement: Data is available on request from the authors.

Conflicts of Interest: The authors declare no conflict of interest.

References

1. Kai, W.; Feng, D. Coal-gas compound dynamic disasters in china: A review. *Process Saf. Environ.* **2020**, *133*, 1–17.
2. Wang, E.; Zhang, G.; Zhang, C.; Li, Z. Research progress and prospect on theory and technology for coal and gas outburst control and protection in china. *J. China Coal Soc.* **2022**, *47*, 297–322.
3. Li, X.; Chen, S.; Wang, S.; Zhao, M.; Liu, H. Study on in situ stress distribution law of the deep mine: Taking linyi mining area as an example. *Adv. Mater. Sci. Eng.* **2021**, *2021*, 5594181. [CrossRef]
4. Li, Z.; Zhang, X.; Wei, Y.; Ali, M. Experimental study of electric potential response characteristics of different lithological samples subject to uniaxial loading. *Rock Mech. Rock Eng.* **2021**, *54*, 397–408. [CrossRef]
5. Yuan, L. Research progress of mining response and disaster prevention and control in deep coal mines. *J. China Coal Soc.* **2021**, *46*, 716–725.
6. Qiu, L.; Liu, Z.; Wang, E.; He, X.; Feng, J.; Li, B. Early-warning of rock burst in coal mine by low-frequency electromagnetic radiation. *Eng. Geol.* **2020**, *279*, 105755. [CrossRef]
7. Niu, Y.; Wang, E.; Li, Z.; Gao, F.; Zhang, Z.; Li, B.; Zhang, X. Identification of coal and gas outburst-hazardous zones by electric potential inversion during mining process in deep coal seam. *Rock Mech. Rock Eng.* **2022**, *55*, 3439–3450. [CrossRef]
8. He, M.C.; Miao, J.L.; Feng, J.L. Rock burst process of limestone and its acoustic emission characteristics under true-triaxial unloading conditions. *Int. J. Rock Mech. Min.* **2010**, *47*, 286–298. [CrossRef]
9. Jia, Z.; Xie, H.; Zhang, R.; Li, C.; Zhang, Z. Acoustic emission characteristics and damage evolution of coal at different depths under triaxial compression. *Rock Mech. Rock Eng.* **2020**, *53*, 2063–2076. [CrossRef]

10. Li, X.; Chen, S.; Liu, S.; Li, Z. Ae waveform characteristics of rock mass under uniaxial loading based on hilbert-huang transform. *J. Cent. South Univ.* **2021**, *28*, 1843–1856. [CrossRef]
11. Dou, L.; Yang, K.; Chi, X. Fracture behavior and acoustic emission characteristics of sandstone samples with inclined precracks. *Int. J. Coal Sci. Technol.* **2020**, *28*, 1843–1856. [CrossRef]
12. Zhang, R.; Liu, J.; Sa, Z.; Wang, Z.; Wang, C. Experimental investigation on multi-fractal characteristics of acoustic emission of coal samples subjected to true triaxial loading-unloading. *Fractals* **2020**, *28*, 2050092. [CrossRef]
13. Feng, X.; Ding, Z.; Ju, Y.; Zhang, Q.; Ali, M. "double peak" of dynamic strengths and acoustic emission responses of coal masses under dynamic loading. *Nat. Resour. Res.* **2022**, *31*, 1705–1720. [CrossRef]
14. Kong, X.; Wang, E.; Hu, S.; Shen, R.; Zhan, T. Fractal characteristics and acoustic emission of coal containing methane in triaxial compression failure. *J. Appl. Geophys.* **2016**, *124*, 139–147. [CrossRef]
15. Li, H.; Qiao, Y.; Shen, R.; He, M.; Cheng, T.; Xiao, Y.; Tang, J. Effect of water on mechanical behavior and acoustic emission response of sandstone during loading process: Phenomenon and mechanism. *Eng. Geol.* **2021**, *294*, 106386. [CrossRef]
16. Wang, X.; Asem, P.; Hu, C.; Labuz, J.F. Microcracking in tensile fracture of a brittle rock. *Eng. Fract. Mech.* **2021**, *251*, 107789. [CrossRef]
17. Jiang, W.; Liu, Y. Study on variation of electrical resistivity under uniaxial pressure environment for rocks. *J. Geol.* **2009**, *33*, 299–302.
18. Li, S.; Xu, X.; Liu, Z.; Yang, W.; Xu, L. Electrical resistivity and acoustic emission response characteristics and damage evolution of sandstone during whole process of uniaxial compression. *Chin. J. Rock Mech. Eng.* **2014**, *33*, 14–23. [CrossRef]
19. Jia, P.; Lei, L.I.; Liu, D.Q.; Wang, X.S.; Wang, D.C. Insight into rock crack propagation from resistivity and ultrasonic wave variation. *Theor. Appl. Fract. Mec.* **2020**, *109*, 102758. [CrossRef]
20. Liu, Q.; Qiu, L.; Zu, Z.; Wei, S.; Cheng, X.; Yin, S. Variation characteristics of apparent resistivity of fractured coal sample in loading process. *J. Xi'an Univ. Sci. Technol.* **2021**, *41*, 731–738.
21. Xu, X.; Liu, B.; Li, S.; Song, J.; Li, M.; Mei, J. The electrical resistivity and acoustic emission response law and damage evolution of limestone in brazilian split test. *Adv. Mater. Sci. Eng.* **2016**, *2016*, 8052972. [CrossRef]
22. Shi, Q.; Qin, Y.; Chen, Y. Relationship between thermal conductivity and chemical structures of chinese coals. *ACS Omega* **2020**, *5*, 18424–18431. [CrossRef] [PubMed]
23. Liu, X.; Wang, L.; Kong, X.; Ma, Z.; Nie, B.; Song, D.; Yang, T. Role of pore irregularity in methane desorption capacity of coking coal. *Fuel* **2022**, *314*, 123037. [CrossRef]
24. Liu, Z.; Wang, W.; Yang, J.; Shi, M.; Ma, T.; Wang, D.; Yang, H. Review and prospect of study on conductive properties of coal and cbm reservoirs. *Prog. Geophys.* **2020**, *35*, 1415–1423.
25. Lin, B.; Zha, W.; Liu, T. Experimental study on molecular structure differences between the tectonic coal and primary coal in pingdingshan coalfield. *Vib. Spectrosc.* **2019**, *103*, 102930. [CrossRef]
26. Chen, P. Direct Current Electric Method Response of Regional Coal and Gas Outburst Danger and Its Application. Doctor Thesis, China University of Mining and Technology, Xuzhou, China, 2013.
27. Wang, X.; Wang, E.; Liu, X.; Zhou, X. Failure mechanism of fractured rock and associated acoustic behaviors under different loading rates. *Eng. Fract. Mech.* **2021**, *247*, 107674. [CrossRef]
28. Qiu, P.; Ning, J.; Wang, J.; Hu, S.; Li, Z. Mitigating rock burst hazard in deep coal mines insight from dredging concentrated stress: A case study. *Tunn. Undergr. Sp. Technol.* **2021**, *115*, 104060. [CrossRef]
29. Chen, X.; Li, W.; Yan, X. Analysis on rock burst danger when fully-mechanized caving coal face passed fault with deep mining. *Saf. Sci.* **2012**, *50*, 645–648. [CrossRef]
30. Lu, C.; Dou, L.; Zhang, N.; Xue, J.; Wang, X.; Liu, H.; Zhang, J. Microseismic frequency-spectrum evolutionary rule of rockburst triggered by roof fall. *Int. J. Rock Mech. Min.* **2013**, *64*, 6–16. [CrossRef]
31. Li, D.; Wang, E.; Kong, X.; Ali, M.; Wang, D. Mechanical behaviors and acoustic emission fractal characteristics of coal specimens with a pre-existing flaw of various inclinations under uniaxial compression. *Int. J. Rock Mech. Min.* **2019**, *116*, 38–51. [CrossRef]
32. Qiu, L.; Zhu, Y.; Song, D.; He, X.; Wang, W.; Liu, Y.; Xiao, Y.; Wei, M.; Yin, S.; Liu, Q. Study on the nonlinear characteristics of emr and ae during coal splitting tests. *Minerals* **2022**, *12*, 108. [CrossRef]

MDPI
St. Alban-Anlage 66
4052 Basel
Switzerland
www.mdpi.com

Sustainability Editorial Office
E-mail: sustainability@mdpi.com
www.mdpi.com/journal/sustainability

www.ingramcontent.com/pod-product-compliance
Lightning Source LLC
Chambersburg PA
CBHW061506220926
51850CB00004B/186
9783036594309